MODERN PLANT BIOLOGY

Johnny Derouen - Spring
1974

RICHARD
RIOW

MODERN PLANT BIOLOGY

Howard J. Dittmer

The University of New Mexico   Albuquerque, New Mexico

VAN NOSTRAND REINHOLD COMPANY

New York   Cincinnati   Toronto   London   Melbourne

To Lois and Debbie

Van Nostrand Reinhold Company Regional Offices:
New York Cincinnati Chicago Millbrae Dallas

Van Nostrand Reinhold Company International Offices:
London Toronto Melbourne

Library of Congress Catalog Card Number: 72-110116

Published by Van Nostrand Reinhold Company
450 West 33rd Street, New York, N.Y. 10001

Published simultaneously in Canada by
Van Nostrand Reinhold Ltd.

15 14 13 12 11 10 9 8 7 6 5 4 3 2 1

Preface

In preparing this text on general botany, I have sifted through the abundance of information resulting from recent research efforts in biological science. A recurring question was, how should this valuable new information be used, and where should it be introduced in the curriculum? An ideal program is one that offers an intelligent balance between the traditional study of botany, stressing the physiology and morphology of the plant kingdom, and the latest developments in the field, particularly those in molecular biology. This is the approach offered here.

An additional objective has been to cover the most recent social and scientific advances regarding the relationship of the plant kingdom to man and the environment. Ecosystems, trophic levels, and energy flow cycles are discussed in great detail, along with the limitations of land, water, and atmosphere for plant life. In addition to a separate chapter on plant products of economic importance, treatment is given to the important economic aspects of plants in the various groups. Also included is material on radiation, energy levels, and the origin of life.

The excitement of botany is found in the relationship of the cells, tissues, and organs to the total organism, and not simply the realization that an organism consists of little more than a collection of atoms and molecules. Therefore, molecular biology is presented as the beginning, not the end, in plant development. Although the first few chapters are written to emphasize the significance of molecular biology, it has been integrated into the discussion of the various organs, organisms, and populations rather than considered as a completely new biological field.

From a pedagogical standpoint the chapters are arranged so that after developing the molecular concepts early, the cell, its components, and the higher plants are studied before the survey of the plant kingdom. The chapters on genetics, evolution, ecology, and economic botany may be more conveniently used where they are now, but they could be readily taken up earlier or later as the instructor wishes. The same is true for those who prefer to start with the lower organisms. Instructors would then start with Chapter 14, on the classification of plants, followed successively by the various chapters on the plant kingdom.

My purpose in writing this text has been to provide the non-major, as well as the professionally oriented students, with a true appreciation of the beginnings of life, how it evolved into the higher organisms, and the increasingly important role of the plant kingdom in the lives of everyone—both the botanist and the layman.

I wish to thank Professors Alan S. Heilman, Burlyn E. Michel, and Patricia L. Venable for valuable suggestions in revision of the early draft. The typing and retyping, technical assistance of photography, and general mechanical help have been provided by Mary Coe, Jennifer Yearout, Margaret Elliott, E. Christine Lassen, Judith Williams, Kathleen Kahn Mahon, Janette Hensley, Lydia Kaniszewski, and Elisabeth Gray. Illustrations and technical assistance have been provided by Jeanne L. Garcia, Dr. Ben Pilcher, Dr. Horace Simms, and Dr. John Beakley. The artists, Carolyn Dalbey Findley and Pamela Kaye Bolton, who executed the illustrations, often under considerable pressure, have, of course, given the text greater fluency and lucidity than is possible solely in the written word. I am grateful for the help of those named above and to the many scientists in the field of botany who have made their contributions earlier.

H.J.D.

Albuquerque, New Mexico

Acknowledgments For Illustrations

The following persons, publishers, and institutions have contributed illustrations from previously published articles or books. Credit for the use of redrawn illustrations is also included in the legends. The copyrights for all such illustrations are retained by the original publishers and authors.

Alexopoulos, C. J., 1962. *Introductory Mycology*. 2nd ed. John Wiley & Sons, New York.

Benson, Lyman, 1957. *Plant Classification*. DC Heath and Co., Boston.

Bonner, James, and J. E. Varner, 1965. *Plant Biochemistry*. Academic Press, New York.

Bonner, James, and A.W. Galston, 1952. *Principles of Plant Physiology*. W. H. Freeman and Co., San Francisco.

Brown, W. H., 1935. *The Plant Kingdom*. Ginn and Co., Boston.

Carolina Biological Supply Company. Elon College, North Carolina.

Chamberlain, C. J. *Elements of Plant Sciences*. McGraw-Hill Book Co., New York.

Chicago Natural History Museum.

Coulter, J. M., and C. J. Chamberlain, 1910. *Morphology of Gymnosperms*. University of Chicago Press, Chicago.

Dittmer, H. J., 1964. *Phylogeny and Form in the Plant Kingdom*. D. Van Nostrand Co., New York.

Dittmer, H. J., E. F. Castetter, and Ora Clark, 1954. *The Ferns and Fern Allies of New Mexico*. University of New Mexico, Albuquerque.

Dittmer, H. J., and R. D. Spensley, 1947. *The Developmental Anatomy of Descurainia Pinnata Ochroleuca*. University of New Mexico, Albuquerque.

Eames, A. J., and L. H. MacDaniels, 1947. *An Introduction to Plant Anatomy*. McGraw-Hill Book Co., New York.

Haupt, A. W., 1953. *Plant Morphology*. McGraw-Hill Book Co., New York.

Hill, J. B., L. O. Overholts, H. W. Popp, and A. R. Grove, Jr., 1960. *Botany, A Textbook for Colleges*. 3rd ed. McGraw-Hill Book Co., New York.

Laboratory Outlines, 1963. W. H. Freeman Co., San Francisco.

Maheswari, P., 1950. *An Introduction to the Embryology of Angiosperms*. McGraw-Hill Book Co., New York.

Martin, G. W., "Identification of several fungi by photograph." Nitragin Corporation, Milwaukee, Wisconsin.

Prescott, G. W., 1954. *How to Know the Fresh-Water Algae*. William C. Brown Co., Dubuque, Iowa.

Radio Corporation of America, New York.

Sager, Ruth, 1960. "Genetic Systems in Chlamydonionas." Science 132:1459-1465.

Sistrom, W. R., 1962. *Microbial Life*. Holt, Rinehart and Winston, New York.

Sinnott, E. J., 1955. *Botany—Principles and Problems* McGraw-Hill Book Co., New York.

Stein, Janet, 1958. "A Morphologic and Genetic Study of *Conium Pectorale*." *Am. J. Botany* 45:664-672.

Stoermer, E. F., H. S. Pankratz, and C. C. Bowen, 1965. "Fine Structure of the Diatom, *Amphiphleura pellicida*." *Am. J. Botany* 52(10):1067-1078.

Swingle, D. B., 1935. *Plant Life*. D. Van Nostrand Co., New York.

Sutton, E., 1961. *Genes, Enzymes, and Inherited Diseases*. Holt, Rinehart and Winston. New York.

The Cell, Life Science Library.

The Earth, 1962. Life Science Library.

U.S. Forest Service (Jerry Riffle, Photograph).

U.S. National Park Service.

Water, 1966. Life Science Library.

Weatherwax, Paul, 1956. *Botany*. McGraw-Hill Book Co., New York.

Weisz, P. B., and M. S. Fuller, 1962. The Science of Botany. McGraw-Hill Book Co., New York.

Wollman, E. L., and F. Jacob, 1956. "Sexuality in Bacteria." *Scientific American,* July.

Contents

1

Introduction

The Science of Botany

Botany is the study of plants, but long before the word "botany" was coined man was interested in plants and the products they provided for him. Man's discovery of plants and their uses probably began with the dawning of the early Stone Age. This was a period of trial and error, with little opportunity for second chances. Very likely the lethal effects of a poisonous mushroom were soon learned, as well as the desirable qualities of many fruits, roots, leaves and stems.

It was at least 10,000 years ago that grains such as wheat, rye, barley and rice were making it possible for large, permanent civilizations to become established in the valleys of the Nile and the Tigris-Euphrates, and in other areas where people were interested in a sedentary life. At about the same time peas and beans were discovered to have excellent keeping qualities in the dried state as seeds, and to be quite edible when fresh.

Plants also have a very ancient history, as well as a bright future, as the source of certain pharmaceuticals. Apparently the early uses were based on the general appearance of the plant. Liverworts, for example, were used in the treatment of liver ailments, bloodroot was considered a cure for anemia, and the acorn in its involucral cup was a fecundity symbol to the Druids.

The early study of medicine was virtually a study of botany. Hippocrates listed 240 useful plants in his medical compendium. In the second century A.D. Apuleius wrote extensively about medicinal herbs, giving the names of the plants in Greek, Latin, Egyptian, Punic, Celtic and Dacian along with each plant's description, its place of growth, its properties and the disease for which it was considered remedial. At the same time Galen, author of a *Materia Medica,* stated that doctors too lazy to acquire a knowledge of botany should quit the profession.

For the next fourteen centuries botany, like the other sciences, had little recorded development. The earlier works of Aristotle and Theophrastus were considered so complete and so authoritative that there was little else to learn and certainly nothing to question. In fact, during the latter part of this fourteen-century span of superstition and inertia, the University statutes at Oxford have recorded that Bachelors and Masters who did not follow Aristotle faithfully were liable to a fine of five shillings for every point of divergence.

By the seventeenth century, curiosity concerning the growth of plants and the tissues of which they are composed became evident. In 1664 a Dr. Merret in England noticed that plant tissues healed if wounded, that the cut ends of plants could be grafted, and that wilted plants could be revived by pouring water on their roots. Since he knew about Harvey's discovery of the circulation of blood, Merret immediately jumped to the conclusion that plants have a similar method of circulation. This started a debate that lasted for two centuries, until translocation and the movement of sap were finally ascertained, although they are still not completely understood.

For many years beginning with the seventeenth century, the major emphasis in botany was in taxonomy, with a good many scientists attempting to develop a satisfactory system for the naming of plants. In the latter part of the seventeenth century Nehemiah Grew wrote *Anatomy of Plants*. It was, however, more of a philosophical treatise than a scientific accomplishment. He was still influenced by the work of Aristotle but his colorful expressions stimulated others to work in the field. In encouraging investigation he stated, "For though a man shall never be able to hit stars by shooting at them; yet he shall come much nearer to them, than another that throws at apples."

Grew had a failing that is not uncommon among students using a microscope for the first time. He made drawings of tissues and organs which were not there and some of his descriptions were more fantasy than observation. Yet his description of the cross section of a root was quite accurate. He described the "skin" (epidermis) covering a cortical body; a ligneous (hard) center; air spaces, which he called pores, in the cortex; and the medullary rays, which he called insertions. He also observed chloroplasts and the green substance they contained, but it was not until fifty years later that Ingen-Housz really identified chlorophyll.

While Grew was busy in England, Marcello Malpighi was doing much the same thing in Italy. Malpighi described the spiral vascular element, the annual rings, and rays. In fact, their work was quite similar and when they discovered this, through publications, each writer offered to withdraw his paper in favor of the other. This magnanimity had its counterpart two centuries later when Darwin and Wallace discovered that they had virtually the same ideas concerning evolution and the survival of organisms. It would be interesting if Grew and Malpighi could come back today and see how many people are now working in the field that they thought was crowded with just the two of them.

The major work in plant physiology really began in the latter part of the eighteenth century when Priestley published his *Experiments and Observations on Different Kinds of Air*. From this point on we began to learn about respiration, photosynthesis and their associated functions in plants and animals. The significance of chlorophyll became known, and nutrient and water relationships were investigated and repeatedly confirmed.

In 1838 Schleiden and Schwann introduced the cell theory; Mendel discovered genetics; hybridization through planned selection became common. Evolution became a popular subject, although its adherents were extremely few until well into the twentieth century.

Botany and the related biological sciences have now entered a new phase of development which is concerned mainly with the interaction of molecules within the organelles of the plant cell. This change in emphasis for the biological sciences is due to the accelerated pace of research aided by such new instruments as the electron microscope (Fig. 1), new techniques, more intensive experimentation, and new interpretations which have added considerably to our knowledge of the finer mechanisms of the cell.

Since the study of biology today leans more heavily toward a

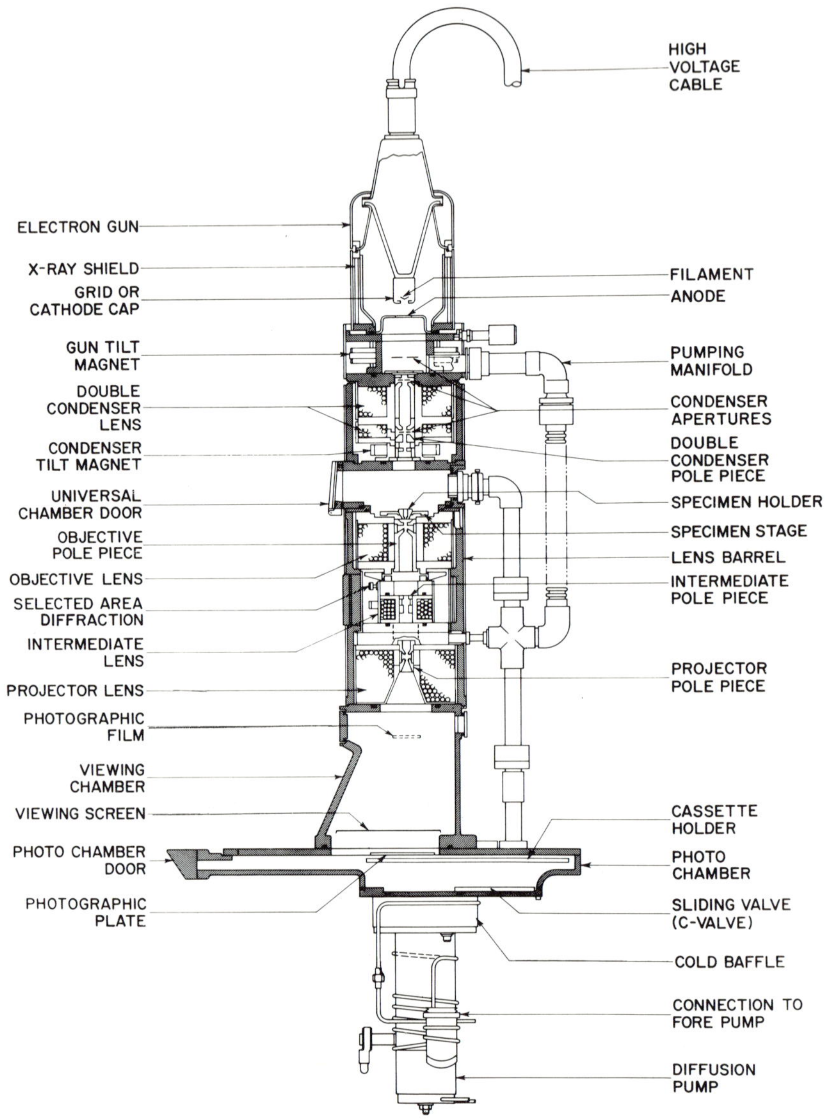

FIGURE 1 Electron microscope. Detail of the complicated inner mechanisms of the electron microscope. (Courtesy Radio Corporation of America.)

functional approach than toward the descriptive, the biological fields have often been renamed. Consequently we see such titles as biosystematics, morphogenesis, embryogenesis, physiological genetics, genecology, developmental cytology, immunoembryology and many others. We are not, however, discarding the earlier titles or their subject content. They are still as important as ever. We are merely broadening our horizons.

The new discoveries in cellular research have led to a new field of molecular biology, and with it greater perspective in interpreting the organisms studied. Splitting of the atom, analysis of the nucleic acids, discovery of the delicate chemical reactions in photosynthesis and respiration, and the building of synthetic organic molecules have all led to Nobel prizes. This has resulted in new research in biochemistry and biophysics in investigations of the minutest fragments of living matter. It is now possible for us to look with greater comprehension into the beginnings of life and perhaps someday we may be able to create life itself by a manipulation of the atoms and molecules of which all organisms are composed. But let us not forget that we are still studying plants and animals, and although we know much more about the governing mechanisms which control every moment in an organism's life, we still have the whole organism to consider. A jigsaw piece does not tell us what the entire puzzle is about until we have put the pieces together.

Life as we see it in the organism, whether plant or animal, is a unique, complex combination of nonliving materials arranged in patterns of chemical activities which give the organism the ability to utilize various foods, to reproduce, and to carry on all the activities we associate with being alive. We know that all matter, whether organic or inorganic, is formed of basic units termed atoms, and that these units may be combined to form a great variety of molecules.

Living systems are composed of very complex molecules. In the organism these molecules are both numerous and variable but all function in a coordinated manner. Generally we consider the viruses the simplest living systems, but even they exceed in size and complexity the largest of the nonliving molecules.

Above the virus level, molecules have become organized into the structure which we term a cell. Not too many years ago we thought of the cell as the smallest functional unit in all organisms. Now we know we must shift our emphasis from the cell as a unit to a study of its individual parts and their coordinated activities. This becomes possible through the use of modern developments in electronics and biochemistry. Formerly our knowledge of the cell was limited to the light microscope, which has a maximum magnification of about 2000 times. The electron microscope now has made it possible to work at magnifications which range up to 200,000 times. The unit of measure with the light microscope (Fig. 2) is the **micron,** of which there are about 25,000 to an inch, but with the added dimension of the electron microscope we speak of **angstroms,** of which there are about 250,000 per inch. Tiny cellular

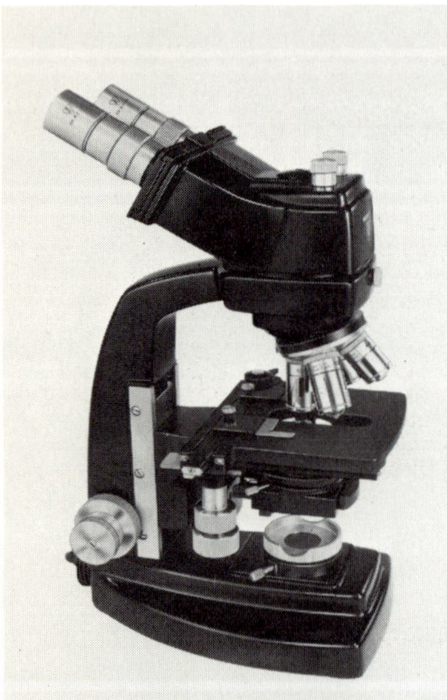

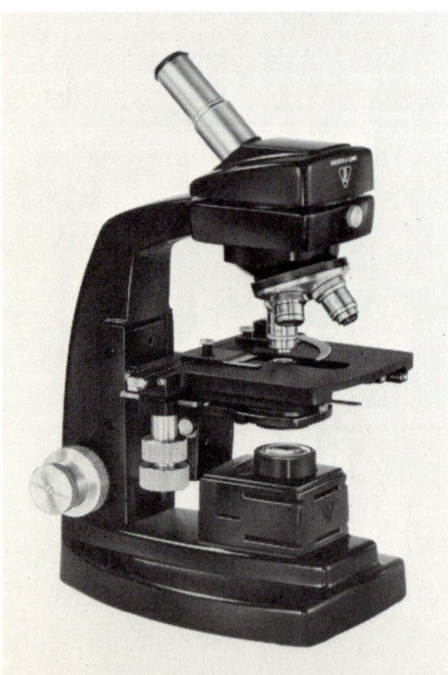

FIGURE 2 Modern light microscopes. A. Binocular with four objectives, mechanical stage and base illuminators. B. A high quality monocular microscope with three objectives, mechanical stage and supplementary light. C. A diagram of the monocular instrument showing the position of the lenses, and the names of the microscope parts. (Courtesy Bausch and Lomb Incorporated, Rochester, New York.)

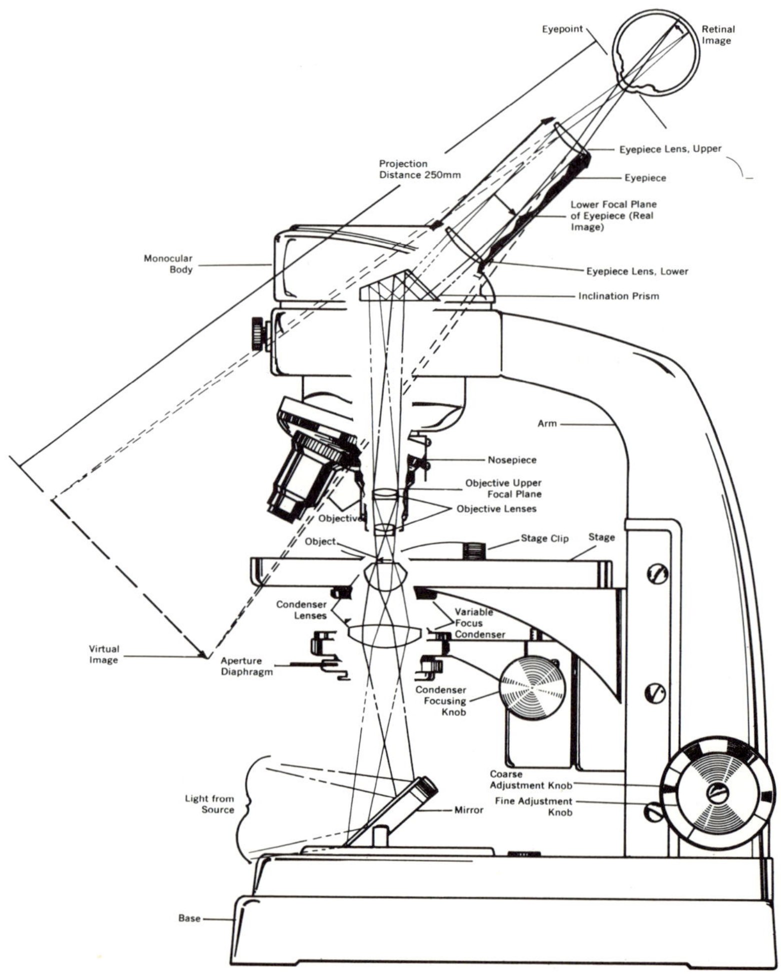

inclusions which formerly appeared as mere specks to the light microscopist are now revealed as organelles with definite structure, many of which are closely associated in a functional system contained within the cell. Furthermore, through the use of various stains and meticulous staining techniques, additional cellular inclusions have been revealed which heretofore had been only suspected.

Units of Measurement

Length

Å = Angstrom unit = 0.1mμ
mμ = millimicron = 0.001 micron
mm = millimeter = 0.1 cm = 25.4 per inch
cm = centimeter = 10 millimeters = 2.5 per inch
m = meter = 100 centimeters = 39.4 inches
k = kilometer = 1000 meters or 1.6 kilometers = 1 mile

Mass

1 kilogram = 1000 grams = 1,000,000 milligrams
1 gram = 1000 milligrams
1 kilogram = 2.2 pounds
1 ounce = 28.35 grams
16 ounces = 1 pound

Volume (wet)

1 liter = 1,000 milliliters = 1.06 quarts = 1000.27 cubic centimeters
ml = milliliter = 0.001 liter

Temperature

Degrees centrigrade (C°) $= \frac{5}{9}(F° - 32)$

Degrees Fahrenheit (F°) $= \frac{9}{5}C° + 32$

With the development of new biochemical methods, chiefly through the use of radioactive agents, it has become possible to "tag" certain atoms and molecules and follow them through the various life processes. Following a single atom through such processes as photosynthesis and respiration has made it possible for us to deduce the mechanisms involved. Formerly, these mechanisms were classified only by the raw materials required and the end products obtained. Although the emphasis has shifted to the atom and the molecule, the cell remains the functional coordinated unit of structure.

The Scientific Fields

While astronomy, geology, physics and chemistry are based on the study of inorganic substances and physical phenomena, biology is the study of living things. The study of biology gives us an appreciation of such natural resources as forests, grasslands, water, soils, animals, and human populations. The breadth of subject is so great that it is used, in part at least, in all major professional areas of medicine, industry, and agriculture. Consequently, biology serves as a foundation for such professions as medicine, dentistry, pharmacy, nursing, agriculture, horticulture, entomology, forestry, and many others. There is, of course, an interdependence of one subject on another. Therefore when we study biology in any depth we must also study such related sciences as chemistry, physics, mathematics, geology, geography, paleontology, psychology, and the social sciences and humanities as well.

Biology has grown to such an extent that the subject matter is divided into many fields for convenience in study. The two major areas, botany and zoology, are known to nearly everyone, but the subdivisions of these sciences are not generally known.

For example, in both botany and zoology we study such basic areas as anatomy, ecology, evolution, genetics, morphology, physiology and taxonomy. In both major fields we also have related areas. In botany we also study bacteriology, economic botany, mycology, phycology, plant pathology, virology, and others.

Divisions of the plant sciences are listed under major groupings but in studying any of these subjects there would be considerable overlapping of content.

Morphology a study of the form and structure of organisms.

Anatomy a study of the internal structure of tissues and organs.

Cytology detailed study of the cells and the subcellular constituents.

Histology the detailed study of the microscopic structure of tissues and organs. Histological technique, the preparation of tissues on slides for microscopic examination, is often studied with histology.

Mycology the study of fungi.

Paleobotany the study of plant fossils and their remains in the strata of the earth's surface.

Phycology the study of algae.

Taxonomy the study of identification and classification of plants. It is often referred to as biosystematics, and includes chromosome studies of the various species.

Physiology the study of function, processes, and chemical activities in the plant and its organs.

Bacteriology now frequently termed microbiology, since it deals with more than just the study of bacteria. Virology, the study of viruses, may therefore be considered a division of microbiology.

Ecology the study of the interrelationships of plants with their environments and other members of the biotic environment. Ecology overlaps plant geography—the world wide distribution of plants.

Economic botany the study of the economic importance of plants, how they may be grown, improved, marketed, stored, and processed. The more important aspects of economic botany include agriculture, horticulture and forestry.

Embryology the study of the early developmental stages of an organism.

Evolution the study of the developmental changes in organisms resulting in changes that may be of such magnitude as to produce new species.

Genetics the study of inheritance or the transmission of characteristic determiners.

Plant pathology the study of plant diseases, their causes, symptoms, and cures. There are probably more people employed in the field of plant pathology than in any other botanical field.

Science and Scientific Knowledge

Before considering the various life processes and the evolution of plants, let us look at the method by which we have come to understand so much about that which is living and how these life processes operate. In man we have a species endowed with the ability to learn new things. He uses the trial-and-error method of the primitive unicellular organisms but also has the capacity of solving problems through thought and logical processes. Everything we have today in a material or artistic sense is due to this capacity for thought, to the powers of communication, the acquisition of knowledge, deductive reasoning, and the desire to learn. Once man began to reason and understand, discovery, learning and knowledge began to occupy most of his time. What then has learning done for us in science?

The Latin word *sciens* means "knowing," and consequently science is often defined simply as knowledge. The French use the term *la science* to encompass all types of learning, while in German *die Wissenschaft,* which means knowledge, is freely used to mean science. In general, we think of a scientist as one who learns by observation and experimentation, accepting as fact only that which can be proved. Perhaps this is too limiting a definition. Many of the things we learn and accept as fact cannot be proved by experimentation immediately, but the information available is so overwhelming that we come to accept it since there is no other explanation which suffices. The comforting thing about accepting knowledge not susceptible to immediate proof, such as knowledge of the manner in which life was created, is that a good scientist is always willing to accept new information and modify his thoughts when other factual material becomes available. Bigotry, prejudice, bias, and dogma should never be acceptable in the thoughts and work of a scientist. Openmindedness, integrity, theory, hypothesis, method, proof, fact, observation and experimentation should figure in every scientist's life and the work he does.

The thing which most distinguishes scientific knowledge from other types is that it is based on rigorous investigation; it is a systematic accumulation of information, supported by experimental evidence. Scientists follow these concepts in whole or in part, depending upon the lines of investigation they follow or the fields in which they work. A scientist may pose a question about nature, then set about collecting all the pertinent evidence, from which he deduces an experimental hypothesis. He then tests his hypothesis by every avenue open to him and from his results forms certain conclusions. He may then accept the results as proof of his original hypothesis, or reject the results as nonconclusive or unsatisfactory. Often he must modify his hypothesis and repeat his experiments with modifications.

Science, then, has its basis in the **scientific method.** The method is not limited to science, for any study in any field may use the same approach to reach an understanding of the facts available. Not all scientists have aptitudes or interests pointed in the same direction but all use the same approach to gather evidence, verify the experimental data and interpret the conclusions. Scientists of different abilities and interests are often associated to propose new problems and interpret the findings of others. Some are imaginative theorists interested in putting new knowledge to work.

Theories formulated through the scientific method suggest ways not only for interpreting natural phenomena but also for applying it. The theory of relativity not only suggested the nuclear processes which keep the sun shining but also made it possible to work out the same nuclear processes for the atomic age.

Physicists and astronomers have expanded our horizons immeasurably, so that dimensions of inconceivable quantity in both time and space are now upon us. With this new scientific influence on our culture technological knowledge has only begun to be felt, but the future is so bright that we may predict a world free of want and ignorance if the human mind can become as unbiased sociologically as it should be scientifically.

The method of science is not in itself profound; actually the steps are quite commonplace, logical and rational. It is a system almost anyone would use in any problem involving several steps. Whether baking a cake or constructing a house, there is a perfectly logical plan to follow. The person who carries out his planned program in the most systematic way invariably achieves the best results. There are many limiting factors which must be considered in most scientific work. A baker, without previous knowledge of the end result, would probably debate within his own mind whether he should use baking powder or yeast as a leavening agent in his mixture. A contractor must consider the nature of the substrata upon which he will set his foundation to erect a building. Weight factors, stress, and strength of materials are variables in this type of work. These are examples of applied science, and most of the principles involved have been worked out before in basic experiments.

Some of the steps involved in the scientific method are given below but the procedures will vary for different problems.

Observation Any problem begins with observation although all things cannot be observed directly. We may not be able to see atoms but we can observe their effects. We may not be able to see the enzymes and their immediate step-by-step breakdown of the substrate, but we can test for their products step by step. Generally we can visualize mentally the problem and, ultimately, the results.

The danger in observation is that we might not view the problem as it stands without bias or prejudice. Too often a person tends to see what he wants to see or what he has been told to see. The beginning student, using the microscope for the first time, often fails to see the intended object and instead mistakes any number of foreign particles for it. Experience in observation is a prime requisite, but more important is the avoidance of unconscious prejudice which may be planted by a preconceived idea, a teacher's instructions, or even an earlier experiment. The scientist, through observation, examines the problem and accumulates the factual data for testing.

Hypothesis Once we have observed a situation we may ask ourselves the question, "What makes it grow like this?" or "How is it possible for a plant to live in this habitat?" We may also ask the question, "Why does the plant grow?" The last question is difficult if not impossible to answer. "Why" questions are, in general, not easily answerable questions in biology. Why are all organisms made up of cells? Why do stars twinkle? Why do the swallows always (or almost always) arrive in Capistrano the same day each year? We can supply answers but the answers to such questions are often based on supposition and purpose which cannot be readily proved.

Thus when we are presented with a problem we can look upon what it is and how it works and by planned study, observation and experiment find satisfactory answers. The scientist, therefore, looks for answers in problems that are testable.

Once the problem has been determined the next step involves postulating an answer. Since most experimental problems require many steps for a solution, the scientist thinks of various techniques he might follow to arrive at a logical answer. This attempted explanation while in the tentative stage is the hypothesis.

Experimentation The follower of the scientific method must now plan the experiment step by step, then perform the experiment, testing each variable as it develops and compare the results with controls which have been methodically prepared. Well prepared experiments are designed so that there is but one variable, such as time, light, or heat; all other factors must remain constant. The control is set up at the same time and is treated in the same way except that there are no variables. This is at least as important as any other phase of the experiment, for it will indicate whether the variable selected in the experiment was a wise choice. This is done not once but many times and when the result is obtained this technique is repeated again and again to be sure the result was no accident. If over a period of years the results are repeatedly confirmed, we may think of the original hypothesis as a **theory;** eventually, if no exceptions are found, the theory may become a **scientific law** or **principle.** Schleiden and Schwann, working independently, observed that the plant and animal tissues they examined were composed of cells. After looking at a great many different tissues they proposed the **cell theory** that all organisms are composed of cells. Now, one hundred years after they made their observations, we have universally accepted their theory as a **cell principle.**

Each experiment is different and the techniques to follow in setting up a proper method are just as different. Some experiments only require a few minutes to perform; others may require years. But once the experiment has been performed, the answer obtained, and the hypothesis proved, we are ready to take our case to court; we have the evidence.

Supplementary Reading

Baker, Herbert G., *Plants and Civilization*. Wadsworth Publishing Company, Inc., Belmont, Calif., 1965.

Bonner, J., *The Molecular Biology of Development*. Oxford University Press, New York, 1965.

Keosian, John, *The Origin of Life*. Reinhold Publishing Corporation, New York, 1964.

Knoblock, I. W. (editor), *Readings in Biological Science* (Part I). Appleton-Century-Crofts, New York, 1967.

Lenhoff, Edward S., *Tools of Biology*. The Macmillan Company, New York, 1966.

Questions for Discussion

1. If you were planning an experiment in biology and another in chemistry, why might it be more difficult to plan the experimental method in biology than in chemistry?
2. If you were an early inhabitant of the earth, how would you probably have selected a plant for medicinal purposes?
3. You are going to plant a new lawn in a yard. What problems will you have to consider to establish a successful lawn?

4. In preparing a scientific experiment, what steps are usually taken?
5. How may the classification of facts concerning a particular problem contribute to its solution?
6. How are the different botanical fields related to each other?
7. What is the significance of a control in setting up a scientific experiment?
8. Prepare a scientific experiment, listing the various steps involved, the order in which the steps are to be performed, and the variables that one is most likely to encounter.
9. How does the "trial-and-error" method of learning differ from the scientific method?
10. What were factors which limited the work of early botanists both mechanically and spiritually?

2

Matter and Energy

Only in the last few hundred years has man become aware that the world is more than a globe created for and built around him. Astronomers now tell us that billions of stars have planetary systems and thousands of these planets are suited to sustain life. It is very likely that man is not the pinnacle of all life and that there are innumerable beings on remote planets of other systems whose intelligence exceeds our own. Our sun is still a comparatively young star and our earth a young planet. If we consider time important in the creation and evolution of life, thousands of planets have had an advantage over ours in the development of species.

The Universe

The universe is made up of two fundamental components, **matter** and **energy,** each of which may be converted into the other under certain conditions. We think of matter as occupying space

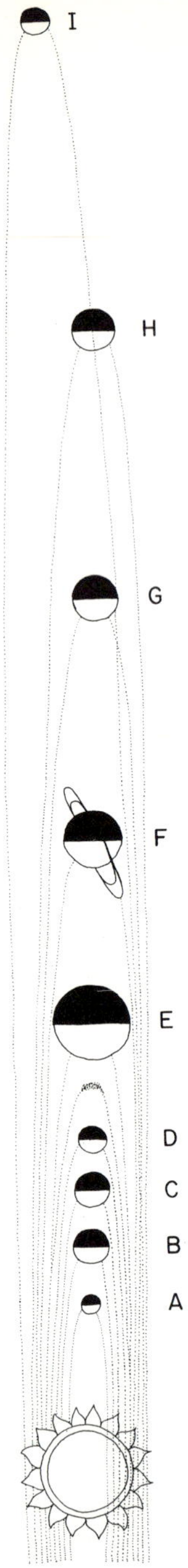

FIGURE 1 Our solar system with the positions of the various planets with respect to the sun. A: Mercury. B: Venus. C: Earth. D: Mars. E: Jupiter. F: Saturn. G: Uranus. H: Neptune. I: Pluto. (After Beiser.)

and having weight, and of energy as the ability to do work. Generally we think of energy as being produced in the transformation of matter, but in the discussions to follow we shall also see how matter is formed from certain types of energy.

All matter seems to be composed of the same elementary substances and all planetary systems seem to be similar in their composition and origin. It follows that conditions which prevail on one planet to create, sustain and evolve life must also be present on others with similar atmospheric, climatic and environmental situations. The race is now on to see who can be the first to actually prove that life does indeed exist on other planets. Neil Armstrong, upon landing on the moon, said it looked very much like our high deserts. Of course it has no vegetation but the moon does have similar basaltic rocks. It is quite probable that if our moon had an atmosphere as dense as the earth's, its maximum and minimum temperatures would be more moderate and life would be possible.

It is generally believed that all galaxies started in much the same manner and with much the same material. Great clouds of dust and gaseous materials are believed to have been present throughout the universe. Gradually these particles were brought together by molecular attraction and gravitational pull. As the molecules joined, cores of molten material were produced. The greatest masses formed suns or stars, and the smaller ones became planets and satellites. Those bodies formed in relatively the same celestial area, held by their orbital pathways, became an organized galaxy with numerous planetary systems. Others were completely individual bodies or formed different galaxies. Although there are other theories describing the origin of the universe, this one has the widest support today. Since most astronomers are of the opinion that all bodies of the universe arose in the same manner, we might consider the earth as a typical example.

All matter on earth, the planets and the moons of our solar system (Fig. 1) is believed, by scientists who study the celestial bodies of the universe, to have come from the gaseous masses and dust clouds left out of the central clouds which formed the sun. Although this star, 864,000 miles in diameter, is considered only a medium sized star among billions in the cosmos, to us it is the source of all substance and energy. It still consists of great masses of gases, and its atoms wander freely, taking part in repeated nuclear fusions of hydrogen atoms into helium atoms. This tremendous activity keeps the sun's core from collapsing and solidifying, and heats not only the sun's enormous envelope of gas but the rest of the solar system as well.

The fusion reactions transform 657 million tons of solar hydrogen into 652.5 million tons of helium ash every second. The remaining atoms are destroyed or there would be no nuclear transformation taking place. Einstein's law, $E = mc^2$, explains this process. In this equation the amount of energy, E, created in the process equals the mass of the destroyed matter, m, multiplied by the square of the velocity of light, c.

The energy produced by the destruction of matter deep within the sun works its way to the sun's surface and then radiates into space. If this nuclear energy were not dissipated into space (and to us) the solar temperature would rise so rapidly the sun would explode. If the tremendous energy released at the sun's core reached the surface all in the form of gamma waves, as originally created, the result would be radiation death wherever life exists throughout our entire solar system. However, the rays are tempered on their outward journey from the sun's core by atoms of

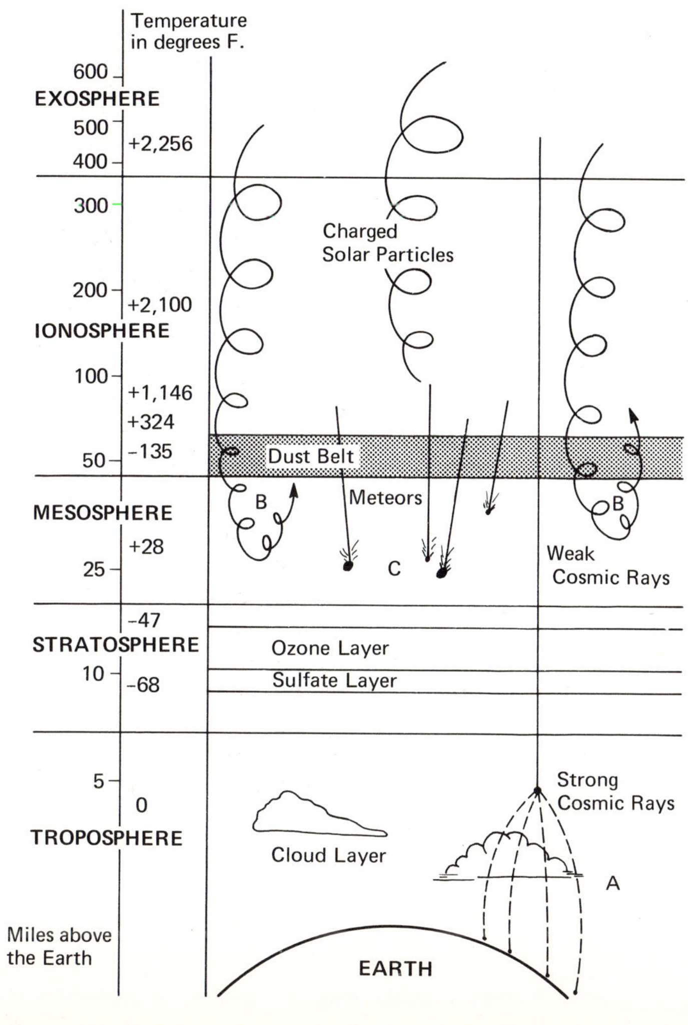

FIGURE 2 The first hundred miles of the earth's atmosphere are made up of a variety of layers which protect us from the lethal radiations of the sun and are partially blocked by the sun's own atmosphere and the magnetosphere some forty thousand miles up. Note the various levels, approximate distances, and temperatures at the different levels. A: strong cosmic rays; B: weak cosmic rays; C: meteors. (Adapted from *The Earth,* Life Nature Library, 1962.)

gas which are part of the sun but play no part in the central fusions. Before these rays of thermonuclear energy can thrust their way through a "shell" some 80,000 miles thick, where the atoms are less tightly packed, they encounter a **photosphere** where the atoms are mostly turned back by the force of the sun's gravitational field. Those that get through form a region of fiery red spray known as the **chromosphere.** The energy that has been thus gentled and reduced by passage through the photosphere and chromosphere is further decreased by the terrestrial atmosphere covering the earth (Fig. 2). Even though most of the atomic energy is lost, the sun's brilliance is so great it cannot be looked upon directly but must be observed through smoked glass or dense filters to prevent blindness.

Although the energy of the sun is created as nuclear energy, the most dangerous of the energy forms which man has identified, it reaches us in the form of **radiant energy** or **light.** It is difficult for us to perceive this form of energy at work, yet it is found throughout the universe. It appears as a mere sparkle from distant stars, but in the form of light, radio waves, x-rays, and infrared and ultraviolet rays, radiant energy floods the earth, making life both possible and pleasant (Fig. 3).

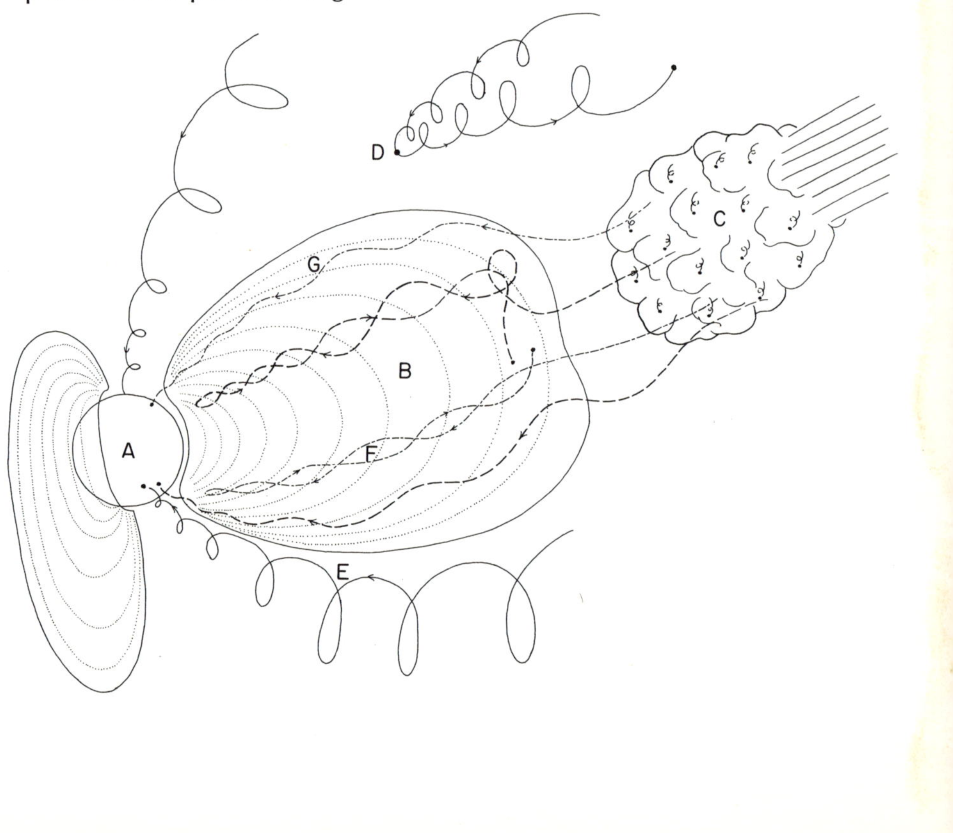

FIGURE 3 The bombardment of the earth (A) by radiation particles from the sun. Protecting the earth from the intense cosmic rays (C) is a great magnetosphere (B), some forty thousand miles thick. Many of the cosmic rays are weak (D) and are deflected by the magnetosphere. Others have greater penetrating powers (E) and reach the earth. The cosmic particles include solar protons (F) and solar electrons (G). (Adapted from *The Earth,* Life Nature Library, 1962.)

The Earth

It is doubtful that the earth was ever a molten ball, but certainly the squeezing of the atoms by gravity and radioactivity did produce a hot central core from which the heat rose to warm the earth's crust to perhaps hundreds of degrees.

At first there was no sea. The primeval ocean was formed when the temperature of the earth's surface fell below the boiling point of water. Water was present on the earth from the beginning but was pent up in the interior rocks. Gradually water vapor was released and rose in great cloud masses that enveloped and darkened the earth.

For perhaps hundreds of years the earth's crust was so hot no moisture could fall without being immediately converted to steam. This helped cool the earth by transferring heat from the rocks to the atmosphere where the heat could be dissipated into space and the vapor reformed into more rain clouds.

For thousands of years the great overhanging cloud masses prevented the sun's rays from reaching the face of the earth in force. Gradually the rocks were cooled until falling rain no longer was converted to steam as it hit the earth but remained to trickle down among the rocks forming rivulets, streams, and lakes. The rains continued to fall, not in a great deluge but probably as it does today. It took perhaps a billion years to form the oceans and these were not as large at first as they are today. As the eons passed water vapor kept rising from the volcanoes and hot rocks below. With normal evaporation this added moisture to the atmosphere and bulk to the seas.

For millions of years the boundaries kept changing. Mountain systems rose and were eroded away by crustal movements and the physical and chemical actions of sun, rain, heat and cold. Very early the continental masses were established and the seas were contained. The inorganic earth was relatively stabilized but life had not yet appeared.

Life probably first occurred about two billion years ago in the environment of the seas as very large molecules capable of reproduction (Table 2-1). These molecules apparently arose in the warm water marshes which contained marsh gas (methane) and ammonia in an oceanic vat of soluble minerals. Molecules containing carbon combined and recombined until the right combination for a stable organic molecule (such as an amino acid) was reached. It seems reasonable to assume that before living organisms were generated

Eras	Periods
Cenozoic Began 60,000 years ago	Recent Pleistocene Pliocene Miocene Oligocene Eocene Paleocene
Mesozoic Began 185,000,000 years ago	Upper Cretaceous Middle Cretaceous Lower Cretaceous Jurassic Triassic
Paleozoic Began 520,000,000 years ago	Permian Pennsylvanian (upper Carboniferous) Mississippian (lower Carboniferous) Devonian Silurian Ordovician Cambrian
Proterozoic Began over 2 billion years ago	Bacteria, primitive algae, and primitive invertebrates were present at this time
Archeozoic Began over 3 billion years ago	No fossils found in these ancient rocks but the first unicells may have arisen at this time

TABLE 2–1 The Eras, Major Periods, and Probable Antiquity of the Earth, Based on the U.S. Geological Survey Chart of 1956.

spontaneously, organic molecules must first have been generated spontaneously. This was probably achieved many times through a process of chemosynthesis in which certain molecules formed organic compounds from energy obtained from inorganic sources. These first independent synthesizing molecules were chemoautotrophic. They derived their carbon from carbon dioxide but obtained their energy from such inorganic compounds as ammonia and those of sulfur and iron to form organic compounds. Macromolecules formed in chemosynthesis of this type and by combinations with other molecules ultimately led the way to the formation of molecules of deoxyribonucleic acid (DNA), the viruses and certain types of bacteria. Through millions of years the gradual changes involved

$$\text{Atoms} \rightarrow \text{Molecules} \rightarrow \text{Macromolecules} \rightarrow \text{Subcellular particles} \rightarrow \text{Cells}$$

The first fragments of life lived by consuming smaller molecules and were classified as heterotrophic organisms. But as the millenia passed certain cells acquired the ability to capture energy from the sunlight and used it to make food out of chemicals dissolved in the sea. These organisms took CO_2 dissolved in the ocean water and combined it with H_2O in the presence of light to produce a simple carbohydrate.

C. B. van Niel, of Stanford University, provided our first understanding of how bacteria can carry on photosynthesis without evolving oxygen. He proposed that light splits water, not carbon dioxide. Light acts on the water, making (H) and (OH) radicals.[1] The (H) then provided the hydrogen with the means to convert carbon dioxide to a carbohydrate. In plants the (OH) was split, making oxygen, but the earliest bacteria apparently lacked the enzymes necessary to catalyze this reaction. In bacteria the (OH) was recombined with (H) from another source to again form water. These were our first autotrophic organisms and may be considered our first plants.

At least as important to later organisms as the initial formation of sugar was the by-product formed in photosynthesis. This was, of course, oxygen. Free oxygen in the air, as we breathe it today, was not present before the green plants manufactured it. It did make possible life on land, as well as in water, and therefore contributed immensely to the evolution of life, for most of the evolution of higher plants and higher animals took place not in the sea, but on land.

All green plants utilize the energy of light in the making of food, and all other forms of life depend on plants directly or indirectly for food. Once solar energy strikes the chlorophyll molecule, it is transformed into chemical energy as food and can be used by us in the metabolic processes of growth, reproduction and work (Fig. 4).

[1] Radicals pass through reactions without change, so that the group (OH) acts much as if it were a single atom. In the following reaction

$$Fe + H_2SO_4 \longrightarrow FeSO_4 + H_2$$

the SO_4 group is the sulfate radical and is passed from one side of the equation to the other without change.

ANIMALS	1 steer	300 rabbits
TOTAL BODY WEIGHT	1,300 lb	1,300 lb
FOOD CONSUMPTION/DAY	16 2/3 lb	66 2/3
DURATION OF 1 TON OF HAY	120 days	30 days
HEAT LOSS/DAY	20,000 kcal	80,000 kcal
GAIN IN WEIGHT/DAY	1 lb	8 lb
GAIN FROM 1 TON OF FOOD	240 lb	240 lb

FIGURE 4 The efficiency of energy conversion. In this illustration the 300 rabbits and the single steer each weigh 1,300 lbs. and each group gained 240 lbs. by eating one ton of hay. However, the rabbits consumed the hay and gained weight in 30 days while it took the steer 120 days, making the rabbits four times as efficient as the steer in food production. (From Kleiber, M., *The Fire of Life,* John Wiley & Sons, New York, 1961.)

Radiant Energy

We usually think of energy as the ability to do work which includes chemical synthesis, action, luminescence, cellular maintenance, and a variety of other functions. The chemical energy derived from a food molecule produced in photosynthesis is converted to an action energy which the cell uses to rearrange the atoms of simple molecules into more complex ones in the building of protoplasm. Every transformation in the cell is associated with energy either in the conversion of it or in its utilization. These conversions result in a decrease of useful available energy, for some energy in the form of heat is always lost.

Energy, therefore, occurs in many forms, of which radioactivity is one. Radioactive materials occur naturally in the environment, but man has recently been adding to this natural radioactivity by artificially producing radioactive atoms on a considerable scale by nuclear fission and nuclear fusion. Since radiation has become so much a part of our lives in recent years it seems pertinent to discuss briefly what it is. The term refers to a wide range of modes of emission, propagation, and absorption of energy. In one way or another energy is always emitted or transferred in radiation. We have already discussed some aspects of cosmic radiation but there are other types. Specifically, they are alpha rays, beta rays, gamma rays, X rays and neutrons. All five give rise to a common phenomenon, **ionization,** when they interact with matter. Ionization is the removal of electrical charges from, or their addition to, electrically neutral atoms and molecules, which then become either negatively or positively charged. Molecules may be split in the process into separate fragments of either charge. These split electrically charged particles are called ions.

If crystals of NaCl are placed in water all of the ions, both positive and negative, become detached and wander about in the solution. If it is a very dilute solution, all ions may be completely free to act as independent particles; if the solution is more highly concentrated, oppositely charged ions seem to attract one another, resulting in a fraction of the total number becoming inactive and acting as neutral particles:

$$\mathrm{NaCl + H_2O \rightarrow Na^+ + Cl^- + (H^+ + OH^-)}.$$

Water, easily the best solvent known, has properties which make it peculiarly fit as part of the environment and part of the substance of life. Substances dissolved in water readily ionize because water molecules act as small magnets attracting the ions of the solute (in this case, Na and Cl), thereby weakening the attraction between them.

Ionization is of great importance to life. Though water may be considered the substrate in many of a cell's ionic reactions, it is also remarkably inert and innocuous. Most substances are unchanged by being dissolved in it. Therefore they may remain in solution in the cell until such time as they may be needed.

Atomic nuclei are relatively complex structures which form the core of atoms. They are made up of positively charged protons and electrically neutral neutrons, both of which are elementary particles of about the same mass as the hydrogen atom. The positively charged nucleus is surrounded by negatively charged electrons which move in orbitals around it. The charge of the electron is of the same magnitude as that of a proton but of opposite sign, so that the number of orbiting electrons in neutral atoms is equal to the number of protons in the nucleus (Fig. 5). These orbiting electrons participate in the formation of chemical bonds.

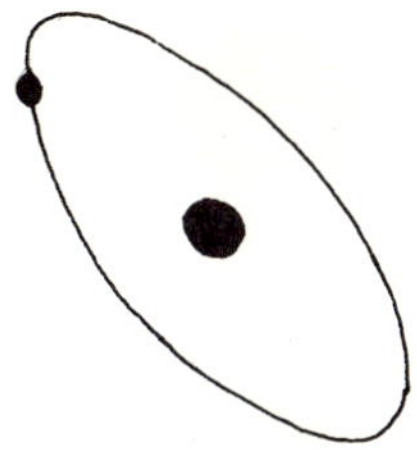

A

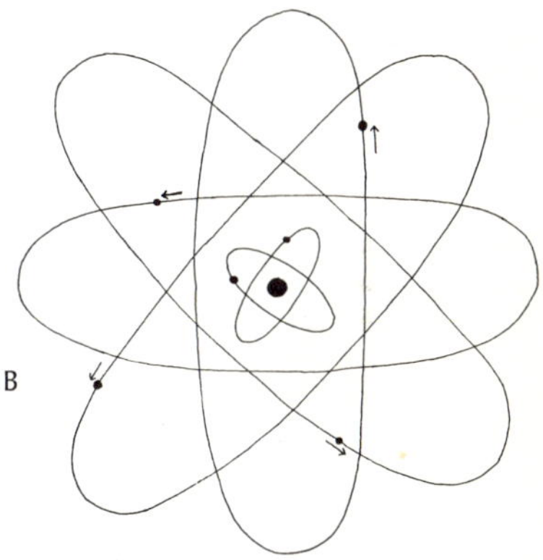

B

FIGURE 5 The atom is visualized as a tiny solar system with electrons traveling around a central nucleus in fixed circular or elliptical orbits. However, an atom may actually be more complex than this. A: hydrogen; B: carbon; C: oxygen.

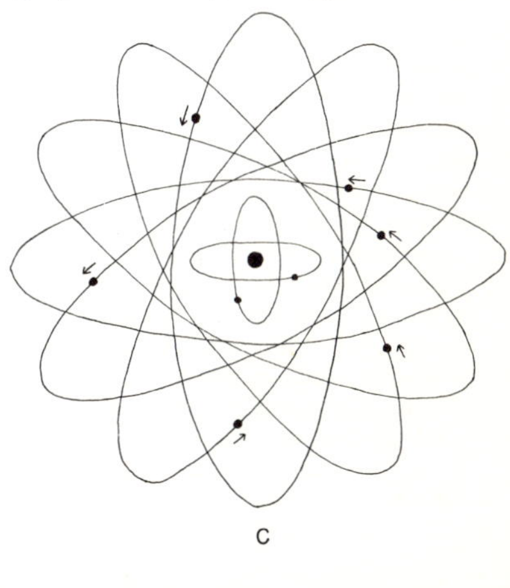

C

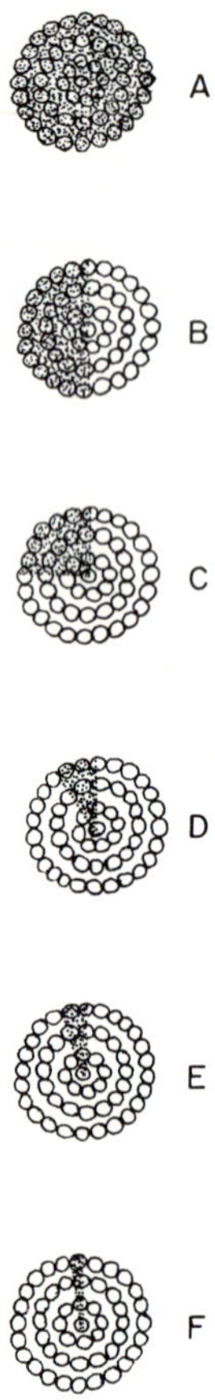

FIGURE 6 The age of once-living matter may be determined by measuring its content of radioactive Carbon-14, since all living things absorb some Carbon-14 from the air when living. This absorption ceases when the organism dies. Carbon-14 has a half-life of 5,568 years, which means that in this period half of the remaining carbon will decay into nitrogen. In the above diagrams the darkened areas indicate the amount of Carbon-14 remaining. If we started with A, in which the tissue has absorbed its Carbon-14 about 27,840 years ago, each succeeding illustration shows the amount remaining after each period of 5,568 years. (Adapted from *The Earth,* Life Nature Library, 1962.)

The chemical properties of atoms are determined by the electrons outside their nuclei. The most specific characteristic is that the electrons move around the nucleus. They are kept in definite orbital paths by electric attraction. The negatively charged electrons are attracted to the positively charged protons in the atomic nucleus. Electrons all carry negative electric charges which cause them to repel each other, but their attraction to the nucleus keeps them in orbit. In general, electrons of an atom are organized into a series of layers or shells, lying one inside the other around a nucleus. Those electrons orbiting in the outer shells are at a higher energy level than those orbiting in inner shells. An electron may jump into an outer shell if additional energy, such as heat or light, is supplied to the atom from outside. Such an electron is said to be in an **excited state.**

The number of protons defines the chemical element to which the atom belongs. A **nuclide** is a type of atom characterized by the number of protons and neutrons contained in its nucleus. In a given element, various nuclides with the same chemical properties may be identified that differ only in the number of neutrons and therefore in the mass of the nucleus. These are the **isotopes** of the element. For example, there are five isotopes of carbon (C^{11}, C^{13}, C^{14}, etc.) and three isotopes of hydrogen (H^1, H^2, H^3). All isotopes of a given element have the same number of electrons and these are chiefly responsible for the element's chemical properties; the isotopes of an element differ only in the neutrons in their nuclei.

With the development of the cyclotron and the nuclear reactor it has become possible to form new isotopes artificially. Some of these isotopes can be made radioactive and, with the aid of an instrument such as the Geiger counter, can be traced in their various pathways and chemical combinations in metabolism.

Nuclear disintegrations of unstable isotopes occur at different rates but with a certain probability per unit of time. The time required for 50 percent of the atoms of a nuclide to disintegrate is spoken of as the half-life and may vary from a fraction of a second to a billion years (Fig. 6). However, the absorption of radioactive substances by organisms is dependent more upon the chemical properties of the substance than on its rate of disintegration.

Types of Radiation

Alpha rays are positively charged particles (helium nuclei) emitted with definite and characteristic kinetic energy by nuclei of some radio-nuclides during disintegration. Alpha rays produce dense ionization in matter but their range, or penetration, is small —usually less than 0.1 mm in water or living tissues.

Beta rays are electrons emitted by nuclei of certain radioactive nuclides. They also produce ionization in the matter through which they pass but their range is greater than that of the alpha rays. Generally this range is no greater than 2 cm and none penetrate deeper than 8 cm in water or living tissue.

Gamma rays are electromagnetic radiation emitted by nuclei of some radioactive nuclides. They ionize matter indirectly through the ejection of high-speed electrons from the material in which they are absorbed. Gamma rays may be ejected to a considerable depth in matter. No definite range can be given since they penetrate any thickness of matter but with decreasing intensity.

X rays are also electromagnetic radiation and interact with matter producing biological effects in the same way as gamma rays. They differ from gamma rays only in that emission is extra-nuclear instead of nuclear. The X rays used for diagnostic medical procedures are less penetrating than gamma-rays from most radioactive nuclei. However, it is possible to generate X rays that are more penetrating than gamma rays from any radioactive nuclei.

Neutrons are constituents of atomic nuclei from which they are ejected during nuclear processes such as fission. Neutrons are uncharged and cannot produce ionization directly. Fast neutrons (of high energy potential) lose energy by collision with nuclei of light atoms, especially those of hydrogen. These nuclei recoil and, being charged, produce ions as they dissipate the energy transferred from the neutron. The transmission of energy from fast neutrons to recoil nuclei can take place at a considerable depth in tissue but, like gamma and X rays, fast neutrons have no definite range.

Cosmic rays of certain types reach the earth from outer space; they consist of a complex group of heavy particles with different energies of galactic and solar origin. The highly energetic fraction of primary radiation interacts with atoms present in the upper atmosphere giving rise to secondary cosmic radiation that is composed of particles and of electromagnetic radiation. Each component of secondary cosmic radiation produces ionization in its own characteristic manner. The low-energy fraction of primary cosmic radiation, trapped by the magnetic field of the earth, becomes part of the inner and outer belts that girdle the globe at two different altitudes.

One comforting thought about radiation from natural sources

is that it has been going on since the beginning of time without sudden and dramatic effects. Generally we think of natural radiation as arising from two sources, cosmic rays and naturally occurring radioactive materials. Such radiation remains relatively constant with time but it does vary from place to place.

The variation in composition of the cosmic rays will be of greater concern to those in space travel than for most of us on earth. The intensity of the cosmic ray exhibits a latitude effect which is due to the prevention, by the earth's magnetic field, of incoming primary particles below a certain critical minimum of energy from reaching the earth's atmosphere. This critical cut-off energy decreases with increasing latitude, with the result that the cosmic ray intensity at the geomagnetic equator is lower than at other latitudes.

Time variations of cosmic ray intensity result from a number of causes including solar flares, the eleven-year solar cycle, the twenty-seven-day lunar period, temperature changes in the upper layers of the atmosphere, barometric pressure changes and air fronts. Another variation associated with the sun is the so-called cosmic ray storms, which are of short duration, lasting from one day to a couple of weeks. These storms and the decreases in intensity which follow are correlated with the number and size of sun spot groups as well as the appearance of solar particles. They therefore follow the eleven-year solar cycle.

The naturally occurring radioactive materials are found throughout the earth's crust and may vary in gamma-ray dosage from something less than cosmic rays to values very much higher. These radioactive materials are the long-lived uranium (U^{238}), thorium (Th^{232}), radium (Ra^{226}), and others including the cosmic-ray-produced isotopes of carbon (C^{14}) and hydrogen (H^3). The natural radionuclides are more concentrated in granitic rocks than in basaltic rocks. Sedimentary rocks such as limestones and sandstones are low in radioactivity but some shales are more active, especially those containing organic matter. Marine deposits are in general more radioactive than non-marine deposits.

Radiation sources may give rise to external and internal irradiation of organisms. In the former radiation reaches the organism from sources outside, while in the latter radiation comes from radioactive materials taken into the organism by ingestion, inhalation, or injection. Alpha and beta radiations are usually more significant in internal exposure since radioactive substances may enter into the metabolism of the organism and become deposited in certain organs rather than being uniformly distributed throughout the tissues. Cosmic rays are predominantly an external source of radiation but naturally occurring radioactive materials may give rise to both external and internal irradiation.

Biological Effects of Radiation

An understanding of the effects of radiation on subcellular and cellular structures is a prerequisite to the understanding of radiation effects on whole organisms, since radiation damage occurs at the lowest level of cellular organization. It is through knowledge of the structure and mode of replication of the macromolecules and other components of the nucleic acids that new insight has been gained in the action of genes in controlling cell activities. Initial experiments indicate that the great majority of genetic changes induced by radiation experimentally are identical with those which occur in nature. However, it has been shown experimentally that radiation may cause a greater number of harmful mutations and, apparently, at a more rapid rate than that occurring naturally. To appreciate the full effect of radiation, each biological system will have to be studied independently. We are apparently on the right track when we think in terms of the sensitive molecule but the actions become more complicated as the organization level rises, and we must think in terms of the sensitive cell, tissue and organ as well.

While a detailed discussion of radioactive materials would be inappropriate here, the significance of radiation in biology is very great. Plants and animals may obtain natural radioactive substances from contaminated water, including public water supplies. This radioactive content may be reduced by filtering systems. Many of the natural radioactive substances in soil and water are absorbed by plants and transferred to man. Since it is known that absorption from natural radiation is much higher in some areas than others, studies of the plant and animal populations in these areas are likely to contribute considerably to the fund of biological knowledge concerning the effects of radioactive materials.

At present radiation is being used in insect control, through the sterilization of male screw worms and other insects. The U. S. Food and Drug Administration has cleared the process of irradiation for wheat and potatoes in human consumption. However, in 1966 Cornell University scientists discovered, when culturing normal carrot cells in irradiated coconut milk, that chromosomes in the carrot cells were damaged although the carrot tissue itself had not been irradiated. They also reported that chromosomes in the roots of vetch and *Tradescantia* were damaged and root growth was inhibited when the plants were grown in a culture medium of irradiated sucrose. The fact that normal growth was inhibited and damaged or altered chromosomes were found indicates that we must use greater care in the irradiation of foodstuffs, particularly those containing sugar.

Some radio-nuclides are produced by the explosion of nuclear weapons and inevitably enter into the food chains through soil, water, and plants. Two of these are cesium and strontium. Both readily enter plants through the absorptive surfaces. When contaminated grass is eaten by cows, the radioactive substances are transferred to the milk and meat—and eventually to humans.

One interesting case was observed by the writer where strontium-90 (Sr^{90}) was absorbed by large trees surrounding an atomic waste disposal pit near one of our atomic energy plants. The pit had been filled with radioactive wastes and covered over for a number of years. Gradually the Sr^{90} moved through the soil by capillarity towards the periphery of the pit, the trees grew larger, with their roots growing toward the pit, and contact was made. The Sr^{90} was absorbed by the roots and transported to the upper portion of the stems, branches, and leaves. The radioactive material was detected by Geiger counters and photographic plates (Fig. 7). The interesting thing about this was that the Sr^{90} moved up the trees and into the branches and leaves only on the pit side of the tree, while the opposite side remained free of any radioactive substances. This accentuates the well-known fact that there is very little lateral conduction in plants. The roots on one side were able to absorb materials on that side for the parts above them, while the roots on the opposite side of the plant were not in Sr^{90} contaminated soil and had none to absorb and conduct. Other effects of radiation and radioactive materials will be taken up in later chapters.

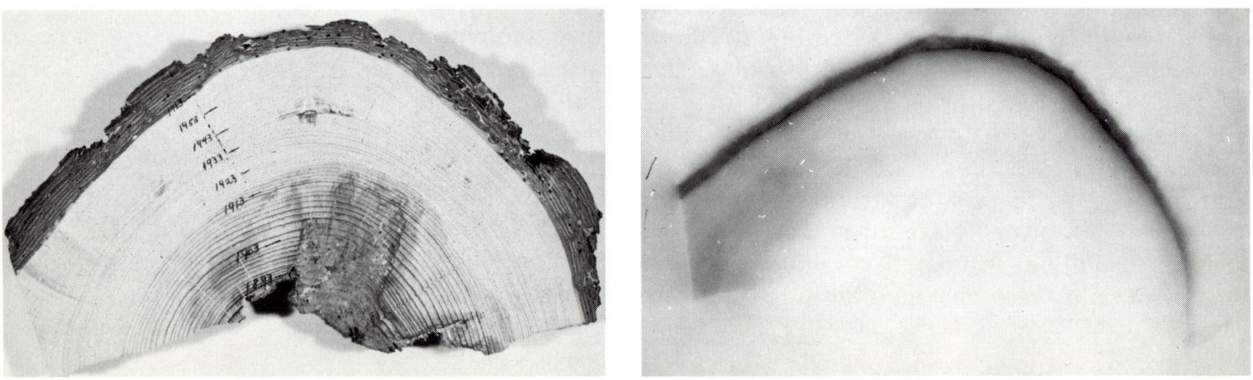

FIGURE 7 Section of western yellow pine tree which has taken up strontium-90 from the soil. A: Note variations in width of annual rings and age of the tree. B: A photographic plate had been placed over the cut stem and exposed to the radioactive strontium. Note that the greatest amount of radioactivity is in the inner bark and the outer xylem rings, although the clouded portion of the plate shows that approximately the outer thirty annual rings have some activity. (Courtesy Los Alamos Biological Laboratories.)

Chemical Energy

Of all the forms of energy—heat, light, electrical, nuclear, and chemical—the last is the most useful and vital to man. It is an extremely versatile form and can be readily controlled whether in the form of an oxidative process or an explosive force.

Structure of the Atom

Chemical energy is created when matter is broken by the separation of its complex bonds. Atoms are formed like tiny solar systems. In the center is the nucleus, which might be compared to the sun, and whirling around the nucleus, which carries a positive charge, are negatively charged electrons. Some atoms are relatively simple, like hydrogen, which has one electron circulating by itself around the nucleus. Helium has two orbiting electrons.

The numbers of protons, neutrons and electrons in an atom account for the differences in the elements which they form. An element, therefore, is composed of similar atoms. Atoms are ordinarily electrically neutral since the number of protons is equal to the number of electrons surrounding them. However, in some reactions electrons may be removed from or added to an atom, giving it a negative or a positive charge. An atom in which such an imbalance of protons to electrons occurs is called an **ion.** Often there are variations of an element. An element with the same nuclear charge but with different numbers of neutrons in the nucleus is an **isotope.** Usually they have the same name, as ${}_6C^{12}$ and ${}_6C^{14}$, where the name carbon is retained, but in hydrogen (which has no neutrons), one isotope is called *deuterium* (one neutron) and another is **tritium** (two neutrons).

The number of protons in the nucleus of the atom is called the **atomic number** of the element. Since the hydrogen atom has but one proton and one electron its atomic number is one; carbon has a nucleus of six protons, six neutrons, and six electrons. Its atomic number is therefore 6.

The number of protons plus the number of neutrons present in an atomic nucleus determines the **atomic mass.** This mass is the same for all protons and is given the arbitrary value of 1. A mass of 1 also characterizes a neutron. However, the mass of an electron is very much less than 1; specifically it is 1,836 times less, which makes its weight practically nil; consequently the total mass of an atom is concentrated almost entirely in its nucleus. Since hydrogen has but one proton, a single electron on the outside, and no neutrons, the atomic mass of hydrogen is 1. At the other end of the atomic scale is uranium. Its nucleus contains 92 protons and 146 neutrons. The atomic mass of uranium is therefore 238, the sum of the protons and neutrons present. Between these extremes of 1 and 238 we find the atomic masses of all other elements according to the specific numbers of protons and neutrons present in the atomic nuclei. Many elements in nature consist largely of one isotope, and the atomic weights of these elements can be more readily fixed in proportion to the number of its protons and neutrons.

Another characteristic of an atom is its **atomic weight.** This is defined as the sum of the weights of its protons and neutrons. Given the atomic number of an atom, the atomic weight minus the atomic number equals the number of neutrons in the nucleus (see Table 2-2).

Element	Symbol	Atomic Number	Atomic Weight
	Major requirements		
Carbon	C	6	12.000
Hydrogen	H	1	1.008
Oxygen	O	8	15.999
Phosphorus	P	15	30.974
Potassium	K	19	29.102
Nitrogen	N	7	14.007
Sulfur	S	16	32.064
Calcium	Ca	20	40.080
Iron	Fe	26	55.847
Magnesium	Mg	12	24.312
	Minor requirements		
Boron	B	5	10.811
Manganese	Mn	25	54.938
Copper	Cu	29	63.540
Zinc	Zn	30	65.370
Cobalt	Co	27	58.933
Molybdenum	Mo	42	95.940
Sodium	Na	11	22.990
Iodine	I	53	126.904
Selenium	Se	34	78.960

Table 2–2 Elements known to be essential for plant growth. Those listed as major are used in fairly high quantity, while the minor elements are used in smaller amounts and in some plants rarely, if ever.

Isotopic atoms differ from normal ones only in physical properties, that is, in mass, not in chemical properties. In naturally occurring elements a small percentage of the atoms present are isotopic. If a naturally occurring sample of an element is weighed, the weight will not be an integral whole number but a slight fraction more. The measured atomic weight of hydrogen is 1.008; that of carbon is 12.011. The added fractions are due to heavier isotopes in the sample. Isotopes are useful in biological studies, for if both isotopic and normal atoms are used in a chemical process, the different mass of the isotopic atoms provides an identifying label through which such atoms may be identified from the normal ones.

The chemical properties of an atom are determined by the number and arrangement of the electrons in its outermost shell. Some atoms are very stable and do not readily react with other atoms. These are atoms in which the outermost electron shell is said to be complete or closed. Elements of helium, neon and a few others are of this type. Atoms of other elements possess incomplete or open outer shells, are less stable chemically, and consequently are more reactive.

When an atom enters into a chemical reaction, its nucleus retains its atomic weight and does not undergo change, but the electrons of the outer shell are increased or decreased in number. In some cases they may be shared with another atom. Chemical reactions involve either the transfer of electrons from one atom to another or the sharing of electrons by two atoms.

Bonds Since it is usually inconvenient to represent the structure of each atom by a diagram of its nucleus and orbiting electrons, the electrons are represented by dots placed around the symbol of the element. The dots represent only the number of electrons in the outer ring of the atom since it is only these electrons which normally take part in a chemical action. Therefore the structural formula for carbon dioxide is $:\underset{\cdot\cdot}{\ddot{O}}::C::\underset{\cdot\cdot}{\ddot{O}}:$ A molecule of oxygen which consists of two atoms has a formula of O : : O. More commonly dashes are used in place of the pair of electrons shown by the dots, as O=O or O=C=O. In either case, for the molecules of O_2 or CO_2, the connecting electrons or dashes are spoken of as the bond.

A sharing of a single pair of electrons between two atoms is illustrated by a single bond; if two or three pairs of electrons are shared between two atoms the linkage is a double or triple bond and is illustrated by double or triple connecting dashes:

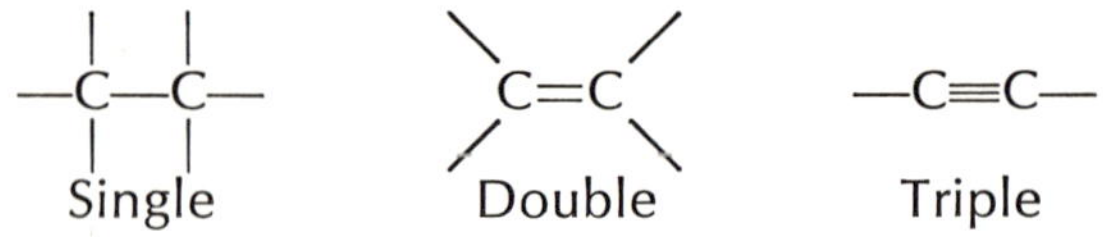

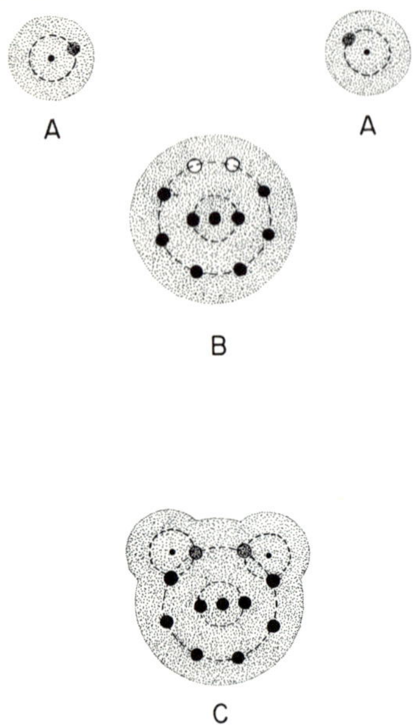

FIGURE 8 The formation of a molecule of water. The two hydrogen atoms (A) each need one more electron to be stable. The larger oxygen atom (B), with six electrons in its outer ring, needs two more to fill its orbit. When the three unstable atoms pool their electrons the result is a stable molecule of water. (From *Water,* Life Science Library, 1966.)

Valence The number of hydrogen atoms one atom of an element can hold in combination is called the valence of that element. Oxygen (O_2), for example, is a bivalent atom since it combines with two hydrogen atoms, H—O—H; nitrogen is a trivalent element and may be written NH_3 or

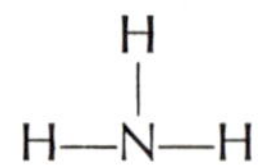

In many cases an element does not form compounds with hydrogen, but since all elements that form compounds of any kind form oxides we can infer the valence of an element from the formula of its oxide. For example, sulfur forms the oxides SO_2 and SO_3. Similarly, other elements have two or more oxides, and thus have two or more valences.

Atoms also display certain physical properties which may be observed in a study of electron behavior. For example, it is the electron which conveys electric currents along a wire. It has also been determined that when electrons move from outer orbit to the inner orbit there is a measurable emission of light, and when electrons move from an inner orbital to an outer orbital there is an absorption of light into the nucleus which contains all the stored energy of the atom. It was observed in the latter part of the nineteenth century that certain elements were able to produce an image on a photographic plate and that this exposure could take place through thick layers of paper and other objects. These are the so-called radioactive elements described earlier.

The Formation of Molecules

The next level of matter is the molecule created when atoms are tightly bonded together. When an atom of oxygen joins with two hydrogen atoms, forming a molecule of water (Fig. 8), the oxygen and hydrogen will lose their identities as gases. Molecules of water will in turn collect together and form raindrops, hail and snowflakes. Atoms of different types are constantly combining to form molecules of entirely different identity.

According to a widely accepted theory, in order to react molecules must collide with one another so that an electronic interaction may occur. This is known as the **collision theory of chemical reactions** and applies to both living and nonliving systems. According to this theory factors which affect the rates of chemical reactions are:

1. The concentrations of the substances involved in the reactions and therefore the likelihood of greater or fewer collisions.
2. The affinity of the reacting substances for each other from an energy viewpoint and therefore increased or decreased reaction.
3. The effect of temperature in promoting reaction.
4. The presence of catalysts to bring about a union.

Nearly all chemical reactions occurring in living cells, and therefore all metabolic processes, are controlled by specific catalysts, the **enzymes.** Their function is not only to bring about the reaction but to accelerate the process greatly.

Energy relationships demonstrated by all chemical reactions exhibit two major properties: (1) The total weight of the products is equal to the total weight of the reactant. (2) Energy is either liberated or absorbed. In most cases at least part of the energy is released as heat.

In many biological systems the reaction cannot occur without free energy to promote the reaction. This works both ways, as we shall see in the discussion of photosynthesis and in other processes. The reaction liberates energy which is picked up in a chemical bond, which then serves as an energy source for further chemical activity, creating self-propelling cycles of chemical action which liberate additional energy.

By far the most important compounds to man, as an initial source of chemical energy, are those containing carbon atoms. All organic substances, carbohydrates, fats, and proteins, whether of plant or animal origin, contain carbon. These substances are the end products of the plant metabolic process in which light energy is converted to chemical energy. Termed **photosynthesis,** this process takes place when water and carbon dioxide, with the aid of chlorophyll and light, are combined, forming a simple carbohydrate molecule, $(CH_2O)_n$, which is then converted into a monosaccharide sugar, $(C_6H_{12}O_6)$. The entire process is described in detail in Chapter 7, but one of the remarkable things about photosynthesis is that the plant is able to use light energy to split a water molecule into hydrogen and oxygen, which are then used in the synthesis of food. The green plant has chlorophyll, which becomes greatly activated when exposed to sunlight. The radiant energy causes the chlorophyll to vibrate rapidly, apparently, acting as a stimulus breaking the water molecule. The simple sugars which are formed are then built into more complex carbohydrates, fats and proteins.

Double sugars ($C_{12}H_{22}O_{11}$) are made from two simple sugars ($C_6H_{12}O_6 + C_6H_{12}O_6$) by the removal of a molecule of H_2O when they are joined. Sucrose is a combination of one molecule of glucose and one of fructose. Another double sugar has the same chemical formula as sucrose but differs in the arrangement of its constituent atoms. Maltose is composed of two molecules of glucose; another double sugar, lactose, is made of one molecule of glucose and one of galactose. The largest carbohydrate molecules are starches and celluloses, which are composed of a large number of simple sugars joined together either in long chains or in branched chains. Fats are also composed of carbon, hydrogen and oxygen but they have much less oxygen in proportion to the carbon and hydrogen. Two carotenoid fats are $C_{40}H_{56}O_2$ and $C_{40}H_{56}$.

Proteins are composed of amino acids, which contain one element not found in either carbohydrates or fats. This substance is nitrogen (N). In every amino acid there is an amino group (NH_2) and a carboxyl group (COOH). There are about seventy amino acids known, but only about twenty different kinds are used in the construction of proteins.

The amino acids (one of which is glycine, NH_2CH_2COOH) are the building blocks of proteins. If you consider each of the twenty different amino acids as a letter in the alphabet and then think of all the words that could be created out of twenty letters in the alphabet, with their thousands of combinations, you can see how twenty different amino acids could build millions of different proteins. One important protein in plants is chlorophyll; the empirical formula for a molecule of chlorophyll *a* is $C_{55}H_{72}O_5N_4Mg$; the structural formula for this molecule is depicted in Chapter 7, page 105.

In this chapter we have considered energy and matter and their use by living things. While energy may not be organic in its origin, it is one of the intangible substances found in all life. Reactions of various types are always possible if sufficient energy is available to trigger the action. Chemical changes involve atomic or molecular rearrangements resulting in the building of numerous compounds as well as their breakdown. Matter forms the fundamental construction units of all inorganic and organic constituents. An organism is able to store and use energy in the organization of these constituents into something living. Once the organism loses this ability it is no longer living.

Supplementary Reading

Augenstein, L. G., R. Mason and H. Quastler, *Advances in Radiation Biology,* Vol I. Academic Press, New York, 1964.

Beiser, Arthur (editor), *The Earth.* Life Nature Library. Time Incorporated, New York, 1962.

Bonner, James, *The Molecular Biology of Development.* Oxford University Press, New York, 1965.

Grobstein, Clifford, *The Strategy of Life.* W. H. Freeman and Company, San Francisco, 1965.

Johnson, Willis H., and W. C. Steere, *This is Life,* Essays in Modern Biology. Holt, Rinehart and Winston, New York, 1962.

McElroy, W. D., *Cell Physiology and Biochemistry,* 2nd ed. Prentice-Hall, Inc., Englewood Cliffs, New Jersey, 1964.

Oparin, A. I., *The Origin of Life,* 3rd ed. Academic Press, New York, 1957.

Radio-Isotopes in Soil–Plant Nutrition Studies. International Atomic Energy Agency, Vienna, 1962.

Questions for Discussion

1. What conditions would one expect to find on another planet for life to exist there?
2. Would you expect or not expect life to be found on another planet? Why?
3. Explain why the first organisms were probably anaerobic rather than aerobic.
4. Explain how atomic energy may well have been the great energizing force in the creation of life and might still become life's destroyer.

5. What are the different types of energy and how do they differ from each other?
6. What are the factors, both organic and inorganic, which prevent life from being completely destroyed by gamma radiation?
7. The planets are essentially spheres in shape. What does this suggest to you about how they were formed? Why not discs or cubes?
8. Classify and describe the different types and sources of energy.
9. What do we mean by nuclear disintegrations?
10. How has the study of radiation been of great benefit to the study of biology?
11. How are the energy needs of the United States being met today in comparison with fifty years ago and before?
12. What is meant by the term "sensitive molecule"? How is it related to mutations?
13. Discuss the economic importance and application of irradiation.
14. Discuss the transformation of energy.
15. What do we mean by "levels of matter"?
16. Why are plants likely to be damaged less by excessive irradiation than animals?

3

The Chemical Nature of Protoplasm

Protoplasm is the living substance of all organisms. Its component ions, molecules and colloidal particles are organized into a great many morphological bodies and functional systems. The individual structures are independently organized into units of specific action. These units in turn make up successively larger and more complicated bodies collectively organized as the functional unit of life—the cell.

Almost all of the elements (105 in all) which make up the earth are present in protoplasm. Many of these elements apparently have no specific function in protoplasm and probably are not used. The most important elements in living systems are those listed in Chapter 2.

The elements found in protoplasm may be present as independent ions. More commonly they are organized into compounds of extremely complex molecular structure. These compounds are usually of three major types: (1) water, (2) inorganic salts, and (3) organic compounds.

Water, the universal solvent, is the simplest but by no means the least important. Most cells, by weight, are at least three-fourths H_2O. Water dissolves more different substances than any other solvent and in the cell it forms a protective medium for the dispersion of numerous organelles. Water also protects the living material against sudden thermal changes, dissolves the waste products of metabolism and assists in their removal. All substances which enter or leave a cell do so with a certain dependence on water. In fact, water is so fundamental in the life processes it is amazing that all organisms are not still aquatic creatures. But many have become terrestrial and have the freedom of this type of environment in spite of the fact that their cells and their physiological activities have remained aquatic.

Solutions, Suspensions and Colloids

In the cell, some of the macromolecules form the insoluble components while others are dissolved in the cellular fluids. The properties of the cellular protoplasm depend on the kinds and quantities of the substances present. Mixture of a substance with water may form a solution, a suspension, or a colloid depending on the size of the dispersed particles. Water may have ions or molecules of a substance dissolved in it, in which case it is a **solution.** In chemical parlance the water is termed a **solvent** and the dissolved substance is a **solute.** Most alkalis, acids and sugars form true solutions when added to water.

Some mixtures have the particles formed from the aggregates of large numbers of molecules suspended in water. If the particles are very large the mixture is termed a **suspension** and must be constantly stirred to keep the particles from settling to the bottom. If the particles are too small to settle out but too big to form a true solution the mixture is a **colloidal** suspension. When the particles dispersed in the liquid are solid we term it a **suspensoid;** if it is a colloid in which a liquid is dispersed within a liquid, such as droplets of oil in water, it is an **emulsoid.** Cream, mayonnaise, and soap are mostly emulsoids, while homogenized milk is both suspensoid and emulsoid.

A colloid usually retains its consistency not only because of the size of the particles but also because the particles usually have the same electric charge and therefore tend to repel each other, keeping the particles uniformly dispersed. Large particles are unaffected by these weak forces and fall to the bottom of the container but particles of colloidal size (1/1000 to 1/10 micron in diameter) may be pushed back and forth or up and down against the

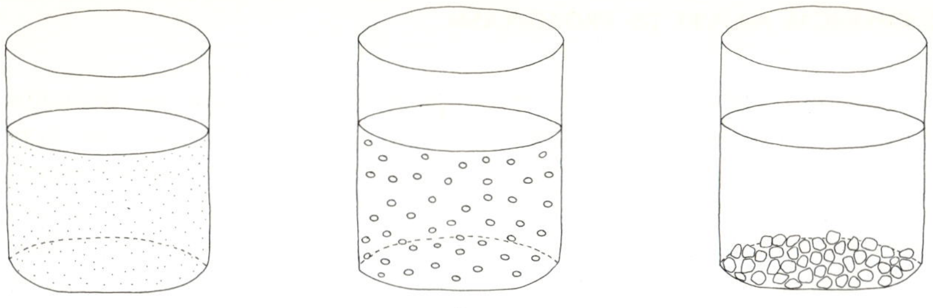

FIGURE 1 Diagrammatic illustration of (A) a solution, (B) a colloid, and (C) a suspension. In A the substance is completely dissolved and uniformly distributed; in B the particles are small enough to remain uniformly distributed by molecular attraction; in C the particles will remain suspended only when the water is agitated, otherwise they settle to the bottom of the container.

forces of gravity. This random movement of suspended particles is termed **Brownian motion** and may be easily observed under the microscope. The movement is due to the intrinsic energy of the particle. Those molecules within the fluid move more freely than those on the surface because surface forces tend to exert an inhibiting influence.

Sols and Gels

Colloids in a liquid condition are also often termed **sols,** but some may form a solid or semisolid and are considered **gels.** If we take a package of gelatin and pour its contents into a bowl, then add the recommended amount of hot water, we seem to have a solution. If we then allow the mixture to cool it becomes a solid. In this case the powdered particles have absorbed the water; the water is the dispersed phase and the gelatin is the continuous phase. If the solid gelatin is now reheated it changes back to a sol. In other words, this is a colloid that becomes a sol when water surrounds the gelatin particles and a gel when the gelatin particles surround the water.

Some colloids may be changed to another state by chemicals instead of heat. Others, like whipping cream, may be changed from an emulsoid to a gel by simply beating it. In some cases, like the gelatin, the colloidal system is reversible but in others, like egg white, it is not. Heat will change egg white from its sol state into a gel but it cannot be changed back to a sol state again.

Other Properties of Colloids

The fact that protoplasm is essentially a colloid makes it a highly efficient mixture for the many chemical activities constantly going on in the cell. Many chemical reactions occur only on a surface and a colloidal system provides this type of medium for the rapid reactions associated with metabolism.

Colloidal systems tend to form boundary membranes between the colloid and a different medium such as air, water, a solid surface, or even a different colloid. This boundary is termed an **interface** and is due to the concentration of particles in a layer or layers where the colloid and the other medium are in contact. The skin-like surface of boiled milk and the surface of a standing custard are interfacial membranes.

Colloids also age. Young, fresh colloidal systems have particles enveloped by **bound** water. The water molecules are adsorbed to the particle surfaces by molecular attraction. Gradually the binding capacity or molecular attraction diminishes and some of the water is lost. The colloid begins to contract as the result of the exudation of water. This may be observed in mayonnaise or mustard left standing open. Action similar to this occurs in living systems of protoplasm and may be a factor contributing to the aging of organisms.

Diffusion and Osmosis

In every living cell there is continuous movement of molecules. Ultimately the molecules in such an environment become evenly distributed, for there is greater movement of molecules from areas of greater concentration to those of lesser concentration than in the opposite direction. This movement is **diffusion.** After even distribution is achieved the molecules continue to move in opposite directions, but now at an equal rate, thus retaining a status of equilibrium.

If we take a small crystal of readily soluble dye like crystal violet and place it in the bottom of a beaker (Fig. 2) we soon note

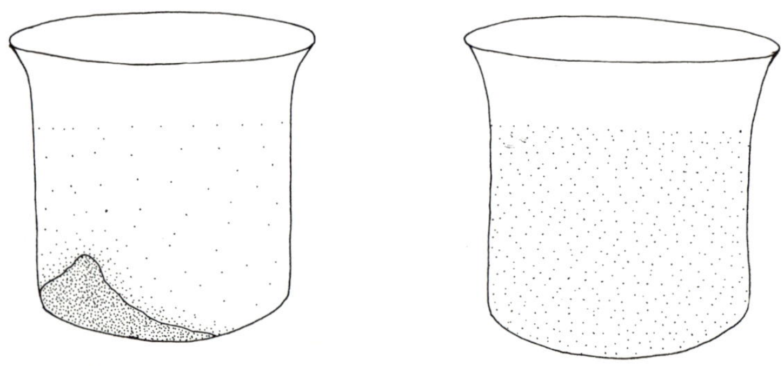

FIGURE 2 A demonstration of diffusion. The molecules of the crystalline material gradually move throughout the beaker on the left as they become dissolved in the water. In the beaker on the right the molecules have all reached equilibrium.

the movement of violet from the crystal towards areas in the water which are colorless. If the container is allowed to remain still, eventually the water will be uniformly colored and the crystal will have disappeared. This is a typical illustration of diffusion.

If a permeable membrane is placed between the area containing the dye crystal and the remainder of the water in the beaker, the dye molecules will pass just as easily from one region to another. This is also diffusion. If molecules of another substance are present which cannot penetrate the membrane, then we have either an impermeable or a differentially permeable membrane (Fig. 3). If water molecules and some substances in solution can penetrate the membrane but others cannot, we speak of the diffusion as **osmosis** (Fig. 4). Living membranes, such as the plasma membrane of the cell, are differentially or selectively permeable, for not all substances can pass through the membrane nor can all substances that penetrate do so at the same rate.

A fairly complete definition of osmosis would be to say that it is the movement of substances in solution from regions of higher concentration to those of lesser concentration through a differentially permeable membrane, each substance moving independently of any other. Assuming that water is in equal concentration on either side of the membrane, there would be an equal bombardment of water molecules on either side with an equal penetration of these molecules to maintain equilibrium. If a cube of starch is placed on one side we will note that the starch remained immobile. Although we could say that the starch is in greater concentration on one side than on the other, the starch does not penetrate because it is insoluble and the definition states that substances must be in solution. Since the starch is insoluble it has not changed the molecular concentration of the solvent and is for all practical purposes a rock lying in the bottom of the beaker.

The situation can be changed by placing a little of the enzyme **diastase** in the starch side of the beaker. Diastase digests the starch, changing it to sugar, which is soluble. Now there is a greater concentration of sugar on one side of the membrane than on the other and consequently there is a bombardment of sugar molecules on the membrane. At first a greater number of sugar molecules are moving from the region of greater concentration to the other side, but eventually equilibrum of sugar and water molecules is reached and the solution is just as sweet on one side as on the other. If we could convert the sugar back to starch, as a plant can do, we would find that it would then be insoluble. It would be equally distributed on both sides of the membrane, but would no longer participate in the osmotic concentration of the solution.

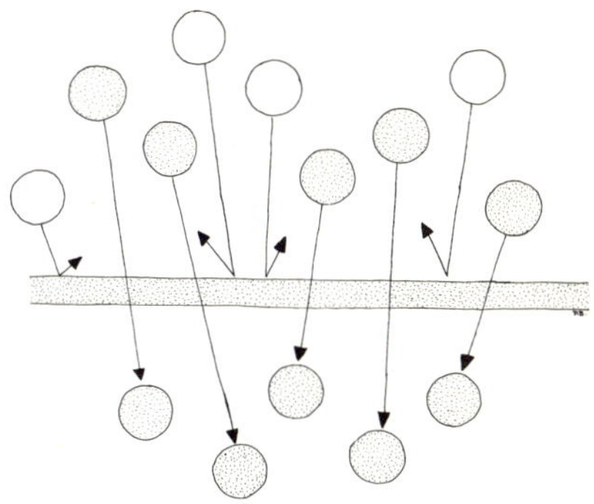

FIGURE 3 Diagram of a selectively permeable membrane. One type of ion passes through readily while the other type does not.

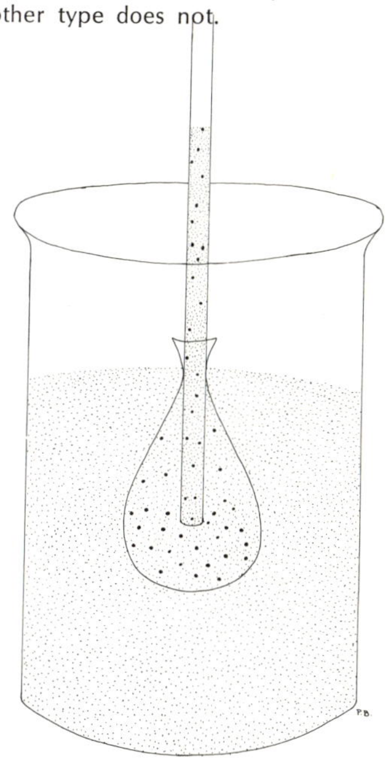

FIGURE 4 A simple demonstration of osmosis. The concentrated substance, such as a salt or molasses, in a membranaceous bag is forced up the tube along with some water which comes in from outside the bag. Water is in greater concentration outside the bag and moves into the bag faster than the molasses can osmose through. If left alone for a long time equilibrium of both substances inside and outside the membrane would be attained.

Since our definition includes the statement that each substance moves independently of any other we might add salt in the form of sodium chloride (NaCl) to the one side which contains the sugar in solution. Assuming that we had 5 percent glucose and 3 percent salt on one side, eventually we would reach equilibrium when each side had 2.5 percent glucose and 1.5 percent salt. The osmotic movement or diffusion of such substances in water, with few if any opposing forces, is spoken of as **passive transport.**

Theories of Osmosis

Several theories have been proposed in explanation of osmosis, any one of which may operate under certain conditions and at certain times.

The **sieve** theory was proposed a good many years ago. Although it could not be proved, its proponents suggested that cell membranes were permeable to those substances which had molecules small enough to penetrate tiny pores in the membranes. Considerable evidence has been found to support this theory through the use of the electron microscope, which shows that the membranes are unequally thickened and have certain sieve-like qualities. However, the pores which appear to be present are so small that little more than water could get through, making it necessary for us to look for other mechanisms to explain osmosis.

A **solution** theory was also suggested which stated that those substances entering and passing through a membrane had to be soluble to the membrane. This theory is sometimes explained on the basis of **enzyme** activity in which the membrane contains enzyme molecules capable of placing in solution certain substances that are insoluble normally. Once the substances have been made soluble by the enzyme they pass through and then may go out of solution on the other side. The solution theorists often accept the enzyme "theory" as a necessary supplement to their explanation of osmosis.

An **electrical** theory has been accepted by most plant physiologists. This may be considered a separate theory, but more commonly it is used also in explanation of the theories suggested above. The assumption is made that an electrical charge is responsible for the movement of different ions (Fig. 5). If within a cell we have a higher concentration of positively charged ions while on the outside there is a greater concentration of negatively charged ions, an electrical potential gradient is established and there will be a movement of molecules across the membrane, eventually establishing equilibrium.

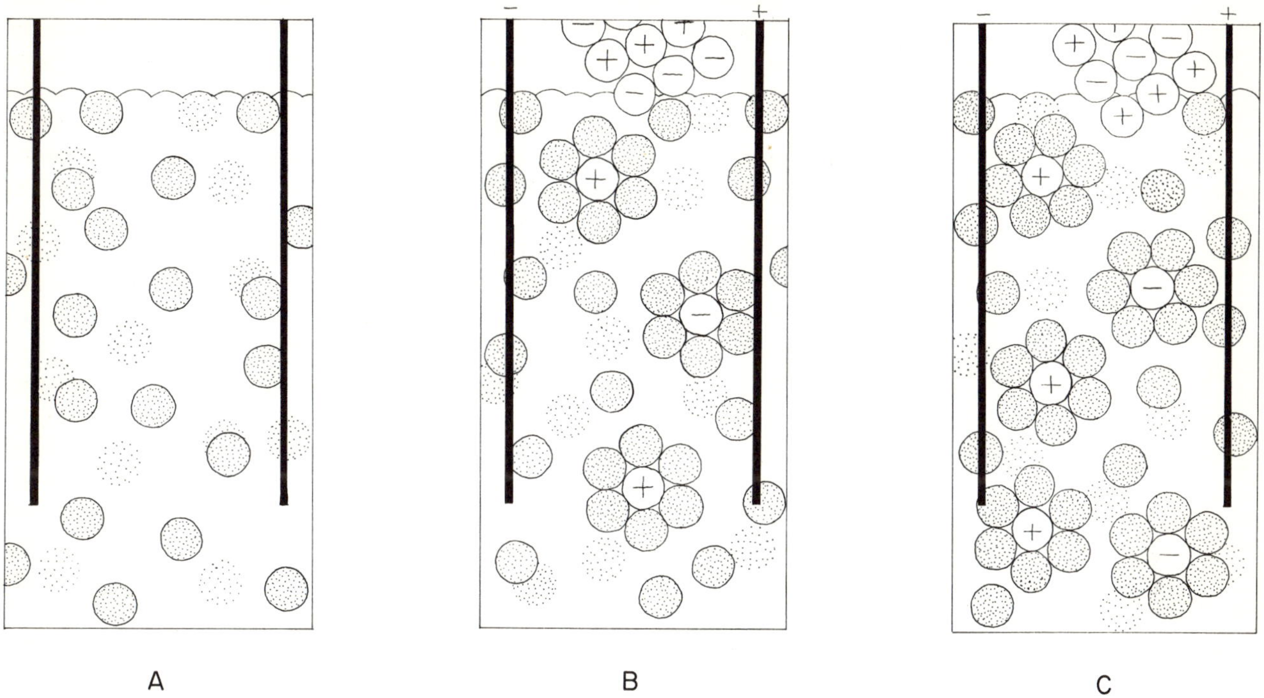

FIGURE 5 Diffusion of ions in solution. In A there is only pure water. In B salt crystals of sodium (positively charged ions) and chlorine (negatively charged ions) are introduced into the water. The negative ions move to the positively charged electrode and the positively charged sodium ions move to the negative terminal. As more salt is added (C), more ions are set in motion. (Adapted from *Water,* Life Science Library, 1966.)

Osmosis in Living Organisms

In the operation of any living cell or osmotic system, true equilibrium of all substances would result in a virtual status quo, but because of the metabolic processes, photosynthesis, respiration, digestion, assimilation, and others, there is a continuous change in the chemical contents of the cells with a continuous movement of at least some substances first in one direction then in another. These changes in the chemical contents and osmotic concentrations keep the organism's cells in an active state and make possible the various life processes.

It has been quite obvious for a long time that the plasma membrane is far from static and that as a living selective structure it controls the direction of flow of compounds. The membrane has the capacity to be permeable to certain compounds from a less concentrated solution outside to a more concentrated solution inside, thus moving against the concentration gradient. Root hair cells, for example, may have a much higher sugar concentration than the surrounding soil, but very little of the sugar passes through the membrane into the soil. The root cells may also have an ionic concentration of potassium higher than that in the soil, yet the cell takes in potassium against the concentration gradient. Other compounds may be selectively passed through the membrane in a similar manner. It is assumed that energy is required to transport materials across the membrane, particularly against the concentration gradient. The mechanism required is an energetic coupling which apparently derives its supply of energy from adenosine triphosphate (ATP). This type of movement involving an energetic coupling system is termed **active transport.**

The theory suggested for energetic coupling transfer proposes that a carrier specific for a certain substance will pick up the compound on one side of the membrane, move it through, and leave it on the other side. The carrier is then free to move back through the membrane again and is ready to pick up another molecule of the same substance. It is unknown what this carrier is, but the affinities of certain substances are sufficiently understood to offer an explanation. Potassium (K) may be coupled with ATP, which then moves across the membrane, releases the potassium, and may move back. It is also suggested that the ATP + K is attacked by an enzyme, adenosine triphosphatase, after it has crossed through the membrane, causing the ATP to lose part of its phosphorus and changing it to adenosine diphosphate (ADP). Once the potassium is freed, the ADP is rephosphorylated by the addition of phosphorus, and the ATP is restored and ready to carry on its transport duties.

Inorganic Salts

All cells contain inorganic salts, such as calcium carbonate or sodium chloride, in limited amounts. Their role in plant and animal metabolism has been well known for many years. Chlorophyll could not be formed if magnesium were not present to become part of the molecule of this important pigment. Similarly **hemoglobin** (a combination of an iron containing a porphyrin termed *heme* and a protein, *globin*) in our own bodies loses its characteristics and our normally healthy red color pales if iron is lacking. Curiously, plants lose their green color and become pale (in plants we call it **chlorosis**) if iron is unavailable. In animals iron is part of the hemoglobin molecule; in plants it acts as a catalyst in chlorophyll formation. In either case, whether it be anemia or chlorosis, the doctor often recommends iron as a medication and it usually brings about a favorable response quickly.

Many of us have experienced extreme fatigue, weakness, and even fainting due to profuse sweating in extremely hot weather. Generally this is due to the loss of salt (NaCl) with the water of perspiration. The simple administration of salt tablets will correct the situation and the body is restored to a normal, well regulated metabolism within a very short time. Plants do not show such dramatic responses to simple medication but they do respond to the same deficiencies and excesses of inorganic salts. Their reactions are simply slower because they do not have a vascular pumping mechanism such as we have to channel all substances in the body fluids throughout the body in a few minutes. Although inorganic salts are present within the protoplasm in minute amounts, they are indispensable to the normal activities of the cell and the whole organism.

Organic Compounds

All living systems must be able to utilize foods and convert them into new protoplasm in the cell. Some of these compounds are broken down to provide the energy for the life processes, others become storehouses of energy for future use, and others are immediately incorporated into the structure of the cell. These organic compounds are classified as **carbohydrates, fats,** and **proteins.** With water and certain inorganic salts these compounds form enzymes, hormones, vitamins, and nucleic acids necessary for the vital activities of the cell, integrating its function in the whole organism.

Carbohydrates

Carbohydrates are the first foods formed in plant cells. They are also the chief storage product and the major constituent of plant cell walls. Molecules of carbohydrates are composed of carbon chains in which every carbon atom maintains a bond with a

hydrogen atom and a hydroxyl (OH) group except those on one end. In a monosaccharide one of the carbon atoms forms a carbonyl (CO) group. If this is one of the two terminal carbon atoms, the monosaccharide is an aldehyde and is called an aldose. If the carbonyl group is not in a terminal position the monosaccharide is a ketone and is called a ketose.
The formulae for two aldoses are:

Glucose

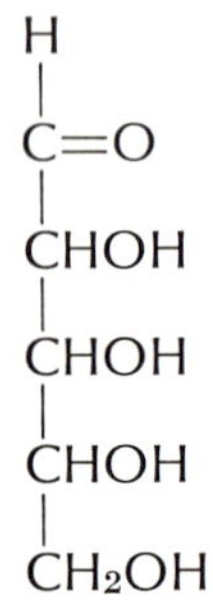

Ribose

The formulae for two ketoses are:

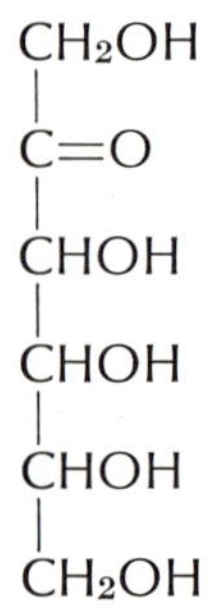

Fructose

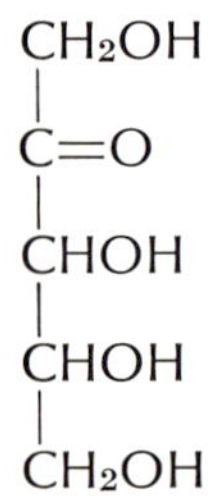

Ribulose

In writing the empirical formula for these monosaccharides we would simply show the number of atoms of each element in the molecule, as $C_6H_{12}O_6$ or $C_5H_{10}O_5$. Most carbohydrates come close to this proportion of hydrogen to oxygen, although the actual number of atoms may be considerably higher and therefore may be represented by the formula $C_nH_{2n}O_n$. Monosaccharide sugars may link together forming disaccharides or trisaccharides or even higher polysaccharides.

Sucrose, perhaps the most common sugar storage product in plants (as in sugar cane and sugar beets), is a disaccharide represented by the formula $C_{12}H_{22}O_{11}$ or by the structural formula:

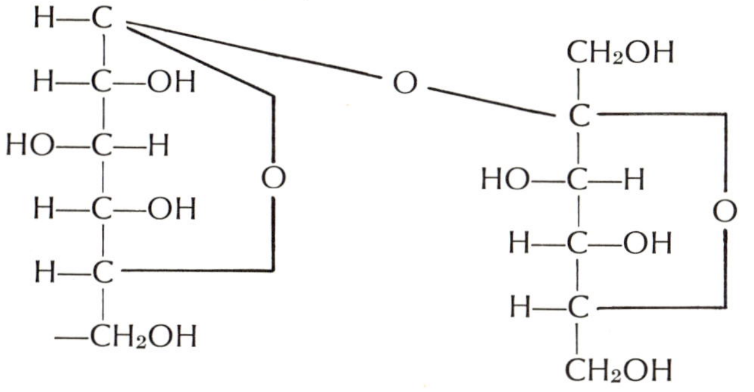

Sucrose

The two most commonly occurring polysaccharides in plants are starch and cellulose. Both may be represented by the empirical formula $(C_6H_{10}O_5)n$ in which *n* represents a number of molecules. It is estimated that approximately 200 to 1000 glucose molecules are linked together forming a starch molecule. Cellulose is similar except that the molecule is made up of about 1500 glucose molecules and the bonding is somewhat different, resulting in a very stable substance.

Fats

Fats (lipids) are composed of molecules containing carbon, hydrogen and oxygen, as are carbohydrates, but the arrangement and proportions of these atoms give fats an altogether different set of characteristics than carbohydrates. Fats have certain common properties, such as their insolubility in water but solubility in certain organic solvents. Fats are, in general, composed of chemically combined **fatty acids** and **glycerol,** the assimilative products of fats in digestion (Fig. 6). Gram for gram, fats have a much higher caloric value than either carbohydrates or proteins. Based on their somewhat variable chemical constitution, fats are classified as: (1) triglycerides, (2) fatty acid derivatives, (3) phospholipids, (4) carotenoids, and (5) sterols and steroids.

Triglycerides (fats and oils) Triglycerides are chiefly made up of saturated or unsaturated fatty acids. The former are found more commonly in animal tissues and are mainly stearic acid, $C_{18}H_{36}O_2$, and palmitic acid, $C_{16}H_{32}O_2$. Unsaturated fatty acids are liquid and include such plant fats as oleic acid, palmitic acid, and linolenic

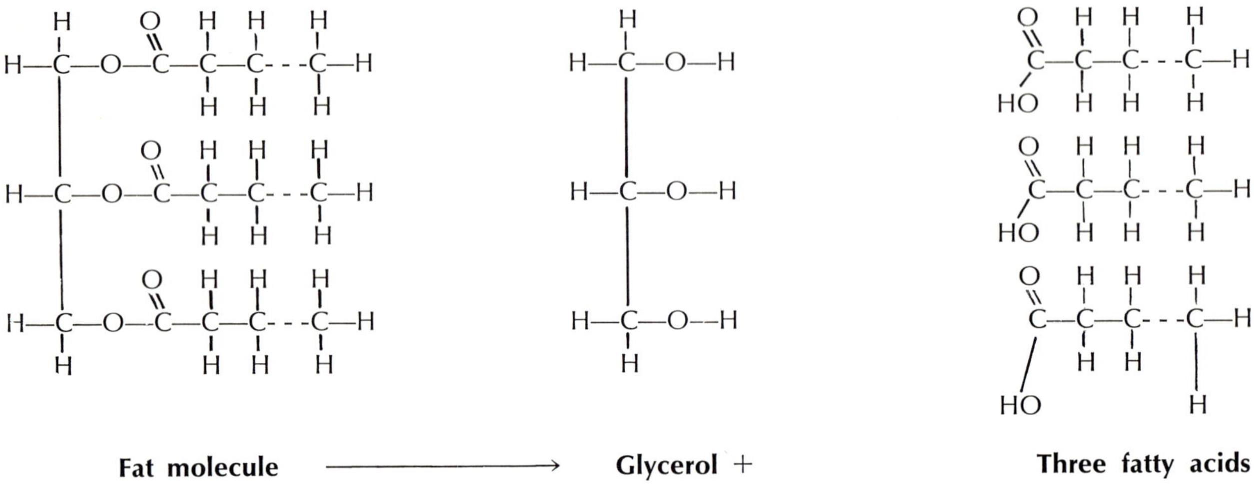

FIGURE 6 A fat molecule and its breakdown to glycerol and three fatty acid molecules in digestion. In this process the water molecule, HOH, is split, with H joining the glycerol and OH joining the fatty acid. Dotted lines indicate that several carbon atoms, each flanked by two hydrogen atoms, may form chains of greater length.

acid. All have one or more double bonds in their molecules of which the formula for oleic acid, $CH_3(CH_2)_7CH{=}CH(CH_2)_7COOH$, is one of the least complex. Unsaturated fats are usually considered oils. They have a melting point close to 0°C, while the melting points of the solid animal saturated fats vary from 30° to 50°C.

Fatty Acid Derivatives (waxes, fatty acid aldehydes[1] and others) Waxes occur in plants as cuticular wax covering the leaves, fruits, seeds, and epidermis of herbaceous stems. The plant waxes contain small amounts of straight-chain ketones[1] and hydrocarbons[2]

[1] Organic molecules which contain the carbonyl group (C=O, a carbon-oxygen double bond) at the beginning or end of their carbon chains are called **aldehydes** (formaldehyde, HCHO, is the simplest). If the carbonyl group is attached to an organic compound other than at the ends it is called a **ketone,** as acetone,

$$CH_3\overset{\overset{\displaystyle O}{\|}}{C}CH_3.$$

[2] Organic compounds containing only carbon and hydrogen as methane (CH_4), or propane (C_3H_8).

(paraffins) in addition to the fatty acid esters[3] which are more typical of animal waxes, such as beeswax. These waxes have very high melting points (60°–80°C).

Phospholipids Phospholipids are comprised of a large number of lipids containing not only C, H, and O, but phosphorus and occasionally nitrogen. The phospholipids are not abundant in plants but they are components of the plasma membranes; phospholipids may also occur with calcium and magnesium salts in plant parts containing these elements. Perhaps the best known phospholipid in plants is lecithin in wheat germ.

Carotenoids Carotenoids are colored lipids in which the carotenes are hydrocarbons with forty carbon atoms, and the xanthophylls are their hydroxyl or ketone derivatives. They are found in plastids. In leaves they are usually masked by the chlorophyll until autumn, when the chlorophyll decomposes. As it fades the color of the carotenoids becomes apparent. In some plants the carotenoids are abundant, such as in the roots of carrots, the fruit of squash, and the petals of many flowers.

Sterols and Steroids Fats of the sterol and steroid type are much more abundant in animals than in plants. Cholesterol is representative of many of the animal sterols; ergosterol, a C_{28}–sterol, is found in plants. Ergosterol received its name from the fungus ergot in which it is found, which parasitizes the grains of rye and other cereals. Ergosterol has also been found in yeast and in other molds. Upon irradiation with ultraviolet light ergosterol is converted by a change in its molecular structure into vitamin D_2, which may be used to relieve the symptoms of rickets. Steroids are also present in some of the basic nitrogenous substances known as alkaloids. Generally alkaloids are considered only as proteinaceous substances. However, a few are linked with steroids and form steroid alkaloids, several of which are found in members of the potato family, the Solanaceae.

Proteins

Proteins are extremely important constituents of living matter. We have known for some time that they are indispensable for life. They are found in the cells of all plants and animals and even in

[3] A combination of an organic acid and an alcohol:
CH_3COOH (acetic acid) + C_2H_5OH (ethyl alcohol) $\longrightarrow$ $CH_3CH_2—O—CO—CH_3$ (ethyl acetate) + H_2O.

viruses, some of which are lacking carbohydrates and fats. Proteins in organisms are of two general types: (1) the soluble proteins, found in the cell sap and body fluids; (2) the insoluble proteins, which are components of the membranes and other structural elements in the cells of many tissues. Plant proteins are of greatest abundance in the nuclei and in the meristems of rapidly growing stem and root tips. Proteins, like carbohydrates and fats, always have carbon, hydrogen, and oxygen in their molecules, but in addition nitrogen is also present. Most proteins contain 50 to 55% carbon, 5 to 7% hydrogen, 20 to 25% oxygen, and 15 to 18% nitrogen. They may also contain phosphorus, sulfur, magnesium, iron, and other elements. The molecules are very long and complex and are made up of linked units called amino acids.

Amino Acids About twenty amino acids of varying molecular structure are found in most proteins. The basic unit is

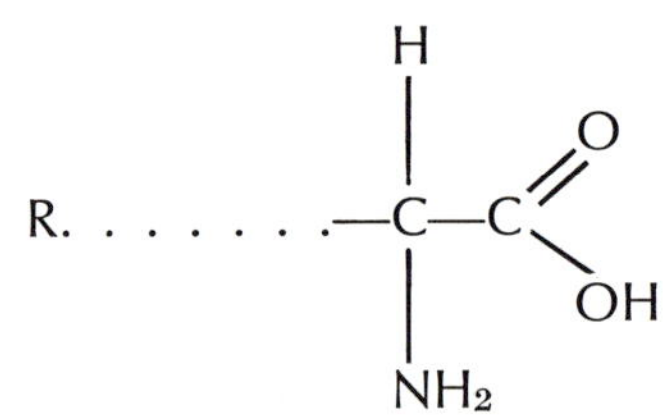

"R" is a symbol for "radical" and implies that many other atoms and molecules may be linked to this group.

Amino acids are identified by the presence of at least one NH_2 group, termed the amino group, which probably was derived from ammonia (NH_3) by the loss of one H atom. The simplest of the amino acids is glycine (Fig. 7).

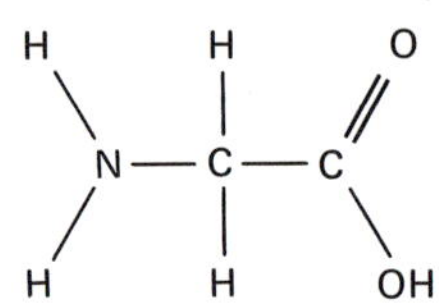

FIGURE 7 The amino acid, glycine.

From this simple framework many complex proteins are built which apparently form the basis of living materials. Since amino acids can combine in a great variety of patterns, thousands of different protein molecules are built. Since we have so many complex protein molecules we can see why we may have such a variety of living organisms. Proteins and their basic amino acids result in a framework on which to build variation and evolution.

If a simple amino acid of the formula $H_2N—CHR—COOH$ is combined with a second by the formation of an acid amide bond, a substance of the formula $H_2N—CHR—CO—NH—CHR—COOH$ is formed. It is called a dipeptide, and the acid amide linkage between the two amino acids is called a peptide bond. Extension of the amino chains to many more members leads to the formation of polypeptides, one of the final steps in protein synthesis. Proteins are generally very unstable substances. They are readily changed by heat or radiation, by acids or bases, or by a number of organic solvents. A protein may be separated into its amino acids through the action of a specific enzyme in a process of hydrolysis, or two amino acids may be combined to form a simple protein by dehydration as shown in Figure 8.

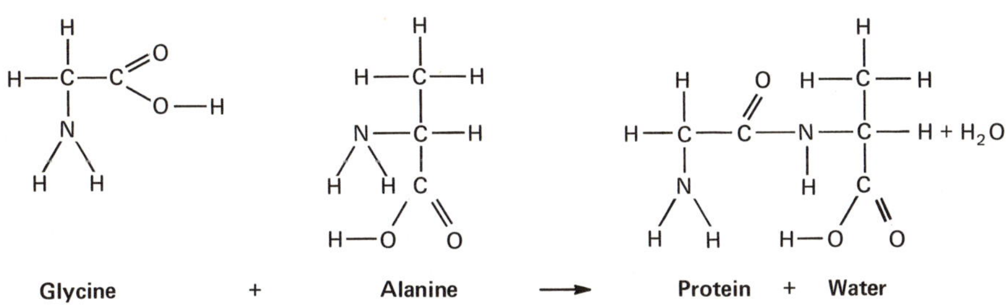

FIGURE 8 Formation of a protein from two amino acids. In digestion the process is reversed.

Nucleoproteins and Nucleic Acids Another type of nitrogen-containing compound forms a molecular skeleton composed of a ring to which other substances are bound, forming much more complex substances. This is a **pyrimidine** and looks like this:

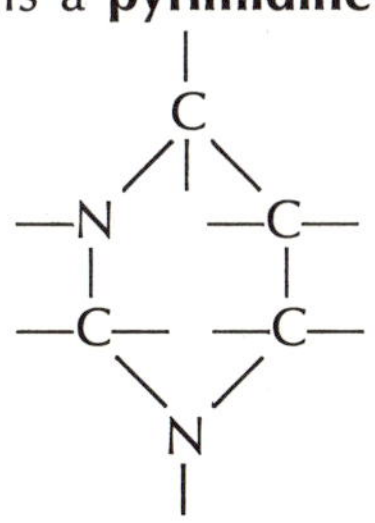

A second, related, compound has a double ring and is called a **purine:**

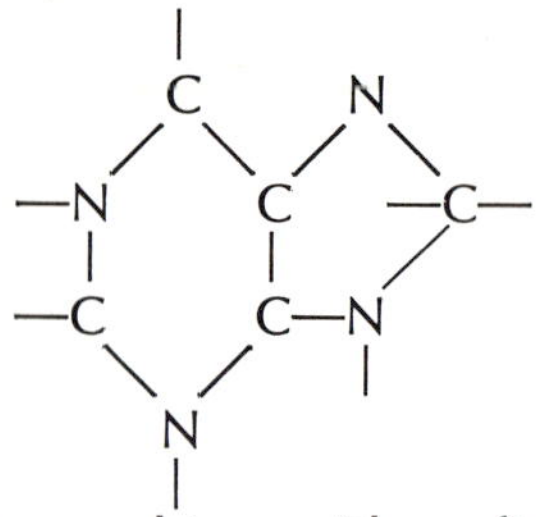

These compounds combine with a five-carbon sugar and a phosphate forming **nucleotides:**

Pyrimidines + sugar + phosphate ⟶ nucleotides
Purines + sugar + phosphate ⟶ nucleotides.

Gradually the complexity of the molecules is increased as the nucleotides are united with other nucleotides to form nucleic acids:

Nucleotides + nucleotides ⟶ nucleic acid.

Finally, it is believed, we may reach a zenith in chemical synthesis when the nucleic acids are combined with proteins to form nucleoproteins:

Nucleic acids + proteins ⟶ nucleoproteins.

Once nucleoproteins were formed the origin of life was possible, for these molecules were sufficiently complex that living systems could be built around them.

The **nucleoproteins** form the largest molecules known to chemistry. They have the ability to reproduce themselves, a characteristic of life apparently possessed by no other type of molecule. It is generally assumed that the nucleic acids present in the nucleus (a few are found in cytoplasm) are bound to the proteins forming the highly complex nucleoprotein molecules. The nucleic acids themselves are composed of a five-carbon sugar, a base, and a phosphate group. The sugar may be either of two types: ribose or deoxyribose. The molecules formed are therefore recognized as either ribonucleotide or deoxyribonucleotide. They form ribonucleic acid (RNA) and deoxyribonucleic acid (DNA), respectively. The two nucleic acids can be differentiated from each other by treating tissue containing the substances with ribonuclease, an enzyme which catalyzes the cleavage of RNA but not DNA. The two nucleic acids may also be separated by incubating tissue in warm dilute alkali. This technique hydrolyzes the RNA to nucleotides, but the DNA is unaffected and may then be precipitated by acidification.

RNA and DNA are arranged as long chains of purine (adenine and guanine) and pyrimidine (cytosine and uracil or thymine) molecules held together by links through phosphate and ribose or deoxyribose sugars (Fig. 9). The order in which the purines and pyrimidines are arranged in the chain determines the sequence of amino acids joined together forming a protein on the ribosome. It is this sequence of amino acids in the macromolecule which gives each protein molecule its specific enzymatic properties and therefore its ability to catalyze certain reactions within the cell. The significance of this sequence has been compared to that of the arrangement of the letters of the alphabet in different words to convey different meanings.

Though it is unknown how protein and nucleic acid combine to form natural nucleoprotein molecules, X-ray analysis gives the striking result that the length of an amino acid residue in the peptide is identical with the length of a nucleotide unit in the nucleic acid molecule. It has therefore been assumed that the nucleotide chains of the nucleic acid have a helical structure like the peptide chains of the protein. It is also believed that the two chains combine by means of hydrogen bonds between their helices.

When DNA is isolated it is found to consist of threadlike particles of very high molecular weights. Although its chemical composition varies greatly, the differences seem to be due mainly to the arrangement of the substances of which all DNA molecules are composed. These substances, adenine, guanine, thymine, and cytosine, are attached to a five-carbon sugar, 2-deoxyribose, to form a deoxynucleoside. The nucleosides are linked together by a phosphate group in such a way that we can conclude DNA is a polynucleotide made up of phosphate ester linkages between the sugars. It has been determined that the DNA molecule is in the form of a double spiral with each spiral made up of alternating sections of phosphate and sugar; the variation in chemical composition of different DNA molecules depends, then, on the arrangement of the four bases: adenine, cytosine, guanine, and thymine (Fig. 10). These bases do not occur in a hit-and-miss, random fashion but in a well organized manner. The adenine (A) is always linked with thymine (T), and cytosine (C) with guanine (G) (see Fig. 13). The amount of A + T content may be much greater than C + G in some molecules, or the reverse may be true. However, the total amount of A to T is nearly always equal as well as the amount of C to G. The number of possible combinations of nucleotides in a DNA molecule is estimated at approximately 10^{200}, giving this molecule all possible types of variation. DNA has been recognized as the transmitting basis for genetic characteristics and

FIGURE 9 A diagrammatic illustration of the DNA double helix or spiral staircase configurations. (After Watson and Crick, Cold Spring Harbor Symposia on Quantitative Biology, Vol. 18, 1953.)

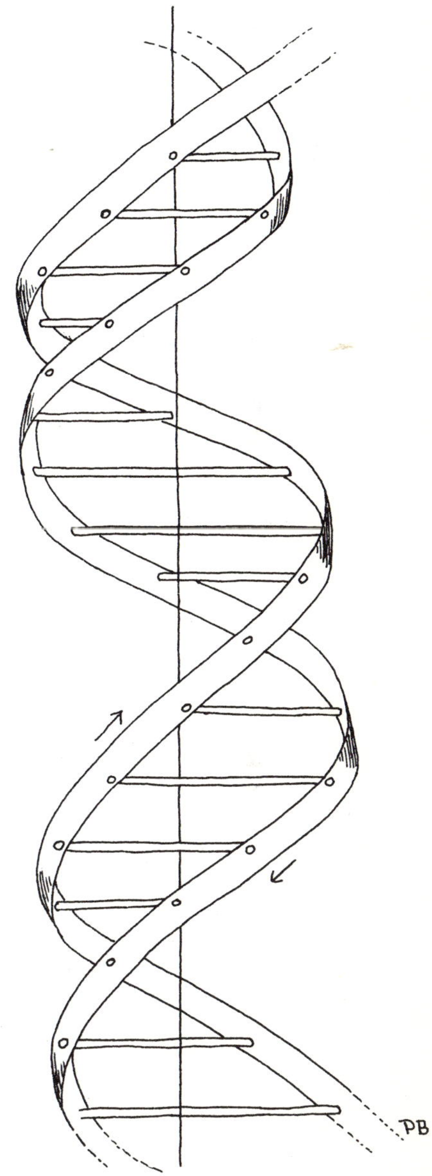

with its wide latitude for variation it may well be the substance responsible not only for variation within the species but also for the creation of new species. DNA is found in virtually all organisms. Considering the billions of years of time available and the numerous molecular combinations possible, DNA may well be the method in organic evolution.

It has been established that genes on the chromosomes, or in the chromatin material of organisms in which chromosomes are not distinctly formed, are DNA. Genes, then, are units of biochemical action and their functions are interpreted on the basis of DNA structure. The different genes in a cell carry different codes and a cell manufactures amino acids and proteins as the DNA code dictates. In addition to being present in nuclear material DNA is also present in the cytoplasm in such organelles as chloroplasts, where it also plays a role in multiplication.

In contrast to DNA, RNA varies somewhat in its base ratios and has a single-stranded helix, instead of the double helix. We recognize three kinds of RNA which are used in protein synthesis:

1. *Messenger RNA,* which transmits the genetic information from the DNA molecule in the nucleus to the cytoplasm.
2. *Ribosomal RNA,* found in the cytoplasmic particles called ribosomes, on which protein synthesis occurs. The ribosomes are associated with the extensive tubular system, the *endoplasmic reticulum,* in the cytoplasm.
3. *Transfer RNA,* a smaller molecule than messenger or ribosomal RNA which acts as an adaptor in protein syntheses. Specific amino acids are formed according to the arrangement of the bases of a specific transfer RNA.

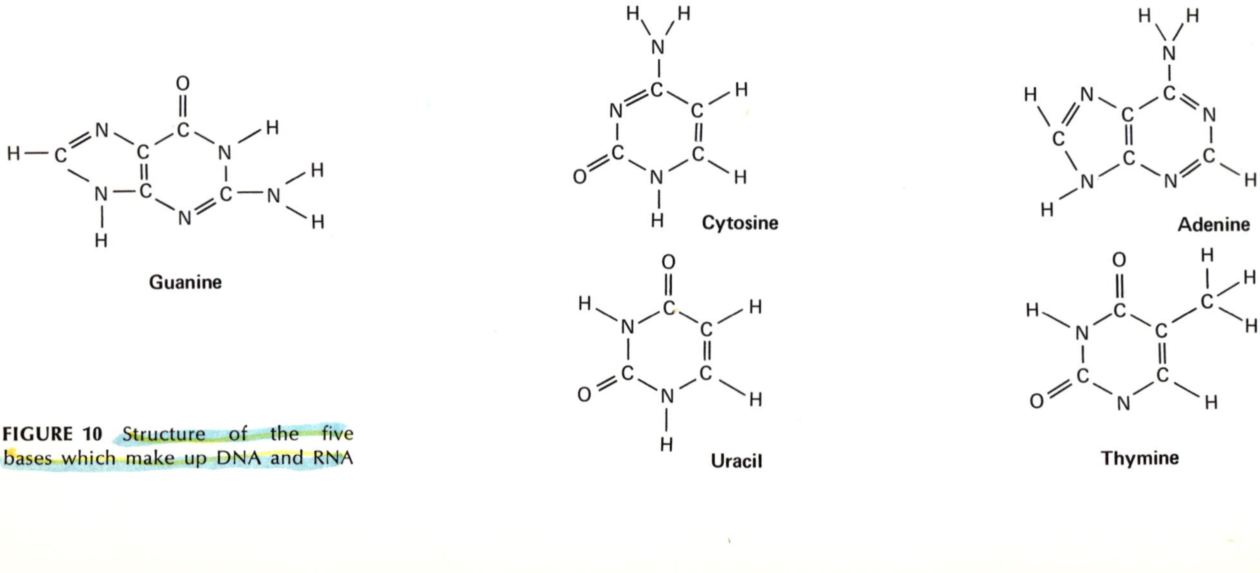

FIGURE 10 Structure of the five bases which make up DNA and RNA

RNA nucleotides are spaced in much the same manner as those of DNA. However, thymine has been replaced by the base uracil (Fig. 11). Thus the four bases are cytosine, guanine, adenine and uracil. Furthermore, the five-carbon sugar which alternates with the phosphorus in the spiral is ribose instead of deoxyribose. The evidence available indicates that DNA acts as the template to form RNA much as it reproduces itself. Because of the similarity of the nucleotides it is possible for the bases of RNA to pair with those of DNA. When RNA is synthesized on the DNA, the cytosine of the RNA can pair with the guanine of the DNA and vice versa, and the thymine which pairs with the adenine of DNA is replaced by uracil. Once formed, much of the RNA leaves the nucleus and moves into the cytoplasm where it functions in protein synthesis.

FIGURE 11 A diagrammatic illustration of the RNA molecule. The base thymine has been replaced by uracil.

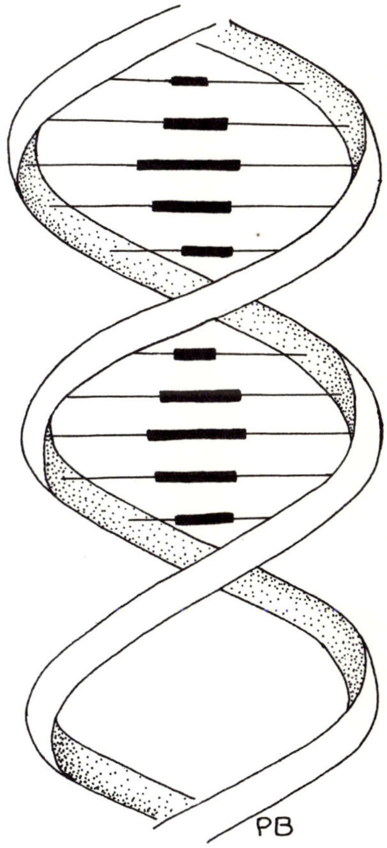

Metabolism

In the activities of all cells there is a continuous formation and building of new protoplasm from the food materials which have been digested and gradually assimilated. At the same time there is a continuous alteration and disintegration of these molecules in other life processes. We speak of those processes in which more complex substances are being built from simpler ones as **anabolism.** Such a process is photosynthesis, in which sugar is formed from carbon dioxide and water. **Catabolism** is the term used where complex substances are being reduced[1] to simpler ones, as in respiration, where sugar is reduced to carbon dioxide and water in the presence of oxygen. Anabolic and catabolic processes collectively

[1] **Oxidation** is a process in which electrons (e^-) are removed from an atom or molecule. The reverse process, the addition of electrons to an atom or molecule, is **reduction.** This may occur when oxygen is not present, as:

$$Fe^{++} \underset{\text{reduction}}{\overset{\text{oxidation}}{\rightleftarrows}} Fe^{+++} + e^-.$$

Fe^{++} is the electron donor which is oxidized to Fe^{+++}. In the reverse reaction Fe^{+++} is the electron acceptor which is reduced to Fe^{++}.

Generally oxidation is a process in which oxygen is added or is a reaction in which hydrogen gives up electrons to oxygen and is thereby oxidized; oxygen takes up the electrons and is thereby reduced:

$$2\ H_2 + O_2 \longrightarrow 2\ H_2O.$$

Oxidative phosphorylation is the process by which ATP is formed using the energy liberated during the passage of electrons through the respiratory chain:

$$\text{ADP} + \text{Inorganic Phosphate} \longrightarrow \text{ATP}.$$

constitute metabolism, which occurs constantly and is regulated to a large degree by the cellular enzymes which form many intermediate products similar in both processes before the end products of each are reached.

Free use of the term *metabolism* is made throughout this book but very little of *anabolism* and *catabolism* since they are both vital processes working continuously but are difficult to distinguish individually in metabolism. For example, digestion is a catabolic process in which carbohydrates, fats, and proteins are reduced to the end products glucose, fatty acids, glycerol, and amino acids to be absorbed and assimilated in the cells. On the other hand, assimilation is an anabolic phase of digestion where these reduced end products are regrouped or assimilated into other food products and used in storage, building new cells and tissues, and carrying on other metabolic activities.

Enzymes

Enzymes are catalytically active proteins. Sometimes simply defined as organic catalysts, enzymes accelerate chemical reactions or initiate the manner in which a certain substrate is to be metabolized. The substrate is a substance which will undergo chemical reaction due to a specific enzyme or series of enzymes. Many reactions are reversible, with the same enzyme responsible for both directions. In these reactions the enzyme would be capable of catalyzing more than one substrate. Since enzymes are not used up in the reaction very small amounts of enzyme are used in the metabolism of large amounts of a substrate. However, greater quantities of the catalyzing agent would hasten the breakdown of the substrate, since this is a surface relationship phenomenon and more enzyme molecules would be in contact with the substrate. The macromolecules of the proteinaceous enzyme are much larger than those of the substrate and each has an irregular configuration in which the catalyst has a precise spot where the substrate molecule can join it (Fig. 12). This relationship has been compared to a lock and key in which the substrate molecule is the key. Once the product molecules are formed from the substrate they break away and the enzyme is prepared to receive the substrate molecule.

Generally enzymes work very rapidly, catalyzing one reaction after another. In some chemical reactions within the organism as many as 50,000 catalytic reactions may occur within a second's time. This accounts for the fact that a small amount of the enzyme

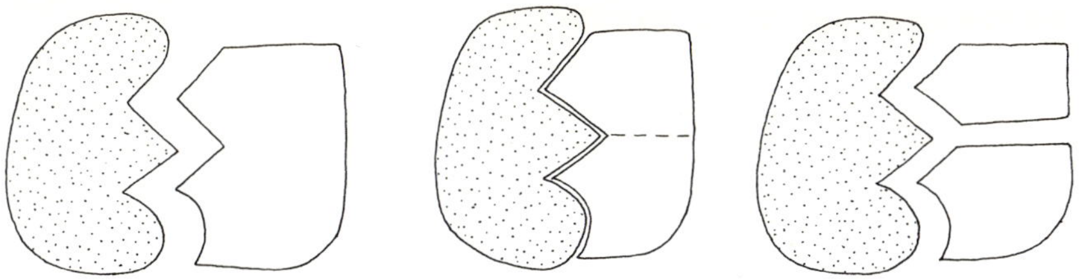

FIGURE 12 Diagrammatic illustration of the enzyme-substrate reaction. The enzyme molecule is stippled. (A) The enzyme is making contact with the substrate; (B) contact is made and the substrate is split into two product molecules and the enzyme molecule is ready for another substrate molecule of the same type. (Buffaloe)

may catalyze a large amount of the substrate. Enzymes are usually localized in the cell; some are found in the mitochondria, others in the plastids, the ribosomes, and so on. Each does its own particular job but all are part of the elaborate scheme of metabolic processes made possible by the numerous enzyme systems.

In the 1950's the main lines of the metabolic pathways of small molecular weight compounds were discovered. These discoveries set the stage for the highly fascinating achievements in molecular biology of the 1960's which showed how the macromolecules always associated with life are formed and function.

Enzyme Synthesis

We know that living matter depends entirely on the mechanisms of enzyme systems which make possible the highly organized sequence of chemical reactions recognized as "life." How these enzymes were made was for many years an unanswerable question but now we know that enzyme synthesis is carried on by ribonucleic acid (RNA). Genetic control of the species is thus made possible through the constancy of enzyme synthesis. In any species the same enzymes are formed from generation to generation, maintaining uniformity in the species. Changes, and thus evolution, do occur over long periods of time. This is brought about by changes in the pattern of enzyme synthesis, which induce mutation.

That enzymes are synthesized from free amino acids was proved by Spiegelman, using radioisotopes, in 1953. He grew a strain of bacteria *(Escherichia coli)* in a medium with C^{14} labeled lactic acid as the only carbon source. These bacteria had only labeled C^{14} in their cells, including the carbon of the proteins which they formed. The radioactive bacteria were then induced to synthesize an enzyme which they did not have before and were given unlabeled amino acids in their medium to feed on. When the enzyme was isolated it was found to be almost completely free of C^{14}, which proved to the investigator that the enzyme now produced by the bacteria must have been made from the unlabeled free amino acids put into the medium at the same time as the enzyme inducer.

Since these first discoveries of enzyme synthesis, much of the enzymatic machinery that makes proteins from amino acids has been analyzed and separated into its components.

These include: ribonucleic acids, enzymes, ATP (adenosine triphosphate , the energy-yielding substrate), cofactors (coenzymes), and the amino acid substrates. In the initial reaction, amino acids are combined with ATP and activate the substances, bringing them to an energy level sufficient for subsequent reactions. They are then attached to the ends of specific ribonucleic acid molecules. Each amino acid has a specific ribonucleic acid which, with the aid of an enzyme, carries the amino acid to the site of protein synthesis on the very small but numerous ribosomes. These facts were accumulated by taking components of the protein synthesizing system from cells and inducing the ribonucleic acids and enzymes to react with radioactive amino acids. Protein was isolated at the end of the reaction and the amount of radioactivity incorporated was a measure of the extent of protein synthesis.

As complicated as this may seem, it may be summarized by:

$$\text{Amino acid} + \text{ATP} \longrightarrow \text{energy (radioactivated amino acid)}$$
$$\text{Radioactivated amino acid} + \text{RNA} + \text{Enzyme} \longrightarrow \text{protein (on the ribosomes).}$$

This is unfortunately not the whole story, for the information every cell receives in carrying on its activities is in its DNA code. When a new cell is formed, it receives from the parent cell all the instructions it needs to live, and to live in a way identical to that of its parent.

In each DNA macromolecule there is a double-stranded chain, with the strands linked in such a way that to each particular pyrimidine in one strand there is joined a purine from the opposite strand and vice versa. At division the two strands separate lengthwise just as we might separate a ladder lengthwise by cutting each rung in the middle (Fig. 13).

After separation each strand builds up its opposite again by assembling the matching purine or pyrimidine molecule along its length. We speak of this as base pairing since both purines and pyrimidines are bases, chemically. DNA directs the protein synthesis on the ribosomes by assembling a molecule of RNA along its (DNA) length which then becomes detached and leaves the cell nucleus to go to the site of protein synthesis, carrying with it the information coded for it in the DNA. This has been demonstrated

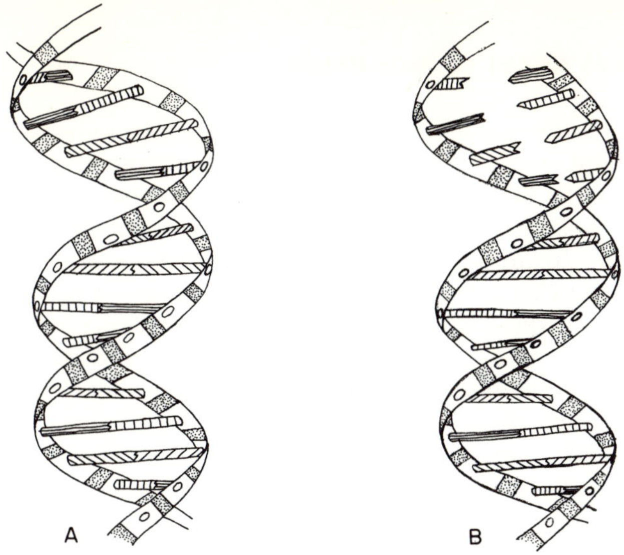

FIGURE 13 Replication of the DNA molecule. A. portion of the molecule with bases joined. B. the bases beginning to split. After complete separation each half will regenerate a new half to replace the one lost in division. (Fraser)

by a virus infection of the bacteria *E. coli*. After the virus DNA enters the bacterium the virus DNA makes its own RNA which then becomes attached to the bacterial ribosomes. The virus is then producing enzymes and using the substance of the bacterium as raw material for growth. This is how a virus infection works.

Additional discussion of the specific enzymes involved in the various plant processes will follow as each is considered in later chapters. Let it be emphasized here, however, that enzyme-substrate reaction involves more substances generally than these two.

In order for certain enzymes to function properly they must be associated with coenzymes. These coenzymes are often specific vitamins such as pantothenic acid, thiamine, or riboflavin, all of the vitamin B complex. In addition to these organic coenzymes, certain inorganic substances are also required for enzyme catalyses and play the same role as coenzymes. These substances include iron, magnesium, zinc, and other elements. Plants form many of the organic coenzymes (vitamins) they use under optimum conditions of growth, while animals obtain most of them in their diet. Thus many deficiency diseases in humans and other animals are due to vitamin deficiencies, which are often reflected in cellular abnormalities resulting from an impairment of the normal protoplasmic machinery. Plants more commonly show deficiencies resulting from a lack of inorganic substances in the soil and consequently suffer an impairment of growth.

We may wonder why so much emphasis is placed on the chemical constituents of the cell when the entire organism seems so much more important. Certainly a multicellular organism is much more than merely the sum of its parts, but it is through the study of the individual parts that we gain an appreciation of the complexity of the organism as well as of the phenomena we associate with life. Much of the work in modern biological research is concerned with the growth and metabolic activities of the cell because this is where all normal mechanisms of growth, as well as those which are abnormal, are initiated. Diseases such as cancer owe their peculiar growth patterns to cellular abnormalities. Therefore, if we can discover the mechanism which triggers this abnormality we are well on our way to finding a cure and a preventative for diseases of this type. Since so much has been learned in recent years concerning the metabolic activity of the cell and its parts, we may look with considerable optimism for greater control of cellular abnormalities in the future.

Supplementary Reading

Allen, J. M. (editor), *Molecular Control of Cellular Activity*. McGraw-Hill, New York, 1962.

Bonnor, D. M. (editor), *Control Mechanisms in Cellular Processes*. Ronald Press, New York, 1961.

Fraenkel-Conrat, H., *Design and Function at the Threshold of Life: the Viruses*. Academic Press, New York, 1962.

Jensen, William A., *The Plant Cell*. Wadsworth Publishing Company, Inc., Belmont, California, 1964.

Loewy, Ariel G., and Philip Siekevitz, *Cell Structure and Function*. Holt, Rinehart and Winston, New York, 1963.

Speakman, J. C., *Molecules*. McGraw-Hill, New York, 1966.

Swanson, Carl, *The Cell,* 2nd ed. Prentice-Hall, Inc., Englewood Cliffs, New Jersey, 1964.

White, Emil H., *Chemical Background for the Biological Sciences*. Prentice-Hall, Inc., Englewood Cliffs, New Jersey, 1964.

Questions for Discussion

Questions applicable to this chapter and the next are at the end of Chapter 4.

4

The Cell

Although our knowledge of the cell has increased greatly in the last fifteen years, the cell has been known as a unit of structure for over three hundred years. It was in 1663 that Robert Hooke, using a microscope primitive by our standards, observed the tiny cubical structures in cork which he termed cells. But it was not until 1838 that Matthias Schleiden, a German botanist, was able to state with confidence that the structural unit of all plant material was the cell. His announcement was followed a year later by a similar statement from his countryman Theodor Schwann, a zoologist: "cells are organisms and entire animals and plants are aggregates of these organisms arranged according to definite laws." This statement has since been recognized as the Schleiden and Schwann cell principle.

Hundreds of scientists have studied the cell as the structural and functional unit of organization and development with the idea that to advance beyond the purely descriptive phase biologists must understand the inner mechanics of the cell. This problem was summarized in 1962 by a French microbiologist, Andre Lwoff, who stated:

> When the living world is considered at the cellular level, one discovers unity. Unity of plan: each cell possesses a nucleus embedded in protoplasm. Unity of function: the metabolism is essentially the same in each cell. Unity of composition: the main macromolecules of all living beings are composed of the same small molecules. For, in order to build the immense diversity of the living systems, nature has made use of a strictly limited number of building blocks. The problem of diversity of structures and functions, the problem of heredity, and the problem of diversification of species have been solved by the elegant use of a small number of building blocks organized into specific macromolecules. . . . Each macromolecule is endowed with a specific function. The machine is built for doing precisely what it does. . . . If the living system did not perform its task, it would not exist. We have simply to learn how it performs its task.

Thus, it becomes obvious that all cells are basically the same, whether plant or animal. But only basically are they the same, for they do vary in size, content, function, and ability to reproduce. The major components of plant and animal cells are listed below. The parts of a typical plant cell are shown in Figure 1.

Major Components of Plant and Animal Cells

1. Nucleus—Present in all but the most primitive organisms
   a. Nucleolus—usually present but variable in number
   b. Chromosomes—in all nucleated cells, variable in number but constant for a given species
   c. Nuclear membrane—in all nucleated cells

2. Cytoplasm—In all nucleated cells, otherwise the cellular contents may be simply termed protoplasm
   a. Endoplasmic reticulum—usually well organized in all cells
   b. Mitochondria—present in all living cells
   c. Ribosomes—present in all living cells
   d. Golgi bodies—usually observed in all cells
   e. Chloroplasts—usually limited to certain nucleated, thin-walled cells of plants
   f. Centrioles—cells of all animals and a few algae and fungi
   g. Lysosomes—cells of animals and possibly some plants

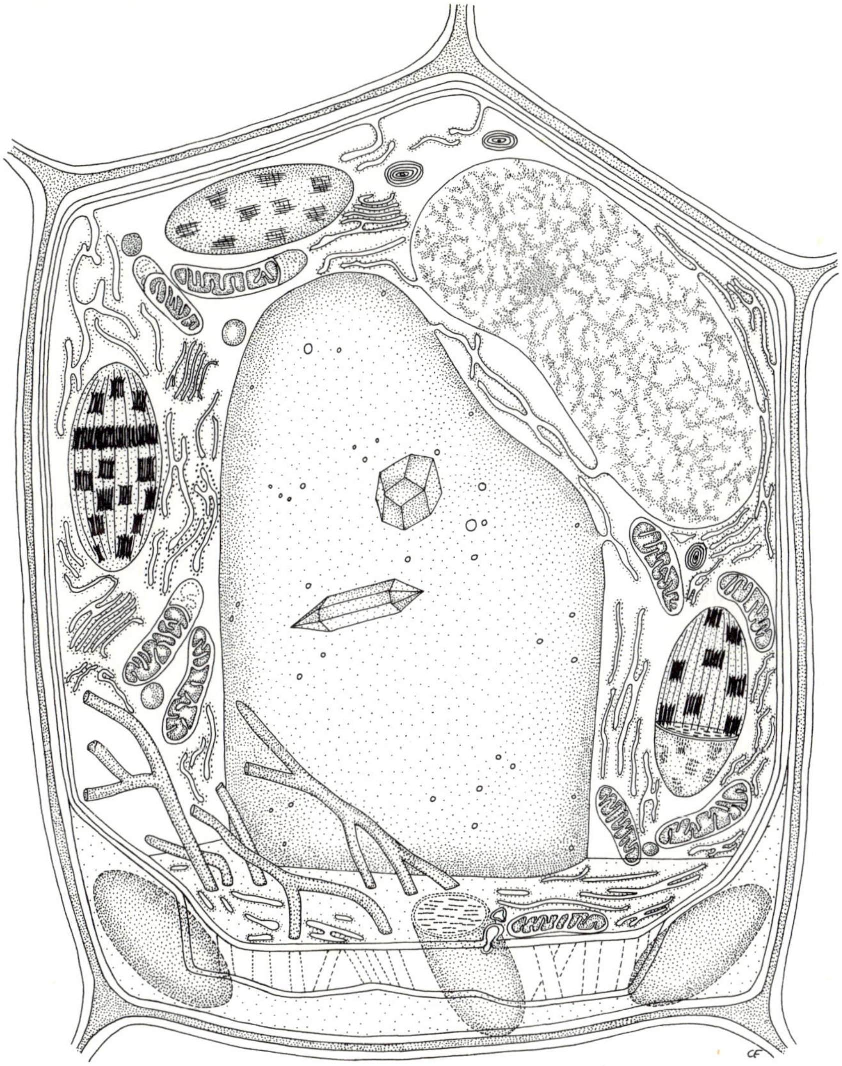

FIGURE 1 A diagrammatic illustration of a plant cell as pictured through an electron microscope. The large central area with a few crystals is the vacuole; at its upper right is the nucleus. The egg-shaped bodies with darkened lamellated areas are plastids. The smaller egg-shaped organelles are mitochondria; the small circular structures are starch grains: At the top of the vacuole and also to its left are golgi bodies. Only a small portion of the endoplasmic reticulum is included near the periphery of the cell. (This drawing and those of the individual organelles in this chapter were redrawn from *The Cell,* Life Science Library, 1964.)

h. Vacuoles—in all cells, more conspicuous in large older cells; include water, fat and other cell sap vacuoles
i. Granules, crystals, starchgrains, pigment and other minute particles in cells of various types

3. Surface Structures
   a. Cell membrane, plasma membrane, or cytoplasmic membrane, bounds all cells
   b. Cell wall—present only on plant cells, composed chiefly of cellulose but other substances as pectin, hemicellulose, and lignin often present

The Cell Wall

During growth, plant cells divide into two; their division is completed by a concentration of protoplasmic material which is the forerunner of a new cell wall. This concentrated substance forms a thin layer called the **middle lamella** and is composed mostly of a pectic compound which cements the new cells together. The primary wall is then formed by the cytoplasm between the cytoplasmic membrane and the intercellular substance. This wall is made up mostly of pectin and hemicellulose microfibrils and expands as the cell grows. In most cells a secondary wall forms in the same manner between the cytoplasmic membrane and the primary wall. It may be thick or thin and may contain a number of substances. In wood cells lignin is deposited in the spaces between the microfibrils of cellulose. The secondary wall is usually unequally thickened, often leaving perforated areas called pits or pit fields which facilitate the translocation of water and dissolved substances from cell to cell. The most extreme thickening is found in the water conducting elements of the xylem and fibers, where the wall may be thickened in rings or as a spiral, appearing like a spring, or with a branching reticulate pattern.

The Cytoplasmic or Plasma Membrane

For many years the nature of the membrane which limits the outer portion of the cytoplasm was conjectural. Some microscopists believed it was only a surface phenomenon, while others considered it a true membrane distinct from the cytoplasm itself. Now through the aid of the electron microscope it has been shown to be a true membrane about 100 Å (angstroms) or 0.01 μ (micron) in thickness, consisting of large protein molecules on its inner and outer surfaces and fat molecules sandwiched between the protein. The protein molecules are capable of folding or unfolding and therefore make it possible for the membrane to expand or contract. The physical and chemical nature of the membrane contributes to its selective permeability, the property of allowing only certain substances to enter or leave the cell. Its degree of permeability may vary from time to time as the molecules in the membrane contract or expand. It has also been shown in some animal cells that the membrane is very much convoluted or undulated, which would increase its absorptive area.

The fatty portion of the membrane is suggested by the fact that fat solvents readily penetrate cells. It has also been shown that some bacterial cells, at least, have enzymes either as a part of the membrane or closely associated with it. This would increase the permeability of the cell to substances that these enzymes (permeases) could act upon. Not all of the permeability functions of the plasma membrane are known, but we have firmly established at least that the membrane is elastic in some cells and fairly rigid in others, and is a living, selectively permeable structure.

The membrane system of a cell is far more than a single plasma membrane within the cell wall and another membrane surrounding the nucleus. There is a labyrinth of membranes, vesicles, and tubules, referred to as the **endoplasmic reticulum** (ER) (Fig. 2), in the cytoplasm which form channels of transportation for the movement of inorganic substances, enzymes, and foods. These compounds may then be selectively channeled to the centers of cellular activity where they carry out their specific activities.

FIGURE 2 The endoplasmic reticulum. Note the flattened, tubular nature of the reticulum and the tiny black dotted ribosomes associated with it.

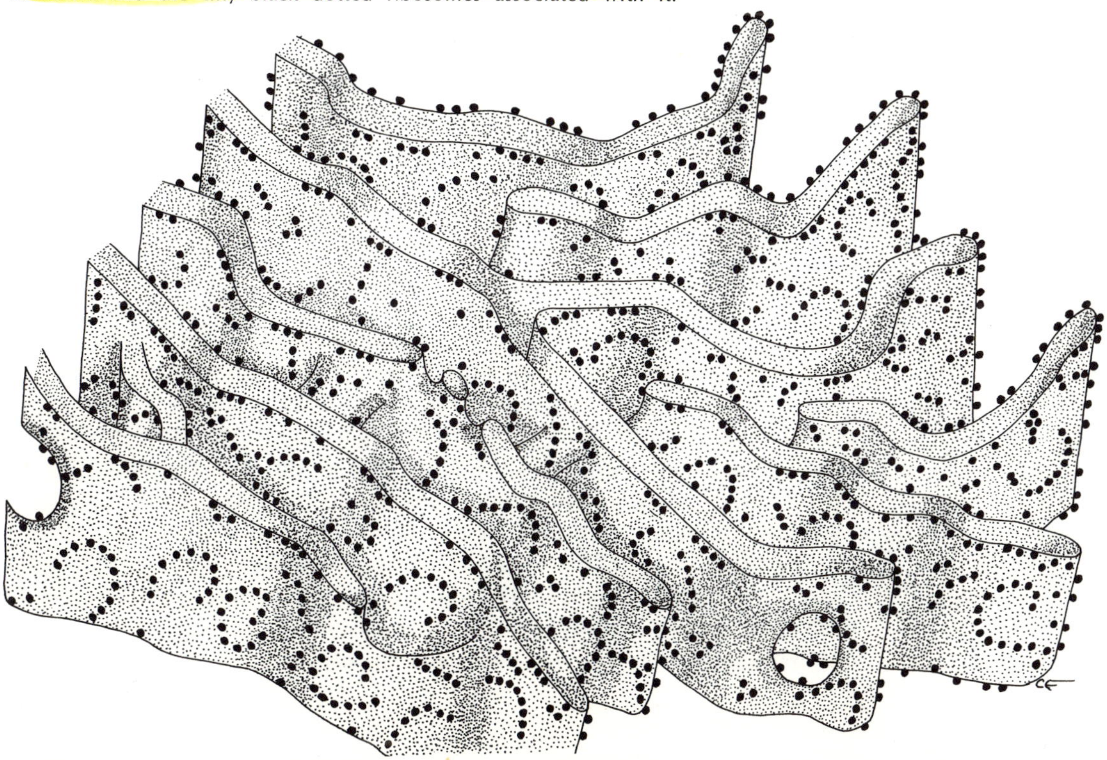

The Endoplasmic Reticulum

The ER is a network of small, flattened channels which form an elaborate system of membranes extending from the nuclear membrane to the plasma membrane. It is apparently an extension of the outer membrane of the nucleus, but provides channels throughout the cytoplasm for the passage of materials and also provides additional surfaces for chemical reactions and as storage depots for synthesized materials. The ER is quite variable in cells of different types and it may be very loosely organized or tightly packed. Since the protoplasmic contents of cells are continually in motion, the ER is constantly shifting in position, breaking up into pieces, then reforming and breaking again.

Ribosomes

Usually associated with the endoplasmic reticulum are tiny granules richly supplied with RNA. These tiny granules, identified as ribosomes (Fig. 2), have a protein forming capacity which is a necessary property if the cell is to grow and function properly. Their association with and proximity to the ER undoubtedly increase the activity of the membranes. It is estimated that ribosomes number as many as 500,000 per cell.

Mitochondria

The energy converting organelles called mitochondria (Fig. 3) may be spherical in shape, but more commonly they are elongated or sausage-shaped. Each mitochondrion is composed of a double lipoprotein membrane which is similar to the nuclear and plasma membranes. The inner membrane is invaginated, forming many folds within the organelle, greatly increasing the surface area of the membranes. Inside the mitochondrion is a liquid phase which aids, as do the membranes, in respiration. In the process of respiration the carbohydrates, fats, and proteins are broken down by specific enzymes outside of the mitochondrion into pyruvic, fatty, and

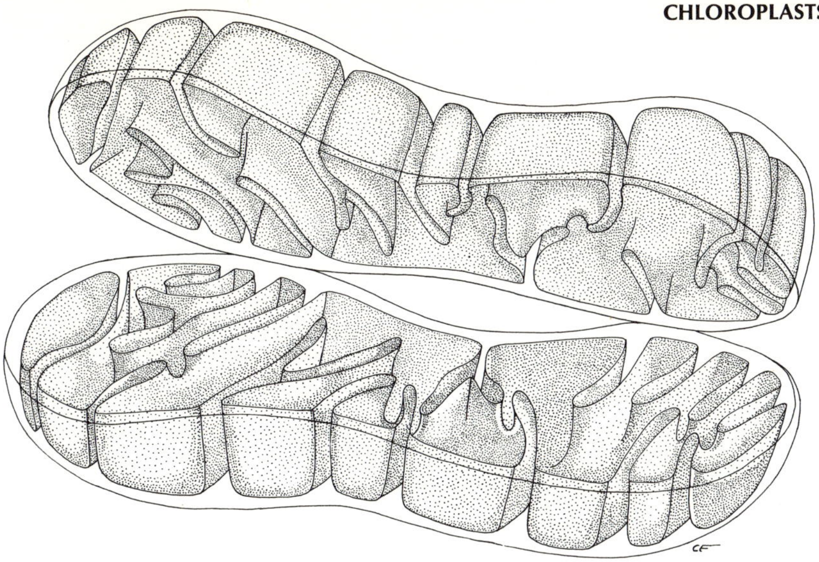

FIGURE 3 A mitochondrion. Note the many invaginations of the membrane within the organelle which greatly facilitate its function.

amino acids. These substances may then diffuse through the mitochondrial membranes into the liquid phase where carbon atoms are removed from them. This is an oxidative process, since oxygen is involved. The continuous metabolism of glucose and other substances in the cell results in a continuous production of carbon dioxide and the utilization of oxygen, releasing water and energy. The latter is not liberated as heat, but is passed along as a new chemical compound and finally as adenosine triphosphate (ATP). The breakdown of the metabolic fuels occurs in the liquid phase of the mitochondrion but ATP production takes place in the membrane where enzymes break down some molecules and reform others until the rearranged atoms become ATP. This is therefore an oxidative phosphorylation reaction,[1] an important part of the exchange of gases in respiration. A single cell may have as many as five hundred mitochondria.

Chloroplasts

Chloroplasts are structures (Fig. 4) unique in and limited to plant cells. They are not found in all plant tissues but are limited to thin-walled cells chiefly in the mesophyll of leaves, in the outer cortical cells of herbaceous stems, in some aerial roots, and in cells

[1]See footnote 1, Chapter 3.

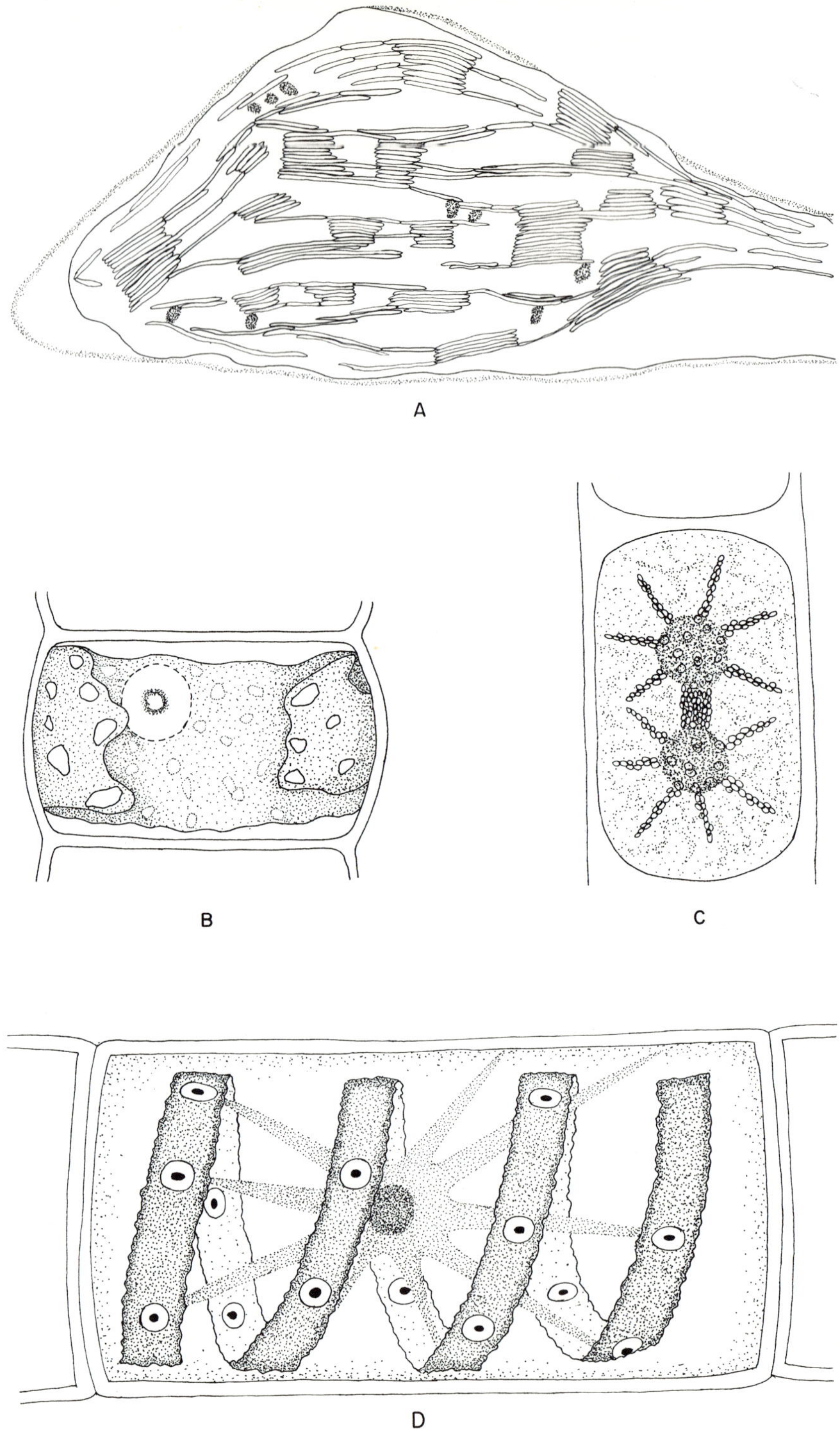

FIGURE 4 Chloroplasts. A: drawing from an electron micrograph of a section through the disc-shaped plastid of spinach. (This drawing is enlarged about 30,000 times while the remainder are drawings from light microscope observations and are enlarged about 300 times.) B: *Ulothrix;* C: *Zygnema;* D: *Spirogyra;* E: *Oedogonium;* F: cross section of the plastid of *Closterium;* G: *Chlorella.*

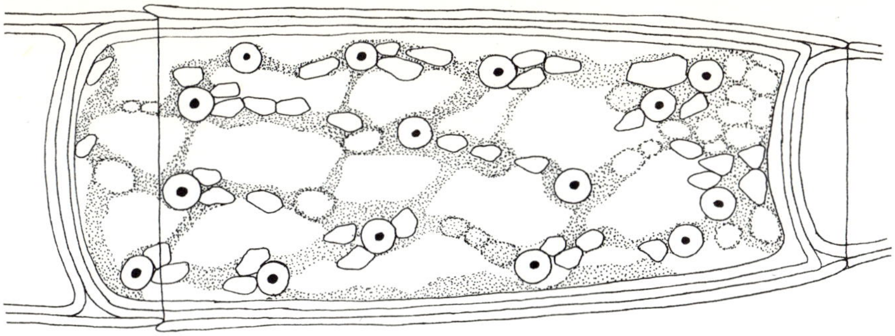

E

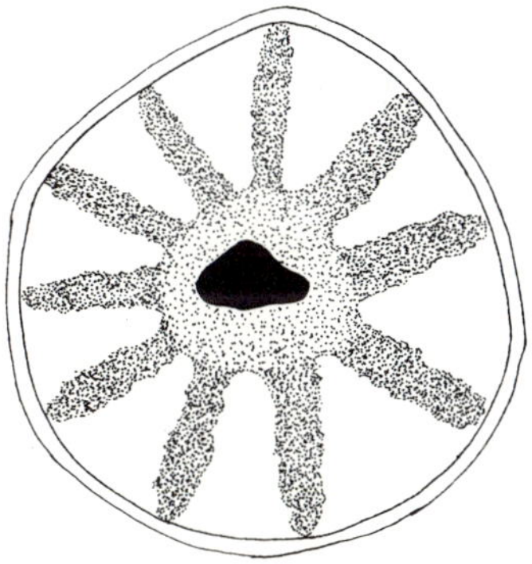

F

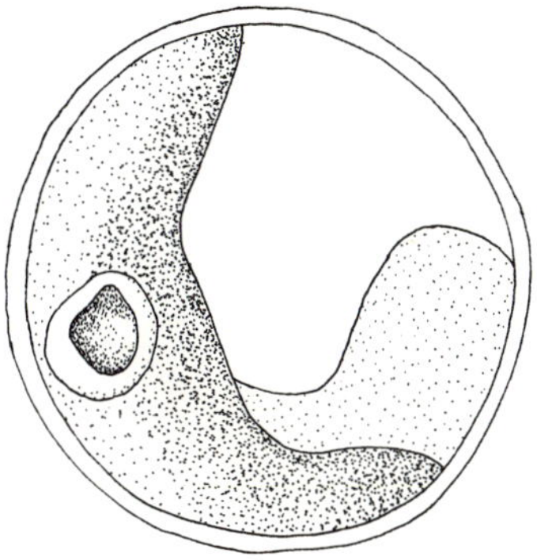

G

of similar types in lower plants including some unicellular organisms. Like mitochondria, chloroplasts may have different shapes; in fact, no other organelle of the cell varies so much in form. In *Spirogyra* the chloroplast may be ribbon-shaped, in *Chlamydomonas* it is cup-shaped, in *Ulothrix* it is curved or bracelet-shaped. In the higher plants it is usually spherical, oval, or disc-shaped and numbers about fifty per cell. Whatever its shape a chloroplast is bound by a limiting membrane and is organized internally into non-lamellated areas, the **stroma,** and small platelike or lamellated areas termed the **grana.** Chlorophyll molecules, fatty substances, and proteins are found within the grana. The individual substances are separated within the lamellated grana like the ingredients of a multidecked sandwich. The arrangement seems to be ideal for absorbing light and utilizing it in photosynthesis. The stroma is the more aqueous portion of the chloroplast, but its content of dissolved salts and enzymes indicates that it too has an important role in food synthesis.

Not all plastids contain chlorophyll. Some plastids may be found in storage tissues, where they may lack grana but store large quantities of carbohydrates, fats, and proteins. Frequently these plastids will form grana and chlorophyll when exposed to light. Other plastids, the chromoplasts, may have very little green chlorophyll but contain pigments such as carotenoids. Plant tissues, such as those in the carrot root or in the fruit of the red pepper, have numerous cells with plastids of this type. Plastids may reproduce by fission, as illustrated in Figure 5.

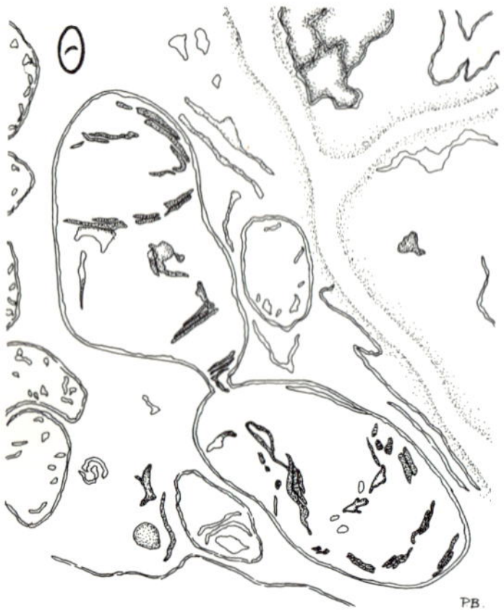

FIGURE 5 Division of a plastid in tobacco. (From an electron micrograph in Bonner and Varner.)

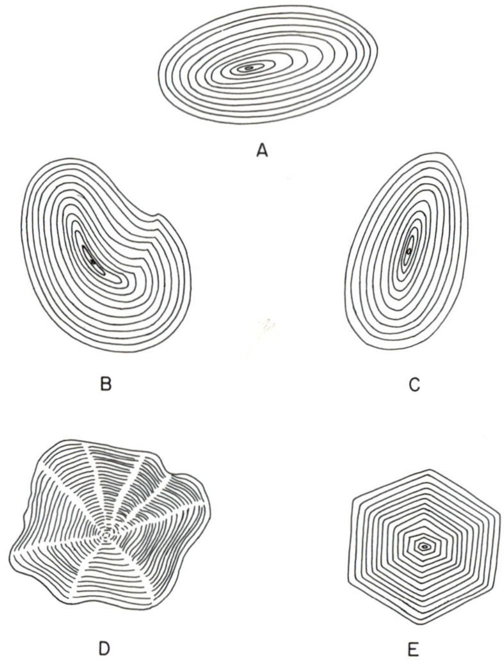

FIGURE 6 Examples of starch grains. A: potato; B: lima bean; C: wheat; D: corn; E: rice. (Adapted from Bonner and Galston.)

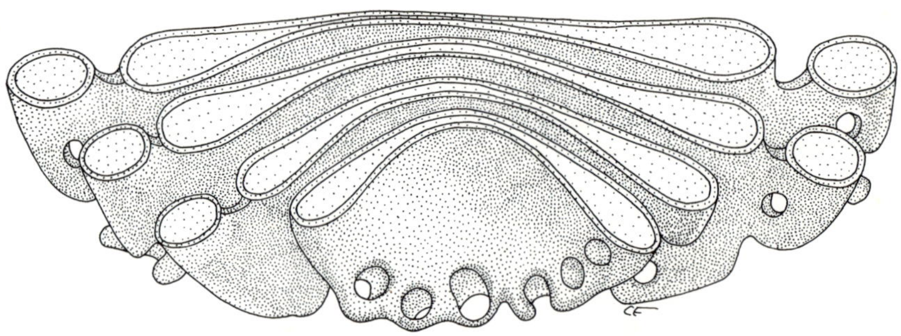

FIGURE 7 A diagram of the golgi apparatus.

Starch Grains

The most common storage product in plants is starch, and this is stored in plastid-like structures which vary in shape according to the species of plant (Fig. 6). In all, the starch is laid down in crystalline layers usually quite easy to see under the light microscope.

Golgi Structures

The Golgi structures (Fig. 7) appear as short channels or plates very much like the ER, but are smaller, discontinuous, and found in relatively compact bundles. It is doubtful that they have a direct relationship with the ER. It has been observed in some cells that with an increase in the rate of metabolism, the Golgi structures enlarge. Various enzymes and other substances collect on the Golgi membranes; these substances are more commonly lipid than proteinaceous. Although their functions have not been definitely established, they are associated with the endoplasmic reticulum and may be responsible for cellular secretions, including substances found in the cell wall.

Vacuoles

In most plant cells, particularly those which are thin-walled and very active metabolically, vacuoles occupy a large portion of the cell (see Fig. 1). They are quite variable in size depending on the amount of cell sap present, the age of the cell, and its activity. Vacuoles probably arise by expansion and coalescence of the spaces within the ER as indicated by the vacuolar membrane, the **tonoplast,** which is similar to that of the ER.

Vacuoles help to maintain the turgor of the cell. Sugars, salts, and other dissolved substances, which form the cell sap in the vacuole, provide a hypertonic fluid which causes water to move into the cell, thus maintaining cell turgidity. In some plants pigments and waste materials, in the form of crystals, are stored in vacuoles.

The Nucleus

Perhaps the most important part of a cell is its nucleus (see Fig. 1). It is recognized as the controlling center since it contains the nucleic acids, which regulate the activities of the cytoplasm, and by its genes and chromosomes determines the character of the whole organism as well. The nucleus is usually spherical in shape and densely granular. The nuclear membrane is made up of two layers, the outside member of which is continuous with the membranes of the ER. The outer nuclear membrane also has pores through which large molecules pass in and out of the nucleus. Staining reveals that the nucleus is made up of a network of fine threads, with irregularly thickened nodules. This material has long been recognized as the chromatin which later forms the chromosomes when the cell and the nucleus are in the process of division. More recently it was shown that the chromatin consists of at least four substances: DNA; RNA; a protein of low molecular weight, called histone; and another protein which can be distinguished from histone. Just how these substances are grouped together to form the chromatin network of a chromosome is as yet unknown. We do know, however, that DNA is the molecule of heredity and the manner in which its atoms are arranged determines the inheritance of the organism.

The nucleolus (there may be several nucleoli in a nucleus) is a deeply stained, rounded structure formed by certain chromosomes and closely associated with them. Its function is not certain, but it is rich in proteins, particularly the substance RNA. It is believed the RNA from the nucleolus passes into the cytoplasm where it joins the ribosomes in protein synthesis.

The role of the nucleus in directing the activities of the cytoplasm has been demonstrated by a German biologist, Max Hammerling, for a green alga, *Acetabularia,* as described in Chapter 16.

Centrioles and Lysosomes

Two additional bodies found in the cells of some organisms, chiefly animals, are centrioles and lysosomes. The centriole is a small granular body located in the cytoplasm near the nucleus or in some cases, within the nucleus. The centrioles function in cell reproduction; in flagellated cells they seem to have some control over the movement of the flagella. Centrioles are found in the cells of all multicellular animals and most unicellular animals; in the plant kingdom they are found only in the cells of some algae and fungi.

Lysosomes are subcellular sacs of approximately the same size as the mitochondria and contain digestive enzymes. The lysosomes are apparently involved in the breakdown of macromolecules and bodies as large as bacteria which may have entered the cell. Lysosomes are known to occur in most animal cells, but they have not been definitely shown to form in the cytoplasm of plant cells.

SUMMARY

We have, in Chapters 3 and 4, described the structural details and functional units of the cell as we see them today through the aid of the light and electron microscopes. We see the cell now as a dynamic and highly integrated structural unit, but our interest from here on will be on the whole organism and how cells become differentiated and specialized in the building of the entire plant.

We have had to examine the separate parts in order to learn what a cell has within it to carry on all of its various activities. As a result we have come to realize that the numerous molecular reactions that occur reflect the integrated structural organization of both nucleus and cytoplasm.

Supplementary Reading

Allen, J. M., *The Molecular Control of Cellular Activity*. McGraw-Hill, New York, 1962.

Brachet, J., The Living Cell, *Scientific American,* September 1961.

Crick, F. H. C., Nucleic Acids, *Scientific American,* September 1957.

Fawcett, Don W., *The Cell, Its Organelles and Inclusions*. W. B. Saunders Co., Philadelphia, 1966.

Fraenkel-Conrat, H., *Design and Function of the Threshold of Life: the Viruses*. Academic Press, New York, 1962.

Giese, A. C., *Cell Physiology,* 2nd ed. W. B. Saunders Co., Philadelphia, 1962.

Hurwitz, J., and J. J. Furth, Messenger RNA, *Scientific American,* February 1962.

Jellinck, P. H., *The Cellular Role of Molecules.* Scott, Foresman & Co., Chicago, 1967.

Jensen, W. A., *The Plant Cell.* Wadsworth Publishing Co., Belmont, Calif., 1964.

Knobloch, I. W., *Readings in Biological Science,* 2nd ed. Appleton-Century-Crofts, New York, 1967.

Kornberg, A., *Enzymatic Synthesis of DNA.* John Wiley & Sons, New York, 1962.

Lehninger, A. L., How Cells Transform Energy. *Scientific American,* September 1961.

Loewy, Ariel G., and Philip Siekevitz, *Cell Structure and Function.* Holt, Rinehart, and Winston, Inc., New York, 1963.

McElroy, W. D., *Cellular Physiology and Biochemistry.* Prentice-Hall, Inc., Englewood Cliffs, N. J., 1961.

Swanson, C. P., *The Cell,* 2nd ed. Prentice-Hall, Inc., Englewood Cliffs, N. J., 1964.

Questions for Discussion

1. Individual cells tend to be spherical, while those within tissues tend to be angular. Explain.
2. Of what advantage is it to protoplasm to be a colloid rather than a solution?
3. Would you expect plant cells to suffer more or less damage from radiation than animal cells? Explain.
4. Diagram a plant cell and indicate the parts which are considered living and those considered non-living.
5. What tissues have the smallest cells? The largest? How is size related to their specific functions?
6. If a tree should suddenly find its cell walls dissolved away how would its form be affected?

7. How is the absorptive surface area within a cell increased?
8. How does the construction of mitochondria and chloroplasts greatly increase their chemical activity?
9. Small cells are said to have a much greater metabolic activity than larger ones. Explain.
10. How are complex carbohydrates and other foods formed from monosaccharides?
11. How would animals be affected if their cells developed cellulose cell walls like plants?
12. How are proteins formed from amino acids?
13. Define nucleoproteins and describe their formation.
14. Describe the action of enzymes in metabolic processes.
15. What is meant by a macromolecule?
16. What do we mean by the "unity of life"?
17. What are the theories which have been developed in explaining the process of osmosis?
18. How may the permeability of the various organelle and cell membranes affect the processes of photosynthesis and respiration?
19. Describe the functions of the ribosomes, Golgi structures, mitochondria, and vacuoles.
20. How are the organelles mentioned in the preceding question morphologically constructed for their particular functions?
21. Under what conditions will dissolved substances, separated by a membrane, pass from one side to the other? What conditions might keep them from passing through?
22. Under what conditions would a greater amount of water pass from one side of a membrane to the other?
23. Apply the action of root hairs in absorption to the above two questions.
24. One method of preserving meat is to salt it. How does this keep it from spoiling?
25. In acid bogs of the North, plant and animal material is often well preserved for many years without being fossilized. Explain.
26. In many storage organs, such as potatoes, large quantities of starch are stored in the cells. How did it get there and what keeps it from leaving the cells?

5

The Plant Body

Thus far we have considered the cell as the basic unit of all organisms. In unicellular organisms the cell carries on all of its metabolic activities independent of any other cell. When combined into a multicellular organism the cells form a highly complex interdependent system which gives the organism many new capabilities, consequently we think of multicellular organisms as possessing several levels of organization contributing to life, growth, reproduction, and environmental association.

Levels of Organization

In Chapters 1 to 4 we have largely considered the initial level of organization. This included the formation of molecules and the molecular structure of the subcellular particles found in cells of various types.

As remarkable as these discoveries concerning molecular biology are, they are only the beginning of our understanding of the integrated activities of the complex multicellular organism. Above the subcellular level of organization is the cell. Each cell is a living unit capable of reproducing itself and carrying on the metabolic functions of the whole organism. Early in their formation cells are isodiametric in shape and densely granular within. Gradually they change into the functional units they are destined to become and are organized into a third level of organization, the **tissue** (Fig. 1).

Each tissue is an integral portion of the whole organism, made up of cells of similar appearance and function. Tissues in turn become organized on a fourth level, that of an **organ.** Such a structure might be a leaf, which is made up of the protective epidermal tissue, the photosynthetic tissues, the vascular conductive tissues, and the mechanical supporting tissues. Each tissue has its own functions but all are coordinated in forming the leaf as a functional organ specifically adapted for the production of food (Fig. 2).

An assemblage of organs all working together as a controlling and coordinating mechanism make up the **organism,** the highest level of independent organization. Each organism is dependent on the activities of all its organs (Fig. 3). If a plant loses all of its leaves and cannot replace them it will die in about one season; if it loses its roots it will die in a few hours or days. If it loses its stem, it will have lost its conductive channels and its growing points for continued growth.

Although we may think of the organism as the pinnacle of biological development, it is possible to think of a final and even higher level of organization. This is an organism population (Fig. 4).

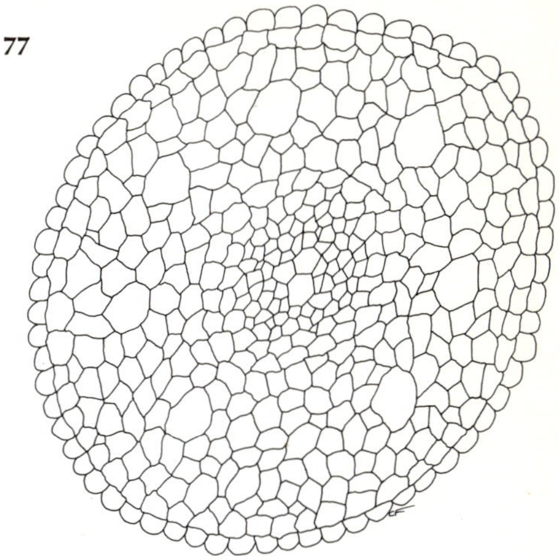

FIGURE 1 Cross section of young sunflower root. All cells are essentially isodiametric, although the central cells are beginning to differentiate into vascular tissue and the outside row is differentiating into the epidermis.

FIGURE 2 A: The fan-shaped leaf of the Washington palm. As the leaf matures the blade splits longitudinally into narrow segments. B: The pinnate leaves of the date palm. They are often eight to ten feet long.

A

B

FIGURE 3 A banana plant with its long leaves, stem, and fruit visible.

FIGURE 4 The mixed population of a rain forest in a subtropical climate. The tall trees in the center are Caribbean pines, with a few live oaks. Below the trees are some rather tall shrubs and in the foreground are low-growing thatched palms.

FIGURE 5 The sparse, mixed vegetation of a desert. In the right foreground is a jumping cholla, next to the cane-like stems of an ocotillo. In the background are palo verde trees and mesquite.

Each organism is a part of a community made up of many species of micro- and macroorganisms, and this community is every bit as much an integrated unit as the subcellular particles which make up the cell (Fig. 5).

In succeeding chapters we will consider the various organs of the plant body and survey the plant kingdom; finally the various plant populations will be described.

Parts of the Plant

It should be obvious to anyone with a reasonable awareness of the organic world that plants, like animals, have great diversity in form and structure and vary from one-celled organisms to large multicellular bodies of great complexity. The body (Fig. 6) of the typical land plant is centered around an axis consisting of (1) the **root,** which grows downward into the soil where it anchors the plant and serves as the absorptive organ for water and minerals (Fig. 7); and (2) the **shoot,** which serves as the aerial organ carrying the green leaves into the light and atmosphere where they carry on photosynthesis (Fig. 8). The shoot matures into the **stem,** which produces leaves at its **nodes.** The intervals between the nodes are called **internodes.** In perennial stems the nodes and internodes gradually disappear as the diameter of the stem increases since leaves are produced only on the actively elongating shoot. Growth in length of plants, unlike that of animals, occurs mainly at the tips (stem and root), with increase in diameter occurring at certain specialized embryonic areas such as the vascular cambium and cork cambium. In addition to leaves the shoot system also produces the flowers, fruits, and seeds, the specialized organs of reproduction in the higher plants.

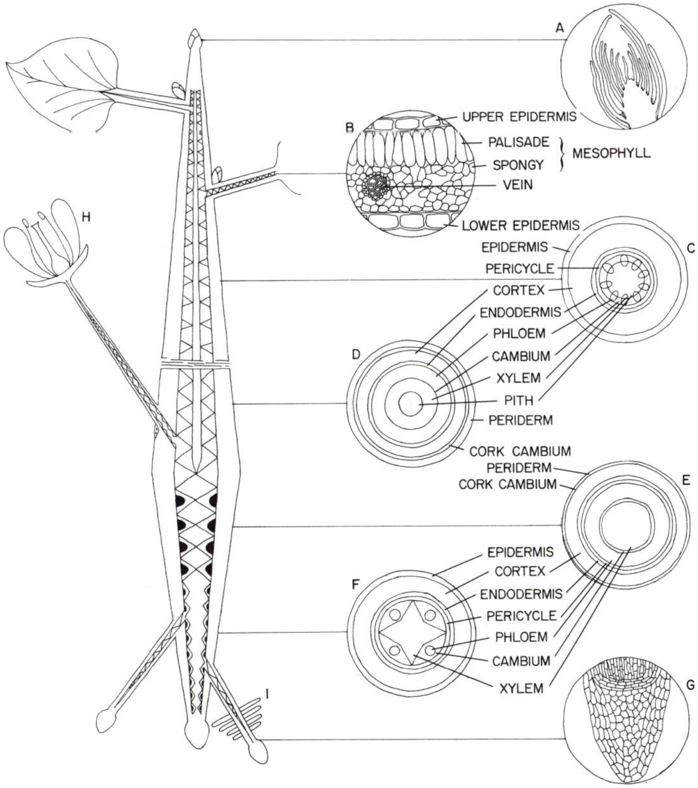

FIGURE 6 Diagrammatic illustration of a plant. A: longitudinal section of the apical bud; B: cross section of a leaf; C: cross section of a one-year-old stem; D: cross section of a two-year-old stem; E: cross section of a two-year-old root; F: cross section of a one-year-old root; G: longitudinal section through a root tip; H: a flower; I: lateral root with root hairs.

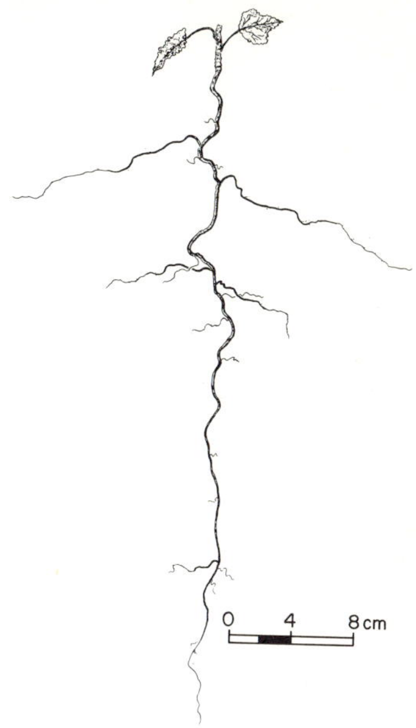

FIGURE 7 The root of a globe mallow plant. Several lateral or secondary roots arise from the long central tap root.

FIGURE 8 A cluster of bamboo plants, which consist mostly of stems with narrow leaves.

The Root

The higher plant begins life as an embryo in the seed, where the main axis consists of the primordial shoot tip and root tip (Fig. 9). As the root emerges from the seed coat it grows downward, regardless of the position of the seed in the soil. The root has a rather tough cap composed of undifferentiated cells which form a cushion over the growing tip. The outer cells of the tip produce new cells and tissues upward as the root elongates and also add new cells to the cap as the cushion cells wear off and are thrust aside. The growing point of cells in the root is the apical meristem, where division of cells occurs rapidly when the environmental conditions for growth are favorable. Farther upward, on the somewhat older portion of the root (usually only a few hundred microns from the tip), the cells are no longer very active in division but are elongating and beginning to differentiate into the more specialized cells and tissues of the mature plant body.

As the primary root penetrates the soil and gains its first foothold, tiny root hairs form as cylindrical protrusions from the outside walls of the epidermal cells. The root hairs absorb water and minerals from the film of moisture on and around the soil particles. Somewhat later, but still within the first few days of growth, lateral roots begin to arise from the main axis of the primary root and their cylindrical bodies add to the initial anchoring and absorptive organ of the young seedling. The structural framework of the subterranean plant part is now established. No further discussion of this organ will be given here. Detailed description of the form, structure, types, and metabolism of roots will be presented in Chapter 9.

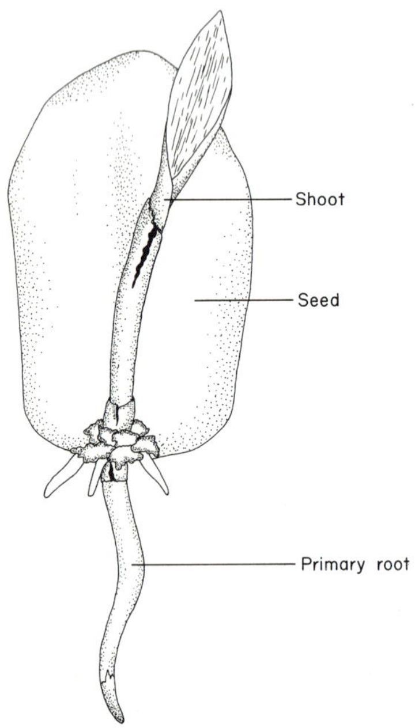

FIGURE 9 An embryo of wheat which has emerged from the seed and is establishing its first root and shoot.

The Stem

The shoot of the developing embryo has a history very similar to that of the root except that it grows upward regardless of the position of the seed. The shoot tip has a number of meristematic cells at its apex which form a dome-shaped mass of rapidly dividing isodiametric cells. As in the root, those outer cells are cubical in shape but towards the base of the apex they gradually elongate and begin to differentiate into the tissues which will form the primary body of the aerial portion of the plant. The maturation and differentiation of cells from this organ will result in conductive, protective, storage, and secondary embryonic tissues.

Although the shoot tissues differentiate in a manner very similar to that of the root, their functions are, in general, quite different. Whereas the root is the primary absorptive organ for the plant, the stem with its leaves is the primary food producing region.

The young stem tip, including the apical region of rapid division and the elongating and differentiating cells, is enclosed by embryonic leaves and protective bud scales (Fig. 10). The entire structure is the bud, really an embryonic twig, which becomes apparent the next year when the growing points actively develop into an elongated stem with leaves and lateral buds in their axils. These lateral or axillary buds will then, the following year, develop into new lateral branches bearing leaves, or in some cases buds may contain embryonic flower buds and develop successively into flowers, fruits, and seeds.

FIGURE 10 Longitudinal section through a bud, showing outer bud scales and inner embryonic foliar leaves.

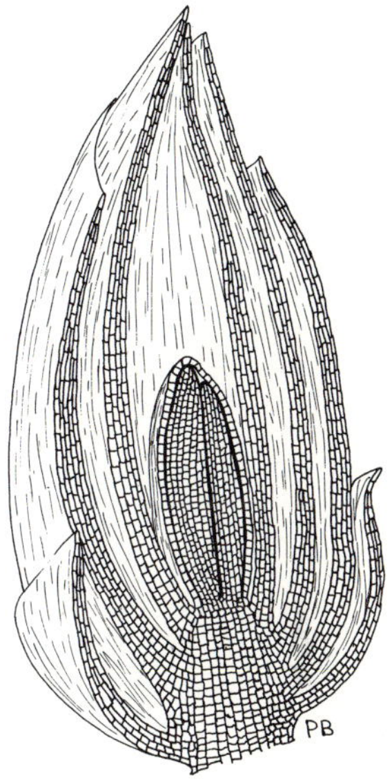

Leaves

The leaf is generally considered as the flattened or expanded extension of the stem (Fig. 11). The expanded portion is the **blade,**

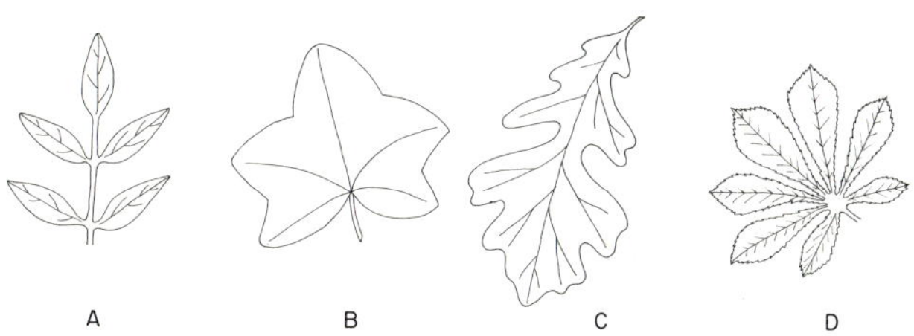

FIGURE 11 Leaf types showing simple and compound. A: portion of pinnately compound leaf of ash; B: palmately lobed and palmately veined leaf of maple; C: pinnately lobed and pinnately veined leaf of oak, D: palmately compound leaf of horse chestnut.

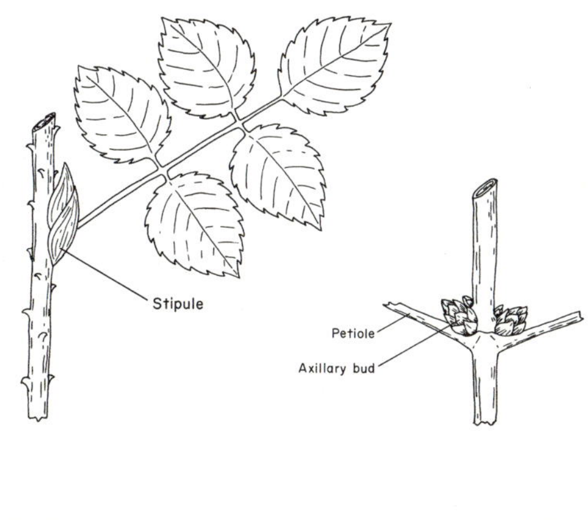

FIGURE 12 A. *Left:* Portion of rose stem showing one pinnately compound leaf with stipules at base of petiole. *Right:* Portion of ash stem showing axillary buds in axis of petioles. (After Hoag)

and the stalk which connects the leaf to the stem is the **petiole.** The petiole is absent in some species, in which case the leaf is said to be sessile. Some leaves also have small spinelike or leaflike appendages termed *stipules* at the base of each leaf. In the axil, where the leaf joins the stem, is the axillary (lateral) bud. In some species there may be more than one bud in this position. The pattern of venation is easy to see on the underside of leaves because the veins usually appear prominently as raised lines or ridges. The larger veins are also somewhat raised above the upper surface in some leaves.

It has been stated above that growth in stems and roots is chiefly apical, with continuous growth from the tips as long as the plant lives. Leaves vary from this manner of growth. They begin to develop from clusters of meristematic cells at their apex, but secondary meristematic groups of cells arise near the margins, adding cells to the length and width of the growing leaf. In the case of monocotyledonous leaves, such as grass, the meristematic region is near the base (an intercalary meristem), where cells are continually added, permitting the leaf to continue its growth from this point rather than at the apex. This makes it possible for us to mow the lawn without seriously injuring the grass.

There are many different modifications of the leaf organ from the expanded type. These include the needles of pine trees, the spines of certain cacti, tendrils of many climbing plants, the thick storage leaves of succulent plants such as the agave, and the thick concentric leaves which make up the bulb of the onion. The functions of the leaf and the various types of leaves and their variations are discussed in Chapters 6 and 7.

Uses of the Vegetative Parts

The root, stem, and leaves are the vegetative organs of the plant, and some plants of economic importance are raised for no other reason than for one or more of these parts. Some roots such as the sweet potato and sugar beet are raised for the storage products they contain. The stems of sugar cane and the tuber (a modified stem) of the Irish potato are used for food, while the stem of flax provides us with a fiber from which linen is made. The leaves of lettuce, cabbage, and spinach are a few of those which provide us food products while several species of agave, pineapple, and banana have tough fibers in their veins which we use in the manufacture of coarse ropes.

Many other products are obtained from the tissues of these vegetative parts including drugs (quinine from the bark of the cinchona tree), lumber from the xylem of numerous woody species, beverages, such as tea from the leaves of *Thea sinensis,* and condiments such as sage and bay. Tannins, dyes, and other products come from these vegetative organs and many will be discussed in Chapter 28.

FIGURE 13 Vegetative reproduction in strawberries. The lateral branches give rise to new plants at their tips.

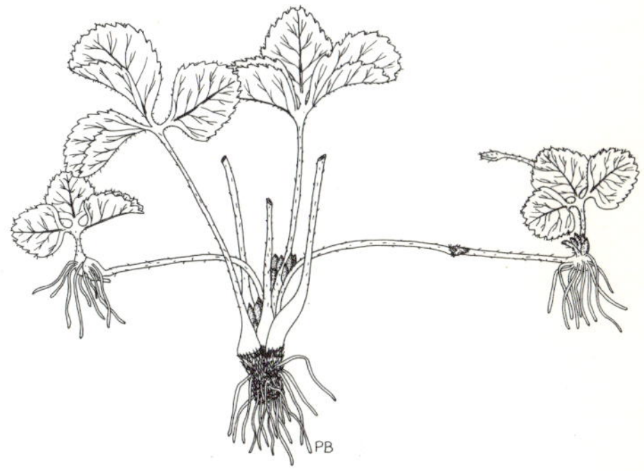

The Flower

Many plants may reproduce by slips, cuttings, or runners (Fig. 13) from the vegetative organs, but sooner or later all of the angiosperms (flowering plants) produce flowers. In the case of annuals, this is in the first year; biennials produce flowers usually in the second year, and perennials often require several years before the first flower is matured (Fig. 14).

The flower (Fig. 15) is the sexual reproductive organ of the plant. While we will not go into the different types in this chapter, it would be well to learn the names of the parts of a typical complete flower as illustrated in Fig. 16.

FIGURE 14 *Agave parryi,* the century plant. It takes from 10 to 20 years for a plant to produce the flower stalk shown here, then the plant dies. Usually before the plant dies small plants are produced around the base. This cluster of plants was probably produced over ten years ago by an earlier plant.

FIGURE 15 *Top:* The flower spike of Bird-of-paradise flower, *Caesalpinia gilliesii.* The slender exerted stalks are the filaments of the stamens. *Bottom:* Flower of a cactus. It has numerous stamens and perianth segments. The ovary is only slightly visible in the lower center of the flower. Cactus flowers rarely stay open for longer than one day.

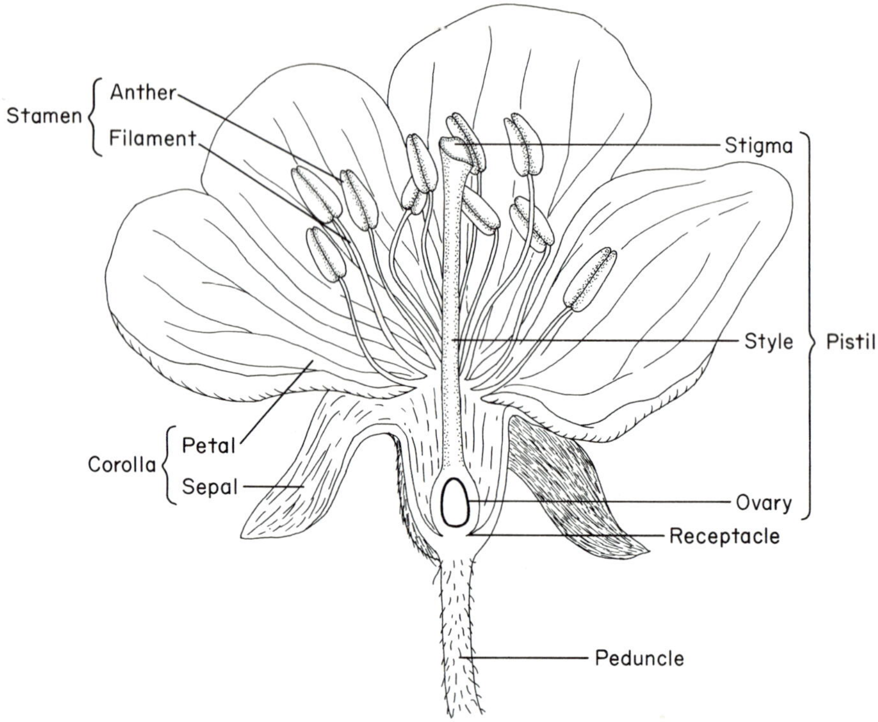

FIGURE 16 Diagram of a complete flower.

Supplementary Reading

Delevoryas, Theodore, *Plant Diversification*. Holt, Rinehart and Winston, New York, 1966.

Esau, Katherine, *Vascular Differentiation in Plants*. Holt, Rinehart and Winston, New York, 1965.

Galston, Arthur W., *The Life of the Green Plant,* 2nd ed. Prentice-Hall, Inc., Englewood Cliffs, N. J., 1964.

Salisbury, F. B., and R. V. Parke, *Vascular Plants: Form and Function*. Wadsworth Publishing Co., Inc., Belmont, Calif., 1964.

Questions for Discussion

1. Compare and contrast the cells, tissues, and organs of a plant with those of an animal.
2. Would it be more advantageous for survival of the species for a plant to be an annual or a perennial?
3. In what ways are plants capable of reproduction?
4. Name some products of economic importance obtained from plant organs.
5. What are the various levels of organization in plant structure?
6. How does the type of growth found in plants contribute to the typical shape we observe in such plants as a pine tree, an oak, and an iris?
7. How does a large population of many species within an area contribute to biotic development?
8. Herbaceous stems have most of their supporting tissues in the form of a cylinder while a young root has most of its thick-walled cells in the center of the structure forming a rodlike organ. Of what advantage is this type of tissue development to the plant parts in the immediate environment in which they grow?

6

The Leaf

One of the most remarkable organs in the world of plants and animals is the green leaf. It has the ability to absorb the energy of sunlight and, through the process of photosynthesis, convert this radiant energy into a form that can be readily utilized. All of the foods the green plant uses are synthesized within it from simple inorganic substances such as carbon dioxide, mineral salts, and water. These, with the sugar products formed through photosynthesis and related chemical conversions, produce the proteins, fats, and complex carbohydrates used in growth by plants as well as all other organisms.

Form and Structure

The substance responsible for the first carbohydrate molecules formed is **chlorophyll,** which is normally a mixture of two related pigments, chlorophyll **a** and chlorophyll **b.** In the vast majority of plants chlorophyll is localized in **plastids,** located in the cytoplasm in nearly all cells of the leaf except those of the veins and the protective epidermal cells. These **chloroplasts** (Fig. 1) are continually moving around in the cytoplasm so that first one, then another is placed nearer the surface of the leaf.

The efficiency of leaves in absorbing light depends to a large extent on their form and structure. The fact that leaves are usually very thin structures makes it possible for light to penetrate nearly all cells. Leaves of dicotyledonous plants have a relatively broad expanded **blade** or **lamina** attached to the stem by a cylindrical **petiole** (Fig. 2). The leaves may be produced singly at a node, in which case they are **alternate** in their arrangement. When two leaves arise at the same node, on opposite sides of the stem, they are **oppositely** arranged and if three or more arise at the same node they are **whorled** (Fig. 3).

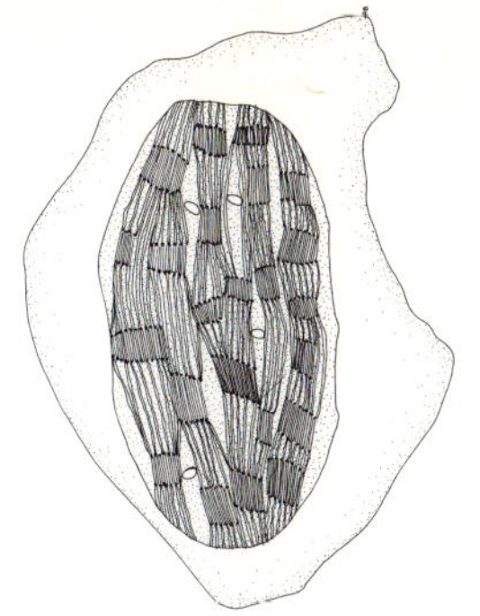

FIGURE 1 Chloroplast of a higher plant. (Redrawn from an electron-micrograph in Rabinowitch and Govindjee, *Scientific American,* July, 1965.)

FIGURE 2 Elm leaf with pinnate venation and short petiole. This is a simple leaf.

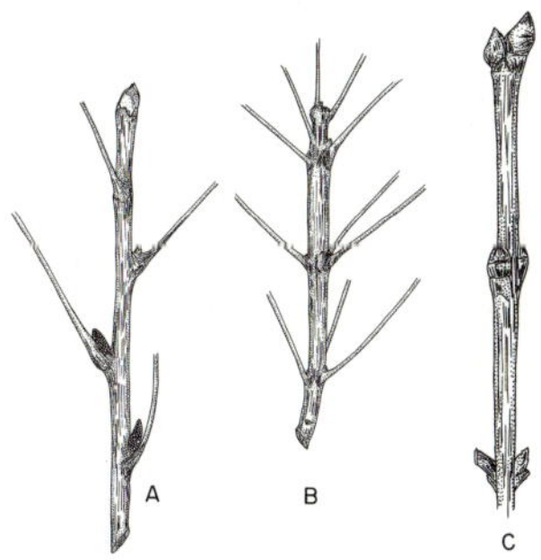

FIGURE 3 Types of leaf arrangement. A: English walnut with alternate arrangement; B: *Catalpa,* whorled; C: Lilac, opposite.

The blades of different species vary considerably in shape. Some are narrowly elongated, such as the willow leaf. The cottonwood leaf is usually deltoid, a catalpa leaf is heart shaped, and an elm may have an ovate leaf. There are many other modifications in leaf shape and many leaves are variously indented. Some have sharp, **serrate** toothlike **margins** (Fig. 4); others have larger and

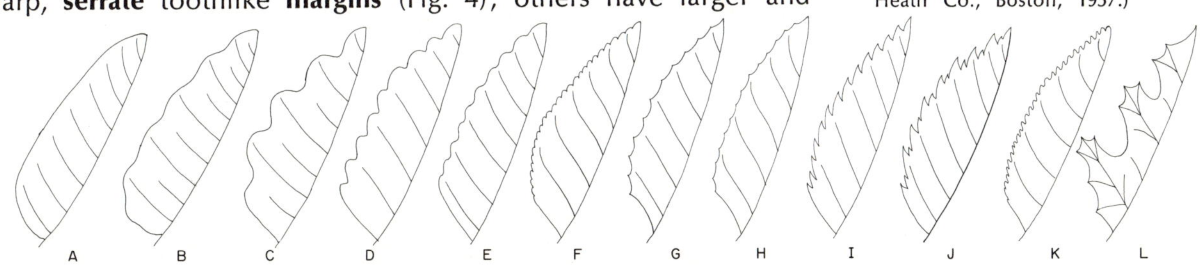

FIGURE 4 Leaf margins: A-entire; B-undulate; C-sinuate; D-crisped; E-crenate; F-crenulate; G-dentate; H-denticulate; I-serrate; J-doubly serrate; K-serrulate; L-pinnately lobed; M-retrorse. (Redrawn in part from Benson, *Plant Classification,* D.C. Heath Co., Boston, 1957.)

FIGURE 5 Palmate venation and lobing in maple leaf.

coarser teeth and are said to be **dentate.** If the teeth points are rounded the margin is **crenate.** Some leaves, such as those of certain species of oak and maple, are **lobed.** In oaks this lobing occurs along the side, extending almost into the midrib, and the leaf is termed **pinnately** lobed, while in maple leaves the lobes extend outward from the union with the petiole and the leaf is therefore **palmately** lobed (Fig. 5). If a leaf is without lobing or indentations of any kind in the margin, it is said to be **entire.**

All of the leaf types described above are **simple** (Figs. 2, 5), since they consist of but one blade, even though they are quite variable in shape. A second type of leaf is **compound** (Fig. 7). We

FIGURE 6 The small simple leaves of the Ocotillo persist for only two or three weeks during moist weather. They fall off during dry weather but may be replaced several times a year if moisture returns. The cluster of flowers at the apex is bright red.

FIGURE 7 The compound leaves of mesquite. The light colored structures which appear fluffy are flower clusters.

determine whether a leaf is simple or compound by the position of the axillary bud, which is found in the axil of a leaf but not in the axil of a **leaflet.** Leaflets are the separate parts of a compound leaf and, for all practical purposes, are leaves except for the fact that they are without the bud in their axils where they join the **rachis,** which is an extension of the petiole. Leaflets also lie in a single flattened plane, while leaves arise in different positions on the stem, and generally at the time of leaf fall the whole leaf drops at once rather than the individual leaflets.

Pinnately compound leaves (Fig. 8) have leaflets arising laterally along the rachis, while in **palmately compound** leaves (Fig. 9)

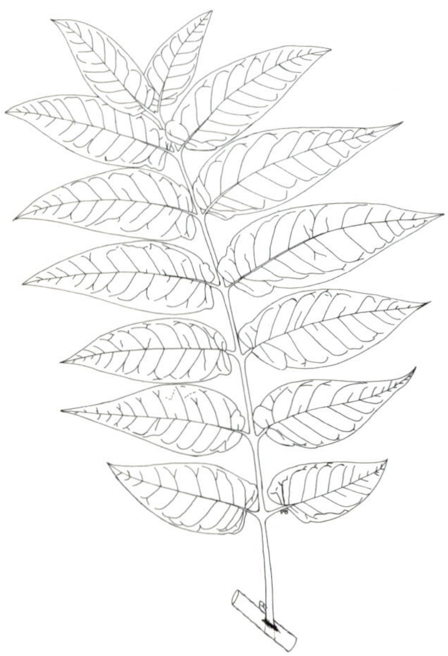

FIGURE 8 Pinnately compound leaf of tree-of-heaven.

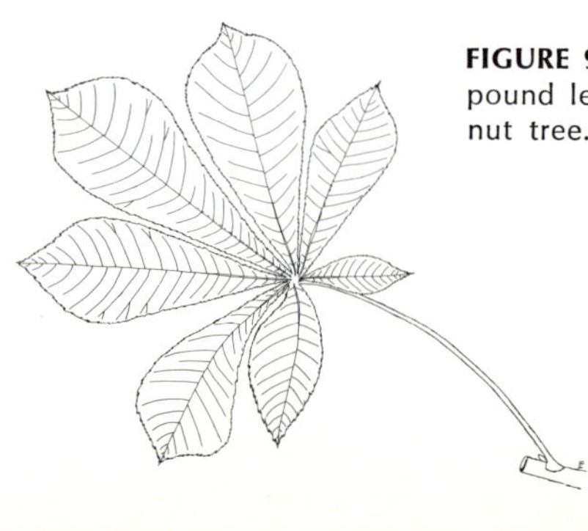

FIGURE 9 Palmately compound leaf of horse chestnut tree.

the leaflets arise from a common point at the tip of the petiole. Honey and black locust trees, box elders and roses produce pinnately compound leaves. Horse chestnut trees and the vine, Virginia creeper, have palmately compound leaves.

In dicotyledonous leaves, arrangement of the veins is **netted.** In this arrangement there are one or more prominent veins, usually in the center of the leaf or in the center of the larger leaf divisions, from which the smaller veins arise, giving rise in turn to the very tiny veinlets, producing a net-like appearance and termed **net venation** (see Fig. 2). When there is a single midrib with secondary veins arising laterally from it, the leaf is pinnately net-veined, as in the elm or oak. When several prominent veins diverge like fingers from the base of the blade, the leaf is palmately netveined, as in the maple (see Fig. 5) and redbud.

Monocotyledonous plants have leaves that are in general elongated and comparatively narrow (Fig. 10). However, there is almost as much diversity in leaf form in monocots as there is in the shape of dicot leaves. Grasses and plants such as *Iris* and lilies have flat, narrow, elongated leaves, while chives and onions have tubular leaves, the leaves of palms (Fig. 11) are large, fanlike and often variously incised and lobed, and *Smilax,* a small vine, has ovate leaves not greatly different in shape from those of an elm tree.

The leaf of a grass plant (Fig. 12) is divided into two parts: (1) the basal portion is the **sheath** which goes around the stem as in corn, where it is prominently displayed; and (2) the apical portion, the **blade,** is the typical leafy part.

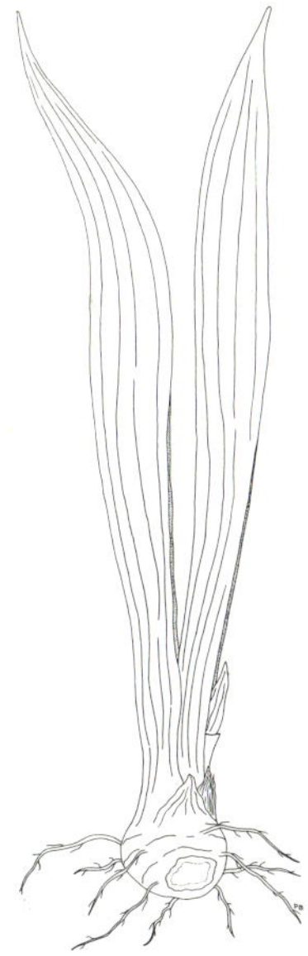

FIGURE 10 *Iris,* a typical monocotyledonous leaf with parallel venation.

FIGURE 11 The Washington palm. A monocotyledon with broad dissected and essentially parallel leaves.

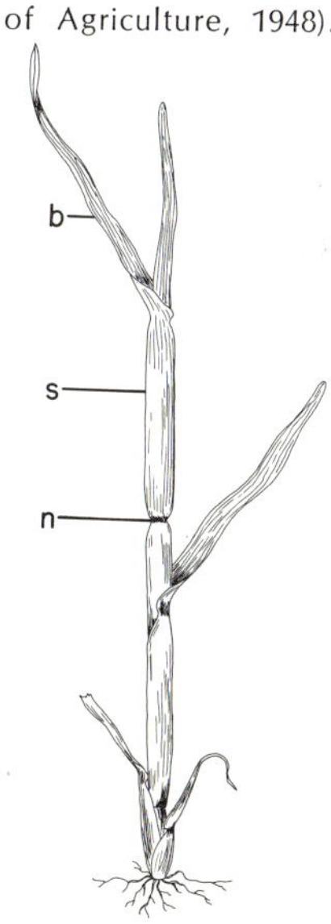

FIGURE 12 Leaf of grass. b–blade, n–node, s–sheath. (From *Grass,* Yearbook of Agriculture, 1948).

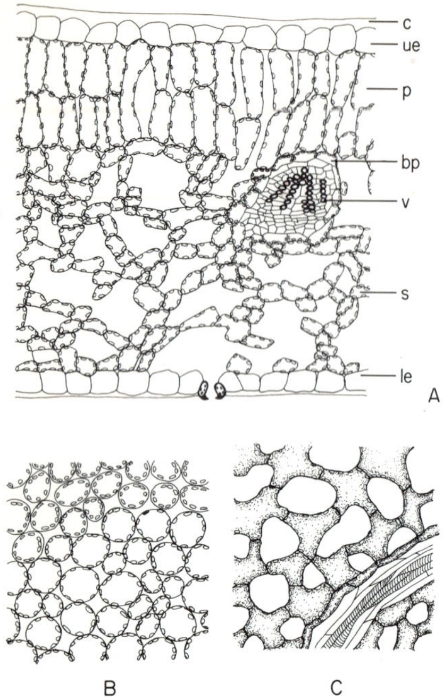

FIGURE 13 A-Cross section of a typical dicotyledonous leaf. c-cutin, ue-upper epidermis, p-palisade parenchyma, v-vein, bp-border parenchyma, s-spongy parenchyma, le-lower epidermis. B-paradermal section of leaf showing palisade cells in cross section. C-paradermal section through spongy layer showing irregularly shaped cells and vein in longitudinal section.

The venation of monocotyledons is usually **parallel,** with veins of about equal size extending side by side through the length of the leaf. These larger veins are joined by small inconspicuous veinlets. In some large monocotyledon leaves, like those of the banana or canna, there is a very prominent midrib but the secondary veins which arise from it run at a tangent and parallel to each other.

The above characteristics are especially important to us in the identification of plants. To the plant, they are important survival characteristics, but when it comes to the metabolic activities associated with leaf structure the one thing most leaves have in common—thinness—is most significant.

The anatomy of the typical dicotyledonous leaf blade (Fig. 13) includes an outer **epidermis,** the tissues enclosed by the epidermis, termed the **mesophyll** (middle leaf), and the network of **veins.**

The Epidermis

The epidermal cells usually appear brick-shaped in cross section, but in face view they are usually irregular in shape with interlocking lobes. The outer wall is cutinized, especially that of the upper epidermis, although in many leaves the lower epidermis may be cutinized as well. The epidermis forms a tight framework over the cells of the mesophyll, protecting these cells from mechanical injury and from drying out.

The epidermis usually contains modified cells other than those making up the integumentary layer. The most important of these are the **guard cells** which border a tiny pore, the **stomate** (Fig. 14).

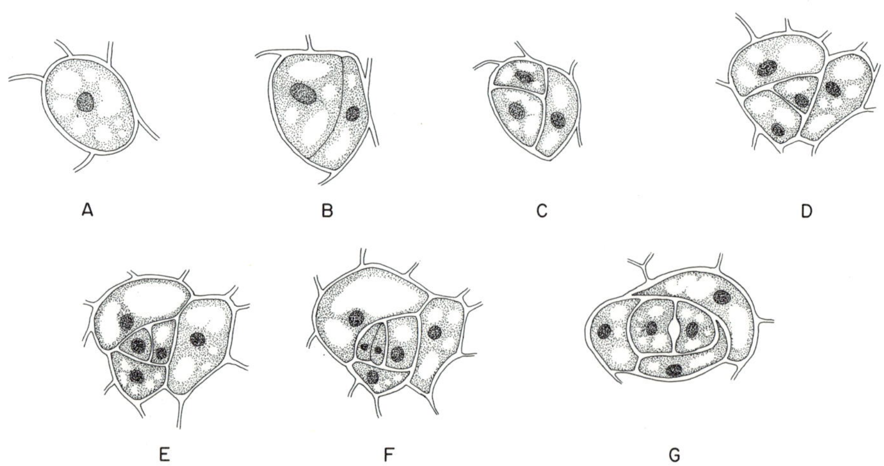

FIGURE 14 Development of a stomate showing division and differentiation of cells in the epidermis to form the pore and guard cells in G.

These modified structures are usually much more abundant on the leaf's lower surface than above, but in different species they may be formed in equal numbers on both surfaces, or may be completely lacking in the upper epidermis or, rarely, in the lower epidermis.

Other modified cells which arise from the epidermis are hair cells and glandular cells (Fig. 15). These may be portions or protrusions of the outer epidermal wall or may develop as multicellular structures from the epidermis. Glandular hairs may be sessile or stalked and are usually sticky to the touch.

The guard cells regulate the opening and closing of the stomates, which in turn regulate the movement of water vapor, oxygen, carbon dioxide, and other gases into and out of the leaf. When the plant has an adequate supply of moisture, the stomates are usually open due to the **turgor** pressure within the guard cells.

FIGURE 15 Epidermal hairs. A-glandular. B-filamentous, septate type. C-stellate type.

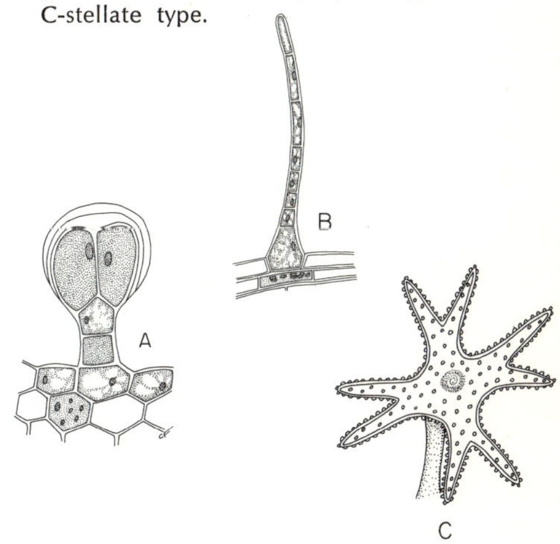

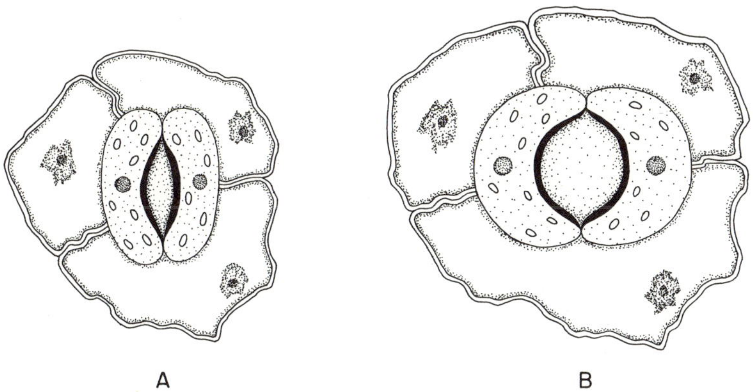

FIGURE 16 Stomates in leaf epidermis. A-partially closed stomate due to decrease in pressure in the guard cells. B-open stomate with expanded guard cells. Note thickened wall on inner portion of guard cells. (After Coulter and Dittmer.)

The latter normally occur in pairs, and together in face view look very much like a tire's inner tube. The inner walls are very much thicker than the outer, so that when turgor pressure is great, the outer portion is pushed outward, towards the surrounding epidermal cells, pulling the inner wall with it like an inflated balloon with unequally thickened walls. This enlarges the pore opening and free diffusion of gases occurs. If wilting should occur the turgor pressure reduces the extension of outer walls of the guard cells, no longer forcing the inner walls apart, the size of the pore diminishes, and there is less diffusion of water vapor and gases.

The opening and closing of the stomates is not as simple as the description given above, where the operation seemed to be merely a water pressure phenomenon. Guard cells often contain chloroplasts and, as a result, manufacture sugar under light conditions. The sugar often diffuses into the cell sap in the guard cells and some of the sugar in the chloroplasts of the guard cells is condensed into starch. Investigations have shown that during the day, when the concentration of sugar increases as the result of photosynthesis, the osmotic pressure of the guard cell's sap increases, water diffuses in from adjacent cells, increasing the turgor pressure of the guard cells, and the stomate opens. At night, or when the rate of photosynthesis is reduced, the amount of sugar is decreased because much of it is transformed into starch. This reduces the osmotic pressure of the cell sap in the guard cells, causing the water to move from the guard cells to the nearby epidermal cells and reducing the turgor pressure on the guard cells, which then become flaccid and close or at least decrease the stomatal opening.

Factors other than water, light, enzymes, and sugar content, such as temperature, may have a bearing on the opening and closing of stomates, but in the final analysis, the construction of guard cells is such that anything which controls the turgor pressure of the cells will have an effect upon the opening and closing of the stomate.

The Mesophyll

Within the two layers of the epidermis lies the greatest mass of tissue of the leaf. It is usually divided into two distinct regions which we term the **palisade parenchyma** and the **spongy parenchyma.** The upper of the two is the palisade, which is made up of elongated cylindrical cells oriented in a position at right angles to the leaf surface. Cells of this layer contain the greatest concentration of chloroplasts, which are located towards the periphery of each cell. Plastids move continuously up and down within the cell, so that first one plastid, then another, is nearer the source of light shining through the upper leaf surface.

Cells of the palisade region (there may be one or more layers of palisade cells) have numerous air spaces between them. If a section is cut parallel to the main axis of the leaf (we call this a paradermal section) the palisade cells appear circular in outline with some cells touching but often with air spaces between them (see Fig. 13B).

The spongy parenchyma layer which lies below the palisade is in close association with the opening of the stomates in the lower epidermis. The cells themselves are often isodiametric, but usually they are variously lobed (see Fig. 13C) and form a loose network of cells which seldom are in contact with each other on all sides.

As a result, large **intercellular spaces** are very prominent in this layer, forming an interconnecting network much like the tunnels of a branching cave system or the ramifying canals of a sponge. As a result of these air spaces ready diffusion of carbon dioxide and oxygen takes place throughout the spongy mesophyll. The gases diffuse readily through the interstices of the palisade layer. Chloroplasts are less numerous in the spongy mesophyll cells, but most cells have them. It is estimated that there are two to four times as many chloroplasts in palisade cells as in those of the spongy region.

An interesting study made several years ago by Turrell revealed the total amount of internal cell surface exposed by cells of the mesophyll. He had an opportunity to count the leaves on a large catalpa tree which had been taken down. He found that there were 26,040 leaves on this tree and after sorting them into different sizes for ease in measuring, calculated that in all, they had 390 square meters of externally exposed surface and that stomates in the epidermis numbered about 1,850,000,000. Turrell then made microscopic sections, both cross and paradermal, and by a mathematical formula based upon the size and shapes of cells and size of the air spaces calculated that the internal surface exposed by the mesophyll cells was about 5,100 square meters, most of this exposed surface being in the palisade layer.

The significance of investigations of this nature becomes apparent when we realize that all substances entering or leaving a leaf go through these surfaces. The large network of intercellular spaces form a reservoir for the carbon dioxide and oxygen which diffuse through the cell surfaces in the processes of respiration and photosynthesis.

Veins

The transportation system of the leaves is composed almost entirely of the anastomosing network of veins. Veins are located in all parts of the leaf, with rarely more than two or three cells separating one small veinlet from another. The conducting elements, which make up the greatest portion of the veins, are xylem and phloem. As in all vascular tissues, the xylem conducts water and minerals to the various cells of the mesophyll and epidermis. The products of photosynthesis move downward in the phloem to the stem, where the conducting elements continue on into the roots and their branches.

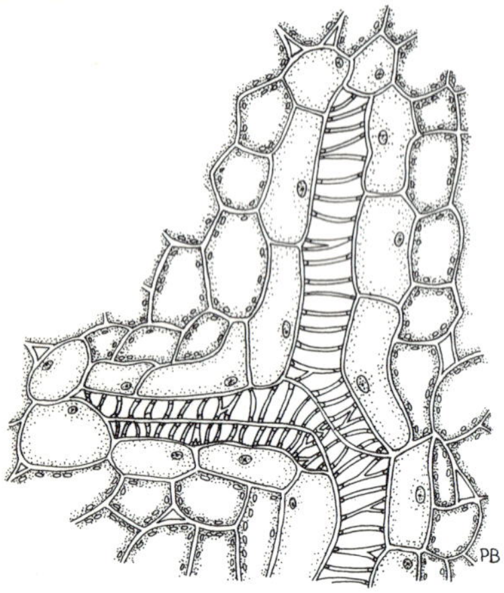

FIGURE 17 Vein ending in the mesophyll of a leaf. The cells adjoined to the tracheids are termed border parenchyma. (Redrawn from W. W. Robbins and J. E. Weier, *An Introduction to Plant Science.)*

Besides xylem and phloem, the large veins may also contain thick-walled fibers, and these may be surrounded by layers of thick-walled and thin-walled parenchyma cells, giving the added function of support to veins of the leaf. As the vein network extends to the more distant areas of the leaf, the veins decrease in size. The ends of the veinlets usually consist of no more than a single water-conducting element, a **tracheid,** but it is surrounded by a layer of thin-walled cells, the **border parenchyma** (Fig. 17), which aids in the diffusion of water from the tracheid to the adjacent cells.

Leaf Fall

It is characteristic of all perennial plants to lose their leaves at various times. This may be seasonal, as in autumn, when the leaves of most deciduous plants fall at about the same time, or it might be a more or less continuing process in evergreens where a few leaves drop at a time after they are several years old and new leaves have already developed on the new shoots.

We often think that leaf fall is associated with frost, and certainly a very early frost will hasten the process, but it is likely that the length of the light or dark period, **photoperiodism,** has a much greater effect in stimulating the production of growth-promoting substances, such as hormones, which initiate the growth of specialized cell layers at the base of the petiole, cutting off the leaf. This specialized zone is known as the **abscission layer** (Fig. 18), and forms at different times in different species. The cells in this layer are usually short, lack lignin, usually have suberin in their walls, and definitely form an area of mechanical weakness. They are protective in that they retard water loss and also protect the stem from the invasion of fungi, bacteria, and even insects.

The actual fall of leaves may be due to a dissolution of middle lamella layers through cells of the abscission layer, or the cells themselves may disintegrate forming a weakened layer through the abscission zone.

It is obvious to most of us that deciduous trees in the temperate zone vary tremendously in the time when leaf fall is accomplished. It is the end of September as this is being written. Honey locust trees have been losing their leaves in our area for about the past two weeks; cottonwood leaves are just beginning to turn yellow; no change seems to have taken place on the leaves of

FIGURE 18 Abscission layer beginning to form at base of petiole. (Courtesy Carolina Biological Supply Company.)

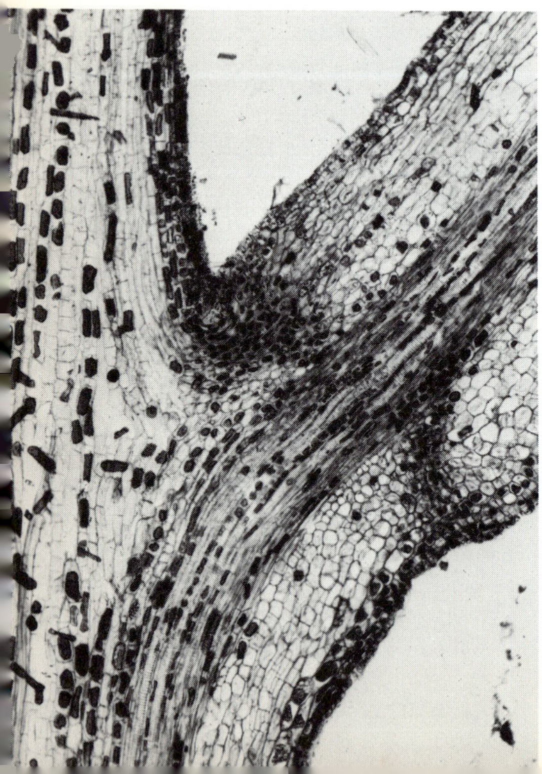

Chinese elms or willow trees. The leaves of my neighbor's American sycamore tree turn brown every year around the first of September, while my European sycamore, less than fifty feet away, doesn't begin to turn color until after the first of October. Everyone could make the same observations on trees in his area. In this section of New Mexico the first frost of the year will come about October 20, yet by that time many species will have lost most of their leaves. Trees with leaves that change color will already have changed; on other species, where the leaves normally remain in a brown, nonfunctional condition, the leaves will continue to remain on the trees for weeks or even months longer.

Leaf Coloration

Fall coloration in leaves of deciduous plants is apparently due to a combination of several factors, both external and internal. Only certain species turn decidedly different colors from the usual green. Many species of maple turn various shades of red, as do sumac and Virginia creeper, as a result of an increase in the anthocyanin pigments and carotenoids. These plants are most colorful when the fall weather has been fair with an abundance of sunshine, when the cells in the leaves have an adequate supply of sugar, and when tannin is present.

The transformation occurs when the green color of chlorophyll fades as a result of decomposition. The carotene and xanthophyll pigments become more prominent and in some species of *Populus,* like valley cottonwood, the leaves become a brilliant yellow and gradually change from green to yellow to orange to brown. In aspens the leaves fall soon after turning yellow, but in the cottonwood they may remain on the tree throughout much of the winter, long after they have turned brown and ceased functioning.

Modified Leaves

The preceding discussion has been mainly about foliage leaves, which are, of course, the most common, but there are certain modified leaves which bear little resemblance to the usual types.

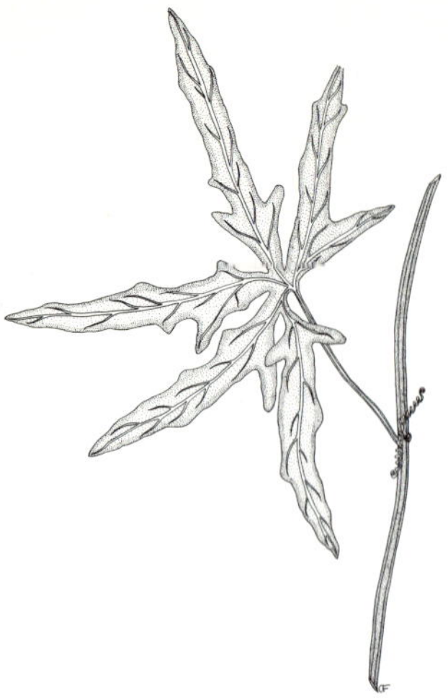

FIGURE 19 Diagram of the gourd plant *Cucurbita digitata* showing tendril which in this case is a modified leaf.

These structures include bud scales, tendrils, thick fleshy storage organs, spines, and the unusual insect-catching organs of some insectivorous plants.

Bud scales are short, thick resinous leaves which form overlapping layers of protective scales over the young growing tip of the stem.

Tendrils (Figs. 19, 20) may be a modification of an entire leaf or of only a leaflet. The tendril is a slender, elongated cylindrical organ which entwines itself around supporting structures, such as a nearby plant or object. Peas, grapes, vetch, *Clematis,* and other plants frequently develop tendrils and grow up on supporting objects.

Some plants such as the century plant *(Agave)* (Fig. 21), *Sedum, Bryophyllum,* and other succulents have thick storage leaves containing water or food materials.

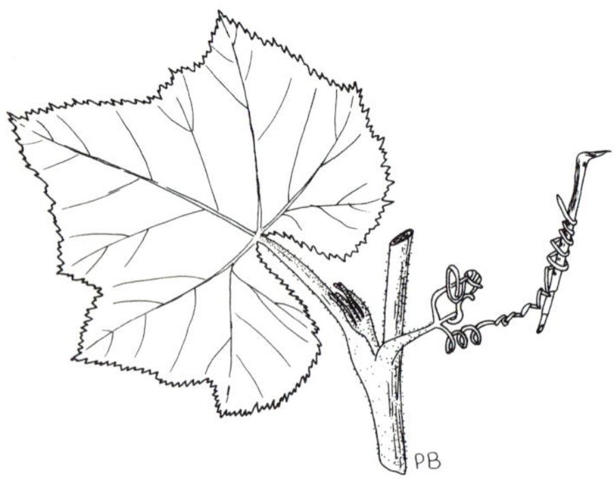

FIGURE 20 Tendril of a squash plant. The leaf on one side of the stem has become modified into a twining tendril. (Brown)

FIGURE 21 The *Agave* or century plant in the lower left of picture has large veins in its fleshy, bayonet-shaped leaves, which are extracted for making coarse rope. The prickly pear cacti in the center of the picture are nearly all stem. The expanded portions are cladophylls, a specialized stem, and the leaves are spines.

Spines (Fig. 22) are found on certain plants, such as cacti, whose leaves have become modified into this type of structure. Some plants, like the black locust, have stipular spines. The stipules, instead of being leaflike, have become spiney.

FIGURE 22 Spines of cacti. The larger spines are in part modified leaves and in some modified bud scales in origin.

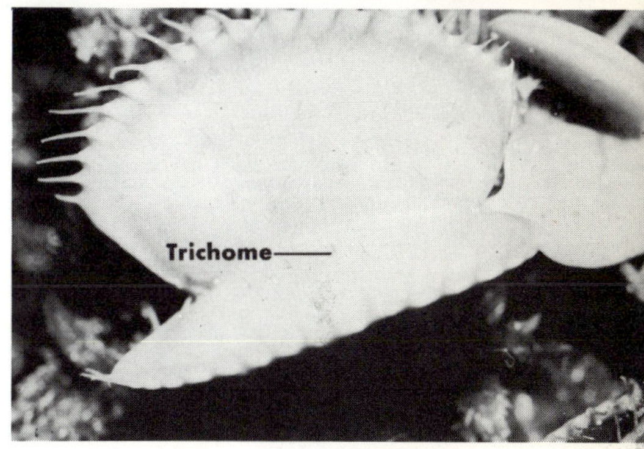

FIGURE 23 Habit photograph of the sundew in foreground and the Venus flytrap just behind it. (Courtesy Carolina Biological Supply Company.)

FIGURE 24 Sundew plant. Closeup of insectivorous leaf showing entrapped ant. Note how the glandular tentacles bend over towards the prey. (Courtesy Carolina Biological Supply Company.)

FIGURE 25 Venus flytrap. Closeup of leaf partially opened. Note the fine skewer-like trichomes which help to hold an entrapped insect. (Courtesy Carolina Biological Supply Company.)

The most unusual of the modifications occur in the insectivorous plants, whose leaves have specialized structures enabling them to catch insects (Fig. 23). Then by means of enzymes the insect is digested and the nitrogenous products of the insect diffuse into the cells of the leaf. These plants contain chlorophyll and therefore also manufacture food as do other green plants.

The sundew plant *(Drosera)* (Fig. 24) is a tiny plant seldom more than three inches in diameter, with a rosette of tiny spoon-shaped leaves with sticky glandular hairs on their surfaces. Insects alighting on the shiny hairs become stuck; gradually the leaf secretes enzymes which begin the digestive process. The Venus-flytrap *(Dionaea)* (Fig. 25) has a hinged, trap-like leaf which snaps shut when an insect lands on the sensitive hairs, causing the two halves of the expanded leaf to close quickly.

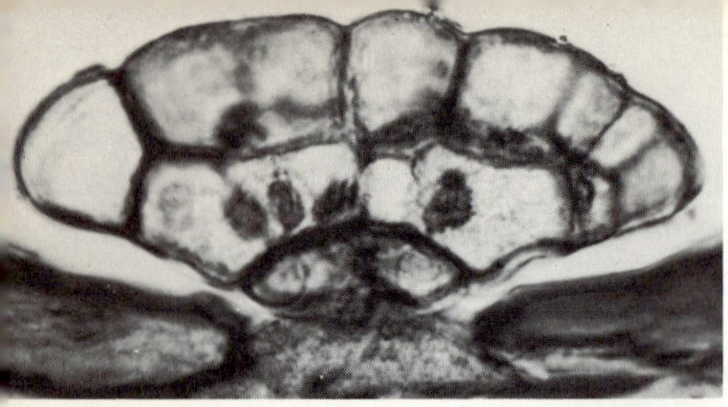

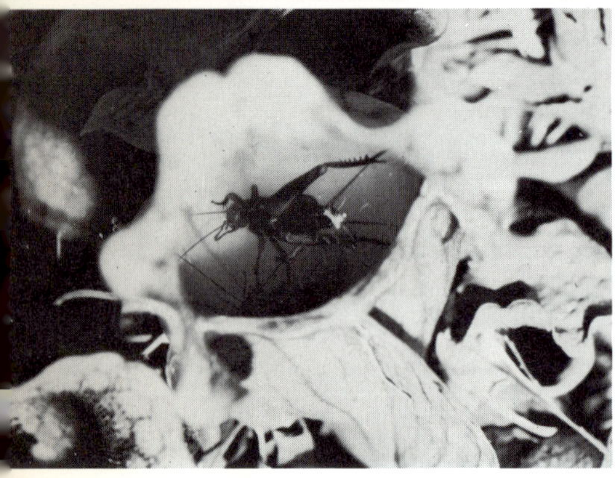

Several different species of pitcher plants are known. These plants have no motile parts, but the leaf forms a reservoir with spines inside and pointed downward. Insects which fall or fly into the reservoir have trouble getting out and eventually are assimilated by the digestive juices secreted by the plant into the reservoir (Fig. 27). The trap leaf of the purple pitcher plant, *Sarracenia purpurea* (Fig. 28), measures about five inches in length when fully mature. William R. West, who has studied this type of plant extensively, describes the leaf as having a basal portion which is small, solid, and round. The central cavity begins about an inch above the base, reaching its maximum diameter about midway up the leaf, thereafter decreasing in size towards the open end. Near the apex is a rolled lip or collar called the nectar roll. The opposite side of the leaf extends upward about an inch farther. This portion is funnel-like and surrounds the collar on three sides. Its surface is covered with nectar glands and strong down-curved hairs. An insect alighting there is directed downward to the next zone, which begins at the mouth of the pitcher. This area has nectar glands but lacks the stiff hairs. It also has slick ridges which again point downwards, allowing the insect a very precarious hold. If an insect slips downward from this zone it reaches the next area which is highly cutinized and very slick, offering virtually no chance to escape. From here the insect slips into the bottom which is called the zone of absorption. The fluid found in the trap is composed of water and digestive juices capable of digesting sugars, starches, and proteins but not chitin. Bacteria are also found in the fluid and undoubtedly aid in the digestion. It is possible to identify the remains of the most common victims. Ants form a large part of the pitcher plant's prey, but larger organisms such as flies, wasps, spiders, and crickets are also trapped.

Insectivorous plants are generally small, although pitcher plants may attain a height of two or three feet; none *are* capable of capturing anything larger than an insect or perhaps a hummingbird or small rodent. Stories of plants capturing human beings are, to say the least, exaggerations.

FIGURE 26 Venus flytrap. Greatly enlarged photomicrograph of a digestive or secretory gland of the modified leaf of this insectivorous plant. (Courtesy Carolina Biological Supply Company.)

FIGURE 27 *Sarracenia purpurea*. A pitcher plant. Note insects within the modified leaf in lower right. (Courtesy Carolina Biological Supply Company.)

FIGURE 28 *Sarracenia purpurea* closeup of the opening in this pitcher plant. Note two crickets, one trapped inside, the other entering. (Courtesy Carolina Biological Supply Company.)

Supplementary reading and questions for discussion on leaves follow Chapter 7.

7

Plant Metabolism

Photosynthesis

The primary function of a leaf or green portion of a plant is **photosynthesis.** This process may be defined as the production of organic compounds by plants from carbon dioxide and water in the presence of light energy. It is estimated that more than five billion tons of organic matter is produced annually through the process of photosynthesis. Obviously photosynthesis is carried out on a grand scale, yet it utilizes two very simple substances which are almost everywhere. Carbon dioxide is present only to the extent of 0.03% in the atmosphere; the other substance, water, is not exceeded by any compound on earth.

We know that photosynthesis is both oxidation and reduction and that in it water is oxidized and carbon dioxide is reduced. In the oxidation of water, oxygen is released. Carbon dioxide is reduced through the utilization of electrons from water oxidation and a simple organic compound is formed. Long before we were able to trace the path of chemical activities in the green leaf during photosynthesis, scientists were aware that some chemical conversion process occurred in plants, through the aid of external substances, to manufacture foods. At first we were limited by our lack of knowledge of the various gases in the atmosphere; we were also unable to analyze the various food substances and identify sugar and starch in the leaf. Once chemists had worked out the techniques for identifying these substances we knew *what* was happening in the leaf, but it was many years before we found out *how* it happened.

Research on Photosynthesis

One of the early experimental scientists who worked with plants was a physician and chemist by the name of Jan Baptista van Helmont (1577–1644), who lived near Brussels. He planted a 5-pound willow branch in a container of dry soil which weighed 200 pounds. This was protected from dust by a cover and watered daily with rainwater. At the end of five years he removed the plant and found it had gained 164 pounds. He also dried the soil and found that it had lost only about two ounces. His conclusion was that the increase in weight in the plant had been gained entirely from the water, even though the tissues in the plant were obviously different from water. It was a logical conclusion at the time, since the atmospheric gases had not yet been discovered.

It was not until 1772 that Joseph Priestley observed that "good air" supported the burning of a candle flame. He also observed that good air supported the breathing of an animal. If a plant was placed under a bell jar with a mouse, the animal lived longer than it would have without the plant. However, this experiment did not always work for him. He apparently failed to realize that the plant could only purify the "noxious" air in *light*. This fact was confirmed in 1779 by Jan Ingenhousz, a Dutch physician. He repeated Priestley's experiments, finally stating that only green plants were effective and that this effectiveness was dependent on the intensity and duration of light.

Additional contributions to the early work in photosynthesis were made by De Saussure in 1804, who showed that green plants purify air by adding oxygen to it and at the same time remove carbon dioxide from air under light conditions. The green color in plants, so essential to the photosynthetic process, was shown to be due to an actual substance by Dutrochet, but it was much later that this green substance was found by Willstatter and Stoll to consist of two green pigments, the chlorophylls, and two yellow pigments, xanthophyll and carotene.

The energetics of photosynthesis was initiated by Engelmann, who recognized that light energy is absorbed by the leaf pigments. In his experiment he projected a spectrum along the lengths of filamentous algae. He then grew oxygen-sensitive bacteria in the culture and observed that they crowded along the portion of the algal filaments which were illuminated with the blue and red rays. This observation of the importance of the blue and red rays in photosynthesis has been confirmed repeatedly by others using different green plants.

The first direct evidence that green plants produce carbohydrates in light was obtained in 1864 by Julius Sachs, who observed the growth of starch grains in the chloroplasts of leaves during the process of photosynthesis. As a result of this work, botanists were able to illustrate the photosynthetic process by the classical equation:

$$6CO_2 + 6H_2O \xrightarrow[\text{green leaf}]{\text{light}} C_6H_{12}O_6 + 6O_2.$$

For many years this description of photosynthesis was used in botany classes around the world, with little new information as to exactly how the process occurred. Many suggested intermediate substances possibly formed before sugar was obtained, but no one proved that they were formed. Gradually some knowledge was accumulated about the quantitative aspects of photosynthesis, including the amount of carbon dioxide and water used, the rate of sugar production, and the amount of oxygen released, but the intermediate chemical steps remained undetected.

Modern Findings

For some time it had been known that sulfur bacteria grow in waters containing hydrogen sulfide. In the presence of light these bacteria convert hydrogen sulfide into sulfur. Van Niel, a professor at Stanford University, used this information to draw an analogy to the process in the green plant. He deduced that H_2O was a reactant in photosynthesis in the green plant just as H_2S was a reactant to the sulfur bacteria. He also concluded, from this information, that the hydrogen which reduces carbon dioxide in photosynthesis is transferred from water and that the oxygen which is released comes from the water and not from carbon dioxide.

The hydrogen involved in this reaction plays a critical role in the entire process of photosynthesis, for it is at this point that light energy is turned into chemical energy. The oxygen formed from the photosynthesis of water at this stage is reformed from the available hydrogen and from some of the oxygen in carbon dioxide as follows:

$$(CO_2)_n + (H_2O)_n \xrightarrow[\text{energy}]{\text{light}} (CH_2O)_n + (O_2)_n + (H_2O)_n.$$

In the 1930's Robin Hill demonstrated that chloroplasts contain an electron transport system. He found that a suspension of illuminated chloroplasts could transfer electrons from water to a suitable acceptor. He successfully isolated chloroplasts from ground-up leaves and found that they would produce oxygen in light if supplied with a reducing substance such as ferric ions. In the presence of light the ferric ions (Fe^{+++}) are reduced by the acceptance of an electron (e^-) and converted to ferrous ions (Fe^{++}) with a release of oxygen.

Information on the intermediate products which form in the process prior to sugar formation was provided by Melvin Calvin of the University of California. He and his co-workers studied extracts of cultures of a unicellular green alga, *Chlorella*. They interrupted the photosynthesis of these algae at various time intervals and analyzed them for the products produced at different times. Using carbon dioxide containing radioactive carbon ($C^{14}O_2$) they discovered that carbon appears first in a three-carbon compound, then later it appears in sugar, more complex carbohydrates, amino acids, and other compounds. Furthermore, they took radioactive sugars, disassembled them, and established the precise location of the carbon atoms in each compound.

The exact location of the whole photosynthetic process was proved by Doctor D. I. Arnon and his co-workers, also at the University of California. They succeeded in removing chloroplasts from cells and causing them to photosynthesize outside the cells. They also discovered that suspensions of chloroplasts isolated from spinach leaves would produce adenosine triphosphate (ATP) from light, adenosine diphosphate (ADP), and inorganic phosphorus.

Chloroplasts, the Site of Photosynthesis

Chloroplasts serve as the site of chemical activities in which an elaborate electron system has developed. The effectiveness of chlorophyll is due largely to the organization of the plastid (Fig. 1). The electron microscope has revealed to us that chloroplasts are constructed of parallel layers of **lamellae** running the length of the plastid. In some places these layers thicken, and when a number of lamellae thicken at about the same place they are termed **grana.** It is in these areas that the chlorophyll is concentrated. Separating the lamellae are nongreen proteinaceous layers termed **stroma,** which contain the enzymes required for CO_2 fixation and other enzymatic processes. Lamellae, which are chemically lipid, probably contain the chlorophyll in single molecular layers which give the plastid considerable surface area for the absorption of light and the action of the chlorophyll molecules.

The entire plastid is surrounded by a membrane that has differentially permeable properties. Thus products involved in photosynthesis enter and finished products move outward. It is also through this membrane that the enzymes and coenzymes concerned with the conversion of CO_2 to sugar enter and leave as the various steps of photosynthesis are accomplished.

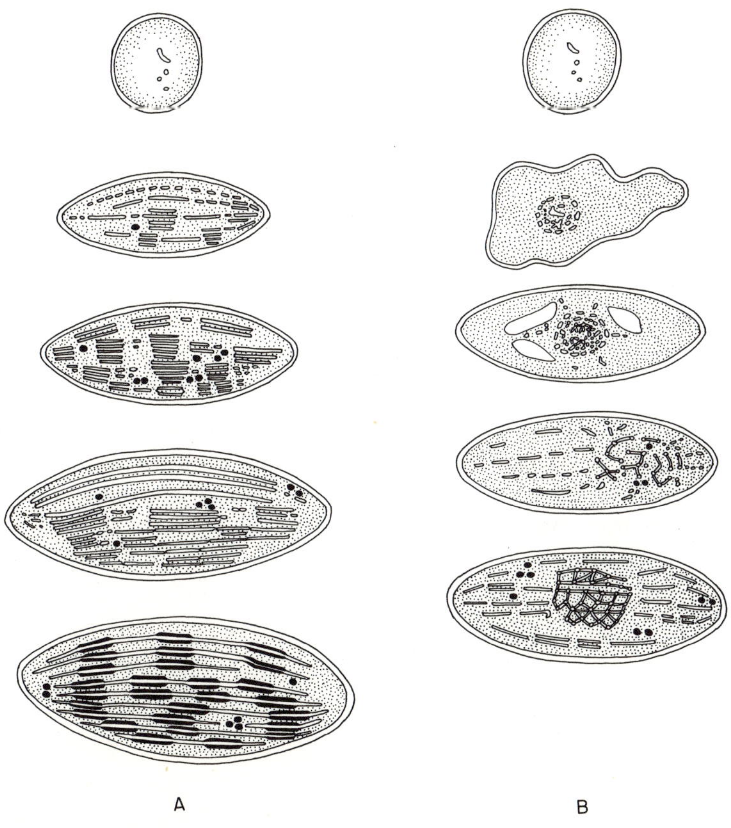

FIGURE 1 Development of a chloroplast. In series A, beginning with the proplastid at the top, the chloroplast is developed in light. In the B series the plastid is developed in the dark. By placing it in the light the plastid in series B can be converted to the normal type. (From Von Wettstein in *Developmental Cytology,* Ronald Press, New York, 1959.)

FIGURE 2 *Coleus* leaves. Note the variation in color pattern. These leaves are mostly red as a result of the predominant anthocyanin pigment. The light areas are yellow where the carotenoid pigments are predominant. Greens are present in the more mottled.

It is possible to isolate the intact chloroplasts from the broken cells by centrifugation and demonstrate that they retain for some time the ability to carry on photosynthesis. While this was successfully done by Arnon and his co-workers, earlier attempts were unsuccessful, probably because the techniques had not been refined sufficiently. These earlier workers were successful, however, in making the isolated chloroplasts produce oxygen in light.

Other Plastids

Higher plant cells often contain other types of plastids which do not have the lamellated structures of the chloroplasts. These **leucoplasts** are colorless and serve as a storage center for a cell's reserve food, such as starch. Leucoplasts seem to have the conversion property of changing sugars to starches, and therefore must have the enzyme systems necessary to accomplish this process. Chromoplasts may be plastids from which chlorophyll has disappeared, or they may never have been green. The chromoplasts usually contain concentrations of yellow, orange, and red pigments which may develop with a decline of the chlorophyll pigments (Fig. 2) in such maturation processes as the ripening of fruits. Tomatoes, for example, are at first green, then become whitish as the chlorophyll fades, and finally become red as the fruit ripens. A series of chemical transformations occurs in this ripening process which is not yet fully understood.

Chlorophylls

Chlorophylls are composed of carbon, hydrogen, oxygen, nitrogen, and magnesium, and are usually formed as two similar but separate pigments, termed chlorophyll *a* and chlorophyll *b*. Chemically they have the empirical formulae of:

chlorophyll *a*: $C_{55}H_{72}O_5N_4Mg$

chlorophyll *b*: $C_{55}H_{70}O_6N_4Mg$.

Other chlorophylls, such as *c*, *d*, *e*, etc., are also present in some groups of lower plants, but *a* and *b* are the principal ones in higher plants.

The properties of chlorophyll are attributed to the way the atoms are arranged in the molecule, shown in Figure 3. Chlorophyll *b* has a —CHO group in place of the —CH_3 group enclosed by the dotted circle. Note the porphyrin rings (I-IV) attached to the magnesium center.

Chlorophyll has as its basic structure the porphyrin ring composed of the four nitrogen-containing rings and their connecting carbon atoms. Chlorophyll is very similar to the hemoglobin of the blood, which is also made up of porphyrin rings, but differs mainly in that an atom of iron is at the center of the ring in hemoglobin rather than magnesium as found in chlorophyll.

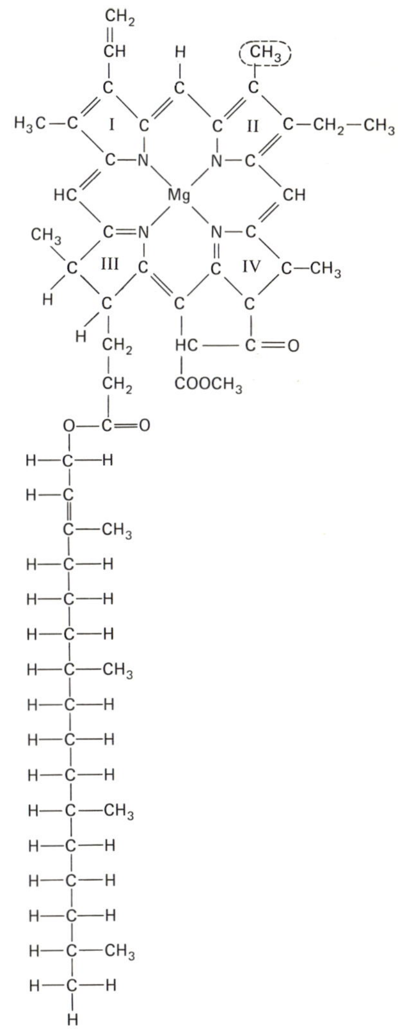

FIGURE 3 Structure of the chlorophyll *a* molecule. Chlorophyll *b* has a —CHO group in place of the —CH_3 group enclosed by the dotted circle. Note the porphyrin rings (I–IV) attached to the magnesium center.

The Source of Energy

The energy which supports the activities of all organisms is derived either directly or indirectly from the sun. We may think of light as a series of waves or rays, termed *photons,* each of which carries a single unit or quantum of energy. To be effective, light must be absorbed, and this absorption varies with the different objects upon which it falls. When the visible light is almost entirely reflected the object is white, and when nearly all of the light is absorbed the object is black. If one thermometer is wrapped in a white cloth and another in black and both are placed in sunlight, the thermometer wrapped in black will have a considerably higher temperature than the other. In objects of color certain wavelengths of light are reflected while others are absorbed. Such objects display the wavelengths of the light which they reflect. If we allow the colors of the spectrum to pass through a green leaf, we note that the green wavelengths are reflected but those of the reds and blues are absorbed. This indicates which of the colors are used in photosynthesis. Although chlorophyll is the major receptor pigment in light absorption in the leaf, the carotenoid pigments are also capable of light absorption. Since they do not participate in photosynthesis, it is assumed that they pass their photoactivated increased energy on to the chlorophylls.

The visible colors of the spectrum may be observed by sending a beam of light through a prism. The colors are due to a graduated series of wavelengths (Fig. 4). When light is observed in this

FIGURE 4 Approximate distribution of colors in the spectrum of sunlight. The numbers refer to the various wavelengths measured in Angstrom units.

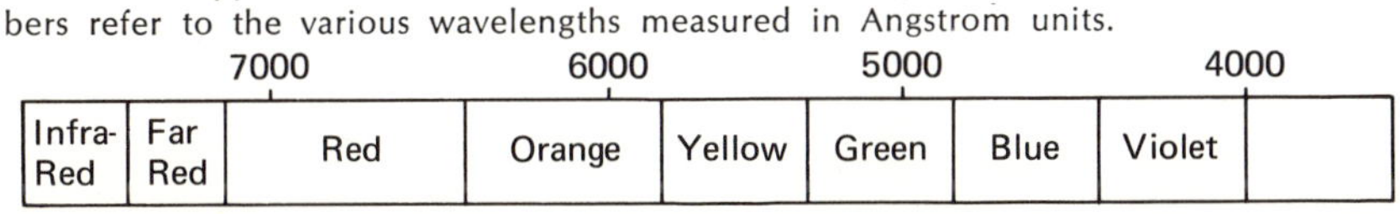

way we note that the range of the visible spectrum is from the blues of approximately 420 millimicrons to the reds at about 670 millimicrons, with the orange, yellows, and greens near the center of the bands. Beyond the effective red rays are the longer wavelengths of far-red and infrared, while at the other end of the spectrum are ultraviolets and X rays.

When light is absorbed by the chlorophyll molecule, it is apparent that the electrons in the atoms of the chlorophyll molecule are moved into an orbit farther from the atomic nucleus, thus absorbing a quantum of light energy. The molecule is then in an excited state. However, this is an unstable situation and as the electrons move rapidly back into their original position, the energy is released. The energy which has excited the chlorophyll molecule brings about a photosynthetic chemical reaction rather than being lost as heat or light. When an electron leaves the chlorophyll molecule a portion of the structure is removed from the molecule. Since electrons have a negative charge, the more electrons removed the more unbalanced electrically the chlorophyll becomes. The chlorophyll then becomes positive (+), which attracts more negative electrons to it, providing a continuous flow of electrons to and from the chlorophyll. During photosynthesis the light energy of a stream of photons is converted to the electrical energy of the electrons. This electrical energy produces ATP in the chloroplasts.

In certain plants, at least, it was discovered that the red light when supplied to plants alone was not as effective as when the

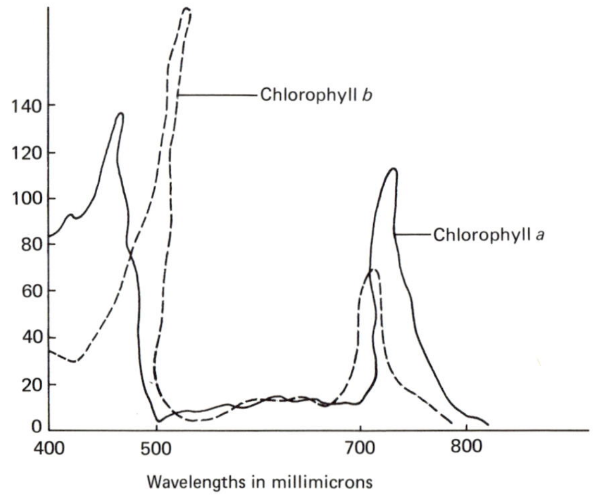

Graph from Allen, M. B., C. S. French, and J. S. Brown, "Native and Extractable Forms of Chlorophyll in Various Algal Groups," 1960. *Symposium on Comparative Biology* Vol. I.

blue lights were supplied simultaneously. This has led to the speculation that different kinds of chlorophyll must be activated at the same time to achieve optimal photosynthesis. The difference in the absorption spectra of chlorophylls *a* and *b* for certain algae is shown in the graph above. It will vary somewhat for other plants.

The minimum light in which a plant can survive but in which it will not grow appreciably because it is balanced by respiratory losses is its **physiological compensation point.** Every plant has a minimum standard compensation point and this point may vary for different species from about two to one hundred footcandles or more.

The light energy when absorbed by chlorophyll initiates a series of reactions in which the electrons of chlorophyll flow to other compounds which accept the chlorophyll electrons. These compounds, known as electron acceptors, have been identified as a lipid soluble material termed plastoquinone, another known as ferrodoxin, which is a water soluble protein containing iron, and a third compound called cytochrome, which is relatively insoluble. Cytochrome may take up or donate electrons; in other words, it may be reversibly oxidized and reduced:

$$\text{cytochrome } Fe^{+++} + e^{-} \underset{\text{oxidation}}{\overset{\text{reduction}}{\rightleftharpoons}} \text{cytochrome } Fe^{++}.$$

Specific kinds of cytochromes may carry electrons from one energy level to another so long as there is a continuous supply of electron-acceptor molecules. This is significant in photosynthesis, for a cytochrome in the reduced (Fe^{++}) condition absorbs different wavelengths of light than the same cytochrome absorbs in the oxidized potential (Fe^{+++}). Changes in the absorption spectra due to cytochromes in the reduced or oxidized state may be altered by the introduction of food or oxygen into the biological electron transport system. Consequently the introduction of light energy raises electrons from a low energy level to a level high enough for them to add carbon and contribute to the formation of carbohydrates.

Plastoquinone is not an iron compound like ferrodoxin and cytochrome, but is related to vitamin K. For some time it had been known that vitamin K was localized in the chloroplasts. Because of its location Arnon and others believed vitamin K had some connection with photosynthesis and it was observed that when it was added to isolated chloroplasts it greatly increased the production of ATP. In the process of photosynthesis vitamin K acts as a catalyst and is manufactured by the plant.

The role of ATP in biological systems is unique. In the mitochondria ATP is made from food, and in the process of oxidation the stored energy of food from the chloroplasts is quickly changed to ATP. Everything which is done inside a plant or animal is dependent on the availability of ATP. Furthermore the ATP of all animals and plants is the same. However, while plants can also produce this substance in both chloroplasts and mitochondria, man and all other animals generate ATP only through respiration in the mitochondria.

In photosynthesis the highly activated electrons are used in another way. They are stored in a compound known as nicotinamide adenine dinucleotide phosphate (NADP). This substance resembles ATP since both have an adenine base, a ribose sugar, and phosphate. NADP and ATP are different in that the latter has a high-energy phosphate bond at the end of its molecule, while NADP has a nicotinamide group with a positive charge at its end. This charge attracts the electron from photosynthesis and the electron acceptor, ferrodoxin, transfers this charge to NADP. The latter then becomes negatively charged, attracting a proton (H), forming NADPH. This substance can then reduce carbon dioxide to a carbohydrate.

In summary this part of the reaction is as follows:

$$H_2O \xrightarrow{\text{light}} [H]\ [OH]$$

$$[H] \longrightarrow NADP \xrightarrow[\text{chloroplast}]{\text{light}} NADPH \rightarrow \text{available for reduction of } CO_2$$

$$ADP + P \longrightarrow ATP$$

Initially the ATP was formed from ADP with the addition of inorganic phosphate. The chloroplasts used light energy to force the additional phosphate group on ADP forming ATP. This process is termed *photophosphorylation,* which distinguishes it from the oxidative phosphorylation occurring in the mitochondria. It is a type of reaction that occurs several times in the photosynthetic cycle, but starts in the early phase of the cycle where light-excited chlorophyll molecule electrons flow to electron acceptors and back to chlorophyll.

In the case of photophosphorylation involving NADP the electron acceptor NADP captures an electron and combines with a hydrogen ion forming NADPH. In this process the electrons from the excited chlorophyll molecules are absorbed by NADP and are not recycled back to chlorophyll but continue on to reduce carbon dioxide in the carbohydrate reaction series.

The Paths of Carbon and Oxygen

In the green plant, the NADPH and ATP produced by photosynthetic electron transport are used primarily in the fixation of CO_2, as shown in the outline of the photosynthetic cycle (Fig. 5).

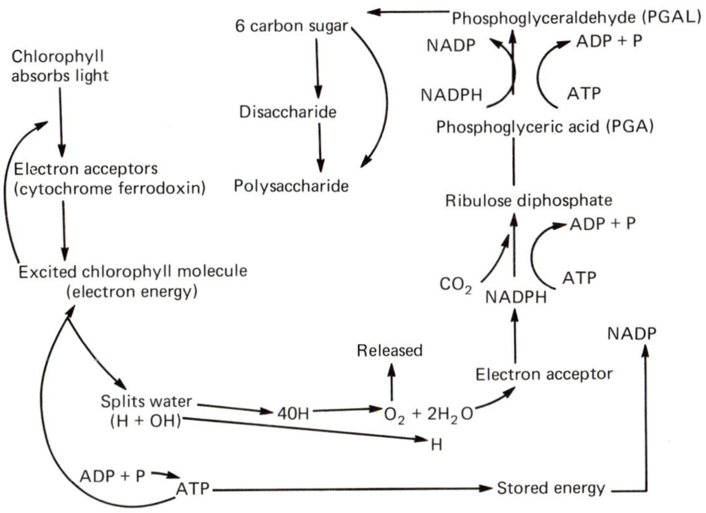

FIGURE 5 An outline of the successive steps in photosynthesis.

Formerly we were limited in our interpretation of photosynthesis to the following equation:

$$(CO_2)_n + (H_2O)_n \xrightarrow[\text{green plant}]{\text{light}} (CH_2O)_n + (O_2)_n$$

This told us very little except for the raw products involved, and the end product, $(CH_2O)_n$, representing a carbohydrate. The intermediate steps involved in the conversion of CO_2 to a carbohydrate remained unknown until the 1940's, and even then were only partially discovered. Scientists were interested in finding out the source of the oxygen which is evolved in the process. Did it come from the ionization of H_2O or CO_2? The answers to this and other questions were achieved through the use of radioactive trace elements in the cells of the leaf. Oxygen was traced by Reuben and his co-workers, who used the heavy isotope of oxygen, O^{18}. They found that when they used O^{18}-labeled water (H_2O^{18}) the oxygen evolved in photosynthesis was O^{18}. When they used O^{18}-labeled carbon dioxide ($CO_2{}^{18}$) only small amounts of labeled oxygen were evolved, showing that the evolved oxygen came from some other source. The summarized equation for this reaction may then be written:

$$(CO_2)_n + (H_2O^{18})_n \rightarrow (CH_2O)_n + (O_2{}^{18})_n$$

Tracing carbon through its pathway proved somewhat more difficult. The intermediate products produced are in very small quantities and the reactions which change the intermediate products to the next step are exceedingly rapid. In less than one second a plant actively engaged in photosynthesis can convert some CO_2 into an organic form.

Calvin and his associates solved the problem through the use of carbon-14 (C^{14}) and paper chromatography. This technique made it possible for them to separate microgram quantities of a complex mixture into its various components. They exposed a portion of a plant or cultures of *Chlorella* to C^{14}-labeled carbon dioxide for periods of time ranging from about 0.5 second up to several minutes. At the end of each exposure time the plant material was killed and extracted with hot alcohol. The paper chromatography technique was then used to sort out the various radioactive components in the plant material. From studies of a number of autoradiographs taken of this material it is clear that after only one minute of photosynthesis radioactive carbon appeared in more than a dozen compounds. By reducing the period of $C^{14}O_2$ exposure fewer components were labeled and when the exposure time was reduced to 0.5 second the only labeled compound was 3-phosphoglyceric acid (3-PGA), the source of two carbon atoms in the three-carbon compound. The experimenters knew that the -COOH group of phosphoglyceric acid comes from carbon dioxide but they did not know what compound in the cell contributed the other two. They then discovered that a five-carbon sugar containing two atoms of phosphorous, ribulose diphosphate, combines with carbon dioxide. This then forms an unstable six-carbon compound which splits, forming two three-carbon molecules of phosphoglyceric acid as follows:

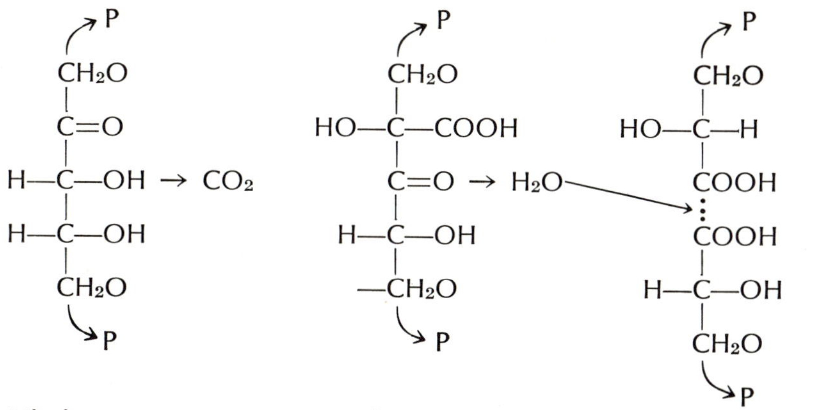

Ribulose diphosphate — 6-carbon intermediate compound — two molecules of phosphoglyceric acid

Photosynthesis is a continuing process and in order for us to understand how it operates continuously it was necessary to learn how ribulose diphosphate (the CO_2 acceptor) was replaced, since it must be regenerated as fast as it is used up in the CO_2-fixing reaction. This process was found to be a series of reactions. First PGA is reduced by H atoms from NADPH to the three-carbon sugar phosphate (triosphosphate) through the action of the enzyme triosphosphate dehydrogenase. Two of the three-carbon triosphosphates then join to form the six-carbon sugar fructose phosphate ($C_6H_{12}O_6+P$). The atoms of the fructose molecules are simply rearranged in producing glucose phosphate ($C_6H_{12}O_6+P$), in which the numbers of atoms remain the same. When glucose and fructose phosphates are combined forming sucrose the phosphate groups split off.

$$C_6H_{12}O_6\textcircled{P}\!\!\diagup\!\textcircled{P} + C_6H_{12}O_6\textcircled{P}\!\!\diagup\!\textcircled{P} \rightarrow C_{12}H_{22}O_{11} + H_2O$$

In the condensation of a polysaccharide such as starch, the phosphate group also breaks away to go back into the complex of food synthesis:

$$C_6H_{12}O_6\textcircled{P}\!\!\diagup\!\textcircled{P}_n \rightarrow (C_6H_{10}O_5)_n + (H_2O)_n.$$

Recycling of the Photosynthetic Groups

There is a continuous supply of energy available for metabolism through the transfer of electrons in the photosynthetic cycle. The oxidation of one molecule of phosphoglyceraldehyde to phosphoglyceric acid produces one new high-energy phosphate bond. This phosphate group may then be transferred to ADP, forming ATP. This compound is the most readily convertible substance in plant and animal metabolism and can be used directly in a large number of biological reactions. In the course of this reaction two hydrogen atoms from phosphoglyceraldehyde may reduce a pyridine nucleotide ($PN + 2[H] \rightarrow PNH_2$). The two hydrogen electrons on this pyridine nucleotide may then move to oxygen by way of electron acceptors and ultimately form water.

Summary of the Carbon Pathway

Carbon reduction is initiated by the reaction of CO_2 with acceptor molecules ribulose diphosphate; with the aid of the enzyme carboxydismutase in the chloroplast the carbon dioxide and ribulose diphosphate combine, forming an unstable six-carbon intermediate compound which breaks down, yielding two molecules of PGA. The enzyme triosphosphate dehydrogenase utilizes NADPH and ATP, reducing PGA and forming the three-carbon sugar glyceraldehyde (PGAL). Some of these compounds may then be used by the plant parts in metabolism at this point. The remainder of the fixed carbon will undergo the series of reactions, resulting in the regeneration of ribulose diphosphate. To continue the process of photosynthesis the carbon reduction cycle must have a continuous supply of NADPH as well as ATP, both of which are supplied by the light reactions of photosynthesis.

Solving the complicated series of reactions was made possible by the use of $C^{14}O_2$, which shows how the three, four, five, six and seven-carbon sugar phosphates are formed, converted, and reversed in their cycle. Radioactivated carbon also reveals the production of double sugars and starches from the six-carbon sugars. C^{14} also is observed as a component of organic acids formed from the sugars through oxidation, as well as fatty acids through reduction, and amino acids by a combination of oxidation and uptake of ammonia. The entire process establishes the sugar molecule as the building block of the plant cell and its constituents.

Sequence of Steps in Photosynthesis

1. A quantum of light is absorbed by chlorophyll.
2. The chlorophyll molecule is activated by light energy.
3. The photoactivated pigment ionizes water.
4. Cytochromes may carry electrons from one energy level to another.
5. Oxygen is released from the hydroxyl radical $2(OH) \rightarrow O_2 + 2H$.
6. The hydrogen is transferred to organic molecules such as nicotinamide adenine dinucleotide phosphate (NADPH).
7. The ATP is used to drive various chemical reactions, including reduction of CO_2 to sugar.
   $$2NADPH + 2H^+ + CO_2 \rightarrow 2NADP + H_2O + CH_2O$$
8. In the reaction, water is again reformed.
9. The radiant energy has been converted to chemical energy.
10. Radioactive carbon shows that an early compound formed is phosphoglyceric acid (PGA), a three-carbon sugar.
11. A five-carbon sugar, ribulose diphosphate, apparently forms before the three-carbon sugar.
12. The five-carbon sugar adds CO_2, then splits into two molecules of PGA.
13. The PGA is reduced to form PGAL by $NADPH_2$ and ATP photoreactions.
14. Two of the three-carbon sugar phosphates combine to form the six-carbon sugar $C_6H_{12}O_6 + P$.

Respiration

Respiration is the process which is going on at all times in all living cells whether plant or animal. It is usually defined as the oxidation of organic substances within living cells with a release of carbon dioxide, water, and energy. In the higher animal respiration is often considered synonymous with breathing. However, breathing is a purely mechanical process involving lungs and related organs which draw air into the respiratory passages, where oxygen and carbon dioxide are exchanged through the moist membranes of the respiratory system. From here the oxygen is carried by means of the animal's circulatory system to the body cells where it is used in cellular respiration. At the same time the carbon dioxide is removed from the body cells by means of osmosis and diffusion and carried back to the respiratory organs and membranes where it is expired.

In plants, and in all animals lacking a specialized breathing apparatus, respiration is fundamentally the same. In all, including the higher animal, respiration involves surface membranes for the diffusion-osmotic process in the exchange of oxygen and carbon dioxide.

Respiration is not a simple gaseous exchange process. It involves the breakdown of energy-containing substances such as carbohydrates, fats and proteins in the cell, and this involves enzyme systems. Dozens of oxidative enzymes are required for the series of reactions in the breakdown of food to release energy, carbon dioxide, and water. In the process the energy liberated is stored temporarily in ATP where it can be used in doing work.

In photosynthesis this substance (ATP) was formed in the chloroplasts, where energy absorbed by light was transferred to chemical energy in the synthesis of ATP and ultimate reduction of carbon dioxide by hydrogen from water. This was a photophosphorylation process, whereas in respiration the ATP is formed in the oxidation of organic compounds in an **oxidative phosphorylation** process. As in photosynthesis, the ATP is used in various cellular activities which oxidize the ATP into ADP and phosphate. The latter becomes a part of the inorganic phosphate in the cell and when energy of the required type is available the ADP and P can be recombined forming ATP (Fig. 6). Adenosine triosphosphate is

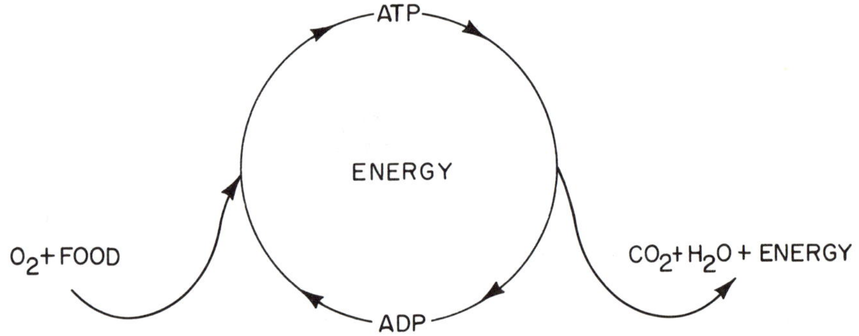

FIGURE 6 The energy reactions involving ATP in the initial phases of respiration.

then available for more chemical conversions. At times of rest a cell may not be respiring and expending energy very rapidly; consequently there may be a buildup of ATP. If the cell is stimulated to do work the metabolic activities are increased and the ATP is quickly changed to ADP and P, which in turn are resynthesized into ATP for more work energy.

The process of respiration is usually summarized in the equation:

$$\underset{\text{sugar}}{C_6H_{12}O_6} + \underset{\text{oxygen}}{6O_2} \rightarrow \underset{\text{carbon dioxide}}{6CO_2} + \underset{\text{water}}{6H_2O} + \underset{\text{energy}}{E}$$

Although the equation shows energy present in the sugar molecule, more energy must be added to initiate the reaction. In a burning process we know that oxygen must be present so that the substrate (in this case sugar) will burn. A match will kindle such an oxidative reaction, but in respiration the heart of activation required for the action of enzymes is supplied by transferring energy from ATP molecules to the glucose molecule. In this process the phosphate is transferred from ATP, along with its energy, to the glucose molecule. Following the union the sugar is activated and available for the next step.

This step is accomplished by enzymes which split the activated glucose molecule into two three-carbon sugars (2-$C_3H_6O_3$). Further oxidation of these sugars involves the removal of hydrogen to a hydrogen acceptor identified as nicotinamide adenine dinucleotide (NAD). No atmospheric or molecular oxygen is involved in this reaction, but as a result of dehydrogenation an intermediate product, pyruvic acid ($C_3H_4O_3$), is formed as below:

$$\underset{\text{3-carbon sugars}}{2C_3H_6O_3} + \underset{\text{hydrogen acceptor}}{NAD} \rightarrow \underset{\text{pyruvic acid}}{C_3H_4O_3} + \underset{\text{hydrogen acceptor}}{NADH_2} + \underset{\text{(energy)}}{ATP}$$

The energy obtained in the process is added to that already in the cell as ATP. This series of reactions involving the change of glucose to pyruvic acid is termed *glycolysis* and in some organisms, such as certain bacteria, it is the only source of energy. Glycolysis may occur in the presence or absence of oxygen, but oxygen is not used in the reaction.

The site of respiration, including the enzymes associated with pyruvic acid oxidation, electron transport, the portion of respiration termed the Krebs cycle, and oxidative phosphorylation, is in the **mitochondria.** These minute organelles have a convoluted internal framework of membranes in which the electron transport system and the enzymes of oxidative phosphorylation are located. Between the convoluted membranes are pockets containing the soluble mitochondrial enzymes.

Since the discovery of the role of mitochondria in respiration, it has been assumed that the electron transport system in higher plant mitochondria is the same as that in animal mitochondria. Evidence has gradually accumulated indicating that this is not true. In animals oxygen utilization is almost completely inhibited by carbon monoxide and cyanide, while in plants these substances may inhibit oxygen utilization in some, may reduce the rate of oxygen utilization in others, and in others may not affect or may even increase the respiration rate. This variation seems not to vary with the different species of plants tested but with the type of cytochromes present and the behavior they exhibit in mitochondrial electron transport.

It has been observed that the rate of substrate oxidation is slow if all of the ADP has been converted to ATP, but if more ADP is provided to the mitochondria the rate of substrate oxidation is greatly accelerated. By examining the different types of cytochromes and identifying those which are insensitive to cyanide and carbon monoxide we are able to note their reaction to the cytochrome electron transport system. The insensitive tissues seem to have two pathways of electron transport, one using cytochrome oxidase, the other using a different type of cytochrome. If the electrons pass through the cytochrome oxidase ATP is produced from ADP and respiration is accelerated, but if the electrons pass through the other cytochrome only heat is produced.

A knowledge of the flow of electrons in the ATP-producing pathway is important to our understanding of the process of respiration as well as that of photosynthesis. While our present information is not complete, great impetus has been given to increased study of the electron transport systems with the discovery that certain cytochromes vary in their reactions apparently in both mitochondria and chloroplasts.

Aerobic Respiration

Respiration in the presence of oxygen, and with the enzymes necessary to complete the breakdown of glucose, is aerobic. In this process there are six molecules of CO_2 and six molecules of H_2O produced for every molecule of glucose used:

$$C_6H_{12}O_6 + 6O_2 \rightarrow 6CO_2 + 6H_2O$$

During the reaction thirty-eight molecules of ATP are produced, which are then made available to the plant. The entire reaction requires a series of complicated steps although it is usually

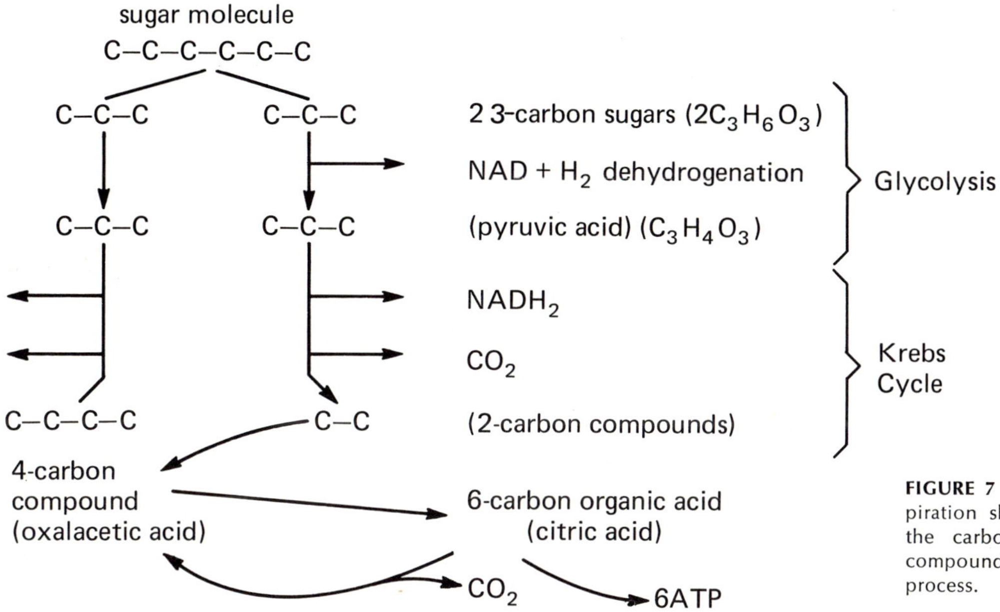

FIGURE 7 General diagram of respiration showing the distribution of the carbon atoms in the various compounds formed and split in the process.

divided into two parts, **glycolysis** and the **Krebs cycle** (Fig. 7). In the first part the glucose molecule is subjected to enzymatic action which converts the six-carbon sugar into three-carbon compounds, the pyruvic acid molecules. In this respect aerobic respiration has the same reactions as anaerobic, but at this point the presence or absence of oxygen will alter pyruvic acid conversion.

When oxygen and the appropriate enzymes are present, a carbon atom is removed from the pyruvic acid and a two-carbon compound is formed. This two-carbon compound is acetyl coenzyme A, and is a compound capable of combining with the four-carbon compound oxaloacetic acid.

During the conversion of glucose to pyruvic acid only a small portion of the total energy present in the glucose is liberated to the cells as ATP; most of the energy remains in the pyruvic acid molecules.

The Krebs tri-carboxylic acid cycle is initiated with the formation of citric acid from acetyl coenzyme A and oxaloacetic acid. Citric acid initiates a cycle in which two carbon dioxide molecules are released and oxaloacetic acid is reformed. During respiration, electrons pass through a series of reactions which take them from high energy levels to lower energy levels (Fig. 8). When at the

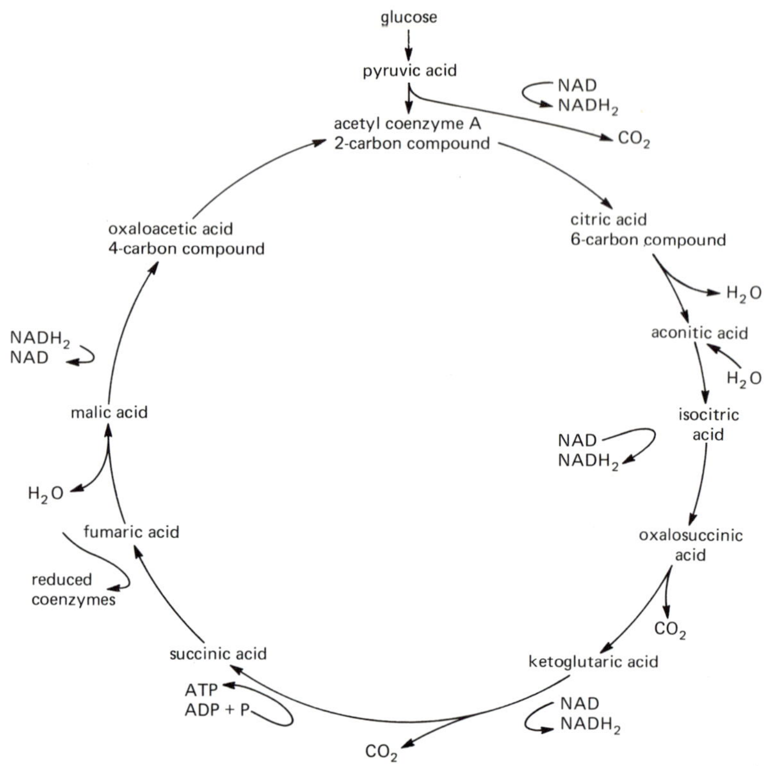

FIGURE 8 An outline of the Krebs or citric acid cycle which shows a series of compounds formed in the oxidation-reduction reactions during aerobic respiration with CO_2 and H_2O as end products. (After R. A. Goldsby, *Cells and Energy*, The MacMillan Co., New York, 1967.)

lowest level, the electron combines with oxygen and H^+ ions forming water. In the intermediate levels the energy released is formed into high-energy phosphate bonds converting ADP to ATP.

A pair of electrons is released at six places in the glycolysis-Krebs cycles. These are accepted by NAD, forming $NADH_2$. In these reactions the electrons move through a series of compounds losing energy and forming ATP. The compounds involved include NAD, a vitamin such as riboflavin, and a series of cytochromes. In the final reactions the H_2 from the $NADH_2$ is combined with oxygen to form water with the aid of the enzyme cytochrome oxidase.

Anaerobic Respiration

Most organisms derive the major portion of their energy from the metabolism of carbohydrates. As we have seen, this involves a breakdown of glucose or similar compound, which has six carbon atoms, into two pyruvic acid molecules each of which has three carbon atoms. This conversion involves the action of at least a dozen enzymes. The process of carbohydrate metabolism to form pyruvic acid occurs in all cells and is not dependent on oxygen. In our muscles and elsewhere it is called glycolysis, while the process in yeast is called **fermentation.**

In fermentation or glycolysis relatively small amounts of energy are liberated, therefore large amounts of carbohydrates are necessary to sustain life under anaerobic conditions.

One coenzyme involved in the process is NAD, which is frequently mentioned as a hydrogen acceptor or donor in various biological oxidation-reduction systems. Another coenzyme necessary for alcoholic fermentation is thiamine pyrophosphate (TPP), which is involved in the decarboxylation (removal of CO_2) of pyruvate to form acetaldehyde in fermentation.

The first step in fermentation is the addition of phosphate to a glucose molecule; soon a second phosphate is added. To make this combination possible two ATP molecules are required and the product is called fructose 1,6-diphosphate (Fig. 9).

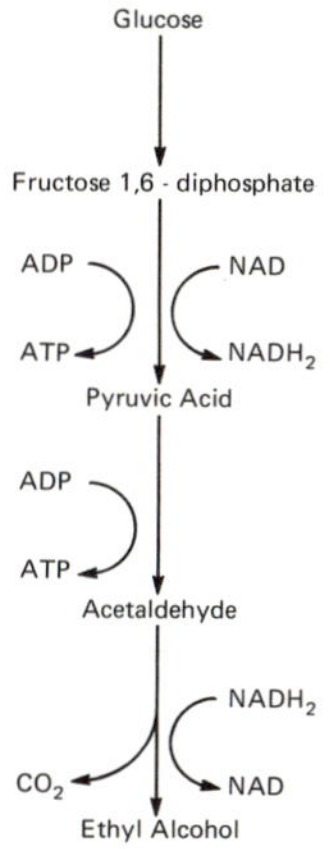

FIGURE 9 An outline of the steps involved in fermentation with alcohol and carbon dioxide as end products.

The second step is the cleavage of the six-carbon diphosphate forming two three-carbon compounds. The third stage results in the formation of two ATP molecules during which NAD is reduced to NADH. For each glucose molecule used, four ATP molecules and two NADH molecules are produced at the expense of two ATP molecules. Consequently there is a net gain of two ATP molecules. An enzyme, enolase, removes water through dehydration from 2-phosphoglyceric acid to yield phosphoenol pyruvic acid. This substance has a high-energy phosphate which yields ATP and pyruvate. In the final stages the acetaldehyde is reduced to ethyl alcohol, and at the same time regenerates NAD from NADH.

Organisms which carry on fermentation are capable of combining hydrogen with pyruvic acid or some organic compound derived from pyruvic acid. The reaction may be summarized:

pyruvic acid		hydrogen acceptor		an alcohol		carbon dioxide		hydrogen acceptor
$C_3H_4O_3$	$+$	$NADH_2$	$\rightarrow$	C_2H_6O	$+$	CO_2	$+$	NAD

This process of alcoholic fermentation is characteristic of some bacteria and yeasts. We take advantage of the reaction in the production of wines and beers, and in the initial phases of distilled liquor production. It is also the process used in the preparation of raised breads, where the evolution of the carbon dioxide gives the bread a light, porous texture.

In the process of fermentation, depending upon the organism involved, various alcohols such as ethanol, isopropanol, butanol, and glycerol may be produced, as well as such organic acids as acetic, lactic, butyric, and propionic.

True Anaerobes The vast majority of plants live as aerobes, but many microorganisms live under environmental conditions with very little free oxygen, or in an atmosphere completely devoid of oxygen. Most of the yeasts and numerous bacteria live in environments of low free-oxygen content, but a few survive in the complete absence of oxygen, and these live much as the first organisms that apparently inhabited the earth lived. They are true anaerobes. In pure anaerobic respiration hydrogen is combined with oxygen from some inorganic compound such as iron, sulfur, or nitrogen as:

$$2H_2O + 5S + 6HNO_3 \rightarrow 3N_2 + 5H_2SO_4 + \text{energy.}$$

In this equation the sulfur serves as the hydrogen acceptor. When combined with sulfur and oxygen the hydrogen forms sulfuric acid. The nitrogen derived from the nitrate (NO_3) escapes into the atmosphere.

The principal organisms living under these conditions are the sulfur and iron-oxidizing bacteria and a few other types like the denitrifying bacteria.

Economic Importance of Respiration

Aside from the actual manufacture of food in photosynthesis, no process is more important in the ripening, storage, and production of new foods than respiration. In the preceding paragraphs a number of products obtained from the fermentation processes, which are partially anaerobic, were mentioned. In addition to those, such foods as sauerkraut, cheeses, and other milk products owe their particular characteristics to the respiratory process.

The storage environment of grains, potatoes, apples and many other fruits, and vegetables must be carefully regulated. If grains containing too much moisture are stored, the rate of respiration increases, reducing the dry weight of the grain. If grains are stored at too high a temperature and in poorly ventilated elevators, there is danger of spontaneous combustion. The rise in temperature is due to the increase in respiration in the seeds as well as the accelerated growth in bacteria and other microorganisms living on the grain.

Apples and many other fruits must be kept at optimum temperature and at optimum humidity. The rate of respiration is approximately doubled for every 18°F rise in temperature up to temperatures of about 100°F. Very few fruits can be stored below 32°F without freezing, but optimum temperatures have been established for most.

If there is too little moisture in the air, fruits will shrivel and lose weight; if the humidity is too high the molds and yeasts grow more readily, rotting the stored fruit. Some fruits, like bananas, can be picked green and ripened in warm storage rooms or allowed to ripen on the grocers' shelves. Many fruits such as cherries and plums have a very high water content and will not keep long under any system we have devised, except to dry or preserve under heat and pressure processes which destroy bacteria and molds. These are just a few of the economic problems with which we must deal as a result of the respiratory processes.

Summary

Photosynthesis and respiration provide us with a carbon turnover, known as the carbon cycle (Fig. 10), which changes sugar into carbon dioxide and back again into sugar. Photosynthesis is the vital process which converts the sun's light energy into food energy, while respiration uses the products of photosynthesis, sugar and oxygen, to convert the stored energy into activity. The cycle goes on and on and from it spring ATP molecules, which store the sun's

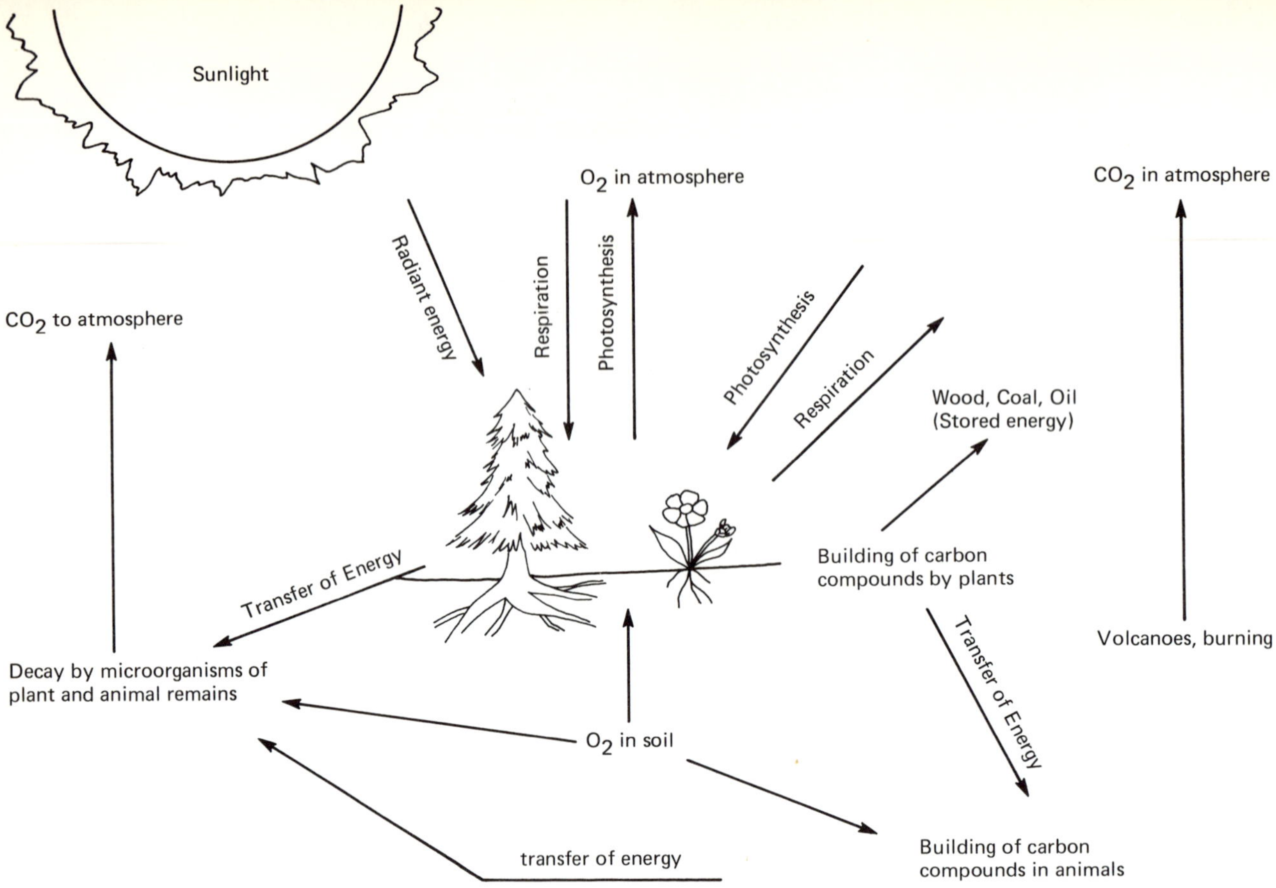

FIGURE 10 The Carbon Cycle. Arrows indicate the exchange of O_2 and CO_2 and the places where these gases are used and released.

energy in a chemical bond until the bond breaks and the energy is released.

At every step along the way specific enzymes are present splitting or adding new molecules and making the next step possible. These regulators of the various metabolic processes are special proteins which speed the various chemical reactions inside the cell. Each enzyme is partially or completely specific and, acting as a true catalyst, is not used up in the chemical reactions it stimulates. It does provide the proper conditions for the energy transfers involved by bringing the proper raw products together.

The main product of photosynthesis, sugar, is a relatively stable substance, but when needed and under the proper conditions it can become highly reactive. The enzyme mechanism which triggers sugar action and makes it available for immediate use is phosphate, provided by ATP. Once the emergency is passed the enzyme systems uncouple the phosphate atoms and leave the sugar again in a stable, nonreactive condition ready for the next moment of activity.

Comparison of Photosynthesis and Respiration

Photosynthesis	Respiration
1. Occurs only in the green cells of plants.	1. Occurs in the living cells of all plants and animals.
2. Occurs only in light.	2. Takes place in all cells under both light and dark conditions.
3. Uses CO_2 and H_2O.	3. Uses food and O_2.
4. Food and O_2 are produced.	4. CO_2 and H_2O are produced.
5. Light energy is trapped in chlorophyll and stored in food as chemical energy.	5. Energy in food is released, temporarily stored in ATP, or lost as heat.
6. Light energy is used producing ATP.	6. Food oxidation produces ATP.
7. Hydrogen transferred from water to NADP is used forming NADPH and then reducing CO_2 to food.	7. Hydrogen is transferred from food to NAD or NADP forming NADH or NADPH and then to O_2 forming H_2O.
8. Light energy is converted to chemical energy.	8. Chemical energy is converted into work or heat energy.
9. Results in an increase in weight.	9. Results in a decrease in weight.
10. All life depends on photosynthesis and its stored products exceed those used in respiration.	10. Respiration is dependent on the products of photosynthesis directly or indirectly.

Water Relations

The role of water in the processes of photosynthesis and respiration has been discussed in some detail, but the importance of this compound cannot be overemphasized. In photosynthesis and respiration ionized water is part of the electron transport system. As a compound, water becomes a solvent which makes it possible for various substances to go into solution and pass through the various cellular membranes in virtually all physiological processes. The movement of all substances in soil or within the organism is made possible by water. The chemical processes of digestion and assimilation depend on water. Water in the cells maintains turgor, and turgor contributes to the rigidity and support of the plant. Water is a major constituent of protoplasm. In many cells it comprises about 98 percent of the total weight of the cell; in others such as dry seeds there will be as little as 10 or 15 percent water.

Most organisms require fresh water for their various metabolic processes. In recent years considerable concern has been expressed about the declining quality of the earth's fresh water as a result of pollution from the expanding human population and increased industrialization. Less than 1 percent of all the water in the world is available to us as fresh water. About 2 percent more is frozen in the polar ice caps. All of the rest, about 97 percent of all the water in the world, is in the oceans and seas, too salty to drink or to use in irrigation for crop plants.

All organisms need water and usually in copious quantities. Much of our food is taken in with water, and the waste materials are removed at least in part by water. Consequently one of the

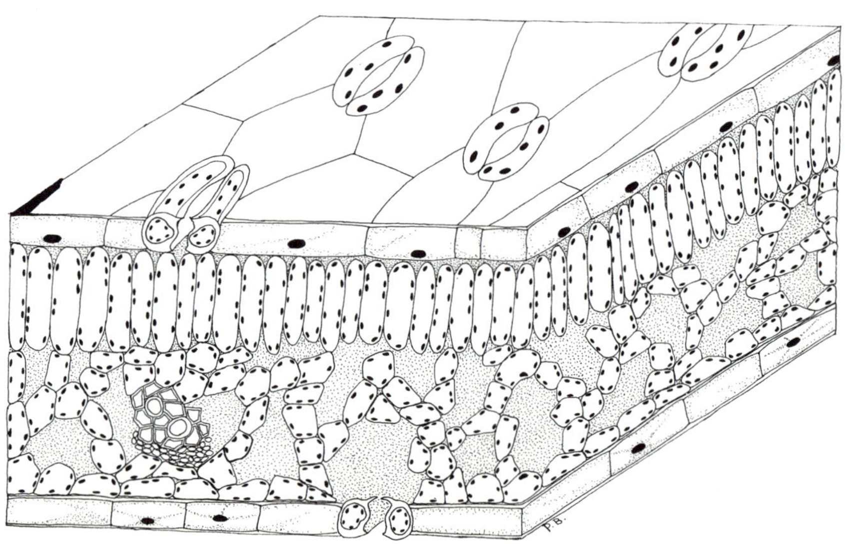

FIGURE 11 Diagrammatic three-dimensional drawing of leaf showing upper epidermis with several stomates, rather compact palisade tissue and a loose spongy layer showing one vein in cross section. The surfaces of the cells exposed to the intercellular spaces lose copious quantities of water in transpiration.

most critical problems confronting any organism is loss of water. No organism has complete control over loss of water, although higher animals generally have an integumentary covering which enables them to retain body water more efficiently than plants.

Transpiration

The loss of water from a plant in the form of water vapor is termed *transpiration*. Most of this water is lost through the open or partially opened stomates and is therefore called **stomatal transpiration.** Some is lost through the cuticle surfaces of leaves and stems and is referred to as **cuticular transpiration.** This type of transpiration accounts for only about 5 to 10 percent of water loss from plants of temperate habitats; the remainder is lost through the stomates.

If a plant could close its stomates (see Fig. 11) completely the loss of water would be greatly retarded, but if it did this it would be virtually impossible for oxygen and carbon dioxide to be exchanged in the respiratory and photosynthetic processes. Therefore, a plant cannot really control the loss of water to any extent, but must adjust itself to limited photosynthesis when the transpiration rate is high and adapt itself to periods of desiccation it cannot control.

Nearly all the water a plant uses is absorbed by the roots from the soil and conducted through the xylem elements to the parts above ground. If a plant is capable of absorbing ten ounces of water per hour and uses one ounce per hour, it has a safety margin of nine ounces. If transpiration exceeds nine ounces per hour then the plant wilts. This is due to loss of water from the cells and subsequent loss of turgor. It is not uncommon to see garden plants wilt in the hot afternoon of a summer day and then gradually recover as the sun goes down. The roots are then able to absorb enough water to overcome the water loss and the plants regain their turgidity. The stomates of most plants are open in the daytime and close at night, which indicates when the greatest loss of water is likely to be. During wilting, stomates will close, but often too late to prevent temporary wilting.

The rate of water loss will vary with different species and with variations in the environment. It has been determined for one apple tree that the rate of water loss was 0.0306 cc per square inch

FIGURE 12 *Fouquieria splendens.* Above. the tip of a stem with apical inflorescence and numerous leaves on the stem. Below. the stems devoid of leaves and the way ocotillo usually looks in its dry native habitat.

of leaf surface per hour. A large apple tree with approximately 100,000 leaves and 5 square inches of leaf surface per leaf would lose 4 gallons of water per hour. It has also been determined that corn requires about 368 pounds of water to produce a pound of dry matter. Studies of date palms in Southern California, which grow in very sandy soils and under high temperatures, have shown that dates need 90 acre-inches of water per year in order to produce a good crop. This is a loss of between 400 and 500 quarts of water per day in midsummer. Since date palms grow in areas of little rainfall, the water must be brought to them by irrigation. Contrast this vapor loss with that of a giant saguaro cactus twelve feet tall growing in the same type of habitat but not supplied with water by irrigation. This plant loses only 0.02 quart per day and never wilts from loss of water.

The U.S. Department of Agriculture states that about 4,000 tons of water are required to produce one man's cereal and meat per year in the United States. The U.S. Bureau of Reclamation stated that plants raised under irrigation require 5,000 to 6,000 tons of water to raise enough food for one individual per year. This is about 20 to 30 percent more than the national average and is due to loss of much water through evaporation in the dry habitats where irrigation is necessary and also to seepage of water from the irrigation ditches before it reaches the fields where the crop plants are growing. Because of the amount of labor involved in maintaining irrigation systems, raising plants under irrigation is a much more costly enterprise than raising crops where farmers may depend on adequate water from rainfall.

Modifications Reducing Water Loss

Through evolution many plants have acquired modifications which have made it possible for them to survive in habitats of high temperatures and low humidity. These adaptations include compact tissues and thick leaves, the development on cacti of spines instead of broad leaves, which greatly reduces the extent of the transpiring surfaces, the shedding of leaves during dry periods and growth of new ones when moisture returns as in *Fouquieria splendens* (Fig. 12), the development of a thick cuticle, especially on the upper leaf surface, and, perhaps most important of all, the formation of a highly viscous protoplasm which retards the loss and movement of water within the plant. This characteristic has been acquired by many plant species inhabiting desert regions.

External Factors Affecting Transpiration

External factors which are important in affecting the rate of transpiration include light, temperature, humidity and wind.

Light As was mentioned earlier, light many increase the rate of transpiration by its effect of keeping the stomates open during the daytime. This may vary during the day as the intensity of the light increases or diminishes. Another factor is that sunlight is absorbed as radiant energy by the leaf. A very small portion of this light energy is used in photosynthesis but the remainder of the light is converted into heat energy, increasing the temperature of the leaf above that of the surrounding air. The increase in temperature results in an increase in vapor pressure within the cells of the leaf, thereby causing an increase in water loss.

Temperature Obviously temperature is governed to a large extent by light, since the increase in vapor pressure within the leaf under increased temperature causes a greater diffusion of water molecules out of the leaf cells, into the intercellular spaces, and out the stomates into the atmosphere where the vapor pressure is less. The evaporation rate of water is doubled with about every 10°C rise in temperature. The cells in the leaf are affected in about the same way because of increased vapor pressure with increased temperature. Plants growing in shade will consequently have less tendency to transpire excessively, since the leaf temperature there will be about the same as the surrounding air, and vapor pressures will be about the same.

Humidity When the humidity is high, such as above 80 percent, the rate of transpiration is relatively low, but as the moisture in the atmosphere diminishes the rate of transpiration increases. This is due to the moisture concentration gradient between the intercellular spaces in the leaf and the outside air involved in the diffusion of water vapor. A decrease in humidity of the air surrounding the leaf would increase the concentration gradient from the leaf to the outside air, thereby increasing transpiration.

In greenhouses it is customary to keep the humidity high by sprinkling the walls and floors, as well as the plants, and also to use vaporizers to reduce the rate of transpiration as much as possible. Furthermore, as the temperature increases within the greenhouse a greater amount of moisture must be added to the atmosphere to maintain the humidity at the same level.

Wind The movement of air over the leaf surfaces tends to remove the moisture-laden air which has just diffused from the stomates. Thus the moisture-vapor gradient is increased from the leaf interior to the outside, increasing transpiration. Some plants have epidermal hairs which help to hold the moist air a little longer; other plants have sunken stomates with a small reservoir formed by invagination of the leaf tissues, which protects the stomatal openings more from direct contact air currents than if the stomates were on a flat leaf surface. Although these adaptations seem to have evolved in many species it is doubtful that they greatly retard water loss. The tendency for some plant leaves to curl when the wind is blowing, or when heat is excessive, probably is more effective since it reduces the amount of leaf surface readily exposed to the wind.

Under some situations a moderate wind may even reduce transpiration. This is possible when the sun is shining and the leaf temperature is higher than that of the surrounding atmosphere. A moderate, cooling wind may then reduce the leaf temperature, which in turn reduces the rate of transpiration.

Closing the Stomatal Aperture

In general it has been indicated that stomates are usually open in the daytime and at least partially closed at night. Research now in progress at the Connecticut Agricultural Experimental Station by Professor Zelitch and his associates has revealed that a number of substances, chiefly organic acids, may be sprayed on leaves, causing the stomatal apertures to close. In the experimental areas this has resulted in a decrease in transpiration to the extent that the water table, from which the plant roots were absorbing water, rose 0.8 inch within about a two to three week period.

Although this may be a good water conservation practice, it does slow down the manufacture of food in the plants and consequently slows growth. Where native plants are growing and water conservation is desirable it seems like a good practice, but where we are raising plants for food or other products, the reduced rate of photosynthesis would result in a decline in crop production.

Economic Importance of Transpiration

Commercially, transpiration is a problem in the raising of plants of economic importance just as much as adverse temperatures, wind, and soils of low fertility. In many vineyards grapes are taken down from the arbor in winter to reduce the drying effects of strong winds. This permits the vines to retain their moisture, which hastens their growth in spring and increases the yield. Farmers who raise winter wheat (planted in fall for spring harvesting) always hope for a good snow cover, not to keep the wheat warm but to keep it from drying out. In many areas shelter belts of trees are planted on the windward side of grain fields to reduce the velocity of wind as it crosses the field, thereby reducing transpiration and increasing yield. Orchards are frequently planted on the leeward side of buildings, hills, or protective structures for the same reason. Frequently a type of strip farming is practiced in many areas. Low crops, like beans, may be planted in long rows about fifty feet wide alternating with rows of sunflowers spaced throughout the field to give protection to the beans.

Although transpiration may seem like an ordinary event in the life of a plant it would be difficult to overestimate its influence on adaptation and survival. Only an aquatic organism which stays submerged does not experience the effects of transpiration.

Guttation

Water is lost from plants in one other way. While water lost from plants in transpiration is in the form of water vapor, in this type of loss the water is in liquid form and is termed the water of guttation. It usually occurs in spring when the soil is saturated with water and the roots are absorbing large quantities of water and building up root pressure. When the humidity is high and there is little or no movement of air currents, small drops of water will form on the tips or margins of some plant leaves. This is especially characteristic of grasses but other plants may guttate water under the same soil, water, and atmospheric conditions.

In guttation water is exuded through modified stomates which seem unable to close. In some cases the water is forced through cell paths or cellular glands, which appear as loosely organized cell openings in the leaf margins. They do not all have the same characteristics, but all such means of water exudation are termed **hydathodes.** In any case guttation is simply a means of eliminating water from the plant when the roots are absorbing more water than the leaves can transpire.

The water of guttation may be distinguished from dew by position and chemical content. Dew collects on objects, including plants, as the result of an atmosphere which is highly humid and relatively warm. As the warm, highly saturated air comes in contact with cool surfaces it condenses, forming droplets of water. On plants this may be on the flat surface, while the water of guttation is always on the margins. Dew is also pure water, while guttated water frequently has mineral salts or sugar in it as the result of osmosing through cells containing these substances and carrying some of them along with it.

Supplementary Reading

Arnon, D. I., The Role of Light in Photosynthesis, *Scientific American,* November 1960.

Bassham, J. A., The Path of Carbon in Photosynthesis. *Scientific American,* June 1962.

Beevers, H., *Respiratory Metabolism in Plants.* Row, Peterson, Evanston, Illinois, 1961.

Bonner, James, and J. E. Varner (editors), *Plant Biochemistry.* Academic Press, New York, 1965.

Calvin, M., Energy Reception and Transfer in Photosynthesis, in *Biophysical Science,* Study Program, J. L. Oncley (editor). John Wiley & Sons, New York, 1959.

French, C. S., Photosynthesis, in *This is Life,* W. H. Johnson and W. C. Steere (editors), Holt, Rinehart and Winston, New York, 1967.

Goldsby, R. A., *Cells and Energy.* The Macmillan Co., New York, 1967.

Jensen, W. A., and L. G. Kavoljian, *Plant Biology Today,* 2nd ed. Wadsworth Publishing Co., Inc., Belmont, Calif., 1966.

Kozlowski, Theodore, *Water Metabolism in Plants.* Harper and Row, New York, 1964.

McElroy, William D., *Cell Physiology and Biochemistry,* 2nd ed. Prentice-Hall, Englewood Cliffs, New Jersey, 1964.

Moment, Gairdner B., *Frontiers of Modern Biology.* Houghton Mifflin, Boston, 1962.

Steward, F. C., *Plants at Work*. Addison-Wesley Publishing Company, Reading, Massachusetts, 1964.

Turrell, F. M., The Area of the Internal Exposed Surface of Dicotyledon Leaves, *American Journal of Botany* 23:255–264, 1936.

Questions for Discussion

1. What advantage is it to a leaf to be broad and thin?
2. Leaves of water plants usually have very loosely organized mesophylls. In what ways is this an adaptation?
3. The upper surface of a leaf is usually darker in color than the lower. What does this indicate to you?
4. Explain the diffusion of O_2 and CO_2 through the stomates in a leaf.
5. In many highly populated cities the CO_2 content of the atmosphere is higher than that in rural areas. Explain the reason for this distribution of atmospheric gases.
6. Some leaves produce alkaloids of economic importance and others produce highly aromatic substances. Where are these formed and stored in the leaf?
7. Leaves which we eat are relatively low in calories. Why should this be true?
8. Why do leaves usually turn yellow in autumn?
9. The atmosphere contains about 80 percent nitrogen, which diffuses through the stomates with O_2 and CO_2. How much nitrogen diffuses back into the atmosphere? Explain.
10. How does the removal of leaves from a tree by insect attack affect the health of the plant?
11. Many leaves of aquatic plants lack stomates. How do these plants obtain O_2 and CO_2?
12. Leaves of tropical forest plants are deep green in color, while those of most desert species are grayish-green. Explain this difference in coloration.
13. Describe the different leaf types and how they may be used in the identification of plants.
14. The fresh weight of plants is usually greater in the morning than in the evening. Explain.
15. What are the advantages of the deciduous habit to a tree? the disadvantages?

16. Epidermal leaf cells are usually brick-shaped in cross section but irregular in shape and interlocked in surface view; this is also largely true of cells of the spongy mesophyll. Of what advantage is this to the leaf?
17. Explain the factors which control the opening and closing of stomates.
18. Why wouldn't it be better for all of the cells of a vein to conduct the same necessary water, food materials, and inorganic substances than for separate elements to conduct these substances?
19. Tree shelter belts have been reasonably effective in increasing the yield of crop plants or orchard trees planted on the leeward side of the shelter belt. Why should this be so?
20. Why do herbaceous flowering plants grow and thrive much better in early spring in deciduous forest areas than in fall?
21. Why are plants growing in dark areas usually pale?
22. Compare and contrast the process of photosynthesis with that of respiration.
23. Would there be any advantages to a plant's having more stomates on the lower than on the upper leaf surface?
24. Describe an experiment demonstrating the process of photosynthesis to show the use of the various elements involved.
25. In areas where the prevailing westerlies are the principal source of wind, on which side of an orchard would you expect the fruit produced to be larger and better? Why?
26. One of the major advances in the food sciences is the storage of fruits, such as apples, and root crops, such as potatoes. What are the factors which must be considered in keeping these foods available for us twelve months of the year?
27. Describe various types of modified leaves.
28. Compare the respiratory process in plants with that in animals.
29. What would happen, and how soon, if the carbon removed from the atmosphere in photosynthesis remained in the plant and animal bodies as an inert substance and was not released by the organisms of decay?
30. Would a chloroplast be a more efficient structure if it were a globular object full of chlorophyll instead of containing numerous lamellated plates? Explain.

8

The Stem

The stem is often the most conspicious part of a plant and yet many people regard it as only a connection between the leaves and roots. The giant redwoods and many other woody plants have their greatest bulk in the stem. Without the extensive growth provided by stems many plants could not survive.

FIGURE 1 Stems make up about 99 percent of the above ground portions of these two cacti. *Left:* Giant saguaro. *Right:* This barrel cactus is fifteen inches in diameter and about two feet tall. In these plants the leaves are modified spines.

FIGURE 2 *Left:* Cottonwood tree with angular branches and leaves on young shoot. *Right:* Joshua tree showing much-branched stems. Rarely do monocotyledonous plants branch as much as this.

Possibly the most important function of stems is giving rise to leaves, but in addition the stem supports the leaves by placing them in favorable positions for the absorption of light, supports flowers and fruits, and aids in the dissemination of seeds. Most stems are extensively supplied with conductive channels, facilitating the upward movement of water and minerals to the upper portions of the plant and the downward movement of food to the roots Some stems store food and water, some are capable of vegetative propagation, and many short-lived stems remain green and carry on photosynthesis.

FIGURE 3 Stem tip. The immediate tip is meristematic, with the cells becoming larger and gradually elongated. The lateral appendages are young leaves with embryonic buds in their axes.

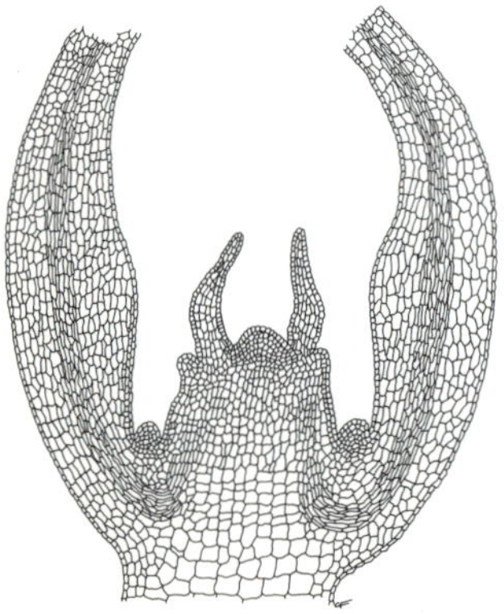

Apical Growth

The stem tip (Fig. 3) is an embryonic region consisting of an active meristematic area at the apex, with regions of elongation and maturation immediately below.

Regions of Growth and Development

The cells of the **apical meristem** are all of about equal size and isodiametric in shape. As these cells divide mitotically in growth, the older ones slightly farther back begin to elongate; for convenience this area is called the **region of elongation.** Just back of this region the cells begin to differentiate into tissues, thus this level is called the **region of differentiation** or, when the tissues are all differentiated, the region of **maturation.** This activity of growth, elongation, and maturation takes place within the upper few millimeters of the stem tip.

Just below the apical leaf growing points, the leaf primordia arise as lateral outgrowths. In this region the internodes are very short with the result that young developing leaves of the second, third, and fourth nodes extend above the apex and reach a stage of differentiation far beyond that of the adjacent tissues in the central axis. Often branch primordia arise in the axils of the third and fourth leaf primordia, initiating an early appearance of lateral buds, but in other species, these growing points may not appear until later.

Buds

Surrounding the stem tip and enclosing the regions mentioned, including the primordial leaves and branch primordia, are bud scales which enclose the entire structure in woody plants. During much of the year, the internal structures are growing slowly and in winter they remain dormant (Fig. 4). This stem tip with its surrounding structures is then a **bud.** If it is at the tip of a branch it is a **terminal** or **apical bud** and if it is in the axil of a leaf it is a **lateral** or **axillary bud.**

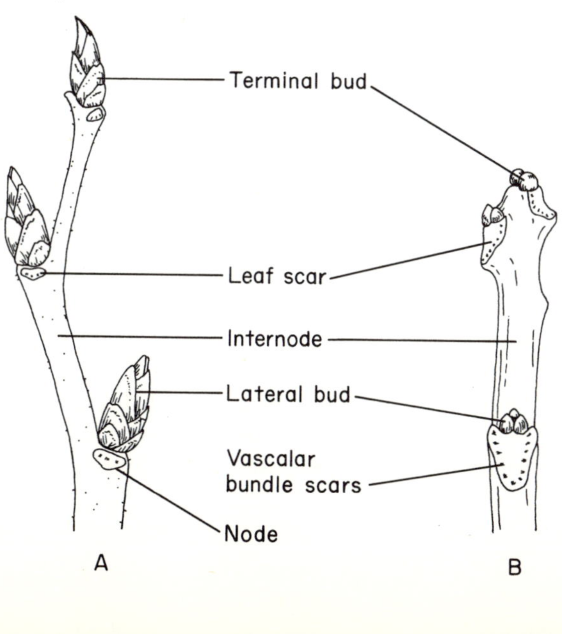

FIGURE 4 Tip of branch as it appears in the winter. Left: American elm; Right: Tree-of-heaven.

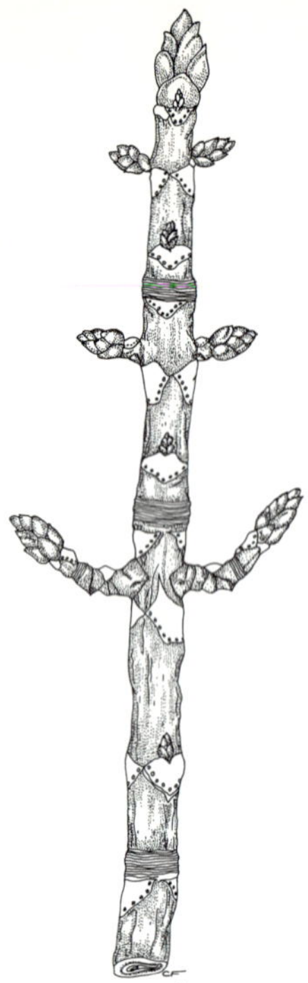

FIGURE 5 A four-year-old twig showing terminal buds at the tip of the main axis and lateral branches. Horizontal rings indicate the end of each year's growth; leaf scars are represented by nonstippled areas; dormant buds still remain on the second and third year old portions; within the leaf scars are vascular bundle scars. (Lab. Outlines.)

In spring, when the terminal buds and lateral buds become active, tips expand with an increase in numbers of cells in the meristem and an increase in size of cells of the differentiating tissues. The bud scales fall off, leaving small ridged scars around the stem. These are the **bud scale scars** and they indicate the position of a new year's growth. The age of most twigs can be determined by the number of terminal bud scale scars left by each year's growth as the twig elongates. The scars remain for several years until they are finally obliterated by the increase in diameter of the stem (Fig. 5).

In autumn, when the leaves drop from the tree, scars are left where the leaves were attached. These are termed **leaf scars.** They are distinctly shaped in different species and are often helpful in identifying species. Within the leaf scar are a number of minute scars left by the breaking of the vascular element at this point. They are the **vascular bundle scars,** and they too are very definite in number, arrangement, and position in each leaf scar for a given species.

Although the discussion thus far has been limited to the growth of vegetative buds and their parts, there are also flower buds inside of which are flower primordia instead of leaf and branch primordia. Flower buds, of course, produce flowers which develop the usual floral parts and ultimately produce fruits and seeds. In a few trees, but more commonly in shrubs, both leaves and flowers are produced in the same bud. These are known as **mixed buds.**

The differential expansion of buds and the branches to which they give rise determine the overall form of higher plants. In most dicotyledonous species the apical bud growth dominates that of the lateral buds (Fig. 6). This is due to an inhibiting substance produced by the apical bud which exerts its effect on the laterals and prevents or delays their development. Removal of the apical bud results in an almost immediate response in growth by one or more of the lateral buds. There is also experimental evidence that in some plants removal of a prominent lateral shoot will stimulate growth of the apical shoot, indicating that movement of the inhibiting auxins may be up as well as down and not confined solely to the main axis.

FIGURE 6 This apple tree was pruned to remove the apical buds and force the laterals to develop into branches. The tree was supported on a trellis which has been removed for the picture. Trees pruned in this way are termed espalied.

Apical Dominance

Different species often exhibit different degrees of **apical dominance.** That is, some plants such as the gymnosperms spruce and Douglas fir have spire or pyramidal outlines and maintain this shape as long as the central apical bud remains alive and active. This type of branching is termed **excurrent** (Fig. 7). Most angiosperms, such as the oak, maple, and elm, have a type of branching described as **deliquescent** (Fig. 8). These trees when young have fairly prominent apical dominance, but after a number of years lateral buds become more active and begin to form large branches which grow upwards at an angle. It soon becomes difficult to recognize which is the original apical point; many forks have formed and the tree spreads out forming a more or less flat top with a canopy effect. Palms have a strong apical dominance but rarely branch, and the leaves are replaced each year at the apex. This type of plant is said to be **columnar** (Fig. 9).

FIGURE 7 Excurrent branching of conifers. Center is a spruce, *Picea engelmanni,* to the right there is a western yellow pine, *Pinus ponderosa.*

FIGURE 8 A Chinese elm showing typical flat-topped deliquescent branching.

FIGURE 9 Columnar manner of growth of the Washington palm.

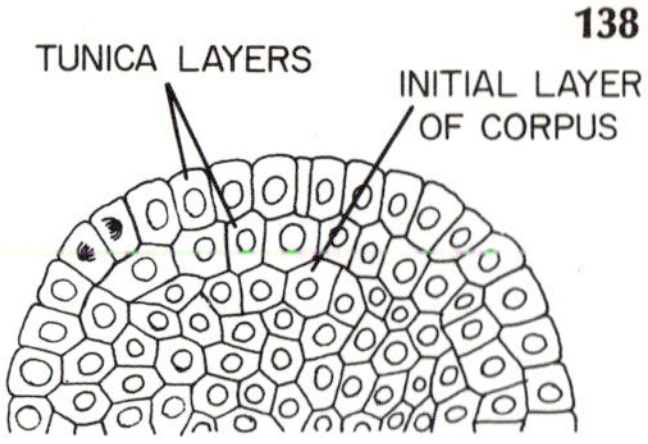

FIGURE 10 Diagram of very young stem tip showing isodiametric cells. Redrawn from Esau, *Plant Anatomy*, 1965. (John Wiley and Sons.)

Primary Meristems

As the stem grows, the youngest cells are those of the apical meristem. As stated earlier, these cells are all similar morphologically and active in cellular division. Gradually the cells begin to differentiate. At first there is a barely discernible difference and it is more a geographic than actually a tissue differentiation. One accepted theory of the origin of the tissues in the stem is the **tunica corpus** concept.

The Tunica

The tunica is defined as the apical external layer or layers of cells that apparently give rise to nothing more than the **dermatogen,** which in turn matures into the **epidermis.** In some stem tips, where the tunica may be several layers thick, primordial leaf cells seem to have their first divisions in this region. If it is true that leaves actually arise in this relatively undifferentiated tissue, then it is more than a tunic.

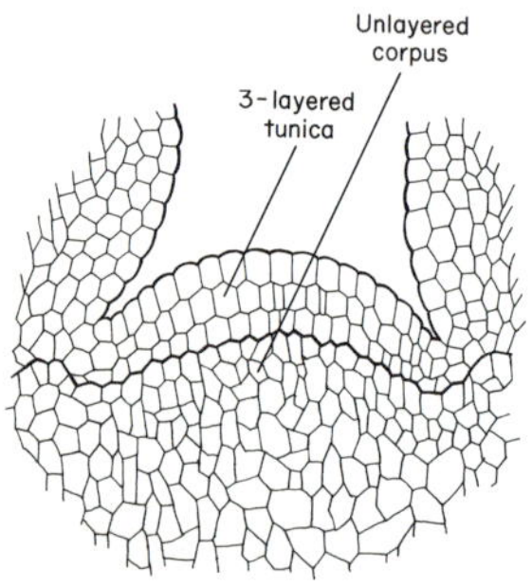

FIGURE 11 Somewhat older stem tip with tunica well differentiated and two lateral leaves beginning to develop. (From Sinnott, *Plant Morphogenesis,* 1960. McGraw-Hill Book Co.)

The Corpus

The corpus occupies the major portion of the stem tip and has cells which divide in all planes. The portion which remains meristematic and occupies the greater portion of the tip is the **ground meristem.** From this portion of the tip the primary tissues differentiating first are the pith, in the very center of the stem, and the cortex, just beneath the epidermis.

In addition to the ground meristem, the corpus also gives rise to procambium. In a cross section of the young stem tip the procambium cells appear to be smaller than those of the surrounding ground meristem but in longitudinal section the procambial cells are longer and often pointed at the ends. The procambium cells give rise to the primary vascular tissues, which consist of (1) the primary xylem, the water conducting cells, (2) the primary phloem, where food is conducted, and (2) the vascular cambium between the xylem and phloem in those species where cambium is differentiated.

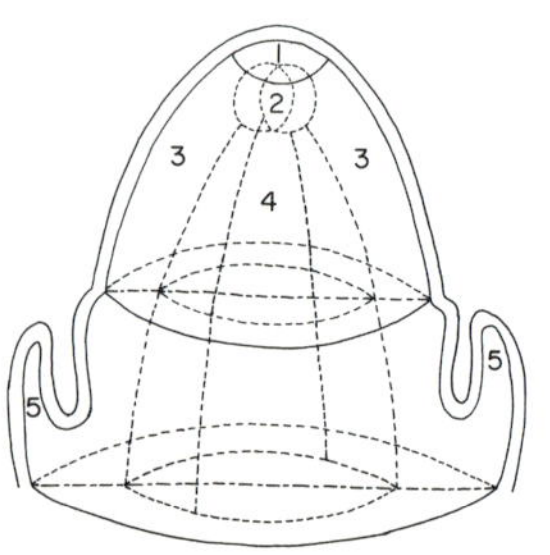

FIGURE 12 Diagram of stem tip of white fir. The apical initials (1) give rise to the other meristem region which collectively may be termed the ground meristem. (2) The mother cells divide slowly but towards the exterior of the group give rise to the (3) peripheral meristem and (4) central meristem; (5) leaf primordia.

These, then, are the primary tissues which have differentiated from the meristem of the shoot tip:

Primary Tissues from the Apical Meristem

Tunica	Dermatogen (Protoderm)	Epidermis
Corpus	Ground Meristem	Cortex Pith
	Procambium	Xylem Phloem Vascular cambium

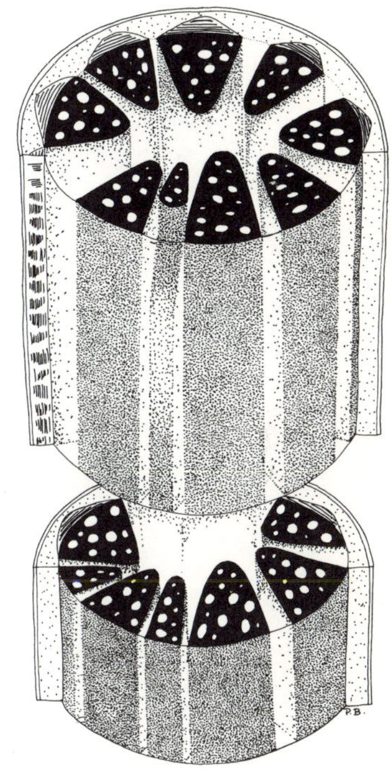

FIGURE 13 Section through a herbaceous stem. The darkened areas are bundles of xylem with triangular regions of phloem just outside the xylem and this in turn surrounded by cortex and an epidermal layer.

If we cut a cross section through a young stem and examine it under the microscope, we can readily see the concentric arrangement of the tissues (Fig. 13). At the outside is the epidermis, a single compact layer of cells endowed with cutinized substances in their outer walls which helps to cut down the loss of water in transpiration. This noncellular layer is called the cuticle. In young stems, as in leaves, stomates are present in the epidermis. Within the epidermis is the cortex, a zone consisting of several to many layers of relatively unspecialized cells in which food may be stored. The cells of the outer layer or layers of the cortex may have slightly thickened walls and be classified as *collenchyma,* if only the corners are thickened, or as *schlerenchyma,* if the lateral walls are thickened all around. Most cells of the cortex, however, are thin-walled and isodiametric and are termed *parenchyma.*

In some species, like milkweed, the cortex may contain secretory cells or canals which contain latex or other resinous substances. Some species may also have elongated thick-walled fibers in the cortical region.

Inside the cortex the primary vascular tissues appear as at least two well differentiated areas; the outer of these is the phloem, which in angiosperms usually consists of sieve tube elements, **companion cells,** thick-walled fibers, and some parenchyma. Sieve tubes are formed from vertical rows of large elongated cells, known as sieve tube members. They arise by division of a procambium cell in which one daughter cell becomes a sieve tube member and the other differentiates, in angiosperms, into the much smaller companion cell (Fig. 14). This cell may divide again, which accounts for the larger number of companion cells adjacent to a single sieve tube. In gymnosperms, companion cells are lacking.

The large sieve tube members at first contain a nucleus and an abundance of cytoplasm; as the cell matures, the nucleus disintegrates but the cytoplasm remains. The name "sieve" comes from the fact that the end walls have numerous pores through which cytoplasmic strands extend, connecting the members of the sieve tube. These end walls are called **sieve plates** and as the cell matures, a deposition of material forms a circular **callose** around each pore.

Fibers which form in the phloem are usually towards the outside of the conducting tubes and are termed **bast** (Fig. 15). Where these fibers occur in abundance, such as in the stems of flax or hemp, they are removed by a retting and combing process and are used in making linen and fine hemp rope.

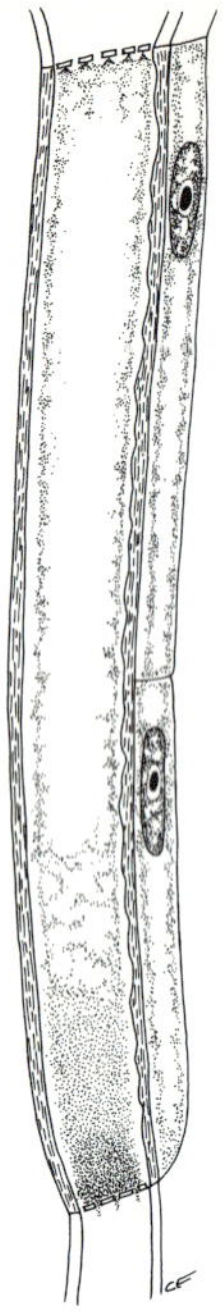

FIGURE 14 Sieve tube with end pores and adjacent companion cells.

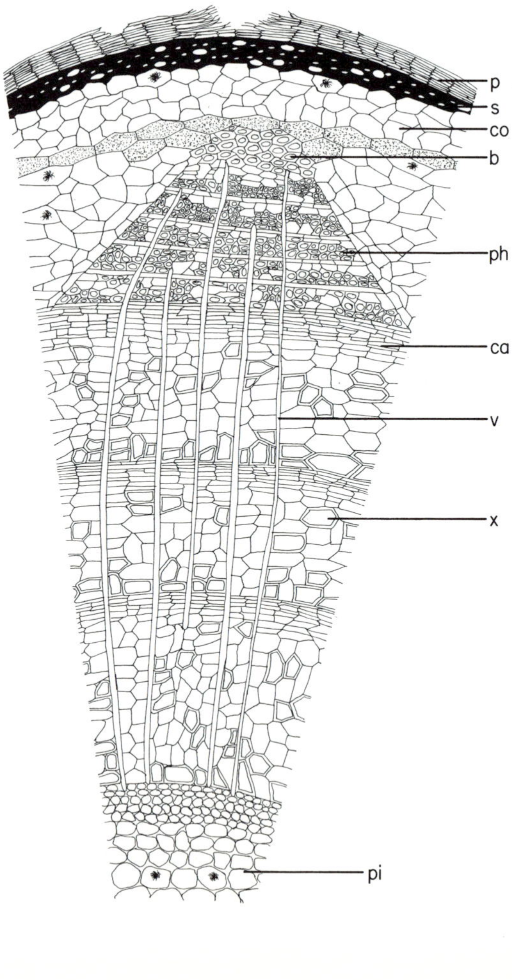

FIGURE 15 Cross section of three year old stem: p-periderm; s-sclerenchyma cells of outer cortex; co-parenchyma cells of inner cortex, stipled cells are in the endodermis; b-bast fibers in outer phloem; ph-phloem; ca-cambial region; r-rays; x-xylem; pi-pith structures in a few cells are crystals.

Xylem

Xylem is located central to the phloem tissue and is in approximately the same radius. Xylem is composed of several different types of cells as well as tubular channels for the conduction of water and minerals in solution. One type of cell is a **tracheid,** an elongated cell with more or less pointed ends. Soon after the tracheid is matured, it loses its protoplasmic contents, and although it is a dead cell, it is capable of functioning as a conducting unit.

The walls of tracheids are considerably thickened, but contain numerous circular or oval **pits** which permit movement of substances from cell to cell. The pit is not, however, simply a hole (Fig. 16). It is a depression within the wall with a **pit membrane** at its base, formed from the primary wall of the tracheid when first formed. **Simple pits** seem to be breaks or depressions in the wall with only the membrane forming a connection between the wall parts. **Bordered pits** have a slightly raised extension of the tracheid (or vessel) wall extending over a portion of the membrane. Where the border of one cell lies opposite the border of another cell, we speak of it as a **full-bordered pit;** if the raised border appears on only one side of the pit membrane, it is **half-bordered.**

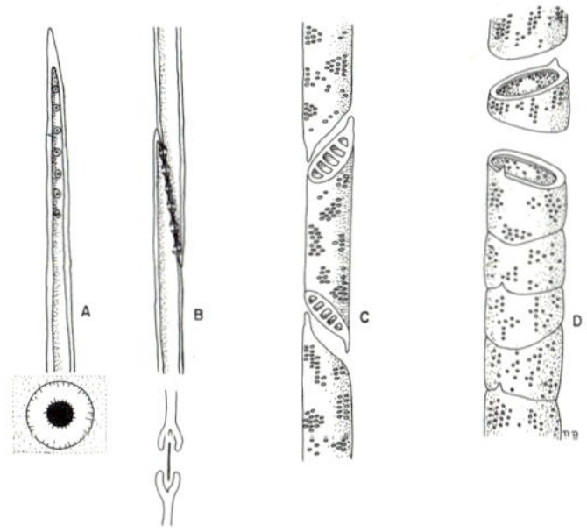

FIGURE 16 Pits in the lateral and end walls of A and B in tracheids, and in the lateral walls of the vessels in C and D.

In addition to pits in the wall, tracheids may also have distinctively thickened walls (Fig. 17). If the thickening is in separate

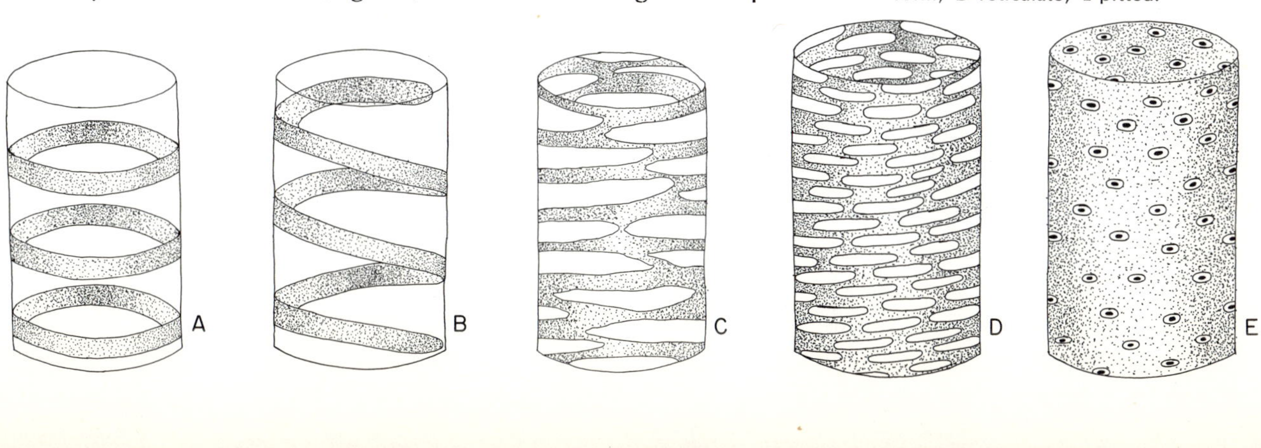

FIGURE 17 Markings in xylem elements. A-annular; B-spiral; C-scalariform; D-reticulate; E-pitted.

rings it is spoken of as **annular;** if the thickening is spring-like we refer to it as **spiral.** Some tracheids have a transverse thickening with interconnecting bars and are called **scalariform;** finally, this thickening may be **reticulate,** when it forms a network. Sometimes the thickening is really a combination of scalariform and reticulate and therefore is termed **scalariform-reticulate.**

Tracheids may be found in all species of vascular plants, including angiosperms, but only these higher plants also have **tracheal tubes (vessels).** A vessel is formed when a number of cells develop end-to-end and form a tube. Vessels have no end walls separating the components of the tube although they may have bars forming a ladder-like network over the ends of some (Fig. 18).

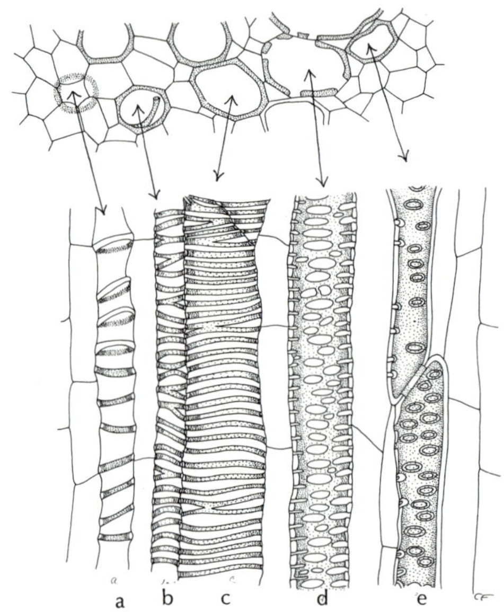

FIGURE 18 Longitudinal and cross sections of xylem elements. showing different types of markings. a-annular, b-spiral, c-scalariform, d-reticulate, e-pitted. (Redrawn from W. H. Freeman Lab. Outlines in Biology.)

As in tracheids, the protoplasts are lost soon after differentiation from the procambial cells and the vessel members become united into a long, nonliving conductive tube. Wall thickening is of the same type as in tracheids, but in vessels the reticulate and scalariform types are much more common.

Xylem fibers are similar to those found in phloem and other parts of the plant (Fig. 19). Xylem parenchyma is also of the same type found in the phloem. Although the parenchyma associated with xylem elements is not usually isodiametric, the cells are thin-walled, contain protoplasm, and function in the storage of food and water with a limited capacity in the conduction of soluble materials.

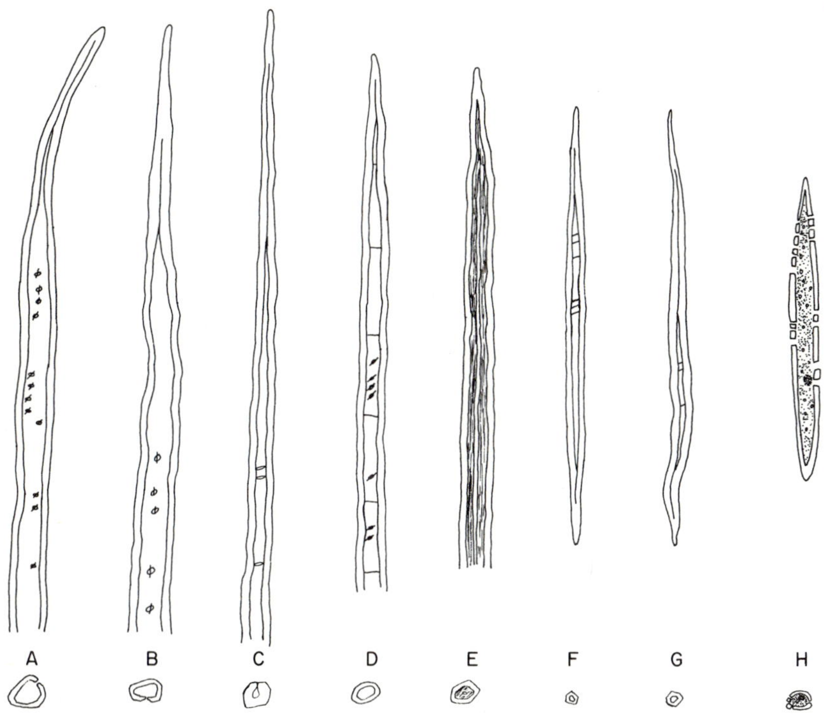

FIGURE 19 Wood fibers, longitudinal and cross sections showing thickness of walls. A-apple, B-tulip tree, C-white oak, D-mahogany, E-red oak, F-hickory, G-*Guaniacum sanctum,* H-*Sassafras.* (Redrawn from Eames, *Introduction to Plant Anatomy,* McGraw Hill Book Co.)

Vascular Bundles

In young herbaceous dicotyledonous stems, as well as in those of monocotyledons, xylem and phloem occur in discrete areas which we term the **vascular bundles.** In monocotyledons as well as most dicotyledons, the bundle consists of xylem towards the inside and phloem towards the outside. However, bundles of monocotyledons also contain fibers and parenchyma and are enclosed by one or more layers of cells termed the **bundle sheath.** This type of vascular bundle is called **closed,** not because it has an encompassing sheath but because it lacks cambium.

Dicotyledonous bundles are considered **open** because they do have a vascular cambium situated between the primary xylem and phloem which will give rise to secondary members of these conductive tissues. In stems that grow rapidly, or live for more than one year, the cambium divides, giving rise to new xylem towards the inside of the cambial layer while it is also adding phloem to the outside. Further discussion of cambial activity will be given in the section on the development of secondary tissues.

The Pith

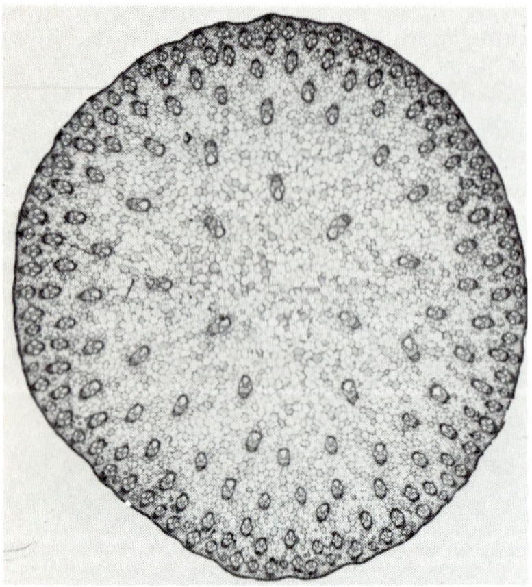

FIGURE 20 Corn stem. Cross section. Note concentration of vascular bundles towards the periphery of the stem. Courtesy Carolina Biological Supply Company.

The center of dicotyledonous stems is occupied by the pith. The cells are usually parenchymatous. These cells store food materials and conduct soluble materials short distances. In some stems where the vascular bundles are not touching each other, cells of the pith grow between them and are considered **pith rays.** As the vascular bundles increase in size as a result of cambial cell division the pith is compressed. After a year or two of growth in woody stems, pith cells are nearly obliterated and the pith ray cells, which are also added by cambial activity, become crowded and narrow.

Monocotyledonous stems owe their vascular system and arrangement of tissues to the development of procambial cells that are "scattered" in the ground meristem instead of being in a concentric ring as found in the meristem of dicotyledonous angiosperms and gymnosperms. The bundles which develop in the stem in a scattered manner from these procambial areas seem to have no definite arrangement (Fig. 20). Actually there is a much greater concentration of bundles towards the outside of the stem than towards the center.

The typical monocotyledonous stem consists of an outer ring of epidermal cells, several rows of collenchyma and sclerenchyma cells, which give considerable strength to the stem, and then the great mass of undifferentiated parenchyma cells within which are found the scattered vascular bundles.

FIGURE 21 The Joshua tree, *Yucca brevifolia.*

In very large monocotyledons, such as palms and **Yuccas** (Fig. 21), which attain tree height, the primary tissues derived from the ground meristem would be inadequate to take care of the water requirements for such a large stem. Plants of this type differentiate a specialized type of cambium within the parenchyma cells, which become active in division, giving rise to new vascular bundles. This type of cambium gives rise to new conductive cells towards the inside and may also add a few more parenchyma cells towards the outside of the stem, giving greater bulk to the stem and increasing its conductive capacity.

A vascular bundle of a monocot (Fig. 22), such as corn, consists of several large vessels of xylem towards the inner portion of the bundles and phloem, containing well formed sieve tubes and companion cells, towards the outside. In addition there may be a number of smaller xylem elements around the larger vessels and a rather conspicuous bundle sheath composed of many thick-walled fibers around the entire bundle. In the vascular bundles of many monocotyledons, especially corn, there is a large air space between

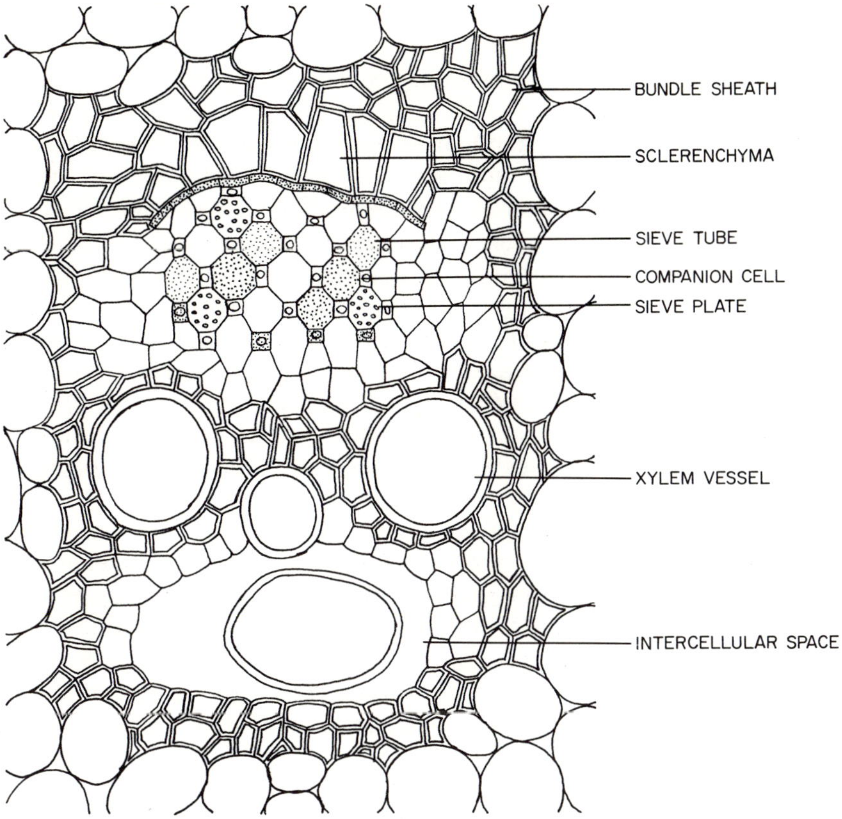

FIGURE 22 Vascular bundle of corn.

the large vessels which apparently is formed from the rapid expansion of the stem. This air space was apparently the early stages of growth of an annular vessel which was simply stretched and broken. In many of these air spaces a wall or portion of it, typical of the type in the large circular vessels, still remains to indicate its origin.

Secondary Growth

The primary stem tissues discussed above have been derived through growth and differentiation of the apical meristems. If this were the only method used by plants in growth there would be little opportunity for increase in diameter except for increase in size of the cells. Fortunately the cambium, which was derived from the procambial strands, is capable of division into new cells which then differentiate into secondary xylem and secondary phloem (Fig. 23).

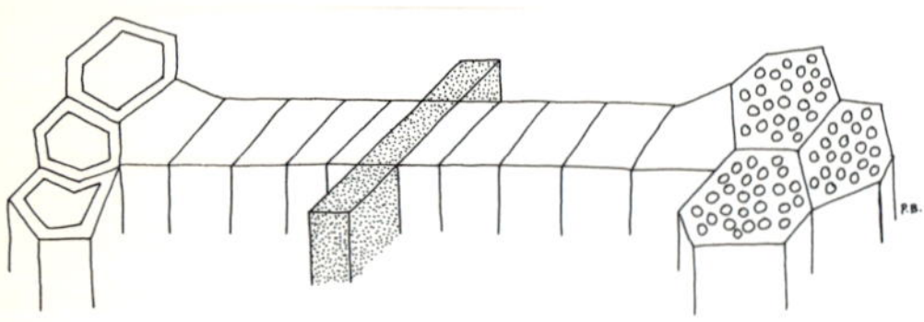

FIGURE 23 The cambium layer is stippled in the center and the tissues it gives rise to are on either side. Xylem is to the left and phloem to the right.

As a result of this type of growth, increase in diameter of a woody stem seems as unlimited as the growth of the apical meristems. The great mass of tissue which one sees in the cross section of an old stem has, then, been derived from this single layer of cells which was formed at the stem tip and now forms a slender cylinder throughout the length of the stem.

When a cambium cell divides, one of the resulting daughter cells continues in its cambium capacity, while the other gradually becomes differentiated and adds to the adjacent tissue. At division the new cell which forms to the inside of the one retaining its cambial capacity becomes xylem; if the new cell which is formed is to the outside of the cell retaining its cambium characteristic, the new daughter cell differentiates into phloem. Some of the new cells formed in this manner differentiate into ray cells, parenchyma, and fibers as well as the new conductive elements of xylem and phloem. In this way the cambium, working in both directions, continues to increase the amount of wood and phloem as the stem grows older, a process that continues year after year and in the case of some giant sequoias has remained active for over three thousand years.

FIGURE 24 Early development of the periderm. The cork cambium is present by the stippled layer in the center. The cells have just divided to give rise to radial rows of cells towards the inside, the phelloderm, and to radial rows of cells to form the phellem on the outside. A young lenticel is forming in the phellem.

If growths of the new secondary tissues were equal, the amount of phloem added outside the cambium would be equal to the xylem added to the inside. However, the amount of secondary phloem produced is less than that of the secondary xylem. Furthermore, the amount of xylem produced persists and eventually forms the great bulk of the plant body while the phloem is pushed farther out and is gradually crushed and sloughed off of the tree as the newer tissues arise within.

The fate of the epidermis and the cortex is the same as that of the older phloem. As the woody stem increases in diameter a new meristematic layer, the cork cambium or **phellogen,** arises immediately beneath the epidermis in the cortex (Fig. 24). The newly formed phellogen begins to divide, as does the vascular cambium, giving rise to rows of a corky tissue, the **phellem,** towards the outside and occasionally adding additional rows of cells towards the inside. These new cells just to the inside of the phellogen are termed **phelloderm.** Collectively the three new tissues are called **periderm** and they, with the old phloem and any cortical cells which might still remain, make up the **bark** (Fig. 25).

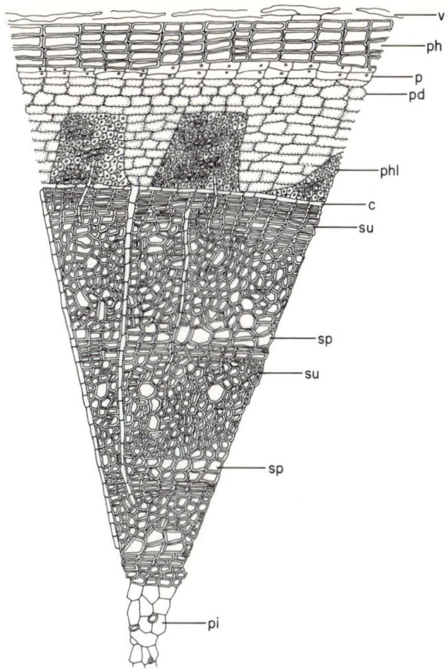

FIGURE 25 Section through a woody stem; r-rhytidome; ph-phellem; p-phellogen; pd-phelloderm; phl-phloem; c-cambium; su-summer wood; sp-spring wood; pi-pith.

The bark of trees is a highly efficient covering since most of its cell walls are heavily impregnated with suberin and are essentially cork. These cells protect the stem from loss of water and in addition form a layer of insulation which protects trees from excessive heat during forest fires.

The bark of different plant species is often quite distinctive (Fig. 26). Some bark peels off in large strips or sheets; that of a

FIGURE 26 The bark of several mature trees. Mulberry has a very uniform and disturbed pattern, the bark of Scotch pine is orange in color on the younger branches and peels off in thin paper-like pieces; Russian olive bark peels off in narrow strips; Valley cottonwood is rather irregularly furrowed and quite coarse; Arizona cypress bark comes off in small thin sections.

FIGURE 27 The irregular bark of ponderosa pine which has formed the rhytidome.

Russian olive is stringy; ponderosa pine bark comes off in small jigsaw shaped pieces (Fig. 27); and that of the shag-barked hickory falls away in long strips.

Wood

Now that we have discussed the origin of tissues which make up the body of the woody plant, we should consider in some detail that part of the stem which means so much to us from an economic standpoint. Examining the cross section of a cut tree shows that the wood is made up of concentric rings. Identification of these rings is quite possible without the aid of a hand lens, but for the most part the cells which make up a ring can be identified only with the aid of a microscope. Rings are usually formed at the rate of one a year and consequently are called **annual rings.**

During spring and early summer, when conditions for growth are at their best, the cells formed are fairly large, but as the season progresses growth becomes less rapid, and the cells formed are much smaller. This differentiation in size of the cells in one annual ring sets it apart visually from the next ring, which will begin with large cells then decrease in size in the latter part of that year. The large cells which were formed in the early part of the growing season are termed **spring wood,** and the smaller ones which make up the outer portion of that annual ring are those which comprise the **summer wood** (Fig. 28). Therefore, in each annual ring we have

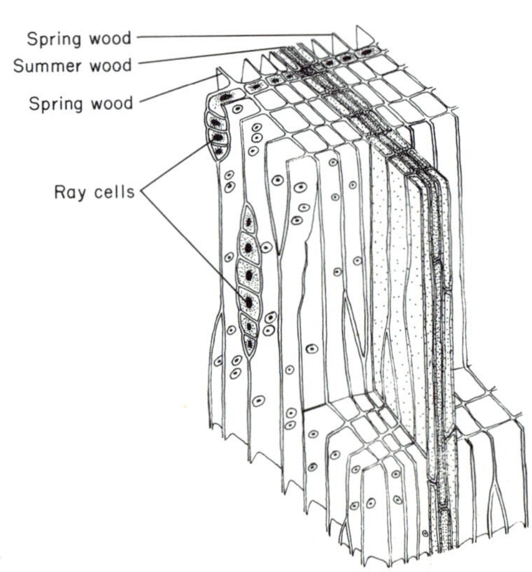

FIGURE 28 Diagrammatic section of wood showing xylem elements (tracheids) in longitudinal section and ray cells in cross section. This portion is of a one year old section with some spring wood showing from the second year's growth.

spring wood and summer wood. Each ring represents one year's growth, so the age of the tree can be determined by counting these rings. Occasionally, when a period of unusual drought occurs in the middle of the growing season, a tree may lay down what appears to be two rings in one year. An expert in **dendrology** (the study of tree rings) can usually distinguish these "false rings" from the true annual ring.

Some years ago the study of dendrology was applied to the dating of Indian ruins in the Southwest and elsewhere. This new science was known as **dendrochronology.** By overlapping the outer tree ring patterns with those of the inner rings taken from logs at different building sites or at lower levels of the same site, it was possible to determine when the most ancient trees were cut and when they were used in building homes. Some Indian ruins have been successfully dated back to about 1200 A.D. in this manner.

Types of Wood

As a tree grows older the functioning of its conductive elements becomes less efficient, primarily through the crystallization or precipitation of substances such as resins, tannins, oils and other materials within the tubes and the gradual occlusion of the apertures. This begins in the xylem elements in the center of the tree and works its way outward as the tree ages. The non-functional xylem formed in this manner is called **heartwood,** while the younger xylem, which is functional, is **sapwood** (see Fig. 25). Each year as a new ring of sapwood is added to the outer mass of xylem a ring of sapwood becomes heartwood. The transformation is not quite as clear-cut as the above statement would imply and it varies in different species. Certainly a tree is several years old before any heartwood forms and it may not be produced at exactly the same time as new sapwood. The time of formation, in other words, is approximate. Heartwood is often darker in color than sapwood, which makes the distinction between the two areas obvious.

Wood is classified in another way which is often of much more significance to us economically than other characteristics. Some woods are classified as **hardwoods** and some as **softwoods.** This distinction is often made even though there is considerable overlap in the actual density, strength, and texture. Hardwoods are angiosperms, in which the xylem conductive elements are composed of both **tracheal tubes** (vessels) and **tracheids.** Wood containing tracheal tubes is said to be **porous,** and if the xylem members in the springwood have many large tracheal tubes which gradually diminish in size to the small tubes in the summer wood we describe the wood as **ring porous.** If there are some large elements diffused with the smaller ones in the summer wood it is classified as **diffuse porous.**

Softwoods are gymnosperms. This type of wood has only tracheids as the conducting elements, along with the usual rays, and since it lacks the tracheal tubes it is classified as **non-porous.** Annual rings are usually just as obvious in gymnosperms as they are in angiosperms and the cells are just as efficient in the translocation of substances in solution.

Wood Qualities

The uses we make of woods are many and we determine their greatest efficiency by experience and the characteristics which we have found in them. Following are some of these characteristics.

1. **Density** refers to a wood's comparative weight per unit of volume. The lightest is balsa, which has a specific gravity of 0.12 while some wood, like South American ironwood, has a specific gravity of 1.3. You will probably recall that water has a specific gravity of 1.0.

2. **Durability** is a reflection of a wood's ability to withstand the attacks of bacteria, fungi, and certain insects. Redwood and swamp cypress are two of our most durable woods.

3. **Hardness** tests are only approximate, but some woods like balsa may be indented by the finger nail while others like oak and hickory require a knife or similar instrument to scratch them.

4. **Porosity** does not necessarily refer to whether a wood contains only tracheids or tracheal tubes, but rather refers to its permeability. Oak, for instance, which is obviously ring porous, is a highly desirable wood for use in tight cooperage for the storage and aging of liquids.

5. **Strength** of a wood may be divided into several categories.

a. Crushing strength is resistance to endwise compression.
b. Tensile strength is the ability to withstand pulling apart.
c. Shearing strength is the ability of the elements to withstand forces tending to slide them by one another.
d. Crossbreaking strength is the ability to sustain a load at right angles to the grain.

6. **Grain, texture,** and **figure** are interrelated characteristics and depend upon the position of the xylem elements, rays, size of annual rings, size of elements, and cut of the wood. A **transverse**

or **cross section** of a log reveals the concentric rings but in longitudinal cuts the grain becomes apparent. **Radial** or **quarter-sawed** wood is taken through the radius of the log. This shows the annual rings in longitudinal view and also shows the rays in longitudinal view at right angles to the grain. This often presents a pleasant pattern. Since only a few boards can be considered quarter-sawed from a single log, this type of pattern is not common and is therefore more valuable. **Tangential** or **plane-sawed** wood is cut from the outside and one board after another is cut in this manner from large logs. This type of longitudinal cut is through spring and summer wood of several rings, giving a pattern that is often described as tent-shaped. Unusual figures in wood may be due to irregularities in growth causing a wavy pattern, by dormant buds which have been grown over, by numerous ray cells, and by knots. Although knots may be considered desirable in such woods as knotty pine which are used for paneling, they are structural defects and lessen a wood's usefulness for mechanical support. The chief economic uses of some of the different woods are listed below.

Hardwoods

1. Oaks (white, red, or black)—cabinet work, veneers, tight cooperage, flooring, trim, furniture, and general construction
2. Maples—flooring, furniture, musical instruments, veneers, implement handles, maple syrup
3. Walnut—cabinet wood, gunstocks, edible nuts
4. Beech—furniture, flooring, woodenware novelties
5. Hickories—axe handles, wheel spokes, hickory and pecan nuts
6. Cherry—furniture, veneers, handles, fruit
7. Ash—tool handles, baseball bats, oars
8. Sycamore—butcher's blocks, musical instruments; especially good when quarter-sawed for veneers, trim, furniture
9. Gums—cheap veneers, veneer cores, wood rollers, gunstocks
10. Poplars—Pulp, boxes, crates, excelsior, fuel

Softwoods

1. Douglas fir—lumber and general heavy construction, plywood; highest quality softwood in the United States
2. Redwood—shingles, timbers, tanks, tight cooperage, caskets
3. Hemlocks—source of tannin, boxes, pulp, cheap construction
4. Spruces—paddles and oars, pulpwoods, interior trim, ornamental planting

5. Firs (white and balsam)—paper pulp, boxes, crates, Christmas trees, resinous blisters on bark form Canada balsam
6. Arborvitaes—shingles, boats, cooperage
7. Pines—white pine used for many years for ship masts

Red pine—crates, pulp
Long-leaf pine—general construction, boxes, crates
Western yellow pine—general construction, boxes, crates
Lodgepole pine—posts, telephone poles, construction lumber
Piñon pine—edible seeds, fuel

8. Cedar—cedar chests, closet linings, pencils, woodenware

Modified Stems

Modifications in form and structure from the usual cylindrical stem body are found in a good many species. Often these modifications are so different from the usual stem type that close anatomical examination is necessary to ascertain that the structures are stems and not roots or leaves, which they may superficially resemble. Externally these organs usually produce leaves and buds and internally they have the tissues and vascular system typical of stems. For the most part these modified stems are storage organs.

Rhizomes

This type (Fig. 29) is a more or less elongated stem that arises from a lateral bud near the base of the main stem axis and extends horizontally. Externally it has nodes and internodes with small, usually scalelike leaves arising from the nodes and with axillary buds arising within the leaf axils. Aerial shoots frequently develop from these buds. Adventitious roots are usually produced along the length of the rhizome. *Canna* and *Iris* have modified stems of this type.

FIGURE 29 Rhizome of an *Iris*.

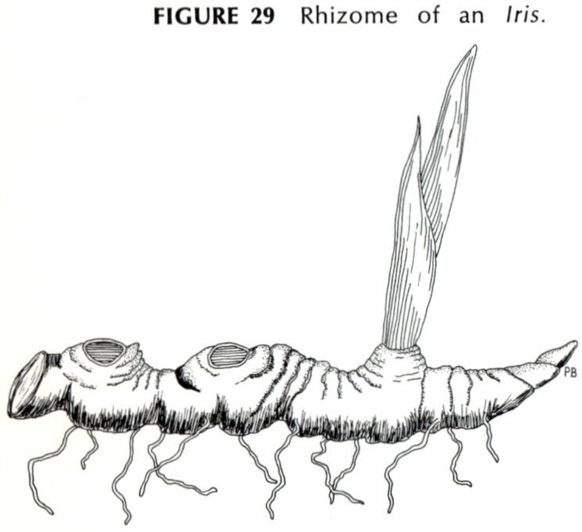

Stolons or Runners

The stolons or runners are very similar to the rhizome but are very slender. In some plants like strawberries they develop above ground while in Kentucky bluegrass they are subterranean. Bermuda grass produces stolons both above and below the soil surface.

Tubers

Perhaps the best known tuber is the Irish potato (Fig. 30). It is really a very short rhizome, greatly increased in diameter. As the subterranean portions of the plant develop, slender rhizomes branching out from the main plant axis enlarge terminally forming the tubers. Externally there is a terminal bud with many lateral buds arranged in spiral manner, typical of the foliage leaf and bud arrangement. Each "eye" is produced at a node and consists of a reduced scale leaf or leaf scar, in the axil of which are one or more rudimentary buds surrounded by tiny scale leaves. The point of attachment to the parent plant is at the end opposite to the terminal bud.

Internal organization of the potato is marked by a very active growth of storage parenchyma, most of which is developed in a central region termed the **medulla.** A second area of cellular activity is the **procambium area,** which lies just outside of the internal phloem and the **cortex.** As the medulla increases in size the vascular elements are pushed outward. To compensate for this active medullary growth, cells of the cortical, perimedullary, and vascular regions divide in three planes. Following cell division in the pith and cortex the external phloem and perimedullary zones become the regions of greatest growth activity.

The vascular cylinder which lies between the cortex and the perimedullary zone is narrow, containing some secondary xylem and phloem in addition to the primary elements. The epidermis of the potato tuber is a short-lived tissue, chiefly because of the rapid growth of the tuber. A phellogen layer is soon formed in or immediately below the epidermis which then builds up a periderm, chiefly of phellem cells around the soft tuber cortical cells. The phellogen remains active throughout the life of the tuber. Potatoes owe their excellent keeping qualities to the phellogen, which forms phellem layers over the outside of a potato if it should be cut or injured in any way.

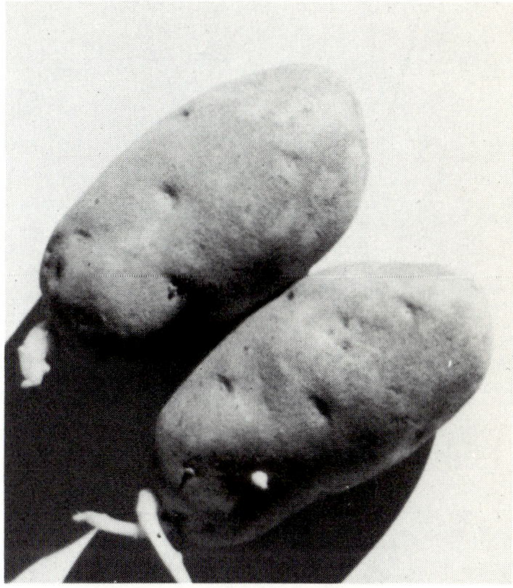

FIGURE 30 Potato tuber; note lateral buds. Apical dominance is demonstrated by these two tubers. Sprouting starts first in the apical bud which may completely inhibit the development of the laterals or merely delay their development.

Bulbs

A bulb (Fig. 31) is essentially a very short stem consisting of a growing point surrounded by thick fleshy bud scales, which are in turn surrounded by thin suberized bud scales. The arrangement of the bulb leaves is concentric so that the older ones surround the younger with the apical meristem lying in the center of the entire structure.

FIGURE 31 Bulb of an onion in longitudinal section showing the large fleshy leaves, and small stippled stem at the base from which both leaves and roots arise.

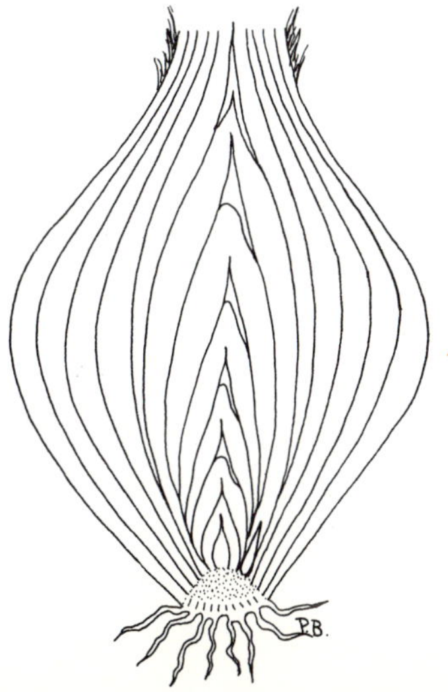

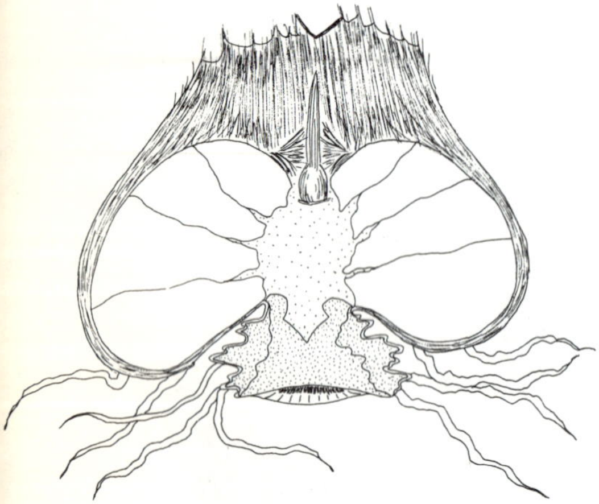

FIGURE 32 *Gladiolus* corm. Note last year's shriveled corm at the base; the new large corm with stippled vascular area in middle from which vascular traces radiate outwards to the nodes and leaves.

In spring one or more axillary buds begin to develop into new bulbs from the old structure. The youngest leaves in the center of the old bulb elongate and become green, providing food for the new bulbs now forming. Typical bulbs are those of onion, lily, tulip, and garlic. The essential anatomical differences between onion and garlic bulbs is that in the former the entire bulb is virtually a mass of fleshy leaves, with very small axillary buds. In garlic the axillary buds are very prominent and become the garlic cloves of commerce. The fleshy leaves of garlic are reduced in size and number, forming scale leaves around the developing cloves.

Corms

The *Gladiolus* (Fig. 32) and *Crocus* have modified stems which fall in the corm category. This type has a short, fleshy, vertical axis with a highly vascular central zone surrounded by a large compact cortical region in which the vascular bundles are scattered. Large numbers of adventitious roots arise from the base of the corm and at its apex tufts of leaves are developed. Externally the corm is surrounded by scale leaves which arise from concentric nodal rings. Buds develop in the axils of these scale leaves and the following year produce new shoots; ultimately new corms are produced in a cluster around the old corm, which then dries up and withers away. In contrast with bulbs corms have much more stem tissue, fewer scale leaves, and longer internodes. A corm differs from a tuber in that it is the enlarged base of the stem rather than the swollen tip of a rhizome. **Taro** (Fig. 33) is an example of a tropical corm used extensively for food.

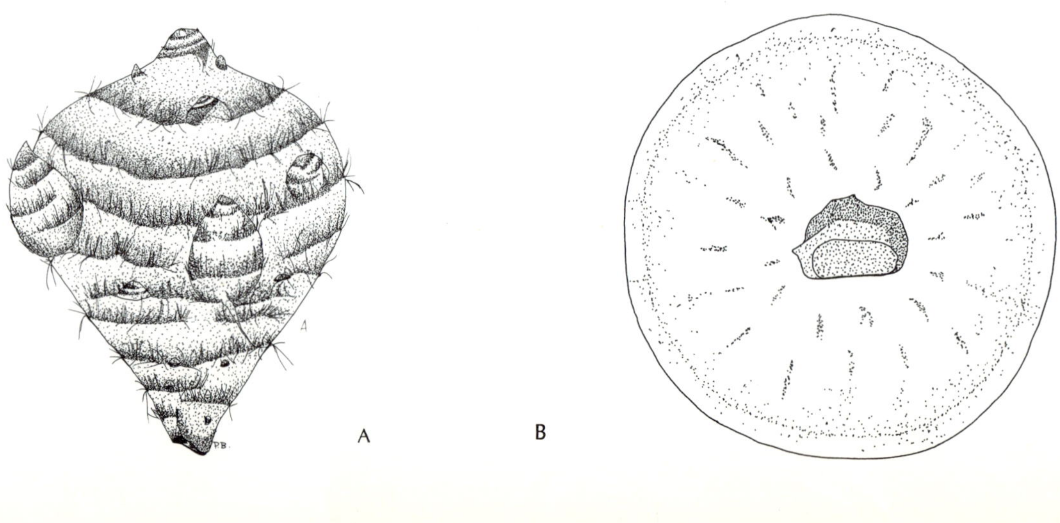

FIGURE 33 *Taro.* A: The corm of an important tropical food plant. Note the nodes, internodes and lateral buds which give this underground organ the characteristics of a stem instead of a root. B: Cross section of the same plant with the internal anatomy of a stem.

Cladophylls

Superficially cladophylls resemble large fleshy leaves. They are primarily photosynthetic in function and generally produce small, scalelike leaves that absciss soon after they appear. Clusters of persistent spines are also produced at the nodes. These nodal spines were at one time considered modified foliage leaves but recently a plant anatomist, after studying the origin of the parts, stated that the spines were really homologous to bud scales. The "hands" or "pads" of cacti are cladophylls (Fig. 34). In addition to the large spines there are numerous tiny spines scattered over the surface of

FIGURE 34 Cladophyll of a cactus, *Opuntia lindheimeri.* The large spines arise at nodes, like buds, and have an origin similar to bud scales.

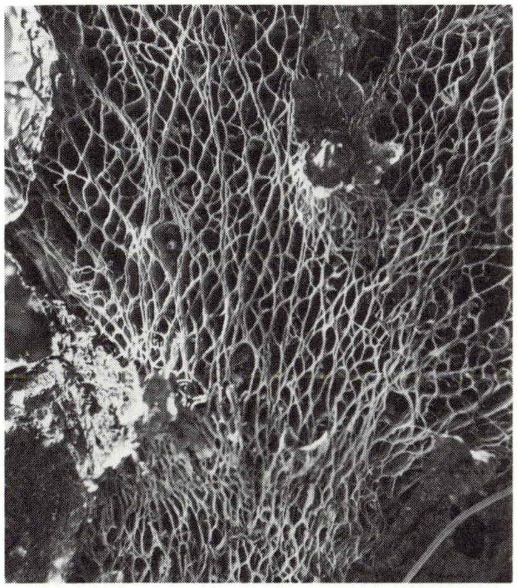

FIGURE 35 The vascular system of one "pad" from the cladophyll of *Opuntia lindheimeri.* (Photo by Ben Pilcher)

most cactus pads. These are of epidermal origin and are, therefore, considered trichomes.

Least conspicuous of the cladophylls are those of *Asparagus* and *Ruscus* (Fig. 36). Those of the former are narrowly cylindrical

FIGURE 36 *Ruscus,* an unusual monocotyledonous plant in which the stem branches are broad and flat like leaves. In the center of the broadened cladophyll is a small leaf and within its axis is the flower.

and look more like ordinary short stems; those of *Ruscus* are leaf-like and arise as laminar stems in the axis of tiny scale leaves.

FIGURE 37 The crown-of-thorns plant. When the picture was taken no leaves were present.

Thorns

Thorns usually arise in the axis of leaves as do ordinary branches, and they may even bear leaves. This origin, plus their unusual modification, places them along with the other stems discussed in the category of modified stems. The honey locust tree produces large branched thorns, often over eighteen inches long, and many of these persist for years along the major branches and trunk of the tree. *Pyracantha* thorns develop at the nodes and are much shorter than those of honey locust but still may be two or three inches long. The crown-of-thorns plant (Fig. 37), a native shrub of our Southwestern deserts, has large branched thorns which make up the greatest amount of growth of the entire plant. The thorns of this plant produce small but numerous leaves during the rainy season.

Movement of Water in Stems

One of the functions of stems mentioned earlier was to form a connecting organ between leaves and roots. In a plant like a dandelion, where the stem is no more than a small cushion of cells between the leaves and tap root, the stem does little more than produce leaves. If, on the other hand, we consider the importance of the stem to the leaves of a redwood tree three hundred feet tall, the stem becomes an amazingly efficient organ in the lifting of water to such heights. How this is done without any type of pumping mechanism is still conjectural. No one theory that has been suggested has as yet proved adequate in explaining the ascent of sap. Possibly this is because we lack the equipment or know-how, and possibly the correct theory has not yet been suggested. The best that can be offered in explanation is the type of tissues that make up a stem and the various forces we understand which can move water at least a small distance.

Over the years plant physiologists have suggested that water ascends through the xylem elements as a result of imbibition, osmosis, capillarity, transpirational pull, root pressure, and cohesion-tension forces. The first three are all operative to a certain degree but are limited to just a few inches under nearly equal pressure systems. **Imbibition** is defined as the absorption of water and swelling of colloidal particles as the result of the adsorption of water molecules onto the internal surface of certain substances. This is the usual way in which water penetrates cell walls and all dead tissues. Once the water has saturated the cell walls it then moves into the protoplasm through the cell membranes by osmosis. Trees which have been felled in wet forests will, after a time, become pretty well saturated with water simply from the imbibition of water where there is very little gravitational pull to prevent the movement of the water. Imbibition is not adequate to explain the movement of water any appreciable distance where there might be an antagonistic force such as gravity.

Osmosis

While osmosis is not a satisfactory explanation for the movement of water to great heights, it is a process which accounts for the movement of water and solutes into and out of the cells. It is the means by which plant cells maintain their water content despite the loss of water from transpiration. When water is abundant the cells are in turgor, but as the turgor pressure falls water moves in by osmosis from adjacent cells. Since the movement of water to the upper levels involves movement through tubes of hollow dead cells, osmosis is not a factor in the upward movement, but it is the force which helps to get the water into the xylem conduits and ultimately helps to move the water out of the tubes to the living cells nearby.

Root Pressure

Root pressure is a force developed in the root cells when there is excess water in the soil and quantities have been moving into the root cells. It can be demonstrated by cutting off the top part of a plant and noting the exudation of water and solutes from the cut end. The pressure can be checked by attaching a manometer to the cut stem. At best only a few atmospheres of pressure have been observed in this manner. Some roots, especially gymnosperms, exhibit virtually no root pressure. When a plant is rapidly transpiring and water is needed in greatest amounts the movement of water from the roots results in loss of pressure. Root pressure might be of some use in forcing air bubbles from stems or reducing water pressure in cells when the soil is saturated, but it is definitely a minor force in raising water to the tops of trees.

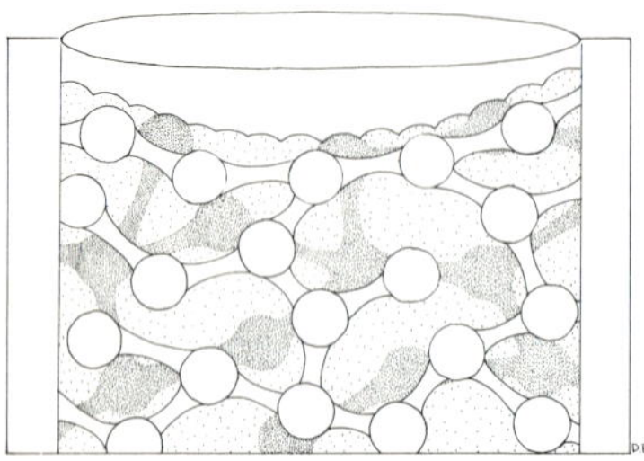

FIGURE 38 The capillary attraction of water. The circular objects are oxygen atoms connected by the elongated hydrogen bonds. The two margins represent a section through a glass tube. Attracted by the oxygen atoms in the glass, the water molecules adhere to the sides of the glass tube pulling the other molecules with them and forming a crescent-shaped surface.

Capillarity

One of the most common explanations for the movement of water not under force or pressure has been capillarity. We have all seen the movement of water in small-bore glass tubes, the smallest caliber tubes permitting water to rise the greatest distance (Fig. 38). Water will also move perhaps two feet in soils of fine texture and much less than that in coarse-textured soils. But the extent of capillary movement of water in the xylem elements is relatively small. In the first place, although xylem elements are small, they are not small enough to permit a rise of many feet by capillarity; secondly, the force (an open tube with air above the water) which permits water in a glass tube to rise by capillarity is lacking in the closed vascular elements of plants. Capillarity is therefore not the answer.

Cohesion-tension Theory

The difficulty in explaining satisfactorily how a tree over three hundred feet tall manages to get water to its uppermost leaves lies in trying to explain the rise of water mechanically in a tube, based on atmospheric pressures. We know the environment is subjected to pressure from the air which at sea level is 14.7 pounds per square inch, arbitrarily considered **one atmosphere** (atm). To raise water to a height of three hundred feet would require a pressure difference from the soil area to the treetop of ten atmospheres, since one atm of pressure can raise a column of water thirty-two feet.

Since this amount of pressure has not yet been measured directly, the theories discussed are efforts to explain this movement. The cohesion-tension theory has the greatest support at present. The basic principle is the attraction of water molecules for each other. In large-bore tubes this attraction is not very obvious, but in the long slender tubes formed by the xylem elements the cohesiveness of the water column has much greater force than the downward pull of gravity. Water in a tube the thickness of a xylem vessel or tracheid has a strong resistance against being broken and as long as the column remains unbroken the water can be moved up.

The movement upward is aided by transpiration, the use of water in photosynthesis, and maintenance of cell turgor. As the water moves out of the xylem tubes to adjacent cells water from below moves upward. Forces which act against the movement are gravity and the resistance of the xylem walls to the water, due partly to friction and partly to the molecular attraction of the walls to the water. It would seem, therefore, since tracheids have end walls while vessels do not, that the latter are more efficient water-conducting channels than a series of tracheids. So far as the actual movement is concerned we are unable to say this is true, since no method of direct measurement has been devised which does not interfere with the conducting system.

The importance of maintaining a continuous cohesive column of water in the tube can be demonstrated in a negative way with cut flowers. Florists usually recommend that flower stems should be cut off underwater a couple inches from the cut end so that air bubbles do not get into the water conducting elements. Air breaks the continuous water column and since water is then unable to move upward the flower soon wilts. If cut flowers are to be kept in water several days it is recommended that a short piece of stem should be cut off the end each day to remove any air bubbles and to insure a continuous supply of water to the upper blossoms, stem, and leaves.

Vegetative Propagation

Grafting

In the raising of fruit trees, numerous shrubs, many shade trees, and some herbaceous plants, no horticultural method is so universally practiced as grafting. It is a process in which a portion of one plant is attached to another. If the operation has been done properly and the two members are compatible a growth union will be accomplished and the graft will be successful.

Plants have less physiological antagonism to grafting than animals. Thus varieties of the same species can be grafted quite easily, different species of the same genus can usually be grafted together, and often the related genera in the same family can be grafted. Only rarely can members of different families be grafted.

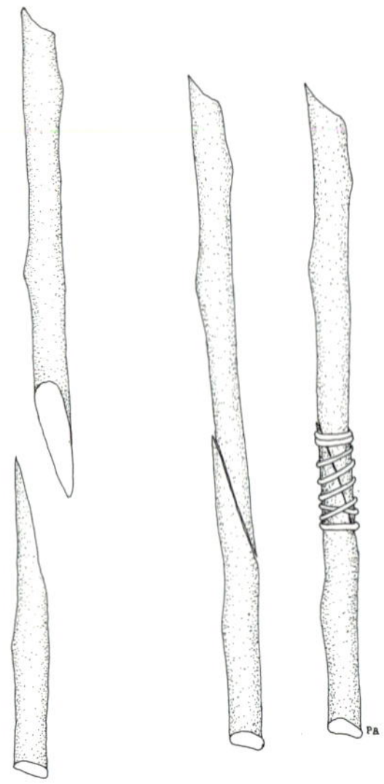

FIGURE 39 Whip graft. The two members are cut at a sharp angle, placed together and securely fastened together.

Several different grafts are practiced depending on the types of plants being used and the size of the parts to be grafted. The part of a plant which is to be transferred is the **scion,** while the rooted recipient member is the **stock.** Care must be taken that the young growing tissue, the cambium, of both members is in as near contact as possible. Although it would be virtually impossible to place the single row of cambial layers of the two members in direct contact every time, they must be close enough to grow together while there is still food left in the cut portions to continue cell division. Once growth has joined the scion with the stock and new connecting vascular elements have been formed at the union the graft may be considered successful.

Whip Grafts are made by cutting both scion and stocks at a sharp angle, placing the two cut ends together, tying with cord, wrapping with cloth impregnated with wax, and supporting if necessary (Fig. 39). Members of such a union must have cut ends of approximately the same size. Cleft grafts may be made using members of the same or different sizes (Fig. 40). **Bridge** grafts are usually made on young orchard or forest trees which have been girdled by rabbits or porcupines which have eaten away the bark and soft tissues. Twigs are taken from the upper portion of the tree and fitted into the girdled area in an attempt to make contact with the cambium layers. All of the above grafts must be bound firmly together and covered with beeswax or a grafting wax to retard the loss of moisture from the exposed surfaces.

A final type of graft, which is probably used more than the others, is the **bud** graft. An axillary or terminal bud is carefully cut from a selected variety and serves as the scion. The bud, along with a shield-shaped bit of bark tissue, is placed in a slitted opening under the bark of the stock, which is then flapped over the bark of the scion leaving only the bud exposed. A number of buds may be placed on the stock base so as to produce several desirable scion branches.

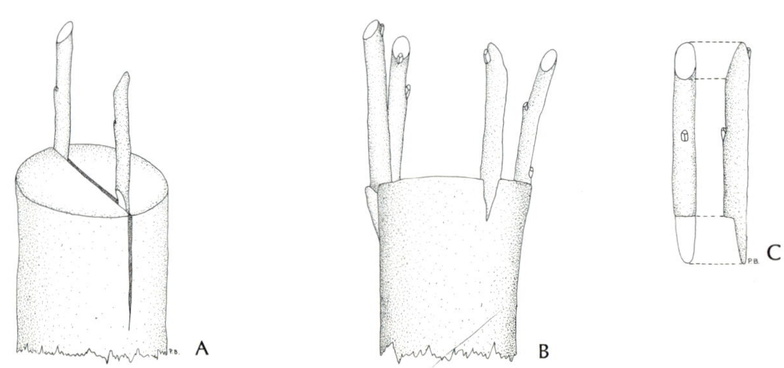

FIGURE 40 A and B are examples of cleft grafting where the scion and stock are of different size. C illustrates how twigs may be cut for placing around the girdled area of young forest trees.

Many people are of the opinion that new varieties are created in this manner. They are not. The type of fruit or flower that will be obtained will be that of the scion. The reasons for grafting plants are (1) to obtain resistance to disease and frost, characteristics usually found in the stock but not always in the scion; (2) propagation of seedless varieties; (3) propagation of plants such as apples which are normally cross-pollinated and therefore would not always produce the same variety from seed; (4) grafting often makes it possible to obtain flowers and fruits earlier; and (5) you can be sure of your results—that is, the fruit or plant top will be that of the scion.

Commercially grafts are made for the above reasons, but many grafts are made simply to illustrate possibilities. Branches of many different varieties of apple trees may be produced on a single wild crab apple rootstock simply by performing a series of grafts. Tomatoes may be grafted onto potatoes but the net result would not be large potatoes on the roots and tomatoes above ground. The plant cannot manufacture enough food for adequate storage organs on both ends. Dwarf varieties of almost any species of apple can be produced by grafting the apple branches onto slow growing quince stock.

There is a movement of solutes such as hormones and other substances between stock and scion. Often this helps to produce hardiness in the two members. The Jimson weed, which produces the drug atropine, can be used as a rootstock for tomatoes, with the result that atropine will be translocated and stored in the tomato fruits.

In different areas of the country stocks of different wild varieties native to that region are usually used, making it possible to grow varieties of a specific scion in that region. The reasons for grafting plants, then, are numerous but let it be emphasized again that we do not obtain new varieties in this way, we merely continue their successful propagation.

Cuttings

Slips are small portions of a stem, root, or leaf which may be employed in the vegetative propagation of plants (Fig. 41). African violets may be propagated by placing a leaf in water or, preferably, moist sand. Roots usually develop from the petiole, stem tissue differentiates, and a new plant is on its way to development. Most cuttings are made from stems in which a twig is cut from a tree

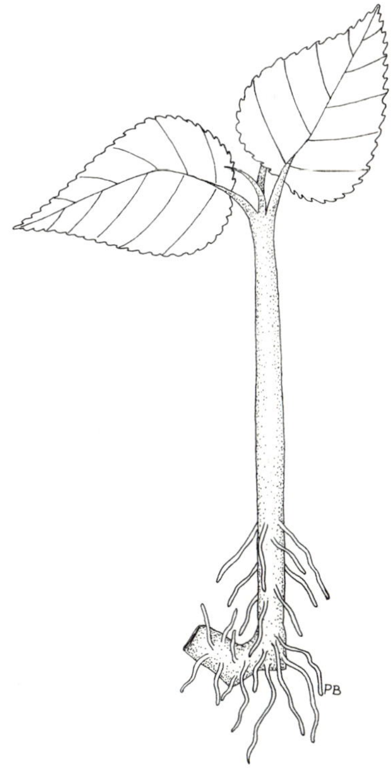

FIGURE 41 *Coleus.* Many plants are propagated by taking a shoot of the plant and allowing it to root in sand or other suitable rooting substrate.

or shrub, usually treated with a rooting compound and placed in moist sand or soil, where rooting occurs. Commercially pineapples, sugar cane, sweet potatoes, cacti, and many flowering plants such as chrysanthemums are propagated in this manner.

The rooting compounds usually employed are growth-stimulating hormones such as indoleacetic acid, indolebutyric acid, and naphthaleneacetic acid. Many hard-to-root plants, such as gymnosperms and some angiosperms, may be stimulated to root using one or more of these compounds.

Layering

A method of asexual reproduction practiced commercially by many growers and occurring naturally in many plants is layering. It consists of bending the branches of a tree or shrub to the ground, covering a portion with soil, and allowing adventitious roots to form on the covered portion. After the roots have become established the distal portion can be cut from the parent plant and the rooted branch can be transplanted. Vines such as grapes, raspberries, and blackberries are frequently propagated in this way.

Variations of the layering technique may be found in mound layering and air layering. In the former the shrub is pruned heavily, particularly with removal of lower branches. Soil is then mounded over the bases of the pruned stems. Roots develop around the bases of these shoots, which are then separated from the mother plant and transplanted. Currants, gooseberries and many other freely branching shrubs are propagated in this manner. Tall plants which cannot be easily bent may be propagated vegetatively by deeply cutting or girdling a branch. This area is then wrapped with wet moss, often treated with a hormone rooting compound, covered with a plastic bag to retard water loss, and left to grow. After roots have developed in the moist moss the branch can be cut off and planted. The house rubber plant *Ficus elastica* is frequently propagated in this way.

Clones

All of the descendants of a single plant which may arise as a cluster around the parent or be propagated from it vegetatively are considered a clone. In nature this may include a good many trees such as aspens, shrubs such as some species of *Artemisia*, and perennials such as the purple nightshade and wild morning glories. Man has continued the propagation of many desirable species or varieties which have arisen by mutation or by slips, cuttings, or grafting from the original plant or its descendants. This includes several varieties of apples, roses, peaches, strawberries, chrysanthemums, and carnations. Many of these plants would produce flowers and seeds but would not breed "true." Dependence of propagation on seed for these varieties would probably mean their loss to horticultural science.

Supplementary Reading

Biddulph, S., and O. Biddulph, The Circulatory System of Plants, *Scientific American,* February, 1959.

Clowes, F. A. L., *Apical Meristems.* Blackwell, Oxford, England, 1961.

Esau, K., *Anatomy of Seed Plants.* John Wiley & Sons, New York, 1960.

Zimmerman, M. H., How Sap Moves in Trees, *Scientific American,* March, 1963.

Questions for Discussion

1. Why is it usually more difficult to count the annual rings on a tree growing in the tropics than on one in a temperate climate?
2. Why are the following considered modified stems: Irish potato, bulb, corm, rhizome, and cladophyll?
3. What causes the "tent pattern" in plane-sawed lumber? Why doesn't quarter-sawed wood show the same pattern?
4. If a sign were nailed to the side of a tree, five feet above the ground, in 1965, how high would the sign be in 1978, if the tree grew four inches taller each year?
5. To what is grain in wood due? What causes knots?
6. How does one distinguish soft wood from hardwood?

7. Why are some softwoods more resistant to decay than most hardwoods?
8. Classify the various tissues found in stems according to their function and also according to the cell type.
9. Observe the bark on six different trees in your area and describe the variations you note in the bark of each.
10. How are solutes conducted laterally in large woody stems?
11. Necessary nutrients may be placed in trees by drilling small holes about three inches deep in the trunk about every four inches around the base of the tree and placing the nutrients in the holes. How does this achieve the desired result? Why wouldn't one hole be adequate? Why not drill all the way to the center of the trunk?
12. Why is cambium considered a lateral meristem?
13. There are several theories on how water and solutes reach the tops of large trees. Which seem like the most logical answers to you? How could you demonstrate them?
14. How does the bud of a tree differ from that of a herbaceous plant?
15. What characteristics do we find in the cells of bark which give them such excellent insulating qualities?
16. How can you show that there is less mechanical tissue in the stems of submerged aquatic plants than in those of terrestrial plants?
17. How could you show that water ascends in xylem elements?
18. Why are gymnosperms much better adapted to survive unharmed in heavy wet snows or ice than angiosperms?
19. In some trees the heartwood is much darker in color than the sapwood. How would you explain this difference?
20. Name some plants in which the stems are leaflike and the leaves stemlike.
21. What explanation can you offer which suggests why the vascular cambium gives rise to xylem on its inside and phloem on the outside?
22. How may one determine the age of a woody twig? A palm tree?
23. List the various primary tissues found in a woody stem; also list the secondary tissues.
24. Describe the various types of markings found on xylem elements and the various types of pits.
25. Discuss the characteristics of wood which make it so variable in color, density, and economic importance.

26. Why is plywood made with the grain of adjacent layers running in different directions?
27. How can you prove that a potato tuber is a modified stem and not a root?
28. How can you determine whether thorns on a plant are modified leaves or modified stems?
29. Why are the fibers obtained from monocotyledonous plants usually much coarser than those obtained from dicotyledonous plants?
30. Why is the grafting of two different varieties of woody plants so universally practiced?
31. In stems the heartwood cells cease to function in conduction but still remain in position, while in animals functionless cells are usually absorbed. How would you account for this difference?

Study for Wed & Thursday

9

The Root

FIGURE 1 Storage root of the gourd plant, *Cucurbita foetidisima.* The central tap root was 4½ feet in circumference and weighed 159 pounds. The lateral root in the foreground was cut off 10 feet from the main axis and weighed an additional 20 pounds.

The primary functions of roots are **anchorage** and **absorption.** In addition many plants are **vegetatively propagated** by roots. The **conductive capacity** of the plant is determined to a large extent by the number of roots arising from the base of the stem, in monocotyledonous plants, and by the diameter of the root provided by secondary thickening and numbers of roots, in plants of other groups. Many plants also develop roots which serve as large **storage organs** (Fig. 1) for the deposition of starch, water, and other substances.

Root Types

Root systems are generally classified as **tap** or **fibrous** depending upon the size and number of the individual roots as well as their origin. Tap roots begin to form in the embryo of the seed and therefore are **seminal roots** (Fig. 2). However, many plants do not form the large tap roots typical of the seminal roots pictured.

Soon after the young plant is formed these plants produce additional roots from the internodes near the base of the stem. These are called **adventitous** roots, since they arise from some place other than the embryonic root of the seed or another root. A large corn plant develops dozens, perhaps hundreds, of these adventitious roots. Since these roots are all slender and of about the same diameter we speak of the root system as **fibrous** (Fig. 3). Most grass plants have root systems of this type. They start out as seminal roots from the seed, gradually many more are added as adventitious roots, and soon the root system is considered fibrous. In a few plants such as *Dahlias* and sweet potatoes roots arise from the stem base and are therefore adventitious in origin, but they become large swollen organs much like tap roots. Large roots such as these formed in clusters around the base of the stem are classified as **fascicled** (Fig. 4).

The word **primary** is used occasionally to indicate the first and only main root of the plant. In the case of single tap root systems the word "primary" is well used, but in a plant like corn two or three roots may arise from the seed before the main body of adventitious roots develops. For this type of plant all roots arising directly from the stem may be considered **main** roots. Secondary roots are those which arise from the main, tap, or primary roots, tertiary roots arise from secondaries, quaternaries from tertiaries, etc.

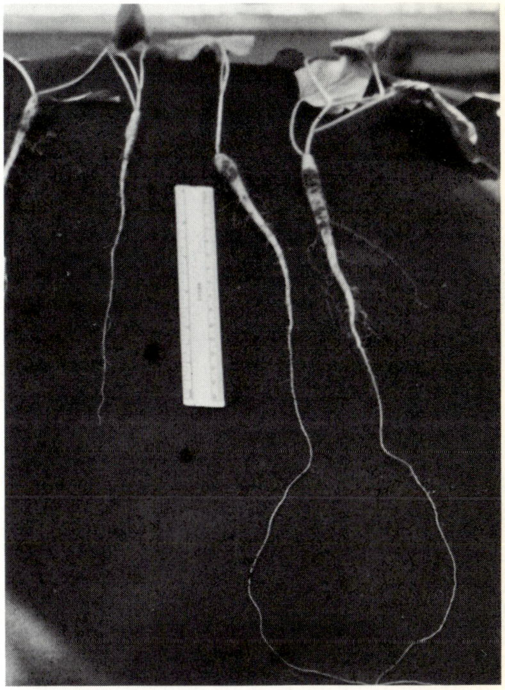

FIGURE 2 The seminal roots of young gourd plants.

FIGURE 3 The fibrous root system of maize. Most of the roots are adventitious in origin.

The roots of most plants are evenly and uniformly spaced within their immediate environment. Each is close to the available moisture and nutrients within the slender cylinder of soil surrounding the root. The most efficient type of root system in this respect is fibrous. Some years ago the author made a study of a winter rye plant *(Secale cereale)* which was raised in a greenhouse in a box containing two cubic feet of soil. Counts were made of all root categories including main, secondary, tertiary, and quaternary divisions. On this one plant, based on careful counts from extensive samplings, it was found that there were approximately 13,800,000 roots and if laid end to end they would have a total length of approximately 385 miles. This tremendous growth took place in four months from the time the seed was planted.

Later, samples were taken of a number of field plants, including winter rye, to see how extensively the roots develop under normal field conditions. Samples three inches in diameter and six inches deep (forty-two cubic inches) were collected. The numbers of roots, lengths, and diameters of the root system in this size sample are listed in the table below.

It should be quite clear by now that a grass, such as winter rye, has a very extensive root system capable of penetrating the interstices of the soil with a very expansive surface area for the absorption of water and soil solutes. If we considered this same sample of soil again and assumed a uniform distribution of roots and root hairs, there would be approximately 150 root portions and about 300,000 root hairs in each cubic inch of soil and their combined surface area, exposed in this **cubic inch** of soil, would be about 30 square inches of root and root hair surface for the absorption of soil substances.

When we mention as many roots and root hairs in a single cubic inch as given here the question arises what room is left for the soil, water, oxygen, minerals, and microorganisms in this same soil volume. The figures in the last column of each table show how much volume was actually occupied by the roots and root hairs. If you calculate for the percentage of roots and root hairs you will see that taking all roots and root hairs together the volume actually occupied by these plant parts was less than one per cent of the total volume. In other words the roots and root hairs grow through the soil spaces, occupying little of the volume but maintaining contact with the soil surfaces over a large area and transfer-

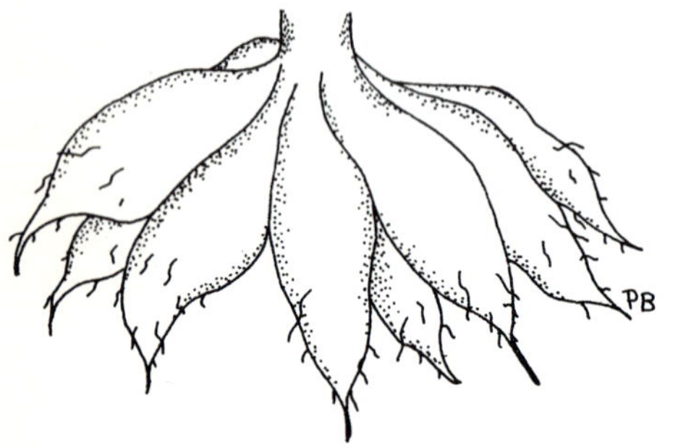

FIGURE 4 The fascicled root system of *Dahlia*.

Table 9–1 *Secale cereale;* winter rye, quantitative analysis of the roots (A) and root hairs (B) in a soil sample 3 inches in diameter and 6 inches deep (a total volume of 42 cubic inches or 688,256 cu mm)

Root categories	Total number of roots	Ave. root length (mm)	Total length of roots (feet)	Ave. diam. of roots (microns)	Total root surface (sq. inches)	Volume in cu mm	Percent of Total Volume occupied by roots
Main*	130	99.0	40	570	35.0	3,300	.49
Secondary	3,670	10.6	130	195	37.0	1,180	.16
Tertiary	2,600	4.3	40	105	5.9	100	.01
	6,400		210		77.9	4,580	.66

*Roots which arise directly from the base of the plant.

Root categories	No. of hairs per mm of root length	Total no. of hairs in millions	Ave. length of root hairs (microns)	Total length of root hairs (feet)	Ave. diam. of hairs (microns)	Total root hair surface (sq. in.)	Volume of hairs in cu mm	Percent of volume occupied by hairs
Main	507	6.4	1,720	37,000	15	889	1,950	.27
Secondary	135	5.3	940	16,500	12	272	550	.08
Tertiary	73	0.8	590	1,600	12	29	50	
Totals		12.5		55,100(10.4 mi)		1,190	2,550	.35

ring soil substances to the plant tissues. Root systems are patterned like most absorptive organs, such as our intestines with their villi and circular folds, for maximum increase in absorptive surfaces with a minimum of volume.

Apical Growth

During primary growth roots develop from an apical meristem which is surrounded by and protected by a cellular and often mucilaginous **root cap** (Fig. 5). Regions of the root apex include, as they do in a stem, the **meristematic tip,** which is quite short but is characterized by isodiametric cells active in cellular division; immediately back of this region and often extending several millimeters is the **region of elongation,** where the cells are considerably elongated. Within the next few millimeters the cells become differentiated into root tissues and form the **region of maturation** (Fig. 6).

FIGURE 5 Root tip showing root cap and developing root hairs.

FIGURE 6 Root tip showing regions of: a-root cap; m-meristem; e-elongation; d-differentiation.

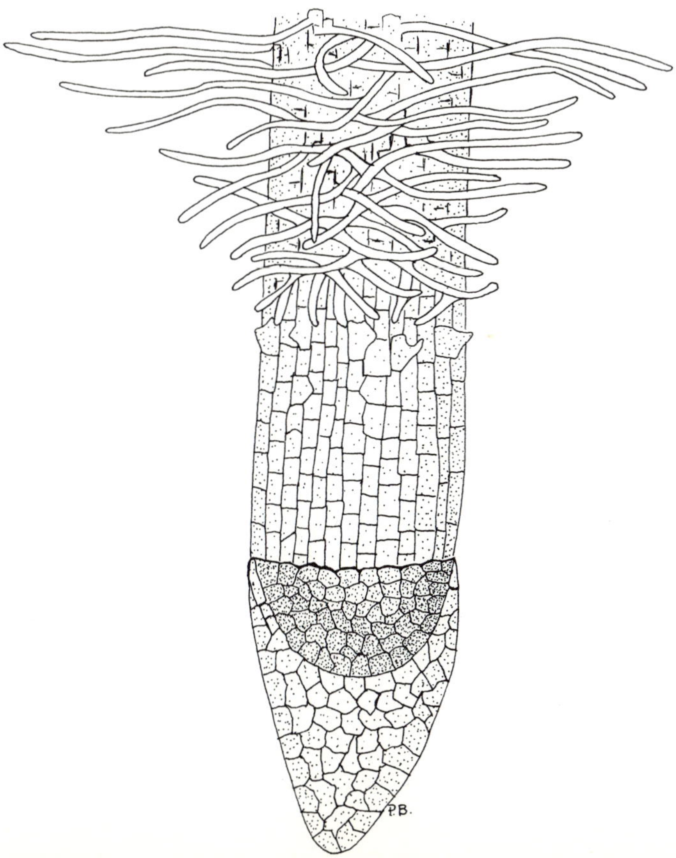

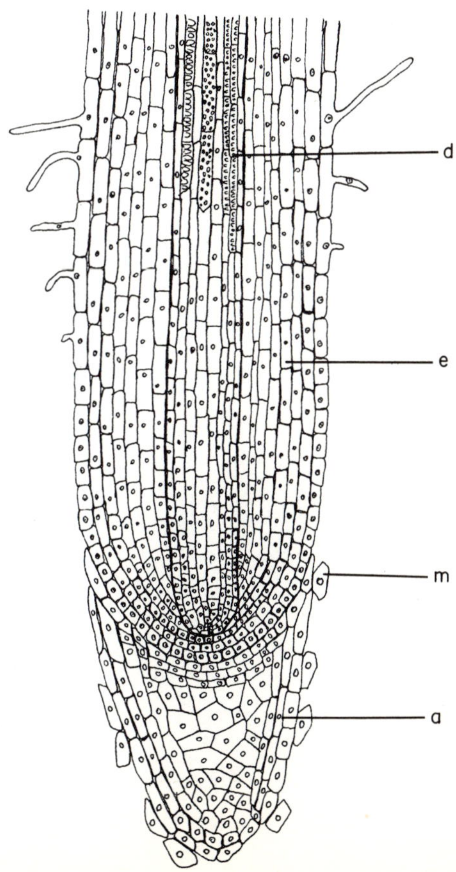

Differentiation of Root Tissues

Three primary meristematic tissues are evident when one examines a cross section of a root tip (Fig. 7). The central body of cells comprise the **procambium** and will differentiate into the cells of the **stele,** which includes cells of the xylem, phloem, and pericycle. Cells forming the outer layer of the primary meristem form the **protoderm,** which differentiates into the epidermis. Many cells of the epidermis, above the region of elongation, will give rise to root hairs, producing the largest area of water and nutrient absorbing surface. Between the procambium and protodermis is the **ground meristem,** composed of isodiametric cells which will become the **cortex.** In most roots, cells in the cortical region occupy the largest volume. The inner row of cells of this cortical region usually differentiates into a thick-walled **endodermis.**

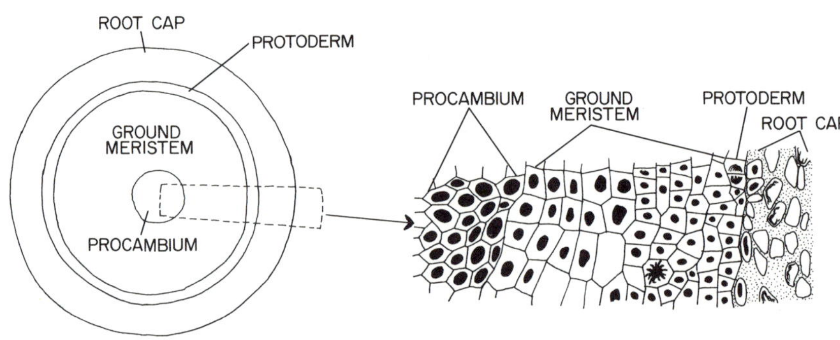

FIGURE 7 A cross section of a young root tip and the appearance of the cells formed in the primary tissues. (redrawn from Robbins, W.W. and T.E. Weier. *An Introduction to Plant Science,* 1950. John Wiley and Sons.

FIGURE 8 Cross section of a young angiosperm root showing isodiametric cells in cortex and prominent diarch xylem. The endodermis is not yet suberized in this young root.

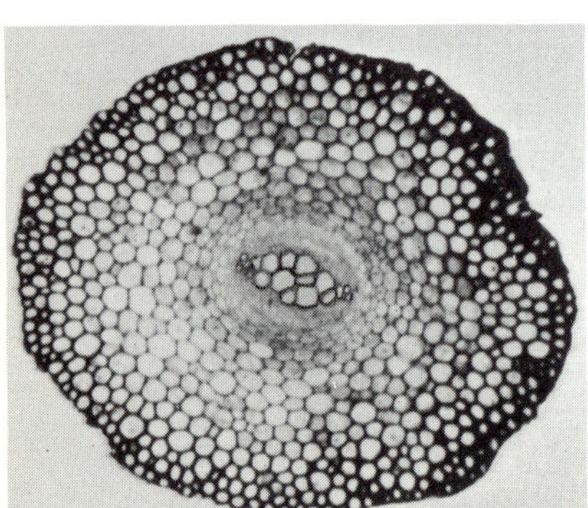

If we were to examine closely a young root in cross, as well as longitudinal, section we would see that cells of the epidermis differentiate first, followed by the cortex, where the cells become large but remain more or less isodiametric as they mature (Fig. 8). Cells of the stele are more compactly arranged than those of the cortex, and are somewhat more variable in their manner of maturation. The first cells to differentiate from the procambium may be those of the phloem, but more commonly a few **protoxylem** elements appear first. Usually the xylem matures in a centripetal manner. That is, the protoxylem cells will be towards the periphery of the xylem elements and the **metaxylem,** which matures later, is in the center. When xylem is matured with the protoxylem external to the position of the metaxylem, the maturation is termed **exarch** (Fig. 9). In roots where the protoxylem is formed towards the inside of the xylem mass and the metaxylem is matured towards the

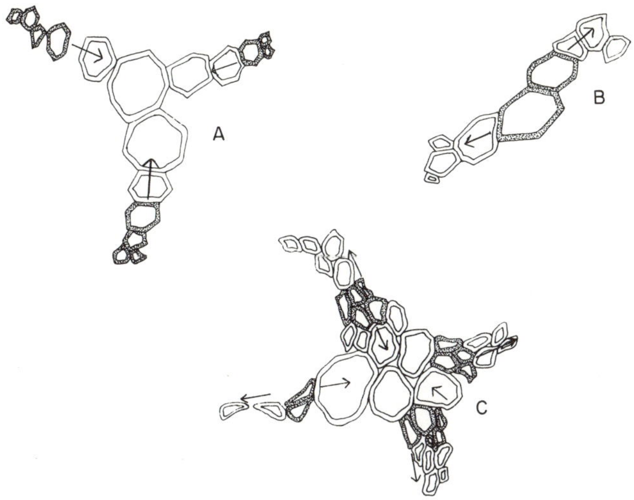

FIGURE 9 Maturation of primary xylem. A- exarch in this triarch stele, B- endarch in a diarch stele and C- mesarch in this tetrarch arrangement. The number of points in the xylem does not necessarily determine the type of maturation.

outside it is **endarch.** In many plants this vascular pattern may persist for the life of the root without additional growth; in others, with cambium, secondary growth soon adds so many secondary elements that the primary xylem elements make up a very small percentage of the xylem tissue and may even be obliterated by the growth of the new tissue (Fig. 10).

Phloem differentiates in much the same manner as the xylem in the exarch root. The outermost primary phloem cells are those of the **protophloem,** while those which differentiate later and within are those of the **metaphloem.**

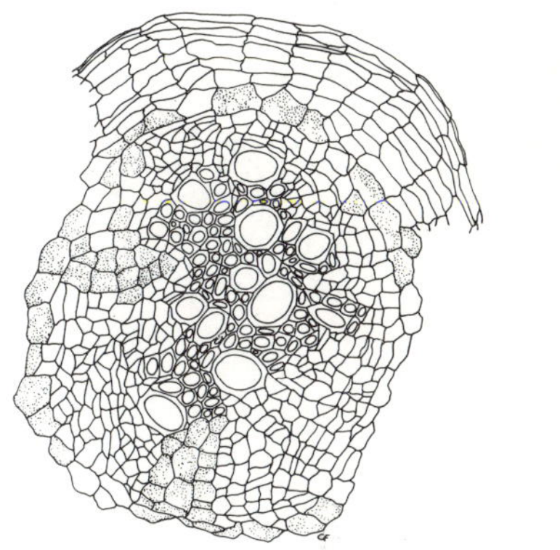

FIGURE 10 Cross sections of the root of the Bird of Paradise plant, *Caesalpinia gilliseii. Above:* a young root with chiefly primary xylem. *Below:* An older root with prominent secondary xylem, and active cambium and developing periderm.

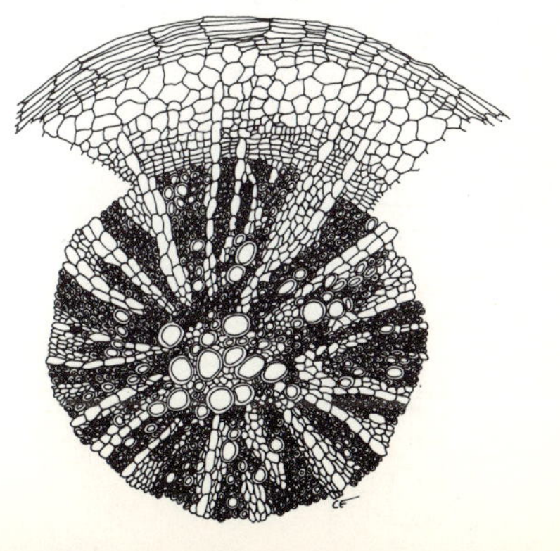

Arrangement of Tissues

The arrangement of tissues in most roots is that of a protostele which has a central xylem and all other tissues external to it. Usually the primary xylem differentiates with arms or points radiating from a common center like the points of a star. Accordingly the xylem may be termed *diarch* when only two points are formed, or *triarch, tetrarch,* or *polyarch* if three, four, or many are developed (Fig. 11).

The primary phloem is usually formed between the points of the xylem (Fig. 12) and may consist of a few well-formed sieve tubes (Fig. 13) with companion cells, or the phloem may be quite extensive in area. The determination of a precise arrangement of alternating strands of phloem and xylem tissue is apparently an inherited factor and is quite constant within the species.

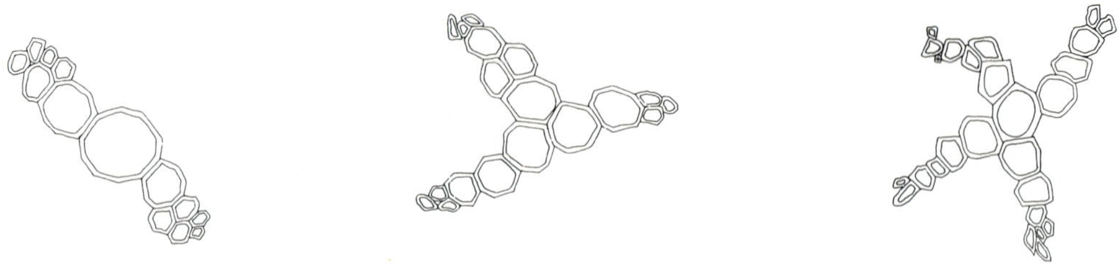

FIGURE 11 Development of xylem. *Left:* Diarch; *center:* Triarch; *right:* Tetrarch.

FIGURE 12 Cross section of a young root with a secondary root beginning to form. ep-epidermis; c-cortex; en-endodermis with some cells heavily suberized; p-pericycle; ph-phloem; x-xylem; sr-secondary root.

FIGURE 13 Formation of sieve plate in root tip of white sweet clover. The light line between the two cells is the middle lamella; the light areas between the stippled dark areas are believed to be developing pores; the elongated and round structures are portions of the endoplasmic reticulum. Drawn from an electron micrograph in Bonner, James and J. E. Varner 1965 *Plant Biochemistry*. Academic Press.

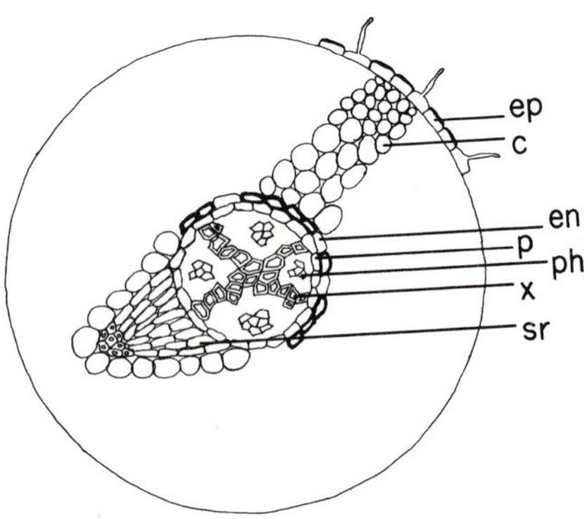

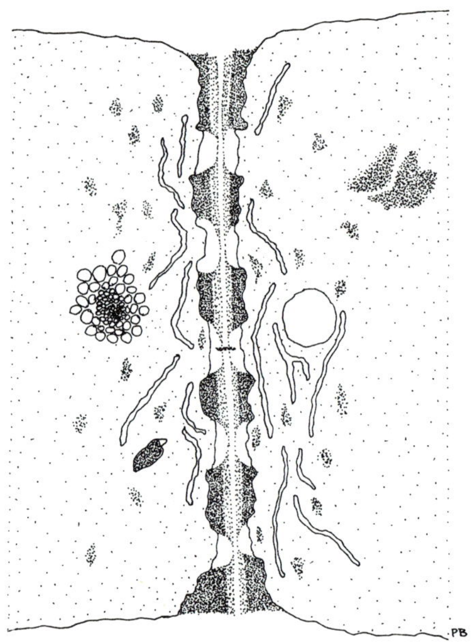

FIGURE 14 Development of a lateral root in vertical section.

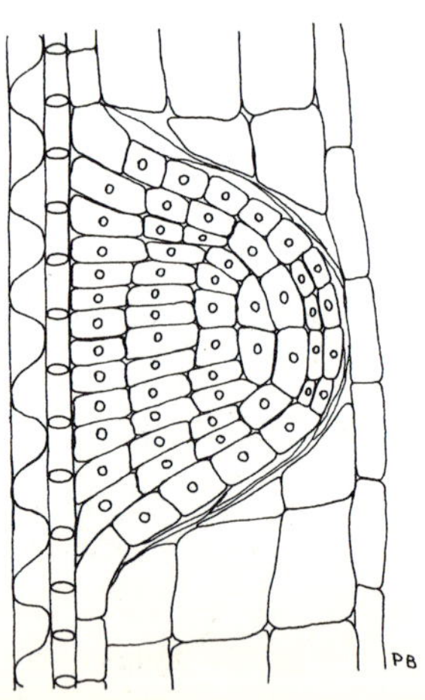

Surrounding the vascular tissue is a differentiated cylinder of cells, usually only one cell-layer thick, known as the **pericycle** which was differentiated from the procambium. The pericycle is composed of parenchyma cells which have retained their capacity for cell division and additional growth even after reaching maturity. In most species the pericycle is an extremely important tissue, for it is in this layer that branch roots arise (Fig. 14). In addition pericycle may also give rise to the cork cambium and also to some cells of the cambium.

The **endodermis** is considered to be the innermost row of cells of the cortex since it differentiated from the ground meristem as did the other cells of the cortex. The endodermal cells are usually quite different in appearance as well as function from other cortical cells. The endodermis cells are characterized by the presence of thin, suberized lamellae, called **Casparian strips** on the radial and transverse walls. They form a seal between the protoplasts and adjacent endodermal cells through which water passes, going into the stelar region, in a directed manner. In young roots these Casparian strips develop first in the endodermal cells which are outside the phloem area; as the root matures all cells of the endodermis develop Casparian strips and the suberization may then cover all walls of the endodermal cells.

Cambium, which is formed between the xylem and phloem in many dicotyledonous roots, forms as it does in the stems. The vascular cambium differentiates in part from the undifferentiated procambium occurring outside the xylem concavities (between the points) and inside the phloem areas, and in part from the pericycle in contact with the xylem points.

FIGURE 15 Cross sections of two roots of *Yucca elata*. A. first year root with central xylem elements, small phloem areas, pericycle, but still lacking suberized walls in the endodermis. B. older root with same tissues but with highly suberized endodermis. *Yucca* is a monocotyledonous plant and consequently does not develop vascular cambium. Compare the diagrams of this plant with the tissues of the dicotyledonous root in Figure 10.

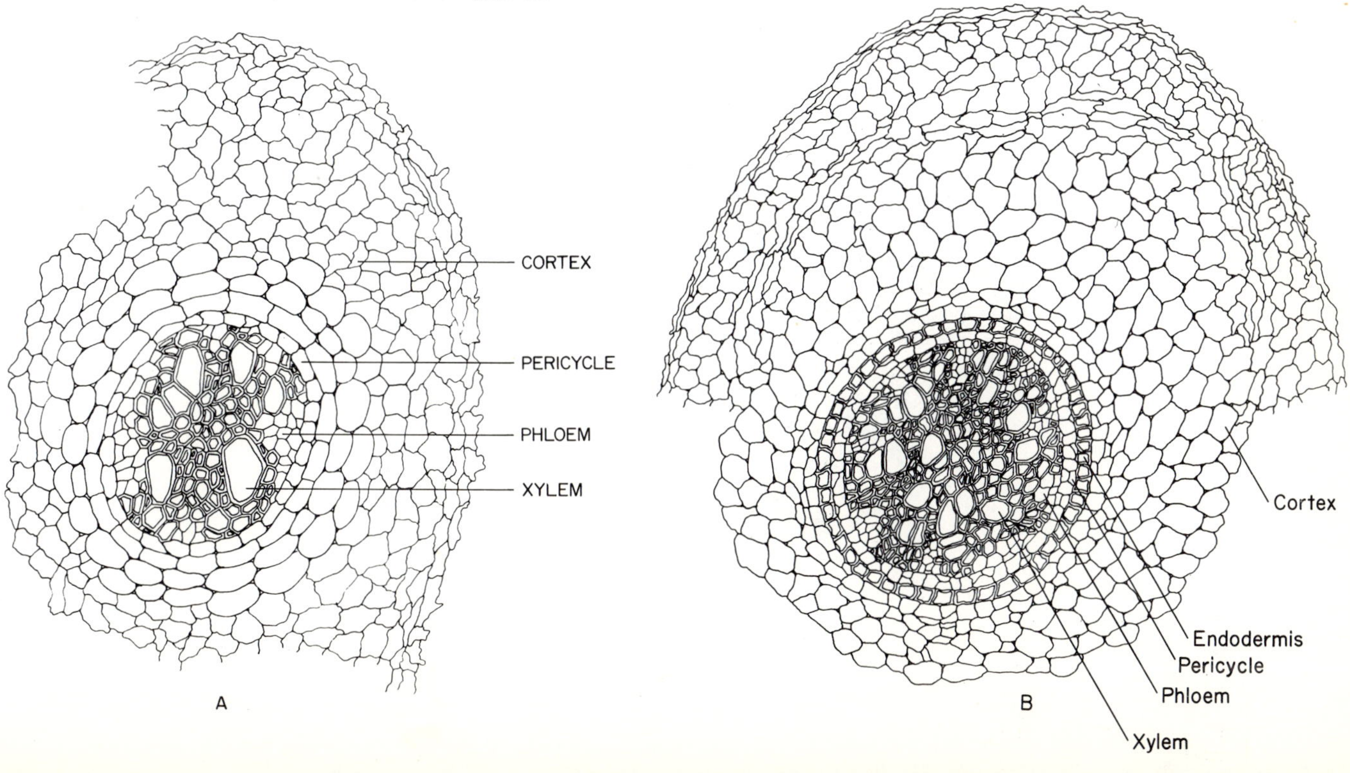

The roots of most woody plants increase in diameter as the result of cambial activity just as in the stem. Eventually the cambium gives rise to the secondary xylem between the xylem points, previously occupied by parenchyma, and the xylem loses its primary radial pattern. Phloem cells are added to the outside of the cambium and soon the phloem areas become united into a cylinder of phloem-conducting elements. As the root ages xylem is laid down in annual rings as in the stem, and the phloem and other tissues are pushed outward by the addition of xylem from the vascular cambium. Consequently an old root has the tissues of, and looks very much like, an old stem in cross section.

FIGURE 16 A portion of the cross section of a root of the Bird of Paradise, *Caesalpinia gilliesii.* This root has a well-developed periderm, a large xylem area but relatively little cortex.

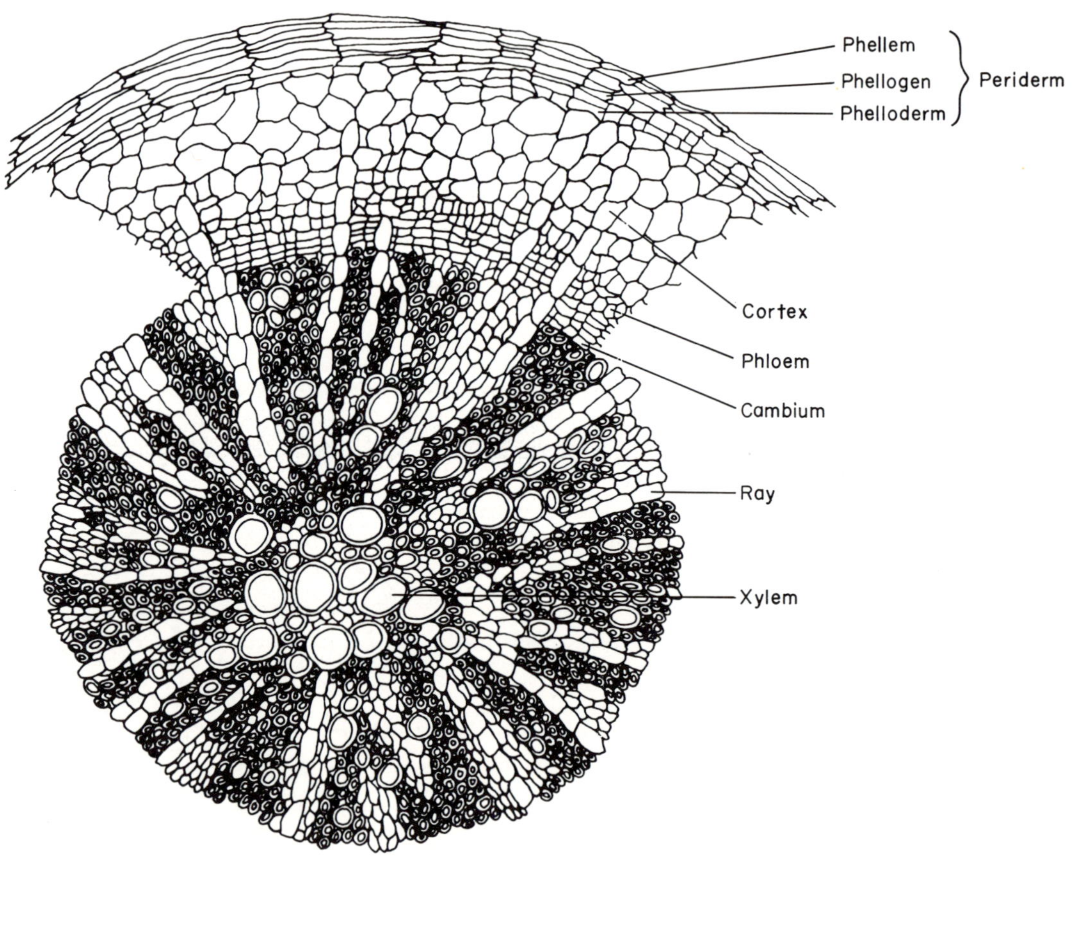

At about the same time that the vascular cambium is becoming active in most roots the cork cambium begins to differentiate from the pericycle. As in the stem a periderm is produced by the cork cambium but in the root it is generally formed entirely internal to the cells of the epidermis, cortex, and endodermis. As a result these cells are soon sloughed off as dead tissues following early growth of the cork cambium.

Lateral Roots

The origin of branch roots is quite deep-seated. One or more pericyclic cells begin to divide forming what at first appears to be multiple layers of pericycle cells (see Fig. 14). Soon a root apical meristem is organized from these cells and tissues of the root become differentiated with vascular elements of the parent root joining the new elements of the young branch root. The new rootlet grows through the peripheral tissues of the parent root, eventually making contact with the soil and penetrating the soil interstices. In some large roots where secondary thickening sloughs away the peripheral tissues, including the pericycle, branch roots may arise in the cambium.

Adventitious roots, which develop from stems, originate internally in much the same manner as branch roots. Usually the root primordium originates in the medullary ray region, between the vascular bundles in young stems, while in older stems the lateral root may develop near the cambium in a vascular ray, or from the cambium itself.

Branch roots do not appear to be as regularly arranged as leaves and stem branches described earlier. This is due to a large extent to the twisting and turning of the parent root as it grows through the soil spaces. However, in water or in uniformly loose soils it is possible to see that roots may have a two-ranked, three-ranked pattern or more. Diarch roots, for example, may have two rows of branch roots and be considered two-ranked, or they may produce four rows and be four-ranked. Generally the branch roots arise near or just outside the xylem points, but they may also form outside the phloem areas. The two-ranked condition is easy to observe in some species of cattails which grow in water or a water-saturated soil.

Root Hairs

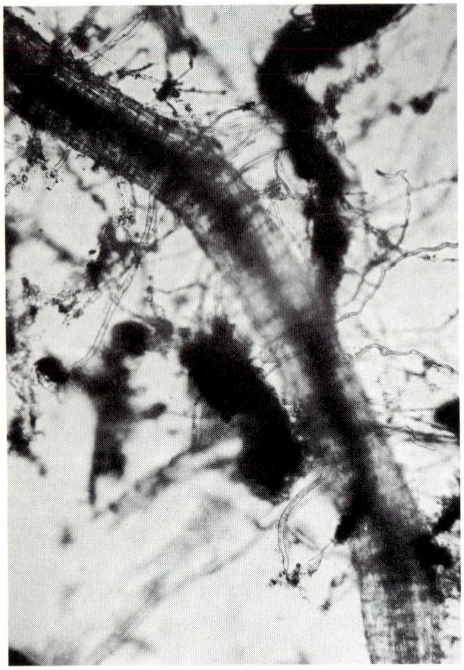

FIGURE 17 Young root of Bermuda grass with root hairs. Numerous soil particles are still attached to the hairs; the very slender branched filaments are ectotrophic fungi. The root is about 150 microns in diameter.

For most plants the root hairs are the premier absorbing organs (Fig. 17). A root hair begins growth as a slight swelling on the outer wall of an epidermal cell and rapidly expands into an elongated cylindrical hair with a rounded end. Size of the hairs is determined largely by genetic factors since it has been observed that different species will have different sized root hairs even when grown in the same soil and water situations. Soybeans, for example, have hairs about two hundred microns in length, while those of oats will be about three thousand. The diameters of the hairs also are quite constant for the species. Some will be as little as five microns in diameter while others may be as much as twenty-five microns, although ten to fifteen microns is the usual size for most plants.

Earlier in this chapter the number of roots on a single rye plant was given as 13,800,000. These roots produced approximately fourteen billion root hairs with a total area of exposed surface of about four thousand square feet. This area plus the surface exposed by the roots (2500 sq. ft.) gave this rye plant growing in two cubic feet of soil an absorbing surface of about sixty-five hundred square feet.

The importance of absorbing surface to a living system is, of course, emphasized by this example (Fig. 18). Yet if one feels this is too much let us be reminded that if all our red blood corpuscles were laid side by side they would cover nearly an acre (43,560 sq. ft.).

Although a root hair is usually described as a short-lived structure it may persist for weeks, months or even into the next year when there is no sloughing of the epidermal tissue. Consequently monocotyledonous plants have root hairs covering most of their roots, while in dicotyledonous tap roots, with rapidly developing secondary tissues, hairs may be found only near the tips.

Root hairs grow by an apical or tip extension and not by a general wall elongation. This manner of growth is rather typical of cell enlargement, not only in tissues of higher plants but also in lower species such as the hyphae of fungi, in which there is a migration of protoplasm from the older portions of the cell to the tip thereby transporting food for the formation of new cell wall as the tip grows forward. Recently it was shown that the cell wall of a root hair consists of cellulose and pectic substances, and that the cellulose layer is extremely thin at the tip.

Through the aid of the electron microscope Dawes and Bowler have shown that the cell wall of the radish root hair consists of two layers of microfibrils. The outer layer originates at the tip and has a reticulate pattern, while the inner microfibrils are parallel and longitudinal, arising farther back from the tip. Apparently due to the expansion of the hair tube, the outer reticulate pattern thins or spreads out and a pectic matrix is deposited between the fibrils of the cellulose meshwork. Minute thin areas in the cell wall, possibly developing pits, are evident. Covering the entire hair is a layer of mucilage with a very delicate cuticle beneath, while just within the cuticle lies an amorphous matrix which both covers and impregnates the meshes of the cellulose fibrils.

The role of root hairs in absorbing substances from their environment is greatly enhanced by the mucilaginous layer. It also makes the root system more efficient in binding soil and possibly attracts bacteria and fungi, which are almost always found in and around the hairs. The microflora and microfauna aid materially in breaking down organic substances in the soil, thus making certain substances soluble for root absorption.

The importance of the absorption surface to the plant can be seen in transplanting nursery stock. For many plants, especially evergreens, the plants must be dug leaving as much of the soil around the roots as possible. This is usually accomplished by "balling" with burlap after digging around the plant. When lifted out of the ground the ball should not be so much as broken, or sufficient numbers of hairs and roots will be broken off to prevent a successful transplant. In plant nurseries the growers frequently dig down around nursery stock to cut off the tips of spreading roots to encourage more root growth near the central axis and make transplanting later easier.

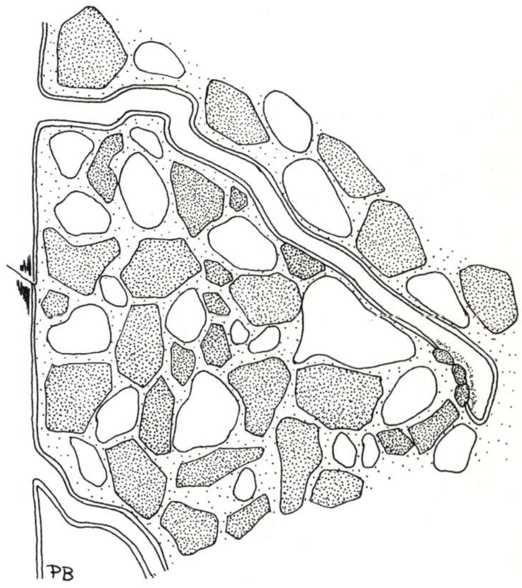

FIGURE 18 Diagrammatic illustration of a root hair growing through the soil interstices.

FIGURE 19 Endotrophic fungi in the root cells of an orchid.

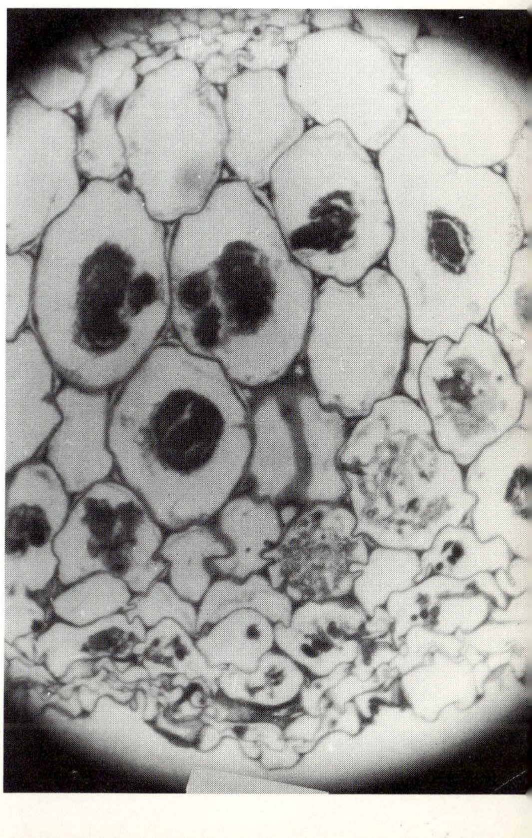

Mycorrhizae

Associated with the roots of many plants are soil fungi which form an intimate relationship contributing to the nutritional requirements of both the host plant and the fungus. The relationship does not take on the characteristic host-parasite problems but is rather a **symbiotic** relationship. The combination of root and fungus is termed *mycorrhiza,* which means fungus-root. Certain endotrophic fungi invade the root cells of many young roots (Fig. 19), causing them to develop in a hypertrophied manner, forming short branched rootlets which greatly increase the absorptive surfaces of the root system. The fungi apparently help in the digestion of some of the organic substances in the soil and in this manner make nitrogen and other elements available to the plant. It is also quite probable that the associated fungus may absorb minerals from the soil and by their intimate relationship permit these substances to diffuse into the root cells while they absorb sugar and other substances from the root cells.

Many roots do not have the abnormal growth described as mycorrhizae, but even though they remain characteristic roots they usually have fungal hyphae growing in very close association with them. Fungi which grow around the roots and root hairs but do not enter the root cells are termed **ectotrophic.** These fungi are beneficial since they help to break down the complex organic substances in the soil, and with their hyphal surfaces in contact with roots and root hairs there can be a ready exchange of nutrient materials.

The roots of many forest trees may be parasitized by certain angiosperm species. This may be a host-parasite relationship, or in some cases the parasite is also green and photosynthetic and merely absorbs some of its nutrients from the host. Figure 20 shows

FIGURE 20 *Juniperis monosperma* root parasitized by *Pedicularis centrantheca*. *Left:* The dark member is the Juniper host and the light colored the parasite. *Center:* parasite tissue invading host's xylem. *Right:* Juniper root is the smaller darkly stained tissue and the *Pedicularis* parasite tissue is to the left. (Photo by Jerry Riffle, U.S. Forest Photograph.)

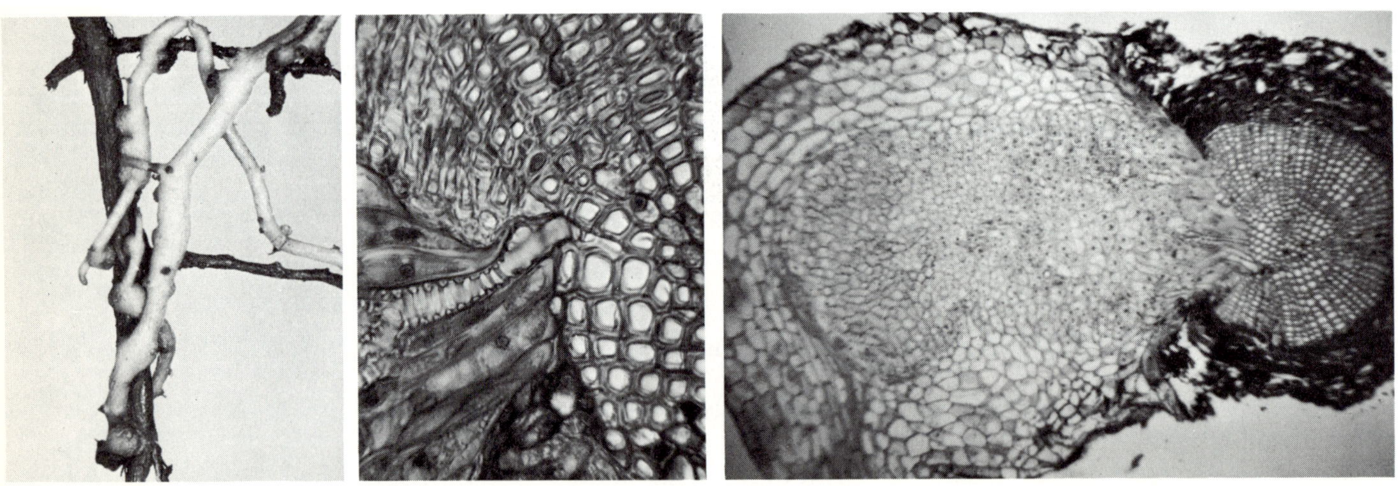

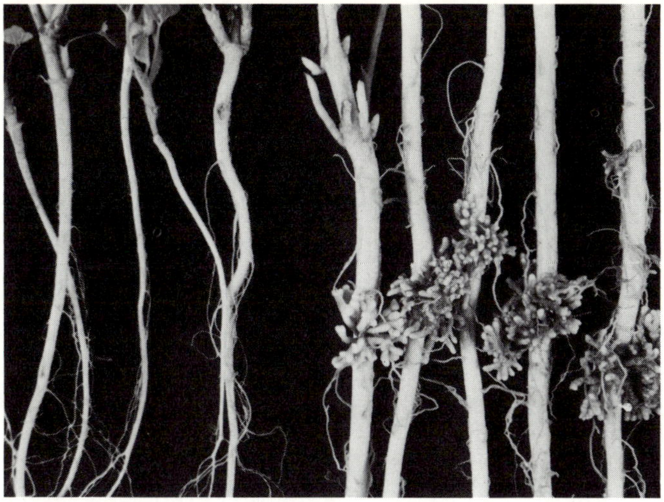

FIGURE 21 Alfalfa roots. The plants developing from uninoculated seed are at left and those with numerous bacterial nodules were grown from inoculated seed. Courtesy of The Nitragin Company, Inc., Milwaukee, Wisconsin.

FIGURE 22 Field of blue lupines. The luxuriously growing plants to the left in the picture were inoculated while those in the right half were not. Courtesy of The Nitragin Company, Inc., Milwaukee, Wisconsin.

an excellent example of parasitism by the angiosperm parasite, *Pedicularis centrantheca,* on *Juniperis monosperma.*

Similar malformations, but of great benefit to the plant, are the hypertrophied nodules developed on the roots of legumes. Seeds of these plants are often coated with nitrogen-fixing bacteria of the genus *Rhizobium,* which causes the roots to develop characteristic nodules (Fig. 21). Fields of legumes grown from inoculated seeds are usually much more productive than those grown from uninoculated seeds (Fig. 22).

Supplementary Reading

Craft, A. S., *Translocation in Plants.* Holt, Rinehart and Winston, New York, 1961.

Esau, K., *Plant Anatomy,* 2nd ed. John Wiley & Sons, New York, 1965.

Galston, A. W., *The Life of the Green Plant.* Prentice-Hall, Inc., Englewood Cliffs, N. J., 1961.

Jensen, W. A., and L. G. Kavaljian, *Plant Biology Today*. Wadsworth Publishing Co., Belmont, California, 1966.

Salisbury, F. B., and R. V. Parke, *Vascular Plants: Form and Function*. Wadsworth Publishing Co., Belmont, California, 1964.

Questions for Discussion

1. Explain why fibrous rooted plants are more efficient in binding soil than tap root systems.
2. Why do fibrous root systems occupy so little of the volume of the total area of the soil they occupy?
3. Sugar content in a root hair is usually higher than that in the surrounding soil medium. Why isn't the dissolved sugar in the root hair diffused into the soil?
4. How is the conductive capacity of roots increased?
5. Root hairs often remain on the older roots of monocotyledonous plants for months or even years, but are usually found only near the tips of dicotyledonous species. Explain.
6. Would you expect roots to have an alternate, opposite, or whorled arrangement similar to that of leaf arrangement?
7. In roots of the same diameter we find that xerophytic plants have a much higher percentage of xylem tissue than hydrophytic plants. Suggest reasons for this difference.
8. Compare and contrast the apical growing points of stems and roots.
9. What adaptations have roots made to increase their absorptive capacity?
10. Would you expect the xylem elements in stems to be shorter or longer than those in roots? Why?
11. What are the embryonic tissues found in dicotyledonous roots? In monocotyledonous roots?
12. How do mycorrhizae benefit roots?
13. How do the hypertrophied growths caused by certain bacteria on leguminous roots benefit the plants?
14. Describe the different types of root systems.
15. Considering the various plant organs such as leaves, stems, and roots, how are these organs modified for greater or less absorption?
16. Why might a root be a better organ for food storage than a stem?

10

Soil and Mineral Nutrition

The environment of roots is far more than an anchoring substrate. In many respects the nature of the plant cover and the successful growth of the plants depend upon the soil conditions as well as climatic factors. The environmental factors that operate through the soil are spoken of as **edaphic factors.**

Soil has been described as that part of the earth's crust which supports plant life. It has also been defined as disintegrated and decomposed rock and humus material. Probably no definition would suffice, for soil is a complex system. Its components are many but they may be divided into two categories: organic and inorganic. The organic matter consists of the remains of living organisms and the soil organisms themselves. The inorganic substances include the minerals, soil air, and soil water and its dissolved substances.

Organic Composition of Soil

Soil development begins with the disintegration of the parent rock material aided by the mechanical weathering processes of freezing, thawing, rain, and wind. Soon simple forms of life establish themselves among the rock particles and help to reduce the particles further by adding their own bodies to the slowly accumulating organic matter as generation after generation dies and is replaced by its own kind and higher forms. As the mantle of disintegrated rock and organic matter thickens, larger plant and animal invaders find the habitat suitable. The final result is an unconsolidated layer rich in minerals, humus, plant, and animal life. This layer varies in its mechanical mixtures and in its chemical content, but it is soil at last. Soil population counts have made for most of the living members inhabiting the upper level of fertile soil. If we consider an acre of soil down to a depth of six inches, which is plough depth, the area of greatest root growth, and the level of highest fertility, we have the following population:

Organism	Approximate amount
Molds	2,000 pounds
Bacteria	1,000 pounds
Actinomycetes	1,000 pounds
Protozoa	200 pounds
Algae	100 pounds
Yeasts	100 pounds
Nematodes	50 pounds
Other worms and insects	1,000 pounds
Weight of an acre of fertile soil 6 inches deep	1,505,000 pounds

Most of these organisms contribute to the building of soil through their ability to decompose almost all types of plant and animal residues. By their action they form humus, which changes the soil character by improving its texture and tilth, by increasing its water holding capacity, and by increasing the acidity in soils which are normally alkaline.

Molds are chiefly aerobic and for the most part they tolerate fairly acid soils, while bacteria do better in soils of less acidity. Bacteria vary from species that are strictly aerobic to some that are strictly anaerobes. They also vary in their growth requirements. Some are **autotrophs** which manufacture their own food from inorganic substances; others, the **heterotrophs,** require suitable organic substrates. The actinomycetes are aerobic and grow best in neutral to slightly alkaline soils; some are capable of living in the high-temperature substrates provided by the decomposition of compost and manure. Protozoans make up much of the microfauna in the soil. They feed to a limited extent as saprozoic organisms on decomposing material but they also eat bacteria. The algae are chiefly bluegreens, and most live near the surface where some light is available. These algae are capable of fixing nitrogen in the soil. Little is known about the soil yeasts, which are comparatively few in number, but their respiratory products, for both the aerobic and anaerobic species, undoubtedly aid in making other substances soluble and available to the higher plants.

The larger forms of life, such as earthworms, nematodes, and insects and their larvae, add their own residue products to the soil and in addition increase the porosity of the soil.

The greatest quantity of organic material in the soil comes from the plants and animals that live in and on the soil. Roots, stems, leaves, fruits, and seeds, along with animal remains, gradually decompose, darkening the color of the soil and increasing its acidity. It would seem that over the centuries this volume of organic matter would become dangerously high and be detrimental rather than desirable. However, the decomposing process caused by bacteria and fungi reduces this organic matter to the soluble end products of respiration, fermentation, and metabolism. The rapid decomposition action of bacteria is called "soil fire," for it is a heat-producing process and if allowed to go unchecked would use up all the organic matter in the soil. Fortunately a regulatory balance is maintained by the rise and fall of the microflora populations as the supply of food for them increases and diminishes.

In areas of little rainfall and shallow soils, such as the great arid regions of the world, the organic matter constitutes only from two to five percent of the soil. In these areas the plant growth is sparse and the animal population low; consequently it takes much longer to build up organic matter. In bogs, marshes, and swamps the organic matter may be over eighty percent, often resulting in highly acid situations which may also decrease the metabolic activities of the soil microorganisms and reduce their populations.

Soil Water

Whether in a liquid, solid, or gaseous state, water performs all functions necessary to sustain life as well as destroy it in some cases. Water permeates every segment of the earth's surface and has been called the universal solvent. All substances, whether organic or inorganic in origin, enter an organism's body in solution. As water falls it percolates downward through the soil environment, placing minerals from the disintegrated rocks in solution and removing accumulated waste materials.

Water moves through the soil through the forces of gravity and molecular attraction. The movement of **gravitational** water is essentially downward as excess water from rain or surface flow percolates through the soil spaces. **Capillary** water may move in any direction; it operates when there is insufficient water for gravitational movement but still enough water to move from soil particle to particle through surface molecular attraction. Plants occasionally wilt when the amount of water in soil is insufficient for roots to absorb and replace that lost in transpiration. Water may also remain in soil when there is too little for capillary movement. In this form it is bound to the soil particles and is termed **hygroscopic** water. This type of water is measured by weighing a given quantity of soil after a plant has wilted in it from inability to take the water away from the soil particles.

Soil pH

In determining the acidity or alkalinity of a solution the concentrations of free H^+ and OH^- ions are measured by means of a pH meter. If the solution has a high concentration of H^+ ions the solution is in the acid range, and if there are more OH^- ions present the solution is in the alkaline range. Pure water is chemically neutral and has been assigned a pH of 7. Readings on the pH meter above 7 indicate that the solution is alkaline, while readings below 7 indicate that the solution is acid. Several pH concentrations are listed below from the entire pH range, which is from 1 to 14.

pH 7 = 10^{-7} or 0.0000001 H^+ (ions) in one liter of solution —this amount is considered neutral.

pH 4 = 10^{-4} or 0.0001 H^+ in one liter of solution, which is quite acid.

pH 10 = 10^{-10} or 0.0000000001 H^+ in one liter of solution, which is quite alkaline.

Alkaline soils are formed as the result of an accumulation of salts such as magnesium, sodium, and calcium in the form of carbonates or oxides. Soils classified as **black alkali** contain sodium, usually in the form of a carbonate (Na_2CO_3). They generally have poor drainage, poor aeration, and low permeability, and are quite toxic to plant growth. To a limited extent the sodium can be counteracted by the application of gypsum ($CaSO_4$), which is acidic in reaction.

White alkali soils, common throughout much of the southwestern United States, have high concentrations of calcium, usually in the form of a carbonate ($CaCO_4$) (Fig. 1). Soils of this type are not as toxic to plants as the black alkali. Most plants will grow in them provided the other necessary nutrients are available.

pH of the soil is far from a stable quantity, since it varies from day to day and from season to season, depending on the availability of water, where the water comes from, the salts carried in it, and the amount of runoff. In arid regions salts from the lower substrates are brought to the surface by capillarity and accumulate there as the water evaporates. Changes in the soil pH can be made within certain limits, but since soil is such a dynamically complex matter the change is seldom permanent. Acid soils are treated with

FIGURE 1 Field of lettuce under irrigation. The light areas in the furrows are calcium carbonate crystals which frequently accumulate on alkaline soils.

applications of calcium (chiefly calcium carbonate) to reduce the acidity, while soils which are highly alkaline may be given applications of superphosphate, or highly organic fertilizers such as manures.

The acidity or alkalinity of the soil, then, is not the limiting factor in plant growth. It is, however, a controlling factor in the nutrient relations in the soil and a contributing factor to the growth and metabolism of the microflora and microfauna which help to make good soil the fertile base it is.

Essential Elements

Previously we have discussed the three major elements which make up the great bulk of the plant body. These are carbon, hydrogen, and oxygen. Carbon is taken into the plant in the form of atmospheric carbon dioxide; water is taken up by the roots and is the source of hydrogen; and oxygen is derived from both the soil and the atmosphere. As important as these substances may be, the plant also requires a variety of other inorganic substances to carry on all metabolic activities.

At least twelve different elements obtained from the soil are considered essential for plant growth. Nitrogen, phosphorus, potassium, sulfur, calcium, iron, and magnesium are considered major elements, since they are used in fairly large quantity. In addition manganese, boron, zinc, molybdenum, and copper are classified as essential minor elements since the amounts required are usually just a few parts per million or even less.

Few soils would have all these elements in a nicely balanced quantity for plant growth, consequently where agricultural plants are raised fertilizers containing one or more of the nutrient elements must be added. The elements which must be added to deficient soils are usually nitrogen, phosphorus, and potassium (potash). To be sold as a fertilizer a product must have a guaranteed analysis of these three elements. Usually it is written 6-4-2, which means the fertilizer contains 6 percent nitrogen, 4 percent phosphorus and 2 percent potassium. Another fertilizer may be 0-52-0; this one has 52 percent phosphorus but is lacking both nitrogen and potash. There are many different types of fertilizers besides these so far as quantity of the various elements is concerned. In nitrogen- or phosphorus-deficient soils, as in desert areas, or where heavy truck cropping is practiced, the fertilizers sold are usually highly fortified with nitrogen and often with phosphorus. Typical ratios are 20-10-0 or 10-20-0. In other areas, where the soils are quite deep and often acid in reaction, the only element that need be added is calcium. This is administered in the form of lime.

Nitrogen

When a plant is deficient in a certain element there are symptoms which the experienced observer will recognize as due to a specific element deficiency. Nitrogen deficiency is usually recognized by a lack of green color in the leaves, which continues to a pronounced yellowing and finally loss of the plant's lower leaves. The plant also has relatively poor vegetative growth, and some plants such as tomatoes develop a purplish or reddish coloration along the leaf veins due to anthocyanin pigments which are produced in unusually great amounts under nitrogen deficiency. Plants, like animals, may also show the effects of excesses. With an excess of nitrogen the vegetative parts become dark green, the leaves become succulent, and the plant in general tends to develop vegetatively but not produce flowers. Nitrogen is taken up by plants in the form of nitrates (NO_3^-) and also in the form of ammonium ions (NH_4^+); however, legumes are not so dependent on nitrogen in these forms since the nitrogen-fixing bacteria, which form the nodules of hypertrophied growth in the roots, enable the plant to use molecular nitrogen (N_2) absorbed by the bacteria.

Phosphorus

Phosphorus is usually added to the soil as a superphosphate or as bone meal, and is taken into the plant in the form of the ions $H_2PO_4^-$ and $HPO_4^=$. Phosphorus is used in many plant processes, particularly in protein building, photosynthesis, respiration, and flowering. A deficiency usually results in stunted growth and a darkening and often a purplish coloring in the leaves as a result of the development of anthocyanin pigments.

Excessive amounts of phosphorus are not easy to detect, since a plant may be exhibiting a deficiency of another element, such as nitrogen, rather than the PO_4 excess. When excessive amounts of superphosphate were recently applied to newly planted lawn plots, the results were excellent for percentage of germination of the seed but the young plants grew slowly and were stunted and chlorotic.

Potassium

Potassium, the third of the three major components of fertilizers, increases root growth, improves drought resistance, reduces respiration thereby retarding energy losses, strengthens stems, and aids in photosynthesis and building of other foods. Potassium acts as an enzyme activator in many cases. It is essential for the action of the enzyme pyruvic kinase, which is necessary in plants to form the compound pyruvic acid used in the normal flow of metabolites in the building phase. When potassium is low the supply of pyruvic acid is limited and abnormal compounds are formed which are not as easily used by plants. Recent research indicates that low potassium levels may restrict the formation of ATP thus limiting the supply of energy needed in reactions such as photosynthesis and respiration. Deficiency symptoms include yellowing of the leaves, which then become brownish around the tips and margins. Growth is restricted and stems are often so weak that the plants readily blow down in winds.

Sulfur

Sulfur is an important component of the proteins in many plants and may be stored in large quantities in onions, garlic, and other plants. Sulfur in the form of $SO_4^{=}$ is normally abundant in most soils, but in soils which are highly alkaline sulfur is not readily available. If elemental sulfur is added to soils with a pH of about 7.5 or more it does not readily go into solution and consequently it remains unavailable. Sulfur is essential to many enzymatic reactions including the oxidation of pyruvic acid. Deficiency symptoms include yellowing of the younger leaves and in severe cases even the older leaves may lose their normal green color.

Calcium

Calcium is the most common element added to soils which are highly acid. In those soils where it is deficient plants may take up abnormally large amounts of magnesium to the point of toxicity. Calcium is especially important to the development of young growing tips, where it forms a compound with pectin. It is found in the middle lamella and apparently influences the permeability of plant membranes in certain mineral ionic exchanges. Calcium deficiency causes a rapid disintegration of the young shoot and root growing tips which is often preceded by the malformation of young leaves. The effects of calcium deficiency are not always easy to determine because of calcium's role in regulating the uptake of other elements.

Magnesium

Magnesium is an important component of the chlorophyll molecule, consequently symptoms of magnesium deficiency include chlorosis of the leaves, which develops upward from the base of the plant, ultimately killing many of the leaves. Magnesium is also needed for the action of many enzymes.

Only during the present century was it found that plants may show deficiencies or even die if deprived of certain elements that are much less abundant in nature than the major elements discussed. These include iron, manganese and several of lesser importance.

Iron

Iron is a part of many important enzymes including the respiratory electron carriers called cytochromes and the oxidative enzymes peroxidase and catalase. In these enzyme systems iron serves in both oxidation and reduction as a reversible reactor. (Fe^{+++}+ electron $\rightleftarrows$ Fe^{++}). Iron acts as a catalyst in chlorophyll formation, consequently deficiency symptoms show immediately in a gradual yellowing of the leaves. The pattern is quite striking since it affects young leaves first in the interveinal areas then gradually works its way to the smaller veins and finally even the largest veins may lose their green color. The administration of iron, in the form of iron sulfate or iron citrate, or in the chelated compounds, to plants suffering from iron chlorosis will usually relieve the symptoms within a few hours if the plants are small.

Manganese

Manganese apparently has a role in the formation of oxygen in photosynthesis. Deficiency symptoms are similar to those of iron in that the leaves become paler; often they have dead brown or gray spots. In the case of manganese deficiency the spotting often leads to loss of the leaves, while iron chlorosis results in loss of color but seldom in leaf loss.

Boron

The exact role of boron is unknown, but a deficiency results in death of meristematic cells. A boron deficiency in fleshy organs of older plants results in the browning and disintegration of internal tissues causing heart-rot in sugar beets and internal browning and rot in apples.

Zinc and Copper

Zinc and copper probably are reversible in their chemical role in the same way that iron reacts. Zinc is found in several dehydrogenase reactions, such as the conversion of ammonium to amino nitrogen; copper aids in the reduction of nitrite to ammonium. Zinc deficiency is exhibited by generalized spotting in interveinal areas, the stunted growth of young leaves, and shortened internodes. Copper deficiency often is manifested by a withering of the tips of young leaves and young shoots even when plenty of water is present.

Molybdenum

Molybdenum apparently has a role in the reduction of nitrates and the fixation of nitrites. The role of molybdenum in the metabolism of nitrogen is greater when molecular nitrogen or nitrate is used as a source of nitrogen for the plant than when ammonium ions are used.

Other Necessary Elements

Many other elements are taken up by plants, and in certain species there seems to be a definite need for some of them. While sodium is toxic to most plants, the saltbush, *Atriplex,* which grows in saline soils, accumulates large quantities of this element and has been shown to require it for successful growth. It is also believed that cobalt is used in some plants, especially legumes, where it aids in the fixation of nitrites in the root nodules. The chloride ion has also been shown to improve the growth of some species, but just what its role is and how much is used is unknown. The difficulty of ascertaining the amount of chloride ions in a soil or solution is due to their frequent presence as a contaminant.

Silicon is frequently taken up by plants and precipitated in the walls. For such a plant as a diatom, where the walls are almost pure silicon, it is obviously necessary; for others it may not have any specific function. Selenium is an element which is toxic to range animals eating plants which have absorbed it from the soil, but to the plant it is not toxic and seems to remain as an inert substance in plant tissue.

Salt Action and Plant Development

Understanding the mineral nutrition of plants involves a knowledge of the interrelations among different plants, their soil environment, and their light affinities as well. Study of one phase of this problem has been sponsored by UNESCO for the past several years, since success would mean the use of more land for agriculture and perhaps the raising of agricultural plants where it has been impossible before. This work has centered around the growing of crop plants in arid regions and also around the salt tolerance of plants, few of which grow well in saline soils. As a result of high concentrations of salts in the soil, the relationship of the plant to its environment is changed. For example, such factors as temperature and light, which have a favorable effect under normal conditions, may exert an unfavorable effect under saline conditions.

Investigations of the tolerance that plants have for saline soils have resulted in a classification of the plants into two groups:

1. **Halophytes** are plants growing in saline habitats. They have become adapted to such habitats due to those features and properties which have arisen during evolution under the influence of conditions prevailing at that time. Halophytes are capable of growing on saline soils containing 0.3 to 20 percent salt, although most grow on soils containing 2 to 6 percent.
2. **Glycophytes** are plants growing in nonsaline soils. They are quite limited in their ability to adapt themselves to salinity since the

conditions under which they evolved did not apparently favor the development of such properties.

There have been many studies on how halophytes are able to regulate their metabolic processes to salinity. These investigations have reached various answers. Different biological properties may allow some halophytes to absorb considerable quantities of salt, accumulating these substances in their tissues and thus regulating their internal osmotic pressure. Halophytes are able to regulate their salt balance in several ways:

1. When excess salt is accumulated it may be secreted by special glands.
2. The salt may be removed by leaf abscission after the leaves have reached their capacity of salt; one species was shown to contain up to 57 percent total salts by dry weight of the plant.
3. Root excretion of salts has also been described for some halophytes.

Because they accumulate salts in great quantity, the halophytes have a high osmotic pressure in their cell sap. As a result of this salt accumulation the root system develops a high suction pressure, which is higher than the osmotic pressure of the soil solutions, enabling the roots to absorb water from the soil. The high osmotic potential in some halophytes is due to an accumulation of organic substances; in others it develops as a result of mineral salts absorbed from the saline substrates.

Usually the total metabolic processes including respiration, photosynthesis, and enzymatic reactions are lower in halophytes than in glycophytes. The former also have protoplasm, which is more viscous, altering their water relations and giving them more drought resistence.

Some interesting observations have also been made on the relation of soil salinity to the sex expression in plants. It has been reported in UNESCO studies that the male trees of poplar, willow, and others usually develop on more saline soils than the females. In monoecious plants there was an increase in the number of male flowers and a decrease in the number of female flowers on plants growing in saline soils.

Numerous experiments have been carried out and many papers written on the growth, development, and yields of plants under saline conditions. Investigations of this type are necessary in order to understand the mechanism of salt action at the different stages of plant development. Data of this nature are used by agronomists and plant growers in evaluating the degree of salt tolerance of crop plants and in breeding new, economically important varieties of salt-tolerant species.

Radioisotopes

In Chapter 2 the uptake of certain radio-nuclides was discussed. It should be emphasized here that radioactive substances which appear naturally or which have been placed in soils are sources of contamination of biological systems. Since it has been shown that certain plants (Western Yellow pine) freely absorb strontinum-90, and other investigators have demonstrated the absorption of thorium-228, radium-224, and additional radioactive substances, plants must be studied extensively to see how this source of radiation may affect us. Many radio-elements are long-lived and therefore constitute a potential hazard by their presence in biological material. From studies of the soil-plant-animal food chains there are sufficient data on the absorption, accumulation, and retention times of radioactive substances in different organs to show that there is a human population contamination problem. At present the amount of radioactive substances released into the atmosphere and falling on vegetation, animals (including man), soil, and water has not reached a dangerous level.

Radiation is not all bad, however. It is quite possible that ionizing radiation could be a useful agent for sterilizing soils for research purposes, since the side effects of the sterilization are often less marked than when other sterilizers are used. Insects and nematodes may be eliminated from soils by moderate doses not affecting the bacteria present, thus making it possible to use radiation as a selective sterilizer. In one study investigators grew clover under sterile conditions and found that the plants grew better when the soil was sterilized by irradiation than when it was sterilized with steam or propylene oxide. However, the plants did grow best in the controls of unsterilized soils.

Studies completed using strontium as the radioactive material show that the rate of downward movement varies with the soil type, the movement of water, the pH, and the clay and humus content. If the soil is quite compact and undisturbed, Sr^{90} placed on the surface will penetrate very slowly and remain in the upper few inches for years. If the soil is plowed to a depth of seven inches or more the Sr^{90} is more deeply placed, and crop plants raised on the soil take up much less of the substance than plants do where the Sr^{90} is on or near the surface. Soil conditions which encourage shallow root development also cause the absorption of Sr^{90} to be increased, while the addition of humus seems to decrease the uptake.

Strontium-90, like stable strontium, is influenced in its absorption by calcium, to which it is related chemically. Normally strontium, unlike calcium, is not found in large amounts in soil. The studies in general show that the uptake of Sr^{90} is greatest in soils of low calcium content; therefore the addition of calcium reduces the uptake of strontium. However, when soils contain adequate calcium for the growth of plants and the soil is essentially calcium-saturated, the addition of more lime has little or no effect on the uptake of strontium. It has also been well established that a plant's ability to absorb Sr^{90} is directly correlated with its ability to absorb calcium. For this reason the concentration of Sr^{90} in biological materials is often reported as a ratio of strontium to calcium.

The role of radioisotopes in making it possible to trace chemical substances through biological systems has been a major break in establishing the field of molecular biology. Without "tagged" or "labeled" atoms we would be unable to trace carbon, hydrogen, and oxygen through the photosynthetic process or through respiration, or to trace the movement of soil nutrients through the plant and determine how plants use these elements in metabolism.

Supplementary Readings

Anderson, A. J., and E. J. Underwood, Trace Element Deserts. *Scientific American,* January, 1959.

Meyer, B. S., D. B. Anderson, and R. H. Bohning, *Introduction to Plant Physiology.* D. Van Nostrand, Princeton, N. J., 1960.

Sutcliffe, J. B., *Mineral Salts Absorption in Plants.* Pergamon Press, New York, 1962.

Questions for Discussion

1. Black alkali is usually toxic to plants, whereas white alkali rarely is. Explain.
2. Plants living in salty marshes often have characteristics similar to those living under extremely arid conditions. How would you explain their similarity in xeromorphy?
3. Compare the major elements required by plants with those required by humans. Discuss the similarities and differences.
4. Sidewalks may be lifted by the tremendous pressure exerted by growing roots. How are roots able to exert this much force?
5. Would you expect the root system of a plant to have greater or less radial symmetry than the aerial portions? Explain.
6. Why do plant roots growing in glazed pots frequently die and rot?
7. In transplanting trees and shrubs it is desirable and even necessary to keep as much soil on the roots as possible. Explain.
8. Describe the differences between topsoil and subsoil.
9. In looking at a cut embankment on a roadside or stream bank we often see several layers of different soil texture and color. How were these layers formed?
10. Why do we think of soil as a dynamic substance?
11. Why can't plants remove all water present in soils?
12. Why, although some necessary elements for plant growth are abundant in soils in which roots are growing, may the plant be unable to absorb them?
13. Why do plants sometimes absorb substances from the soil when they are unable to use them?
14. Why is it better to use superphosphate as a fertilizer in highly calcareous soils than to use bone meal?
15. Describe the form in which such substances as nitrogen, potassium, and sulfur are absorbed, and the function they have in plant growth.
16. What may cause fluctuations in the soil pH?
17. How may halophytic plants maintain their salt balance?
18. How may a human being get an excess of strontium-90 in his system even though he has never been directly exposed to it?
19. Plants capable of growing in highly calcareous soils are not injured or benefited by the application of lime. Explain.
20. What do we mean by deficiency symptoms in plants? Give examples.

11

The Flower, Fruit, and Seed

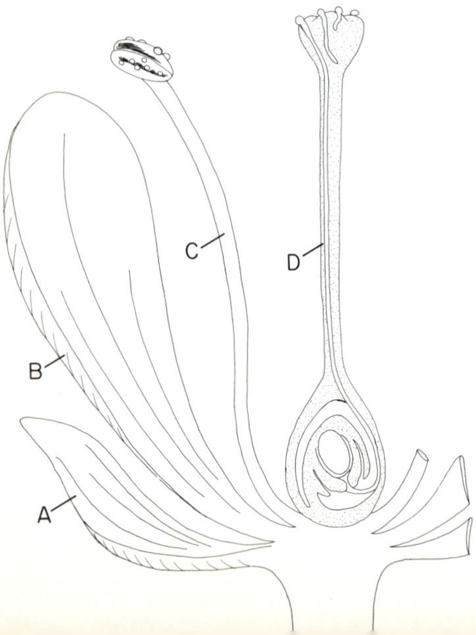

FIGURE 1 A typical complete flower showing A-sepal, B-petal, C-stamen, D-pistil (including ovary, style, stigma). The flower is classified as superior since the ovary is above the attachment of the other floral parts. Germinating pollen grains are shown on the stigma, including one which has completed its pathway to the micropyle of the ovule.

Roots, stems, and leaves are all reproductive organs in that a new plant may be produced from them, at least in certain species. Many crop plants such as sugar cane, sweet potatoes, and white potatoes are propagated mainly by planting sections of the stem or roots. If this were the only method plants had of reproducing in the wild state there would be little variation. A plant produced by cuttings has the same genetic constitution as its parent. Variation and the development of new species in angiosperms take place through the flower, the organ of sexual reproduction. Within the flower seeds are produced and a protective covering in the form of a fruit develops. The flower is the structure which is most important to us in the identification of angiosperm species, and biologically it contributes greatly to evolution in the flowering species.

The Flower

A typical flower (Fig 1) consists of four sets of appendages; the outer set is made up of more or less greenish, leaflike organs, the **sepals.** Collectively they constitute the **calyx.** The next set of appendages, lying within the ring of sepals and usually completely covered by them in the bud, is the **corolla,** each segment of which is a **petal.** Collectively the calyx and corolla make up the **perianth.** Flowers composed of sepals and petals of similar size and shape, which are therefore radial in symmetry, are termed **regular** or **actinomorphic** (Fig. 2). Those in which some of the petals are of unequal size and shape and which exhibit bilateral symmetry are **irregular** or **zygomorphic** (Fig. 3). Lillies, roses and magnolias are regular flowers; plants having irregular flowers include peas, snapdragons, and orchids (Fig. 4). The irregular condition is considered advanced and the regular type of flower is primitive. The two types of flowers are found among both monocotyledons and dicotyledons.

FIGURE 2 Regular flower of cactus showing its radial symmetry, numerous stamens and petals and central pistil.

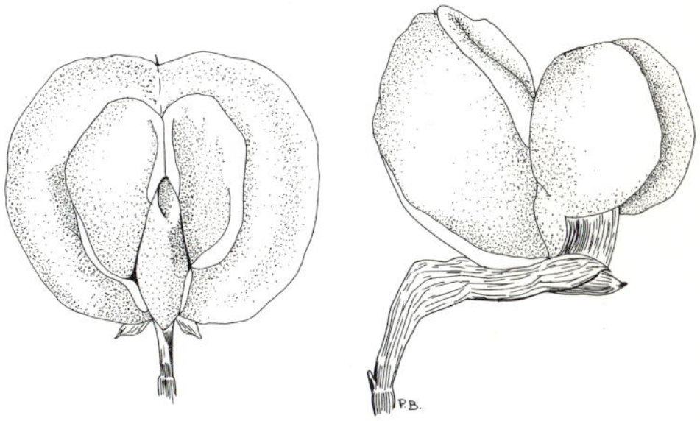

FIGURE 3 Irregular flower of sweet pea showing bilateral symmetry due largely to different size and arrangement of petals. *Left:* Ventral view. *Right:* Lateral view.

FIGURE 4 Irregular flowers of the *Cattleya* orchid.

The next set of appendages, lying just within the petals and often alternating with them, is the ring of male organs, the **stamens.** Located in the center of the flower is the **pistil** or pistils, the female organ of an angiosperm. Flowers having both stamens and pistils are said to be **perfect;** those lacking either stamens or pistils are **imperfect.** Many trees, such as cottonwoods (Fig. 5), willows, and herbaceous plants like corn, have separate staminate and pistillate flowers. If the plant has flowers of only one sex, then the species is **dioecious;** if the staminate and pistillate flowers appear on the same plant the species is **monoecious.**

FIGURE 5 Female or pistillate inflorescence of cottonwood.

FIGURE 6 Male or staminate flowers of the cottonwood tree.

Flowers having all four sets of appendages (violets and roses, for example) are classified as **complete,** and those which lack one or more of the floral sets of appendages (as do begonia, elm, and corn) are **incomplete.** The floral appendages are borne on a receptacle which is the enlarged tip of the flower stalk, the **peduncle.** Those plants which have compound **inflorescences,** i.e., two or more flowers in a cluster, have an individual stalk or **pedicel** for each flower (Fig. 7).

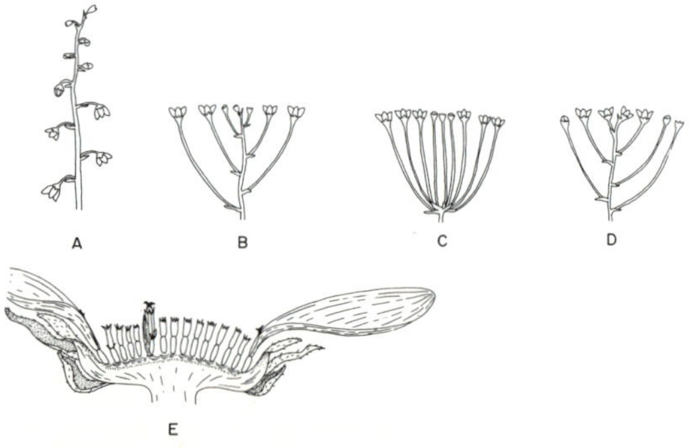

FIGURE 7 Types of inflorescences. A-a raceme, the flowers are on short pedicels; in a spike the flowers are sessile; B-a corymb, the flowers form a flat-topped cluster with the oldest flowers rimming the centrally located buds; C-an umbel, the pedicels are of equal length and appear to arise from the same place. D-a cyme, differs mostly from a corymb by having the oldest flowers in the center and rimmed by buds. E-a head, the flowers are sessile on a common receptacle; the central flowers are usually tubular and regular while the outer ray flowers are irregular.

Compound inflorescences have been given a variety of names which are quite useful in the identification of plants. Probably the most compact inflorescence is that termed a **head,** in which there are many florets borne on a common receptacle. Sunflowers, dandelions, asters and many other members of the Compositae are examples of this type of inflorescence. Another is an **umbel.** In this type the individual flowers are borne on pedicels of equal length arising from about the same place. Examples are dill, carrots, parsnips, and other members of the Umbelliferae family. In some members of this group the pedicels which arise in umbellate fashion also give rise to small floral umbels. This is a compound umbel.

Many grasses produce a much-branched inflorescence which is termed a **panicle** (Fig. 8); other plants have a main axis from which arise lateral individual branches bearing one or more flowers.

FIGURE 8 The panicled inflorescence of Yucca.

This is a **raceme,** and if the flowers are sessile on the main axis it is termed a **spike.** There are other variations of the racemose type, such as the **catkins** of a cottonwood tree, in which the inflorescence is pendulous and often has only male flowers arising from the central axis.

There are two other types of flat-topped inflorescences which superficially resemble an umbel but whose pedicels are of various lengths, the ones which arise lower on the peduncle being much longer and producing the outer rings of the inflorescence. In a **cyme** the first and oldest flowers are near the center of the cluster and the youngest flowers are towards the outside, arising from the longer pedicels. A **corymb** is similar but in its case the older flowers are towards the outside of the cluster and the youngest are in the center.

The Perianth

Since most flowers have conspicuous and colorful petals, the petals are more commonly observed than any other part of the flower. However, in some flowers such as lilies and amaryllis the petals and sepals are of the same size and same color. The only method of determining which structures belong to the calyx and which comprise the corolla is to note that three of the appendages alternate with and arise just outside of the inner three appendages. The exterior ones are, of course, the calyx and the inner the corolla. In some flowers, like marsh marigold and Virgin's Bower, the sepals are the more conspicuous and the petals are very small or lacking. A few plants, such as poinsettias and bougainvilleas, have clusters of very small and inconspicuous flowers surrounded by large and colorful bracts, which are modified leaves.

The Pistil

The pistil is usually made up of a stigma, style, and ovary. The ovary contains the ovules and, subsequently, the seeds. An ovary situated on a more or less convex receptacle, with the other floral parts arising at or below the base of the ovary, is said to be **superior,** or in other words the flower is **hypogynous** (Fig. 9). If the ovary is surrounded by receptacle tissue and the other floral parts arise from the receptacle tissue at the apex of the ovary or above it the ovary is **inferior,** or the other floral parts are said to be

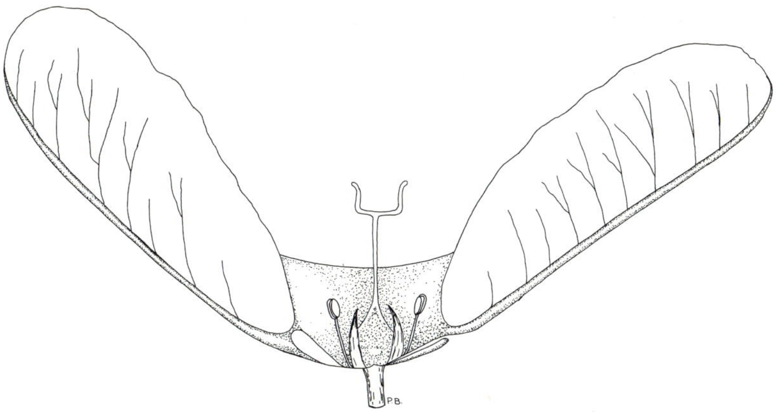

FIGURE 9 Superior flower of maple. Note central pistil with stamens and sepals arising at the base of the ovary.

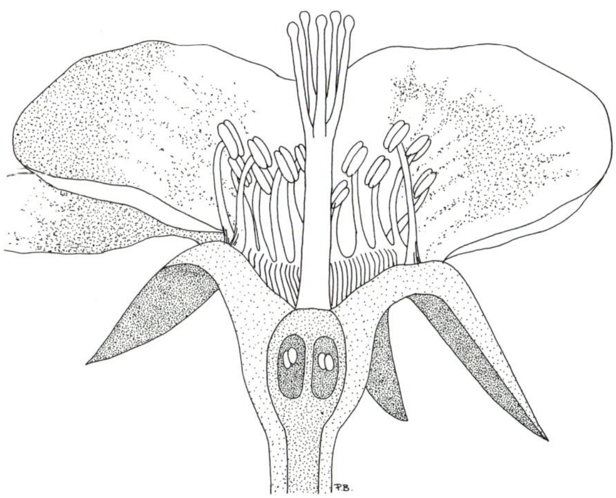

FIGURE 10 Inferior flower. Note attachment of stamens in a ring above the position of the ovary.

epigynous (Fig. 10). Floral parts with an intermediate position in respect to the ovary are **perigynous.**

A pistil is composed of one or more divisions, termed **carpels;** if the pistil has only one cavity, as in a peapod, it is simple; if it is composed of two or more carpels, as seen in the cross section of an orange, the pistil is compound. The carpels may be thin and leaflike or fleshy. On their inner surface, usually near the margins, the **ovules** are produced. They are attached to the ovary wall along the area termed **placenta** by a short stalk, the **funiculus.** If the ovules arise on the inside of the ovary wall they have a **parietal placentation;** if the ovules arise along a central axis they have an

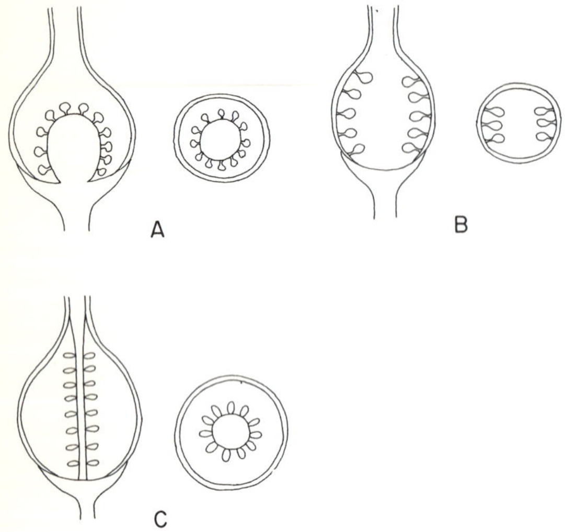

FIGURE 11 Three types of placentation in longitudinal and cross section. A-free central; B-parietal; C-axial.

axile placentation; in some flowers the placenta may form on a central protuberance at the base of the ovary, giving the attached ovules a **free central placentation** (Fig. 11).

Along with the **ovary,** which makes up the principal portion of the pistil, the **stigma** and **style** complete the pistillate portion of the flower. The style may be very long, such as the fine hairlike structures developing from the ovaries on an ear of corn and frequently called corn silk, or the style may be very short or lacking. The stigma is at the apex of the style, or if the latter is lacking, the stigma may appear as a tiny cushion on the upper end of the ovary. The stigma is usually slightly larger in diameter than the style and may be disclike, branched, or feathery, and rough, smooth, or viscid on its surface.

The Stamens

The stamens, which are the pollen-producing organs of a flower, are usually made up of a slender stalk, the **filament,** and the enlarged apical pollen sac, the **anther.** In cross section the anther can be seen to be composed of four lobes, each of which is termed a **microsporangium** (Fig. 12). In young anthers the central

FIGURE 12 Cross section of anther showing four microsporangial areas in which the pollen is formed.

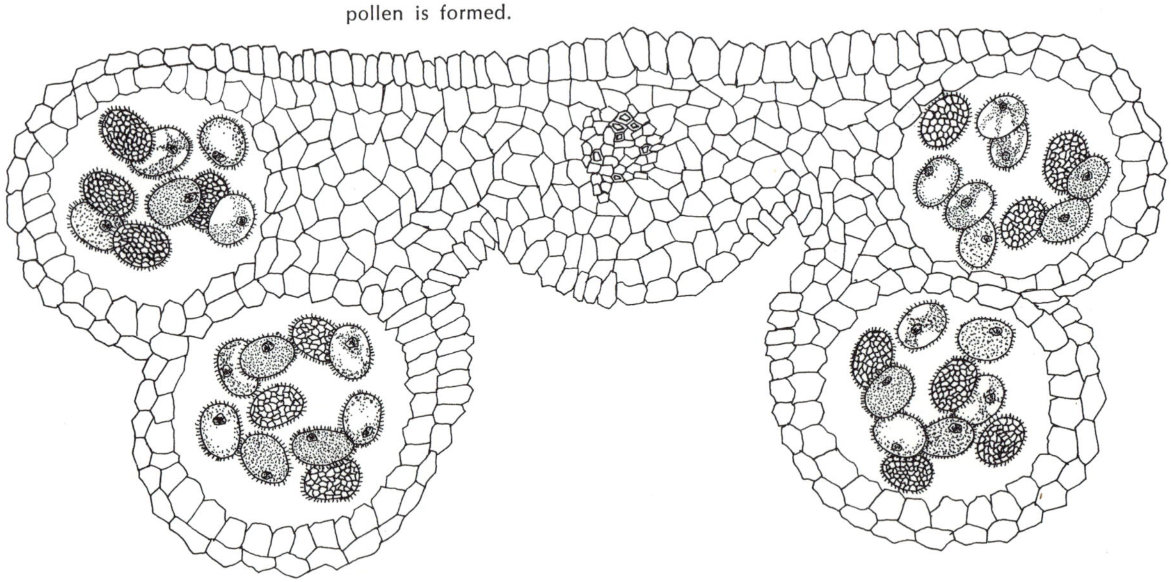

cells of each microsporangium have an abundance of cytoplasm and large nuclei. These cells differentiate into the **microspore mother cells.** They are at first angular in section and tightly packed together, but as the anther increases in size the mother cells separate and become spherical.

With continued growth of the anther the microspore mother cells divide, forming tetrads of **microspores.** In this division the number of chromosomes is halved so that each spore has half the number of chromosomes found in the microspore mother cell. Each microspore mother cell has $2n$, or the diploid number of chromosomes, therefore each microspore has n number of chromosomes and is haploid. This type of division is called **meiosis** and is discussed briefly in connection with megaspore formation below, and again in detail in Chapter 13.

The completion of development for a mature pollen grain occurs when the pollen nucleus undergoes division forming a **tube cell,** which will initiate the development of a **pollen tube** and a **generative cell** (Fig. 13). The latter may then divide again, resulting

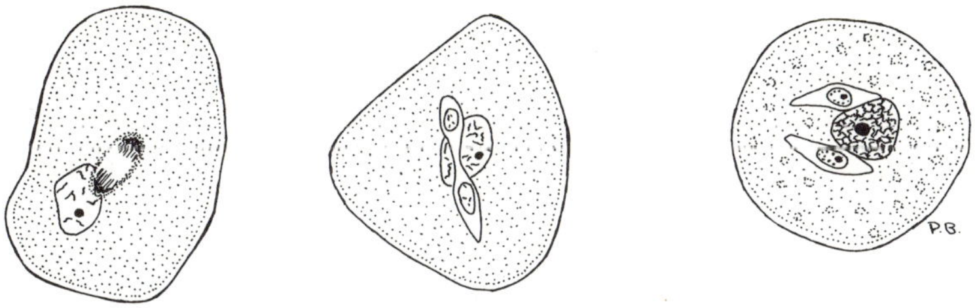

FIGURE 13 Development of the male gametophyte in an angiosperm. *Left:* Tube nucleus and initial division to form two sperms. *Center:* Division almost complete. *Right:* Mature microspore with tube nucleus and two sperms.

in the formation of two **sperms** before the pollen is shed; or the final division may be delayed until after the pollen has been shed and has reached the stigma. The small bit of tissue (including tube cell and sperms) within the microscope wall is known as **gametophyte tissue.** Therefore a mature pollen grain is a **microgametophyte.** As each pollen grain matures, it develops a two-layered wall consisting of an outer **exine** and an inner **intine.** The former is usually ornamented with various sculptured lines frequently quite unique for each species and useful in the identification of plant species.

When the pollen grains are mature they usually fill the microsporangial chambers on each side of the anther so completely that the two lobes on each side merge as one large pollen sac. When the anther walls open the grains are freed and may then blow away or be carried by other agents to a receptive stigma where the pollen grain may germinate and the male gametophyte may fully develop. Usually the pollen grains are free from one another, but in some plants, such as orchids, the spores cling together in a waxy or mucilaginous mass, the **pollinium,** and are shed in this manner.

Agencies Responsible for Pollination

1. Wind	4. Birds and other animals
2. Insects	5. Man
3. Gravity	6. Water

The accomplishment of pollination is not only of preeminent importance to the propagation of plants, it is also of great significance to us. In order that we may have hybridized corn, for example, we must completely control the transfer of pollen to ensure the development of the most productive hybrid. Apple trees may be wind pollinated, but apple growers cannot always depend on the wind at the right time so they usually raise bees in their orchards to ensure pollination. Hummingbirds, moths, and many other organisms are also necessary for the pollination of many flowers. Man has become a very important agency in the transfer of pollen, since by the union of the proper varieties we have developed numerous plants of greater economic importance. In fact, some varieties, such as double petunias, would cease to exist in one year if man disappeared, and certainly would not have been developed without him.

Some flowers are self-pollinated. These are usually enclosed, like sweet peas, giving the wind little opportunity to blow the pollen to another plant. Other flowers, like those of apples and cherries, are open, allowing the wind to carry the pollen around and achieve cross-pollination, and therefore making cross-fertilization possible. Insects may visit any type of flower, but for closed types which may be self-incompatible insects are virtually the only possibility for fertilization. The red clover seed crop is absolutely dependent on the bee for transfer of the pollen since the small leguminous flowers have pistils completely enclosed by the united petals.

Development of the Ovule

The ovules of angiosperms develop as rounded protuberances within the ovary. These ovules will eventually become seeds, but first there are many steps required before the ovule will produce an egg which can be fertilized. The inner portion of the young ovule consists of a cluster of cells termed the **megasporangium** or **nucellus.** It is almost completely surrounded by one to several layers of cells, the **integument,** which will eventually form the seed coat. A small opening, the **micropyle,** is left where the integument does not enclose the megasporangium completely.

The basal or attached end of the ovule is termed the **chalaza** and may ultimately be near the micropyle if the ovule is inverted, or it may be opposite to the micropyle if the ovule is erect. Ovules therefore exhibit several variations in respect to their form and orientation (Fig. 14). They may be erect **(orthotropous);** or the

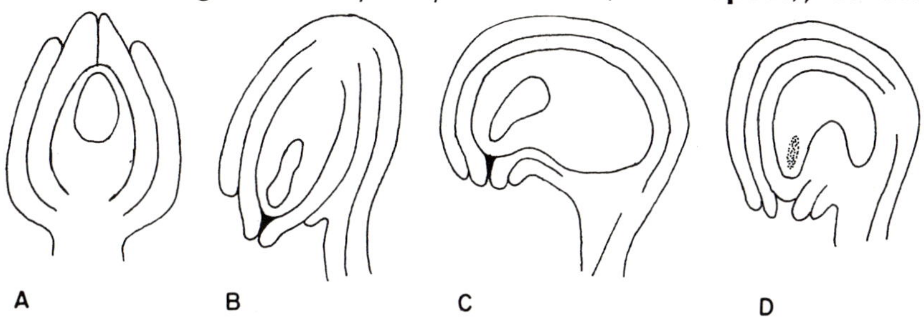

FIGURE 14 Types of ovules in longitudinal section. A-atropous (orthotropous), B-anatropous, C-campylotropous, D-amphitropous.

funiculus may be adnate to the ovule for about half its length, with the micropyle at right angles to the funicular attachment, in which case the ovule is **amphitropous;** or they may have the funiculus attached near the middle of the ovule and be **campylotropous;** or the ovules may appear completely inverted and be **anatropous.**

Development of the Megaspore

Within the megasporangium a single large **megaspore mother cell** is differentiated. This cell, like the microspore mother cells, has the diploid number of chromosomes and like them carries the factors of inheritance.

In each human body cell there are twenty-three pairs of chromosomes, but in each of our sex cells (sperms and eggs) there are just twenty-three individual chromosomes. These sex cells have not received just any twenty-three chromosomes; each has received one member from each of the twenty-three pairs found in our body cells. The same situation is found in all animals and plants. The number of pairs of chromosomes is often different from the twenty-three we have, but the sex cells of each species will always (or almost always) have half the number found in the body cells.

After the megaspore mother cell is mature, it divides forming four **megaspores.** During this division the number of chromosomes contained in the mother cell is halved. The newly formed megaspores lie in a single line within the megasporangium. Usually the three nearest the micropyle disintegrate and disappear while the fourth, nearest the food supply, towards the opposite end, persists. This megaspore now greatly enlarges giving rise to an elongated or oval structure we call the **embryo sac** or **megagametophyte.**

While the embryo sac is enlarging the megaspore nucleus divides, giving rise to eight nuclei. The divisions here are equal, so

FIGURE 15 Development of the 8-nucleate stage in the embryo sac of an angiosperm ovula. A-Megaspore mother cell. B, C-Meiosis occurs in formation of four megaspores. C, D-Three megaspores disintegrate; E, F, G-Division of remaining megaspore to form the 8-nucleate embryo sac.

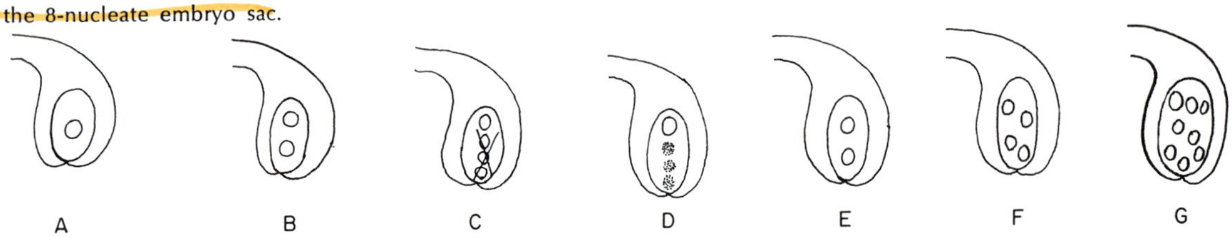

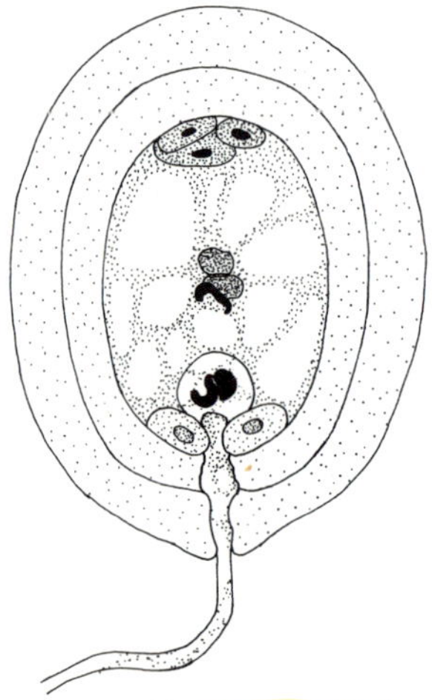

FIGURE 16 Mature ovule showing outer integumentary layers and embryo sac with its three antipodal cells at the chalazal end, two central polar nuclei, two synergids at micropylar end and one egg. The pollen tube has entered the micropyle and one sperm has united with the egg while the second sperm has reached the polar nuclei just before they have fused to form the fusion nucleus.

far as the number of chromosomes is concerned, therefore each of the eight nuclei has the same number of chromosomes as the original megaspore nucleus.

As the embryo sac continues to enlarge the cells of the megasporangium and the integument divide and enlarge greatly increasing the size of the developing ovule. Within the embryo sac the eight nuclei are rather loosely arranged into two groups of four, with four near the chalazal end and four near the micropylar end. One nucleus from each of the two groups moves to the center of the embryo sac and these are now known as the **polar nuclei.** These two nuclei will soon unite forming the **fusion nucleus.** The three nuclei remaining at the chalazal end of the embryo sac are called the **antipodals;** in most flowers these nuclei soon disappear. Of the three nuclei remaining at the micropylar end one in the more central position and nearest the micropyle is the **egg** and the other two, partially surrounding the egg, are the **synergids.** All of the nuclei except the polar nuclei are enclosed by small quantities of cytoplasm and individual membranes, and may therefore be considered cells rather than naked nuclei (Fig. 16).

With the final orientation of the cells and nuclei in the embryo sac, the enlargement of the megasporangium and integuments, the ovule is mature and ready for fertilization.

Variations in the Embryo Sac

While the usual pattern of development of the female gametophyte is as described above in the great majority of angiosperms, the minority of species which are different vary in a number of ways. The greatest variation is in the chromosome numbers found in the fusion nucleus and in the number of nuclei formed in the embryo sac of different species.

In the lily (Fig. 17) the megaspores are formed as usual but none degenerate. Instead, three of them migrate to the chalazal end of the megagametophyte and one remains at the micropylar end. This megaspore divides forming four nuclei, each with the same number of chromosomes as the original megaspore. These four become the egg cell in its usual position with two synergids, while the fourth migrates to the center and becomes one of the two polar nuclei. After the three megaspores have migrated to the chalazal end of the embryo sac they undergo simultaneous division. This should result in six nuclei, each with a haploid set of chromosomes, but after the chromosomes have divided three sets,

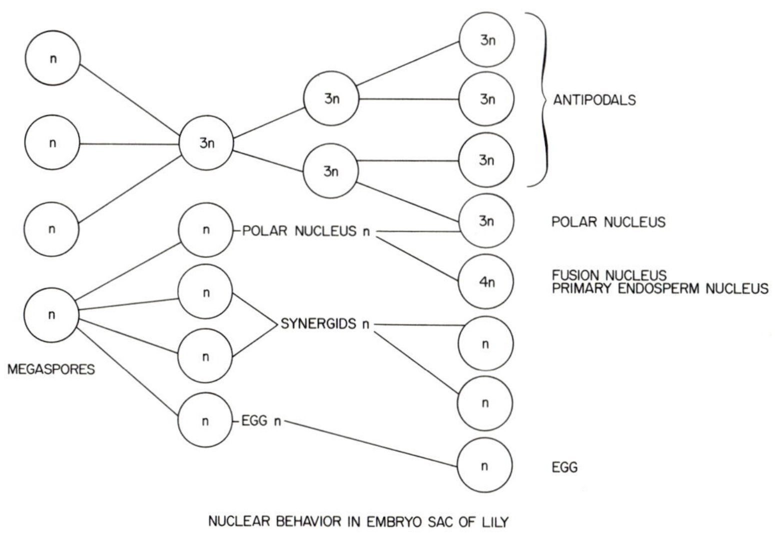

FIGURE 17 Nuclear behavior in the embryo sac of a lily. In this case none of the megaspores degenerate. Three of them fuse to form a $3n$ nucleus and then by two mitotic divisions form three $3n$ antipodals and one $3n$ polar nucleus. The fourth megaspore divides mitotically to form four nuclei, one of which becomes the egg, two synergids, and one polar nucleus. At the time of fertilization the $3n$ polar nucleus unites with the n polar nucleus to form a $4n$ fusion nucleus and when the second sperm unites with it the newly formed primary endosperm nucleus has $5n$ chromosomes. The embryo has $2n$ since it is the result of an n egg and n sperm.

one from each of the megaspores, become entangled and form one large 3*n* nucleus and the three sets from the other half of the divisions form a second large 3*n* nucleus.

The net result of the simultaneous division of these three megaspores is two daughter nuclei, each with triploid sets of chromosomes. These two nuclei then divide again, forming four triploid nuclei in the chalazal end of the embryo sac. Three of these remain as antipodals and eventually disappear. The fourth migrates to the center and becomes the second polar nucleus. In the lily megagametophyte, at maturity, there is therefore a 3*n* polar nucleus and an *n* polar nucleus, which will fuse somewhat later to form a 4*n* fusion nucleus.

There are many other deviations from the typical type and from the lily pattern, including a few in which only one polar nucleus is formed. In still others the number of nuclei in the embryo sac may vary from only 2 to as many as 128 or even more (Fig. 18).

FIGURE 18 Some variations in the number of nuclei in embryo sacs. In all the egg is stippled at the base; the number of polar nuclei is indicated in the center with a darkened cell representing the second sperm. The antipodals vary from none to many at the chalazal end and the synergids vary from none to two. Redrawn from Maheshwari, P. 1950 *The Introduction to the Embryology of Angiosperms*. McGraw-Hill. A:*Oenothera*; B: *Butomopsis*; C: *Polygonum*; D: *Fritillaria*; E: *Plumbagella*; F: *Penzea*; G: *Plumbago*; H:*Acalypha*; I: *Peperomia*; J: *Peperomia hispidula*.

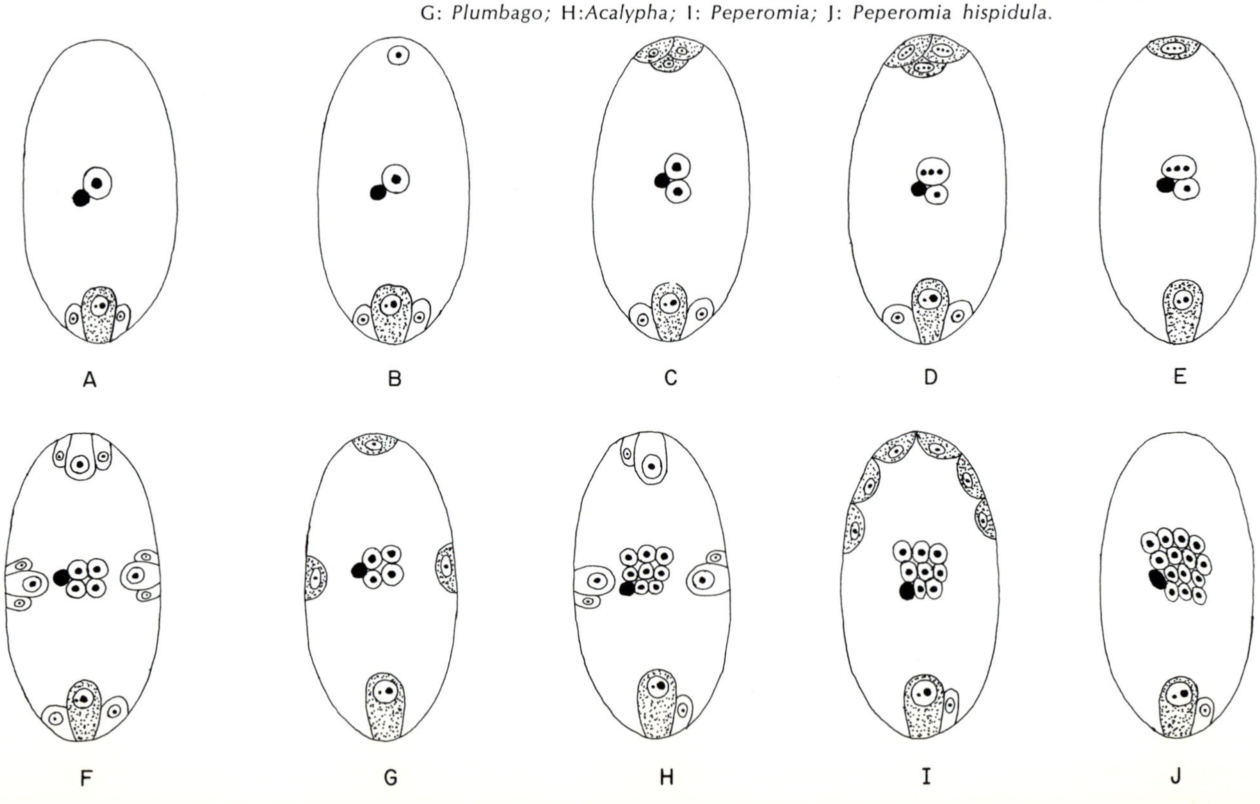

Growth of the Pollen Tube

Following pollination the pollen grains lodged on the stigma germinate and a pollen tube begins to grow downward through the style and through the ovary to the ovule. The ovules in the ovary must be reached by pollen tubes germinated from the many pollen grains lodged on the stigma and style if the ovules are eventually to become seeds. As each pollen tube digests its way down the style the two gametes move down the tube. In some cases they are preceded by the pollen tube nucleus; in others the tube nucleus may follow the gametes; and in others the tube nucleus may remain in the pollen grain and apparently take no part in the growth of the tube. Once the tube has grown through the micropyle and into the embryo sac the male gametes follow and the stage is set for fertilization.

Fertilization

The one male gamete unites with the egg to form a zygote and the second male gamete travels a little farther into the embryo sac and unites with the fusion nucleus. Investigators have reported that the two male gametes sometimes vary in size, the larger uniting with the polar nuclei (fusion nucleus) and the smaller uniting with the egg at fertilization. Since two sperms are involved in unions with the egg and the fusion nucleus, the process is termed **double fertilization.**

Following fertilization the zygote continues on to become the embryo. In most plants this is a diploid structure, but the endosperm which develops as a result of the second fertilization may have a number of different sets of chromosomes. The usual primary endosperm nucleus in angiosperms is triploid, in the lily it is pentaploid, and in still other species it has different numbers. The other nuclei in the embryo sac, namely the synergids and antipodals, usually disintegrate very early and play no part in subsequent events in development of the embryo and endosperm.

Apomixis

Occasionally an embryo develops from an unfertilized egg. This is known as **parthenogenesis** and may be the result of development from a haploid egg, in which case the resulting plant will be sterile. Or an egg may have formed in the embryo sac without a reduction in number of chromosomes. A plant developing from a diploid but unfertilized egg would be developing parthenogenetically and in this case the resulting plant would probably be fertile.

The term *apomixis* refers to a type of reproduction in which no sexual fusion occurs, but in which structures usually concerned in sexual reproduction are involved. Occasionally we note that tissues adjacent to the embryo sac, such as the nucellus, may give rise to new plants. Man has successfully used hormones to stimulate the development of fruit without the formation of embryos (or seeds). It has also been shown that hormones applied to the stigma will occasionally initiate the formation of an embryo without fertilization.

Development of the Seed

Embryonic development begins soon after the zygote is formed, and usually shortly after the primary endosperm nucleus begins to divide forming stored food for the embryo. Within the angiosperms there is considerable variation in embryogeny but one of the most common types is found in the mustard family.

In plants such as *Capsella,* shepherd's purse, and *Descurainia,* tansy mustard, embryonic development begins with a series of transverse divisions of the zygote to form a chain of proembryo cells or six to eight **suspensor cells,** as they are usually called (Fig. 19). The most apical of these cells divides to form an octant of

FIGURE 19 Development of the embryo of *Capsella;* A-F, early stages showing transverse and vertical divisions of the proembryo; G, H, I show embryo forming at apex of suspensor cells and great enlargement of basal suspensor cell; J, K, embryo beginning to differentiate w into cotyledons and hypocotyl; L, Embryo with cotyledons folded over as it does within the seed. Redrawn from Maheshwari, P. 1950 *Introduction to the Embryology of Angiosperms.*

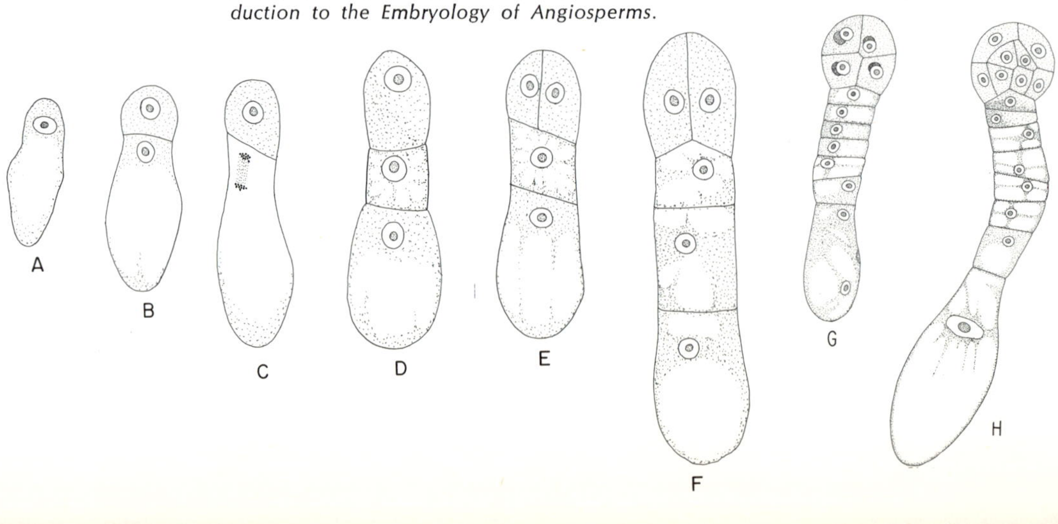

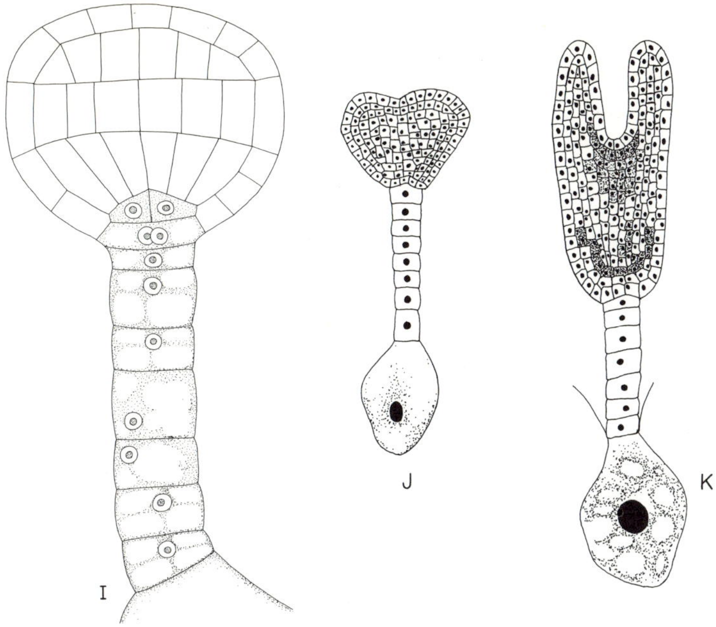

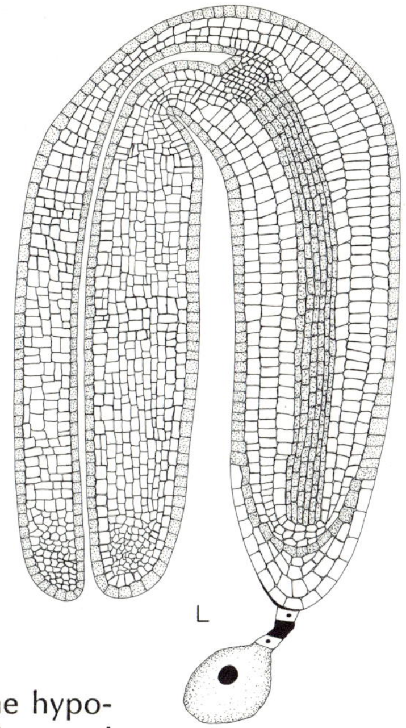

cells in two tiers of four each. The root and the bulk of the hypocotyl are derived from the basal quadrant, and both cotyledons and stem are derived from the apical quadrant.

When the embryo is approximately 100 microns long, the cotyledonary bulges are distinct and have reached a height of 25 microns above the apex. The embryo at this time has a very prominent hypocotyl along with its conspicuous cotyledons. This condition persists as the embryo elongates, the cotyledons reaching the end of the embryo sac and curving back to fill the embryo sac almost completely by the time the seed is mature.

In many dicotyledonous plants, such as those of the mustard family, the endosperm is found in a free nuclear condition in which cell walls are not formed around the nuclei in early embryonic stages. However, by the time the embryo is 100 microns long the endosperm completely fills the remainder of the embryo sac and cell walls have formed. The embryo growing within the seed digests and absorbs the endosperm, so that by the time the seed is mature very little endosperm tissue is left.

Variations in the development of the embryo occur among different species, but the fundamental pattern is essentially that described above. Even in monocotyledons the early stages are very similar, but the embryo becomes organized into a single large cotyledon derived from the cells at its distal end, with the main axis (hypocotyl and epicotyl) in a lateral position.

Seeds of monocotyledons usually have large quantities of endosperm, and the embryos are comparatively small; however, the embryo of an orchid consists of but a few cells and the endosperm is practically nonexistent. Many dicotyledons, such as the castor bean, also have large quantities of endosperm and small embryos, but others like peas and lima beans, through the action of digesting enzymes, absorb the endosperm into their cotyledons as the seed develops and consequently the interior of the seed is virtually all embryo.

The embryo and the endosperm, if present, are covered by a protective coat that arises from the integument or integuments of the ovule and is known as the **testa** or **seed coat.** In some seeds the nucellus may remain as a protective and nutritive membrane within the seed coat or be fused with it. More commonly the nucellus is completely destroyed in the development of the seed. A few seeds have a persistent nucellus which gives rise to a special nutritive tissue, the **perisperm,** but for the most part it soon disappears.

The Seed

The seed of angiosperms therefore consists of a diploid embryo; an endosperm which is usually triploid; and a seed coat developed from the ovule's integuments, which are diploid.

Types of Angiosperm Seeds

The lima (Fig. 20) (or navy, pinto, kidney, etc.) bean when mature has a rather thick and translucent-to-opaque seed coat.

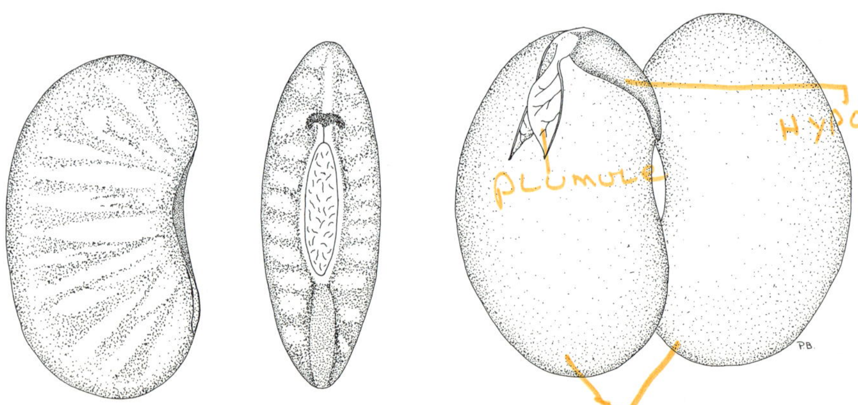

FIGURE 20 Lima bean seed. Left: lateral view with seed coat, Center: ventral view with the large scar showing, the *hilum,* and small stippled area indicating the scar left by the *micropyle. Right:* seed coat has been removed allowing the two large cotyledons to fall apart exposing the rootlike hypocotyl and the leaflike plumule or epicotyl.

Within the notch of the bean is a rather large oval scar left where the seed broke off from the funiculus. This scar is called the **hilum.** Quite near it is a very small opening where the pollen tube entered the ovule and this opening is known as the **micropyle.**

When the bean germinates the tip of the hypocotyl, the radicle, emerges and becomes established in the soil as the primary root (Fig. 21). As the hypocotyl elongates the cotyledons are carried

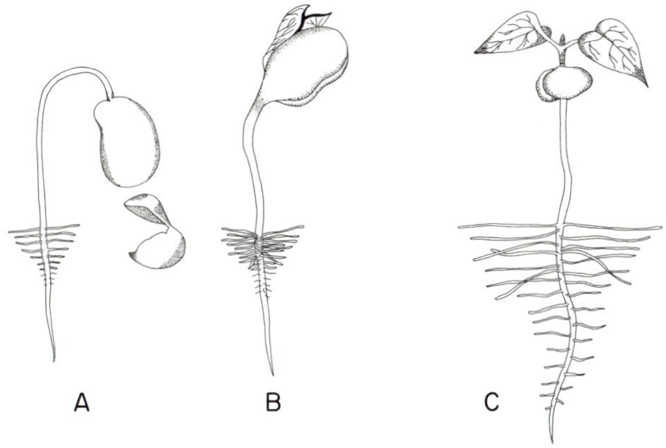

FIGURE 21 Germination of bean seed. Note the dropping of seed coat in A; the emergence of the first leaves in B; in C the first leaves have formed and the cotyledons which were carried above ground are beginning to wither.

a few inches above the ground, leaving the hypocotyl to form not only the root system but also the basal part of the stem. After the hypocotyl has become partially established the epicotyl begins to elongate and form the first true leaves. Soon after the first leaves are established the cotyledons wither and fall from the bean stem.

The castor bean (Fig. 22) is often used in germination studies to demonstrate a different behavior in dicotyledons. The castor bean seed is the one from which we extract castor oil. It is not in the legume family in which peas, lima beans, etc., are classified. At one end of the castor bean seed is a large spongy structure, the **caruncle,** derived from the seed coat. Beneath the caruncle are the hilum and micropyle and along one side is a well marked raphe.

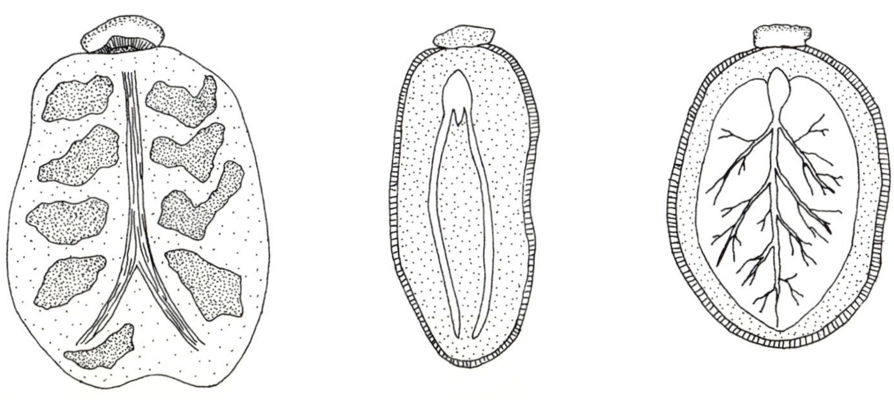

FIGURE 22 Castor Bean. A-External view of seed coat showing markings which are quite variable in pattern. The small raised portion at the apex is a bibulous structure termed the caruncle; B-Section through the seed showing the two thin cotyledons, epicotyl and hypocotyl; C-section through seed showing a cotyledon in flat view. The cotyledons are white and leaflike.

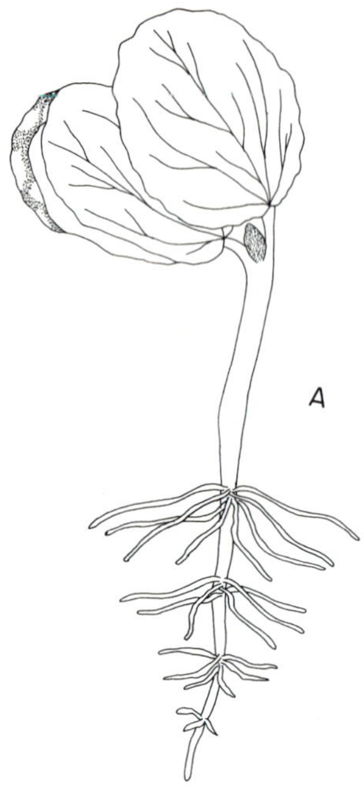

When the castor bean seed germinates, the hypocotyl with its radicle emerges first and becomes established (Fig. 23). The food which is stored in the endosperm is not used in this plant until the embryo begins to germinate. The cotyledons are leafy and membranous; once they have emerged into the light, chlorophyll develops and the cotyledons become photosynthetic. The epicotyl soon develops the shoot and the typical palmate leaves. The cotyledons often remain on the plant stem for several weeks before dropping.

A third type of dicotyledonous seed, the pea, has the same parts as the typical leguminous bean but upon germination the cotyledons remain in the ground (Fig. 24). The shoot and roots become established very quickly with relatively little elongation of the hypocotyl.

Monocotyledonous seeds also vary in certain structural features and in the manner of germination. The seed of a grass plant, such as corn, is really a fruit since the ovary wall is the outside covering. A longitudinal section of a grain (Fig. 25) discloses a rather large cotyledon **(scutellum)** with the main axis of the embryo

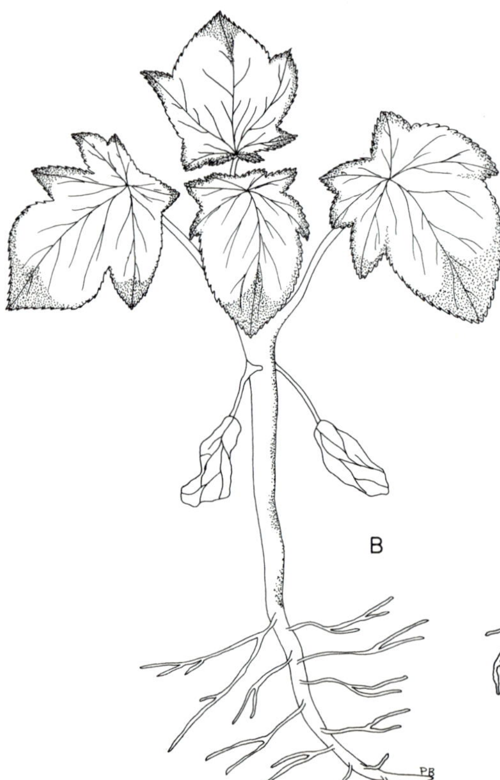

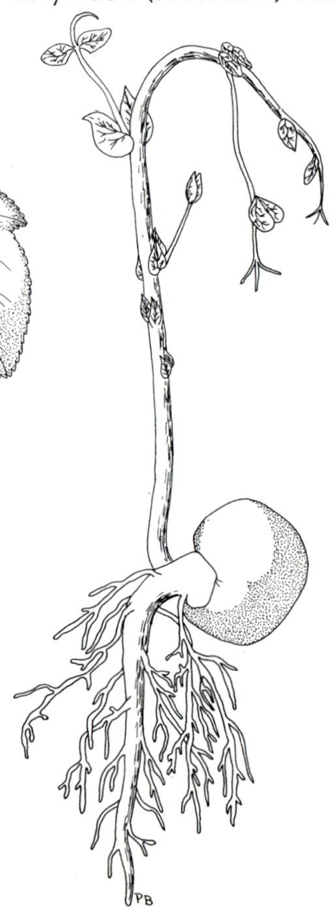

FIGURE 23 Castor bean seedlings. A-Note large leaflike cotyledons and small epicotyl. B-The typical leaves have formed and the cotyledons have withered.

FIGURE 24 Pea seedling. In this plant the cotyledons and seed coat remain below ground.

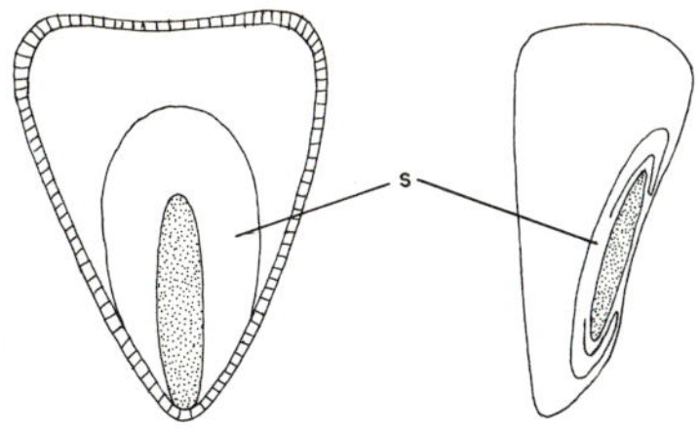

FIGURE 25 Corn grain. *Left:* Ventral view showing embryo. *Right:* Lateral view showing embryo at right; the remainder of the seed is endosperm. The pericarp is not included on the right. (s-scutellum)

attached to its side. This portion of the embryo consists of a very much shortened epicotyl or shoot tip covered by a protective cap termed the **coleoptile;** opposite this structure is the tiny hypocotyl or root tip covered by its protective sheath, the **coleorhiza.** At the time of germination (Fig. 26) both tips elongate, the primary roots

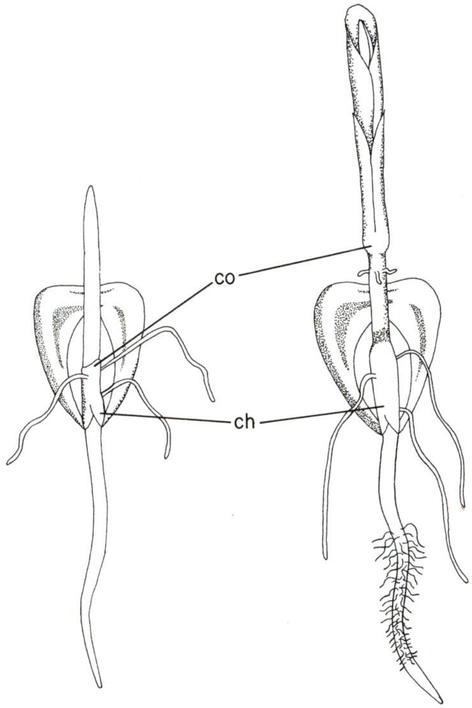

FIGURE 26 Stages in the germination of corn. Note the early development of the adventitious prop roots just above the seed. co-coleoptile; ch-coleorhiza.

becoming established in the soil and the shoot forming the initial leaf. Both coleoptile and coleorhiza elongate for a while but soon are penetrated by the growing stem and root tips and eventually are left behind and wear away. The scutellum remains in the ground with the endosperm, the fused pericarp and seed coat, and is absorbed with the endosperm as the main axis grows.

Storage Products of Seeds

In most seeds the foods are stored in the embryo or the endosperm or both. Of all the principal plant foods the most important come from the endosperm and embryo of grains. Although grains are technically termed fruits the biggest part of the grain is seed. More people in the world depend on the rice grain than depend on any other plant food. Second to rice is wheat, and third is corn. Other seeds which rank in the top twelve most important plant foods are beans, soybeans, and coconuts.

The most common food stored in the seed is a carbohydrate. Usually this is in the form of a starch but some seeds such as sweet corn, peas, and some nuts store quantities of sugar. One of the reasons we prefer to eat sweet corn and peas when they are young and fresh is that they have about 35 percent sugar of total carbohydrates stored but as the seeds mature this sugar gradually changes to a starch and the seeds are then less flavorful.

Fats are often stored in the embryo of a seed; in some, fats are present in such quantity as to be the principal storage product. Economically seeds which provide us with oils include castor beans, corn, flax, peanuts, cotton, soybeans, and many others.

Proteins are found to a lesser extent in seeds than are the other two major foods, but many strict vegetarians obtain enough proteins from seed and other plant sources to construct a healthy diet. Most of the protein in wheat is in the germ (embryo), and many of the legumes store fairly large quantities of proteins in the embryo as a reserve food. Nuts also usually have fairly large amounts of proteins stored in the cotyledons.

Dissemination of Seeds

Many agencies help in the dispersal of seeds and their transfer to other areas. The most obvious is wind. Seeds (and fruits) often have modified appendages derived from the seed coat or fruit coat (Fig. 27) which make it easy for wind to move them considerable distances. Winged seeds and fruits include those of maple, ash, elm, and some pines. Epidermal hairs form tuftlike parachutes on the milkweed, cottonwood, and others. Dandelions and lettuce have very tiny fruits which produce small umbrella-like parachutes; long plumed, persistent styles are characteristic of the fruits of *Anemone* and *Clematis*. Some seeds, such as those of ground cherry, have a large, inflated membranous envelope surrounding them, and then there are many seeds so small, so light, or so flat—

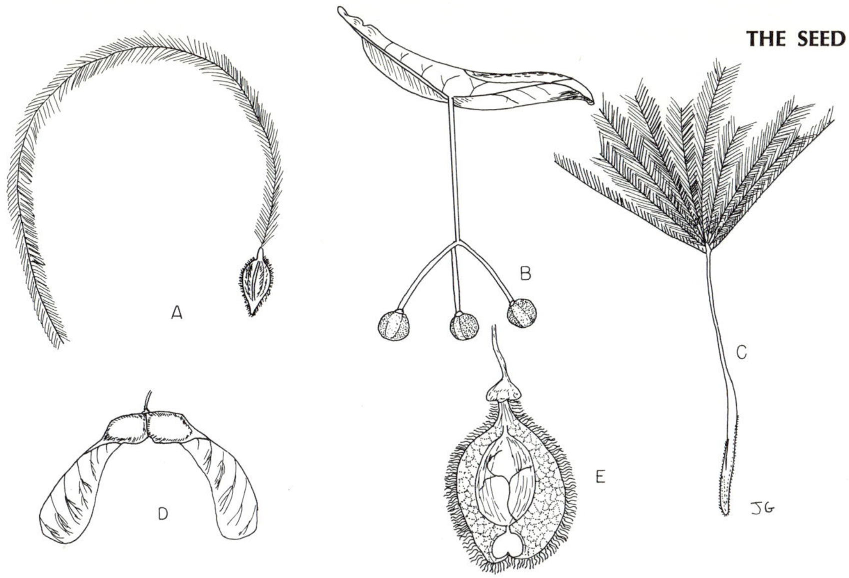

FIGURE 27 Modifications in seeds and fruits for wind dissemination. A-plumed style of *Clematis,* leafy bract of *Tilia;* C-parachute on style of dandelion; D-extended fruit coat of maple; E-extended fruit coat of elm. After Weatherwax, P. 1956 *Botany.* McGraw-Hill.

such as those of mustard, orchids, and yuccas—that wind can easily transfer them.

Water may also transfer the seeds of many plants such as the coconut, with its highly impervious fruit wall, and the tiny seeds of sedges which are enclosed in membranous envelopes. Animals, including man, transfer numerous seeds which have spines, awns, or barbs that stick to fur or clothing and are carried to new areas where they may be picked off and dropped. Other seeds are transported because they are edible or have an edible fruit coat. Squirrels and many other animals carry nuts, apples, and other fruits away for storage or immediate use. Some seeds and fruits, such as grains, pass through the digestive tract and remain viable. Some, like those of mistletoe, have a viscid covering and are carried on the feet of birds, or even of insects, and other animals.

Then there are explosive fruits like the squirting cucumber, the touch-me-not, and many legumes. This type of fruit breaks open or unwinds with sufficient force that the seeds are propelled a distance of several feet. Given a period of many years, a species with only this method of dissemination could cover a large area.

Factors Affecting Germination

Seeds, once mature, will usually germinate when conditions both externally and internally are favorable. Externally the conditions necessary for germination are: (1) an adequate supply of water, (2) an optimum temperature, (3) an adequate supply of oxygen, (4) presence or absence of light for a few species. Some seeds will germinate when they are submerged in water; others will not, probably because too much water usually means too little oxygen. Many of our garden vegetable seeds, such as radish, lettuce, and peas, are classified as cool weather plants and will germinate when the soil temperature is in the thirties or forties, Fahrenheit. Other plants like zinnias, cotton, tomatoes, and corn germinate best when the soil temperature is in the fifties or even above.

Internal conditions affecting germination include such factors as a seed coat impervious to water or oxygen, a seed coat so resistent that the embryo cannot crack it open, an immature embryo, insufficient stored food, or too short a period of dormancy. In nature the seed coat is gradually worn away by the chemical substances in the soil, but we frequently can hasten germination by filing, cracking, or soaking the seeds in water or chemicals to hasten the entrance of water and oxygen. The so-called dormancy or after-ripening period required by some seeds may be a period of weeks or several years and can be affected by changes in temperature.

Quality of Seeds

Quality in seeds is usually recognized by their percentage of germination. Grains are usually at their peak percentage of germination for about one year. Then they gradually decline in percentage of germination so that by the time they are five years old the number of viable seeds left is quite low. These seeds could still be used for animal food, but for use in industries which require fermentation products, as in the manufacture of alcoholic beverages, seeds with a high percentage of germination are necessary. Stories that seeds remained viable after storage for thousands of years in the Egyptian pyramids are without foundation. Generally seeds in their first year or two of maturity are at their highest percentage of viability. Many seeds will not live longer than that, but we do have authentic records of the large lotus seeds surviving for over two hundred years.

Purity of Seeds

For a planter, one of the important considerations about seeds concerns their percentage of purity. For example, if you are planting a Kentucky bluegrass lawn you must have about 99 percent purity in bluegrass seeds, with less than one percent extraneous seeds. A ratio of 90–10 might be much cheaper, but it will cost much more in weeding and care to develop a good lawn. One store visited recently offered five pounds of grass seed for ninety-eight cents. The fine print on the label disclosed the contents as 87.5 percent annual and perennial rye, 1.5 percent Kentucky bluegrass seed; 5 percent meadow fescue, small amounts of red top and clover and "less than 5 percent weed seed."

Anyone planting this seed would have a rough, irregularly textured lawn with numerous weeds to dig out, and then would have to replant much of the lawn the next year because of the big percentage of annual grass seed it contained. Instead of buying a grass seed mixture at twenty cents a pound it is really much cheaper to pay seventy-five cents for a pure, high-quality, guaranteed seed. Always read the fine print on the label; the law requires that the purchaser be informed of the content.

Fruit

A fruit is a ripened ovary, often with closely associated floral parts. While we may usually think of the fruit as a juicy, succulent structure forming after pollination and fertilization, it is variable and not always a clearly differentiated organ. Botanically a fruit is merely the housing for the ovules or seeds. In some cases such as that of the grain (caryopsis), the housing wall is no more than a membrane which is fused with the seed coat. In other plants such as the coconut palm the ovary wall is differentiated into several layers and may be several inches thick.

The three distinct layers of the pericarp, when present, are termed endocarp, mesocarp, and exocarp. A peach, for example, has a very hard, bony endocarp surrounding the centrally located seed; the rich, juicy yellow or orange pulp of the peach is mesocarp, and the thin, fuzzy skin is the exocarp. Fruits having these structural features, like olives, cherries, plums, and apricots, are classified as **drupes.**

The most obvious thing about the growth of the fruit is that it involves a very large increase in volume. The increase in size is due partly to increase in number of cells and partly to increase in cell size. Some fruits also have a considerable increase in the number and size of intercellular spaces. In general, growth by cell division predominates in the early stages, and growth by cell expansion and increase in intercellular spaces in the later stages. In the tomato and many others, cell multiplication ceases soon after the fruit is formed and all subsequent enlargement is the result of cell expansion.

Shape of the fruit is usually predetermined, but in many plants differences in shape arise as the result of unequal growth rates so that the length-breadth ratio changes continuously during growth. Factors such as unequal amounts of water and nutrition, unilateral light, pressure or gravitational forces, and heredity may affect size and shape.

Internally fruit size and shape may be determined by the number of seeds or ovules fertilized. In pears, the earlier the seeds abort in their development, the more elongated is the fruit. Occasionally, as the result of frost at blossom time, the seeds are killed at a very early stage and the fruits that develop are sausage-shaped rather than pyriform. A similar situation exists in strawberries, where the fruit will develop well only in the vicinity of developing seeds that have been fertilized. In flowers where there has been poor or unequal pollination the strawberry fruit is often quite distorted.

FIGURE 28 A-Section through the flower of a strawberry. In B the central receptacle has enlarged to form the fleshy fruit, the small pistils have become achenes and are now much farther apart as the result of receptacle expansion; the stamens and petals have dropped off and the sepals remain as a protective cup around the base.

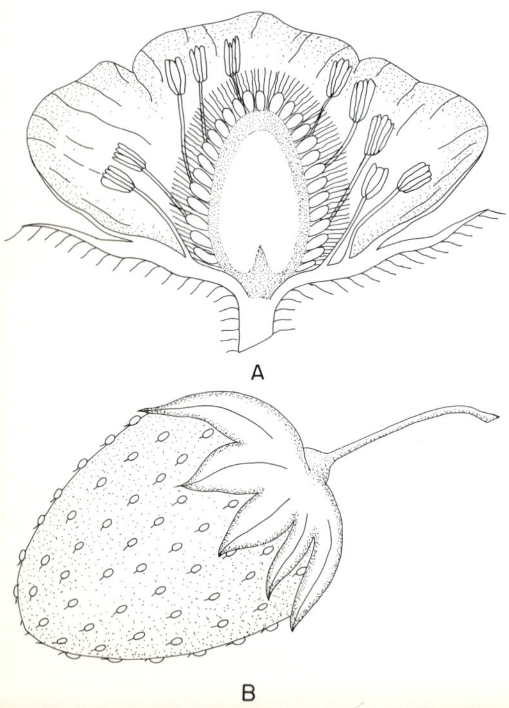

Development of the fruit is quite different from that of the stem and root. In these vegetative organs apical meristems largely determine the shape and extent of growth. A fruit has no apical meristem, but it is the opinion of many botanists that enlargement is influenced by two very active internal meristems in the developing seed. These are the growing embryo and the endosperm. These tissues produce auxins which apparently diffuse out of the seed coat and directly stimulate the growth of the surrounding fruit tissues. Synthetic and extracted auxins have been used successfully for a number of years on developing fruits to increase the growth rate and size of fruit. It is of special value where the developing floral organs are partially sterile, where there has been inadequate pollination, or where the seed content and growth rate of the fruit are below normal.

The various parts of the fruit can be traced back to those of the ovary and to other floral organs which enlarge to become part of the fruit (Fig. 28). Some of the different fruits are listed below.

I. **Dry fruits:** These are often divided into two categories, dehiscent and indehiscent, depending upon whether or not they split open along regular lines at maturity to release the seeds.

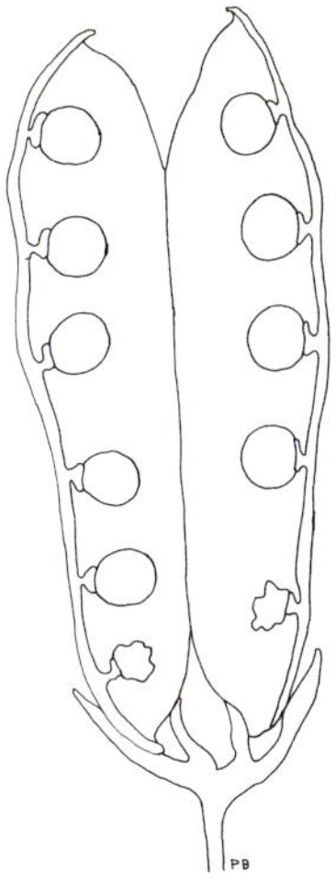

FIGURE 29 Legume, a fruit typical of the Leguminosae family. It has one carpel but splits open along two sutures.

FIGURE 30 An unusual legume is the screwbean produced by *Prosopis odorata*.

A. **Dehiscent**

1. *Legume*—a simple fruit composed of one carpel which at maturity dehisces along two lines (sutures). Examples are beans, peas, and locust trees. (Fig. 29–30)
2. *Follicle*—a simple fruit of one carpel which splits open along one side only. Examples are milkweed, catalpa, and magnolia.
3. *Capsule*—a compound fruit composed of two or more united carpels (Fig. 31). Septicidal capsules, like the *Azalea and Rhododendron,* split along the line separating the carpels. Loculicidal capsules like *Iris* and *Lilium* split at the locules (weakened areas in walls away from the septa). *Poricidal* capsules like the poppy (Fig. 33) and water lotus have pores at the apex of the capsule through which the seeds are liberated.
4. *Silique*—a compound fruit of two carpels which separate at maturity. This type of fruit is formed by members of the mustard (Cruciferae) family, such as radish, cabbage, tansy mustard, and shepherd's purse.

B. **Indehiscent fruits**

1. *Achene*—a small fruit formed by the complete fusion of two carpels with a single seed in the center. In an achene the pericarp is free from the seed coat and joined only by the funiculus at the narrow end of the fruit to the placenta. Examples are chiefly members of the Compositae family such as sunflower, lettuce, and dandelions.

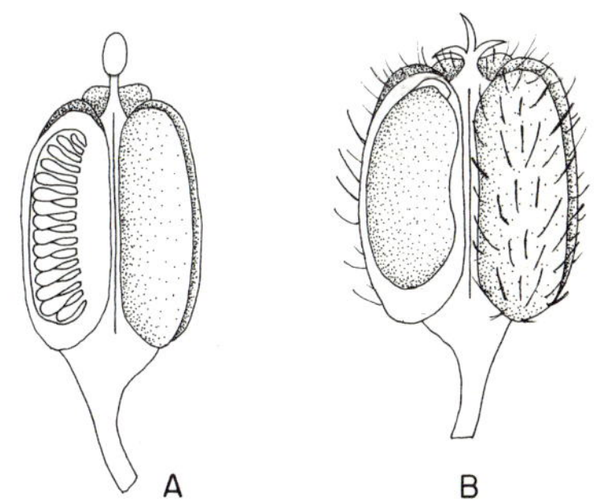

FIGURE 31 Capsules
A. Yucca capsule, with rows of flat seeds in the one opened carpel. B. Castor bean with a single seed exposed in the open carpel.

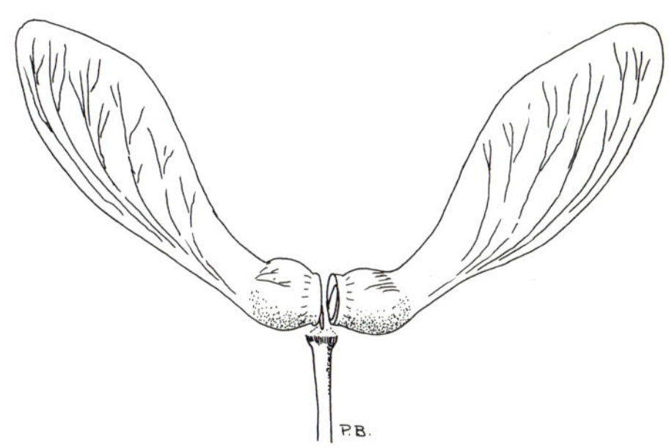

FIGURE 32 Samara of maple.

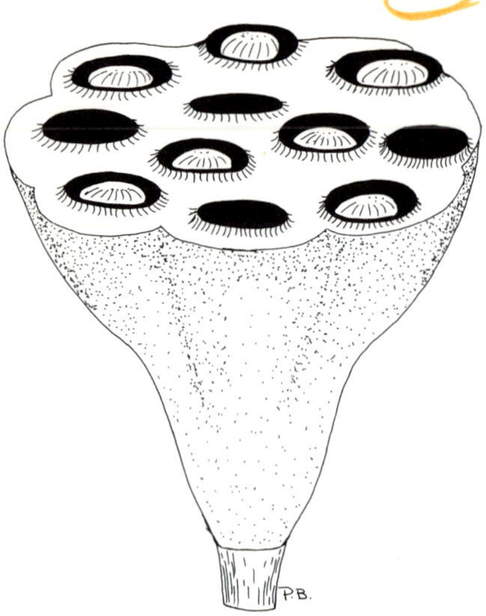

FIGURE 33 Poricidal capsule of poppy.

FIGURE 34 An acorn is an example of a nut. The cup, within which the ovary enlarges, is an involucre formed from the coalescence of bracts.

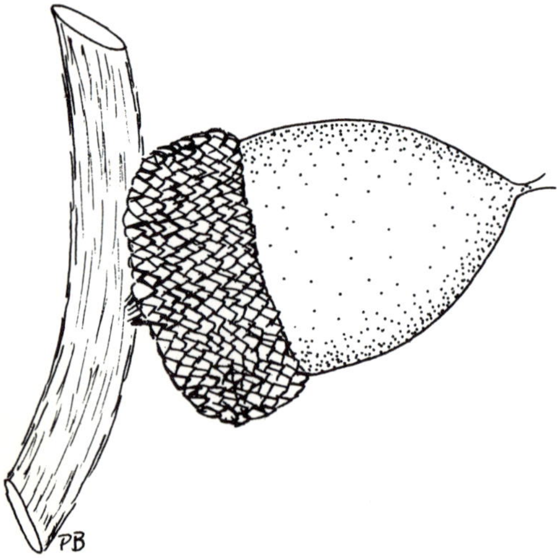

2. *Grain or Caryopsis*—in this fruit the pericarp is fused with the seed coat all the way around. Cereals and all other grasses are examples.
3. *Samara or key*—this is a one- or two-seeded fruit in which the pericarp forms a wing (Fig. 32). Examples include the elm, in which the pericarp is more or less circular, the ash, in which the pericarp is elongated, and the maple, where the entire structure is somewhat key-shaped.
4. *Nut*. The botanical definition of a nut requires that this fruit be one-seeded, with a hard, stony pericarp and associated floral bracts (collectively termed involucre) that fuse to form a cap in the case of an acorn (Fig. 34), a bur or prickly involucre in chestnuts, and a husk in the walnut. Brazil nuts are seeds which form in a large poricidal capsule; an almond is really a seed produced in a drupe; and a peanut is a seed produced in a legume. Therefore these seeds are not botanically classified as nuts.

II. **Fleshy fruits**

A. **Drupe.** As described above, the drupe has a single seed with the pericarp divided into three layers. In addition to the examples listed, avocados, mangos, and apricots are also drupes. (Figs. 35, 36)

B. **Pome.** This type of fruit is often called a false fruit because the major part is developed from the base of the floral tube and the upper portion of the receptacle. The thin outer portion of the core is the ovary wall. Most pomes are many-seeded and five-carpeled. Examples are apples, pears, quince, and many other members of the rose family.

C. **Berry.** This type of fruit is developed from a compound ovary in which the endocarp and mesocarp are very juicy and the exocarp is skinlike. Tomatoes, grapes, dates, and cranberries are examples. Another type of berry is a hesperidium, in which only the endocarp is juicy while the mesocarp and exocarp fuse to form a rind. Citrus fruits like the orange, lemon, and grapefruit are typical examples. A third type of berry is a pepo, in which the mesocarp and endocarp form the fleshy part, and the exocarp is fused in part with receptacle tissue to form a rind. Members of the gourd family such as watermelon, cucumber, and squashes are examples. (Figs. 37, 38, 39)

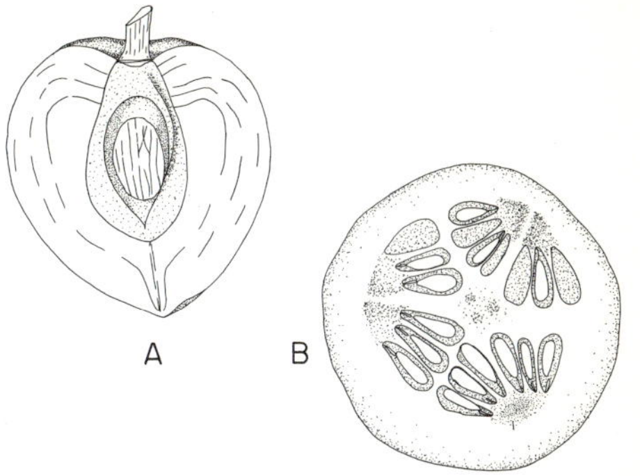

FIGURE 35 Avocado. This soft fruit has a large central seed surrounded by a hard endocarp.

FIGURE 36 A-Drupe of coconut; B-cross section of cucumbers-a pepo type berry.

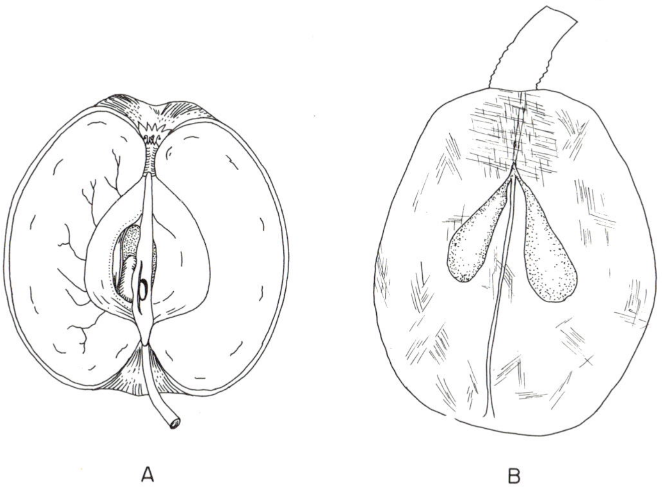

FIGURE 37 A-longitudinal section of an apple, a typical pome fruit; B-longitudinal section of a grape, a true berry.

FIGURE 38 Fruit of the papaya-a berry.

FIGURE 39 *Left:* the hesperidium of a tangerine; *right:* the hesperidium of a lemon.

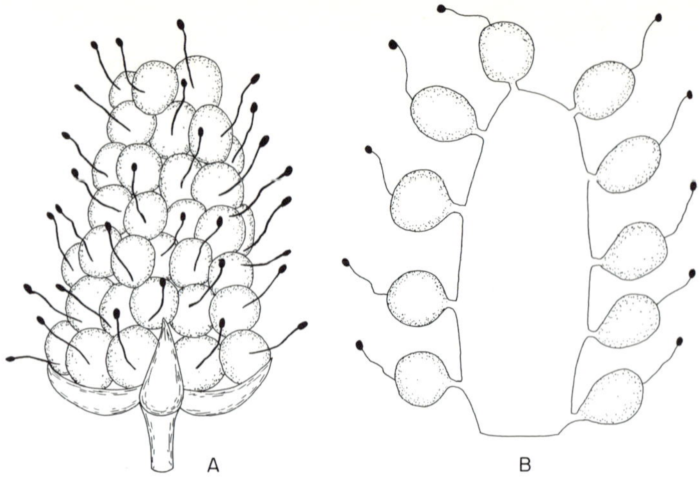

FIGURE 40 A-fruit of raspberry; B-same in vertical section.

FIGURE 41 Multiple fruit of pineapple; diagram shows fruit with fused ovaries.

D. **Aggregate fruits** are formed from many pistils in a single flower. Raspberries (Fig. 40) and blackberries have clusters of drupelets on a small but common receptacle. The strawberry has many achenes developing on a large receptacle. These are a few examples of aggregate fruits.

E. **Multiple fruits** have a number of fruits, usually nutlets, fused together to form a single fruit. In each case it is the fusion of floral parts from a number of flowers on a receptacle. Pineapples (Fig. 41) and mulberries are typical. Another type of multiple fruit is a syconium, like the fig, in which small flowers are produced on the inner wall of the receptacle. A Smyrna fig has pistillate flowers growing on the inside of a bulbous receptacle which must be fertilized by a tiny wasp. The wasp enters the small opening at the apex of the receptacle to lay its eggs. Prior to entering the Smyrna fig the wasp has visited the staminate flowers of the caprifig, where it has become covered with pollen. From the flowers of the caprifig the wasp carries the pollen to the pistillate flowers of the Smyrna fig, making it possible for fertilization to occur and the fruit to develop.

Economic Products of Flower, Fruit, and Seed

The importance of the cereals has been mentioned insofar as food is concerned. Many of our cereal foods are made from the bran layer, which consists of the pericarp, testa, and nucellus of the grain. Patent or white flour is made from the endosperm, and germ is from the embryo. In addition, oils are obtained from the grains and other seeds. The importance of our many fruits for food is obvious, but the many other products include spices like caraway, anise, and celery from the seed and tiny fruits; clove oil from dried flower buds; an insecticide from the flowers of pyrethrum; vanilla from the pod of the vanilla orchid; opium from the capsule of the poppy; and perfumes from the petals of many flowers such as tuberose, rose, gardenia, jasmine, and lavender.

Supplementary Readings

Dittmer, H. J., and R. D. Spensley, *The Developmental Anatomy of Descurainia Pinnata Ochroleuca.* University of New Mexico Press, Albuquerque, 1947.

Hillman, W. S., *The Physiology of Flowering.* Holt, Rinehart, and Winston, New York, 1962.

Mason, H. L., The Concept of the Flower and the Theory of Homology, *Madroño* 14:81–95, 1957.

Salisbury, F. B., *The Flowering Process.* Pergamon Press, Oxford, England, 1963.

Wardlaw, C. W., *Embryogenesis in Plants.* John Wiley & Sons, New York, 1955.

Wilson, C. L., and T. Just, The Morphology of the Flower, *Botanical Review,* 5:97–131, 1939.

Questions for Discussion

1. What are the advantages and disadvantages of wind pollination to plants?
2. Clover is usually pollinated by bees. What would be the future of clover crops if the bees were all killed by an insecticide?
3. How are such plant seeds as those of sandburrs and cockleburrs disseminated? How about the seeds of apples? Of dandelions?

4. How may plants produced from seeds resulting from cross pollination differ from those produced from clones?
5. Why do apple growers usually keep bees in the orchard even though apples are usually wind-pollinated?
6. Are flowers with long, narrow tubular corollas more likely to be pollinated by bees or by butterflies and moths?
7. Night-blooming flowers are usually white, while those of day-blooming species are usually more brilliantly colored. What would this indicate to you?
8. What do we mean by double fertilization?
9. Of what advantage is it to the plant to have seeds which undergo a dormancy period before germinating?
10. What conditions are necessary for the germination of seeds?
11. Flower stalks often elongate rather rapidly after pollination and fertilization. Of what advantage is this to the plant?
12. In many compound inflorescences the flowers mature at different times. Of what advantage is this to the plant?
13. How does filing a small spot on a seed hasten germination?
14. Is it more likely that regular or irregular flowers would be insect-pollinated? Explain.
15. How would plants developed parthenogenetically differ from those developing from a fertilized egg?
16. In what ways is the development of a plant embryo similar to that of an animal? How is it different?
17. How may seeds be stimulated to germinate?
18. Why are young edible seeds often more tasty than mature ones?
19. How are seeds and fruits especially modified for dissemination?
20. In buying seeds for a lawn why are purity and percentage of germination important factors to consider?
21. What are some of the products of economic importance we obtain from fruits and seeds?
22. Are most tree seeds adapted for dissemination by wind or by barbed appendages? Explain how this adaptation is an advantage to dissemination of the species.
23. Describe variations in the germination of different seeds.
24. How are some flowers especially adapted for insect pollination?
25. Why will some seeds such as water lotus germinate when submerged in water while others will not?

12

Growth

Growth has been defined as an irreversible increase in size. However, this definition by no means explains the complex processes involving assimilation, the formation of new protoplasm and new walls, and the development of tissues and organs. In growth two general processes are involved: (1) **morphogenesis,** which is the development of external form of new organs such as roots, stems, and leaves, and (2) **differentiation,** defined as the gradual change into the various tissues and functions acquired by cells and tissues from their common origins in the meristems.

Growth in plants and animals is similar in that both are based on the division and differentiation of cells, but they are different in that animals grow relatively uniformly all over, while vegetative growth in the higher plants occurs chiefly from apical meristems in the shoot and root tips and from lateral meristems, namely the vascular cambium and cork cambium which add girth to the stem and root.

Mitosis and Cell Division

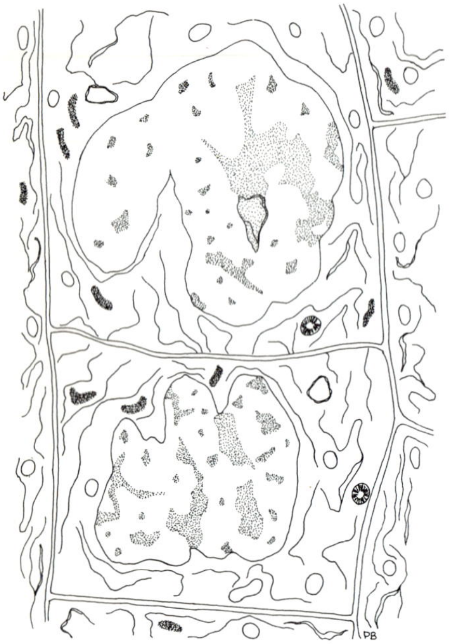

FIGURE 1 Daughter cells in onion root tip. The large central bodies are young nuclei forming. The dark stippled bodies outside the nucleus are golgi bodies; one mitochondrion appears in cross section in each cell. Portions of a forming endoplasmic reticulum appear as slender lines. Drawn from an electron micrograph in Bonner, James and J. E. Varner 1965. *Plant Biochemistry*. Academic Press.

One of the most important factors in growth is the division of cells, increasing the number of cells as well as achieving an increase in size of the organism. As aging occurs there is a reduction in cellular division, but in some tissues it continues throughout the life of the organism. Throughout the organic world cell division is essentially the same whether we are considering animals or plants. In all organisms the division is centered about the nucleus or nuclear material, but it is well to remember that the cytoplasm also undergoes changes that cannot be ignored in the division process (Fig. 1).

Plants have their greatest activity in cell division, in spore production, in the stem and root tips, and in the lateral meristems of the cambium, cork cambium, and pericycle.

Phases in Cell Division

In order to see the progressive steps which occur in mitosis tissues are cut by means of a microtome to a thinness of ten to fifteen microns and stained with a variety of dyes to achieve contrast in the cellular organelles. For convenience in discussing the entire process cell division is generally thought of as having five phases, but it must be emphasized that the process is continuous (Fig. 2). Essentially mitosis is division of the nucleus and its chromosome material, but in most cases it is accompanied by new wall formation and therefore cell division.

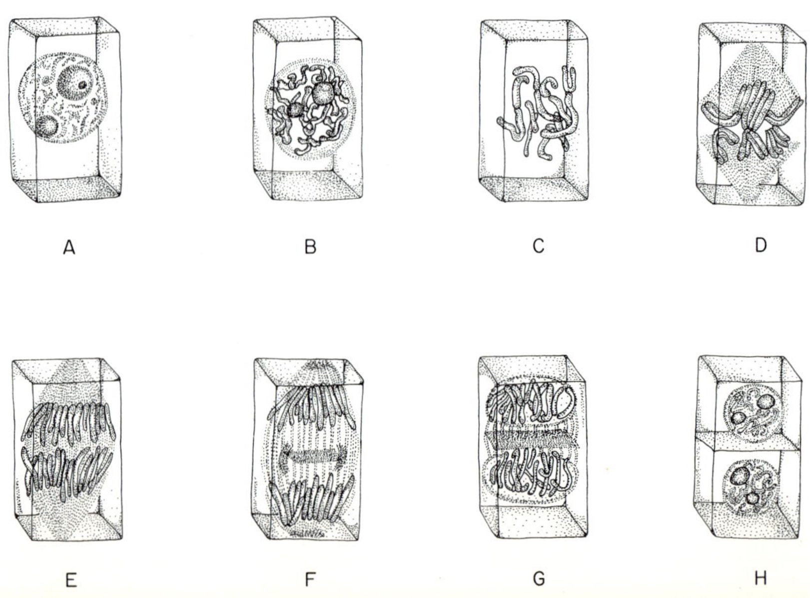

FIGURE 2 Mitosis. Selected stages in equal nuclear and cell division in growth of the onion root tip. A-interphase, B-early prophase, C-late prophase, D-metaphase, E-anaphase, F-early telophase, G-late telophase, H-Daughter cells. Redrawn from *Laboratory Outlines in Biology*. (W. H. Freeman and Company. 1963.)

Interphase This is frequently erroneously termed the resting stage. Actually it is a time of great metabolic activity; the cell is getting larger and reaching a stage of division. During this time the nucleus has gotten much larger in comparison to nondividing cells. Large molecules of nucleic acids and proteins are being synthesized just prior to division. At this time the nucleoli and the fine network of chromatin, composed of the nucleoproteins, which stain readily, are clearly visible but the individual chromosomes are not identifiable. In late interphase the chromatin begins to form long, thin threads.

Prophase The cell is in prophase when the chromosomes become visibly distinct as long, thin threads. Each chromosome is longitudinally doubled at this stage and consists of two identical strands termed chromatids. This process of halving was once described as a "splitting" of the long threads, but we now know it to be a replication of the parent chromosome through a DNA synthesis in the nuclear material. In late prophase the coiled chromatids become increasingly visible, probably as a result of water loss, making the chromatids denser and more readily stained. With this loss of water the chromatids shorten, the coils decrease in number, and the paired chromatids become prominent. In late prophase the nucleoli become smaller and ultimately are no longer visible. The nuclear membrane is also disintegrating at this time and it soon disappears.

Metaphase Up to this point the steps described involve only mitosis, but with the disappearance of the nuclear membrane a new structure, the **spindle,** begins to form in the cytoplasm. This structure consists chemically of long-chain protein molecules oriented between the two poles (opposite ends) of the cell. The spindle strands are fine tubules rather than fibers and are formed from a portion of the cytoplasmic material. Chemical analysis has shown that about 15 percent of the cytoplasmic proteins become spindle. Once the spindle is formed the chromosomes move toward the spindle tubules and become attached to them by organelles centrally located on each chromosome, called the **centromeres.** The chromosomes at this time are in the center of the cell, oriented in a flat platelike position.

Anaphase With the chromosomes in a central position and attached to the spindle by their centromeres, the chromatids separate. One chromatid from each chromosome moves along the spindle in opposite directions to the pole. The centromeres are a controlling factor in chromatid movement; without them the chromosomes cannot be oriented on the spindle and the chromatids cannot separate. Like all phases, anaphase may be divided into early, middle, and late depending on the position of the chromatids in relation to the equatorial plane and the poles. We may consider anaphase as ended when the chromosomes, which can now be recognized as individual chromatids or chromosomes, reach the poles and form densely packed clusters.

Telophase Soon after the chromosomes reach the poles, a nuclear membrane begins to form, the chromosomes uncoil forming slender threads again, and the nucleoli gradually appear. All of the stages of nuclear behavior as we saw them in prophase are now re-enacted in reverse order as the nucleus begins to take on the characteristics of interphase. While the nucleus is reforming, a new cell wall is formed. At first this wall, which began to appear in anaphase, is formed within the broad central area of the spindle fibers and is called a **cell plate,** but it soon develops beyond this area and forms a dividing wall between the developing daughter cells. By this time the spindle fibers have disappeared, the cytoplasm is becoming reorganized, and the nucleus is enlarging.

Controlling Factors

Many factors have some control in cell division. Age, health, and type of tissue have been mentioned. It is also believed that growth-stimulating substances exert a strong effect not only in initiating the division but also on the rate of division. Temperature has been demonstrated as a controlling force in the division of the terminal hair cells of *Tradescantia.* If the temperature is 10°C it takes the cell 135 minutes to complete division, but if the temperature is 45°C it takes only 45 minutes. It has also been demonstrated that withholding food from certain unicellular organisms will force them into a state of starvation during which they will not divide. Bacteria, under favorable conditions, will divide every 15 to 20 minutes and yeast cells may bud and reproduce even faster. Observations of root tip cells indicate that a cell will divide when about 20 hours old, but this is a quite variable factor since some plants grow better under cool temperatures while others grow much faster when the soil temperature is above 24°C.

Importance of Mitosis

The significance of mitosis and cell division is apparent when we realize that all cells, those of our bodies as well as those of a tree, have been formed by this process. Mitosis also ensures a continuous succession of cells endowed with the same characteristics as the original. Organized growth would not be possible if the cells forming an individual or a species were not similar in quality or did not transmit these qualities to their successors.

The nucleus is the control center of the cell. Since it is made up of chromosomes composed of intricate patterns of nucleic acids and proteins, of a specific type for each species, we have in the nuclei the regulators of metabolic activities and the determiners for the structural characteristics of that species and for the individual organism. The remarkable thing about mitosis is that the process occurs with such exactitude. The longitudinal duplication of the chromosomes into identical chromatids occurs time after time for millions of generations, with each pair of resulting daughter cells exactly like its parent down to the minutest degree of chromatin material.

Unanswered Questions

There are many things we still do not understand about mitosis. For example, why do cells divide? What causes them to differentiate into different tissues? Why do they get old and die?

In unicellular organisms we can demonstrate that size is important to division. If they are kept small by removing tiny bits of tissue they will not divide. If the nucleus is removed they do not divide. Ordinarily when one of these organisms reaches adulthood and achieves maximum size it divides. Yet in a multicellular organism size is a predetermined factor. Plants and animals reach a size characteristic of that species, and increase in size no longer takes place. Mitosis still continues in most tissues but only at a pace adequate to repair and replace worn-out cells, and even that pace of division is reduced with old age.

For most cells division is not only a growth or replacement factor; it is also a stimulus to rejuvenescence. The individual cell has itself reached maturity, and then by division it can start the process of youth, maturity, and aging all over again. As each new cell grows new substance is added, new protoplasm is replacing old, and the cell is a new, efficient metabolic unit.

We have discussed cell division as a duplicating process in which the daughter cells are exactly like the parent, but what about those in the multicellular organisms which arise in the same duplicating manner but differentiate into quite different cells, both in appearance and function? The nucleic acids are the same and the

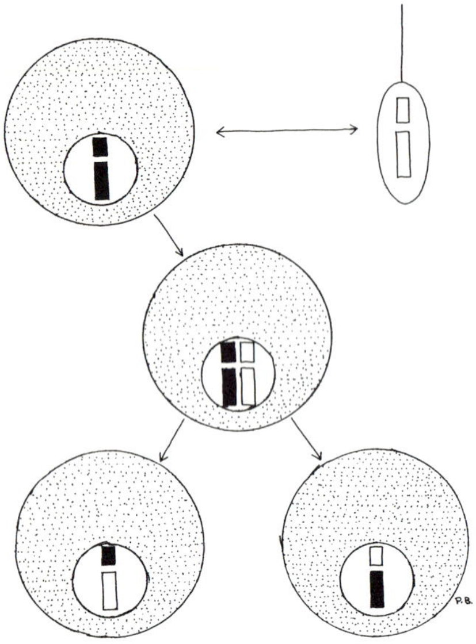

FIGURE 3 An illustration of sexual reproduction. The union of egg and sperm at top results in a combination of unlike chromosomes in the new individual, center. The eggs produced by the new individual have a different genetic constitution from either of the original sex cells.

chromosome number remains the same, but some coded information provided by the DNA molecules exercises control over the cell and directs its destiny. The actual potential for differentiation is determined within the cell long before any visible change takes place; once these chemical alterations have occurred the cell's future is committed (Fig. 3). In the plant, embryonic cells of the tip meristems and cambium all are virtually the same, but within a day or two many of them begin to differentiate into cells and tissues which are radically different from adjacent tissues which had the same origin. For example, cells of the vascular cambium, which are small, thin-walled cells appearing brick-shaped in cross section, divide and ultimately differentiate into xylem cells and phloem cells. Yet the cambium itself always remains as a continuous cylinder of cells separating the thick walled water-conducting cells it has produced from its inner face from the thin-walled food-conducting cells it has produced by division from its outer face. The coded information the chromosomes carried in the inner cell following replication from the mitotic division of the cambium was the determining factor which caused this cell to become a water-conducting tube (or in some cases to remain a generalized cell without specialization). What chemical changes or rearrangement of the DNA molecules takes place to initiate this type of specialization we have not yet determined. However, it seems that the cells which are formed in this manner are committed to become xylem cells long before they begin to take on the morphology and function of xylem cells.

Growth in all organisms which start as the result of the sexual process begins with a fertilized egg. In both plants and animals the cells that develop initially from the fertilized egg are undifferentiated; if this were all that occurred the organism would end up as a shapeless mass. But soon after the initial mass of cells is formed, regions of cellular differentiation begin and the embryo is in the process of development. The pattern of growth, as determined by the organism's genetic constitution, is by now becoming chemically if not physically established. For the higher animal the embryo will reach a fetal stage and finally be borne as a cub, a puppy, or an infant. The embryo of a plant will differentiate root, stem, and leaf primordia and ultimately germinate from the seed as a seedling oak, pine, aster, or whatever its genes and DNA molecules have destined. From the above few sentences it must be clear that there is unity and uniformity in all life; yet when we look at the species around us it is abundantly clear that variation is also a characteristic of life.

Environmental Effects on Growth

Internal forces are not the only ones which operate in controlling the growth of organisms. Many external factors have a very salient effect in promoting or limiting the growth of the organism. Plants probably have a closer relationship to the environment than animals, which can at least move about and protect themselves from climatic changes in the environment to a limited degree. Plants depend upon the hardiness of their individual cells and tissues to survive in the environment in which they are growing. If they are adapted to their particular habitat their survival is assured, but palms planted in a northern climate would not last through the winter, nor would cranberries from the acid soils in the North survive long in the warm, alkaline soils of the Southwest.

External factors which affect the growth of plants include minerals and their availability; soil texture and its pH; light, its intensity, and duration; moisture both in the soil and in the air; and temperature.

Soil

Although soil has been discussed in detail in Chapter 10, it should be emphasized here that the texture of the soil controls to a large extent the amount of water which it contains and how freely the water may move by capillarity from one part to another. Soil texture also has an influence on the pattern of root development and the availability of nutrient elements. A soil's acidity or alkalinity is determined to a large extent by the parent rock material from which the soil particles were formed, by the amount of organic material found in the soil, and by the degree of decomposition achieved by soil microorganisms.

Light

No other external factor has as profound an influence on the growth of the green plant as light. For this reason it is well to consider several aspects of light again. Plants vary in ability to absorb the different wave lengths of light and in growth response to the different colors of the spectrum.

The pigments chlorophyll, carotene, xanthophyll, anthocyanin, and phytochrome are quite selective in the colors as well as in the amount of light they seem capable of absorbing or reflecting. In the case of anthocyanin and phytochrome there seems to be unusual sensitivity in the absorption of the visible spectral rays. Phytochrome through its absorption of red and far-red light seems to act as a "tuning switch" in controlling many growth responses in plants.

Experiments have demonstrated that plants are most sensitive to **red** light, having a wavelength of about 6600 angstroms, and grow well when exposed to this color. In recent years a peculiar aspect of the red rays was noted when plants were exposed to **far-red light,** which has a wavelength of about 7300 angstroms, and is next to **infrared** in the spectrum (Fig. 4). Plants which normally

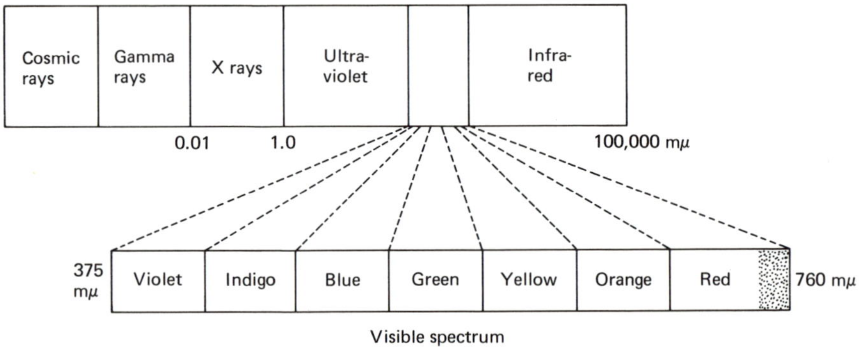

FIGURE 4 The spectrum of radiant energy. The lower box has the colors of the visible spectrum which are located between the ultra violet and infrared. The stippled area in red is the far red which begins at about 730 mu (7300 angstroms). One millimicron (mu) is one-millionth of a millimeter.

responded well by leaf expansion, deep green coloration, and shortened petioles when exposed to the normal red wavelengths were inhibited in their growth responses when the red wavelengths were immediately followed by far-red wavelengths. However, if the far-red is followed by red, then the plant resumes its normal growth activities. In fact, this can be repeated with alternating wavelengths of red and far-red and in each experiment the plant's response will be to that of the last type of wavelength received (Figs. 5, 6).

The manner in which the red rays are absorbed has been suggested by S. B. Hendricks and H. A. Borthwick of the United States Department of Agriculture, who have done considerable research in this field. They are of the opinion that the receptor pigment, **phytochrome,** exists in two forms, **R-phytochrome** and **F-phytochrome.** R-phytochrome accumulates in the plant in darkness, but becomes converted into F-phytochrome in light. F-phytochrome is

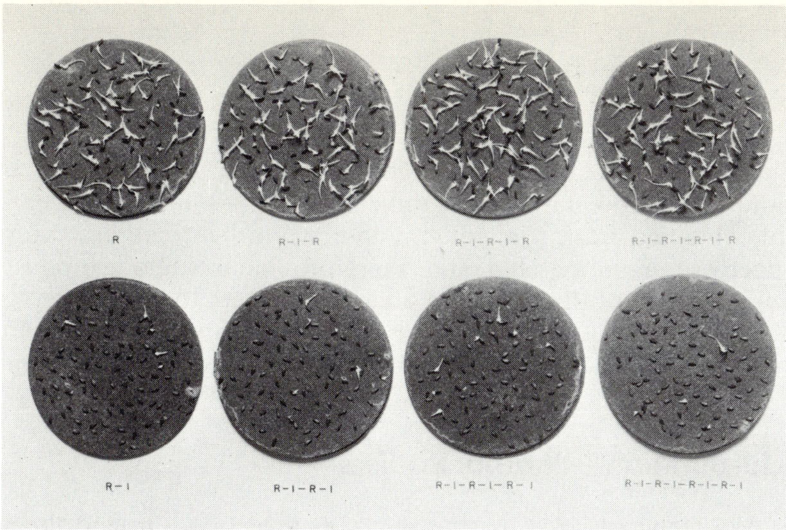

FIGURE 5 The effect of red and infrared wavelengths on seed germination. In the top row all seeds received red wavelengths as the last exposure. In the bottom row of petri dishes all seeds received infrared as the last exposure. Legends under the lots indicate the sequence and number of exposures. Red irradiation periods were for one minute, infrared exposures were for four minutes. Temperature during the one-half hour required to complete the treatment was 27°C. During all other times the temperature was 19°C. (Courtesy Agricultural Research Service, Plant Industry Station, U.S. Department of Agriculture).

then converted back to R-phytochrome. Since the reaction is reversible, red light exposure will induce the usual growth developmental responses.

$$\text{R-phytochrome} \underset{\text{far-red light (darkness)}}{\overset{\text{red light}}{\rightleftarrows}} \text{F-phytochrome} \rightarrow$$

developmental responses

A pigment has been extracted from plant tissues which shows the property of reversibility in the presence of red and far-red light. This seems to be a protein which is blue in color, but additional information must still be accumulated about its chemical makeup. It has been identified as the phytochrome that seems to influence the gibberellin system of the plant, which causes elongation of

FIGURE 6 Reversal of flowering in Kalanchoe plants. The plant on the left is the control and was given short-day conditions. The center plant received the same short-day period but the dark period was interrupted by one minute of red light exposure, which prevented it from flowering. The plant at the right was subjected to the same short-day conditions and had its dark period interrupted by one minute of red light followed by one minute of far-red. The far-red rays nullified the effects of the red rays, consequently the plant bloomed normally. (Courtesy Agricultural Research Service, Plant Industry Station, U.S. Department of Agriculture.)

stems. One effect of far-red light apparently is to increase the amount of gibberellin thus stimulating elongation, while the red light, obtained from sunlight or other white light sources, causes a decrease in gibberellin and therefore inhibits elongation. Although we know that gibberellin can be used to hasten growth responses, chiefly in elongation, it has not been proved as yet that gibberellin production does change under these light conditions.

Light-duration—Photoperiodism

We frequently discuss the effect of light intensity and the effect of the various wavelengths of light on specific growth responses. However, the length of the light period (or really the length of the dark period) may have even more striking effects upon the vegetative growth and flowering of plants. Most species have a certain critical minimum or maximum period of continuous darkness which will force them into bloom or delay this phase of their life cycle.

Plants of temperate regions are divided into three categories based upon their response to the light period **(photoperiod).** They are: (1) **long-day plants,** (2) **short-day plants,** and (3) **day-neutral plants.**

Long-day plants bloom under continuous light periods of over twelve hours. This is a very general statement, for some plants have a much more critical response than simply "over twelve hours." For some the light period must be continuous for fourteen, fifteen, or even more hours in order for them to blossom and produce seeds. Experimentally it can be shown that long-day plants will

FIGURE 7 Flowering responses of *Petunia hybrida* var. Ballerina. The plants were photographed after 64 days on 8, 9, 10, 12, 14 and 16 hour photoperiods from left to right. Note the greater amount of vegetative growth in the three to the left and the increased flowering on those to the right. (Courtesy Agriculture Research Service, Plant Industry Station, U.S. Department of Agriculture.)

FIGURE 8 *Hibiscus* plants grown under 8 hour exposure (left) and 16 hour photoperiod (right). (Courtesy Industry Station, U.S. Department of Agriculture)

FIGURE 9 *Poinsettia* grown under 8 hour photoperiod at left and 16 hour period at the right. This is an excellent example of a short day flowering plant. (Courtesy Agricultural Research Service, Plant Industry Station, U.S. Department of Agriculture)

bloom under continuous illumination, but in nature this light period does not exist except for a few weeks each year in the polar regions (Figs. 7, 8, 9).

Most of our summer flowering plants, such as sunflower, hollyhock, daisy, iris, day-lily, and sweet clover, are long-day varieties.

Short-day plants comprise a group that flower when the light period is short, "less than twelve hours." For many this may mean a light period of eight or even fewer hours of continuous illumination. Plants of this day length remain vegetative during the long light periods of summer and do not produce flowers until the days begin to shorten in fall. Typical examples are poinsettias, cyclamens, chrysanthemums, and some varieties of marigolds, along with many others. Chrysanthemums comprise a very interesting group because they have so many different varieties with different responses. A few are long-day in their flowering habits, but most are strictly short-day. If you look through chrysanthemum catalogues from some of the specialist growers you will notice that the blooming period has been established for a given latitude. Some varieties will be listed as blooming September 15, others October 5, 15, or 20, some December 5, 10, 15, and so on. Obviously those that are listed as flowering late in the fall or early winter will have to be protected from frost long before they are scheduled to bloom. If this is practical a person may plant these varieties out-of-doors, but for the most part these plants will have to be raised in the greenhouse to ensure flowering.

Some plants, such as the cockleburr, are normally short-day bloomers. For this plant, however, a continuous dark period of eight and a half hours is enough to cause it to bloom. In other words, even though the cockleburr is normally a short-day plant, it will bloom under a period of fifteen and a half hours of continuous light. Day-neutral plants are not responsive to day length and often will bloom under either long or short days. Most varieties of tomatoes and cotton, some marigold varieties, and dandelions fall into this category.

The delicate response mechanism which influences a plant to flower or not can be demonstrated in certain experiments. For example if a short-day plant is given a cycle of eight hours of light and sixteen hours of darkness it will flower. If, however, a short exposure of light is given during the dark period—as little as one minute of light from a 25-watt bulb—flowering will be inhibited. It appears, therefore, that an uninterrupted period of darkness is more critical to short-day plants than the total light in a twenty-four-hour period. The same thing is true for long-day plants. That is, an interruption of the dark period, which is long enough to prevent them from flowering, will then promote flowering.

Again it seems that the phytochrome pigment has something to do with the biological clock mechanism which induces flowering. If short-day plants have their long nights interrupted by red light they do not flower, but if this red light is followed immediately by far-red wavelengths the effect of the red is canceled and the short-day plant will blossom. Similar effects have been observed in the long-day plants with the same predictable results. Since it has been previously shown how red and far-red affect the phytochrome pigment, this substance is involved in photoperiodism.

The economic importance of a plant's response to the photoperiod is well-known. A florist must have his poinsettias in bloom a few days before Christmas or he will be out of business. Lilies must be ready for the Easter season. The florist can depend on the coming of Christmas at the same date each year, but because of the wide range of Easter dates much more care must be taken in the raising of lilies to ensure having blooms at the correct time. If one wanted to raise certain plants for seed in the summer above the Arctic Circle, with the light period nearly twenty four hours long, he would, of course, plant long-day varieties. Short-day tobacco varieties are raised in the Quebec, Canada, region in summer since the leaves are the desired product. If the tobacco plants were long-day varieties they would soon flower at the Quebec latitude with much less vegetative growth.

Temperature

An organism living as close to its environment as a plant, depending solely upon its tissues for protection, and upon the fortuitous position in which it is growing, is obviously quite responsive to and dependent on temperature variations. Native plants are, in general, geared to the seasons. Plants frequently go into dormancy periods under periods of unfavorable environmental conditions such as drought or frost, but with the return of moisture and warm weather they return to a vigorous state of growth. For many plants dormancy means the loss of leaves, and dieback of the stem to the roots. For others such as some cone-bearing trees dormancy may mean only a cessation of growth for the duration of the unfavorable period.

Seeds may have their dormancy period broken by being placed in a cold refrigerator, often below freezing, for a twenty-four-hour period for some, to several weeks for others. Otherwise their period of dormancy will continue until favorable environmental factors for germination are present.

Resistance to frost injury by many plants has been described as due to: (1) resistance of the proteins to coagulation when the cells are relatively dry; (2) increased or decreased permeability of the cell membrane to water, resulting in ice formation in the intercellular spaces rather than within the cell; and (3) decreased viscosity of the protoplasm, making it more resistant to mechanical injury by ice crystals. Drought hardiness and resistance to excessive heat are found in desert plants. Nearly the same changes which take place in frost resistant plants occur in drought-resistant protoplasm. The protoplasm becomes more viscous as a result of the accumulation of mucilages, resins, oils, gums, and alkaloids. These substances seem to keep the water in a relatively "bound" condition, which reduces the free movement of water from cell to cell, reducing also the rate of transpiration.

Temperature obviously affects growth in many ways through its effects upon such metabolic processes as digestion, translocation, respiration, transpiration, and photosynthesis. Many plants following a period of low temperatures, which breaks the dormancy period, then resume vegetative growth. Peonies do well in cold winter climates where the tuberous roots are in frozen soil for weeks or months. In milder climates where the soil temperatures fail to reach sub-freezing they do not do so well. Most bulbs, tubers, and other underground stems undergo a dormancy period, if not freezing temperatures, for a short time.

One of the reasons potatoes do so well in Idaho, Maine, and other cool, high-latitude states is that the long day provides plenty of opportunity for photosynthesis to occur and starch to be stored in the tubers. In these areas the nights are cool, which reduces the rate of respiration thus decreasing the amount of food used in this process. In warmer areas the rate of respiration is much greater, which uses up much of the stored food, thereby limiting the size of the tubers.

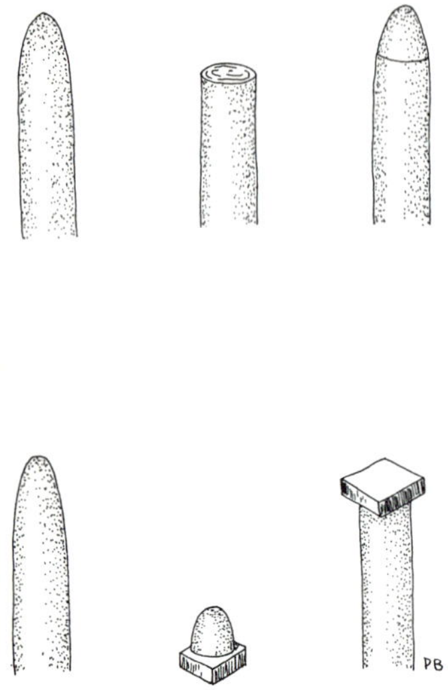

FIGURE 10 Oat coleoptile. *Top:* The tip has been cut off, then replaced, restoring the source of auxin. *Bottom:* The tip is removed and placed on an agar block. Following diffusion of the auxin the block is placed on the decapitated end restoring auxin to the cut portion. (Bonner and Galston.)

Hormones—The Growth Regulators

Plants, like animals, are influenced by certain chemical substances which have a remarkable effect upon the growth, flowering, and movement of plant organs. These substances are generally called growth regulators or **hormones.** They are organic compounds produced in minute amounts in one part of the plant, often influencing the growth in other parts of the plant. Hormones are being continually produced in the areas of their origin, a necessary process since hormones are used up during metabolic processes.

Growth substances were suspected long before they were identified. Over a hundred years ago Charles Darwin, the first proponent of the theory of natural selection, observed that young grass seedlings could be prevented from bending in the direction of light by covering the very tip of the young shoot (now identified as the **coleoptile**) with an opaque cap. Darwin interpreted this inability of the seedling to grow towards the light when covered as a regional reception in which only the tip was able to perceive light. He thought that a response in the tip to light was transmitted downward, causing the curvature.

Auxin

The classical experiments performed by Frits Went about forty years ago helped to interpret this phenomenon further and identified the substance involved as a hormone. This hormone was later given the name **auxin.** Went and others noted that if the tip of an oat seedling was removed, growth was inhibited and within a few hours was completely halted (Fig. 10). If the removed tip was replaced, growth of the stem continued as before. Then an experiment was performed in which the tip was placed on a block of agar for several hours and the block, without the tip, was placed on the decapitated stem. Again growth was renewed. Furthermore, if a block of agar which has been impregnated with the substance from the tip is placed on one side of the tip the stem will renew growth with a curvature away from the side on which the block was placed (Fig. 11). This is caused by the rapid growth of cells on the block side, resulting in a curvature in the opposite direction. Hundreds of experiments of this type involving oats and many other plants have shown us the widespread effect auxins have on the growth of plants. The first auxin determined was indoleacetic acid, but since then a number of related organic compounds have been isolated and their effects observed.

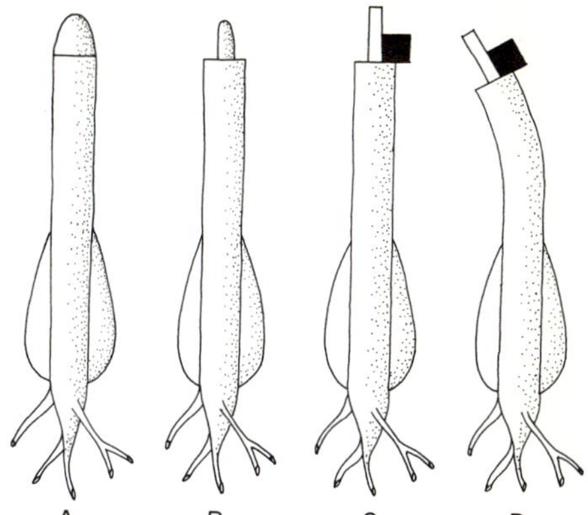

FIGURE 11 Oat seedlings with the coleoptile tip removed in A, then placed on a gelatin block. The block impregnated with the hormone is then placed on the tip of the shoot (C); as the hormone diffuses into the shoot cells it stimulates the cells to more rapid growth on that side causing the curvature in the opposite direction. (Bonner and Galston.)

Auxin is produced in the shoot tip and young leaves. It is transported down the stem independently of the vascular system. It has long been known that this substance is essential for elongation of the stem and the protective coleoptile sheath, but just exactly how it operates has not been definitely determined.

Auxin also exhibits an inhibitory effect on the development of lateral buds. Although auxins are formed in the apical meristematic zone, which includes the lateral buds, these buds generally remain inactive while the terminal shoot continues to grow. If the apical bud is cut off, the lateral buds begin to open and send out branch shoots. This apical growth and the inhibitory effect shown when the terminal bud is present is called apical dominance (see Figs. 3 and 4 in Chapter 8). Trees may be pruned and trained to grow in various shapes by gardeners who wish to take advantage of the manner of branching when the apex of a plant is removed.

Tropisms

We know that there are certain growth movements in plants which we term tropisms. These tropic growth movements are regulated in some manner by auxins.

Geotropism Geotropism, the response to gravity, can be readily observed in roots which are positively geotropic, or, in other words, grow downward in response to gravity. Stems are normally negatively geotropic and grow upward (Fig. 12). If a stem

FIGURE 12 Illustration of a geotropic response. The roots grow downward in a positive geotropic response and the stem tip has curved upward in a negative response to gravity. (Bonner and Galston.)

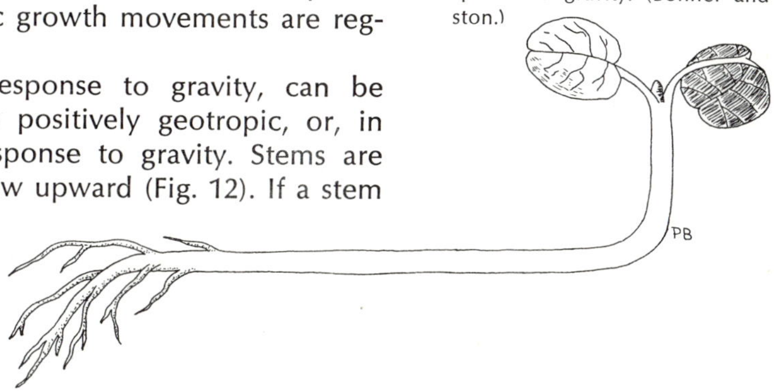

is placed in a horizontal position there is an accumulation of the growth regulator on the lower side of the stem. The auxin accelerates the growth of the cells on the lower side, causing the stem to grow in an upward curvature. For those stems which might be positively geotropic, and especially roots, which begin their development horizontally and then begin to grow downward, auxin accumulates on the lower side of the stem or root. For the most part, however, auxin inhibits the growth of roots even when found in extremely small quantities. Since auxin is toxic to roots in even smaller amounts than would stimulate stem curvature, the positive geotropic response of roots has been explained as a migration of auxin to the lower side of horizontally oriented roots. The auxin inhibits cell growth, causing the root to turn downward, as the cells above grow at their normal rate.

Phototropism Perhaps a more readily observable tropic response is that of the aerial portion of the plant in relation to light. This is termed **phototropism.** It has been demonstrated many times that most plants will grow in the direction of a light source. Their leaves will often orient themselves at right angles to the direction of the light source, a position which gives them maximum surface area for light absorption (Fig. 13). We also know, through the use of light filters, that plants bend only in response to the blue and violet colors of the spectrum. In other words, the short wavelengths initiate a **phototropic** response while the long wavelengths (red) are mainly a stimulus for photosynthesis. The short wavelengths apparently affect the distribution of auxin, which stimulates the cells to grow or elongate more rapidly on the dark side of the stem (away from the light source), causing it to curve towards the light. Auxin is believed to migrate to the unlighted side, or at least to remain in usual concentration on that side, while the hormone migrates from the lighted side, or deteriorates there (Fig. 14).

FIGURE 13 Note the mosaic pattern formed by the leaves of *Cecropia* to place the maximum amount of leaf surface in a favorable light position.

Thigmotropism Unequal growth development of cells on opposite sides of a stem or tendril may also be due to contact, a tropic response exhibited by some plants and called **thigmotropism.** Plants such as gourds, grapes, and peas have tendrils which coil around an object giving support to the climbing plant. The stems of some vines have a similar thigmotropic response. The entire stem gradually grows in coil after coil (termed **circumnutation,** often around a supporting object. Members of the morning glory family, the Convolvulaceae, and strangling figs are examples of this type (Fig. 15).

Hydrotropism Hydrotropism is a movement in response to water. Lateral roots develop much more abundantly in soil strata

which have copious supplies of moisture than in drier strata (Fig. 16). It is doubtful that this could be considered a true tropic response. The growth of roots in response to soil nutrient supplies has been termed **chemotropism,** but again the growth is more of a metabolic response to more adequate supplies of nutrients than a tropic reaction. Increased growth occurs in both cases but differential cell growth rates on opposite sides of the roots probably do not occur in most cases.

FIGURE 14 This ivy vine shows several tropic responses. The stem which grows appressed to the wall is negatively phototropic and positively thigmotropic. The fine light colored stalks growing from the vine stems and away from the wall are actually leaf petioles; the leaf blades have fallen off but grew outward and towards the light showing a positive phototropic response.

FIGURE 15 Monstera deliciosa growing as a vine over the branches of a mahogany tree.

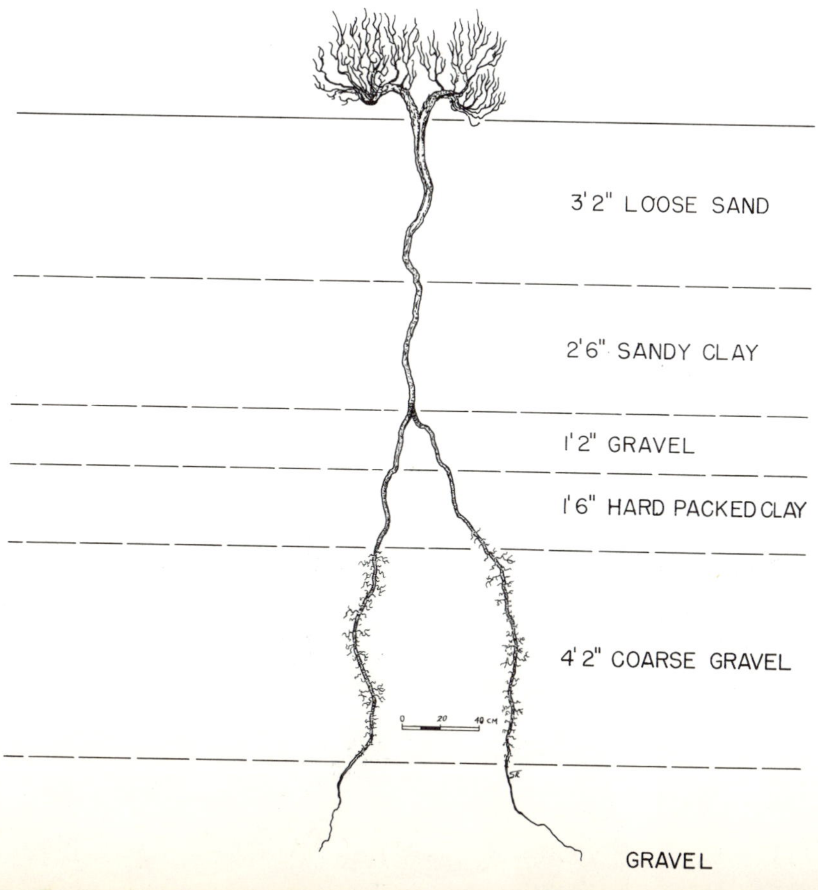

FIGURE 16 A diagram of *Chrysothamnus nauseosus* growing along an arroyo. Note the copious development of lateral roots in the coarse gravel layer which was well supplied with water.

Nastic movements Many species exhibit a type of movement which varies from the tropic responses described above although the movement may be initiated by similar external stimuli. One nastic movement, **photonasty,** is a movement brought about by a plant's response to light. The most obvious are those of many legumes such as honey locust, clover, *Mimosa pudica,* and others in which leaves or leaflets close from the usual horizontal position to an upright position or droop down against the stem. In other cases the leaves fold against the petiole or stem. These are called "sleep" movements and usually occur in the evening generally before dark. Many flowers such as the morning glory, crocus, and California poppy close at night and open in the morning. A different species of morning glory, the moon flower, exhibits its photonastic response by opening in the evening and remaining that way all night, then closing during the day.

Thermonasty is the response to temperature. It is often difficult to determine with some species whether the movement is due to light or temperature stimuli, but many flowers open readily with a rise in temperature and close when the temperature drops. The opening and closure may be the result of turgor changes on the upper and lower sides of leaf blades, petioles, or flower parts.

Perhaps the most remarkable nastic movement is that of **thigmonasty** exhibited by the sensitive plant, *Mimosa pudica*. The leaf is bipinnately compound and at the base of each leaflet, the secondary petiolules, and the petiole are small swollen regions termed **pulvini** which change in turgor with the slightest stimulus. If one of the terminal leaflets is touched it folds upward against the leaflet on the opposite side which has also folded up. Each leaflet along this petiolule then folds upward, the entire petiolule droops and finally the supporting petiole drops downward (Fig. 17). If the

FIGURE 17 *Mimosa pudica,* the sensitive plant. *Left:* These two plants are in typical daytime position. *Right:* The plant on the right has been lightly touched on the leaf tips, causing the leaflets to close and the leaves to droop. Within about fifteen minutes the plant on the right restored its position as in the top picture.

stimulus is strong enough the petiole may drop immediately. Recovery usually takes from fifteen minutes to an hour depending upon the health and age of the plant and the humidity. In the leaflet water leaves the cells on the upper side of the pulvinus, decreasing their turgor pressure and causing the leaflets to fold upward. The pulvini at the base of the petiolules and petiole lose water from the cells on their lower side, which causes the leaf to drop. The water which moves out of the parenchyma cells on one side may pass into intercellular spaces or move into some of the cells on the opposite side, increasing their turgor and aiding the movement.

FIGURE 18 The effect of gibberellic acid. This variety of dwarf Marvel peas was administered a few drops of gibberellic acid in the seedling stage on the shoot apex in the plant at left while the one on the right was untreated.

Gibberellin

Towards the close of the nineteenth century Japanese rice farmers noticed that certain rice plants in their rice paddies were much more elongated than the others. These were observed for some time with the hope that the elongated plants would prove to be the beginning of a new and vigorous strain. Unfortunately they rarely lived to maturity and seldom flowered.

A Formosan botanist in 1926 discovered that seedlings of the elongated type were infected with a fungus, *Gibberella fujikuori*, of the Ascomycete group. It was found that healthy plants could be infected with this fungus. If the fungus was grown in artificial nutrient media a substance was released into the media which when transferred to healthy plants caused growth symptoms similar to the fungus itself. This substance was named **gibberellin** in 1938, following successful isolation and identification of the growth-promoting substance by Japanese scientists.

It was not until about 1950 that gibberellin became known to the Western world, but by 1955 a British chemical laboratory had confirmed the earlier oriental work and had isolated a similar but slightly different substance known as **gibberellic acid.** This substance, like the first, causes hyperelongation in the internodes (Fig. 18), accompanied by a diminution in leaf area, and in long-day plants it stimulates flowering.

We have worked with gibberellic acid in our own laboratories using orchids as test plants. When the seeds were treated we were able to obtain orchids that looked like two-year-old plants in one year. To the orchid grower this is a great boon, for carrying orchids for seven or more years before getting the first bloom is a large investment. As yet we have been unable to get these large commercial orchids to bloom earlier than the required seven years but the work is still in its early stages and orchids are short-day plants on which the gibberellins seem to have little effect in flowering. Our early results in getting a relatively large plant in two years may be deceiving. It is like a young teenager who grows very rapidly for a time; the hormones of maturation do not necessarily accompany an increase in height.

Gibberellin-like substances have recently been isolated from the vegetative parts and young seeds of bean plants. When applied to rosette-type plants, gibberellin causes elongation of the stem to an aerial type and when applied to celery and similar plants it hastens elongation of the petiole, but thus far it has been shown to hasten maturation and the reproduction processes only in certain long-day plants.

Kinins

Another substance with hormone-like properties, **kinetin,** was recently discovered. It is a derivative of **adenine,** one of the components of the nucleic acid DNA compounds. Whereas auxin tends to promote cell enlargement, the kinins stimulate mitotic activity in plant parts and stimulate cell enlargement in leaves where auxins have little effect. It is quite possible that these two growth regulators complement each other in developmental processes. While auxins in the apical bud inhibit the growth of lateral buds, kinins promote their development.

Florigens

Although much work has been done and thousands of articles have been written on the controlling mechanisms of flowering, hormones directly responsible for this activity have not been isolated. Over thirty years ago the name *florigens* was given to these suspected hormones but thus far they have eluded chemical analysis. Since growth regulators have been found in so many plant activities it would be surprising indeed if there were not some flowering regulators directly concerned with the reproductive phase.

Cell Differentiation

How the cell differentiates into the specific functional unit it is to become has been explained on the basis of the coded information DNA has transmitted to it. When we look at an apical meristem we see cells of the same size and shape which have all come from a single cell by mitotic division. All must have the same genetic constitution but some will differentiate into quite different cells such as xylem, phloem, and a dozen different other types. (Fig. 19). Some will remain as parenchyma cells, while others will remain undifferentiated with the power of continual division and differentiation.

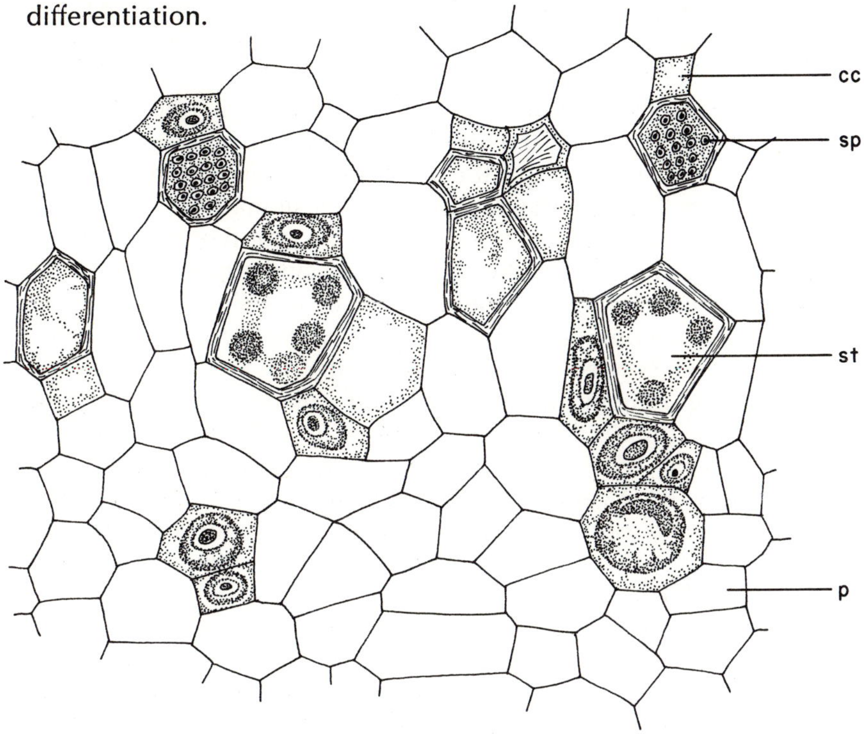

FIGURE 19 Cross section of phloem showing several sieve tubes, two with the sieve plates and some with companion cells. The unstippled cells are phloem parenchyma. cc-companion cell; sp-sieve plate; st-sieve tube; p-parenchyma. Redrawn from *Laboratory Outlines in Biology*. W. H. Freeman and Co.

What is this information that each cell must have in order to carry out its role in the higher organisms with its many tissues, organs, and functions? We must assume that each cell has the potential of becoming any type of differentiated unit capable of any type of function. James Bonner has suggested that each cell has repressant genes which prevent it from being anything but undifferentiated until that cell finds itself in a certain position. For example in the development of a shoot apical meristem all cells are very much alike for about the first half millimeter of its length or so, depending upon the species. Those cells in the outside layer, which, of course, have come from the same source as the internal ones, begin to differentiate into epidermis. These cells were no longer repressed. Cells near the center of the apex begin to form thickened walls and to lose their protoplasm as they gradually become xylem elements. Other cells which are covered by the epidermis

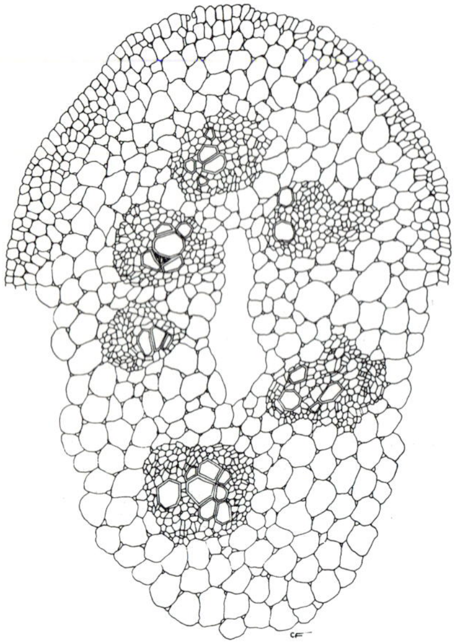

FIGURE 20 A portion of the storage root of *Cucurbita palmata* with small vascular areas arising in the large cortical region.

on the outside and vascular tissue near the center of the shoot remain as undifferentiated parenchyma cells and make up the cortex. But in some species with massive storage roots or modified storage stems some of these cortical cells are many cell widths away from the water- and food-conducting elements. A few of these cells, at strategic places within the large cortex, become meristematic and within a short time form small bundles of xylem and phloem which then contribute to the conducting capacity of the root or stem (Fig. 20).

As long as the cell remains within its normal population it seems to remain just what it is, but removal to a new environment, or injury, or perhaps the pressure of cell overpopulation will give the cell a stimulus to become meristematic.

A cortical parenchyma cell will become embryonic if it is removed, placed by itself in the embryo sac, and supplied with the nutrients for development. Finding itself in this situation the cortical cell will become an embryo and develop into the epicotyl, hypocotyl, and cotyledons of the seedling. In fact, in the early stages of development any cell could be isolated and turned into an embryonic unit capable of developing cells for the whole organism. Many cells which become too highly specialized are committed, by their genetic constitution, very early to develop into specific cells and once the molecules have become arranged for this commitment the destiny of the cells is determined and there is no turning back.

Again, what causes the cells to become one kind of tissue or organ rather than becoming something else or remaining undifferentiated? At present we do not know what the arrangement of the bases is for a cell primordium to commit itself to develop into a flower or to remain vegetative and become a branch with leaves. We are sure that there is such a coding; a number of arrangements for AT, CG within the spiral staircase have been ascertained for certain amino acids. However, this still remains one of biology's mysteries. We do know that we can inhibit the differentiation of cells which ordinarily would form flowers for extended periods under the proper light periods, which was discussed earlier. It must be that these cells have certain flowering genes which need only the stimulus to start building the flowering hormones which will then direct certain cells to develop pistils, others to form stamens, and others to remain as modified sterile floral appendages, the sepals and petals. Probably the floral genes were repressed, as Bonner states, waiting only for the correct length of the dark period, the proper temperature, or a specific cell pH to derepress these genes and get adenine, thymine, cytosine, and guanine into their proper positions on the spiral staircase to initiate flower production. We would like to know how and when a specific cell has its bases arranged into a stamen code, and why it did not become a pistil instead. The commitment of cells, as determined by its genetic code, is one of the most fascinating facets of biology. We have not broken the secret yet, but there are research investigators intensively engaged in the work of decoding. Eventually we will know what A is to T and C is to G and what happens when they are reversed, multiplied, or rearranged in the various sequences possible.

Supplementary Reading

Bonner, James, On the Mechanics of Auxin-induced Growth, *Plant Growth Regulation,* Ames, Iowa. Iowa State University Press, 1961.

Bonner, James, The Next New Biology, *Plant Science Bulletin No. 3,* 1–7, 1965.

Butler, W. L., and R. J. Downs, Light and Plant Development, *Scientific American,* December, 1960.

Hillman, W. S., *The Physiology of Flowering*. Holt, Rinehart and Winston, New York, 1962.

Krebs, H. A. and H. L. Kornberg, *Energy Transformations in Living Matter*. Springer, Berlin, 1957.

Ray, P. M. *The Living Plant*. Holt, Rinehart and Winston, Inc., New York, 1963.

Sinnott, E. W., *Plant Morphogenesis*. McGraw-Hill Book Co., Inc., New York, 1960.

Stowe, B. B., and T. Yamaki, Gibberellins: Stimulants of Plant Growth, *Science,* Vol. 129, pp. 807–816, 1959.

Sussex, I. M., Plant Morphogenesis, *This is Life,* W. H. Johnson and W. C. Steere (eds.). Holt, Rinehart and Winston, New York, 1962.

Sweeney, B. M., Biological Clocks in Plants, *Annual Review of Plant Physiology,* Vol. 14, pp. 411–440, 1963.

Veen, R. Vander, and G. Meijer, *Light and Plant Growth*. Macmillan, New York, 1959.

Thimann, K. V. Plant Growth, *Fundamental Aspects of Normal and Malignant Growth,* W. Nowinski (ed.). Elsevier, Amsterdam, 1960.

Wardlaw, C. W., *Morphogenesis in Plants,* Methuen, London, 1952.

Went, F. W., Plant Growth and Plant Hormones, *This is Life,* W. H. Johnson and W. C. Steere (eds.). Holt, Rinehart and Winston, New York, 1962.

Questions for Discussion

1. Why are proteins considered growth or building foods, while carbohydrates and fats are described as energy foods?
2. How are tagged radioisotopes used in tracing the steps in photosynthesis and related functions?
3. Describe practical illustrations which show the importance of a knowledge of photoperiodism.
4. What cellular changes occur in a stem tip which has been placed in a horizontal position and is responding in a negatively geotropic manner?
5. What are the principal differences in the growth of plants and animals?
6. What types of movement do we find in plants? How is each accomplished?
7. In embryonic development the cells in the embryos of plants are undifferentiated and in a uniform mass like those of animals. What causes certain cells to begin to differentiate into a root tip, while others differentiate into a stem tip?
8. Why do many plants absorb soil substances that they will not utilize?
9. How do the tissues and organs of plants and animals differ? In what ways are they the same?
10. Growth is defined in different ways. Give some thought to composing a definition for growth and be prepared to defend your answer.
11. Since all cells in a plant are apparently identical insofar as their chromosomes and genetic factors are concerned, what

explanation can be offered of how they manage to differentiate into different cell types and tissues?

12. The size of flowers and fruits on a plant part is often increased by partially girdling the base of a branch or the main trunk of a tree. Why should this increase the size of the flowers and fruits on outer portions of the branch?
13. Why cannot weed-killing sprays be used in a vegetable garden?
14. Why will native tropical plants, when transplanted to temperate regions, often fail to bloom?
15. Where on earth would you expect to find the greatest number of short-day species?
16. Describe different types of tropisms.
17. What do we mean by sleep movements in plants? Give examples.
18. What determines whether a certain bud will develop into a branch, a flower, or both?
19. In bud growth are all cells essentially the same morphologically?
20. Some individuals believe that plants should be fertilized only with nutrients from organic sources. As far as a plant is concerned are nutrients from organic sources better than those from inorganic sources? Explain.
21. Why do plants from the legume family contain more proteins in their cells than most other plants?
22. How do the far-red lightwaves of the spectrum affect growth in most plants?
23. Why would a tobacco grower be more interested in raising a short-day variety of tobacco under long-day conditions than in raising the same variety under short-day conditions?
24. What are some examples of cool-weather plants? Of warm-weather ones?
25. How would you explain the fact that many plants grow faster at night than in the daytime, when we know light is necessary for all green plants?
26. At what stage in mitosis could one most conveniently count the chromosomes in a cell?
27. If you were trying to determine the age of a tree by counting its annual rings, at what level of the tree would you make the count?
28. What are the chemical and physical factors which tend to modify plant growth?
29. How could you stimulate a plant to grow taller?
30. In what respects is a mammal a more definite and highly organized individual than an oak tree?

13

Genetics

The link that every generation has with the past and with succeeding generations is through the germ plasm by way of the hereditary determiners termed the **genes.** Consequently the study of heredity is genetics.

When one examines the eggs and sperms of animals and plants he sees certain superficial morphological differences, but upon closer study of the nuclei he finds very little difference in them physiologically. However, the hereditary factors carried on the chromosomes often differ greatly from each other, each controlling the inheritance of one or more characteristics. Often many of the characteristics inherited by an offspring will be maternal, while in other cases the paternal characteristics are more obvious. The differences in molecular content of the genes and chromosomes is a critical one, for it accounts for the fact that "like begets like" and at the same time allows for variation within the species and provides for the great differences between plants and animals and their subordinate groups.

None of us would expect elm trees to have children or even dandelions, nor would we expect dogs to give birth to cats, or roses to produce seed that will develop into carnations; yet when we look at the similarities in the genetic material we find that the

DNA, which is at least the basic material of all genes, is composed of the same chemical substances. As stated in Chapter 3, DNA is composed of a deoxyribose sugar, a phosphate group, and four bases, adenine, thymine, cytosine, and guanine. This is the same in both animals and plants whether of low or lofty evolutionary status. What, then, makes some 2,000,000 species of organisms different from each other?

The DNA Molecule and Genetic Code

From the evidence now available it is apparent that the way the four bases are arranged in the DNA molecule determines the characteristics of the organism. We refer to this as the DNA code. The differences in the arrangement of adenine, thymine, cytosine, and guanine in the DNA molecule result in different characteristics just as the letters in a word make different meanings when rearranged. The word EMIT means one thing, but if the letters are rearranged we have ITEM, TIME, or MITE. Each letter retains its own identity but in different orders the combination has different meanings. E, for example, is always E, but if it is placed in a different position with respect to M, I and T the code is changed and the end result quite different. Or take something much more complicated, like the keyboard of a typewriter. A typist can strike the keys in perfect order to type one correct word after another, but if the keys were rearranged they would produce words which were entirely different and probably meaningless. If only *one* key were altered from the original arrangement, the results would be the same as those from the original set except that now and then a word or two would be misspelled, which would, of course, cause variations in the word structure.

Perhaps this analogy is farfetched, but to get variation in organisms we must have variations in the arrangement of adenine, thymine, cytosine, and guanine (A, T, C, G). Since A and T are always linked together and C is always linked with G we have a somewhat different structural arrangement than simply a scrambling of the letters as in the word ITEM above. Order then becomes of paramount importance. However, although C, G and A, T are always linked they may be in reverse order such as G, C and T, A. In one series in the molecule we might have AT, AT, GC, TA, GC, or we may have CG, AT, GC, TA, GC. Each arrangement has an entirely different code and the results produced by them would be entirely different. Considering the length of a chromosome, the length of the numerous DNA molecules on chromosomes, and the number of AT, GC possibilities in the DNA molecule, the number of possible variations is an extremely large figure which could easily account for the genetic variations in the species now living, as well as all those which have lived in ages past.

In humans the individual chromosomes vary in length from about 1/10,000 to 1/2,000 of an inch when they reach their maximum degree of contraction at metaphase. In an onion the chromosomes are somewhat larger. Of course, chromosomes also vary slightly in length in other plants and animals. The number of genes may also vary on each chromosome, but the number of genes arranged in a linear chain on each chromosome is about 10,000. Each gene has a specific place or locus on a chromosome and each gene has specific chemical and physical properties. Genes also have the ability to reproduce themselves exactly. It is this genetic material which determines the appearance of the organism and is consequently termed its **genotype.** Generation after generation the genes, with their own specific content of DNA, are transmitted on the chromosomes from offspring to offspring. The morphological and physiological potentials of the individual are carried by these genes, which cause the cells to divide and differentiate into the visible characteristics which we term the **phenotype.** The different genes carry different codes and a cell manufactures amino acids which combine to form proteins as the code dictates. The types of proteins formed in an organism determine its phenotype.

Meiosis

The number of chromosomes in every species is consistently the same in all body cells as a result of the replication of the chromosomes in mitosis. Each body cell, therefore, has a full complement of chromosomes, which is spoken of as the diploid number, or as **2n.** Actually we may think of this as a double complement since the chromosomes are present in **homologous pairs.** For example, maize has twenty chromosomes, which may be more correctly spoken of as ten pairs. Similarly garden peas have fourteen chromosomes or seven pairs in each of their body cells. By homologous pairs we mean that the factors of inheritance are the same on each member of a given pair of homologous chromosomes. In peas both members of a single pair will carry a factor for flower color. This factor may be for exactly the same color, such as white, or one of the chromosomes may carry a factor for white while its opposite carries a factor for blue. In other words the homologous chromosomes carried factors for the same characteristic (color) but not necessarily to the same degree. When factors are in contrasting pairs we call them **alleles.** Therefore, the factor for blue is the allele of white and the factor for white is the allele of blue.

If all types of cell division were exactly the same, with the chromosome complement remaining the same, growth could continue in an uncomplicated way. In plants where many species are reproduced by cuttings or other methods of vegetative propagation, mitosis is the only type of cell division necessary for continuation of the species. But in all higher organisms reproduction by a sexual process inevitably occurs. This involves the union of an egg cell with a sperm. If the egg and sperm each had the same number of chromosomes as the body cells the resulting zygote would have double the number, and if mitosis occurred as usual the new product of fertilization would have twice as many chromosomes as its parents. For example, if the egg and sperm of maize plants each had 10 pair of chromosomes the zygote would have twenty pair or forty chromosomes and if this continued the next generation would have 80, then 160, 320, etc. We would soon reach an impossible figure if this continued generation after generation. Fortunately most organisms are provided with a different type of nuclear division to maintain uniformity in the number of chromosomes in the species. This division is called **meiosis.** Generally it takes place in the origin of the sex cells although in plants, especially, it takes place several divisions prior to the formation of the sex cells. Cells arising by meiosis (reduction division) have half the number of chromosomes as the diploid cells and are therefore termed **haploid** or **monoploid.** The basic number is expressed as n and represents one member of each pair of homologous chromosomes carried in the parent cells. Thus each egg and each sperm, which are products of this type of division, has a complement of haploid chromosomes and when they are combined at fertilization the resulting zygote has **2n** or the diploid number. In all subsequent divisions resulting in the maturation of the organism the number of chromosomes remains constant.

The events of division characterizing meiosis are much the same as those in mitosis, but prophase is extended over a longer period of time and the end result is really four daughter cells instead of two. In most plants reduction division usually occurs with division of the spore mother cell in the following manner (Fig. 1).

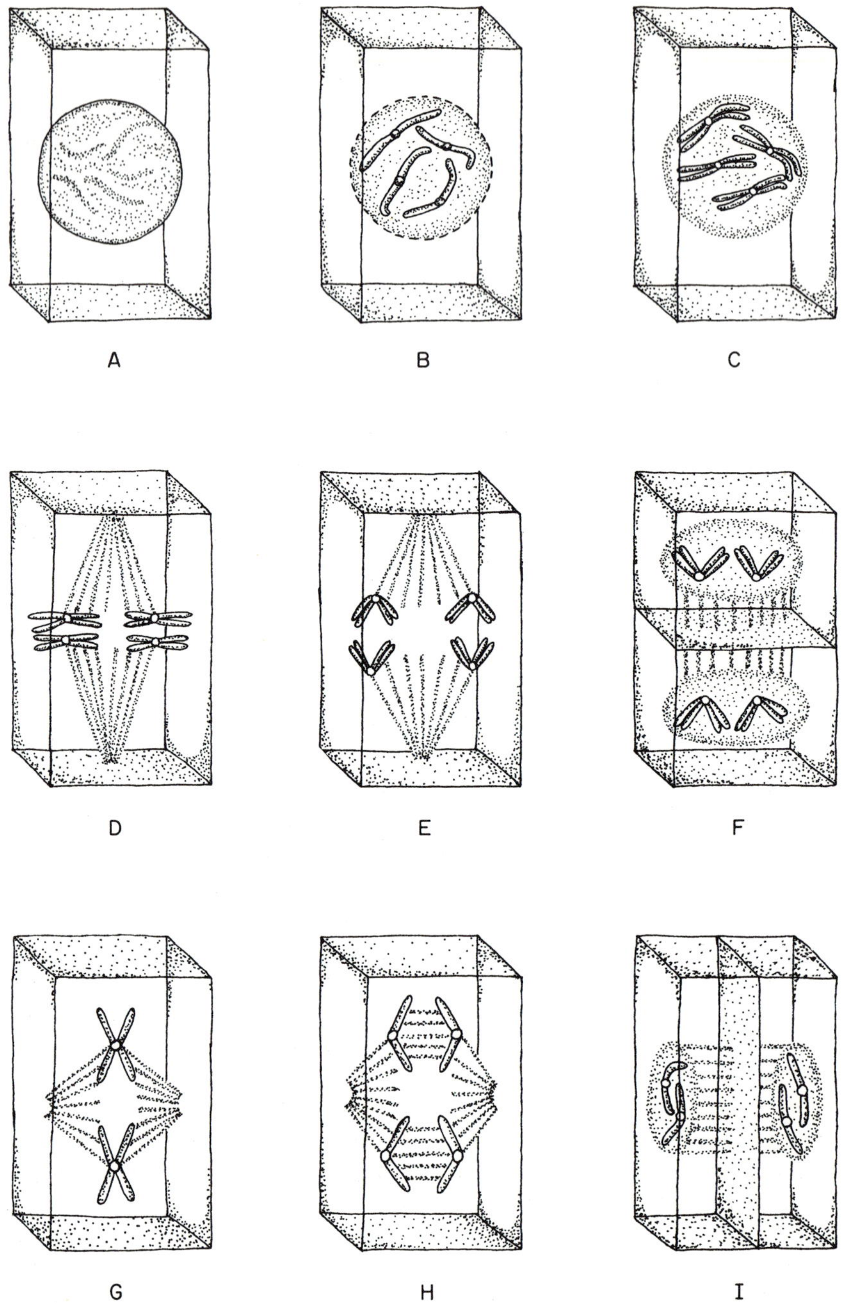

FIGURE 1 Meiosis. A-Interphase; B-early prophase; C-late prophase; D-first metaphase; E-first anaphase; F-first telophase; G-second metaphase with spindle fibers and chromosomes oriented at right angles to the first metaphase; H-second anaphase; I-second telophase. The end result would be four daughter cells instead of just two as illustrated here. Note reduction in number of chromosomes in daughter cells of I. Note also the opposite positions of the new cell plates formed in F and I.

Prophase begins with the development of the chromosomes into thin threadlike structures. In this initial phase it is often easy to determine the number of chromosomes characteristic of the species. The chromosomes at this time are arranged in pairs, identical in size, morphology, and position of the centromeres. Each chromosome is already divided into a pair of **chromatids** and as far as can be determined the DNA of each chromosome has already undergone replication. The pairing of the chromosomes at this time has been described as zipperlike because the pairing is so perfectly blended between the two members. The four strands which appear in each unit are termed a **bivalent.** Soon the chromosomes separate longitudinally and the four chromatids of the bivalent are visible. However, the longitudinal separation is not complete along the entire length, for the bivalent is still held together at various points and the centromeres are not yet split. The points of contact are known as **chiasmata;** there may be several along the length of the bivalent. At this time there may be an exchange of genetic material since the chromatids are in close contact. An exchange of chromosome material in this way is known as crossover and will be discussed later.

Metaphase begins with the disappearance of the nuclear membrane and the appearance of the spindle. Each bivalent chromosome consists of two chromatids and possesses two independent and undivided centromeres, which are arranged above and below the equator of the cell rather than parallel to it as in mitosis. It is also obvious in this stage of meiosis that the pairs of homologous chromosomes align themselves independently of any other chromosomes in the center of the cells.

Anaphase begins when the chromosomes of each bivalent begin to separate and move toward opposite poles of the cell. The centromeres of each chromosome seem to repel each other, resulting in the centromeres' moving toward opposite poles, and the arms of the chromatids become pointed toward the equatorial region. The bivalents are no longer held by their chiasmata but sister chromatids are still held by their centromeres.

Telophase begins when the chromosomes have reached the poles and in some organisms a nuclear membrane begins to form around the chromosomes, ultimately completing the division into two cells. However, in most plants meiosis is really only half over when the nuclear material reaches a telophase at this time. Generally in late anaphase, which we may refer to as I, the chromosomes align themselves into a second metaphase (II) and here the centromeres divide for the first time, separating the sister chromatids, which move in an opposite direction to new poles which we might term east and west if we consider the original poles as north and south. The chromatids are now individually called chromosomes. This results in a second telophase and therefore four daughter cells in all, each with half the number of chromosomes as the original cell with which we started. The chromosomes really only divided once during the first anaphase and their final separation in Anaphase II completed meiosis. In the higher plants and many of lower status the spore mother cell gives rise to a tetrad of spores in this way, the tetrad remaining for some time as a sphere with four quarters.

Meiosis, then, is the way the genetic material is maintained in a specific quantity for the species. We speak freely of the number of chromosomes and their behavior in division. Now we know that a chromosome is a mass of nucleic acids specifically organized and coded with the factors of inheritance for the species.

Nucleic Acids

Experiments with bacteria and other organisms have fully established DNA as a genetic carrier, although in a few viruses such as the tobacco mosaic virus RNA is substituted for DNA. Ingenious techniques have been developed to show the presence and quantity of DNA and RNA in the nucleus and cells. A staining technique known as the Feulgen reaction proved to be a specific stain for DNA. When properly used it stains only this nucleic acid in the cell, making it possible to determine the exact location of DNA. All animal and plant cell nuclei examined in this way have proved to be rich in DNA, while the cytoplasm does not take the stain, or takes it only very slightly. It has been observed that during mitosis chromosomes are deeply stained, while the nuclei of nondividing cells are nearly uniformly stained by the stain on the chromosomes. This establishes the location of the DNA in precisely the part of the cell which we believe to harbor the genes.

Another staining technique was developed which required the use of two dyes, methyl green and pyronin. Methyl green was found to stain DNA and pyronin to stain RNA. This technique, therefore, made it possible to distinguish between these two nucleic acids.

It was also found that a spectrophotometer could be used to detect the nucleic acids. The bases in the nucleic acids intensely absorb ultraviolet light having a wavelength of 260 $m\mu$. No other components in the cell absorb at this wavelength. Since these acids are mainly in DNA and RNA a maximum absorption of 260 mμ indicates the presence of the nucleic acids. Although this method does not differentiate between DNA and RNA, it measures the total quantity of nucleic acids. Therefore if the total quantity of nucleic acids is determined and one measures the amount of DNA by the Feulgen method, the difference will be mainly the amount of RNA remaining.

Repeated experiments using these techniques gave consistent results: DNA is always localized in the nucleus and most of the RNA is in the cytoplasm, although some is in the nucleus. This information led to the discoveries of the functions of these two nucleic acids. It was determined through staining of the chromosomes that these bodies are composed of nucleic acids, chiefly DNA, and that DNA is the substance of which genes are composed. Individuals who have studied fruit fly chromosomes estimate that there are about 1000 genes on the X chromosome of its sex cells and that the average length of a gene on this chromosome is about .005 microns. Furthermore, DNA has the power of self-duplication and the ability to produce RNA. For example, if one strand of DNA has a base sequence series of adenine, thymine, adenine, cytosine, guanine, thymine, it will form a complementary strand of thymine, adenine, thymine, guanine, cytosine and adenine in duplication.

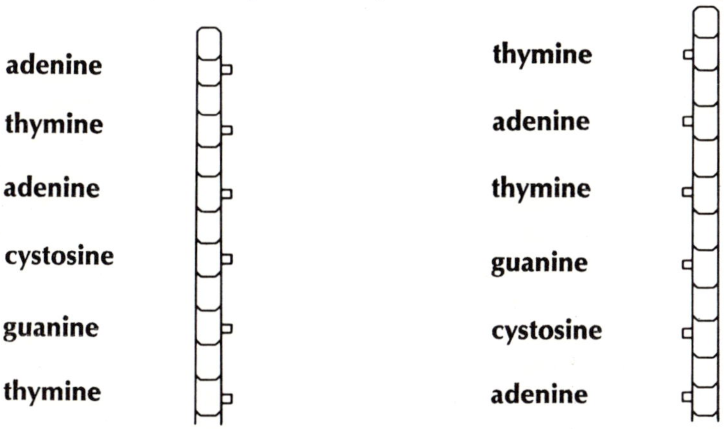

Base sequence **Complementary strand**

Furthermore, the same strand of DNA also serves as a basis for making the messenger RNA, which will move to the ribosomes in the cytoplasm where it stimulates the synthesis of various amino acids in the formation of proteins. The messenger RNA also provides for the arrangement of the amino acids in the polypeptide chain in the building of different proteins.

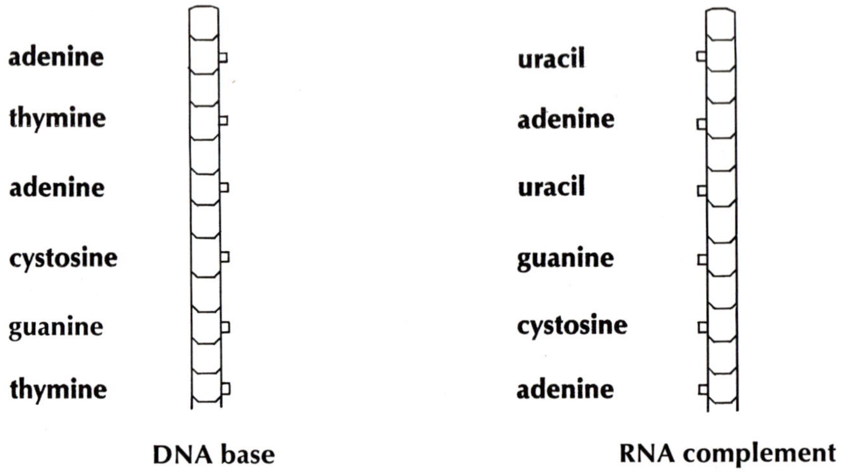

Formation of an RNA Molecule from a Base Strand of DNA

At present the experimental evidence indicates that the type of amino acids produced depends upon the sequence of bases in a strand of DNA, which in turn determines the sequence of bases in the RNA molecule. For example, if the DNA base sequence is T, G, G the RNA sequence will be A, C, C and the amino acid formed is histidine. If the DNA base sequence is C, A, A the RNA base sequence will be G, U, U, and the amino acid is valine. Since proteins are the building substances of all life the knowledge of how proteins are formed initially in the production of amino acids is paramount to our understanding of gene action and the principles of inheritance.

The remainder of this chapter covers the historical aspects of the science of genetics and some of the types of inheritance man has observed. Much of this information was known long before DNA was established as the carrier of genetic information. We still have much to learn and much suspected information to confirm, but the classic work of many Nobel Prize winners in this century has brought new meaning to the work of Mendel and other celebrated geneticists who followed him.

Gregor Mendel and Dominant-Recessiveness

Gregor Mendel (1822–1884), of Austria, was reared on a small farm where he developed a great interest in plants through the help of his father. He learned at an early age how to hybridize and graft plants. Following his early training in preparatory school Mendel completed two years of college and then entered a monastery in Brünn. Here he was encouraged to continue his interests in gardening and hybridization. As a result he began keeping records of the numerous crosses he made, the characteristics he observed and the pedigrees he obtained. Although Mendel used various plants in his studies, he is best known for his crosses with garden peas. He published the results of his work in a rather obscure Austrian journal, *Verhandlungen des Naturforschenden Vereins*.

Mendel's work followed scientific methods with such simple precision that it probably was looked upon by the few who read it as lacking the sophistication which characterized most early scientific papers. Actually, his results seemed so simple because he followed the course of a single hereditary character through each of his crosses and did not look at all characters simultaneously. In all, Mendel traced the transmission of seven inherited characteristics in garden peas. We know now that peas have seven pairs of chromosomes, and the seven characteristics Mendel selected are now known to be the result of genes on these chromosomes in which each pair has one of these characters. This was a most fortunate distribution since it provided a complete, independent assortment.

Mendel examined each individual plant generation after generation and followed each of the seven individual characteristics as **independent units.** This made it possible for him to determine the regular mathematical relationships which he used, and we still use, in predicting how the genes of parents are united in the sexual process, how the genes are then segregated in the development of the gametes, and then how they are recombined in the subsequent generations.

On April 18, 1876, Gregor Mendel wrote a very lengthy letter to Carl Nageli, a well known botanist and evolutionist. Excerpts from this letter show the thought and care Mendel used in his experiments.

> In 1859 I obtained a very fertile descendant with large, tasty seeds from a first generation hybrid. Since the following year its progeny retained the desirable characteristics and were uniform, the variety was cultivated in our vegetable garden and many plants were raised every year up to 1865. The parental plants were *bcDg* and *BCdG:*
>
> | | |
> |---|---|
> | *B*—albumen yellow | *b*—albumen green |
> | *C*—seed-coat grayish brown | *c*—seed-coat white |
> | *D*—pod inflated | *d*—pod constricted |
> | *G*—axis long | *g*—axis short |
>
> The hybrid just mentioned was *BcDG.*

It is obvious from this brief description that Mendel recognized the paired nature of the characteristics (genes) and that the materials of heredity were organized as unit particles. Mendel also recognized specific ratios resulting from his crosses. Another paragraph from his letter to Nageli emphasized this point.

> My experiments with single characters all lead to the same result: that from the seeds of the hybrids, plants are obtained half of which in turn carry the hybrid character *(Aa),* and the other half, however, receive the parental characters *A* and *a* in equal amounts. Thus, on the average, among four plants two have the hybrid character *Aa,* one the parental character *A,* and the other the parental character *a.* Therefore $2Aa + A + a$ or $A + 2Aa + a$ is the empirical simple developmental series for two differentiating characters. Likewise it was shown in an empirical manner that, if two or three differentiating characters are combined in the hybrid, the developmental series is a combination of two or three simple series.

Dominant-Recessiveness was also recognized by Mendel and he chose to recognize it arbitrarily through the use of letters.

> The dominant characters are designated as *A, B, C, D, E, F, G* and as concerns their dual meaning please refer to p. 11. The recessive characters are designated as, *a, b, c, d, e, f, g;* these should remain constant in the next generation. Therefore, from those seeds which stem from plants with the recessive characters only, identical plants are expected (as regards the characters studied).

In this paragraph Mendel shows how true-breeding offspring will be obtained if the parents carry only the recessive characters. Of even greater interest is his last phrase in parentheses, "(as regards the characters studied)." Mendel described only his work and the specific characters he studied. He was a careful, thoughtful scientist and refused to generalize in a qualitative manner but described the quantitative results of his specific experiments. He also showed that he appreciated the importance of repetition in experiments, which is so important in verifying results. He stated:

> By repeated experiments it was proved that if plants with green seeds are fertilized by those with yellow seeds, the albumen of the resulting seeds has lost the green color and has taken up the yellow color. The same is true for the shape of the seed. Plants with angular seeds, if fertilized by those with round or rounded seeds, produce round or rounded seeds. Thus, due to the changes induced in the color and shape of the seeds by fertilization with foreign pollen, it is possible to recognize the constitution of the fertilizing pollen.
>
> Let *B* designate yellow color; *b* green color of the albumen.
>
> Let *A* designate round shape; *a*, angular shape of the seeds.
>
> If flowers of such plants as produce green and angular seeds by self-fertilization are fertilized with foreign pollen, and if the seeds remain green and angular, then the pollen of the donor plant was, as regards the two characters . . . *ab*. If the shape of the seed was changed, the pollen was taken from . . . *Ab*. If the color of the seeds is changed, the pollen was taken from aB; if both shape and color is changed, the pollen was taken from AB.
>
> If the hypothesis that hybrids form as many types of pollen cells as there are possible constant combination types is correct, plants of the makeup
>
> *AB* produce pollen of type *AB*
> *ABb* " " " " *AB* and *Ab*
> *AaB* " " " " *AB* and *aB*
> *AaBb* " " " " *AB, Ab, aB,* and *ab*
>
> Fertilization of ovules occurs:
>
> I ovules *ab* with pollen *AB*
> II ovules *ab* " " *AB* and *aB*
> III ovules *ab* " " *AB* and *aB*
> IV ovules *ab* " " *AB, Ab, aB,* and *ab*
>
> The following varieties may be obtained from this fertilization:
>
> I *AaBb*
> II *AaBb and aaBb*
> III *AaBb* and *aaBb*
> IV *AaBb, Aabb, aaBb, and aabb*

If the different types of pollen are produced in equal numbers these should be in:

I all seeds round and yellow;
II one half round and yellow; one half angular and yellow
III one half round and yellow; one half angular and yellow
IV one quarter round and yellow; one quarter round and green; one quarter angular and yellow; one quarter angular and green.

In the concluding paragraph of his letter Mendel revealed a bit of modesty when he stated, "Every day, from spring to fall, one's interest is refreshed daily, and the care which must be given to one's wards is thus amply repaid. In addition, if I should, by my experiments, succeed in hastening the solution of these problems, I should be doubly happy."

Gregor Mendel's investigations were forgotten for thirty-four years, but in 1900 three scientists working independently discovered Mendel's writings. These men, De Vries in Holland, Correns in Germany, and Tschermak in Austria, gave credit to Mendel, recognizing for the first time experiments which have become the cornerstone of all modern work in plant and animal genetics.

Genetic Ratios

The selection by Mendel of the letters *A* to *G* to represent the seven pairs of characters he studied in peas was an arbitrary choice and can as well be used by anyone. However, for pedogogical convenience today we usually use the first letter of the dominant character to represent the dominant gene and use the same letter in lower case to represent its homologue if it is recessive. For example, Mendel used *A* to designate round seed shape and *a* for angular shape. We now refer to these seed-coat characteristics as smooth and wrinkled, and consequently express smooth as *S*, since it is dominant, and wrinkled as *s*.

Furthermore, if a parent has two factors for smooth it is *SS*, or for wrinkled, *ss*. Each is **homozygous** for its specific character; if the plant should have the genotype *Ss* it is **heterozygous.** Each pair of letters represents a pair of genes on a pair of **homologous** chromosomes whether *SS*, *Ss*, or *ss*. "Homologous" therefore defines a pair of chromosomes for the same thing but not necessarily to the same degree. They both represent factors for seed coat but one letter stands for smooth and the other for wrinkled. Two factors forming a contrasting pair, as *Ss*, are termed alleles; therefore, *S* is the allele of *s*. A character conditioned by a single pair of alleles is termed a **unit character.** Obviously a cross between parents both with genotypes of *ss* could only result in wrinkle-seeded offspring. Similarly if both parents were *SS* all offspring would have smooth phenotypes. If we make a cross between two homozygous parents (P_1) as below:

	Smooth		**Wrinkled**
$\mathbf{P_1}$	**SS**		**ss**
	$\mathbf{F_1}$	**Ss**	

we find that all offspring, (F_1) the first filial generation, are all smooth. Their phenotypes are like the smooth parent but their genotypes are different.

If we then cross members of the F_1 with each other the progeny will be of all types.

		Possible genotypes in sperm	
		S	*s*
Possible genotypes in egg	*S*	*SS* Smooth	*Ss* Smooth
	s	*Ss* Smooth	*ss* *Wrinkled*

We have, therefore, variation in the progeny not only in the genotype but also in the phenotype expressed as a ratio of 3:1. In this cross we have two different phenotypes, smooth and wrinkled, but we have three genotypes, *SS, Ss,* and *ss*.

A cross involving a single pair of factors is spoken of as a monohybrid, while one involving two pairs is a dihybrid. Another characteristic Mendel observed was that some pea seeds were yellow in color and some were green. He also made numerous crosses in which he traced the pattern of its inheritance and found that yellow was dominant to green. If we again use the smooth and wrinkled seed coat characteristic along with the yellow-green color we can demonstrate a dihybrid cross:

P_1 generation:	*SSYY* Smooth yellow	*ssyy* Dwarf green	*Genotypes* *Phenotypes*
	SY ×	*sy*	*Gametes possible*
F_1 generation:	*SsYy* *Smooth yellow*		*Genotype* *Phenotype*

This cross shows two pairs of homologous chromosomes in each parent. The gametes produced contain one member of each of the pairs of homologous chromosomes from that parent. At fertilization the haploid complement of chromosomes in the gametes become diploid in the zygote. Each parent is homozygous for its characteristics but the F_1 are heterozygous and smooth yellow in phenotype like the one parent.

For simple crosses such as in monohybrids and even in dihybrids, where the genotypes of the parents do not produce many different kinds of gametes and therefore not many different progeny, no special technique is required in working out the problem. When two or more pairs of chromosomes in each parent are involved, and they are heterozygous, then the number of progeny and their characteristics can be more easily predicted in a checkerboard arrangement as follows:

F_1 Smooth yellow *SsYy* × Smooth yellow *SsYy*

Possible genotypes in egg		*Possible genotypes in sperm* *SY*	*Sy*	*sY*	*sy*
	SY	*SSYY* Smooth yellow	*SSYy* Smooth yellow	*SsYY* Smooth yellow	*SsYy* Smooth yellow
	Sy	*SSYy* Smooth yellow	*SSyy* Smooth green	*SsYy* Smooth yellow	*Ssyy* Smooth green
	sY	*SsYY* Smooth yellow	*SsYy* Smooth yellow	*ssYY* Wrinkled yellow	*ssYy* Wrinkled yellow
	sy	*SsYy* Smooth yellow	*Ssyy* Smooth green	*ssYy* Wrinkled yellow	*ssyy* Wrinkled green

The ratio in this cross between two completely heterozygous parents in which dominance and recessiveness are involved is 9:3:3:1:

9 Smooth yellow
3 Smooth green
3 Wrinkled yellow
1 Wrinkled green.

As indicated before, Mendel studied each factor separately and described them as **unit characters.** Furthermore he noticed that each character could be inherited **independent** of any other. Therefore, a plant with smooth seed coats could have green or yellow seeds and a plant with wrinkled seeds could have green or yellow seeds. In all of his crosses, each of the seven factors studied was independent of all others. He noticed that smooth seed coat appeared three times more frequently than wrinkled and that yellow color appeared three times more frequently than green. If you will examine the 9:3:3:1 ratio above you will note that there are twelve yellow to every four green seeds and twelve smooth to every four wrinkled. In dominant-recessive problems, involving two pairs of factors, these are the ratios which prevail.

Blending and Incomplete Dominance

Since Mendel's results were published we have found that not all things which are inherited are transmitted on a dominant-recessive basis, nor is each character the result of a single pair of genes. Some factors exhibit a blending or incomplete dominance, some have multiple factors, others are cumulative, some factors are linked with others on the chromosomes, and some are linked on the sex-determining chromosomes.

In the problems we have discussed thus far the heterozygote was always identical in appearance to the homozygous dominant. There are, however, cases where the heterozygote is intermediate and therefore exhibits a **blending** between the two parents. A number of flower species show this type of inheritance.

A cross between a red-flowered four o'clock and a white-flowered one produces an F_1 in which all progeny are pink.

	Red		White	
P_1	*RR*	×	*rr*	
gametes	*R*		*r*	
F_1		*Rr*	Pink	

If the F_1 are self-pollinated the results are as follows:

F_1	*Rr*	×	*Rr*	*Genotypes*
	Pink		Pink	*Phenotypes*

Male gametes

		R	*r*
Female gametes	*R*	RR Red	Rr Pink
	r	Rr Pink	rr White

A second pair of factors may be added, resulting in an entirely different ratio from the typical 9:3:3:1 obtained from two heterozygous parents exhibiting dominant-recessive traits.

In radishes the root shape may be long, oval or round and the flower color may be red, purple or white. If we cross a long-rooted red-flowered radish with a round-rooted white-flowered one we obtain oval-rooted purple-flowered progeny.

P_1 generation:	*LLRR* Long red		*llrr* Round white	*Genotypes* *Phenotypes*
	LR	×	*lr*	*Gametes possible*
F_1 generation:		*LlRr* Oval purple		*Genotype* *Phenotype*
	LlRr	×	LlRr	

Possible genotypes in sperm

		LR	*Lr*	*lR*	*lr*
Possible genotypes in egg F_2	*LR*	*LLRR* Long red	*LLRr* Long purple	*LlRR* Oval red	*LlRr* Oval purple
	Lr	*LLRr* Long purple	*LLrr* Long white	*LlRr* Oval purple	*Llrr* Oval white
	lR	*LlRR* Oval red	*LlRr* Oval purple	*llRR* Round red	*llRr* Round purple
	lr	*LlRr* Oval purple	*Llrr* Oval white	*llRr* Round purple	*llrr* Round white

The offspring in the F_2 are as follows:

Phenotypes	*Ratio*	*Genotypes*
Long red	1	*LLRR*
Long purple	2	*LLRr*
Long white	1	*LLrr*
Oval red	2	*LlRR*
Oval purple	4	*LlRr*
Oval white	2	*Llrr*
Round red	1	*llRR*
Round purple	2	*llRr*
Round white	1	*llrr*

All genetics problems are worked in the same way, but the interpretation of the results depends upon the types of inheritance involved. The dominant-recessive type of inheritance is usually the easiest to work and more known inherited factors probably have this type of inheritance than any other. However, the quality of the genes for a certain character and the number of genes for this character have a modifying effect in many different ways.

A slight modification of the incomplete dominance problem above is found in wheat. Red seed color is dependent upon the presence of two dominants, *R* and *B*; white seed color is dependent upon the presence of both recessives in the homozygous state. Progeny varying from these factors are brown.

	Red		White
Parents	*RRBB*	×	*rrbb*
Gametes:	*RB*		*rb*
F_1		*RrBb*	Red
P_2:	*RrBb*	×	*RrBb*

Possible genotypes in sperm

		RB	*Rb*	*rB*	*rb*
	RB	*RRBB* Red	*RRBb* Red	*RrBB* Red	*RrBb* Red
Possible genotypes in egg	*Rb*	*RRBb* Red	*RRbb* Brown	*RrBb* Red	*Rrbb* Brown
	rB	*RrBB* Red	*RrBb* Red	*rrBB* Brown	*rrBb* Brown
	rb	*RrBb* Red	*Rrbb* Brown	*rrBb* Brown	*rrbb* White

This type of modification is not common. It involves an interaction between two pairs of factors and is like the typical 9:3:3:1 ratio except that the 3 + 3 (Brown) are combined as two unrecognizable genotypes. We call this **incomplete dominant epistasis.** Epistasis is described as the effect one pair of factors has on another pair. In this case the red color can result only if a dominant is present in **both** pairs, and a complete lack of color prevails if neither is present. However, if a dominant is present in only one pair the recessive pair exerts a modifying effect in the development of the full dominant color.

Linkage

In our discussion of Mendelism it was mentioned that the genes separate as unit characters and are inherited independently of each other. It was also demonstrated that the members of a pair of homologous chromosomes separate at reduction division but may appear in the gamete with any of the members of other homologous pairs. Therefore if we have a chromosome complement of *AaBb* the possible combinations in the gametes are *AB*, *Ab*, *aB*, and *ab*. Furthermore, they appear in equal numbers. In genetics the way to determine the types of gametes produced in a plant exhibiting dominant traits is to cross it with the known double recessive. This is termed a *backcross* and in a dominant-recessive problem in which one parent is completely heterozygous the backcross would result in a ratio of 1:1:1:1.

In a cross involving sweet peas in which blue flower color is dominant to red and long pollen grain is dominant to round pollen, the expected ratio of 1:1:1:1 was not obtained.

	Blue long	Red round
P_1	*BBLL*	*bbll*
F_1	*BL*	*bl*

BbLl Blue long

The F_1 was then backcrossed to the double recessive.

BbLl × *bbll*
Blue long Red round

Possible gentoypes in sperm

		bl
Possible genotypes in egg	*BL*	*BbLl* (Blue long)
	Bl	*Bbll* (Blue round)
	bL	*bbLl* (Red long)
	bl	*bbll* (Red round)

This cross should then result in a ratio of 1:1:1:1, as illustrated. However, the results were:

Blue long	43.7%
Blue round	6.3%
Red long	6.3%
Red round	43.7%

The offspring with the highest percentages, blue long and red round, are the parental types. It is assumed that when characters such as these appear in large percentages they are on the same chromosome, i.e., they are linked. If this is true then why do we not find them appearing together 100% of the time? We have found that there is an occasional exchange of genes between chromosomes of homologous pairs and that this exchange occurs with regularity. The exchange is called **crossover** and it occurs during meiosis. As has been mentioned earlier, there is a great deal of coiling of the chromatids around one another during prophase. In some cases a pair of chromatids may break at the corresponding place on each and when they rejoin they become attached to the opposite members. We therefore have new chromatids which are exactly alike in appearance but there has been an exchange of parts.

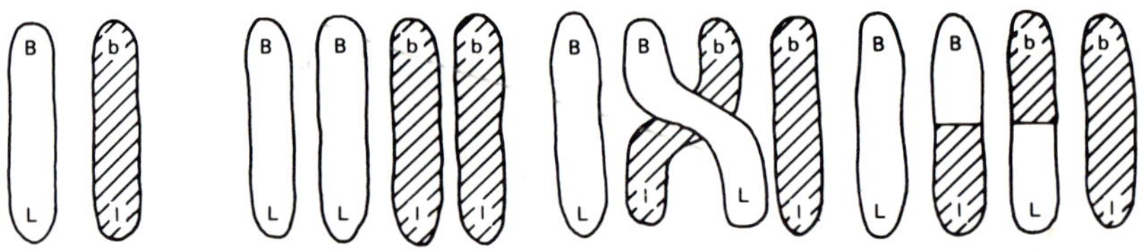

Homologous Chromosomes at synapsis | Formation of chromatids | Coiling of chromatids | Chromatids break and separate, joining with portions of other chromatids

The above cross should have resulted in 50 percent blue long and 50 percent red round but due to crossing over blue long and red round remained linked in 87.4 percent of the crosses and the genes for blue round and red long crossed over in 12.6 percent of the cases. The *BL* and *bl* are **non-crossover** factors and *Bl* and *bL* are crossover factors.

A similar situation is found in corn, where tallness is due to a dominant factor *(T)* and dwarfness to its recessive allele *(t)*, while normal leaves are due to a dominant factor *(N)* and crinkly leaves to its recessive allele *(n)*.

In a cross between a homozygous tall crinkly-leaved plant and a dwarf, normal-leaved plant the F_1 were all tall normal with the genotype of *TtNn*. If we backcross the F_1 with the double recessive we obtain

Tall crinkly-leaved plants	41%
Tall normal-leaved plants	9%
Dwarf crinkly-leaved plants	9%
Dwarf normal-leaved plants	41%

Again, if we had a straight dominant recessive cross of *TtNn* × *ttnn* the resulting ratio should have been 1:1:1:1. Obviously this is another example of linkage, with the genes for tall and crinkly on the same chromosome as well as the genes for dwarf and normal linked on the other homologous chromosome. The genes for tall normal leaf *(TN)* and dwarf crinkly *(tn)* crossed over in 18 percent of the progeny obtained.

A simpler type of linkage, sex-linkage, is quite common in humans and in many other animals. It probably exists also in many plants but since plants are usually bisexual the characteristic is less obvious. In organisms where sex is determined by a pair of sex chromosomes such as XX for female and XY for male, any factor carried on the Y chromosome would be inherited 100 percent in the male progeny. Factors carried on an X chromosome would be transmitted to the female progeny from the male parent in 100 percent of the cases. If the factor is carried on one of the female parent's X chromosomes then 50 percent of both male and female progeny would receive the factor.

In plants as well as animals there are some genes which have a lethal effect. This may be exhibited in early embryonic development, or its effect may not be observed until the organism has lived in apparent good health for years. One lethal factor in plants is **albinism,** in which the factor or factors necessary for the development of chlorophyll are absent. In corn, if the factors **LL** or **Ll** are present the developing corn plant is green with well developed chlorophyll; if the seed carries the factors **ll** the corn plant will grow from the seed but will remain white. After the stored food is used up in germination the seedling will die from starvation. This condition is occasionally observed in other plants, such as maple tree seedlings, as well as corn.

Multiple Factoral Inheritance

Several sets of alleles can produce more or less equal and cumulative effects on the same character. This type of inheritance was also observed in wheat kernels. A cross between two varieties of wheat, one of which had very dark red kernels and the other white kernels, resulted in an F_1 of intermediate shade. This cross probably involved two pairs of factors as follows:

	Deep Red		White
Parents:	$R_1R_1R_2R_2$		$r_1r_1r_2r_2$
Gametes:	R_1R_2		r_1r_2
F_1		$R_1r_2R_1r_2$	Intermediate shade of red

Then if the F_1 were self-pollinated the results would be:

Possible genotypes in sperm

Possible genotypes in egg		R_1R_2	R_1r_2	r_1R_2	r_1r_2
	R_1R_2	$R_1R_1R_2R_2$ Deep red	$R_1R_1R_2r_2$ Dark red	$R_1r_1R_2R_2$ Dark red	$R_1r_1R_2r_2$ Medium red
	R_1r_2	$R_1R_1R_2r_2$ Dark red	$R_1R_1r_2r_2$ Medium red	$R_1r_1R_2r_2$ Medium red	$R_1r_1r_2r_2$ Light red
	r_1R_2	$R_1r_1R_2R_2$ Dark red	$R_1r_1R_2r_2$ Medium red	$r_1r_1R_2R_2$ Medium red	$r_1r_1R_2r_2$ Light red
	r_1r_2	$R_1r_1R_2r_2$ Medium red	$R_1r_1r_2r_2$ Light red	$r_1r_1R_2r_2$ Light red	$r_1r_1r_2r_2$ White

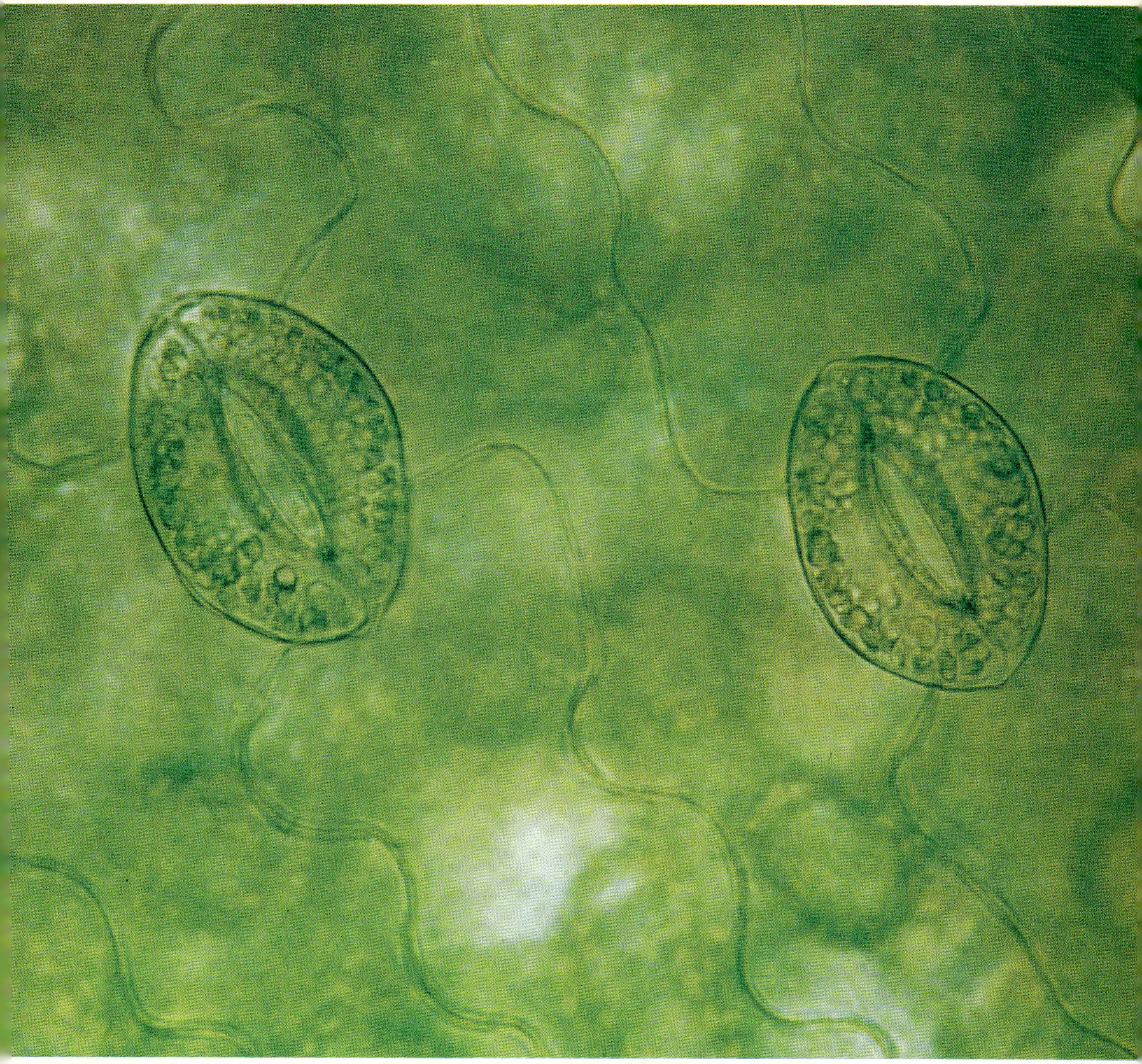

1. Stomates (p. 90).

2. Date grove (p. 126)

3. Cross section of Tilia stem (p. 140)

4. Sequoia grove (p. 133)

5. Honey locust thorns (p. 156)

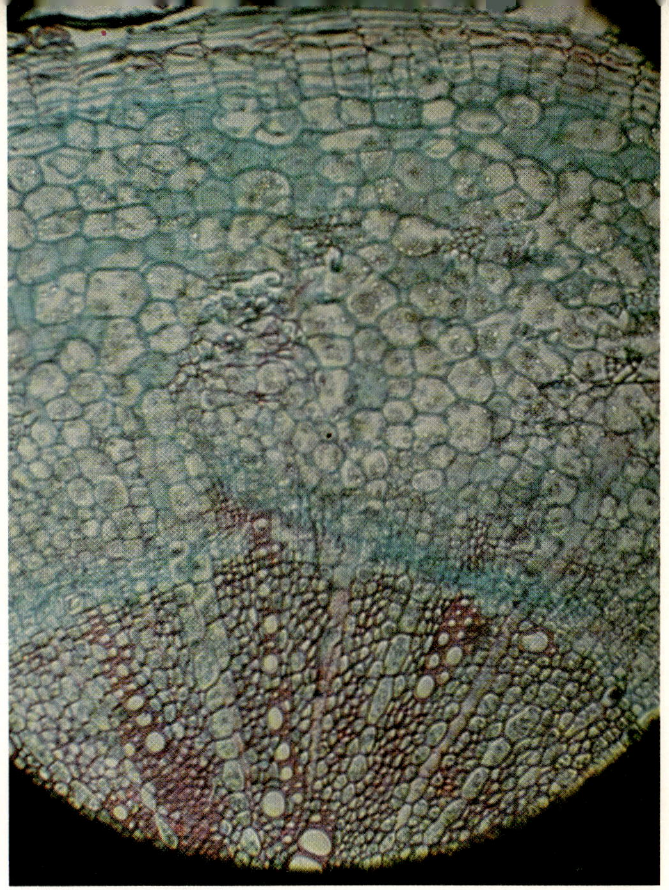

6. Bird of Paradise root (p. 174)

7. Chlorotic cottonwood tree given iron sulfate (pp. 28, 189)

A

B

C

D

E

8.

A. *Erigeron* (p. 199)

B. *Yucca elata* (p. 199)

C. *Geranium richardii* (p. 198)

D. Tulip cluster (p. 197)

E. Tulip (p. 197)

9. A. Dandelion showing method of seed dispersal (p. 216)

B. *Clematis* showing seed dispersal (p. 216)

10. Pineapple (p. 224)

11. Algae:

A. Nostoc (p. 308)

B. Nemalion, carpogonium, carpospores (p. 314)

C. Fresh water habitat (p. 321)

D. Marine habitat (p. 345)

E. Marine habitat—low tide (p. 345)

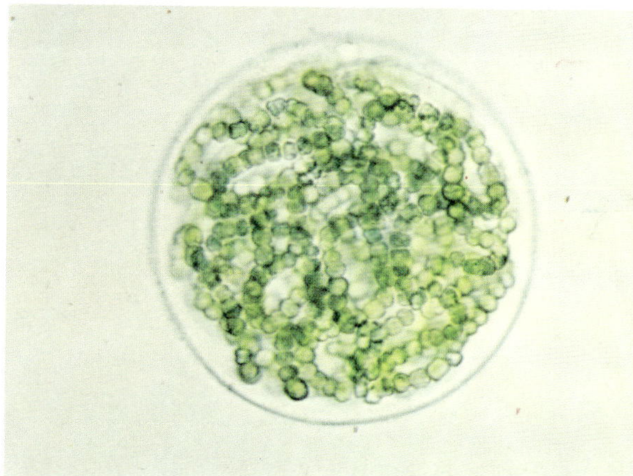

A

B

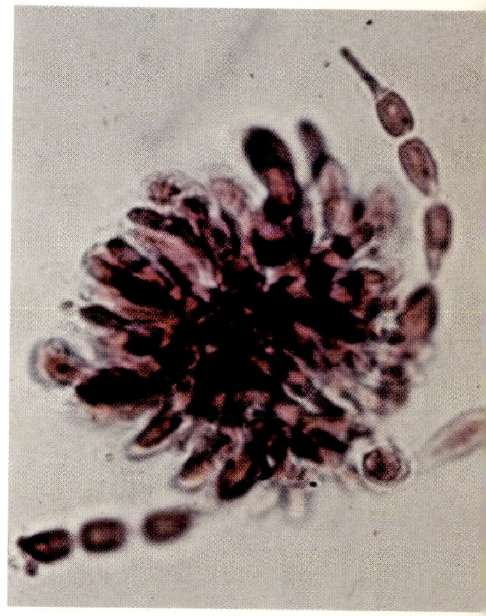

C

D

E

12. Green Algae: A. Zygnema (p. 331)

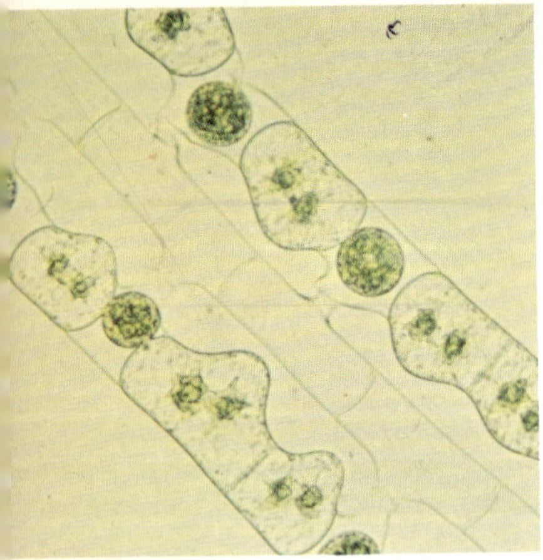

B. Desmid—Microsterias (p. 332)

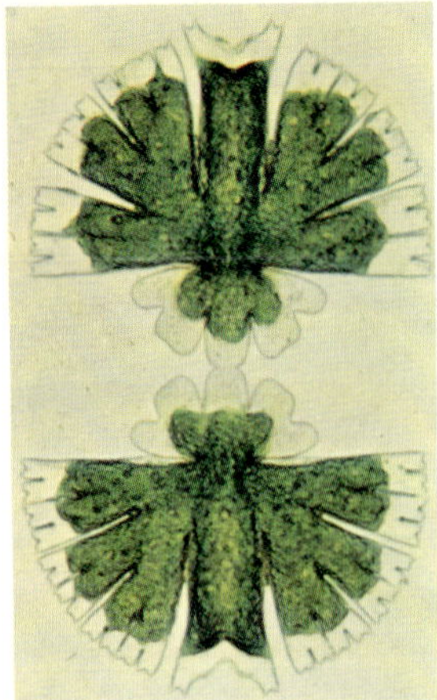

C. Pediastrum (p. 334)

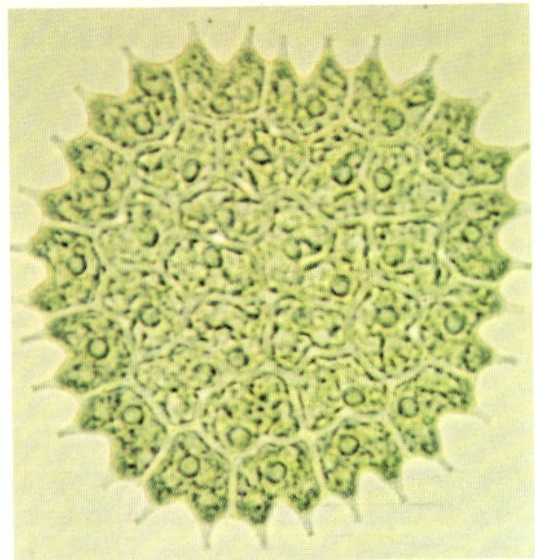

D. A marine green alga-*Codium* (p. 321)

13. Diatom:

A. Valve view (p. 343)

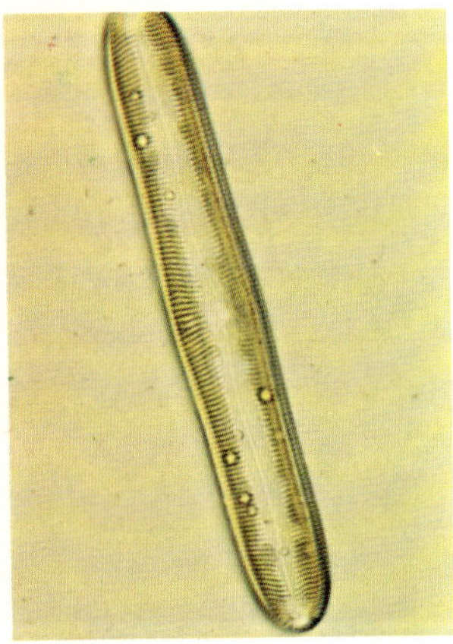

B. Girdle view (p. 343)

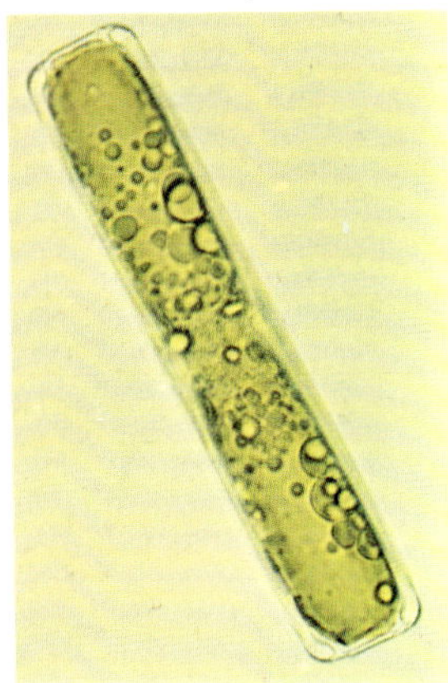

14. Brown algae:

A. *Macrocystis* (p. 348)

B. The elk alga-*Pelagophycus* (p

C. *Nereocystis* (p. 348)

D. *Fucus* (p. 349)

15. Fungi: A. Inkycap-*Coprinus* (p. 380)

B. *Cantharellus* (p. 380)

C. *Fomes* (p. 411)

D. *Hydnum*-corallike fungus (p. 411)

E. *Peziza* (p. 402)

F. Lichens (p. 413)

16. Vertical view of *Marchantia* stomate (pp. 421, 424)

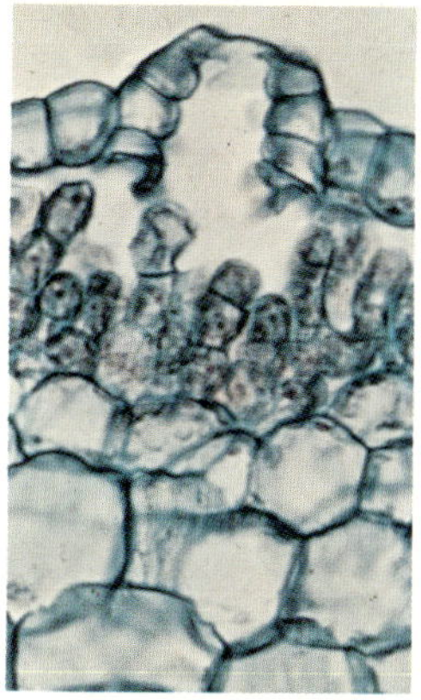

17. Equisetum (p. 471)

18. Ferns:
A. *Pteridium* in forest (p. 476)

B. *Adiantum* (pp. 476, 483)

19. Gymnosperms: A. Cycad (p. 566)

B. Pine forest (p. 505)

C. Germinating pine seed (p. 511)

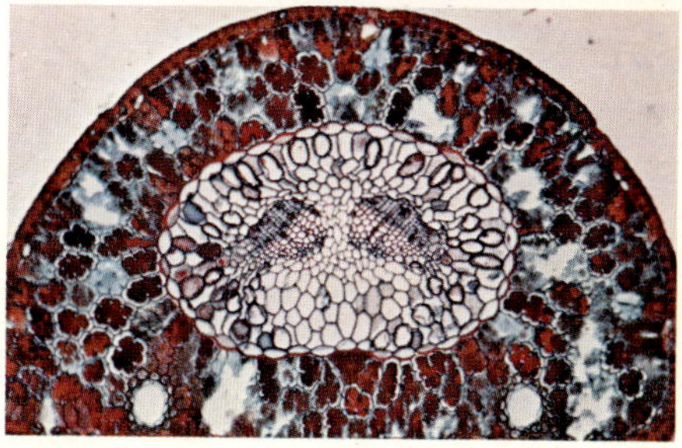

D. Cross section of a pine leaf (p. 506)

20. Habitats: A. Xerosere (p. 566)

B. Hydrosere (p. 565)

C. Aspen and Engelman spruce (p. 576)

21. A. Desert (p. 570)

B. Dwarfed spruce at timberline (pp. 574, 592)

C. Tropical forest in Jamaica

A

B

C

D

E

F

22. Plants of economic value:

A. Broccoli (p. 571);
B. Chili pepper (p. 605);
C. Grapefruit (p. 606);
D. Cotton (p. 613);
E. Papaya (p. 606);
F. Banana and coconut plantation (p. 606).

In this cross, which seems to be the result of two pairs of genes responsible for various grades of color, the capital letters stand for pigment-producing alleles and would produce the following phenotypes:

Genotypes	*Phenotypes*	*Ratio*
$R_1R_1R_2R_2$	Deep red shade	1
$R_1R_1R_2r_2$		
$R_1r_1R_2R_2$		
$R_1R_1R_2r_2$	Dark red	4
$R_1r_1R_2R_2$		
$R_1r_1R_2r_2$		
$R_1r_1R_2r_2$		
$R_1r_1R_2r_2$	Medium red	6
$R_1r_1R_2r_2$		
$R_1R_1r_2r_2$		
$r_1r_1R_2R_2$		
$R_1r_1r_2r_2$		
$r_1r_1R_2r_2$		
$r_1r_1R_2r_2$	Light red	4
$R_1r_1r_2r_2$		
$r_1r_1r_2r_2$		
	White	1

There are many other characteristics inherited as multiple factors, including flower shape in tobacco, fruit size in tomatoes, number of rows of grain on an ear of corn, flower color variations in many species, and numerous examples in humans, such as skin color and probably intelligence.

Characteristics which are inherited as the result of multiple factors usually are dependent upon more than two pairs. In another cross involving a variety of wheat there was only one white-kerneled individual in every sixty-four F_2 progeny. This would be a cross in which each parent has three pairs of factors for the same trait and might be written *RRSSTT* × *rrsstt*. A cross between these homozygous parents would result in a heterozygous progeny of *RrSsTt*. This genotype would result in eight different types of gametes: *RST, RSt, RsT, Rst, rST, rSt, rsT,* and *rst*. If these were placed on one side of a checkerboard and the eight gametes of similar composition from the second F_1 *parent* were placed on the upper side of the square, we would have a sixty-four-square checkerboard. The phenotypes would be one deep red, sixty-two intermediates of varying shades and one white. If this trihybrid cross were of three pairs of dominant-recessive traits, instead of multiple, each for a different characteristic, heterozygous parents would produce progeny in the ratio of 27:9:9:9:3:3:3:1.

This by no means exhausts the different types of inheritance. Many other variations are known, not only of different modes of inheritance, but variations following the types illustrated above. Students interested in genetics will find abundant reading material in this relatively new and fascinating biological field in the many genetics books available.

Selection

The practical aspects of genetics have been known for a much longer time than the mode of inheritance of the genetical factors. For hundreds of years observing farmers have been selecting their best plants for seed for next year's crop as well as their best farm animals for continued expansion of the flock or herd. This type of selection is known as **phenotypic selection.** While it is in general a logical practice to follow, the results are not always as good as expected.

The logical procedure in planting apple trees would be to plant seeds from the largest, tastiest apples. This could be an unfortunate decision since apples are cross-pollinated (the pollen coming from a different plant) and the plant contributing the pollen might have been a scrubby crabapple. The progeny resulting from this cross might well bear fruit of intermediate quality far less desirable than the fruit selected. Other plants in which cross-pollination is common can produce the same unfortunate results.

Hybrid corn, on the other hand, is produced by crossing rather nondescript parents, which in combination result in seeds that will produce larger ears, bigger grains, and often an extra ear on the plant. This is often considered **hybrid vigor,** in which the desirable qualities of the two parents are combined in the progeny. Unfortunately this cross must be made by hand each year to produce the seed, for the large ears and seeds of the hybrid plant cannot be depended upon to produce corn of similar quality.

Inbreeding

The choice of plants to produce crops of good quality or quantity is now more commonly based on **genotypic selection.**

Once the quality of the genes has been determined by the phenotypes they produce, continuation of the character can be achieved. In plants this is often done by **self-pollination** and the line is therefore kept free of outside pollen by inbreeding. In some cases inbreeding in plants is accomplished by mating closely related individuals, as with animals where selected strains are crossed.

In years past it was the belief of many that inbreeding was dangerous and led to inferior stock. This can happen, but inbreeding does not create weaknesses or defects. If they are present as recessives in the genotypes, then inbreeding will bring them out because inbreeding leads to homozygosity. Therefore, if deleterious genes are present in heterozygous individuals they will soon be brought to light in inbreeding. However, many lines of animals and plants have been inbred for numerous generations, establishing a **pure line** which has proved very desirable.

Outbreeding

Outbreeding is equally successful where the individuals are carefully selected. This involves crosses between individuals which are different but which are compatible in the production of sturdy, disease resistant, high quality stock. Plants produced in this way are **hybrids,** and have the quality we term **hybrid vigor.** Increase in size, yield, growth and other characteristics are often obtained by such crosses.

Crosses of this nature have resulted in hybrid corn, wheat, chrysanthemums, and many others. Two pure line strains of the same species have often been successfully crossed to produce an even more desirable product.

One example of the characteristics obtained in a new variety of tobacco where the varieties Broadleaf and Sumatra were crossed is shown in Table 13-1.

Broadleaf	Sumatra	New Variety Round Tip
large leaf	small leaf	large leaf
narrow tip	round tip	round tip
drooping leaf	upright leaf	upright leaf
leaves close	leaves apart	leaves close
texture coarse	texture fine	texture intermediate
leaves thick	leaves thin	leaves intermediate

flavor good but strong	flavor none	flavor intermediate
yield high	yield low	yield high
18 leaves per plant	20 leaves per plant	26 leaves per plant
fair root system	fair root system	wonderful root system

TABLE 13-1 Characteristics of a hybrid variety of tobacco obtained by crossing Broadleaf and Sumatra varieties (E. M. East and D. F. Jones, *Journal of Heredity* 12:51, 1921).

It is obvious from the contents of Table 13-1 that the new variety of Round Tip tobacco is the product of several types of inheritance. Some of the characters are the result of dominant-recessiveness and some of blending, some may be multiple factors and in general the hybrid is the result of outbreeding in producing a plant with hybrid vigor.

Mutations

Mutations are changes in genes potentially capable of being transmitted. Variations from the normal genetic constitution in many species have been known for years but the manner in which they occurred has been subject to conjecture until relatively recently. Since mutants appeared without any known cause they were termed "spontaneous." Their appearance could not be correlated with any external stimuli and no prediction could be made about frequency or type of change.

Various agents such as certain chemicals, radiation by X rays and infrared rays, and radioactive substances have been used to induce mutations. The results have been generally uncertain since the new mutant genes appear infrequently and are mostly recessive. They therefore remain hidden until a fortuitous cross is made bringing the recessive mutant genes together to exhibit their effect in a new phenotype.

Inversion

Muller was successful in producing induced mutants through the use of X rays on the fruit fly, *Drosophila*. He was able, through his experiments, to obtain a predictable rate of mutation which took it out of the spontaneous category. The effect of the radiation was to produce a lethal factor as the result of **inversion** of the genes on a region of a chromosome. If the normal order of genes on the chromosome was a b c d e f g the inversion, resulting in a double break in the chromosome, caused the genes on a segment of the chromosome to be reversed a b e d c f g, as follows:

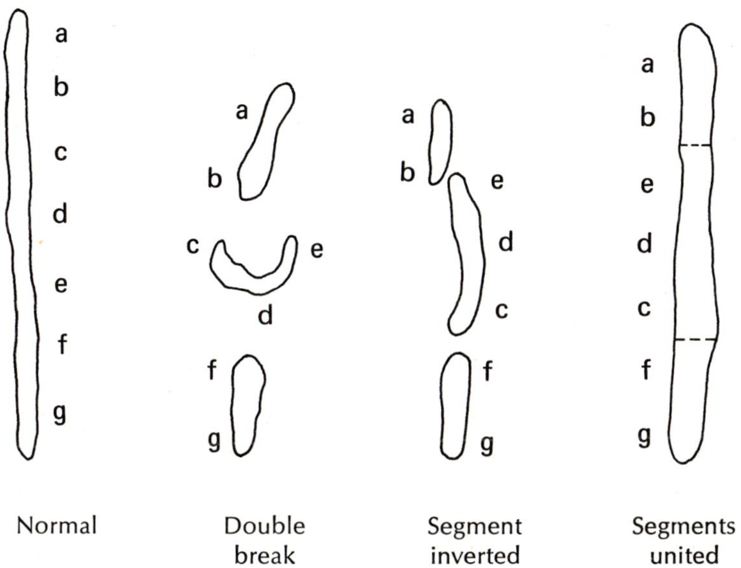

Muller found that this condition occurred on the X chromosome and male fruit flies inheriting this chromosome from their mothers died. The daughter fruit flies received this chromosome in a 1:1 ratio but were unaffected since their other X chromosome inhibited the development of the lethal characteristic.

This experiment provided geneticists with a means of experimentally producing mutants. Since then it has been found that radiation will induce gene mutations of various types, not all of which are lethal.

Normally chromosomes occur in pairs in the somatic cells of the individual, but in the course of genetic research it has been observed that variations in the numbers of chromosomes, structure, loss of parts, and arrangement occur. These are **chromosomal aberrations** and they frequently result in changes in the predictable Mendelian ratios, in linkage groups, and often in the appearance of the individual.

Translocation

Inversion, discussed above, is an aberration. **Translocation** is also a chromosomal aberration, in which a piece of one chromosome breaks off and becomes attached to another chromosome.

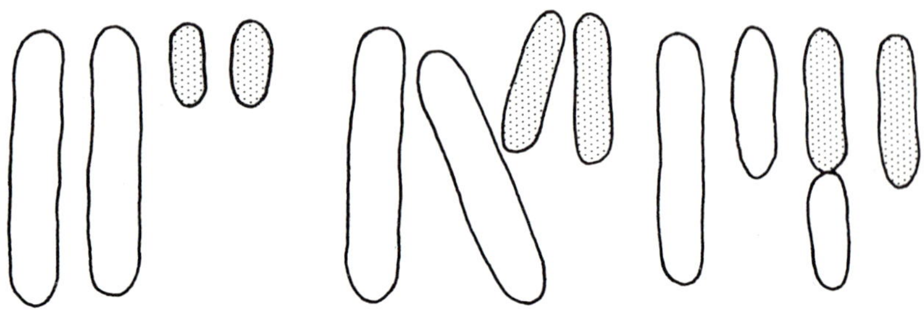

Two pairs of homologous chromosomes | Contact by two members of different pairs | Portion of one chromosome becomes attached to other.

This diagrammatic presentation of translocation illustrates how a segment may be lost from one pair of homologous chromosomes and picked up by another. The translocation probably occurs at prophase when the chromosomes are coiled about each other.

Reciprocal translocations are the exchange of parts by nonhomologous chromosomes. This is, of course, a type of **crossing-over**, except that crossing-over is the exchange of materials between the members of a homologus pair and reciprocal translocation or **segmental interchange,** as it is sometimes called, is an exchange by chromosomes belonging to different pairs.

Polyploidy

Another type of chromosomal aberration occurs when entire sets of chromosomes are involved. We have been emphasizing the fact that the somatic cells of organisms have diploid sets of chromosomes and that they give rise to haploid gametes by meiosis. Occasionally we find individuals that are **polyploid** as a result of whole sets of chromosomes being duplicated. If an individual has three of each kind of chromosome the cells are **triploid,** if four it is **tetraploid,** if five **pentaploid.** Considerable work has been done in recent years in the counting of chromosomes for different species. Many of these investigations have revealed polyploids in different species of certain genera showing how some of these species may have evolved and how they are related.

Polyploids may arise through a doubling of the chromosome number during some phase of the life cycle. This may occur as a result of failure of normal meiosis in megasporogenesis or microsporogenesis. This may result in gametes that are diploid. A fusion between two such gametes would result in a tetraploid individual, or if a diploid gamete fused with a normal haploid gamete the new individual would be triploid. Doubling of chromosomes may also occur in the zygote or in vegetative cells, such as those of a bud during mitosis.

Polyploidy may occur spontaneously or be induced artificially by such chemicals as colchicine and by radiation.

The sunflower, genus, *Helianthus,* has been studied extensively at the University of Indiana by Heiser and his students. The annual species of this genus were all found to be diploid with each species having seventeen pairs of chromosomes. Perennial species of the same genus may be diploid, triploid, tetraploid, or hexaploid. Hybridization occurs readily between most of the species with hybrids artificially produced showing many characteristics of other established species. In a cross between *H. decapitatus* (tetraploid) and *H. annuus* (diploid), a hybrid was produced which has many of the characteristics of the species *H. multiflorus* (triploid). See Table 13-2.

In a cross between two diploid species, *H. annuus* and *H. petiolaris,* Heiser obtained a hybrid which was usually sterile but with many intermediate characters. The hybrid had ten pairs, two chains and one ring of chromosomes indicating that the chromosomes from the two parents were not completely homologous. The chains and ring of chromosomes indicate that translocations and inversions had occurred. Much more complete data on the hybridization and chromosome counts of *Helianthus* species (about fifty in number) are obtainable in the other research papers of Charles B. Heiser, Dale M. Smith, W. C. Martin, A. T. Guard, Ray Jackson, and Sarah Clevenger.

Polyploids arising from a duplication of the chromosome complement within the same species are known as **autopolyploids**. Tetraploids of this type may thus have four identical chromosomes of each kind and at reduction division two of each kind are found in each gamete, which is then a diploid gamete. Fertilization between this type of gamete and a normal haploid one would result

Morphological Characteristic	Helianthus decapitatus (Tetraploid, 2n = 68)	H. multiflorus (Triploid) 2n = 51	H. annuus (diploid) 2n = 34
Stem	essentially glabrous	slightly hispid	hispid
Leaves arrangement	alternate at middle node or above	alternate below middle node	alternate except for lower most nodes
width	3.5-10 cm	8-15 cm.	10-30 cm.
petioles	long-winged	short-winged	little or not at all winged
Disk diameter	1.2-2.5 cm	2.0-3.0 cm.	2.5-25.0 cm.
Phyllaries			
shape	linear lanceolate	broadly lanceolate	ovate
width	2 to rarely 4 mm	4-6 mm	4-15 mm
length	greatly exceeding disk, reflexed	slightly exceeding disk, erect	usually equaling or slightly exceeding disk, erect
Rays			
number	8-12	20 or more	15 to many
color	pale yellow	yellow	usually orange-yellow
Middle cusp of chaff	usually short-acute	short-acute to acuminate	attenuate

Table 13-2 Comparison of selected morphological features of *Helianthus decapitatus, H. multiflorus* and *H. annuus* (from Charles B. Heiser, Jr., and Dale M. Smith, The Origin of Helianthus multiflorus, *American Journal of Botany* 47:860-865, 1960)

in a triploid individual or if fertilization occurred between two diploid gametes a tetraploid individual would be the result. Although autopolyploids may produce fewer viable offspring than the normal diploid individuals, they often are larger, hardier, and more productive than the diploids. Since in autopolyploids there is merely a duplication of the chromosomes, the genetic constitution is not altered except in quantity. Plants or animals not having serious deleterious genes would very likely continue in a very productive manner even though they became autopolyploids. It was pointed out in Chapter 11 that polyploids of various numbers are quite common in certain nuclei of the embryo sacs of higher plants.

Allopolyploids arise from the hybridization of two species or genera. For example, the sperm and egg united in fertilization may each have the haploid number but they remain as individual chromosomes in the zygote since they do not have a homologue with which to pair. Subsequently these chromosomes may duplicate, resulting in a new individual with, for all practical purposes, a diploid number of chromosomes but actually twice as many as each of its parents, assuming each parent had the same number. Then if the individual is fertile it will produce gametes by the usual reduction process, or if it produces diploid gametes and they are united with diploid gametes we have allotetraploids.

Although many allopolyploids are sterile, it is in this way that fertile new species may arise. Considering the countless millions of crosses that occur, it is quite conceivable that among them are many compatible unions.

Radiation

The success we have had in inducing chromosome aberrations has been due largely to radioisotope techniques applied in agricultural research. The production of improved agricultural and ornamental plant varieties requires many years of selection and breeding to obtain the desired characteristics. Radiation has helped to provide a great variety of genetic changes, many of which are economic improvements.

Resistance to disease has been obtained in some species by using radiation-altered stocks. Some improvements in yield have been obtained by irradiation of seeds. The "Somilac" bush navy bean was found to produce about 30 percent more beans per acre than the parent stock and in addition required fewer days from planting to harvesting. Similar results were obtained by irradiation of peanut seeds, which produced a crop of higher yield and greater disease resistance.

Favorable genetic changes are only a small percentage of all changes achieved by radiation. Selection and breeding are still necessary, but we have the means for variation at our disposal.

Recently a study of the effect of radiation on an obnoxious weed, *Tribulus terrestris,* commonly called goat head or burnut (Fig. 2), was completed. Irradiated seeds consistently produced plants bearing larger seeds with longer spines than the parent stocks. The effect desired was to produce a spineless or lethal stock, for the plant grows in pastures and in clover and alfalfa fields. When cattle eat the hay some of the spiney fruits get into the animals' mouths, causing serious sores and ulcers. The spines

A

FIGURE 2 *Tribulus terrestris*. The shoot in A has fruits developed from non-irradiated seeds. In B the fruit is one developed from irradiated seeds. Note larger spines in B. Each segment separates from the fruit into small goathead-like units.

B

also puncture many bicycle tires. Our results indicated that irradiation of the goat head seed is good for the plant but not for humanity.

Radiation has also been used successfully to produce somatic mutations in plants such as fruit trees, which can then be propagated by cuttings or by grafting. Two such beneficial mutations have been reported in peaches of the Fairhaven variety. One branch on one tree had had its genes altered to the extent that it would produce fruit which ripened approximately ten days earlier than normal. A branch on a different tree was genetically induced by radiation to ripen its fruit three weeks *later* than normal. These two mutations may thus increase the peach season by more than a month. Length of season is an important factor to growers of soft, perishable fruits.

Numerous experiments have been conducted with irradiated plant material as well as with animals. Disease-resistant strains have been reported in experiments with wheat, oats and flax. High-yielding forms of cereal grasses with stronger stems that suffer less wind damage have been reported. The elimination of a toxic factor in some otherwise useful forage plants has also been achieved through radiation.

Each year many new and successful genetic changes are described in the scientific literature. The genes form a continuous stream in the germplasm of each species, but the little alterations we can induce in the nucleic acids and the phenotypic results obtained make genetics a very moving and fascinating field.

Supplementary Reading

Bateson, W., *Mendel's Principles of Heredity*. Cambridge University Press, 1909.

Beadle, G. W., Biochemical Genetics, *Chemical Review* 37:15–96, 1945.

Cook, S. A., *Reproduction, Heredity, and Sexuality*. Wadsworth Publishing Co., Belmont, California, 1964.

Crick, F. H. C., The Genetic Code, *Scientific American* 207:66–75, 1962.

Levine, R. P., *Genetics*. Holt, Rinehart and Winston, Inc., New York, 1962.

Moment, G. (editor), *Frontiers of Modern Biology.* Houghton Mifflin Co., Boston, 1962.

Moore, John A., *Heredity and Development.* Oxford University Press, New York, 1963.

Muller, H. J., Artificial Transmutation of the Gene, *Science* 66:84–87, 1927.

Punnett, R. C., The Early Days of Genetics, *Heredity* 4:1–10, 1950.

Sinnott, E. W., L. C. Dunn, and Th. Dobzhansky, *Principles of Genetics.* McGraw-Hill Book Co., New York, 1958.

Questions for Discussion

1. Why are unicellular organisms and primitive species used more extensively than higher organisms to study mutation rates?
2. Many plant and animal species of economic importance have been greatly improved genetically. Could we do the same thing for human species?
3. How could you test the effect of radiation on plant species in future generations?
4. Why does hybrid corn, raised so extensively today, usually fail to breed true?
5. Why may haploid plants produce abnormal gametes?
6. Why does heterosis tend to disappear with inbreeding?
7. Why are hybrid roses always grafted to other root stocks after the new hybrids are developed?
8. What evidences do we have of cytoplasmic inheritance?
9. In garden peas let smooth seed coat be represented by *S* and wrinkled by *s;* let tallness be represented by *T* and dwarfness by *t.* What will the progeny of the following crosses be like?

$$SSTT \times sstt \qquad SsTt \times sstt$$
$$SsTt \times SsTt \qquad SsTT \times ssTt$$
$$SStt \times ssTT \qquad ssTt \times SsTt$$

10. How would the F_2 generation in a cross involving two linked genes be different from a similar pair which are independently assorted?

11. Why is a plant of an F_1 generation, whose genotype has not been established, backcrossed with the double recessive?
12. How would one know that flower color in a given species is the result of incomplete dominance?
13. What is meant by inversion? translocation? sex linkage?
14. Compare and contrast meiosis with mitosis.
15. If one can readily recognize the phenotype of an organism, which is based on its genotype, why do we not know what the genotype of the organism is?
16. Where does crossing-over probably take place in the origin of the sex cells?
17. How do we know that DNA and RNA are essentially the same in all plants and animals?
18. Would you suspect that Mendel observed characters in garden peas other than those he wrote so extensively about? Explain.
19. What are the major points considered in "Mendelism"?
20. Distinguish between the term homologous, homozygous, and heterozygous.
21. In the following ratios what types of inheritance are involved?

1:2:1	15:1
3:1	9:6:1
9:3:3:1	50:50
48:4:3:49	1:4:6:4:1
1:2:1:2:4:2:1:2:1	1:62:1

22. Why is inbreeding occasionally considered undesirable?
23. Would an organism whose somatic cells have forty-four chromosomes as a diploid number be a more complex organism than one in which the diploid number is sixteen?
24. How may mutations be induced? What probably stimulated mutations to occur in ages past?
25. What do we mean by chromosomal aberrations?
26. Study Table 13-2. What types of inheritance are probably involved in the different characteristics listed for these sunflowers and the origin of *Helianthus multiflorus?*
27. What are some of the benefits we have derived from induced mutations and hybridization in plants?
28. Why are some hybrids sterile?
29. How could such plants as seedless grapes originate and be continued in propagation if they do not produce seeds?
30. What do we mean by the DNA code?

14

The Classification of Plants

We have been considering, in most of the preceding chapters, the higher plants and their morphology, physiology, and manner of growth. However, there are thousands of other species, classified in about a dozen major groups, each with certain outstanding characteristics which set it apart. The branch of botany which deals with the classification of organisms is **taxonomy** or **systematic botany.** Taxonomy is probably the oldest of the recognized biological fields, and yet it probably has been as changeable as any of the areas devoted to the study of plants and animals.

For the most part the naming of organisms has been done for the sake of convenience, since any attempt to classify organisms according to a true evolutionary relationship has resulted in inconsistencies and a great many unavoidable errors. This has not prevented man from giving organisms names any more than it has kept him from giving names to almost anything he uses or comes in contact with.

Early Naming Systems

The naming of plants (and animals) undoubtedly goes back to prehistoric times. Early man used many plants for food, shelter, and, probably not much later, clothing. He probably found that some plants were quite tasty, others unpalatable, and still others poisonous. Surely when a type of communication was established one of the first things man must have introduced into his vocabulary was the naming of the various plants and animals he could use or had reason to avoid. He must also have noticed certain similarities in groups of plants which gave him the courage to try new ones on the basis of pleasant experiences with other plants of similar appearance.

Probably the earliest type of systematic classification was based on the uses man found for plants: those which could be used for food; those from which fiber could be extracted for making ropes and clothing; wood which could be used in making weapons; plants which he believed had medicinal value, and last but not least those which he felt had certain supernatural powers and could be used in his religion. Many of the early names which were used were passed down through oral communication for centuries until they became a part of history through the records of the early scribes. Some of these names survive today as common names, and others have been latinized and appear as part of our scientific vocabulary in the taxonomy of organisms.

Theophrastus

In the early days of classification writers were at a loss for names, for the species they encountered were far more numerous than the descriptive words available for names. This led to the use of cumbersome phrases, sentences, and even entire paragraphs to name individual species. Theophrastus (370–285 B.C.) made one of the first attempts to classify all plants known in his time. In his system he definitely attempted to classify plants on the basis of form and longevity. His major groupings were of trees, shrubs, undershrubs, and herbs. The herbs were also classified into annuals, biennials, and perennials. In addition he described certain flower characteristics and his descriptions were used later by other early systematists but usually in a different way.

Tournefort and Ray

The classification systems of Theophrastus, and his contemporary Aristotle, persisted for centuries with little modification and virtually no improvement. By the seventeenth and eighteenth centuries a number of botanists became interested in taxonomy and brought forth a number of publications on systematic botany. Tournefort in 1700 continued the scheme established by Theophrastus in dividing flowering plants into two categories, trees and herbs. He then subdivided these groups into flowers with petals and those without. Flowers were further emphasized in this system, since Tournefort divided them into "petals separate" and "petals united." About the same time John Ray, an English botanist, classified about 18,000 species, in which he recognized the differences between dicotyledonous and monocotyledonous plants.

Linnaeus and Binomial Nomenclature

There were others who preceded Tournefort and Ray and many who followed with suggested innovations in the classification and naming of organisms, but the dominant figure in systematic botany was Carolus Linnaeus (1707–1778).

Linnaeus classified all flowering plants into twenty-three different **classes** based on the number and arrangement of stamens, and then subdivided the **classes** into **orders** on the basis of the styles in each group. A twenty-fourth class included all plants without flowers, namely, ferns, mosses, liverworts, algae, and fungi.

Linnaeus traveled very extensively throughout the European area visiting the major herbaria, arboreta, and gardens. He often worked for months in different countries and so impressed the royal rulers of the different kingdoms that his work was readily supported financially. Between 1735 and 1758 Linnaeus published a number of lists classifying animals and plants. These were the most complete lists published up to that time, but more important was his influence in establishing standards which were recognized and adopted by taxonomists all over the world. Although Linnaeus was not the first to use two names for each organism, he is generally credited with throwing aside the cumbersome descriptions employed by most of his predecessors.

Linnaeus introduced a brief, graphic method of formal description for each species of plant or animal. Descriptions were written in Latin, in his day and still today the only international language of science. Each organism was given a double name and this system of **binomial nomenclature** has been employed ever since. The first of the two names is the *genus* and the second is the specific epithet; this with the genus name is the *species,* as *Rosa fendleri.* Essentially organisms of the same species are fertile although, in plants more than in animals, there may be some cross-fertilization and production of fertile offspring by hybridizing different species within the same genus.

A specific epithet may be used only once within a given genus, but it may be used many times with other generic names. The generic name is always capitalized and can be used only once as a valid name among all organisms. The name of a species, like that of a genus, is selected by the individual who writes and publishes a description of the organism. The species name should be descriptive of some characteristic color, a morphological feature, habitat, use, or some other readily recognized feature. Specific epithets often used include the following:

Color	**Form**	**Locality**
alba	lentiformis	arizonicum
aurea	linearifolius	desertorum
cardinalis	erecta	aquitalis
rubrum	recurvous	missouriensis
viridis	filiformis	nevadensis

Epidermal Surface	**Use**	**Named after Someone**
canescens	catharticus	coulteri
glandulosus	eremiticus	nuttallii
hirsuta	edulis	grayi
lanata	officinale	fendleri
stellatum	usitatissimum	jonesii

Difficult to Determine
ambiguus
confusa
desperatus
obscura

There are hundreds of other names, many of which give the reader a clue to the species identity, such as odortatus, superbus, spectabilis, elegans, triloba, arida, succulentus, and diversicolor.

Expansion of Linnaeus' System

In the more than 350,000 plant species classified, more encompassing categories were needed to indicate similarities and relationships among the different organisms. The name selected for the category just above the genus level is **family.** A family may be unique and contain only one genus, but more commonly it has several. One of our common vegetable families, the **Cruciferae,** the mustard family, has such species as the radish, *Raphanus sativus,* the turnip, *Brassica rapa,* cabbage *Brassica oleracea,* and watercress, *Rorippa nasturtium-aquaticum.*

The family Cruciferae has certain affinities with several other families, including the poppy family, **Papaveraceae,** and the caper family, the **Capparidaceae,** and consequently is grouped with them in the **order Papaverales.** This order, along with about thirty others, is in the subclass **Dicotyledonae.** The latter, along with the subclass **Monocotyledonae,** is in the class **Angiospermae,** in which all flowering plants are categorized. The Angiospermae, along with the two classes Gymnospermae and Filicinae, is in the subphylum Peteropsida. The latter and three other subphyla are in the **Tracheophyta,** the phylum of vascular plants. This phylum and only one other phylum, the Bryophyta, are in the subkingdom **Embryophyta** which with its companion subkingdom, the Thallophyta, comprise the Plant Kingdom.

Plant Kingdom

Subkingdom Embryophyta
Phylum Tracheophyta
Subphylum Pteropsida
Class Angiospermae
Subclass Dicotyledonae
Order Papaverales
Family Cruciferae
Genus *Raphanus*
Species *sativus*

FIGURE 1 Summary of the classification of a radish. In addition to the scientific name there are many common variety names such as globe, icicle, crimson giant, long scarlet, and sparkler, each different from the others.

There are other subcategories such as suborder, subfamily, and subgenus which could be mentioned, but the various subspecies and variety epithets probably have more meaning. Everyone is familiar with the great variety of roses which have been developed to such an extent that they are not only listed as grandifloras, multifloras, hybrid teas, and floribundas but also many are given individual names such as Peace, Herbert Hoover, Show Girl, Countess Vandal, etc.

Less similar in appearance but just as closely related are some of the vegetables we listed earlier:

Brassica oleracea var. caulo-rapa—kohlrabi
" " " *acephala*—collard, kale
" " " *botrytis*—broccoli, cauliflower
" " " *gemmifera*—brussels sprouts
" " " *capitata*—cabbage

Few species have so many varieties, and most of these are man-induced and have been developed to a very high degree for a particular purpose. Few of these plants could survive in nature, if left to propagate alone, but with our help they are doing quite well.

Recent Systems of Classification

While the above categories are in wide use in present day systems of taxonomy, they are often given different designations by other writers.

Engler and Bessey

For close to one hundred years, in the nineteenth century and early twentieth centuries, the dominant system of classification was that proposed by Engler. In this system the entire plant kingdom was divided into four major **divisions** as outlined below.

DIVISION I: THALLOPHYTA—thallus plants
- Subdivision 1. **Algae**
- Subdivision 2. **Fungi**

DIVISION II: BRYOPHYTA
- Class 1. **Hepaticae**—liverworts
- Class 2. **Musci**—mosses

DIVISION III: PTERIDOPHYTA
- Class 1. **Lycopodineae**—club mosses
- Class 2. **Equisetineae**—horsetails
- Class 3. **Filicinae**—ferns

DIVISION IV: SPERMATOPHYTA—seed plants
- Class 1. **Gymnospermae**—confiners and allies
- Class 2. **Angiospermae**—flowering plants
  - Subclass 1. *Dicotyledonae*
  - Subclass 2. *Monocotyledonae*

The Engler system was based mainly upon the reproductive organs, but it failed to take into consideration the great diversity of the organisms in the Thallophyta. Most botanists are of the opinion that it also placed too much emphasis on the "seed habit" (separating the Spermatophyta from the Pteridophyta) since seeds were apparently developed many times in Peteridophytean groups according to the fossil evidence. While the Engler system was reasonably convenient, other systematists, including Charles E. Bessey, sought something better. Bessey's system, outlined below, was not widely adopted. It probably would have been, except for some unfortunate terminology used by him and the fact that the Engler system was so deeply entrenched.

Tippo

In 1942 Tippo proposed a classification system for the plant kingdom in which he divided all plants into two subkingdoms and these into phyla. Succeeding categories were divided as before into classes, orders, families, genera, and species.

Attempting to show the great diversity of organisms in the subkingdom Thallophyta, Tippo divided this group into ten phyla. These phyla, formerly considered classes, comprise a very heterogeneous group which seem to have had several origins and followed several lines of development, and in some instances apparently gave rise to higher organisms.

The second subkingdom, the Embryophyta, dominates the plant world today. The plants themselves are larger and therefore much more conspicuous, but, more than that, they have come to occupy most of the terrestrial habitats and are our most familiar plants.

Perhaps the greatest cause of opposition to Tippo's system of classification has been his abandoning of the word "division" and the use of the word "phylum." The use of the word "division" had become a tradition in botanical circles, and it has been continued by members of taxonomic societies in a recent meeting of the Botanical Congress on the International Rules of Binomial Nomenclature. As a result a number of botanical books have been written using the same categories as Tippo, in some cases with slight modification, but substituting "division" for "phylum." Some books use "division" as a subkingdom designation, others use it as a substitute for the word "phylum," and some use it as a class category. In any case the word "division" no longer has the same status it held in Engler's system.

C.E. Bessey System of Classification for the Angiosperm Families The figures in parentheses indicate the number of orders or families in each group.

Class, Alternifoliae (Monocotyledoneae)
- Subclass—Strobiloideae
  - Orders (5)
    - Families (30)
- Subclass—Cotyloideae
  - Orders (3)
    - Families (17)

Class, Oppositifoliae (Dicotyledoneae)
- Subclass—Strobiloideae
  - Superorder—Apopetalae—Polycarpellatae
    - Orders (7)
      - Families (104)
  - Superorder—Sympetalae—Polycarpellatae
    - Orders (3)
      - Families (25)
- Subclass Cotyloideae
  - Superorder—Apopetalae
    - Orders (7)
      - Families (82)
  - Superorder—Sympetalae
    - Orders (3)
      - Families (22)

The System Used in this Book

For the sake of uniformity in biological circles, we will retain the use of the word "phylum" in this book and will follow the Tippo system in classifying members of the plant kingdom as outlined below. There are inconsistencies in all systems of classification, since many organisms are placed as the result of subjective opinion. If it were possible to classify all organisms according to a known ancestry, or "blood relationship," we could be much more objective. The most important thing is for the student to know the characteristics of the plants in the different groups and where these fit into the phylogeny of the plant kingdom, rather than to be overly concerned about whether the group must be designated as a division, a phylum, or a class.

Plant Kingdom

Subkingdom THALLOPHYTA

THE ALGAE

Phylum Cyanophyta
  Class Cyanophyceae

Phylum Chlorophyta
  Class Chlorophyceae
  Class Charophyceae

Phylum Euglenophyta
  Class Euglenophyceae

Phylum Pyrrophyta
  Class Dinophyceae

Phylum Chrysophyta
  Class Xanthophyceae
  Class Chrysophyceae
  Class Bacillariophyceae

Phylum Phaeophyta
  Class Phaeophyceae

Phylum Rhodophyta
  Class Rhodophyceae

THE FUNGI

Phylum Schizomycophyta
  Class Schizomycetae
  Class Actinomycetae

Phylum Myxomycophyta
  Class Myxomycetae
  Class Acrasieae

  Class Plasmodiophoreae

Phylum Eumycophyta
  Class Phycomycetae
  Class Ascomycetae
  Class Basidiomycetae
  Class Deuteromycetae

Subkingdom EMBRYOPHYTA

Phylum Bryophyta
  Class Hepaticae
  Class Anthocerotae
  Class Musci

Phylum Tracheophyta
  Subphylum Psilopsida
    Class Psilophytineae
  Subphylum Lycopsida
    Class Lycopodineae
  Subphylum Sphenopsida
    Class Equisetineae
  Subphylum Peteropsida
    Class Filicineae
    Class Gymnospermae
      Subclass Cycadophytae
      Subclass Coniferophytae
    Class Angiosperm
      Subclass Dicotyledonae
      Subclass Monocotyledonae

FIGURE 2 This yucca is commonly called soaptree, palmilla, and Spanish bayonet. In scientific terminology it is in the following categories:
Subkingdom—Embryophyta
Phylum—Tracheophyta
Subphylum—Pteropsida
Class—Angiospermae
Subclass—Monocotyledonae
Order—Liliales
Family—Liliaceae
Genus—*Yucca*
species—*elata*

The above outline of the various major groups does not necessarily follow the order in which they may have evolved. In many cases it may, but in such a heterogeneous group as the algae, with seven distinct phyla, there were undoubtedly parallel lines of evolution. We will therefore consider the algae in three different chapters, based on the characteristics which would indicate their relationships and the three major lines of development which they seem to have followed. This will place the Rhodophyta in the same chapter with the Cyanophyta, the Chlorophyta with the Euglenophyta, and the Phaeophyta with the Pyrrophyta and Chrysophyta.

The Thallophyta

The **thallus** plants comprise over 100,000 species, ranging in size from microscopic, unicellular plants to some over one hundred feet long (Fig. 3). It is often easier to describe plants in this

FIGURE 3 The thallophyte *Nereocystis,* a member of the brown algae. The long, curving stipes with their terminal floats and blades are anchored by a large holdfast more than ten feet below the ocean surface.

group on the basis of what they do *not* have, since they lack true leaves, roots, stems, flowers, fruits, and seeds, than to say what they have. Thallus plants have a vegetative body which may consist of only one cell; or it may be differentiated into distinctly formed colonies; or differentiated into a leaflike organ, the **blade** or **lamina,** a stemlike structure, the **stipe,** and rootlike structures termed **rhizoids.** The sex organs and sporangia are usually unicellular and superficial in position. Occasionally the reproductive organs are borne imbedded within protective tissues of the thallus, but they are not surrounded by specialized, multicellular jackets of sterile cells as in the higher groups.

Reproduction in the Algae

Asexual Reproduction In the simpler plants such as the blue-green algae, and primitive members of other groups, asexual reproduction is the only mode known. This is usually accomplished by fission in which the individual cells simply split by a furrowing of the protoplasm separating the cell into two parts. In some algae mitotic division accompanies cell division, but in the simpler plants nuclei are lacking. In those plants where spores are produced fission is usually the type of division which is accompanied by the germination of the spore.

Sexual Reproduction Sexual reproduction is usually accomplished by the union of sex cells **(gametes).** However, in some species there may be a transfer of nuclei between different unspecialized cells resulting in a nuclear fusion. Plants producing gametes which are identical in appearance are said to be **isogamous** (like gametes). This is considered the simplest type of sexual reproduction since the gametes have no distinguishing morphological characteristics. It is apparent, however, that there are certain genetic or physiological differences since gametes from different plants usually unite to accomplish fertilization. Those plants which look alike and produce isogametes, yet where the gametes of two different plants are required for fertilization, we term **heterothallic** and the two plants are often designated as + and − rather than male and female. Plants which are alike but in which gametes from the same plant may unite are termed **homothallic.**

A more advanced form of sexual reproduction is recognized in plants where the gametes are similar in appearance but differ in size. The gametes produced here are **heterogamous** or **anisogamous.** Usually both kinds of gametes are flagellated; the larger is considered the female and termed the **megagamete** and the smaller, male, is called the **microgamete.**

The most advanced type of sexual reproduction is found in the higher thallophytes and all embryophytes. This is **oogamy.** Oogamous plants produce large nonmotile eggs and small sperms. The sperms are usually flagellated in the Thallophytes and Bryophytes and even in the more primitive Tracheophytes.

Reproduction in plants is usually accompanied by alternating stages of haploid (*n*, number of chromosomes) cells or plants and diploid (2*n*, number of chromosomes) plants, the result of a sexual fusion. In the simpler algae the filament is haploid, as are the gametes, but union results in a diploid zygote. Reduction division usually occurs in the zygote and upon germination the zygote produces a haploid filament or haploid spores which then form the filament:

$$\overset{n}{\text{Filament}} \longrightarrow \overset{n}{\text{gametes}} \longrightarrow \overset{2n}{\text{zygote}} \xrightarrow{\text{R D}} \overset{n}{\text{spores.}}$$

Some species have a delayed meiotic phase in which the zygote germinates and forms a second plant which bears sporangia. These structures give rise to spores by reduction division, which in turn form the gamete-producing plants. Plants which form the gametes are **gametophytes** and those which form spores are **sporophytes.** The latter are usually diploid and the former haploid:

$$\overset{n}{\text{Gametophyte}} \longrightarrow \overset{n}{\text{gametes}} \longrightarrow \overset{2n}{\text{zygote}} \longrightarrow \overset{2n}{\text{sporophyte}} \xrightarrow{\text{R D}} \overset{n}{\text{spores.}}$$

We therefore have an **alternation of generation** in which a sexual plant, the gametophyte, alternates with an asexual plant, the sporophyte. In many of the algae having this type of reproductive cycle, both the asexual and sexual plants are identical in appearance and we speak of the cycle as **isomorphic.** In many higher algae, the two stages are quite different and therefore may be termed **heteromorphic.**

The Algae and Fungi

The subkingdom Thallophyta is divided into two groups, the algae and fungi (including the bacteria), based largely on the differences in their metabolic capabilities as well as their dominant morphologic features.

The algae are **photoautotrophic** plants and therefore are able to manufacture their own food through some photosynthetic process. The pigments used in the process are the chlorophylls, which vary along with some of the other pigments in the various algal groups. In fact, pigmentation is a major characteristic in separating the various algal groups into the three groups presented in Chapters 15, 16, and 17.

Bacteria and fungi generally lack internally formed photosynthetic pigments and therefore must depend on external sources for food and nutrients. Most members of this very large group are therefore **heterotrophic** and live as **saprophytes** on decayed organic matter, or live as **parasites** upon or within the cells of living plants and animals. There is a small group of bacteria which contain a green pigment, **bacteriochlorophyll,** which is similar in function to the chlorophyll of green plants but varies slightly from it chemically. It is from organisms of this type that we believe oxygen first diffused into the atmosphere. Some of these pigmented bacteria, when grown anaerobically in the presence of light, absorb hydrogen sulfide and carbon dioxide, produce an organic compound, and release oxygen.

Bacteria, in general, fall into the unicellular category although some are colonial. Fungi are usually multicellular and vary considerably in the types of reproductive organs and sporangia they produce.

The Embryophyta

The thallus plants have been categorized into ten different phyla, but the plants which produce embryos are all found within two phyla: the **Bryophyta** and the **Tracheophyta.**

Variation in Plants

In the chapters to follow it will be noted that there is a gradually increasing complexity in the vegetative and reproductive bodies of the organisms. In the earlier chapters diversity will be much more prominent than advancement. Gradually lines of evolution become established and plants of increasing complexity are evolved. Evolution works in this way. Of the many different body forms developed some proved to have excellent survival value and persisted. Others were unable to adapt themselves to the environment and eventually perished. The survivors continue to undergo changes, through genetic variations, giving rise to new species and new evolutionary lines.

The mechanisms of evolution including polyploidy, mutation, and hybridization are considered in detail in the chapters on Genetics (13) and Evolution (26), and are also referred to as the occasion arises throughout the text. It should be kept in mind that survival and perpetuation of the species depend on random genetic variations which turn out to be beneficial to the species and which are therefore the controlling factors in a species' adaptation to the existing environmental condition. Success and survival in any given environment depend on a fortuitous genetic constitution which happens to be advantageous to its bearers.

Supplementary Reading

Bailey, L. H., *How Plants Get Their Names*. The Macmillan Co., New York, 1933.

Benson, L., *Plant Classification*. D. C. Heath and Co., Boston, 1957.

Boughey, A. S., *Population and Environmental Biology*. Dickenson Publishing Co. Inc., Belmont, California, 1967.

Cove, E. L., *Plant Taxonomy*. Prentice-Hall, Inc., Englewood Cliffs, N. J., 1955.

Cronquist, A., The Division and Classes of Plants. *Bot. Rev.* 26(4) 425–482, 1960.

Questions for Discussion

1. Many people have difficulty remembering scientific names. What, then, is the value of having a latinized system of binomial nomenclature?
2. Why did some of the early systems which classified higher plants as trees, shrubs, and herbs prove inadequate?
3. Will all organisms eventually be completely classified, with no more to be named? Give reasons for your answer.
4. What are the advantages in using a phylum system of terminology in the classification of both animals and plants?
5. What is the advantage in using the floral parts of higher plants, rather than stems, leaves, or roots, in classifying them?
6. The early botanists were chiefly taxonomists. Why would you expect this to be a logical development?
7. Several brown algae are over one hundred feet long, yet do not have the mechanical or conductive tissues of terrestrial plants. How do you account for this?
8. What are the characteristics of the different plant groups which cause us to put them in different phyla?
9. Why was it necessary for the development of systematic biology that a system of binomial nomenclature become established?
10. Why was it wise to select Latin as the international language of science terminology?
11. Why were the best systems of classification devised based mostly on the sexual plant parts?
12. Examine the plant names in a taxonomy book and note how many species names you can recognize in their latinized forms. A knowledge of a romance language is a great help.
13. Why isn't it possible to arrange all plants in a perfect evolutionary series?
14. If one is familiar with plants it is usually rather easy to place a given plant in its correct family even though one has never seen it before. Why is this true?
15. Would it be possible to identify plants by their roots? Explain.
16. The identification of plants, animals, rocks, minerals and almost everything else is based on "keys" of specific characteristics. Prepare a key for the identification of the following items: A rocking chair, straight chair, table, granite boulder, limestone, iron ore, redwood tree, oak, ash, aster, sunflower, dog, cat, rat, pig, watermelon, root of sweet potato, palm tree and milk carton.

15

The Non-flagellated Algae

Phylum Cyanophyta
Phylum Rhodophyta

The blue-green and the red algae seem to have few characteristics in common when superficially examined, but on closer inspection we see that they have very definite affinities. They are the only algal phyla in which flagellated cells of any kind are lacking. They both have **phycobilin** pigments, which are not present in other groups. The phycobilins are water-soluble, proteinaceous pigments. The blue phycobilin is **phycocyanin,** the dominant pigment of the blue-green algae, and the red phycobilin is **phycoerythrin,** the dominant pigment of the red algae. Both groups have the type of chlorophyll designated as *a* and in addition the red algae also have a related chlorophyll designated as *d*.

The food storage reserves are also similar chemically. The blue-greens store most of their food as a **cyanophycean starch,** while the carbohydrate reserve of the red algae is a related polysaccharide termed **floridean starch.**

These are important features which are basic to the classification of algae. They are also unique to these groups and therefore indicate to us a relationship (Table 15-1) The differences are so

Table 15-1 Comparison of Blue-green Algae and Red Algae

	Blue-green	**Red**
Habitat	Fresh, brakish, marine water, some in soil	Mostly warm marine water
Body plan	Unicellular, filamentous, or colonial	Rarely unicellular, mostly large multicellular or filamentous
Cellular organization		
Wall	Distinct, mostly cellulose	Distinct, cellulose
Gelatin	Present in many in and around walls	Often present in large quantities
Nucleus	Nuclear material present but membrane lacking	Nucleus often very small but present
Pigments	Phycocyanin, chlorophyll a, phycoerythrin, carotenoids	Phycoerythrin, phycocyanin chlorophyll *a, d,* carotenoids
Plastid	Lacking	Distinct
Reserve food	Cyanophycean starch	Floridean starch
Sexual reproduction	Lacking	Isogamy to oogamy
Asexual reproduction	Fission, hormogonia	Fragmentation
Types of spores	Akinetes, endospores, "heterocysts"	Carpospores, tetraspores
Flagellated Cells	Lacking	Lacking
Economic importance	Frequent pollutants in water; some may be toxic; some blue-green algae fix nitrogen in the soil	Agar as food base and bacteriological media

great, however, that we may assume that while the red algae have many similarities to the blue-green algae they have evolved far beyond them. For the most part blue-green algae are small, are often unicellular, and are frequently found in definite colonies but have very little differentiation in cellular or tissue structure.

The cells of blue-greens lack organized nuclei and plastids. Furthermore they do not produce gametes of any kind, nor do they have any sporangia beyond a single-celled type which may contain one or a few small spores. No sexuality of any type is known in the blue-greens, but that found in the red algae is so different from the types found in other algae that we believe sex must have arisen separately and after the red algae had branched off from very early, remote blue-greens or from their primitive ancestors.

The red algae have nucleated cells, produce non-flagellated gametes, and often have sporangia which produce asexual spores. In reproduction the red algae are usually sexual while the blue-greens can only reproduce asexually.

Phylum Cyanophyta, Blue-green Algae

CLASS CYANOPHYCEAE

Habitat

The blue-green algae are found in fresh as well as salt water habitats; they are frequently found in fertile soils, occasionally several feet beneath the surface; they live in the waters of hot springs and in the frigid waters of the arctic. They make up a major portion of the free-floating plankton of lakes and ponds, and are continually found in pools and city reservoirs where their decomposing remains pollute the drinking water.

Some blue-green algae live in association with other plants and animals as epiphytes or as symbionts. A few of those in soil probably live as saprophytes and a few found in the digestive tracts of animals may live a passive parasitic existence. Some living in the cells of coelenterates or protozoans are able to continue manufacturing food, some of which diffuses into the cells of the host.

Antiquity

No fossils of blue-green algae are known from the oldest geological era, the Archeozoic, but graphite deposits indicate the

Reproduce by fission

presence of organisms whose remains were decomposed by other microorganisms. Some algal impressions and limestone "ball" deposits appear to have been of blue-green algal origin. Some deposits in lake bottoms, termed **marl,** were possibly formed by aquatic plants that absorbed carbon dioxide in water, and precipitated the calcium and magnesium salts. This precipitation occurs in waters which are highly charged with these salts. Normally calcium is held in solution by carbon dioxide, but if the CO_2 is removed the salts are precipitated on the plant part abstracting the gas. If the salts are not in excess the precipitate may be due to the conversion of calcium bicarbonate into the normal carbonate ($CaCO_3$) by the oxygen liberated from the plant in photosynthesis. The calcium salts are deposited on the outside of the plant as minute crystals. This phenomenon can be observed in most mineral springs and streams today. It is also the way, along with evaporation, that minerals are deposited and accumulated around hot springs and geysers.

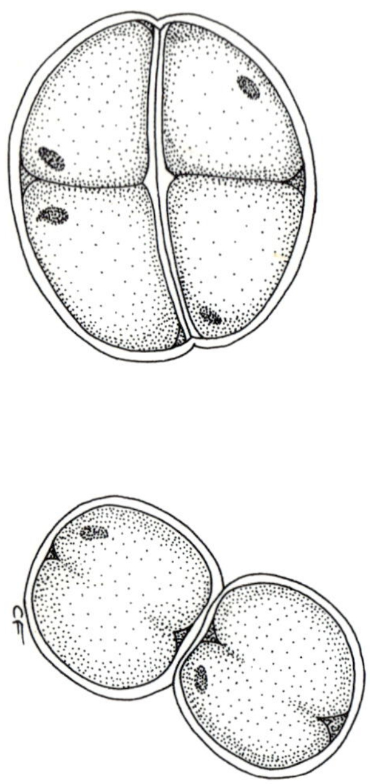

FIGURE 1 *Chroococcus,* a unicellular blue-green, in the process of division.

Morphology

Blue-green algae may occur as unicellular organisms, as filaments, or as colonies which have been developed from cellular division in one, two, or three planes. The individual cells in a filament or colony may be all alike, or they may vary in size and to a slight degree in shape and function.

The distinguishing feature separating the Cyanophyceae into two orders, the Coccogonales and the Hormogonales, is based upon the cell or colony shape, and particularly whether or not the plant is filamentous. The members of the order Coccogonales are all spherical in shape and are either unicellular or in colonies of a few cells as in *Chroococcus* (Fig. 1) and *Gloeocapsa* (Fig. 2) or in large flat rectangular plates *(Merismopedia)* (Fig. 3) or in variously shaped masses like *Microcystis* and *Coelosphaerium.*

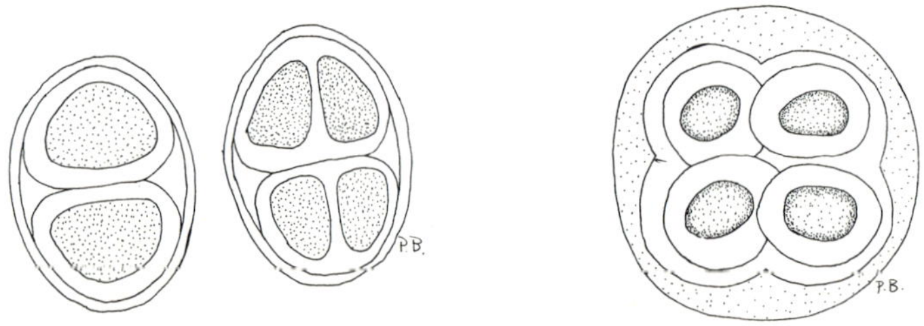

FIGURE 2 *Gloeocapsa,* note the gelatinous sheath which each cell forms around itself. Ultimately at the four-celled stage each cell has three sheaths around it.

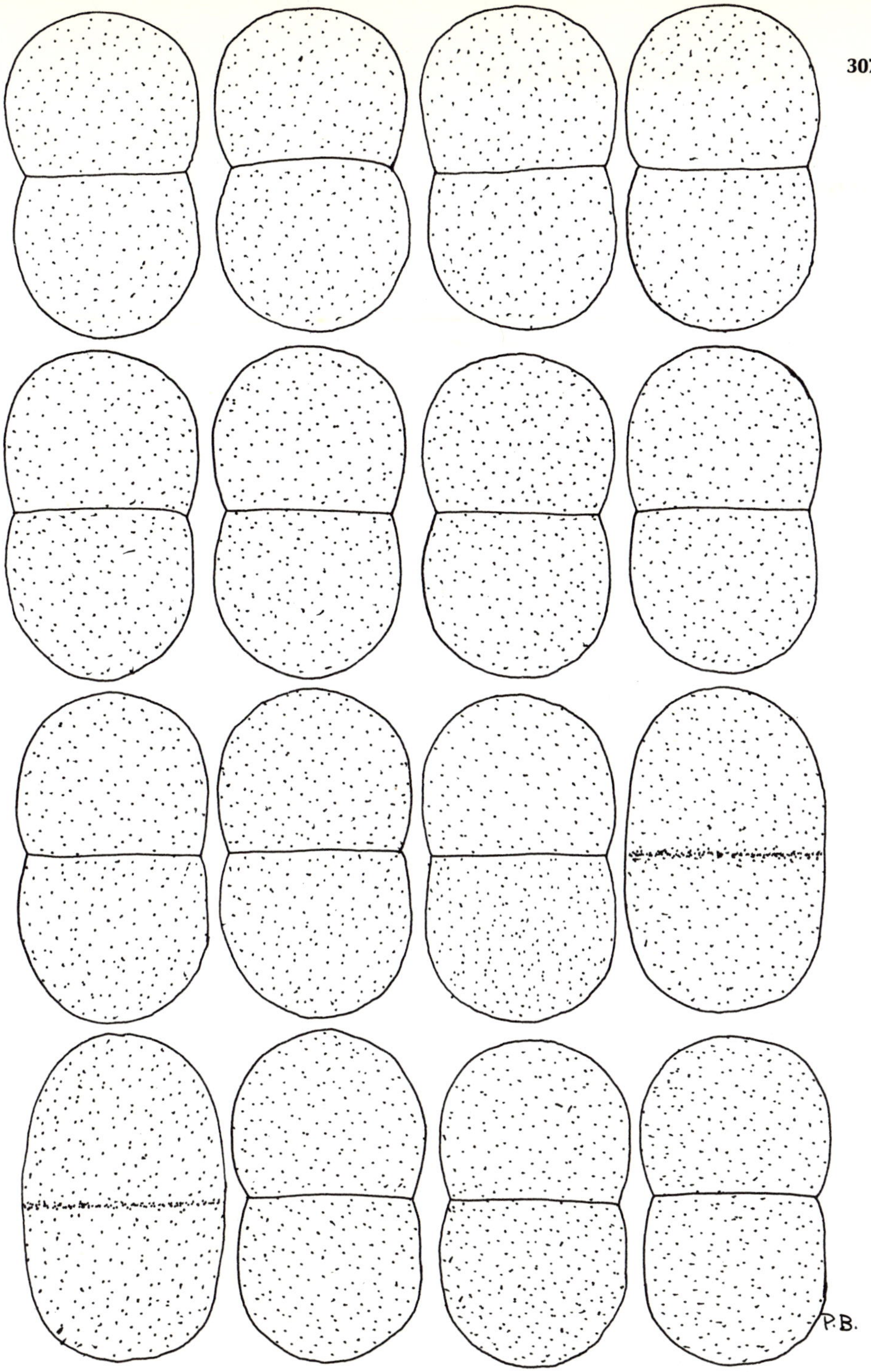

FIGURE 3 *Merismopedia*-a large, flat colonial blue-green in which division occurs in two planes.

FIGURE 4 *Oscillatoria.*

The members of the order Hormogonales are generally larger and filamentous. Reproduction in this group often occurs by a breaking of the filament (a **trichome**) which then continues cell division, adding new cells to the ends of the broken segments. The short sections resulting from the breaking of the trichome are termed **hormogonia.**

The individual species may form filaments which are free from each other, like *Oscillatoria* (Fig. 4); they may have a heavy sheath, like *Lyngbya;* they may form in beadlike chains within gelatinous balls, like *Nostoc* (Fig. 5); they may show a type of false branching like *Tolypothrix* or *Scytonema* (Fig. 6); they may produce branches (true branching) by division in two planes as in *Stigonema;* they may be curved, like *Arthrospira* and *Spirulina;* they may form gelatinous colonies of various shapes, like some species of *Anabaena* (Fig. 7), *Rivularia* (Fig. 8), and *Nostoc;* or they may have filaments which show a certain amount of differentiation with spores, heterocysts, and larger cells at one end and a gradual reduction in size of cells towards the opposite end as in *Gloeotrichia* (Fig. 9) and *Rivularia* (Fig. 10).

FIGURE 5 *Nostoc*-Beadlike cells in filament with heterocysts.

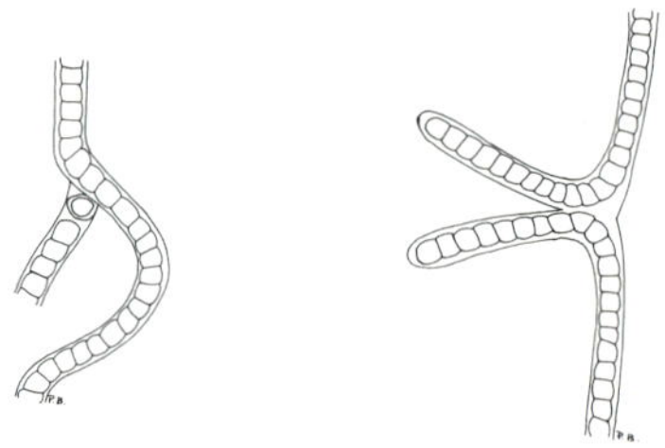

FIGURE 6 *Scytonema.* False branching may occur at the heterocyst or spore or a break in the filament.

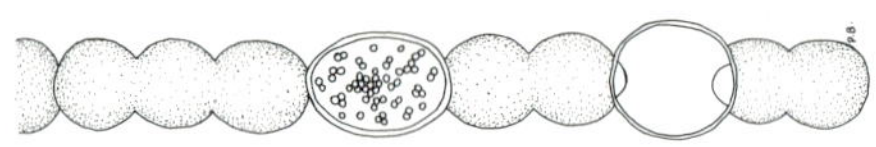

FIGURE 7 *Anabaena.* Species may or may not be in gelatinous colonies. A spore and heterocyst are included.

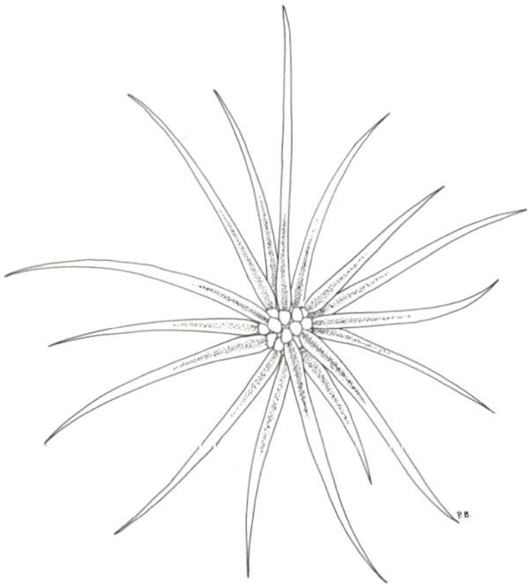

FIGURE 8 Colony of *Rivularia.* There are many more filaments in the typical colony. Note tapering filaments and central spores.

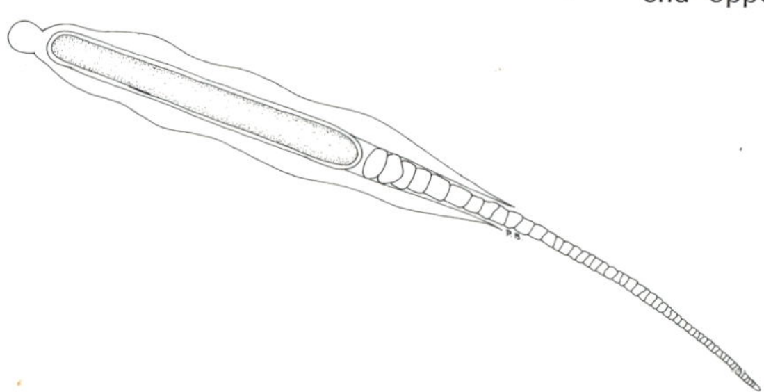

FIGURE 9 *Gloeotrichia.* Not elongated spore and terminal heterocyst.

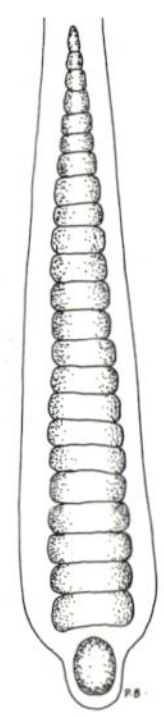

FIGURE 10 *Rivularia,* showing a gradual reduction in size of cells at the end opposite the spore.

Cellular Organization

As indicated earlier, the simplicity of the blue-green algae is exemplified by the lack of differentiation of protoplasmic organelles within the cell. This is true to the extent that nuclear material is not contained within a nuclear membrane, nucleoli do not appear to be differentiated, and cell division is apparently not preceded by mitosis. Cytological investigations of some of the members have disclosed some differentiation of cellular contents similar to that of the chromatin contents of higher plants. This substance has been identified as DNA, RNA, and a protein. Although this chromatin-staining material cannot be said to be a nucleus, the isolation of the radicals of nucleic acids indicates that only the nuclear membrane is lacking; the functional parts of a nucleus at least are present.

The electron microscope has shown that much more is present in the blue-green algal cell than had ever before been identified or even suspected. Note, for example, the drawings made from electron micrographs of *Symploca* (Fig. 11) and *Nostoc* (Fig. 12)

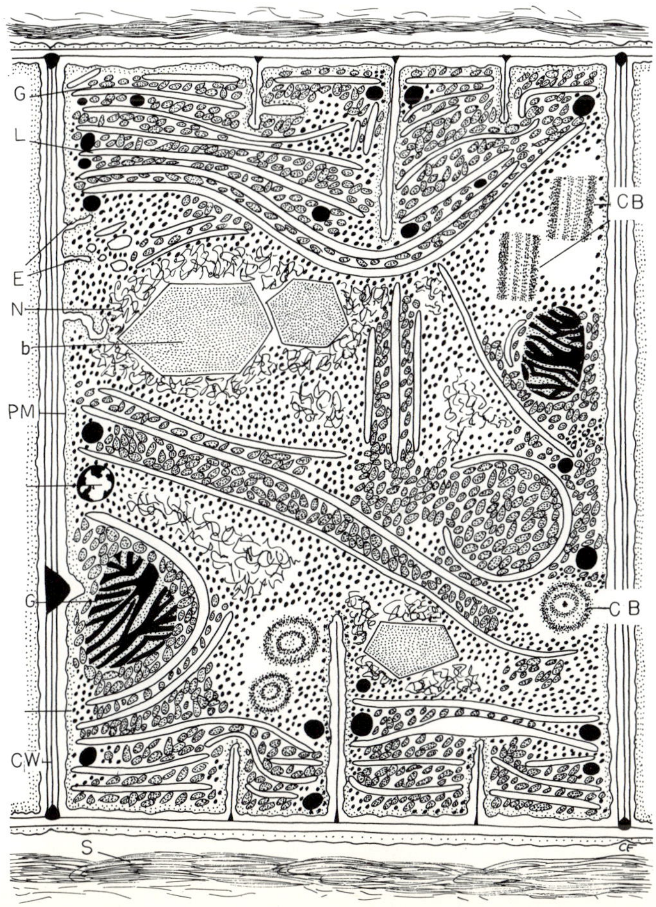

FIGURE 11 Diagram of an electron micrograph of *Symploca muscorum*.

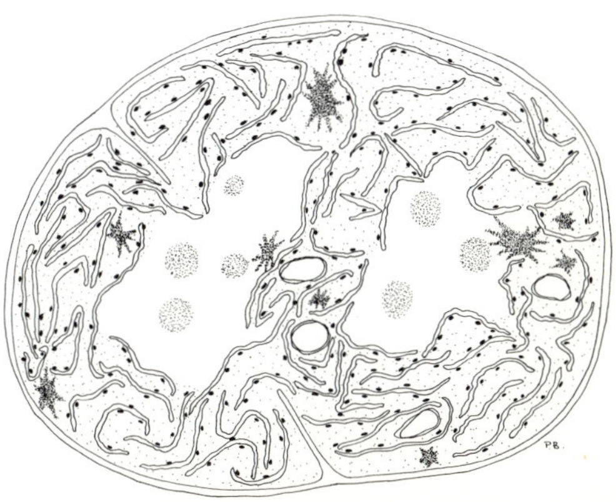

FIGURE 12 A diagram of an electron micrograph of a cell of *Nostoc*. Note invagination of cell wall in early stage of fission; elongated structures are lamellae. Small dark bodies are ribosomes; heavily irregular stippled areas are structured granules; spherical bodies are vacuolelike inclusions; lightly stippled spherical structures are cylindrical bodies in cross section. Drawing made from an electron micrograph in Bonner, James and J. E. Varner. 1965 *Plant Biochemistry*. Academic Press.

cells. The same general cellular contents of higher plants are present but they seem to vary more in quantity and are organized somewhat differently. Compare these drawings with those of *Chlamydomonas* of the Chlorophyta, Fig. 3 (p. 324), Chapter 16. Note the similarity of contents but note also the tendency of the lamellated structures and other organelles to be more concentrated towards the periphery of the cell in *Nostoc* and the greater concentration of cellular material in the chloroplast of *Chlamydomonas.*

The pigments in blue-greens seem to be more highly concentrated in the granular material near the periphery of the cells, but they are not contained in plastids. The phycobilin pigment, **phycocyanin,** is usually the most prominent although in a few species, like those of the genus *Trichodesmium,* the red pigment **phycoerythrin** may predominate. The red hue of the Red Sea is apparently caused by a seasonal abundance of *Trichodesmium.* In addition to these pigments chlorophyll *a* and some carotenoid pigments are usually present. Two of these carotenoid pigments, **myxoxanthin** and **myxoxanthophyll,** have been found only in blue-greens.

Photosynthesis occurs in the same manner as in other plants, but there is a difference in the wavelengths of light which are absorbed. Blue-green algae with the typical color absorb more of the red rays, while those which are reddish, like the Red Algae, absorb more of the blue rays. Yellow, orange, and even green wavelengths may be absorbed depending upon the abundance of the pigments associated with chlorophyll *a.* Although these associated pigments, the carotenoids and phycobilins, are not directly involved in the manufacture of sugar, they do absorb some light and transfer some of the energy obtained from light to the photosynthetic process.

Food Reserves

The storage product produced by blue-greens is starchlike, but it is more like the animal starch, glycogen. It is usually termed cyanophycean starch and often has protein in chemical combination, making it a glycoprotein. Some blue-greens, at least, have proteins which contain diaminopimelic acid. This substance is an amino acid found also in some bacteria but never in higher plants or animals. This is, of course, another reason why we believe bacteria and blue-greens are reasonably closely related or came from common ancestors. Some blue-greens may also store some food in the form of oil droplets, but it is less commonly stored than cyanophycean starch.

Nitrogen Fixation

One of the unique characteristics now recognized in certain members of the Cyanophyta is their ability to "fix" the free elementary nitrogen of the atmosphere. Early reports to this effect were received with skepticism, since free-living, nitrogen-fixing bacteria were often found living in the gelatinous sheaths around the algal filaments. Studies now completed of some blue-greens growing in bacteria-free cultures show that atmospheric nitrogen can be used directly by these algae living in both soil and water. The gaseous nitrogen is incorporated into their amino acids and proteins. Following death of the plants the nitrogenous materials in their bodies are released by the decay organisms forming ammonia, which is then transformed into nitrates by the nitrifying bacteria.

The Cell Wall and Gelatinous Matrix

The cell wall of blue-greens is a conspicuous portion of the cell. It is made up of cellulose fibrils imbedded in pectic components and some hemicelluloses. This wall is a quite rigid structure usually enclosed in a gelatinous or mucilaginous sheath. The sheath may be thick or thin and closely associated with the individual plants. In colonies the individual sheaths may become confluent to such an extent that the cells or filaments appear to be imbedded in a gelatinous mass without distinguishing individuality. Usually the entire colony is limited in its size, but in some species of *Anabaena* and others the cells and gelatinous mass may grow to be several inches thick and several feet long. The sheath increases the water-holding and water-absorbing capacity of the organism and undoubtedly favors survival under desiccating conditions.

Some blue-greens such as *Oscillatoria* are capable of a type of movement which is more or less a type of oscillation. It is believed that this is due to rhythmic waves of alternate expansions and contractions moving along the length of the filament, or that the movement is caused by the secretion of gelatinous material through minute pores in the cell walls.

Reproduction

We have indicated that there is very little differentiation of cells in blue-greens. Aside from slight differences in the size of cells in some species, the only real differentiation occurs in the formation of **spores** and **heterocysts.** The latter are thick-walled cells formed within the filament or at the end. Heterocysts do not germinate as spores but may help to break up the filaments into hormogonia. Spores termed **akinetes** are formed singly or in groups of two or three in the filament. Akinetes contain the entire protoplasmic contents of a cell; in addition the wall of the original cell makes up the outermost portion of the wall of the spore. **Endospores** are produced within the parent cell, but the original wall remains free of the spores, each spore developing its own wall.

While spores may serve as overwintering structures or give added resistence to desiccation, it is upon germination of the spore that new growth is initiated. From this point on, all cells are produced by a process of cell division termed **fission.** This is started by furrows, which begin on opposite sides of the cell, then continue to develop inwardly until they meet, separating the cell into two parts. The two daughter cells may remain together in the filament or separate as in the unicellular species.

Reproduction in the blue-greens, then, is by fragmentation, as in the breaking up of the filaments to produce hormogonia, by the production of spores, or by cell division. In all cases new growth and division is accomplished by fission.

(1) Cell → fission → cells
(2) Filament → hormogonia → fission → filament
(3) Filament → spores → fission → filament

Phylum Rhodophyta, the Red Algae

CLASS RHODOPHYCEAE

The red algae number about four hundred genera, in which are classified about thirty-five hundred species. They are mostly red or reddish in color because of the abundance of the phycoerythrin pigment, but a few of them, especially those living in fresh water, are olive green in color. The red algae are multicellular, but grow in a variety of shapes and sizes from microscopic to several feet long.

Habitat

Red algae are adapted to grow at greater depths in oceanic waters than any other photosynthetic organism, since the blue wavelengths of light penetrate deeper than the red rays and red algae can absorb greater amounts of this quality of light. Some of these algae occur over three hundred feet beneath the ocean's surface, but most are found in much shallower waters. Most of the species are attached by rhizoidal holdfasts; a few are free-floating.

Cellular Organization

In addition to the similarity in the types of pigments found in both red and blue-green algae, the cell walls are rather similar in that both are largely composed of cellulose in the inner layers and an abundance of pectic compounds in the outer layers. At the time of cell division the crosswall, which develops separating the daughter cells, may be formed with a large central pore providing a continuous connection of cytoplasm between cells. Or small pits may be formed, allowing the penetration of thin cytoplasmic strands, the **plasmodesmata,** which may serve as pathways for the translocation of nutritive material.

Food Reserves

The reserve food is called **floridean starch.** It is similar to cyanophycean starch and glycogen, but is not identical to either. The red algae may also form a sugar that is similar to galactose and a very soluble fat-related substance, glycerol.

Reproduction

The reproductive organs of the red algae are unique to this particular group and are the most elaborate among all algae. The male gametes are nonflagellate, tiny, spherical bodies which for the most part float passively in the water, where they make contact with the female organs by chance. The tiny spherical bodies are usually termed **antheridia,** and arise from a male branch on the thallus or filament. The spherical antheridium produces but a single sperm; these are usually termed **spermatia** since they lack flagella.

The female organ is somewhat more complex and may have a variable number of parts in different species. The simplest ones, as seen in *Nemalion* (Fig. 13), are borne on a short branch termed the **procarp,** which has a swollen cell at the apex, the **carpogonium** (Fig. 14), from which is extended an elongated tip, the **trichogyne.**

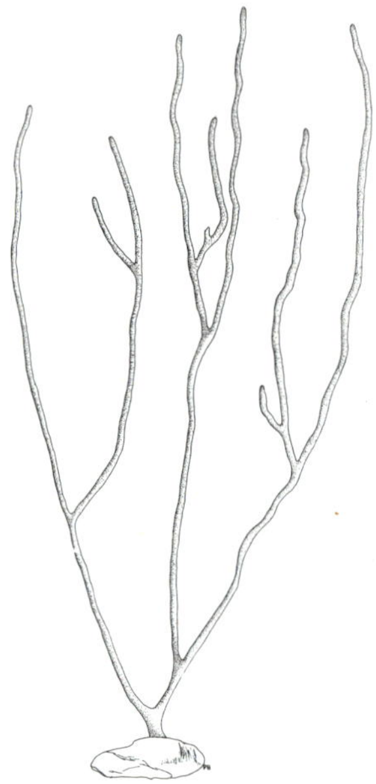

FIGURE 13 *Nemalion.* Showing wormlike filaments.

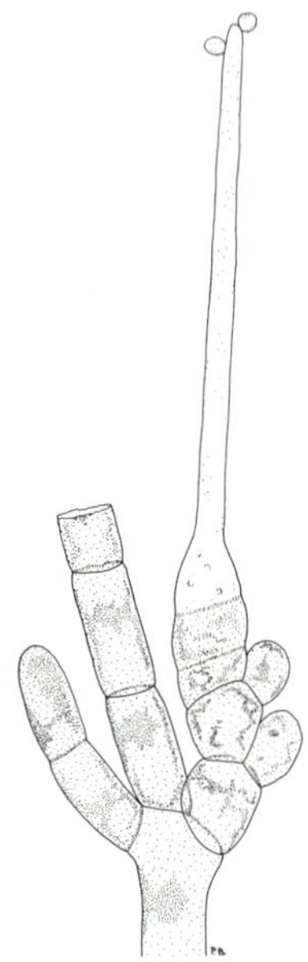

FIGURE 14 *Nemalion.* Showing procarp with swollen carpogonium and elongated trichogyne. Two male cells are adhering to the trichogyne apex.

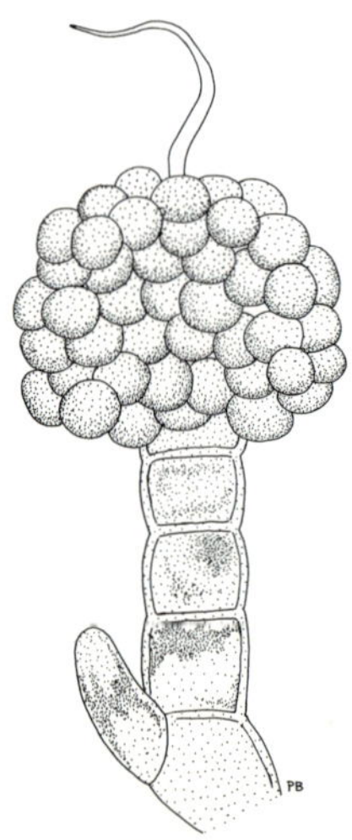

FIGURE 15 Loosely arranged cystocarp of *Nemalion,* showing clusters of carpospores on hidden gonimoblasts, withering trichogyne, and typical monosiphonous branch.

In a plant like *Polysiphonia,* the procarp consists of several cells including a large **pericentral cell,** which divides to form a cluster of four or more cells, one of which becomes the carpogonium. In addition the pericentral cell divides into **auxiliary** cells which form a loose enclosure around the carpogonial cell.

Male cells floating in the water make contact with the trichogyne. The contact walls of the male cell and the trichogyne dissolve away, allowing the contents of the male cell to enter the tube of the trichogyne and migrate to the carpogonium where it unites with the female nucleus. Soon after fertilization the zygote nucleus undergoes meiosis, and then a number of mitotic cell divisions occur, producing many short haploid filaments, the **gonimoblasts.** In *Nemalion* (Fig. 15) these are comparatively few and loosely organized, but in *Polysiphonia* the gonimoblasts are borne within an urn-shaped structure, the **pericarp** (Fig. 16), formed from cells derived from the original pericentral cell. The filaments of gonimoblasts produce **carpospores** at their apex which are budded off and float away as they mature. The entire structure of procarp, gonimoblasts, carpospores, and pericarp, in the case of *Polysiphonia,* forms the **cystocarp.**

In *Nemalion,* as mentioned above, reduction division occurs in the germination of the zygote, thereby producing haploid carpospores, but in *Polysiphonia* reduction division does not occur there and the carpospores produced in the cystocarp are diploid.

In the case of *Nemalion* the carpospores give rise to the gametophytic plants, but in *Polysiphonia* where the carpospores are diploid they give rise to an asexual plant, the **tetrasporophyte** (Fig. 17). This looks very much like the gametophytes but it produces sporangia, which are termed **tetrasporangia** since they produce **tetraspores.** They are formed by reduction division and consequently the gametophytic plants which form from the tetraspores are haploid. An outline of the life cycles of the more primitive *Nemalion* and the more advanced *Polysiphonia* is given below. Other red algae are illustrated in Figures 18, 19, and 20.

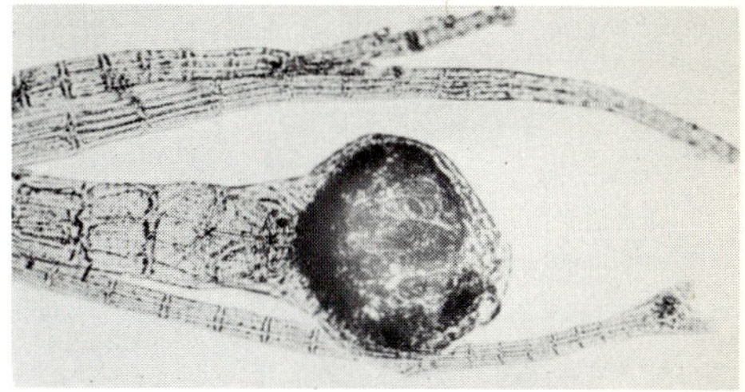

FIGURE 16 *Polysiphonia,* urn-shaped cystocarp (pericarp) with enclosed carpospores.

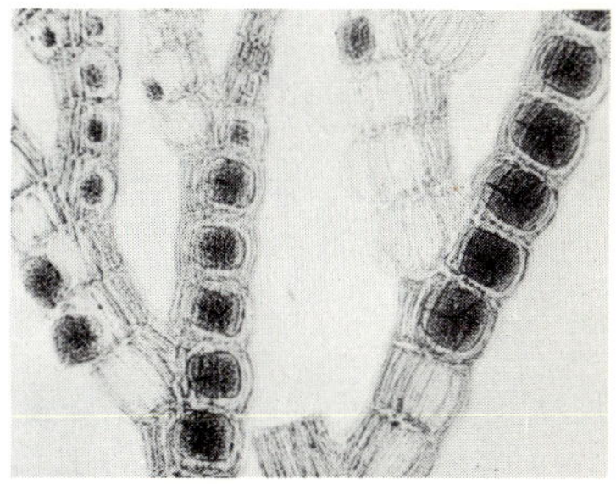

FIGURE 17 Asexual plant of *Polysiphonia* showing tetrasporangia and young tetraspores.

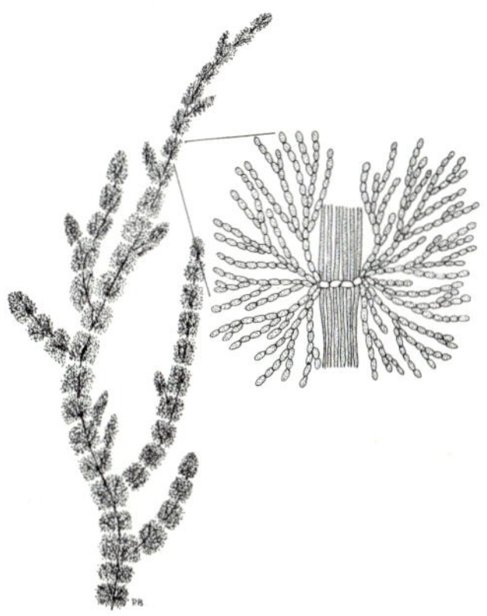

FIGURE 18 *Batrachospermum.* The diagram on the left is of an entire thallus and the one on the right shows a very small segment of the very much branched thallus.

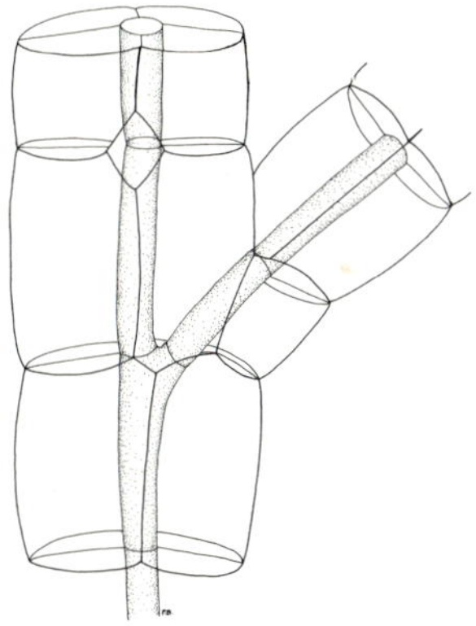

FIGURE 19 Short section of the filamentous thallus of *Polysiphonia* showing the typical polysiphonous or multiseriate filaments. Compare with filaments of *Nemalion* in figures 14 & 15.

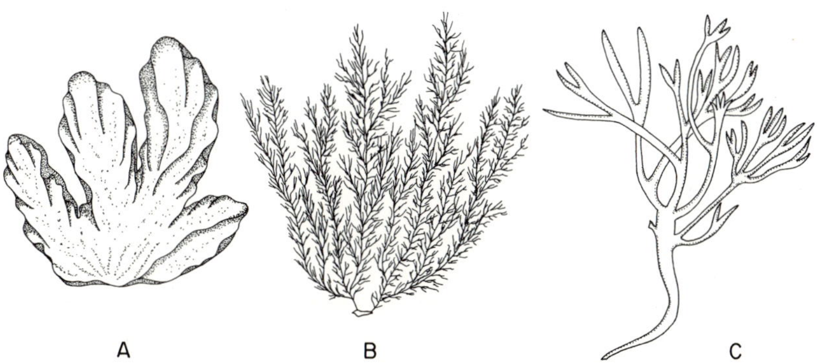

FIGURE 20 Several red algae. A-*Porphyra,* B-*Polysiphonia,* C-*Gelidium.*

Phylogenetic Relationships

The Cyanophyta and Rhodophyta are unique among the algal groups in that they lack flagellated cells of any type. They are also similar to each other in that both have similar wall structure, and have pigments similar in type but varying in quantity between the two groups. Both store starch variants as reserve foods, and both may have protoplasmic connections between cells. They differ chiefly in the lack of organized nuclei and plastids in vegetative cells of the blue-green algae and the fact that red algae have developed a type of sexual reproduction. It would seem, on the basis of these characteristics, that the two phyla are more closely related than other algae although the Rhodophyta have succeeded in evolving a much more complex thallus.

The Life Cycles of Typical Red Algae

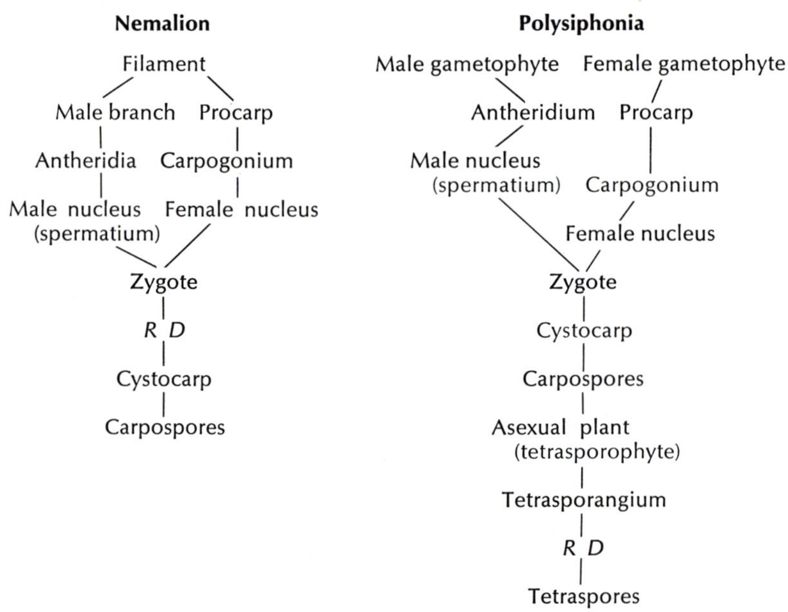

Economic Importance

Blue-green and red algae are both of some economic importance, but reds are much more important for direct usage. Some of the gelatinous species, including *Nostoc,* are used in making soups, but for the most part blue-greens are part of the plankton eaten by fish which we and other large animals at the top of the food chains eat, thus getting the blue-green protoplasm indirectly (Fig. 21).

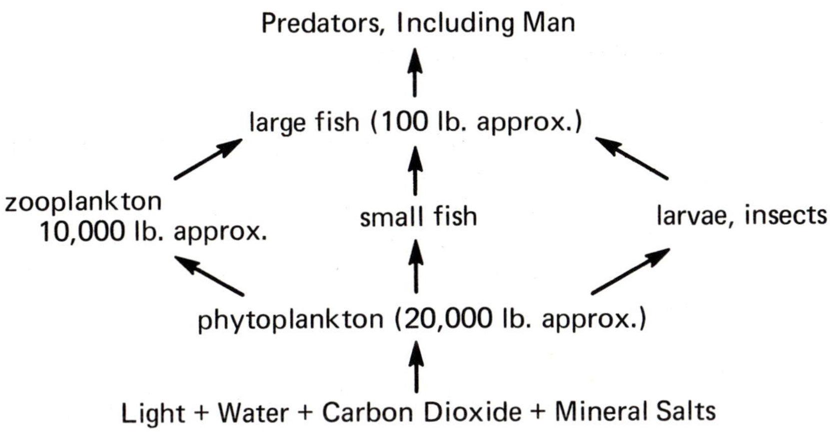

FIGURE 21 An aquatic food chain in which microorganisms utilize the basic inorganic substances to form food which is then transferred to larger organisms above.

More commonly blue-greens are offending polluters of city water supplies. Excessive growth of these organisms in lakes causes the formation of malodorous compounds of hydrogen sulfide and phosporus, nitogenous compounds, and certain other offensive volatile gases which reduce the number of vacationers at resort areas.

A few blue-greens have proved to be poisonous. Some species of *Microcystis* and *Anabaena* have apparently been responsible for the death of cattle after they drank water containing these organisms. Laboratory tests made on mice, rats, and guinea pigs resulted in low blood pressure, respiratory difficulty, and death in 20 to 180 minutes following the drinking of water containing species of the above genera. The active principal in *Microcystis* is described as a potent neurotoxin. Allergic symptoms have also been observed in some human beings after swimming in lakes inhabited by blue-greens such as *Oscillatoria*. The symptoms include dermititis, conjunctivitis, blocked nasal passages, and bronchial asthma.

A few red algae including *Chondrus crispus* (Irish moss) and *Rodymenia palmata* (dulse) are eaten, but most, including these two species and species of *Gelidium* (Fig. 20) and *Porphyra,* are used in processing agar. This product is used universally as a base for bacteriological media, pharmaceuticals, laxatives, and cosmetics, and as a stabilizer in prepared foods such as puddings, pie fillings, mayonnaise, chocolate milk, jellied fish, jellied consomme, and similar products.

Supplementary readings on the algae are at the end of Chapter 17.

Questions for Discussion

1. Blue-green algae are found almost everywhere. What explanation can you offer for this fact?
2. Blue-green algae often live in hot springs and in subsurface habitats. What does this suggest about their physiology?
3. Why is it desirable to classify algae according to their pigmentation?
4. How has the electron microscope aided our study of the algae?
5. In what ways do the red algae show similarity to the blue-greens?
6. In what ways are the blue-green algae similar to bacteria?
7. Describe the economic importance of red algae.
8. Most marine algae are found on rocky seacoasts rather than sandy beaches. Explain.
9. All algal groups have species living in the sea but comparatively few in fresh water. What explanation can be offered for this fact?
10. What do we mean by an alternation of generations?

16

The Green Algae

Phylum Euglenophyta
Phylum Chlorophyta

The Euglenophyta and the Chlorophyta are similar in pigmentation. Both have chlorophyll *a* and *b,* carotene, and several xanthophylls which are essentially the same. However, the morphological differences are many.

Phylum Euglenophyta

The Euglenophyta are comprised of over four hundred species. They have many characteristics which can be used in describing them as animals or as plants. Some of the species lack chlorophyll and are considered saprophytic or holozoic. It is doubtful, however, that euglenoids ingest solid food particles, even though some seem to be capable of it. The photoautotrophic members, at least, have chlorophyll in distinct plastids which move about freely in the protoplasm. The food reserves are largely those of an insoluble carbohydrate, **paramylum,** closely related to starch. This discussion of the Euglenophyta is largely a description of the best known genus in the phylum, *Euglena* (Fig. 1).

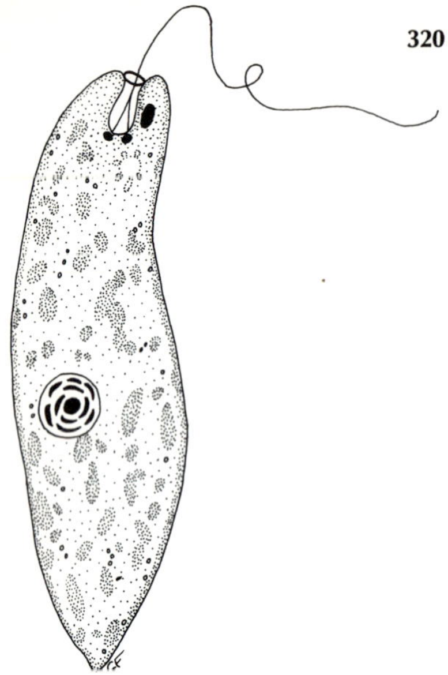

FIGURE 1 *Euglena.* Note the reservoir at the anterior end within which the flagellum arises. At the basal bifurcation of the flagellum are two blepharoplasts. To their right is a larger eye spot. Just below the right blepharophast a contractile vacuole is forming. There is a centrally located nucleus, many stippled chloroplasts, and some very tiny food reserves.

FIGURE 2 Diagram of a flagellum based on an electron micrograph. Note the nine paired external fibrils surrounding the central pair.

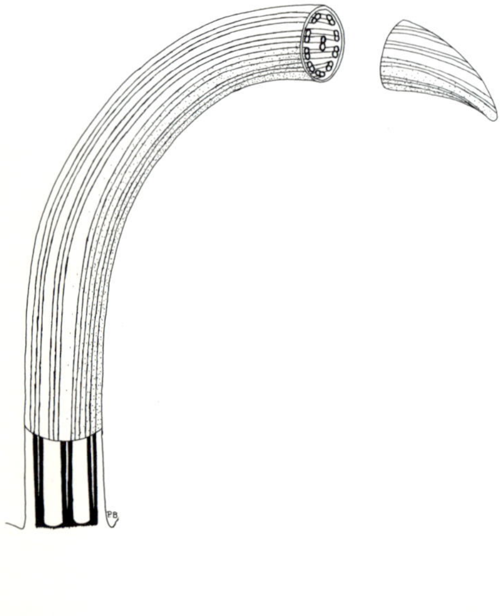

Body Organization

The protoplast of the euglenoids is surrounded by a plasma membrane, the **periplast** or **pellicle.** At the anterior end the euglenoids have a distinct gullet formed by invagination at the tip of the cell. The large opening created by invagination is called the **reservoir;** its anterior opening is the **cytostome** and the narrow passage connecting the cytostome to the reservoir is the **cytopharynx.** Usually within the protoplast at the base of the reservoir one or two pulsating **contractile vacuoles** can be distinguished. These structures maintain water balance in the cell by discharging excess quantities of water which may accumulate in the cell. In the protoplast near the gullet there is usually a small pigmented area termed the eyespot or **stigma,** which may be light sensitive.

Near the center of the cell is a prominent nucleus containing numerous chromatin granules, which become organized into definite chromosomes at the time of division. It is doubtful that any form of sexual reproduction occurs in the euglenoids. Multiplication is accomplished by mitosis followed by longitudinal cell division. This may occur when the organisms are active or after they have come to rest. Division begins at the anterior end, new reservoirs being formed before the simple euglenoid has split into two.

Flagella

The distinguishing feature among all flagellates is the flagellum. Usually the number borne by the members of any group is constant, but in some there is variability in number as well as type and size. The euglenoids may have one, two, or three flagella and these may be of the same length or unequal. Flagella of all organisms, except those of bacteria, are made up of a central core of two strands surrounded by a group of nine peripheral strands (Fig. 2). Usually the two central strands are longer and extend out beyond the outer ring of fibrils. Flagella which appear smooth, without lateral fibrils, are termed **whiplash.** Those with lateral fibrils are the **tinsel** type. Flagellates in the Chlorophyta have the whiplash type, the euglenoids have the tinsel type, and some flagellates described in the next chapter may have both kinds.

In most flagellates the flagellum or flagella arise in a depression at the anterior end of the body or have a lateral attachment. In the euglenoids and flagellates of green algae and perhaps others as well, the flagella are connected to a **neuromotor** apparatus consisting of a small granular body, the **blepharoplast,** at the base of each flagellum. The blepharoplast is connected to a descending fiber, the **rhizoplast,** which is apparently connected to a centrosome that lies within or just outside the nucleus. This mechanism is believed responsible for directing the activities of the flagellum.

Phylum Chlorophyta

One of the two largest phyla of the algal groups, Chlorophyta, is comprised of approximately 5,800 species, the great majority of which are in the class Chlorophyceae.

Habitat

Most of the green algae live in fresh water, but many inhabit brackish waters, and others are marine. Some species grow in or on moist soil, some live epiphytically on larger algae, in lichens, or on higher plants, others live epizoically on aquatic animals, and a few live within the cells of protozoans and coelenterates.

Body Organization

The green algae may be unicellular, filamentous, colonial, or differentiated into a flat or tubular thallus. The prominent green pigments are contained within chloroplasts. These may be small and discoid, plate-like, ribbon-shaped, cup-shaped, or developed into numerous net-like or reticulate connecting strands. Most of the species have one or more **pyrenoids** located within the chloroplast. A pyrenoid is a circular organelle consisting of a central proteinaceous core surrounded by minute platelets of starch. These structures are storage centers for starch accumulation, but algae without them apparently store starch in leucoplasts in much the same manner as higher plants.

Most green algae have a distinct cell wall which consists of an inner layer composed almost wholly of cellulose and an outer layer made up chiefly of pectose. Pectin is secreted by the protoplast and filters through the microfibrils of the cellulose layer. Many green algae secrete gelatinous sheaths around themselves, but this is much more common in the Cyanophyta than it is in the Chlorophyta.

Nucleus

Members of the Chlorophyta have distinct nuclei with one or more nucleoli, and a chromatic network which becomes organized into chromosomes at the time of division. Most species are uninucleate, particularly in the vegetative cells, but some are multinucleate and therefore the thallus or filament is termed **coenocytic.** Some of the species become multinucleate just before the formation of gametes or spores, or just prior to cell division; in other species there is a gradual increase in the number of nuclei within the vegetative cells throughout the life of the plants.

Cell Division

Cell division in the algae, as mentioned earlier, is usually by a transverse furrowing of the protoplast, accompanied by the secretion of new wall material within the deepening furrow. Very rarely is there cell division by means of a cell plate as seen in higher plants.

Usually division of the protoplast in algae with single chloroplasts is accompanied by division of the plastid so that each daughter cell receives a daughter plastid as well as a daughter nucleus. In other green algae with numerous plastids the plastids are simply divided between the two daughter cells, about half of the plastids going to each daughter cell. Multiplication of the plastids then occurs in the daughter cells.

Asexual Reproduction

Vegetative reproduction in the green algae may occur much like that in the filamentous blue-greens by fragmentation of the filament through the aid of external forces or dead cells, or by the formation of a variety of spores. The most common type of spore is the **zoospore,** which many algae form upon germination of the zygote; some may produce them in specialized cells termed **zoosporangia.** Zoospores are thin-walled, flagellated cells which may have chlorophyll and therefore be photosynthetic before they settle down to grow into a new filament, thallus, or colony. **Aplanospores** are similar in form, origin, and function to zoospores but they lack flagella and are essentially nonmotile. Both of the above spore types may be produced singly in their sporangia, or many may be produced in a sporangium. Some green algae produce resting spores with greatly thickened walls like the akinetes of the blue-greens.

Sexual Reproduction

Most green algae are capable of sexual reproduction, although in a few the production of gametes or the fusion of cellular contents of any type have never been observed. Sexual reproduction may be accomplished by the fusion of the protoplast of cells, by the union of well organized isogametes or heterogametes, or by the union of a nonmotile egg with a motile sperm (oogamy) as described under the Thallophyta in Chapter 14. Some of the green algae form gametes from the contents of undifferentiated vegetative calls. Others develop specialized cells termed **gametangia** for the production of isogametes. In the green algae where distinct female gametes are produced, the cellular organ is termed an **oogonium** and the male gamete-producing organ is the **antheridium.**

CLASS CHLOROPHYCEAE

Order Volvocales

Members of this order are the only ones in the phylum Chlorophyta with flagellated, actively motile vegetative cells. The individual plants vary from unicellular to relatively large motile colonies. They all have about the same type of vegetative cells, but there is a considerable variation in the number of cells per colony and they advance through a developmental series in reproduction from isogamy, through heterogamy, to oogamy in the most advanced species. There are approximately five hundred species, found chiefly in fresh water habitats.

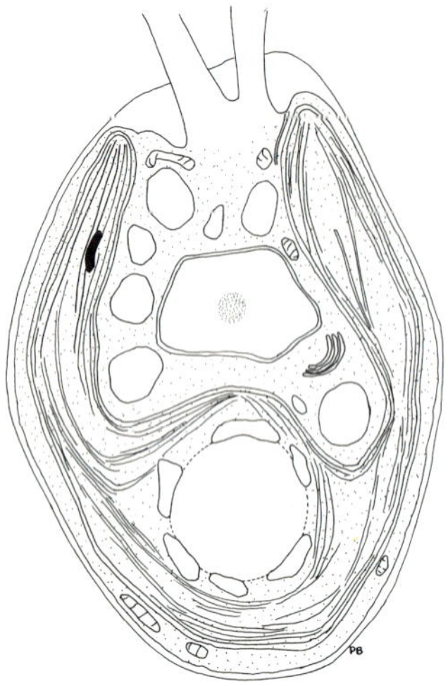

FIGURE 3 *Chlamydomonas*. A diagram of an electron micrograph showing centrally located nucleus. The stippled area within the nucleus is the nucleolus. The large circular area towards the base is a pyrenoid with tiny starch plates surrounding it. Towards the upper portion are contractile vacuoles in formation. As they enlarge some will coalesce with others. The 3 curved rod-shaped lamellated structures near the center is the golgi apparatus. Two flagella are at the anterior end and an eyespot is present to the left.

Chlamydomonas (Fig. 3) is usually described as the most primitive member of present-day green algae. It is also considered the type of organism which may be ancestral to one or more of the flagellate lines of evolution. Support for this opinion is found in part in the small size and simplicity of this tiny unicellular organism and also in its many methods of reproduction. Some species may simply split in two in a process of simple mitotic division and fission. In other species the protoplasts may undergo several divisions forming a number of cells imbedded in a gelatinous matrix. This is termed a **palmella** stage since algalogists first thought them to be species of a genus **Palmella.** After escaping from the cell wall the individual cells develop flagella, become motile, and swim away independently. Some species of *Chlamydomonas* develop gametes which are all alike (Fig. 4); others are heterogamous,

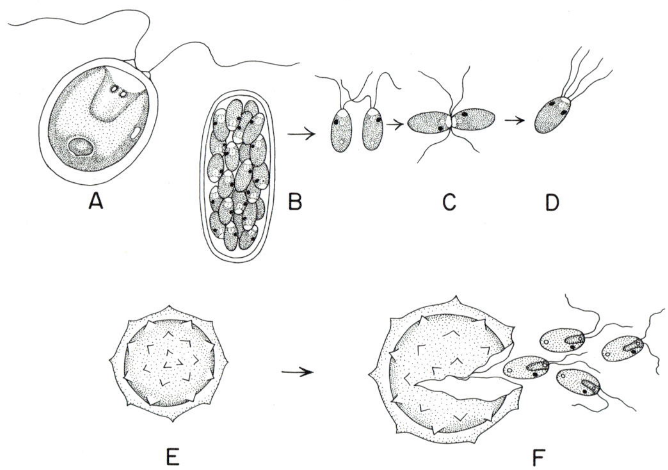

FIGURE 4 *Chlamydomonas* life cycle of one species showing A-adult; B-formation of isogametes; C-union of gametes at anterior ends; D,E-stages of zygote; F-liberation of four daughter cells. Adapted from *Laboratory Outlines in Biology*. W. H. Freeman and Company.

since the gametes are of different sizes; in still other species some cells will produce as many as thirty-two biflagellate gametes while other cells form a single large non-flagellated gamete. This type would then be oogamous.

Sexual reproduction has been shown to be associated with nutrition in certain species. Trainor found that in distilled water cultures with ammonium nitrate added at .03 percent zygote yield was 2 percent, but when the nitrogen level was reduced to .003 percent the zygote yield was increased to 82 percent. It was also found that three-day-old cultures produced a much larger number of zygotes than older cultures. Furthermore, some *Chlamydomonas* cultures were shown to produce about 32 percent zygotes under continuous illumination but would produce 79 percent zygotes under twelve-hour alternating periods of light and dark. The lowest intensity of illumination under which zygotes were produced was eighty footcandles with a gradual increase in numbers up to six hundred footcandles; beyond this intensity sexuality declined.

Temperature also proved to be a factor in sexuality. A zygote yield by as much as 70 percent could be obtained under continuous illumination if the temperature of the medium was raised from 22°C to 27°C. Higher temperatures offset the effects of illumination. Experiments of this type show how external forces in the environment may affect sex.

It has also been demonstrated that vegetative cells in some species will differentiate into gametes but they will not mate unless they are of different strains, which for convenience have been designated as + and −. When the proper conditions prevail, a number of individuals will clump together, then swim away in pairs attached at their flagellated ends. The walls dissolve at this point and the protoplasts emerge and unite in the water. The zygote produced often secretes a wall about itself and may remain dormant for extended periods when conditions are unfavorable for growth. When conditions are favorable the zygote undergoes meiotic division, forming four to thirty-two daughter cells within the old wall. Dissolution of the wall frees the daughter cells to swim away independently.

Investigations of such simple organisms as *Chlamydomonas* provide us with much information on variations in organisms and contribute to our knowledge of how the sexual process is basic to the process of evolution. Although *Chlamydomonas* may maintain itself asexually through numerous generations without change, the introduction of the sexual process initiates morphological and physiological changes which provide numerous combinations and recombinations of chromosome material basic to genetic variation.

The most prominent characteristic of *Chlamydomonas* and of the other members of the Volvocales, which ties them together in this order, is the similarity of their vegetative cells. Each vegetative cell has a single large chloroplast which is rather irregularly cup-shaped. Within the chloroplast are one or two pyrenoids. Located in the open end of the plastid, towards the anterior end of the cell, is a rather large nucleus containing one or more nucleoli. Near the anterior end of the cell there is also a tiny eyespot and in this same area one usually finds one or two contractile vacuoles. The flagella, which are of equal length, arise at the anterior end of the cell where they are connected to the neuromotor mechanism described earlier.

Other intermediate members of the Volvocales include *Gonium*

(Fig. 5), in which the number of cells in a colony vary from four to sixteen; *Pandorina,* with sixteen to thirty-two cells in a spheroid ball; *Eudorina,* in which the cells in the colony number from sixteen to sixty-four; and *Pleodorina,* in which the colony cells may number as many as 128.

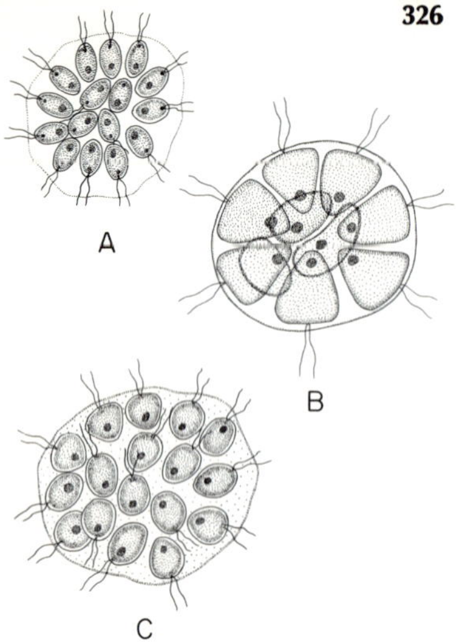

FIGURE 5 A-*Gonium,* B-*Pandorina,* C-*Eudorina.*

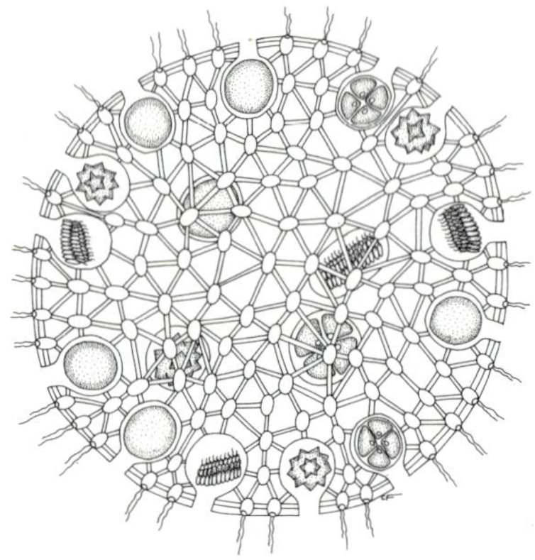

FIGURE 6 *Volvox.* Diagram of large plant showing packets of sperm; large nonmotile eggs; irregular zygotes; formation of daughter colonies; and individual vegetative cells with their prominent cytoplasmic connections.

Members of the genus *Volvox* (Fig. 6) have the largest and most complex colonies of the entire order. They are often over one millimeter in diameter, and a single colony may have over fifty thousand cells. All of the species are oogamous, they have cytoplasmic connections between the cells, and daughter colonies are produced within cells in the posterior half of the colony. Yet the same type of cell with its cup-shaped, pyrenoid nucleus and equal flagella is found here as in *Chlamydomonas.*

The large, nonmotile eggs are formed individually in cells which lose their flagella and increase greatly in size. Sperms are produced in clusters from individual cells in the same way. After sperms are formed they swim away as a sperm colony for a short time before breaking up individually. The sperms then swim to eggs, generally of another colony, where they enter through the gelatinous envelope of the colony and fuse with the eggs. After fertilization the zygote develops a thick spinescent wall and remains there within the gelatinous sheath of the oogonial cell until it is freed by dissolution of the parent wall.

Germination of the zygote is accompanied by meiosis followed by a series of mitotic divisions in which a hundred or more cells are produced, forming a small colony. Succeeding reproductive phases are asexual for a number of generations, each new daughter colony becoming progressively larger until the maximum size for the species is reached.

Order Ulotrichales

Ulotrichales is one of the largest and most diverse orders, from the standpoint of body organization, in the green algae. Members vary from unicellular or small colonial types to the filamentous forms most characteristic of the order. In general they have band or U-shaped plastids, with one or more pyrenoids, and are mostly uninucleate.

Ulothrix (Fig. 7) has a simple unbranched filament which grows attached by means of a holdfast cell in fresh water lakes and streams. The chloroplast curves around within the cubical cells resembling a bracelet.

Ulothrix reproduces asexually by the formation of one to eight zoospores in each vegetative cell of the filament. These zoospores are quadriflagellate and uninucleate, and contain contractile vacuoles and an eyespot at the anterior end. The zoospores swim about for a short while after liberation from the parent cell, then settle on a substrate and divide transversely, forming a basal holdfast cell and an apical cell which continues to divide to form the filament. Occasionally aplanospores are formed instead of the typical zoospores.

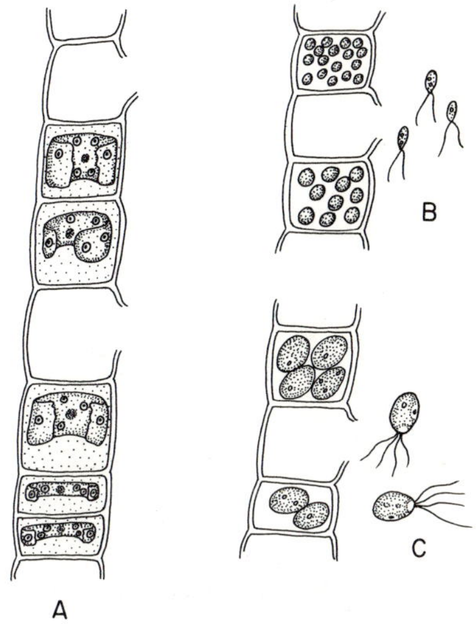

FIGURE 7 Portion of filaments of *Ulothrix*. A-Cells with braclet-shaped chloroplasts. B-cells with gametes. C-cells with zoospores.

Members of this genus are isogamous forms in which gametes are produced by divisions of the protoplast forming sixteen, thirty-two, or sixty-four biflagellate gametes. These break out of the parent cells, swim about, and ultimately fuse with gametes from a different filament. The resulting zygote secretes a thick wall about itself and undergoes a resting period before germination. The protoplast of the germinating zygote undergoes meiosis and forms four to sixteen daughter protoplasts, which then mature as zoospores or aplanospores. These spores then give rise to new filaments.

In many isogamous algae, as in *Ulothrix,* the production of gametes or zoospores is governed by environmental factors. For example, if the water is slightly acid or neutral the cells will produce zoospores; if it is slightly alkaline the cells will form gametes. Similarly, changes in temperature will be more favorable for the development of one type of reproductive organ than another. It has also been observed, in different genera and species, that seasonal variation with changes in length of the light period, which may be correlated with pH, has a pronounced effect. Nutrients, such as the amount of nitrogen, also may affect the cell's development of gametes or spores. Consequently one finds that nearly all cells in a filament will be doing the same thing at about the same time.

Included in the order Ulotrichales are a number of beautifully branched filamentous algae such as *Draparnaldia* (Fig. 8), *Stigeoclonium* (Fig. 9), *Coleochaete* (Fig. 10), *Protococcus* (Fig. 11), and *Sphaeroplea* (Fig. 12).

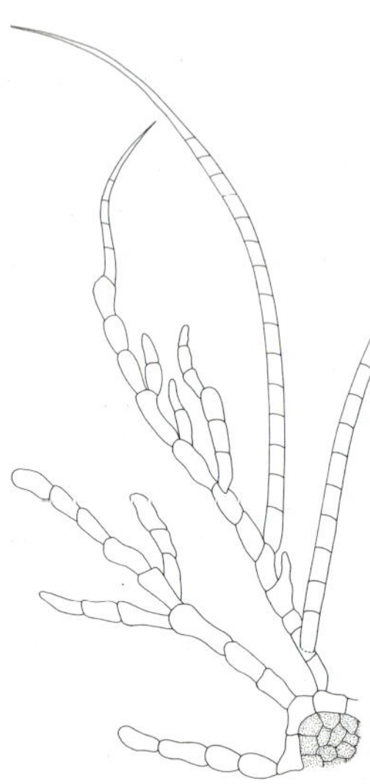

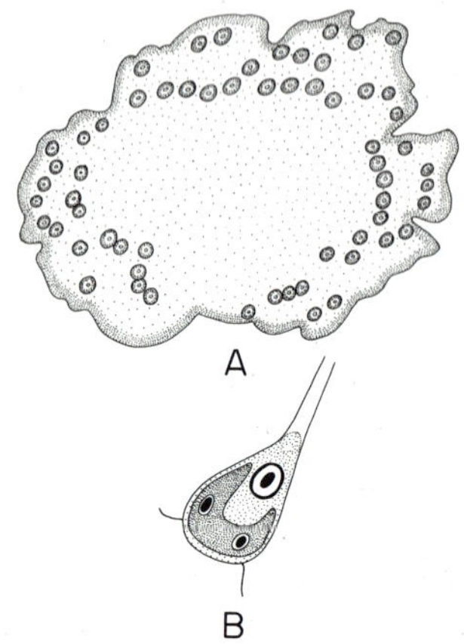

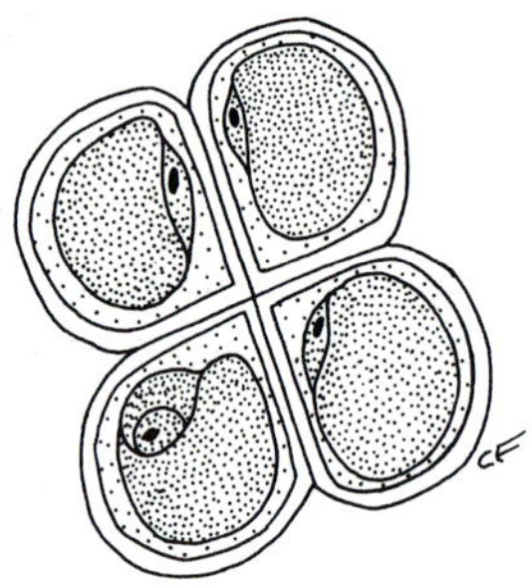

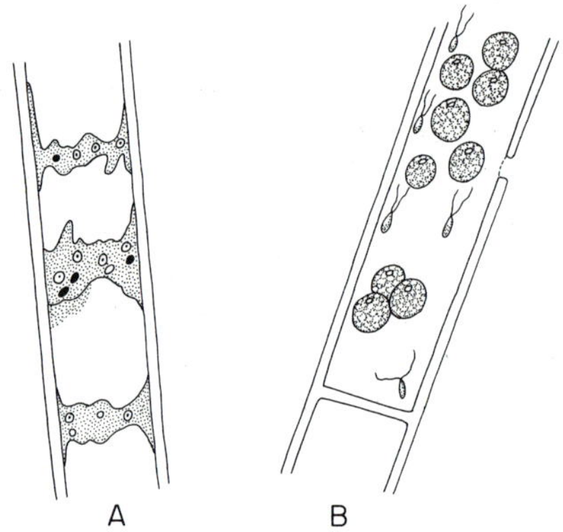

FIGURE 8 *Draparnaldia* showing profuse branching of small filaments from main axis.

FIGURE 9 *Stigeoclonium,* a green alga with long attenuated branches arising from a basal disc.

FIGURE 10 *Coleochaete.* A-Thallus with reproductive bodies. B-Enlarged ooganium with trichogyne.

FIGURE 11 *Protococcus.* The chloroplast is curved encircling the partially hidden nucleus.

FIGURE 12 *Sphaeroplea.* A-Vegetative cell with several annular plastids. B-Portion of cell with eggs and sperm, which have entered the oogonial cell.

Order Ulvales

Ulvales is a relatively small order comprised mostly of marine species. In cellular organization they are similar to the Ulotrichales and probably were derived from them. The zoospores are quadriflagellate and the gametes biflagellate, whether isogamous or heterogamous.

Ulva (Fig. 13) is a heterothallic plant which has an isomorphic alternation of generations. The diploid sporophytic plant produces quadriflagellate zoospores by meiosis. The haploid zoospores grow into the gametophytic plants, which are similar in appearance to sporophytes. In most species the gametes produced are of the same size and appearance, but in a few species one plant, which we may designate as the female gametophyte, produces larger biflagellate gametes than those produced by the other sexual thallus. These plants would be, therefore, heterogamous. Fusion of the gametes occurs in the water and the zygote gives rise to the sporophytic (asexual) plant.

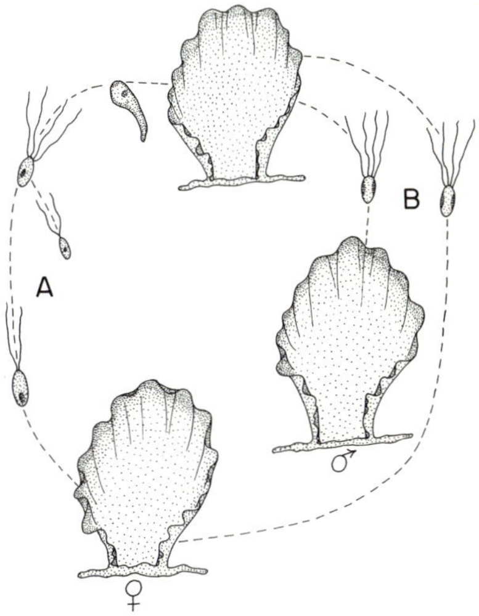

FIGURE 13 *Ulva* showing life cycle of isomorphic plants. A-gametes and fusion to form asexual plant. B-zoospores produced by meiosis from the asexual plant. The isomorphic sexual plants are heterothallic.

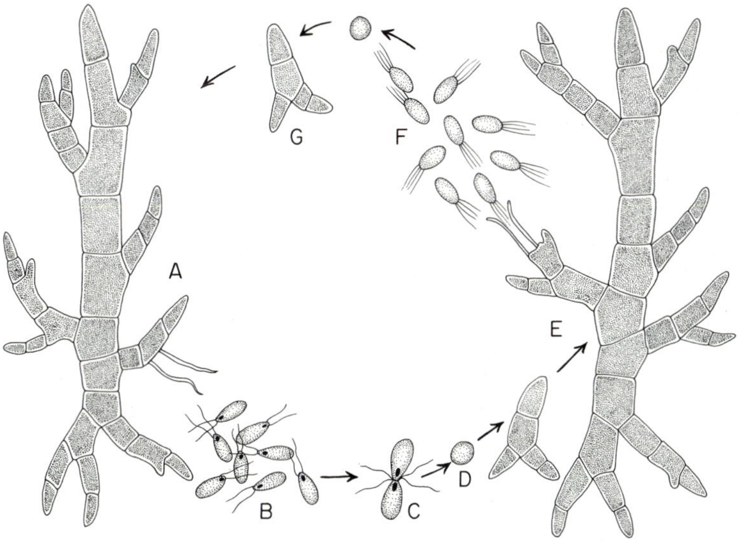

FIGURE 14 *Cladophora* life cycle showing A-sexual, haploid plant; B-gametes; C-fision; D-zygote; E-asexual, diploid plant; F-zoospores produced by meiosis; G-germination of zoospores to form young sexual plant. Redrawn from *Laboratory Outlines in Biology*, W. H. Freeman

Order Oedogoniales

Oedogonium (Fig. 15) is one of our most commonly studied algae of the oogamous type. The filaments are unbranched and the individual cells are uninucleate, with much-branched, reticulate chloroplasts and many pyrenoids. It has a rather unique manner of

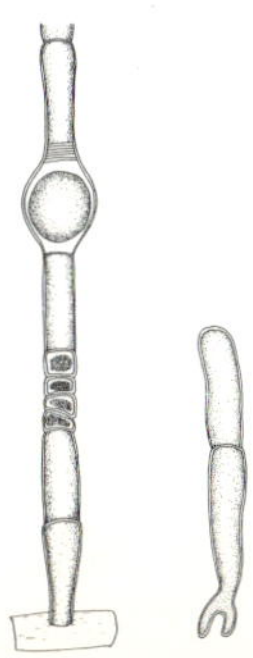

FIGURE 15 *Oedogonium*. Filament showing (*left*) oogonium with single large egg, four young antheridia and holdfast cell. *Right:* young filament following germination.

FIGURE 16 Portion of *Oedogonium* filament with oogonium and three dwarf males attached.

growth in which one or a few cells within the filament divide much more frequently than others and in their division a cell wall ring is left around one end of the cell. The cell then elongates and may then divide again, leaving another ring next to the first. This may continue for a number of divisions with a ring left at the cell end each time following elongation. In addition to this type of intercalary division the *Oedogonium* filaments also grow by elongation and division of the apical cell.

Asexually the various species of this genus may reproduce by fragmentation or by akinetes, which may appear in chains of ten to forty in enlarged cells resembling oogonia, or multiflagellate zoospores may arise singly in zoosporangial cells at the terminus of a filament. The zoospores are spherical or, more commonly, egg-shaped and have a ring of flagella near their pointed ends.

Order Conjugales

The members of the order Conjugales are quite distinct from the other orders of green algae because of their lack of flagellated cells of any type, their variations in chloroplasts, and their method of sexual reproduction by conjugation.

Spirogyra (Fig. 17) is the best-known and most widely studied genus of the order. It is usually easily identified in water by its fine silky texture, and consequently it is called "water-silk." *Spirogyra* is an unbranched filament whose individual cells have one or more ribbonlike chloroplasts spiraling from one end to the other. Growth may occur by the formation of crosswalls and elongation of cells within the filament or at the terminus. Vegetative propagation may occur by the breakup of filaments, with a continued growth and division of cells in the filament.

FIGURE 17 A-*Spirogyra*. Photomicrograph showing stages of conjugation. B-Drawing shows completion of conjugation in which some cells remained vegetative since no contact was made with cells of another filament.

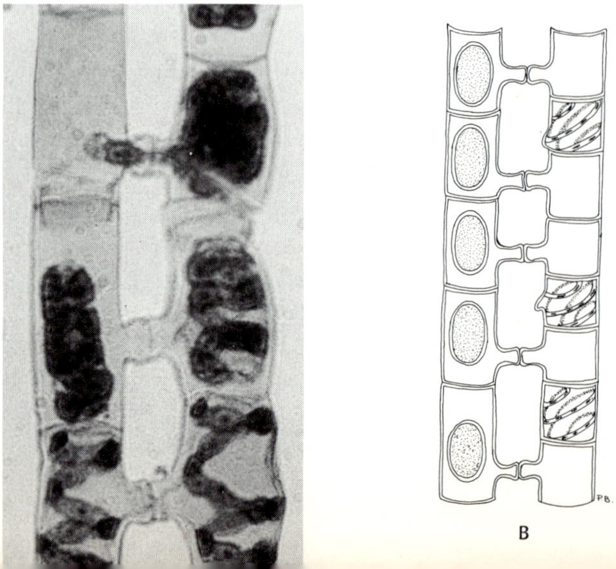
A B

Conjugation in the filamentous members of the Conjugales occurs in several ways, but in all cases it involves the rounding up of the entire protoplasts, following disintegration of the chloroplasts and nuclei, forming large amoeboid gametes. All cells in the filament are capable of forming gametes but it apparently does not happen until conjugation tubes from the cells of two different filaments make contact. This is frequently seen in conjugating *Spirogyra* where many cells in the filaments have made contact and the gametes are formed. Interspersed between the conjugating cells there are often a few which failed to make contact and in these the ribbon-shaped chloroplast is still very much in evidence (Fig. 18).

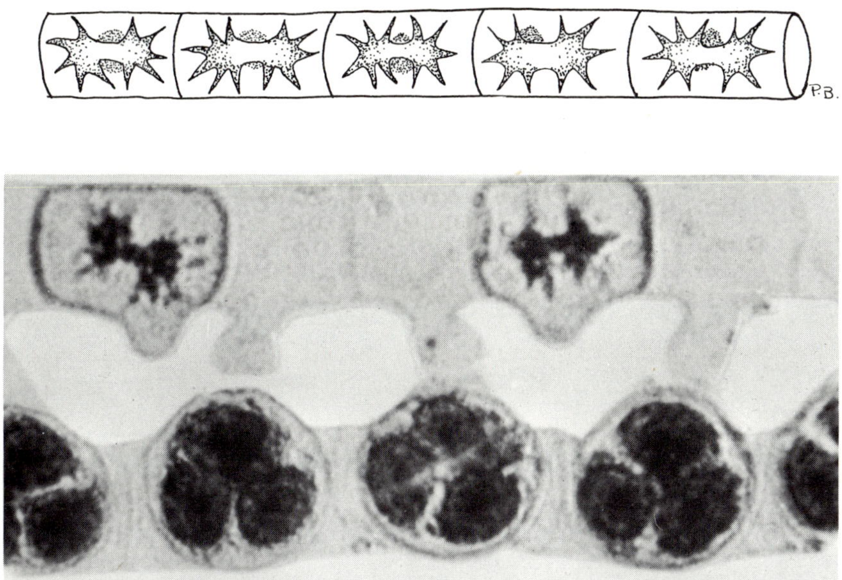

FIGURE 18 *Zygnema*. Star-shaped plastids in cylindrical cells. In the photomicrograph, zygotes have formed and reduction division has occurred showing formation of four daughter protoplasts. Two cells in the upper filament failed to make contact with cells of opposite filament and remained vegetative.

Generally all gametes will move into cells of another filament, leaving behind a filament of empty cells and creating a filament of cells full of zygotes. Although we classify these plants as isogamous, we might think of the filament from which the gametes moved out as male and the recipient filament as female. However, in some species instead of this **scalariform conjugation** we have **lateral conjugation** in which the contents of one cell will move into the cell next to it in the same filament. This would then indicate that some cells in the same filament are male and some are female. But in other species and especially in the genus *Mougeotia* in this order, the contents of the cells of two filaments move out into conjugation tubes and unite in the tube between the empty cells. In this case both would be acting as males. We must therefore conclude it would be better not to think of maleness or femaleness in the Conjugales.

In all members of the Conjugales germination of the diploid zygote is preceded by meiosis, resulting in four haploid nuclei. Three of these disintegrate; the fourth remains and produces a new filament.

Many species of the Conjugales are unicellular and many consist of two symmetrical halves joined by a narrow isthmus as illustrated in Figures 20 and 21.

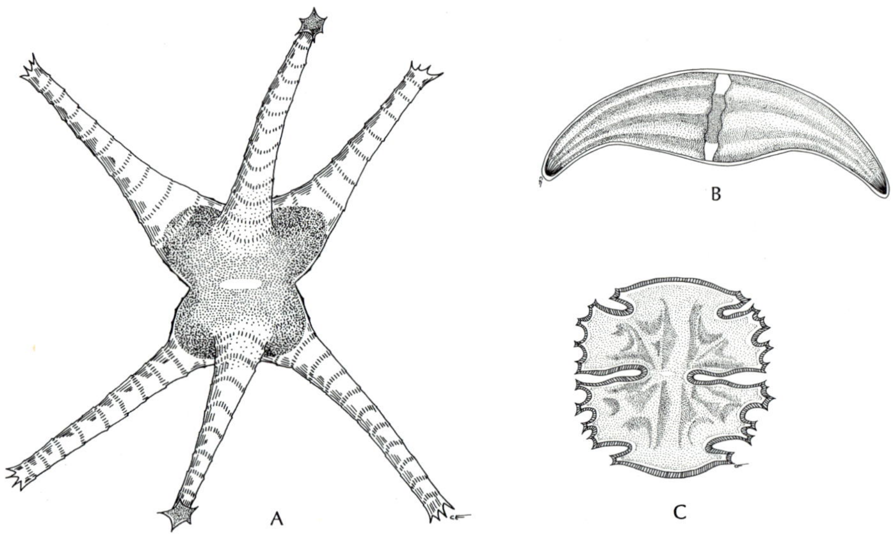

FIGURE 19 A-*Staurastrum*. B-*Closterium*. C-*Microsterias*. All desmids have their cellular bodies divided into two equal halves. From *Laboratory Outlines in Biology*, W. H. Freeman and Company.

Order Chlorococcales

The members of Chlorococcales also vary from unicellular to large colonial, and probably have greater variation in cell shape than any other group. *Chlorococcum* (Fig. 20) has fairly large unicells with a large U-shaped or cup-shaped chloroplast. The cells are uninucleate until just before reproduction. In some cases the protoplast divides into a number of (usually sixteen) fragments, each with a tiny nucleus. These mature into biflagellate zoospores,

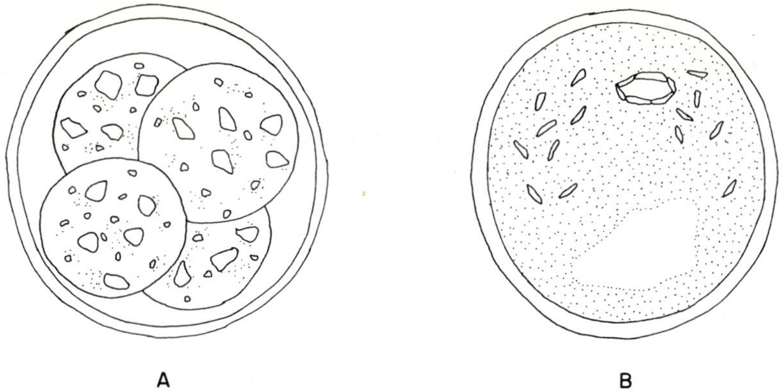

FIGURE 20 *Chlorococcum*. A-Formation of four daughter protoplasts; B-vegetative cell with unstippled nucleus at base; pyrenoid with surrounding starch plates near top. Most cell contents are hidden by the large chloroplast which occupies most of cell.

similar in structure to the vegetative cells of *Chlamydomonas*. The zoospores swim out through an opening in the parent wall, gradually resorb their flagella, and grow into the mature *Chlorococcum* cells. Isogametes are produced in the same way as zoospores, only in larger numbers. Union takes place in the water and upon germination the zygote undergoes reduction division to form four zoospores, each capable of growing into mature plants.

Chlorella (Fig. 21) is a very small member of this order. It has been studied extensively by scientists of the United States Air Force and Navy, as a possible source of food, as a source of oxygen, and as a remover of carbon dioxide in closed space ships and submarines. The plants are only about four microns in diameter, about the size of a large bacterium, and one quart of a moderately thin suspension of *Chlorella* contains about twenty billion cells of the alga. In young cultures new growth results in a fairly high protein content, but as the culture gets older fats and carbohydrates predominate. It has also been found that a quarter pound of *Chlorella* will supply the daily minimal human vitamin requirement except for ascorbic acid.

Chlorella generally reproduces by cell division, but may produce aplanospores. No gametes or zoospores are produced by the members of this genus. Note the electron micrograph drawing of *Chlorella* in Figures 22 and 23 for detail of the cellular contents of this very small alga.

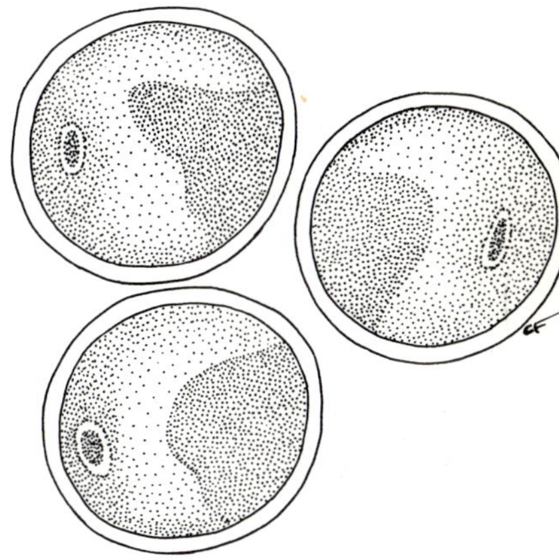

FIGURE 21 *Chlorella*. Vegetative cells showing large U-shaped plastid and single pyrenoid in each. These cells are only about 4 microns in diameter.

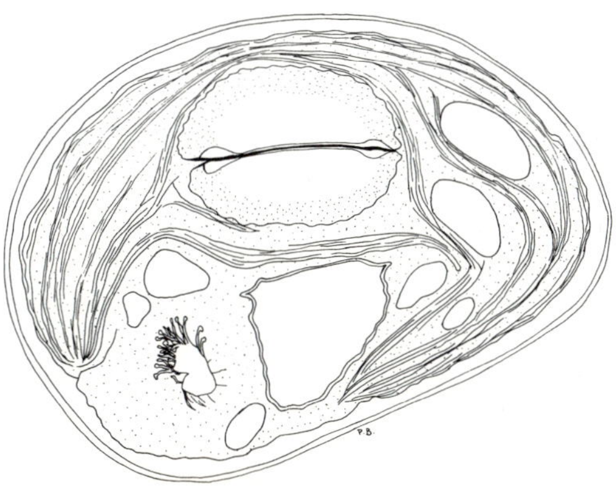

FIGURE 22 Diagram of an electron micrograph of *Chlorella* from Bonner, James and J. E. Varner, 1965, *Plant Biochemistry*. Academic Press.

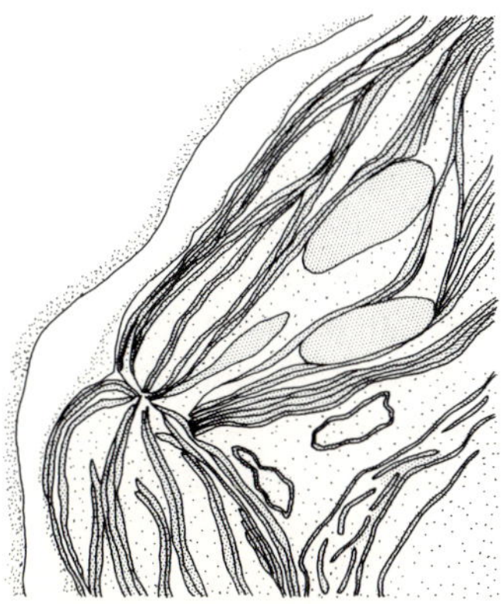

FIGURE 23 Chloroplast division in *Chlorella* showing only a portion of the cell. Drawn from an electron micrograph in Bonner, James and J. E. Varner. 1965. *Plant Biochemistry*. Academic Press.

Other members of the Chlorococcales include *Scenedesmus* (Fig. 24), *Pediastrum* (Fig. 25), and *Hydrodictyon* (Fig. 26).

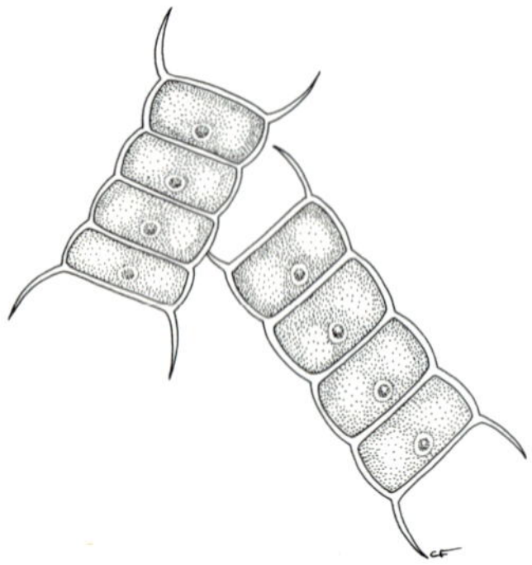

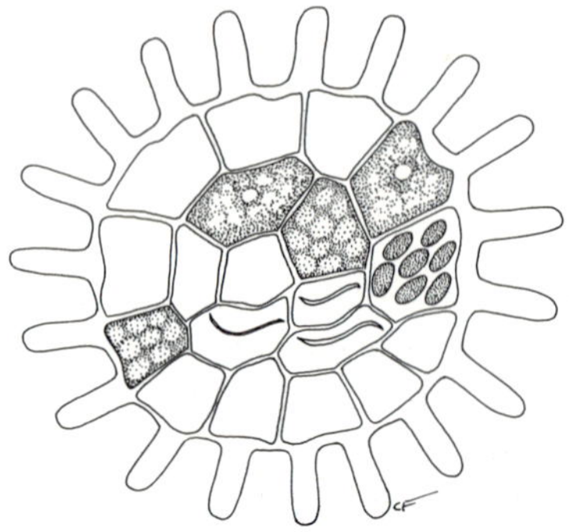

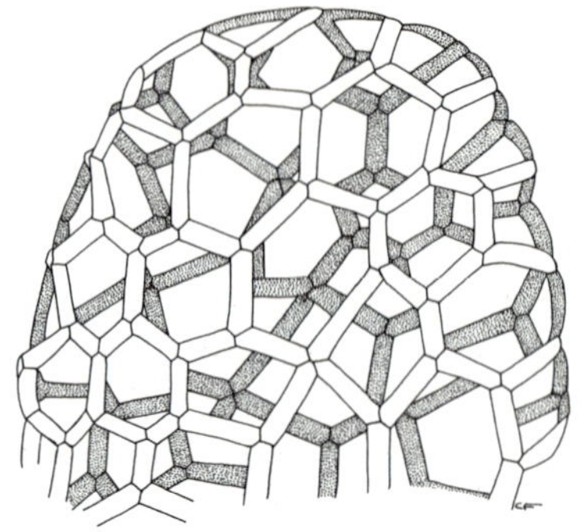

FIGURE 24 *Scenedesmus*. Two plants of typical spinescent 4-celled colonies.
FIGURE 25 *Pediastrum*. A few vegetative cells stippled in and one cell showing gametes.
FIGURE 26 *Hydrodictyon*. Portion of the thallus of the "net alga" showing typical form of cellular connections. From *Laboratory Outlines in Biology,* W. H. Freeman and Company.

Order Dasycladales

The species of the seven genera in Dasycladales all live in warm marine waters attached to pilings and rocks in relatively shallow offshore habitats.

The genus *Acetabularia* (Fig. 27), commonly called mermaid's wine glass, has been studied rather extensively in cytological work because of its very large stemlike cell and the behavior of its nucleus.

FIGURE 27 *Acetabularia,* the mermaid's wineglass.

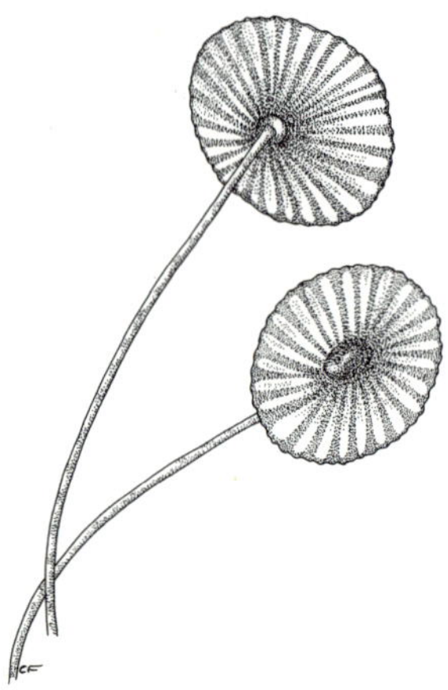

The germinating zygote elongates into a short tube with a differentiated rhizoidal system at its basal end. The nucleus remains in this base as the upper portion of the cell elongates. After a period of growth a transverse whorl of sterile hairs forms just below the apex, followed by a whorl of gametangia at the apex which gives the entire structure the appearance of a slender wine glass.

Throughout the period of growth the zygote nucleus remains near the base of the central stipe, but with the formation of the gametangia it enlarges and divides into a great many nuclei which are carried by the streaming cytoplasm to the thallus apex, forming the gametangial rays. Now the multinucleate protoplasts of the gametangial rays divide into uninucleate protoplasts with enclosing walls termed **cysts.** These cysts enlarge and undergo a series of divisions of which the last is meiotic to form biflagellate pyriform isogametes. The gametes leave the cysts through a lid opening and fuse in the water to form the zygote. The latter germinates directly into the diploid plant.

Acetabularia has been studied by Hammerling and others in an attempt to learn more about the role of the nucleus in determining the growth of the cell. Using two different species with different ray characteristics researchers found that when the upper end of a stalk of one species (without its nucleus) was grafted to the lower of a different species (with its nucleus), the whorled appendages that developed were like those of the basal species. Furthermore, if a graft is performed between two species but the nuclei of both are retained, the cap that develops is intermediate between the two species. This type of study demonstrates the role of the nucleus as a determiner in tissue development even though the differentiation which takes place is in the cytoplasm, rather remotely removed from the nucleus.

Summary

The characteristic which most closely relates the Euglenophyta to the Chlorophyta is the pigmentation. They both have chlorophyll *a* and *b,* carotene and several xanthophylls which are essentially the same. However, there are distinct differences.

The cell walls of the members of the Chlorophyta are of cellulose, and in the flagellate members of this phylum the wall is quite firm and rigid. The euglenoids have a very pliable pellicle covering the cell which enables the organisms to change their shapes quite freely.

Evolutionary relationship between these two phyla is indicated by the flagella, although the types of flagella produced are different. It is probable that the tinsel type of flagellum was derived from the whiplash type. If this is true, then some of the flagellates in the Chlorophyta are more primitive than the Euglenophyta. Both phyla probably have an ancestral origin in some primitive green organisms with characteristics not greatly different from those of the living *Chlamydomonas*. Some of these organisms retained their whiplash flagella, developed cellulose walls, and took on the characteristics of the primitive Chlorophyta leading to the development of other green algae. Some failed to develop the wall and their flagella became modified into the tinsel type; some species discontinued the production of chlorophyll, became saprophytic or holozoic, and became the ancestors of the early animal line.

Both botanists and zoologists include the euglenoids in their discussions of evolutionary development, not because each is convinced the euglenoids are definitely plant or animal, but because these organisms have ancestral qualities common to both animals and plants. Protozoologists include the members of the Volvocales along with the euglenoids and other flagellates in the phylum Protozoa, while in botany the different types of flagellates are separated into different phyla based largely on the types of pigmentation found in the different species.

Supplementary readings on the algae are at the end of Chapter 17.

Questions for Discussion

1. Young cultures of algae, such as *Chlorella,* growing under optimum conditions, may contain nearly 10 percent nitrogen, while old cultures have considerably less. What explanation can you suggest for this change in nitrogen content?
2. Why does the arrested synthesis of protein molecules result in a greater accumulation of amino acids?
3. Why would one expect phosphorus content in algae to increase when they are illuminated rather than kept in darkness?
4. How may the aquatic medium during seasonal changes affect the permeability of the algal membranes?
5. Describe the different types of chloroplasts found in the green algae.
6. Of what advantage is it to an alga to be imbedded in a gelatinous envelope?
7. In what ways are members of the Euglenophyta similar to certain Protozoans?
8. Why is it more likely that certain algae will be used to supply oxygen and food in space ships than higher plants?
9. To make algae more useful as food for humans what type of treatment will they have to receive?
10. What are the essential differences between a multicellular individual and a colony of individual cells?

11. What causes some cells in a filament to produce gametes while later in the season the cells of the same filament will produce zoospores?
12. How has *Acetabularia* contributed to our knowledge of nuclear behavior?
13. What types of reproduction are found in the green algae?
14. Zoospores are usually positively phototropic. Of what advantage is this to the plant?

17

The Yellow-Brown Algae

Phylum Pyrrophyta
Phylum Chrysophyta
Phylum Phaeophyta

Phylum Pyrrophyta

Pyrrophyta is a phylum of approximately 135 genera and over a thousand species mostly of unicellular and colonial algae. Commonly called the dinoflagellates, the members are often classified with the protozoans. Some of the members are uniflagellate or lack flagella entirely, but most are biflagellate. The majority of dinoflagellates have a cell wall made up of regularly arranged, platelike segments composed principally of cellulose. Most of the organisms have a central groove or girdle surrounding the body and separating it into approximately equal halves. Within this groove arise two flagella of unequal length. The tinsel flagellum remains within the groove while the other, the whiplash type, is usually out of the groove and is probably of greater use in motility. The flagella have a neuromotor apparatus similar to that of the other flagellates.

Most of the species have chlorophyll in their plastids, which may be rod-shaped, discoid, or band-shaped. The dominant pigment is xanthophyll, a pigment closely related to fucoxanthin, the prominent pigment of the brown algae. Their reserve foods are stored chiefly as starch and oils, although other substances may be produced by them.

Reproduction

Reproduction in dinoflagellates is chiefly asexual, by simple cell division or by the production of zoospores or aplanospores which escape from the parent wall and then secrete their own wall following liberation.

Sexual reproduction occurs rarely but in one genus, *Ceratium*, (Fig. 1A) it has been described as a type of conjugation similar to that in the conjugales. In *Gymnodinium* (Fig. 1B) the protoplast of a cell divides to form four gametes which are liberated by rupture of the parent wall. Gametes then fuse in the water, forming a zygote which, following a resting period, divides meiotically to form four naked protoplasts. These cells then secrete their own walls typical of the species.

FIGURE 1 A-*Ceratium*, B-*Gymnodinium*, C-*Gonyaulax*.

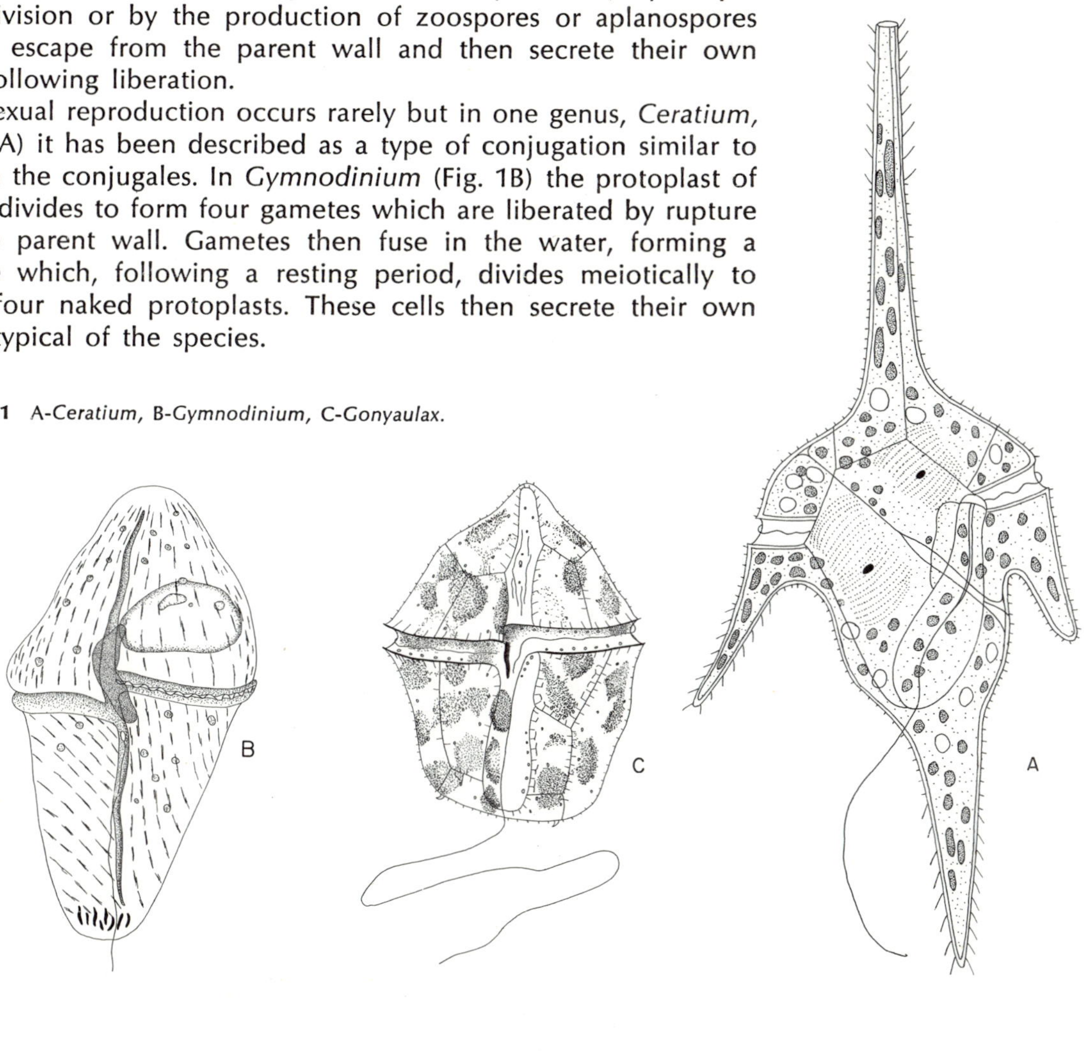

FIGURE 2 *Vaucheria geminata* showing coenocytic filament and stalked sex organ. Four spherical oogonia surround the central curved tubular antheridium.

Several genera including *Gymnodinium* and *Gonyaulax* (Fig. 1C) have acquired a certain notoriety because they are associated with a marine situation termed **Red Tide.** In the warm coastal waters off our Gulf, South Atlantic, and Pacific shores, numerous fish have died and been cast upon the beaches. This has been a seasonal disease and reaches epidemic proportions in certain years. Where these dinoflagellates are in large numbers the sea turns reddish and at this time thousands of fish die. As yet no satisfactory control has been found but many of our marine laboratories have been devoting much time and money to finding a specific cause for the disease and a possible control. It is certain that the dinoflagellates cause the death of the fish, but whether they carry a disease or produce a toxic substance themselves is still unknown.

Phylum Chrysophyta

There are over six thousand species of Chrysophyta classified thus far and most are unicellular species of diatoms. Since there is so much variation in the morphology of members of the group it is convenient to consider them in three classes. Although superficially they seem to be quite unrelated, relationship is indicated by their similarities in pigmentation, chiefly carotenes and xanthophylls; their food reserves, a carbohydrate or oil but never starch; and flagellate similarities.

CLASS XANTHOPHYCEAE

Most of the members of Xanthophyceae are yellowish-green, although green may predominate in a few. Food is stored in the form of oil and **leucosin,** an insoluble white substance of proteinaceous chemical composition. In some the cell wall is composed of overlapping segments which are chemically made up of a pectic compound and occasionally some cellulose. In a few the wall may be impregnated with silica.

One of the best known members is *Vaucheria* (Figs. 2, 3), which

FIGURE 3 *Vaucheria sessilis;* antheridium on left liberating sperm; oogonium on right has within it a fertilized egg.

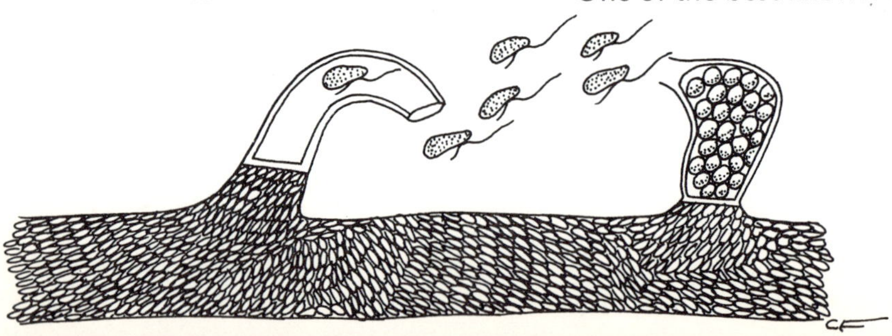

was classified in the Chlorophyta until about a decade ago. It is usually dark green in color, but occasionally the carotenoids predominate. Its food reserves are stored as an oil rather than starch as it was once believed, when its was classified in the green algae. Its single large multinucleate zoospore is produced in the terminal zoosporangium. However, instead of having a uniform distribution of equal flagella over its surface it has pairs of unequal length.

The filaments of *Vaucheria* are coenocytic (lack crosswalls), with numerous nuclei and tiny disc-shaped plastids. Crosswalls usually are found where the sex organs arise and at the terminus where a zoosporangium is found, but not elsewhere.

Reproduction in Vaucheria

The specialized sex organs are rather unique in this genus. Oogonia may arise as enlarged protuberances on the filament or as globular appendages on a gametophore. Each oogonium contains one large nonmotile egg. The antheridia arise as protuberances from the filament but are tubular in appearance and frequently coiled. The sperms have two unequal flagella and are liberated from the tips of the tubular antheridia.

Entrance to the oogonia, usually one in another filament, is gained through a pore in the oogonium. Following fertilization the zygote breaks from the oogonial wall and after a period of dormancy germinates to become a new coenocytic filament.

FIGURE 4 *Dinobryon*-portion of colonial thallus showing protoplasts and bell-shaped lorica.

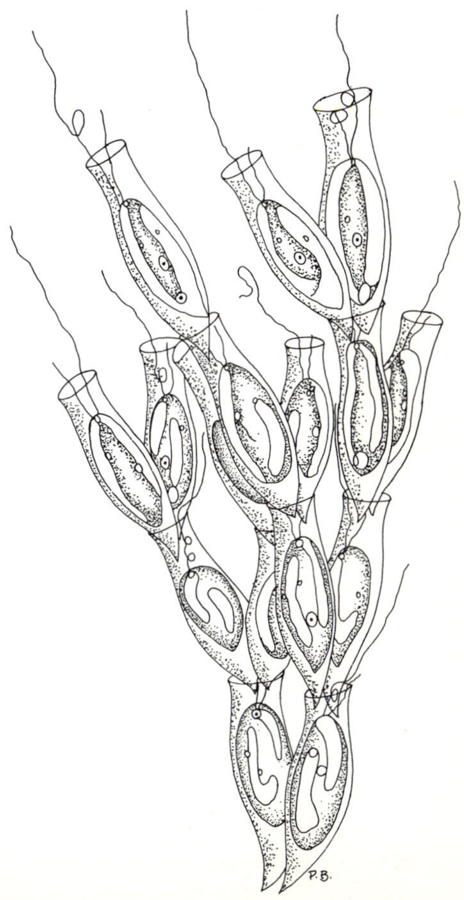

CLASS CHRYSOPHYCEAE

Often called the golden brown algae, the Chrysophyceae are a small class of mostly unicellular, biflagellate (rarely uniflagellate) organisms that are occasionally placed with the Protozoans by zoologists.

Dinobryon (Fig. 4) is a colonial organism in which the cells form dendroid colonies by longitudinal division of the protoplasts. Each protoplast forms in a bell-shaped envelope, termed the **lorica,** and upon division two new cells emerge from the parent lorica and develop coverings of their own. After the colonies have produced a series of these inverted bell-shaped cells they break up and each segment begins the formation of new attached colonies. Isogamy has been recently reported in this genus.

CLASS BACILLARIOPHYCEAE

Commonly called diatoms, the members of this class number about 5,500 different species and probably are more numerous than any other plants on earth. They grow in virtually every habitat, aquatic or terrestrial, that will support life. The order **Centrales,** whose members are radially symmetrical, are found most abundantly in the sea, while the members of the order **Pennales,** which are bilaterally symmetrical, are most abundant in fresh water habitats (Fig. 5).

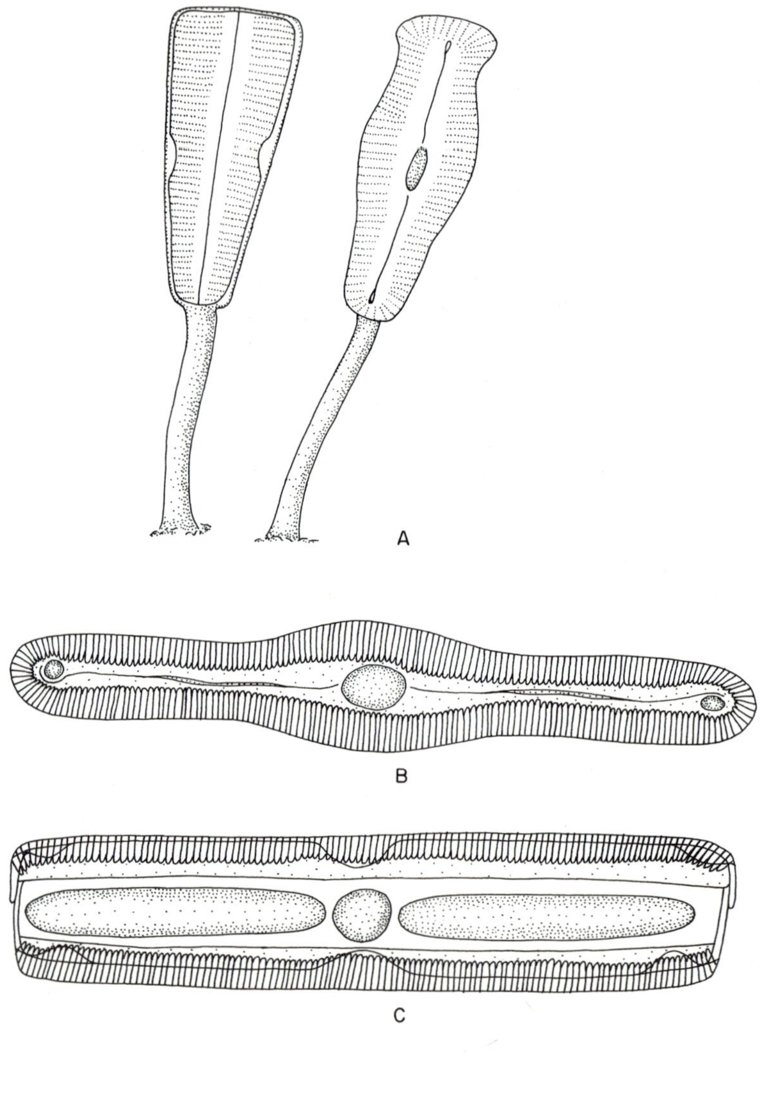

FIGURE 5 Diatoms. A-Stalked diatom shows girdle and valve views. B-Valve view. C-Girdle view both of *Pinnularia*.

Each diatom consists of two halves **(valves).** The lower half, the **hypotheca,** is the inner portion, and the portion above is the **epitheca.** The epitheca fits over the lower as the top of a box or the lid of a petri dish fits over its lower half. Uniting the two halves is a connecting band which encircles the cell.

The diatoms in the Pennales are all bilaterally symmetrical and capable of a jerky type of movement which is due to a streaming of cytoplasm in a central seam or raphe extending from one end of the valve to the other but interrupted in the middle by one or two **central nodules.** Cytoplasm streams along the outer fissure in the raphe from a **polar nodule** to a central nodule, then moves inside and along an inner fissure to the opposite end. The mechanism is very similar to that of the treads of a tank, but the movement is not smooth and involves a good deal of forward and backward motion.

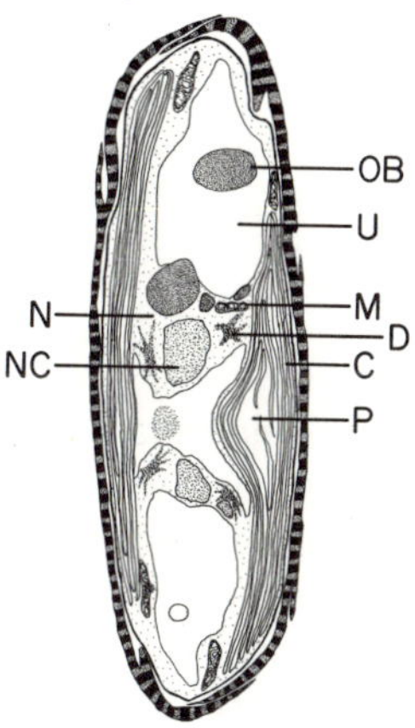

A

FIGURE 6 A diatom, *Amphipleura pellucida,* drawn from an electron micrograph, girdle view longitudinal section showing enclosed lamellated chloroplast (C) on either side; D-dictyosomes; M-mitochondria; N-nucleus; NC-nucleolus; OB-oil bodies; P-pyrenoid. There is considerable duplication of materials throughout the cell. From E. F. Stoermer, H. S. Pankratz, and C. C. Bowen, 1965. Fine Structure of the Diatom, *Amphipleura pellicida. American Journal of Botany.* 52(10) 1067–1078, p. 1068.

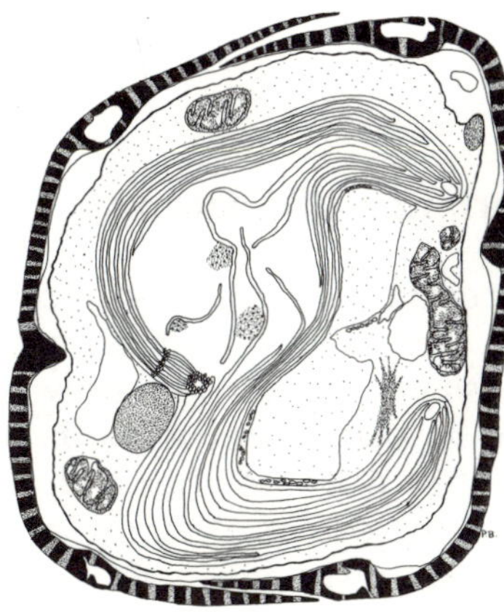

FIGURE 7 Cross section of *A. pellucida* showing particularly the large pyrenoid within the chloroplast. Note lamellae of chloroplast which have penetrated the pyrenoid.

In the Pennales (Figs. 6, 7) there is a centrally located nucleus and one or two elongated chloroplasts. The Centrales have small, numerous discoid plastids. In most diatoms the green chlorophyll is overshadowed by varying amounts of carotenoid pigments, especially fucoxanthin.

Reproduction in Diatoms

Diatoms reproduce by a distinctive type of cell division. The nuclei and chloroplasts divide and half of the contents of the parent cell remains with the epitheca and half with the hypotheca

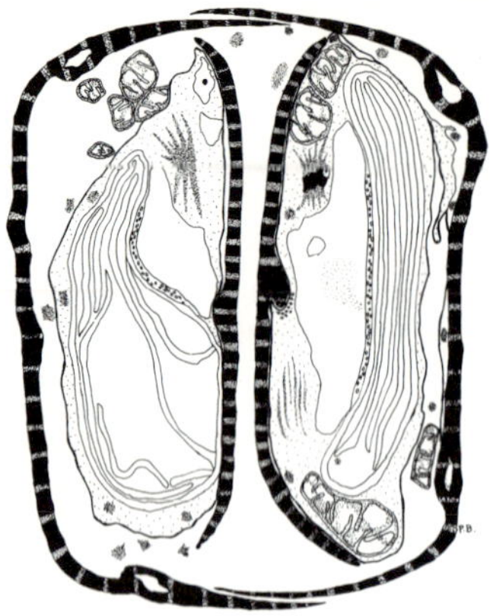

FIGURE 8 Drawing from an electron micrograph of a diatom showing division; note that the nucleus and chloroplast have divided and new walls are forming within the old. Same reference source as in Fig. 17–6.

as the two valves separate. Each half then secretes a new wall within the old (Fig. 8). As a result diatoms become gradually smaller and smaller as this type of division continues generation after generation. Eventually some of the progeny become extremely small. When this happens **auxospores** are formed. This may be accomplished by a shedding of the valves with an accompanying increase in size of the protoplast, which then secretes two new valves restoring the cell to the size of that species before division. It has also been shown that prior to shedding the valves, some species form four or more flagellated **microspores.** These, when liberated, swarm around a protoplast of an oogamous nature which had not divided. Fertilization occurs and the zygote develops into an auxospore.

Several common members of the Pennales are *Navicula, Pinnularia,* and *Pleurosigma.* Three frequently observed members of the Centrales are *Coscinodiscus* (Fig. 9), *Triceratium,* and *Melosira.*

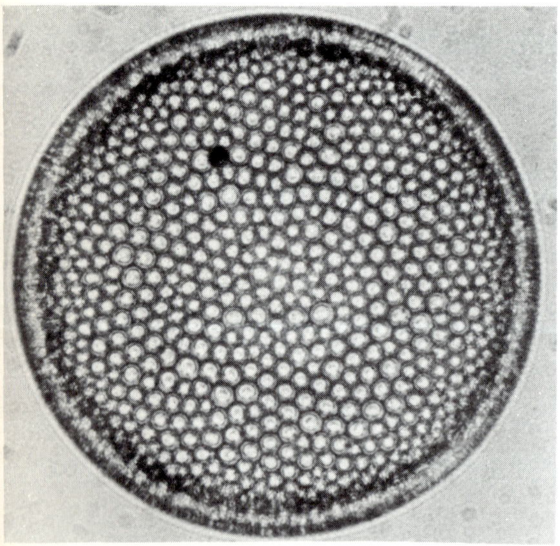

FIGURE 9 *Coscinodiscus,* a member of the centrales.

Importance of Diatoms

Economically diatoms include some of our most important plants. Their highly siliceous walls are practically incapable of decomposition and consequently have accumulated in vast deposits known as **diatomaceous earth.** This is mined and used as an ingredient in silver polishes and other products requiring a very fine abrasive ingredient. Diatoms are extremely light in weight and are used in making lightweight concrete, as an insulator in high temperature furnaces, and as a filter for certain types of acids, oils, and other solvents.

Indirectly, diatoms furnish food for fish and other aquatic organisms, which store the diatom oils and vitamins in their livers. These organisms in turn are eaten by larger fish, and the oils and vitamins are transferred to the livers of halibut, cod, sharks, and other fish and extracted by us for our own use.

In one study it was found that it takes about 400 billion diatoms to sustain a humpback whale for just a few hours. It was disclosed that the diatoms are eaten by small crustaceans, each of which may have over 130,000 diatoms in its stomach; the crustaceans are eaten by herring, each of which may have over 5000 in its stomach at one time; and a humpback whale requires about a ton of herring (about 5000) in its stomach to feel comfortably full.

Phylum Phaeophyta

The largest of all plants classed in the Thallophyta are members of the brown algae. The vast majority are marine, with only a few living in fresh water habitats. Most of them live in the cold-water latitudes, unlike the red algae, which seem better acclimated to warm ocean habitats. Many of the brown algae (Fig. 10) live in relatively deep coastal waters beyond the low tide zones; others are able to withstand long periods of desiccation and live in the intertidal zones as rockweeds.

The members of the Phaeophyta vary in color from olive green to various shades of brown depending on the amount of chlorophyll and the xanthophyll pigment, **fucoxanthin,** present in their plastids. Usually there are a number of plastids present in a cell, but a few browns have but one per cell. Pyrenoids are lacking. The reserve foods are stored in the form of a complex carbohydrate, **laminarin,** or as **mannitol,** a hexahydric alcohol, or in the form of oil globules, but never as starch. The cell walls are composed principally of cellulose and frequently are covered on the outer surface with a gelatinous coating of **algin.** In addition to the plastids, the protoplast contains a centrally located vacuole and a single nucleus with its membrane, nucleolus, and chromatic network.

FIGURE 10 A kelp bed, chiefly *Nereocystis,* along the shores of Vancouver Island, British Columbia.

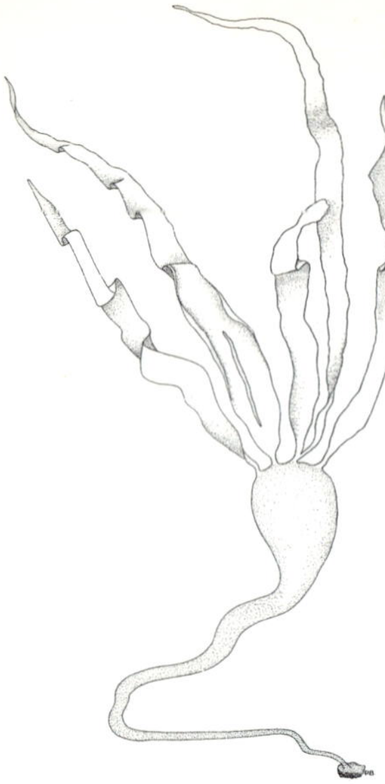

FIGURE 11 A brown alga, *Nereocystis,* showing basal holdfast, stipe, bulbous float at stipe apex, and blades.

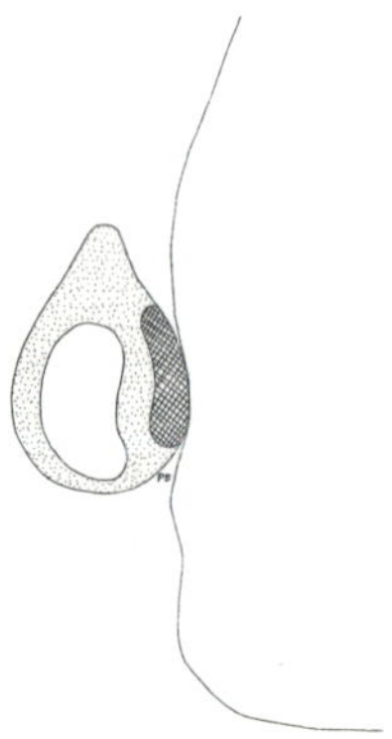

FIGURE 12 Typical appearance of a male gamete (also female in some species) and zoospore in the brown algae.

Morphology

Most members of the phylum have a well-organized thallus differentiated into a holdfast, a stipe, and leaflike photosynthetic organs termed the **lamina** or **blades** (Fig. 11). The thallus of the smaller species is often made up of branched uniseriate filaments very much like many of the green algae. Growth in the larger brown algae is usually initiated by a single apical cell or by a row of apical meristematic cells. Many of the large kelps have an active meristematic region at the juncture of the stipe and blade, or at the base of the blade, which adds new tissue to the distal portions.

The gametes and zoospores, when present, are pyriform (Fig. 12) and bear two laterally oriented flagella which are generally unequal in length. Like the green algae, some brown algae do not have a distinct alternation of generations but most members of the Phaeophyta have an alternation of generations in which haploid gametophytic plants alternate with diploid sporophytic plants in the life cycle. This alternation of generations may be isomorphic or heteromorphic.

The phylum Phaeophyta is composed of a single class, the **Phaeophyceae,** comprised of about two hundred genera and approximately a thousand species. For convenience the class is divided into six orders of which only three will be discussed in detail here. The orders are:

1. Ectocarpales
2. Laminariales
3. Fucales

ORDER ECTOCARPALES

Although this order is comprised of about sixty genera and three hundred species, we will discuss *Ectocarpus* as the typical and principal genus.

Ectocarpus (Fig. 13) looks very much like a small tuft of fine feathers growing on the rocks in shallow coastal waters. It is generally light brown in color and only a few inches tall, but very copiously branched. Occasionally it is found in freshwater streams, especially those that flow directly into the sea, but most species live in the ocean.

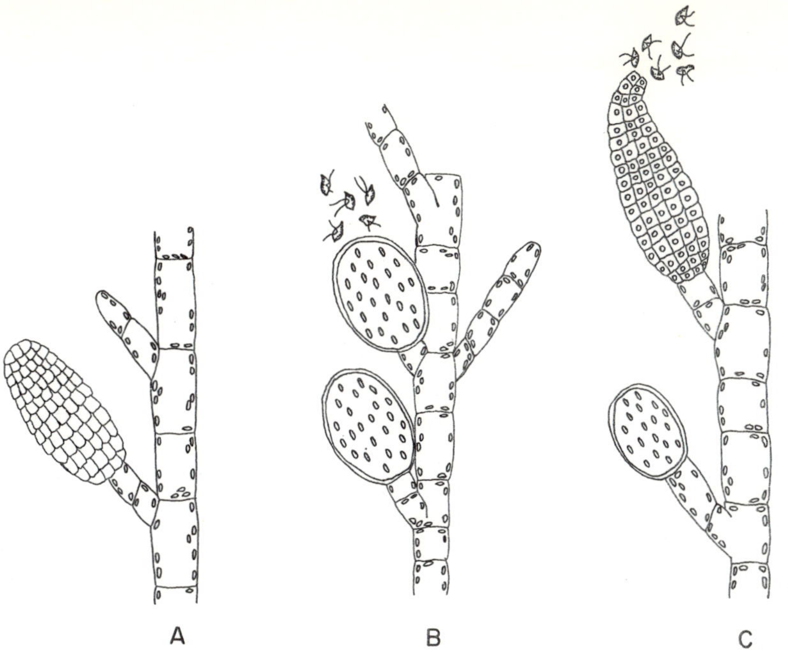

FIGURE 13 *Ectocarpus.* A branch of the gametophytic plant with gametangium. B Branch of sporophytic plant with two unilocular zoosporangia. C Sporophytic plant with a unilocular zoosporangium and a plurilocular zoosporangium near top of branch.

Ectocarpus has an isomorphic alternation of generations in which the gametophyte and sporophyte plants are similar in every respect except the type of reproductive organs they bear. The gametophytic plants produce **plurilocular gametangia,** each of which contains several hundred cubical cells. Each cubical cell gives rise to a single pyriform biflagellate gamete. After their release the gametes swim about and fuse with a gamete from another gametophytic *Ectocarpus* plant. Although the sex cells are isogamous, the plants are distinctly heterothallic.

The zygote germinates directly into the sporophytic plant without meiosis. The diploid plant, at maturity, produces two kinds of sporangia. One type, the **unilocular zoosporangium,** is globose in form and contains a single nucleus in its rather large locule. This nucleus undergoes a series of divisions, one of which is meiotic, producing thirty-two to sixty-four zoospores. These haploid zoospores are morphologically similar to gametes and upon germination give rise to the gametophytic plants.

On other branches of the same sporophytic plants **plurilocular zoosporangia** are produced. These are long, slender zoosporangia with numerous locules. Each cubical locule produces a single zoospore which looks like those of the unilocular type except that they are not produced by reduction division and consequently are diploid. These zoospores germinate directly into the sporophytic plants and consequently complete a short cycle in the life of *Ectocarpus.*

ORDER LAMINARIALES

This is a small order with only about thirty genera and about a hundred species, but it is a spectacular group from the standpoint of size. A number of kelps have sporophytic plants that are over a hundred feet in total length. In their customary habitats, the cold waters of the North Atlantic and the North Pacific and other cold waters of the Southern Hemisphere, they occupy thousands of square miles of coastal waters.

Most kelps live out beyond the low-tide mark where they are anchored by large dichotomously branched holdfasts. Many of the larger species live in waters about fifty feet deep, but with their long stipes the blades manage to reach the surface where they may float out and photosynthesize. Many of the larger species are broken by the waves, but their basal and intercalary meristems keep giving rise to new tissue, replacing that which is worn and torn away.

An extreme in heteromorphism is exhibited by kelps. The sporophytic diploid plant may be many dozens of feet in length, but the gametophytic haploid plants consist of only a few cells.

The Life Cycles of the Isomorphic *Ectocarpus*
and a Heteromorphic Kelp

Ectocarpus

Sexual Plants (*n*)
Gametangia (*n*)
Isogametes (*n*)
Zygote (2*n*)
Asexual plant (2*n*)

Unilocular (2*n*) — Zoosporangia — R D — Zoospores (*n*)
Plurilocular (2*n*) — Zoosporangia — Zoospores (2*n*)

Heteromorphic kelp

Male plant (*n*) — Antheridia (*n*) — Sperm (n)
Female plant (*n*) — Oogonia (*n*) — Egg (n)

Zygote (2*n*)
Asexual plant (2*n*)
Zoosporangia (2*n*)
R D
Zoospores (*n*)

These plants are dioecious. The male gametophyte consists of a minute, branched filament with a few antheridia formed at the ends of the short branches. Each antheridium produces one biflagellate sperm. The female is also a short filament of a few cells, one or more of which become rounded into an oogonium containing a single nonmotile egg. The eggs may be fertilized in the water or a sperm may enter a pore in the oogonium, converting the egg into a zygote.

Soon after the zygote is formed it begins to divide into a filament of cells which continue growth and differentiation into the large plants they are destined to become. Studies have shown that some kelps, at least, will not produce sex cells unless the temperature of the water is below 15°C, which probably accounts for the abundance of large brown algae in colder waters.

Most of the sporophytic plants produce clusters of club-shaped zoosporangia on the blades (Fig. 14). When these clusters are localized they are called **sori,** a word borrowed from fern terminology. Growing in among the zoosporangia one often sees sterile hairs, called **paraphyses,** which may be of some aid in directing the zoospores, by setting up pathways, to the outer portion of the blade. The zoospores, when released, swim actively about by means of their two laterally attached flagella and soon settle down and germinate into male and female plants.

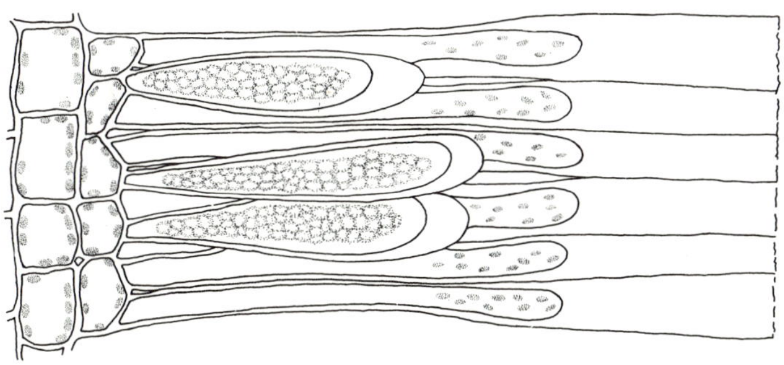

FIGURE 14 Portion of a blade of *Laminaria* showing three zoosporangia and several paraphyses.

ORDER FUCALES

The plants in the order Fucales are much smaller than those in the Laminariales, but they make up a large segment of the vegetative population of the intertidal zone as rockweeds (Fig. 15).

FIGURE 15 *Fucus,* a rockweed growing in the intertidal zone. Picture was taken at low tide.

Fucus is the best known and most widely studied genus in the order. The thallus is conspicuously dichotomously branched, dark brown in color and almost as tough as leather. The plants seldom reach over a foot in length and are firmly anchored by holdfasts. The thallus has a prominently branched midrib along which inflated air bladders arise. The internal organization is much like that of a kelp, including a central spongy pith region surrounded by cortical cells and protective epidermal layers.

Near the tips of branches small raised pustule-like nodules appear. These enclose the reproductive cavities termed **conceptacles.** Within the conceptacles the oogonia and antheridia are produced. In some species only oogonia are formed in a conceptacle, while other conceptacles on the same plant will have only antheridia. These plants are therefore monoecious as are those in which both oogonia and antheridia are found in the same conceptacles. Other species of *Fucus* are dioecious (Figs. 16, 17) and have only one type of sex organ on a plant.

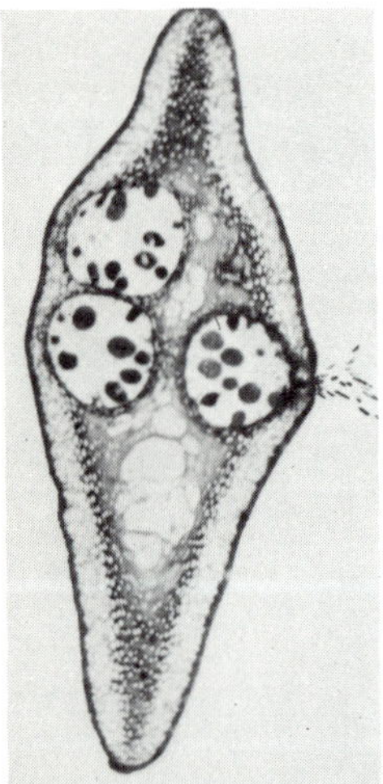

FIGURE 16 Cross section of thallus of *Fucus* with oogonial conceptacles.

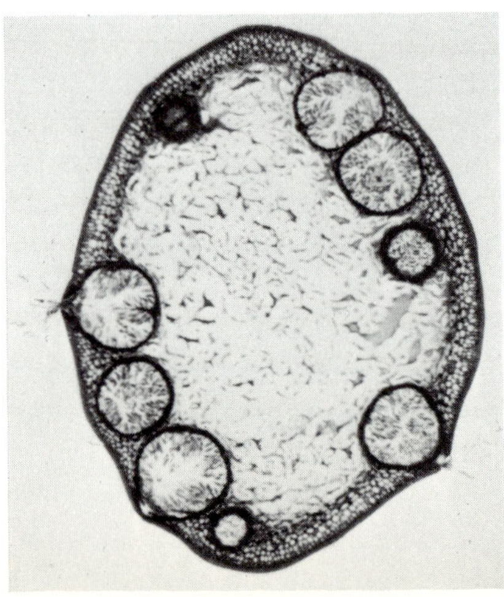

FIGURE 17 Cross section of thallus of *Fucus* showing conceptacle with antheridial branches.

An oogonium (Fig. 18) at maturity contains eight large nonmotile eggs. The antheridia (Fig. 19) are quite small and usually form sixty-four laterally biflagellate sperms. Generally fertilization takes place away from the thallus in the water after the eggs have floated out of the conceptacles (Fig. 20).

FIGURE 18 Diagram of a greatly enlarged oogonium with paraphyses.
FIGURE 19 Diagram of antheridial branch. Antheridia are uniformly stippled.
FIGURE 20 Diagram of *Fucus* egg and surrounding sperm. The latter cause the egg to spin in the water before entrance is gained by one sperm.

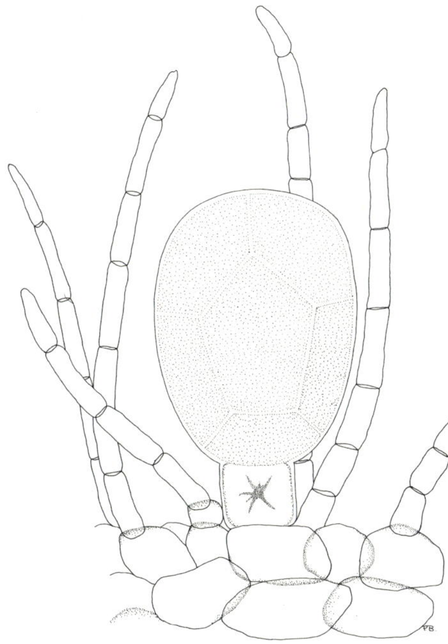

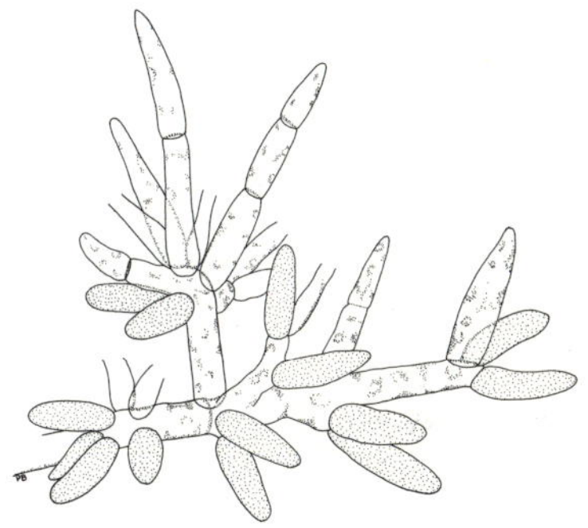

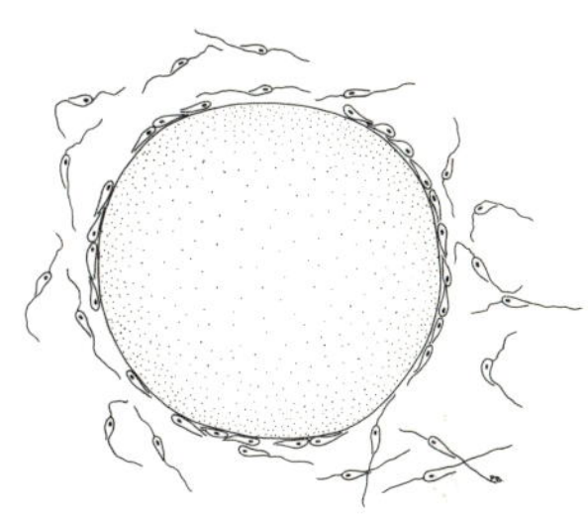

The zygote germinates almost immediately and a new sporophytic plant is formed. In *Fucus* the gametophytic generation is represented solely by the gametes. Some algalogists are of the opinion that the origin of the gametes is much like that in the embryo sacs of the higher plants. In the development of the female reproductive organ of *Fucus* the initial growth is considered a **megasporangium.** The nucleus within divides twice, including a meiotic division, to form four haploid nuclei which are regarded as **megaspores.** At this time there may be a brief pause before the next division, forming eight nuclei which mature into eight eggs with the megasporangium developing into the oogonium.

Development of the male sex organ may be explained in the same way, the initial male cell forming as a **microsporangium.** Meiosis results in the formation of four nuclei considered as microspores. After a brief period of dormancy these cells undergo successive mitotic divisions, giving rise to the sixty-four sperms.

This is a type of gametogenesis found in the vascular plants and in *Fucus*. It is an example of analogy in parallel evolution rather than the beginning of microsporogenesis and megasporogenesis for the plant kingdom. We do not believe that the brown algae have given rise to primitive vascular plants, but that the latter might have been derived from some ancestral plants with vegetative characteristics similar to *Fucus*. This primitive plant was considered to have been a siphonaceous green alga rather than a brown. In any event the manner of origin of the sex cells is not the connecting link.

FIGURE 21 *Sargassum* showing globular floats and blades.

Other members of the Fucales include *Sargassum* (Fig. 21), a free-floating brown alga known as gulf weed which is very abundant in the Gulf of Mexico, the South Atlantic Ocean, and elsewhere. *Ascophyllum* grows to several feet in overall length and is fairly common along our north Atlantic coast. *Pelvetia* is a rather large rockweed which grows on rocks in the intertidal zone in many areas including the California coast.

Phylogenetic Relationships

Several characteristics stand out when an attempt is made to link the three phyla discussed in this chapter. The dominance of the xanthophyll pigments, particularly the presence of fucoxanthin, indicates that these groups are related even though their common ancestry may be far removed. The presence of pyriform zoospores and pyriform gametes in the great majority of the species as well as the presence of the two laterally attached flagella on the motile spores and gametes also are unique to these groups. Furthermore the flagella are usually of the two types in each group: one is of the whiplash type and the other is of the tinsel type. The reserve foods are similar to the extent that they are usually variations of a starch-like product and an oil. While these characteristics do not bring these three phyla together in a beautiful linear evolutionary series, they do show similarities that bring them together in a yellow-green or green-brown series just as the blue-green algae have much in common in their pigmented relationship with the red algae, and as the Euglenophyta and Chlorophyta show relationship in the possession of the same types of chlorophyll.

The Pyrrophyta, the Chrysophyta, and the Phaeophyta probably all arose from very early flagellate ancestors, with the Pyrrophyta remaining very much the same. The Chrysophyta, on the other hand, seem to have developed along two or three different lines with considerable emphasis on the unicellular habit in the members of the Chrysophyceae and especially in the Bacillariophyceae. The greatest variation in body form was achieved by the Xanthophyceae, where the unicellular forms became modified into a siphonaceous series with the evolution of plants like *Botrydium* and *Vaucheria*.

The Phaeophyta probably had as humble an origin as any of the others but somehow the genes of the brown algae and their mutating nucleic acids initiated a greater development in thallus structure so that the browns were able to go on to much greater complexity in body organization. Like the lines of evolution in the green algae, the trend in evolution in the brown algae was increasing complexity in body organization, reduction in motility, isogamy to oogamy, and isomorphism to heteromorphism, with increasing emphasis on a large diploid sporophytic plant. This includes the members of Fucales, where we may consider the gametophyte in the alternation of generations as reduced to one cell. This is really not very different from the great differences in size between the gametophytes and sporophytes of the Laminariales.

Supplementary Reading

Chapman, V. J., *Seaweeds and Their Uses.* Methuen, London, 1950.

Dawson, E. Y., *How to Know the Seaweeds.* Wm. C. Brown Co., Dubuque, Iowa, 1956.

Dittmer, H. J., *Phylogeny and Form in the Plant Kingdom.* D. Van Nostrand, Princeton, New Jersey, 1964.

Fritsch, F. E., *Structure and Reproduction of the Algae,* Vols. 1 and II. Cambridge University Press. London, 1935, 1945.

Hammerling, J., and W. Gunter, Uber den-Wuchsmodus von Acetabularia, *Zeitschr. Naturforsch.,* 136(7):449–454, 1958.

Hardy, Alister C., *The Open Sea. Its Natural History: The World of Plankton*. Houghton Mifflin, Boston, 1956.

Lewin, R. A., *Physiology and Biochemistry of Algae*. Academic Press, New York, 1962.

Moore, Hilary B., *Marine Ecology*. Wiley, New York, 1958.

Prescott, G. W., *Algae of the Western Great Lakes Area*. Cranbrook Institute, Bloomfield Hills, Michigan, 1957.

Prescott, G. W., *How to Know the Fresh-Water Algae*. Wm. C. Brown Co., Dubuque, Iowa, 1954.

Smith, G. M., *Freshwater Algae of the United States*. McGraw-Hill, New York, 1950.

Smith, G. M., *Cryptogamic Botany,* Vol I. McGraw-Hill, New York, 1955.

Tiffany, L. H., *Algae, the Grass of Many Waters*. Charles C. Thomas, Springfield, Illinois, 1935.

Tilden, J. E., *The Algae and Their Life Relations*. University of Minnesota Press, Minneapolis, 1935.

Trainor, Francis R., Control of Sexuality in *Chlamydomonas chylamydogama, American Journal of Botany* 45(8):621–626, 1958.

Walford, L. A., *Living Resources of the Sea. Ronald* Press, New York, 1958.

Questions for Discussion

1. What evidences do we have that some members of the Chrysophyta are a very ancient group?
2. Red tide organisms are believed to cause the death of thousands of tons of fish at certain seasons. How do you believe the death of the fish has been accomplished by these unicellular organisms?
3. Why is it that diatoms do not readily decay like most other algae?
4. Why is it that asexual reproduction by cellular division may occur for many generations and then begins to slow down and may cease altogether if the colony is not restored by some type of nuclear fusion?

5. What characteristics do the members of the three phyla of yellow-brown algae have in common?
6. Describe some algae with isomorphic alternations of generations and some which are heteromorphic.
7. Some brown algae are over one hundred feet in length yet they have no mechanical tissue. How are they able to support themselves?
8. Suggest reasons why red algae live mostly in warm seas while most brown algae live in much colder waters in the higher latitudes.
9. Why are the Fucales generally considered to be the most advanced order of brown algae?
10. Some brown algae such as *Egregia* live in coastal waters where they are continually buffeted by heavy seas. What does this suggest about their metabolic requirements?
11. Describe the various lines of evolutionary development which seem to have occurred among the different algal groups.
12. Describe the economic importance of brown algae.
13. Why is it possible for red algae to live at much greater depths in the ocean than green or brown algae?
14. How is it possible for many species of fish to live at much greater depths in the ocean than algae or other plants?
15. Describe the different types of zoospores found in the algal groups. What do they all have in common?

18

The Primitive Achlorophyllous Microorganisms

Phylum Schizomycophyta
Phylum Myxomycophyta

Microbiology is the study of organisms which are so minute it is necessary to use a microscope, accessory instruments, and a variety of techniques to observe them. The organisms usually included in this microscopic world are algae, some fungi (including yeasts), protozoans, bacteria, rickettsiae, viruses, and slime molds. This chapter will be devoted to a description of the last four groups, but various members of the other groups are often considered in the study of microorganisms because of their minute size and frequent similarities in their reproductive habits.

Microorganisms carry on the same fundamental activities within their cells as do the multicellular organisms. While the unicellular organisms may be structurally simple, their physiological activities include the utilization of food and energy, the growth and development of new protoplasm, and the ability to reproduce.

In general, these organisms are often divided into two basic categories:

1. The **Monera** include those organisms which lack an organized nucleus, such as the blue-green algae, bacteria, rickettsiae, and viruses.

2. The **Protista** do have true nuclei. The ancient members of this group were probably all unicellular, but a more liberal classification of modern forms also includes the nucleated algae, slime molds, fungi, and the protozoa.

Of the monerans, we have already considered the **Cyanophyta** in Chapter 15. Now our attention is directed to the **Schizomycophyta,** the **Myxomycophyta,** and related groups. Organisms in the Schizomycophyta have genes which apparently are present as minute concentrated clumps of DNA protein, but they are not organized into chromosomes as found in other organisms; they do, however, have similar genetic potentials. The cells of these organisms also do not appear to have vacuoles. Some are photosynthetic, although they do not have organized chloroplasts. Their photosynthetic pigment is contained in a unit resembling the single layer of a granum in a chloroplast. These units may be termed **chromatophores,** and several may be found in some bacterial cells.

It is the opinion of most microbiologists that the earliest organisms consisted of little more than an aggregate of nucleic acids and proteins surrounded by a layer of proteins and possibly other nutrients very much resembling the present day viruses. The early protoviruses were apparently capable of securing their metabolic substances from the simple organic compounds created by chance in the prehistoric swamps and oceans. Modern viruses no longer are capable of such independent existence but obtain their foods from the living tissues and enzymes of a host they infect. It is quite possible that present-day viruses are degenerate forms of higher microorganisms which evolve as the result of a ready availability of nutrient substances within the cells of certain hosts.

The early microbial organisms probably multiplied faster than the simple organic substances could be produced. It was then that these primitive organisms developed the enzymes and catalytic ability to use simpler substances and unite them into new substances for growth. This type of synthesis requires energy and it was probably accomplished very simply by a rearrangement of the chemical bonds of microbial nutrients. This was not easily done in the absence of atmospheric oxygen, but with the passage of a billion or more years of time the simple chemical rearrangements which did occur provided some energy and possibly released carbon dioxide as a by-product.

Now with the formation of another compound (CO_2), organisms eventually evolved which possessed a pigment similar to chlorophyll. These organisms were able to use both the simple organic and inorganic substances available but with the new pigment they were able to secure new energy from sunlight. Soon primitive photosynthetic organisms such as the bacteriochlorophyllous types appeared. They produced carbohydrates and free oxygen from the water and carbon dioxide now available. Once these forms became widespread life utilizing oxygen evolved. Energy was secured at least in part by direct oxidation. By this time bacteria and many other microorganisms capable of using different energy sources evolved.

Phylum Schizomycophyta: the Bacteria

As the name Schizomycophyta implies, bacteria are "fission fungi." They are not the only fission organisms, but it is the only mode of cell division known for them. Bacteria vary from about two tenths of a micron to ten or more microns long. The average size for them is about one to three microns, whereas the cells of most organisms run about ten times this figure in average size.

Morphology

Bacterial cells come in a variety of shapes, of which three principal types are obvious (Fig. 1). These are spherical, rodlike,

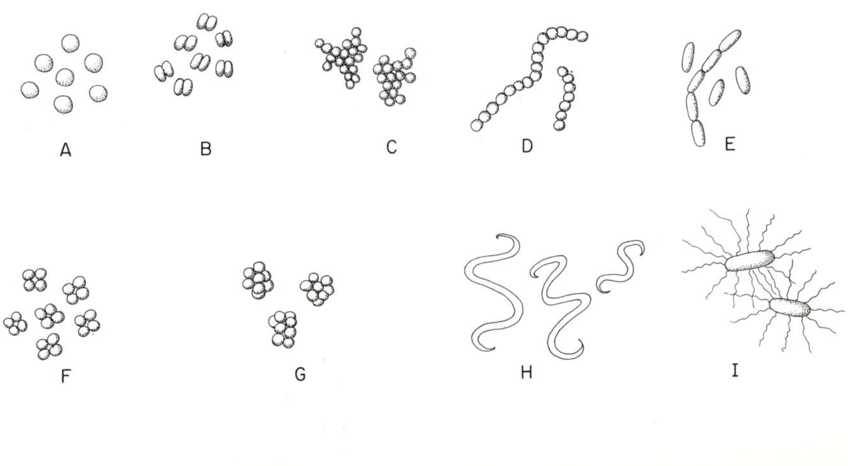

FIGURE 1 Bacteria, showing some of the possible forms: A-Coccus, B-Diplocococcus, C-Staphylococcus, D-Streptococcus, E-Bacillus and Streptobacillus, F-Tetracoccus, G-Sarcina, H-Spirillum, I-Flagellated.

and spiral. The rodlike types are termed **bacilli** (sing. **bacillus**). They may be straight and cylindrical, or long ellipsoids, or occasionally curved. They may be naked or encapsulated and some are found in elongated chains. A single-celled type which is covered with a waxy capsule is the tuberculus bacillus; one which appears in chains is the causative organism of anthrax.

Spherical organisms are the **coccus** types. They are not all perfect spheres, with the result that it is often difficult if not impossible to distinguish between rounded rodlike bacteria and oval cocci. When two coccus cells normally stay together and grow in this paired condition they are termed **diplococcus;** if they are in fours they are **tetracoccus;** if in chains they are **streptococcus;** if in grapelike clusters they are **staphylococcus;** and if in eight-celled, cuboidal colonies they are of the **sarcina** type.

An organism which grows in a spiral is termed a **spirillum** form. Some are shaped like a corkscrew, others are less tightly coiled. Other species have only part of a spiral turn and are comma-shaped. These are termed **vibrios.**

Chemical Composition

The major component of bacteria, like that of other organisms, is water, which makes up about 75 to 85 percent of the total weight. Most of the amino acids have been found in the bacterial proteins which make up about fifty percent of the total dry weight. One amino acid, diaminopimelic acid (DAP), is found only in bacteria and blue green algae. It is a portion of the macromolecular polypeptide structure which gives rigidity to the cell wall.

Carbohydrates are present in the cell walls and capsules, usually as polysaccharides. The cell walls of **gram-positive** bacteria have thirty-five to sixty percent carbohydrates while the walls of **gram-negative** species contain only fifteen to twenty percent carbohydrates. Other carbohydrates, often stored as a glycogen, are in the cytoplasm.

Fats are found in the cell wall; also, as in other organisms, the cytoplasmic membrane is a lipo-protein layer. The lipid content of some bacteria, such as the acidfast tuberculosis bacillus, may run as high as 40 percent.

RNA and DNA are also present in the proteinaceous nucleic acids of bacterial cells. Other substances usually present include sugars, organic acids, nucleotides, vitamins, enzymes, coenzymes, and phosphate esters.

The Bacterial Cell

The protoplast (that portion of the cell within the cell wall) performs most of the metabolic activities of all living cells. This is the site of the respiratory processes, synthesis of proteins, including enzymes, and nucleic acids. Whether or not bacteria have discrete chromosomes is still a debatable question, but no one questions the existence of a genetic mechanism associated with DNA in the bacterial cell.

Covering the cell walls of most, if not all, bacteria are capsules of varying thickness and composition. They may be membranaceous and termed **microcapsules,** or the capsules may be thick and of firm structure, or a loose slime.

Flagella are present on many bacteria. These structures are extremely thin (0.01 to 0.05 micron), but may be as much as 70 microns long. The flagella are composed of an elastic fibrous protein and apparently arise from a granule within the cytoplasmic membrane. Electron photomicrographs show a single bacterial flagellum as composed of two or three fibrils twisted helically around each other.

Endospores are formed within the cells of certain bacteria in all of the major groups. They are highly resistant bodies and extremely high temperatures are required to kill them. Normally a single cell produces only one endospore and within the different species these may be spherical or oval. The endospore may be considerably larger than the usual diameter of the cell, or it may be contained within the usual size of the protoplast.

Staining Techniques

Morphology alone, or the growth of bacteria in various sugar solutions and other substances, rarely is enough for adequate identification. Various dyes have been used over the past hundred years and have not only helped us to see bacteria more clearly but have brought out characteristics which otherwise remain hidden. Two outstanding techniques are used extensively which make it possible for us to further categorize all bacteria. These are the gram stain and the acidfast procedure.

The gram stain divides bacteria into two great groups, the gram-positive and the gram-negative. In this technique the bacteria are stained with crystal violet, which is then followed with an iodine solution that acts as a mordant and helps to fix the dye in gram-positive organisms. After the stain is applied alcohol is used to decolorize organisms that do not readily take the stain, and a contrasting dye, such as safranin, is applied. Organisms that take the purple primary stain are gram-positive and those which lose the stain when decolorized with alcohol, and perhaps take the safranin stain, are gram-negative.

Antibiotics have also been found to vary in their effectiveness against gram-positive and gram-negative bacteria because of the sensitivity of the organisms to the antibiotics. Consequently antibiotic sensitivity tests are frequently run on bacteria isolated from clinical materials. These tests are most frequently done by placing small paper discs containing antibiotics on the surface of an agar plate inoculated with the organism to be tested. The diameter of the zone of inhibited growth around the paper disc is a measure of the antibiotic's effectiveness against the bacteria in the agar.

The acidfast stain such as is used in the Ziehl-Neelsen procedure has carbol fuchsin as the primary stain. The bacterial smear is covered with this solution and heated gently for about five minutes. After heating, the slide material is decolorized in a solution of 1 percent sulfuric acid or 3 percent hydrochloric acid in 95 percent ethyl alcohol. The material is then subjected to a counterstain of contrasting color, such as methylene blue. Acidfast bacteria will retain the red stain and all other organisms will be decolorized by the acid and stained with the blue.

As in all techniques, there is some gradation of individual organisms in responding to the dyes. Some species, for example, may have both gram-positive and gram-negative individuals and may therefore be termed **gram-variable.** In any case bacteria are so sensitive and so difficult to identify that the staining procedures must be carefully standardized to obtain consistent and reliable results.

Resistance of Bacteria

Among all organisms bacteria are, in general, more resistant to extremes in temperature and tolerance to the acidity or alkalinity of the medium than any other group. Even in their active vegetative state many of them are able to withstand subzero and boiling temperatures for short periods, but their powers of resistance are greatly enhanced by the formation of spores. Complete sterilization requires boiling under pressure or the thorough application of strong germicidal fluids to every place bacteria may have lodged. The strength of the germicide required to kill bacteria in their spore forms is often more than ten times that needed to kill them in their ordinary vegetative phase.

Conjugation

The usual method of reproduction in bacteria is simple fission, probably accompanied by an equal division of the internal organelles of the protoplast. It was suggested years ago that there might be a form of sexual reproduction in bacteria, but it had never been observed directly until 1944, when Lederberg and Tatum demonstrated that a sexual type of recombination could occur in *Escherichia coli* (Fig. 2). They found two mutant strains

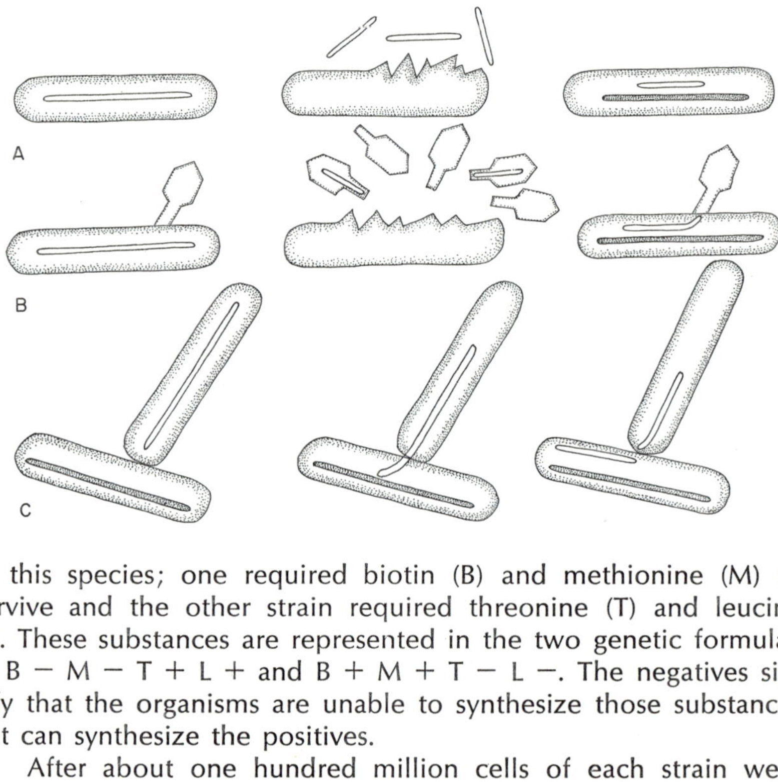

FIGURE 2 Methods of genetic exchange in bacteria. In A the exchange is accidental following a breakup of the bacterial wall and entrance of nuclear material into another bacterium. B. Exchange of material by virus carriers. The prismatic figures are viruses, two of which have picked up genetic material from the infected bacterium and the one on the right has discharged the genetic material it carried into another bacterium. C. Conjugation where two bacteria make direct contact and one (the male) liberates a portion of its genetic material into a second, which we may call the female. Experiments have shown that the amount of material exchanged is largely determined by the length of time the bacteria remain in contact. (From Wollman, E. L. and F. Jacob, 1956, Sexuality in Bacteria. *Sci. Amer.* July) p. 6.

of this species; one required biotin (B) and methionine (M) to survive and the other strain required threonine (T) and leucine (L). These substances are represented in the two genetic formulae as B − M − T + L + and B + M + T − L −. The negatives signify that the organisms are unable to synthesize those substances but can synthesize the positives.

After about one hundred million cells of each strain were mixed, they were incubated and plated on a synthetic medium that contained none of the four growth factors. Neither strain could grow alone on this medium but when the mixed culture was plated a number of colonies appeared. These were found to have the genetic composition B + M + T + L +. The hybrids had the positives of both parental strains. Since these experiments, conjugation has been photographed in *E. coli* and it has been demonstrated that genetic material passes from one bacterial donor (male) to the other (female) when they are in contact. The conjugants do not fuse completely but separate after a short period of attachment.

Grouping of Bacteria

ORDER PSEUDOMONADALES

The bacteria classified in this order are generally considered the most primitive now living. It is also believed that the first cellular organisms may have been of this type. Today there are pseudomonads belonging to all three basic forms and many of them are flagellated. Included also are the chemosynthetic and photosynthetic bacteria. To the latter belong the purple nonsulfur and the green sulfur bacteria. The chemosynthetic types are the sulfur bacteria, the iron bacteria, the hydrogen bacteria, and the nitrifying bacteria.

Photoautotrophic bacteria use light as the primary energy source and a chlorophyll to absorb the light energy. The purple bacteria are thus able to form a simple carbohydrate using raw materials from organic sources, carbon dioxide, and hydrogen as shown below.

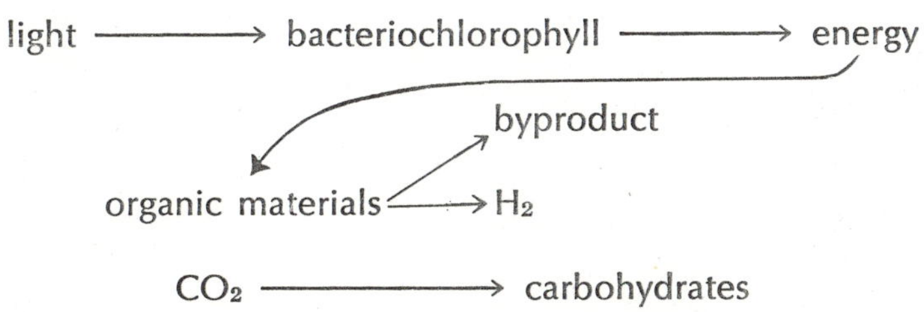

In this reaction oxygen is not used, but some species may accomplish the same reaction using oxygen rather than light and therefore be considered chemautotrophs.

Sulfur bacteria may absorb hydrogen sulfide or molecular sulfur and combine these substances with oxygen:

$$(1)\quad 2H_2S + O_2 \longrightarrow 2S + 2H_2O$$

or

$$(2)\quad 4S + 6O_2 + 4H_2O \longrightarrow 4H_2SO_4.$$

Iron bacteria oxidize soluble iron compounds and gain energy by converting them into insoluble substances:

$$Fe^{++}\ \text{(soluble)} \xrightarrow{\text{oxidation}} e^- + Fe^{+++}\ \text{(insoluble)} + E\ \text{(energy)}.$$

Among the most important of the present-day bacteria are the nitrifying bacteria. One type uses ammonia and produces nitrite ions, and the second type uses the nitrites and converts them into nitrate ions, in which form they become soluble and are absorbed by higher plants which build proteins.

$$\underset{\text{Ammonia}}{2NH_4} + 4O_2 \xrightarrow{\text{oxidation}} \underset{\text{Nitrite}}{2NO_2} + 4H_2O + E$$

$$\underset{\text{Nitrite}}{2NO_2} + O_2 \xrightarrow{\text{oxidation}} \underset{\text{Nitrate}}{2NO_3} + E$$

$$CO_2 + H_2O + E \longrightarrow \text{organic compounds}$$

These reactions are an integral part of the entire nitrogen cycle, shown below, which also includes free elemental nitrogen fixation by bacteria in the soil and in the leguminous roots (Fig. 3).

A

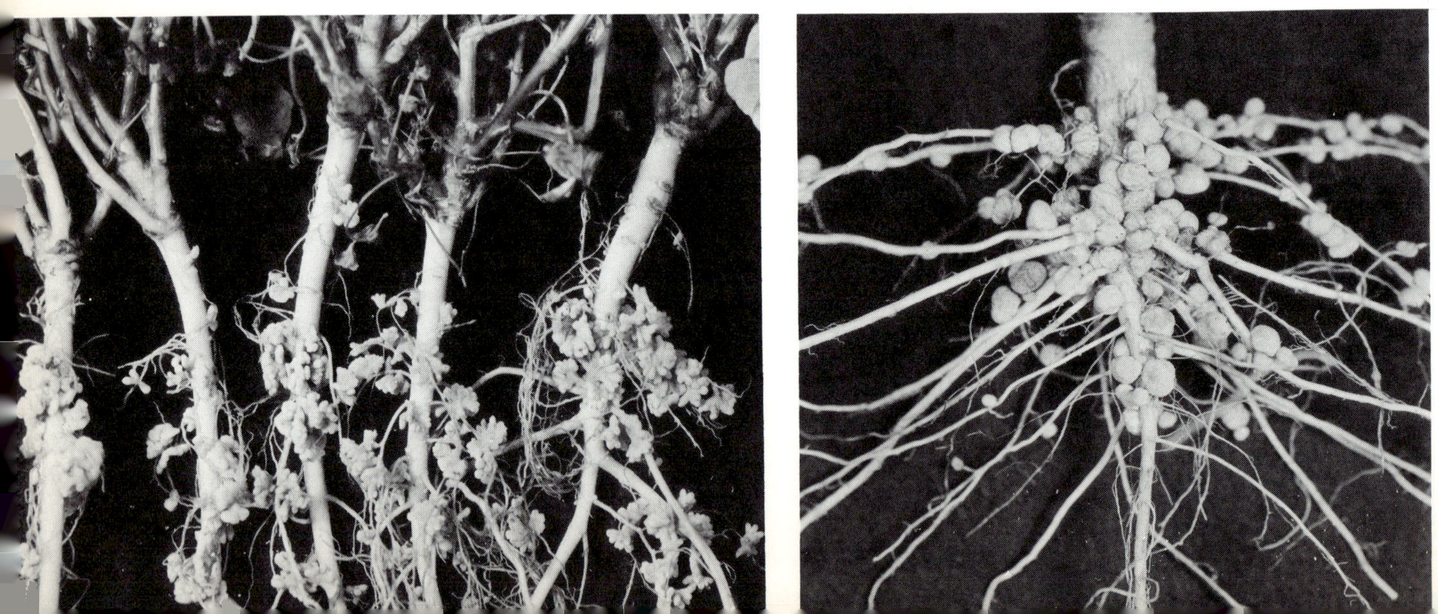

B

FIGURE 3 Roots developed from inoculated seeds. A-Button clover. B-Soy bean. Note the difference in shape of the hypertrophied nodules caused by the growth of the nitrogen-fixing bacteria. (Courtesy Nitragen Corp., Milwaukee, Wis.)

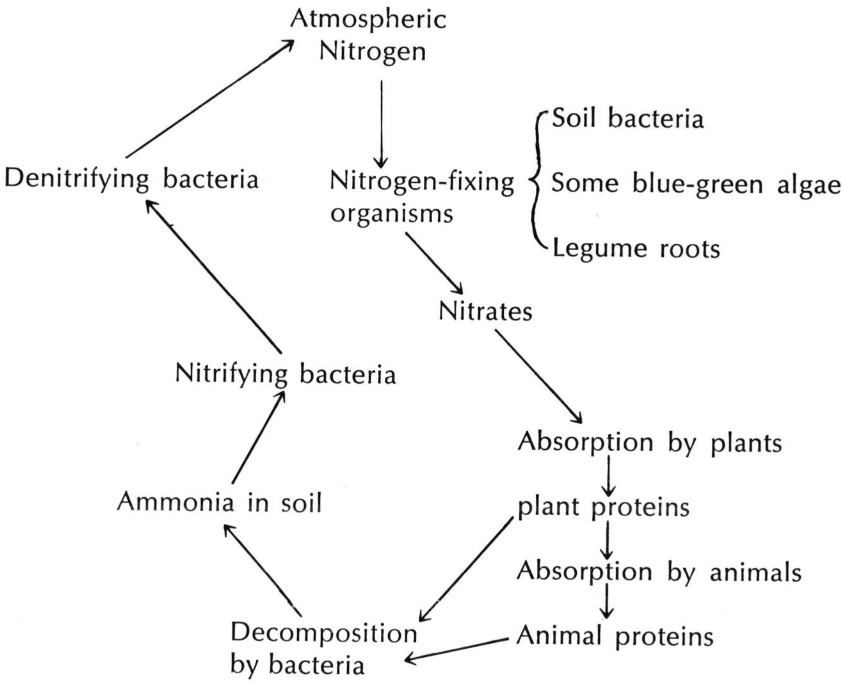

The Nitrogen Cycle

While this is a complete cycling of nitrogen, it is not in itself a complete building process. Through the action of rain and movement of soil water, some nitrates are being continually leached out of the soil and carried away. Furthermore the denitrifying bacteria break up the soil nitrates and free the inert elemental nitrogen to the atmosphere. Due to these steady losses, the system would inevitably run itself down were there not some agency for the reconversion of atmospheric nitrogen into soil nitrates. This is the role of the nitrogen-fixing bacteria outlined in the cycle above.

ORDER EUBACTERIALES: TRUE BACTERIA

Most bacteria studied in connection with disease, fermentation, or sludge disposal, or merely free-living types, are members of this extremely large group. All structural forms appear somewhere in the order, and many have been studied extensively. The

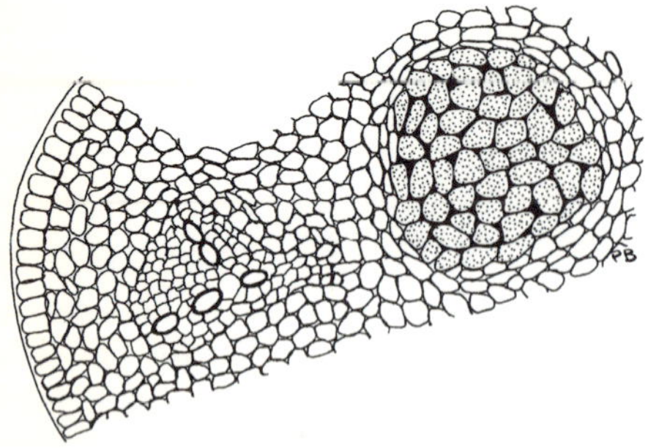

FIGURE 4 Internal structure of bacterial nodule in root cells. Vascular tissue of the root is to the left and hypertrophied cells to the right.

general comments made about bacteria apply largely to members of the Eubacteriales. They are not acidfast, and a few are motile by means of flagella.

Included in this order is the genus *Azotobacter,* whose members are aerobic and utilize atmospheric nitrogen. They are all free-living. *Rhizobium* is also capable of fixing free atmospheric nitrogen when growing symbiotically on the roots of legumes (Figs. 3 and 4).

Escherichia is a common inhabitant of the intestinal tract. *Aerobacter* is found in the intestine and also may grow on plant material. *Proteus* hydrolyzes urea and is frequently found in feces and decomposing matter. *Salmonella* includes the organisms which cause typhoid and paratyphoid fevers. *Brucella* causes undulant fever, and *Pasteurella* is the causative organism of bubonic plague. *Lactobacillus* is responsible in part for the fermenting of dairy and plant products. *Diplococcus* has species which cause pneumonia and gonorrhea. One anaerobic genus, *Clostridium,* is associated with the spoiling of food, as is *Staphylococcus. Clostridium* also causes the acute food poisoning, botulism. These are by no means all of the types or all of the disease producers in the Eubacteriales but they do indicate the great diversity characteristic of the members of the order.

ORDER SPIROCHAETALES

Although generally considered spiral bacteria, the members of this order have occasionally been placed in the phylum Protozoa of the animal kingdom. The cells are quite flexible, making it possible for the organisms to move by a spinning, whirling, or lashing motion. Some have terminal flagella, but these structures are apparently not used in locomotion by the spirochetes.

Most of the members are saprophytic or parasitic but the best known, throughout the world, is the causative organism of syphilis, *Treponema pallidum.*

ORDER ACTINOMYCETALES

The members of this order are frequently called the branching bacteria, since most of them form branched filamentous colonies which superficially resemble the mycelia of some fungi. The hyphal branches, which are never more than 1.5 microns in diameter, often bud off conidia like spores (Fig. 5) of some molds. The members of the Actinomycetales are generally considered soil organisms. Some species produce antibiotic secretions which have been of inestimable value to mankind since their discovery some twenty years ago.

Most of the species which form antibiotic substances are in the genus *Streptomyces*. From its species a good many specific drugs are derived including streptomycin, chloromycin, terramycin, and aureomycin.

FIGURE 5 *Streptomyces*. Portion of the thallus with two branches showing production of conidia.

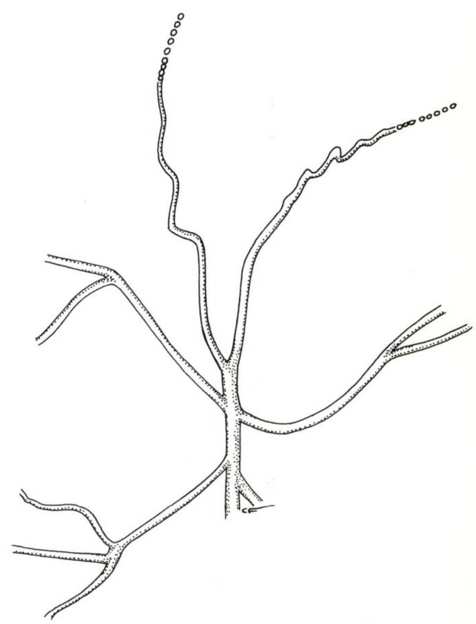

ORDERS MYXOBACTERIALES AND BEGGIATOALES

These two orders are of somewhat doubtful status. Each may be considered a distinct order, but the members of the Myxobacteriales, often called the slime bacteria, are sometimes placed with the slime molds because of their slimy secretions and their tendency to swarm and to form a fruiting body similar to the Myxomycophyta.

The Beggiatoales, known as the gliding bacteria, have a gliding or oscillating movement similar to that found in certain blue-green algae. For this reason, along with other cellular characteristics, some students of the protista have placed them with the blue-green algae as colorless cyanophytes.

Rickettsiae

Rickettsiae are, like the viruses they resemble, all parasitic. They range in size from about 0.2 to 0.5 micron in diameter and from 0.3 to 1.0 micron in length. This is about the size of a large virus. Although they react to antibiotics like true bacteria, they do not stain as readily.

The first rickettsiae were discovered in 1909 by Ricketts in his investigations of Rocky Mountain spotted fever. The group was named by da Rocha Lima in 1916 in honor of Ricketts, who died in 1910 from typhus fever, which is caused by another species of the same genus *(Rickettsia prowazeki)*.

Viruses

Between 1881 and 1885, Louis Pasteur was conducting rabies experiments in his laboratory. From his investigations he was able to show that the causative agent was transmitted in the saliva of dogs. He also was able to demonstrate that healthy dogs became rabid in the usual incubation period if they were innoculated with a small amount of macerated brain tissue from rabid dogs. Pasteur was, however, unable to find any bacteria in the dog's brain tissue or in his cultures which could cause hydrophobia. He, therefore, ultimately concluded that the causative agent was ultramicroscopic.

Similar observations were made by Iwanowski in 1892, when he was studying mosaic disease of tobacco. He was able to show that the disease was contagious by placing the juice of infected leaves on healthy plants and obtaining the typical symptoms of dark green and yellow patterns on the leaves and shoots of infected plants. Iwanowski also found that he could filter the infected juice through unglazed porcelain filters and still it would infect healthy tobacco plants. But when he heated contaminated plant juice to the boiling point of water and then tried to infect healthy plants with the juice, he was unsuccessful in transmitting the disease. Since bacteria are readily filtered out of solutions when dripped through unglazed porcelain filters, Iwanowski concluded that the infectious tobacco mosaic agent was smaller than bacteria and suggested that it might be a toxic bacterial product.

These classic experiments pointed to an unknown substance which was called a "living fluid contagion" or **virus.** For a number of years the names "ultramicroscopic virus" and "filterable virus" were used. It was really only with the development of the electron microscope and the intensive studies of the nucleic acids that we became able to see and know what a virus is like. Viruses range in size from 10 millimicrons to more than 300 millimicrons. (1 millimicron = 0.001 micron. There are 1,000 microns in a millimeter and about 25 millimeters in an inch.) Some viruses are rod-shaped, some spherical, oval, or long narrow cylinders, and some look like tiny tadpoles.

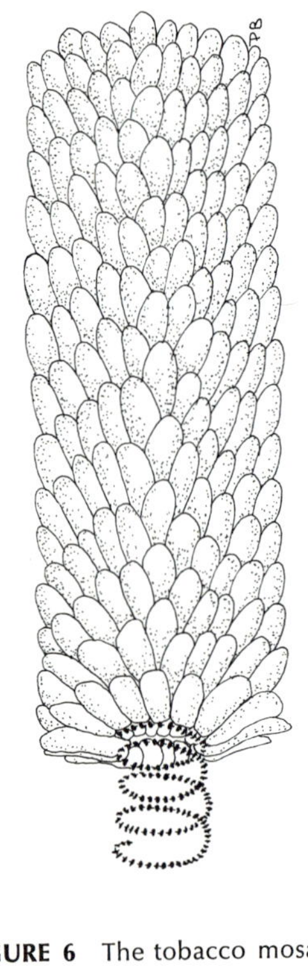

FIGURE 6 The tobacco mosaic virus.

The tobacco mosaic virus (Fig. 6) is a double helix of protein with a strand of RNA in the middle. It has been shown that this virus, which has been studied more extensively than any other, can be crystallized and that this substance is infectious. It was shown that the outer protein layer can be separated from the RNA and that the latter is the infectious portion. The RNA induces the plant to produce more of the same RNA and also to form new protein coats around the RNA strand, thus forming new tobacco mosaic viruses.

Most animal viruses contain DNA and RNA, but RNA seems to be the only nucleic acid in plant viruses. Apparently RNA is therefore a carrier of genetic information in certain viruses, as well as a protein-forming apparatus.

Although some scientists believe that the development of viruses is a fairly recent occurrence, since they are specific obligate parasites that grow only in living host tissue, others think that they were among the first organisms. Those who support the latter theory point out that the primordial organisms probably consisted of nucleic acids and protein, the composition of the viruses we know, although their metabolic activities are different.

The protovirus obtained its chemical components from the pools of organic compounds which had formed by chance in the primeval seas and swamps. These protoviruses built their bodies from the chance organization of the nucleic acids into replicating forms and the ability to produce proteins. These early viruses were millions of years in the making.

The viruses apparently became heterotrophic when large organisms evolved. Thus the viruses grew in an intracellular habitat which was a stable environment. Now the virus is dependent upon its host's cells in which to live, but it has retained the ability to direct its own reproduction and metabolic activities.

Since the invasion of a host cell by virus DNA or RNA changes the normal activity of a cell from producing new cells to producing the infecting virus, the genetic status of the cell is altered. Such cells show abnormalities, and when enough are infected and reproduce as abnormal cells, the host is diseased. In recent years much support has been given to the study of viruses in connection with cancer, since this disease is the result of the rapid growth of abnormal cells.

Some diseases known to be caused by viruses include the following:

Plant diseases		**Animal diseases**
Mosaic of:	alfalfa	Measles
	beans	*Herpes simplex*
	cucumbers	Psittacosis
	potatoes	Some Tumors
	tobacco	Cow and Small Pox
	tomatoes	Polyomyelitis

Yellows in Aster		Hepatitis
Leaf curl in:	cotton	Yellow fever
	sugar beets	Rabies
	tobacco	Some acute respiratory diseases

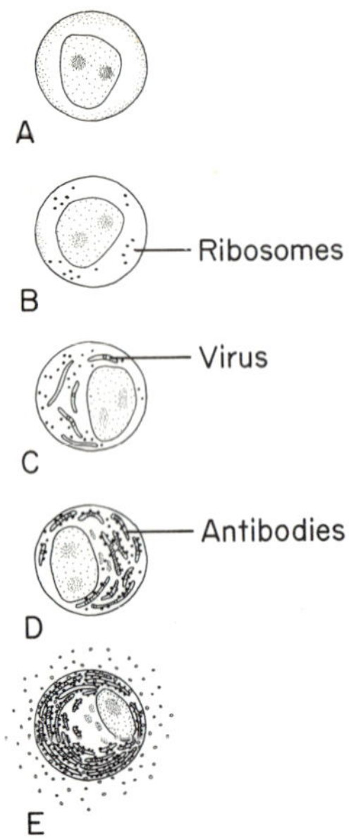

FIGURE 7 Illustration of antibody formation. A-normal young cell. B-a few ribosomes are shown forming outside of the nucleus. C-an antigen (such as a virus) has been introduced which caused the cell to begin to form antibodies. In D and E the endoplasmic reticulum has greatly increased in number along with the antibodies. Redrawn from Nossal, G.J.V. 1964. How Cells Make Antibodies. *Scientific American,* December.

Relationship of Bacteria to Man

Disease Organisms

Disease is usually described as any departure from the typical condition of good health. Here we are concerned with **infectious** diseases, those caused by microorganisms such as bacteria, rickettsiae, viruses, or other types. Other diseases are noninfectious.

Infection results in damage to cells or tissue, which upsets their physicochemical behavior. This usually involves inflammation, fever, and pain, accompanied by changes in cellular metabolism and the accumulation of intermediate metabolic products which are usually toxic. The invading organism or substance is an **antigen** which the cell then combats by the formation of **antibodies** (Fig. 7).

Organisms which cause disease are often termed **pathogens.** Those organisms which have the capacity to produce disease are **virulent.** This quality varies among microbial strains from those that are highly virulent to those that are avirulent. A few representative diseases, and the types of organisms which cause them, are listed below.

Bacterial diseases	Viral diseases	Rickettsial diseases
Bubonic Plague	Yellow fever	Typhus
Tularemia	Encephalitis	Rocky Mountain spotted fever
Typhoid fever	Tobacco mosaic	

Economic Effects of Microorganisms

Food Products We are all aware of food spoilage caused by bacterial growth, as well as spoilage caused by other organisms.

Years ago about the only convenient method available of preventing the decomposition of food materials was desiccation. Fruits, seeds, and other products, including meat, were spread out or hung in the sun to remove as much water as possible. This eliminated the moist medium necessary for bacterial growth. This method is still used extensively both individually and commercially. Later it was found that salting or sugaring the food product increased the osmotic concentration of the medium beyond the limits of survival for most microorganisms; this method was used and still is used for the preservation of many products.

Now we use any number of techniques to preserve food. The methods used are:

1. Controlling the amount of moisture.
2. Use of chemical preservatives to alter pH.
3. Storing foods at low temperature.
4. Use of high temperatures to destroy microorganisms.
5. Use of radiation to destroy microorganisms.

Think about the effects on our economy in the food industry alone if we no longer had the microorganisms of decay. Plums would not have to be changed to prunes. We could remove all refrigerators from homes and supermarkets. All canned and bottled foods could be kept open. Meat and other products could be left completely exposed. Of course, flies might become troublesome, but then there would be no danger of their or other insects' carrying disease organisms. We undoubtedly would no longer be here either, since we need and support an extensive bacterial population in our intestinal tracts. Perhaps the most useful bacteria are the decomposers. Without these organisms, the earth would be many times its present volume due to the accumulation of the remains of all plants and animals over the past two billion years (if there had been organisms to continue the evolutionary lines).

A great many foods are altered by the action of microorganisms, particularly those which have a fermentation reaction. Cucumbers pickled in brine owe their change in color, texture, and flavor to a *Lactobacillus*. Cabbage becomes sauerkraut through the action of several lactobacilli. Fermented milk products include buttermilk, yogurt, butter, and cheese. In some cases molds are also involved and the bacteria usually are species of *Lactobacillus* and *Streptococcus,* such as S. *lactis* and S. *cremoris*.

Industrial Products

Fibers obtained from such plants as flax and hemp are separated and combed from the stems following partial decomposition of the soft tissues by bacteria, especially species of *Clostridium*. Other products include alcohols, acetones, citric acid, and cured leather and tobacco.

Phylum Myxomycophyta

The slime molds have characteristics of both plants and animals and consequently are placed in a separate phylum in the plant kingdom, or they may be placed in the class Sarcodina of the phylum Protozoa in the animal kingdom. During its plant phase the slime mold becomes quiescent and produces sporangia which superficially resemble those of other fungi. Its animal-like phase is the **plasmodium,** which may be several inches in extent and is multinucleate and often multicellular.

The phylum is usually divided into three classes, of which the class **Myxomyceteae** has by far the largest individuals.

CLASS MYXOMYCETEAE

The slime molds of this group usually grow in a habitat abundantly supplied with dead organic matter, in places of fairly high humidity and relatively cool temperatures. In wooded areas they occur on and under the sides of fallen timber.

The plasmodium (Fig. 8) is often very colorful depending to a large extent on the mineral composition of the substrate on

FIGURE 8 Plasmodium of the slime mold *Mucilago crustacea,* formerly known as *M. spongiosa,* growing under a log in spruce forest at about 8500 feet elevation. This specimen was over a foot long and dichotomously branched. Identified by G. W. Martin.

which it is growing. The protoplast is fluid in some portions and quite gelatinous in others. Usually it is fan-shaped, but it may vary quite a bit just as an amoeba might when flowing in different directions.

The plasmodium usually has veinlike ridges running through the slime mass. These ridges are concentrated cytoplasmic streams which fade and reappear as the organism moves about. The greatest streaming within the plasmodium is within these veinlike ridges, although the direction of movement of the entire plasmodium is not closely correlated with the direction of movement in these denser areas. Streaming in the latter is usually in one direction for a minute or so, then the flow reverses itself for a similar length of time. Streaming in the veins is probably associated with metabolism more than with movement from place to place. The motive force in a plasmodial strand is thought to be generated through the interaction of a contractile protein with ATP. Both substances have been found in the plasmodium of some myxomycetes.

As long as conditions are favorable for vegetative development, the plasmodium continues to increase in bulk by assimilative processes, growth of the cytoplasm, and repeated divisions of the nuclei. The nuclei divide simultaneously throughout the plasmodium and require about twenty to forty minutes to complete the process. Before entering into the fruiting stage a mature plasmodium will contain thousands of nuclei, all derived from successive mitotic divisions of the original zygote nucleus and its successors.

The plasmodium of some slime molds will occasionally become hardened and remain in this dormant condition for some time. This hardened mass is termed the **sclerotium.** When suitable conditions for slime mold growth return, the plasmodium grows out of the sclerotial cover and again becomes active.

Myxomycetes vary a good deal in the type of fruiting body **(fructification)** (Fig. 9) they form, but all are small. Usually they

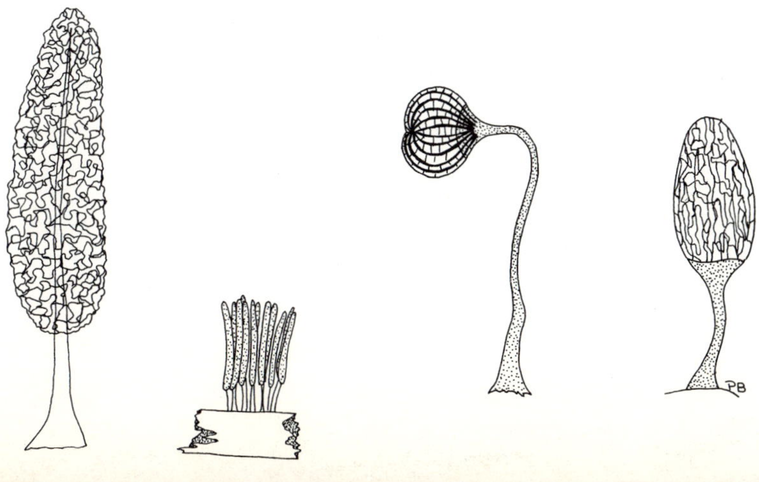

FIGURE 9 The fructifications of several myxomycetes. The capillitia is the network of fibrils within the sporangia.

resemble very small mushrooms or puffballs, but in some the sporangia are not so well defined. Shortly before forming the fructification the plasmodium migrates to a relatively dry area of the substrate. Numerous tiny sporangia begin to form above the drying plasmodium as uninucleate sections of the plasmodium become surrounded by walls, forming large numbers of spores. As the sporangia develop each is surrounded by a wall, the **peridium.**

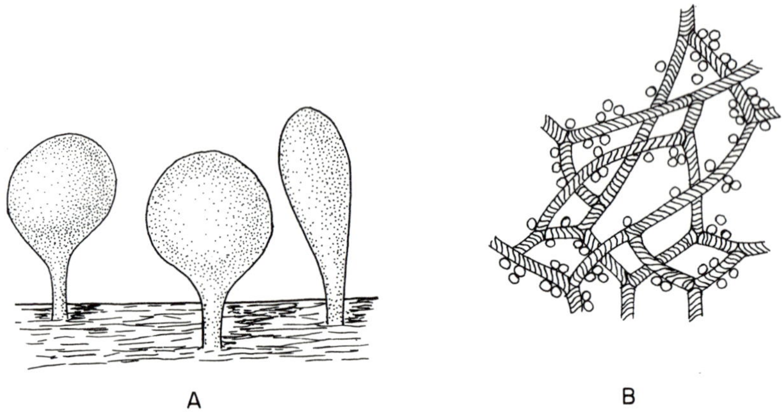

FIGURE 10 A. Sporangial stage of the slime mold *Hemitrichia* B. The anastomosing fibril capillitia with spores, both greatly enlarged in the drawing.

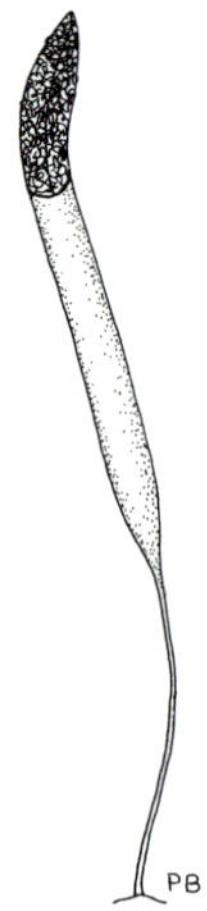

FIGURE 11 *Stemonitis.* One of the tallest of the slime molds. The upper portion of the sporangial wall is cut away to show the capillitia.

Within the sporangium the protoplast forms hairlike filaments, the **capillitia,** which become arranged in a distinct network, characteristic of the species. When the spores are mature the peridium disintegrates, and the capillitium expands and carries the spores somewhat above the base of the fructification, where air currents disperse them to new areas where they may germinate.

The typical life cycle of a slime mold is as follows:

spore ⟶ motile spore stage ⟶ fusion in pairs ⟶ zygote (flagellate or amoeboid) ⟶ plasmodium ⟶ fructification ⟶ sporangium $\xrightarrow{\text{meiosis}}$ spore.

CLASS ACRASIEAE

The members of Acrasieae are few in number, but at least one genus, *Dictyostelium,* has been studied extensively. Although they resemble the myxomycetes their plasmodial stage is a cellular colony composed of hundreds of uninucleate amoeboid cells which retain their wall boundaries. This structure is a **pseudoplasmodium.**

The pseudoplasmodium (Fig. 12) is a part of the reproductive phase. After it is formed it may move about for a time, and then it orients itself vertically and begins to differentiate into a stalk and fruiting tip. Some of the amoeboid cells form the stalk and others migrate to the tip to form a bulbous sporangium. Still others differentiate within to form the spores, which become surrounded by a resistant wall and often with a bit of slime.

FIGURE 12 The life cycle of *Dictyostelium discordeum.* The plasmodial mass is forming in (1). After it is formed the amoeboid cells orient themselves vertically and gradually develop into the fruiting stage. The numbers represent time, in hours, in the life of *Dictyostelium.* (Redrawn from J. T. Bonner.)

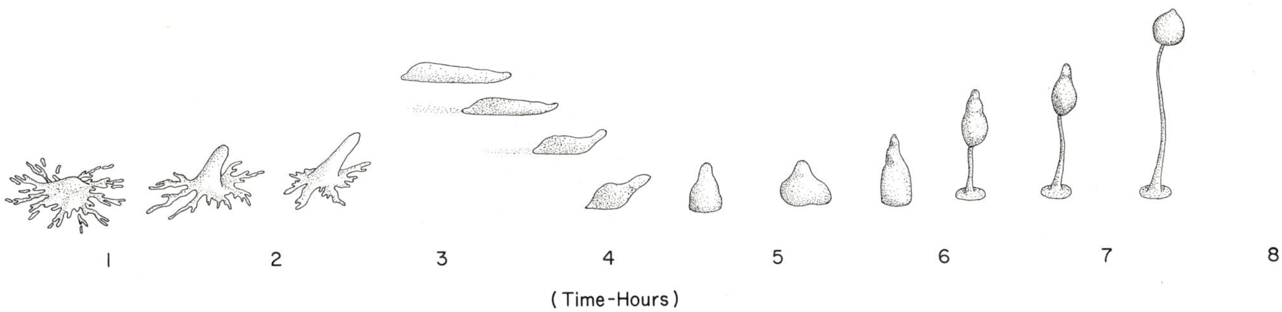

Following their liberation the spores germinate into single non-flagellate myxamoebae, which may then greatly increase in number by division. When the food supply diminishes the myxamoebae cluster together to form the loosely organized cellular pseudoplasmodium.

Independent investigations by J. T. Bonner and K. B. Raper have shown that as the pseudoplasmodium is formed in *Dictyostelium* the larger cells assume an anterior position in the migrating cell mass, while the smaller cells are at the posterior end. Experiments have shown that the pseudoplasmodium can be cut in sections, then grafted together in different positions, and in each case the larger cells migrate to the anterior end. It was also shown that this positioning of cells was correlated with speed of movement. The larger cells were more rapid in their movement while the slower small cells remained posteriorly. Furthermore, spores forming from the anterior cells were larger than the others.

In *Dictyostelium* a hormonal substance, **acrasin,** is secreted by some of the moving amoebae and it seems to attract other amoebae in the vicinity and bring them together forming the pseudoplasmodium.

CLASS PLASMODIOPHOREAE

While the members of the other two classes of the Myxomycophyta are mostly saprophytic and engulfers of bacteria and other microorganisms, the members of the Plasmodiophoreae are mostly parasitic on higher plants. In this group there is a plasmodium, but a fructification is not formed. Sexual reproduction is achieved by the union of motile isogametes which have been formed by successive divisions of the plasmodium. The swarm cells in this group have two anteriorly attached flagella (Fig. 13).

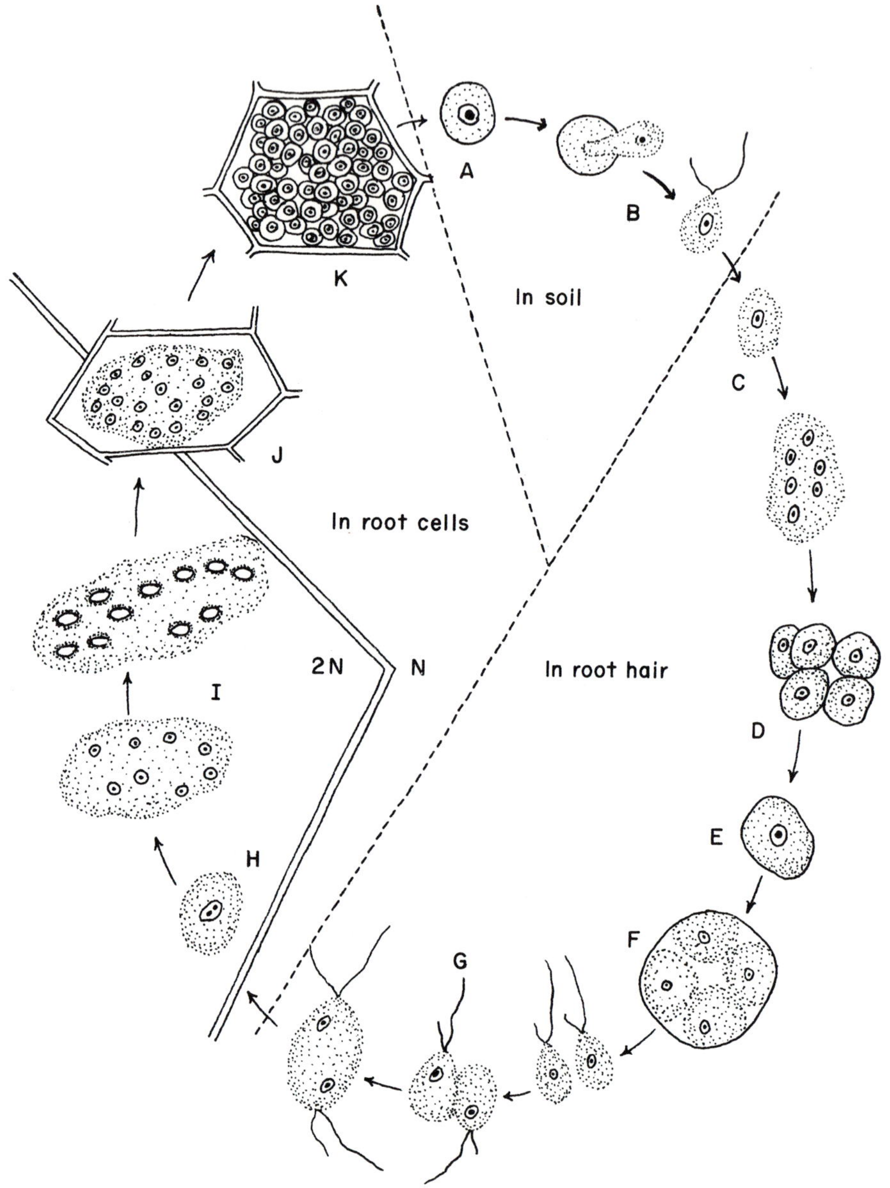

FIGURE 13 The life cycle of *Plasmodiophora brassicae*. A. The resting spore germinates to form a swarm spore C, which loses its flagella and forms a myxamoeba, the nuclei divide many times to form a haploid plasmodium which then differentiates into a number of gametangia D. Each gametangium E, forms gametes F, which fuse G to form a zygote H. The zygote develops into a diploid plasmodium I whose nuclei undergo meiosis to form a haploid plasmodium J. The nuclei in the host's cells then develop into resting spores K which may then be liberated into the soil. Note haploid and diploid phases and areas of development of the different stages. (After Alexopoulos)

One of the most economically important species is *Plasmodiophora brassica,* a parasite of cabbage. Penetration of host tissue usually takes place when the swarm cells make contact with a susceptible host. Once contact has been made with cabbage tissue the swarm cell loses its flagella, forms a myxamoeba stage, and enters the host tissue by a partial dissolution of the host wall. Within the cabbage cell the myxamoeba grows into a plasmodial stage by increase in size of the protoplasm and successive divisions of the nuclei. This multinucleate plasmodium breaks up into as many protoplasts as there are nuclei, and each differentiates into a gametangium. The latter continues to undergo nuclear divisions until the gametangium separates into uninucleate, spindle-shaped isogametes. Fusion of the gametes takes place within the host's cells. After the zygote is formed it takes on amoeboid characteristics and penetrates deeper into the host's tissues.

The zygote amoebae may then form a plasmodium in the host tissue, causing hypertrophy and hyperplasia of the tissue. The life cycle continues as before, new gametes and new resting spores being formed. The disease, club root of cabbage, is spread by the resting spores which are liberated into the soil by disintegration of the root tissue.

Occasionally the members of this class are placed with the true fungi as a very primitive order but they, like another small class of slime organisms, the **Labyrinthuleae,** seem to have more in common with the myxomycetes. *Labyrinthula* is a genus known to cause severe damage to a marine grass which is an important food source of aquatic wildfowl.

Relationships of the Myxomycophyta

While the three classes of slime molds considered here have much in common, their early origins may have been quite different. Both the Myxomyceteae and the Plasmodiophoreae are apparently of flagellate origin although the latter may have arisen from very primitive members of the Phycomycetes. The Acrasieae, which are non-flagellate, may have had their origin from free-living amoebae of the soil. It is generally believed that the amoebae themselves came from flagellates but there are many amoebae which have no flagellate stage and in which no form of sexual reproduction has been observed. It seems safe to assume that the points of divergence were very early in the evolution of these groups and each class became and remained distinct throughout its history.

Supplementary Reading

Bacteria

Carpenter, P. L., *Microbiology*. W. B. Saunders Co., Philadelphia, 1961.

Clinton, C. E., *Introduction to the Bacteria,* 2nd ed. McGraw-Hill, New York, 1958.

Grant, Madeline P., *Microbiology and Human Progress.* Holt, Rinehart, and Winston, New York, 1953.

Sistrom, W. R., *Microbial Life.* Holt, Rinehart, and Winston, New York, 1962.

Van Niel, C. B., Comparative Biochemistry of Photosynthesis, *American Scientist* 37 (3) 371–383, 1949.

Waksman, S. A., and H. A. Lechavalier, *Actinomycetes and Their Antibiotics.* Williams and Wilkins, Baltimore, 1953.

Slime Molds

Alexopoulos, C. J., *Introductory Mycology,* 2nd ed. John Wiley & Sons, New York, 1962.

Bonner, J. T., *The Cellular Slime Molds.* Princeton University Press, Princeton, N.J., 2nd ed. 1967.

Bonner, J. T., Differentiation in Social Amoebae, *Scientific American,* December, 1959.

Gray, W. D., *The Relation of Fungi to Human Affairs.* Holt, Rinehart and Winston, New York, 1959.

MacBride, T. H., and G. W. Martin, *The Myxomycetes.* Macmillan, New York, 1934.

Moscona, A. A., Tissues from Disassociated Cells, *Scientific American,* May, 1959.

Smith, G. M., *Cryptogamic Botany,* Vol. I. McGraw-Hill, New York, 1955.

Questions for Discussion

1. Vaccines have been developed for most bacterial infections but very few successful ones to combat viruses have been discovered. Why should there be this difference?
2. What methods are used to control bacterial and viral infections?
3. How were viruses discovered?
4. What has the electron microscope revealed to us about bacteria that was not previously known?
5. Bacteria are in general more infectious to animals, and fungi are more infectious to plants? What does this suggest about their physiology?
6. What do bluegreen algae and bacteria have in common?
7. Describe some bacteria which are beneficial to man.
8. What do we mean by a mosaic virus infection?
9. Why must we be so careful in cultivating bacteria, even those which are noninfectious?
10. Why are members of the actinomycetes of great benefit to us while many other bacteria are harmful?
11. Why are bacteria considered very primitive organisms?
12. Describe the different types of metabolism found in bacteria.
13. What methods are used to preserve foods from bacterial infestation?
14. What would happen if all bacteria were suddenly removed from the earth?
15. Why cannot plants be treated with vaccines and serum injections as animals can?
16. How does the continuous use of certain antibiotic substances cause even more virulent strains of bacteria to develop?
17. How does drying meat preserve it from bacterial decay?
18. What methods do bacteria employ in producing the harmful effects of their infection?
19. What methods are used in identifying bacteria?
20. Should slime molds be considered plants or animals? Explain.
21. Many slime molds are wood-decaying organisms. At what phase in their life cycle do they do the greatest damage?
22. What methods are used in identifying slime molds?
23. How does the plasmodial stage of a slime mold differ from an amoeba?
24. Describe growth and reproduction in *Dictyostelium*.
25. How do some members of the Myxomycophyta invade plant tissue and cause its decomposition?
26. How does a virus resemble the DNA molecule?

19

The True Fungi

Phylum Eumycophyta
Class Phycomycetae
Class Ascomycetae
Class Basidiomycetae
Lichens

FIGURE 1 A few mushrooms of the genus *Lactarius* growing in a spruce forest. The cuplike caps are from three to four inches in diameter.
FIGURE 2 The giant puffball *Calvatia gigantea*. This specimen is about 12 inches wide in diameter.
FIGURE 3 Gill type mushrooms growing on the trunk of a large spruce tree.

Most of us are rarely aware of the widespread presence of fungi. In the spring and fall we notice a few mushrooms (Fig. 1) and puffballs (Fig. 2) in the lawn or forests, but they seem to be just oddities in the landscape that are with us for a short while. Occasionally we notice unusual growths on the sides of trees (Fig. 3), or a powdery covering on the leaves, and perhaps we have been

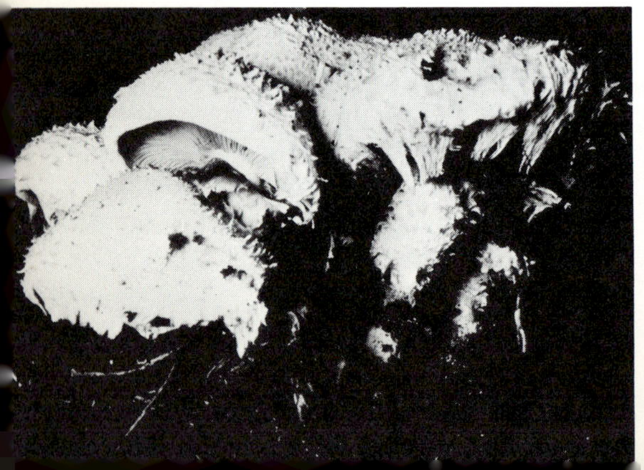

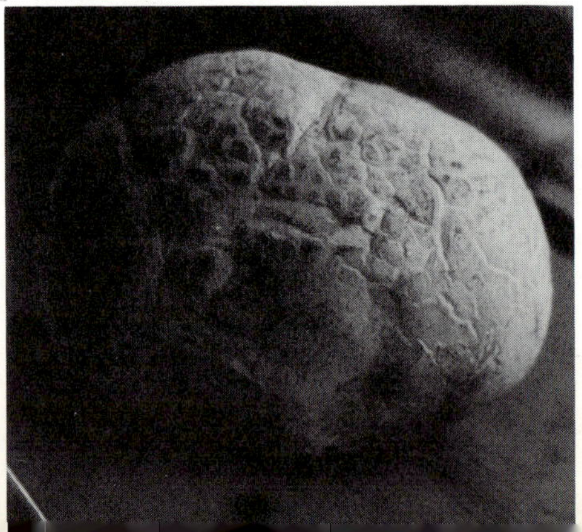

surprised by the mysterious death of a tree or shrub which seemed to be in perfect health a short time before. But seldom do we think of fungi as causing migrations of millions of people, or as a limiting force in the economy of a country, or as the cause of famine for millions almost every year.

In the United States, where we control plant and animal diseases probably more successfully than almost any other country, the average annual loss in field crops (wheat, corn, etc.) between 1951 and 1960 is listed as $1,890,836,000; for forage crops, pasture, and range grasses the loss was $808,701,000; and for vegetable crops it was $290,389,000. These figures do not include the losses in lumber, fruit trees, ornamentals, seed crops, or small fruit crops such as strawberries.

One of the major duties of the thousands of individuals employed by our federal and state departments of agriculture is the control and eradication of plant diseases. We have hundreds of quarantine stations at ports of entry and thousands of research workers trying to reduce the numbers and the effects of disease organisms. Research is continually being done to develop effective fungicides, bacteriocides, and insecticides to control insect vectors. Other investigators are studying the genetics of pathogens, the development of host resistance to fungus invasion, and the effectiveness of crop rotations on soil-borne diseases. Dozens of other projects are underway in an attempt to give us greater returns from our crops and allow the parasites less.

The United States owes much of its Irish population to the fungus which causes Late Blight of Potato. Ireland, as well as several other European countries, had for many years raised the white potato as its principal food crop. Then in 1845 and for several years following, the fungus *Phytophthora infestans* (Fig. 4) destroyed

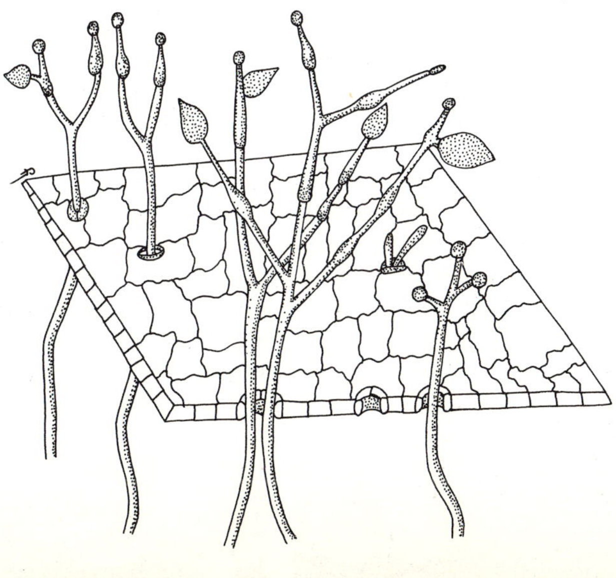

FIGURE 4 *Phytophthora infestans.* Conidiophores growing through the lower epidermis of a potato leaf.

virtually the entire crop. The potato famine resulted in starvation, riots, looting, and murder. Many individuals from Ireland and western Europe, with ambition and an adventurous spirit of survival, migrated to the United States during this stark period.

Major alterations in the economy of many countries and in their agricultural emphasis have also been experienced. The rubber tree *Hevea brasiliensis,* a native of South America, was parasitized by a defoliating leaf spot fungus, making rubber production expensive and hazardous. The fungus was apparently left behind when rubber trees were taken to Malaya. Now most rubber is produced in the Far East.

A similar and compensating move occurred with coffee. Coffee was a major crop in Ceylon until about 1880. Shortly before this time it was observed that a rust was causing defoliation of the coffee trees. After a few defoliations the trees died. Following numerous attempts to control the disease the coffee trees were replaced with tea plants and the coffee industry was moved to Brazil.

The United States cereal crops have suffered severely from rust fungi which have infested fields of wheat, corn, and other cereals. In 1916 alone, it was estimated the United States and Canada lost 300,000,000 bushels of wheat. It was found that the disease had as an alternate host the American barberry. Since this plant is of little value, the government sought to control the rust by removing all American barberry bushes from the wheat belt areas. It was a major enterprise covering thousands of square miles, but it was effectively carried out. We still have wheat rust, since it can be propagated from wheat plant to wheat plant by means of one spore stage, but the rust rarely is able to complete its long cycle now.

One of our most important nut and hardwood lumber trees, the American chestnut, was almost completely wiped out early in this century by the chestnut blight. In recent years the American elm has been suffering a similar experience. The Dutch elm disease is caused by a fungus that produces necrosis of phloem tissue. In some cities of the Midwest and the East, American elm trees have been completely eliminated by this disease. Towns that were once pleasantly shaded by the towering canopies of spreading elms either have nothing left or are faced with the problem of cutting down dozens of dead, leafless trees still standing seventy-five feet tall. Removal of the dead trees is an expensive, major problem and the planting of new trees, while not difficult, is a restorative measure that will not bring shade and beauty to the town again for many years. Obviously the new plantings will have to be some species other than the American elm.

Plant diseases have always been with us. They influence what we eat, what we wear, where we live, and what industries are established in a given locality. Our annual losses from plant diseases amount to about 3.25 billion dollars a year. Fortunately we are an overproducing nation in proportion to our population. Where the population density is high and crop production barely meets the yearly consumption, a serious infestation of a major crop means famine. Control of plant diseases is an absolute necessity for developing countries in order to maintain and advance themselves.

Although the parasitic nature of fungi is always a major consideration, most fungi are innocuous and thousands are beneficial. A number are enjoyed directly as food; others are cultured in milk processing to give the desired flavor and texture to some of our finest cheeses. Fungi are now used in the production of effective and valuable drugs, and also in the fermentation processes in producing bakery goods and alcoholic beverages.

The Structure of Fungi

A few fungi are essentially unicellular, but the majority are composed of microscopic filaments which may be colorless or variously tinted. They never contain chlorophyll. Most fungi have a vegetative phase that is made up of numerous filaments termed **hyphae.** The massed hyphae produced by a single fungus are called a **mycelium.**

Although the mycelia of most fungi are quite similar, the structure of the fruiting bodies is enormously varied. The more primitive species often have no particularly characteristic form and may appear superficially as a downy body with little specialization. Many of the more advanced fungi have their hyphae differentiated into distinct caps and stipes, others form large puffballs, and some produce cuplike fructifications in which tiny sporangia form like eggs in a nest. It is this "fruiting" body which serves as the most useful characteristic for the identification of fungi.

To a large extent the dissemination of fungi is through the medium of spores produced in or upon the variously shaped fruiting bodies. Relatively few fungi have explosive sporangia capable of propelling spores, and even these have very limited power. Spores must depend, for the most part, on wind, water, insect carriers, birds, and other animals, including man, to carry them from place to place. The distance spores may be carried by air currents is virtually limitless. As a result of their tiny size they go undetected even with the closest inspection, and their chances of survival under extremes of temperature, desiccation, and other environmental adversities have been heightened by their resistant and thick cell walls. Perhaps the most important of all aspects of fungus survival is found in the great production of spores. An individual mushroom or puffball may produce spores by the millions, billions, or even trillions. Few of these spores germinate and grow to maturity, but because of the large number produced, the chances of species survival are great.

Phylum Eumycophyta

The phylum Eumycophyta is generally divided into four classes based on the characteristics of the hyphae in the mycelium, the reproductive organs, both asexual and sexual, the fructification, and the spore types.

Classification of the True Fungi

1. *Phycomycetae*. The mycelial hyphae in this class are usually **nonseptate,** i.e., lack crosswalls, and all possess this vegetative structure except for a few of the most primitive species which are essentially unicellular. Sexual reproduction usually consists of a fusion of multinucleate male and female cells. Frequently the asexual spores are flagellate and are really zoospores.

2. *Ascomycetae*. Here the mycelial hyphae are **septate.** The most important characteristic is the formation of a saclike structure, the **ascus,** in which **ascospores** are produced as the culmination of sexual reproduction.

3. *Basidiomycetae*. In this group the hyphae are also septate, but with certain differences in their nuclear behavior. The diagnostic structure is a clublike growth termed the **basidium.** The **basidiospores** are borne on small stalks, the **sterigmata,** at the apex or along the side of the basidium. Two major groups in this class,

the rusts and smuts, have highly modified basidia but they are homologous in origin to the typical basidium.

4. *Deuteromycetae.* The members of this class are called the **fungi imperfecti** or imperfect fungi, since their sexual stage has never been observed and their complete life cycles are unknown. Many fungi which at one time were placed in this class, such as *Penicillium,* have been moved to one of the three preceding classes when their complete life cycles were finally established. Most of the species now classed as deuteromycetes will eventually be moved to the ascomycetes or basidiomycetes.

CLASS PHYCOMYCETAE

There is considerable diversity of form in the members of this class, but the most prominent members are bread molds and water molds. A number of water molds infect fish and other aquatic animals. Bread molds commonly grow as saprophytes or parasites on plants, but they occasionally grow on animal tissues.

The manner of reproduction of the phycomycetes varies from vegetative propagation to the production of gametangia of a well developed type. Spores are produced on special hyphal branches termed **sporophores;** in some, well organized sporangial bodies are formed.

Spores are often produced on elongated hyphal sporophores simply by being pinched off at the tips as the sporophore grows. During rapid growth, chains of these spores are forming constantly (Fig. 5). When spores arise in this manner they are said to have an **exogenous** origin. Spores which are produced within an enclosure such as the ascus of an ascomycete or the sporangium of a bread mold have an **endogenous** origin. When the sporophore produces sporangia it is termed a **sporangiophore.**

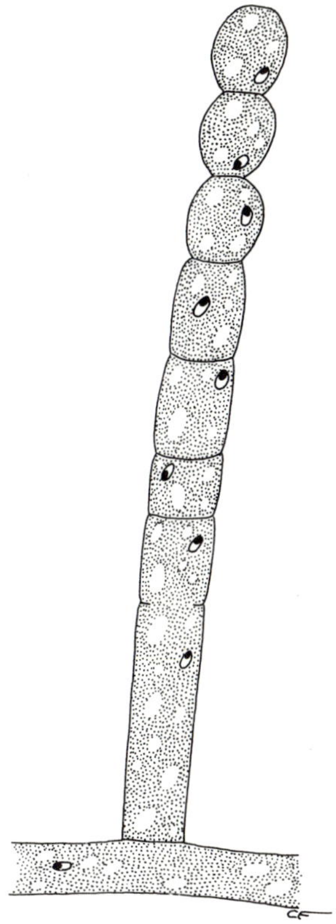

FIGURE 5 Development of a conidial branch with conidia forming at the apex.

In some members of the group such as *Phytophthora* and *Plasmopara,* aerial sporangia are formed and cut off from the parent hyphae. These sporangia are often considered spores in themselves and may be termed **conidia.** The hyphae which bear them are therefore called **conidiophores.** Another genus, *Peronospora,* in this group produces conidia in the usual manner, but upon germination the spores develop **germ tubes** that penetrate the host tissue and bring about new infections.

In the phycomycetes the spores may function either as zoosporangia or as true conidia from which a germ tube, typical of the higher fungi, is produced. In some species the conidia might produce either. At temperatures up to 18°C the conidia of *Phytophthora infestans* usually tend to form zoospores, while at higher temperatures the conidia more commonly form germ tubes upon germination.

Saprolegnia

One of the most common watermolds, *Saprolegnia,* frequently infects aquatic insects or fish. Once the fungus has started to grow in the host, the mycelium develops as a mass of much-branched hyphae which soon forms a furry coat around the insect or fish. At the tips of the modified hyphae, growing out from the host's body, zoosporangia begin to differentiate. Within these zoosporangia biflagellated vermiform zoospores are formed. These are liberated at maturity through an opening in the apex of the sporangium. This is the usual mode of asexual reproduction in water molds, but the shape of the zoosporangia and the manner of discharge of the zoospores in other genera are often quite different from those of *Saprolegnia.*

Sexual reproduction in this genus superficially resembles that of *Vaucheria* of the Chrysophyta. The sex organs of the water mold are cut off by conspicuous, thick crosswalls from modified hyphae which are more highly differentiated than those which produce the zoosporangia.

The male sex organ, the **antheridium,** is a much curved filament that ultimately comes in contact with the female organ, the **oogonium** (Fig. 6). There it releases the male nuclei through fertilization tubes into the oogonium. The oogonium develops as a branch on a short hyphal filament, becoming spherical and containing from one to twenty eggs. After fertilization, which is accomplished by a number of antheridial nuclei migrating into the

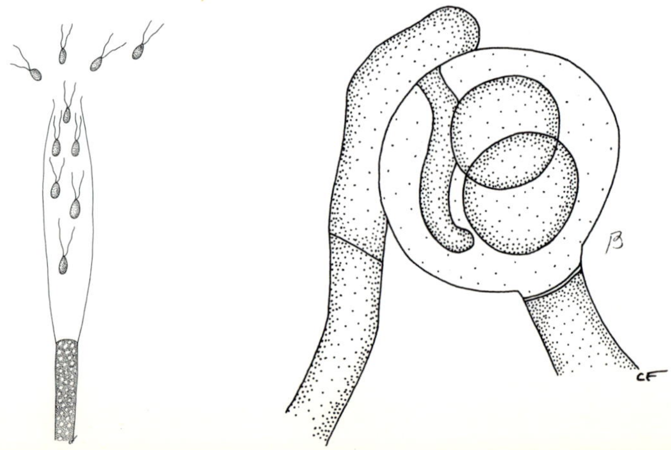

FIGURE 6 A: *Saprolegnia,* zoosporangium releasing zoospores apically. B: *Saprolegnia* showing oogonium with two eggs; elongated antheridium on outside of oogonium, and fertilization tube entering oogonium from antheridium.

oogonium and fusing with the eggs, the diploid cells develop into smooth, thick-walled zygotes. The oogonium generally contains five to ten zygotes, but a few species produce only one zygote in an oogonium. After a resting period the zygote germinates and meiosis occurs. A specialized hypha, the germ tube, emerges from the zygote, and generally forms a zoosporangium at its tip as a result of migration of protoplasm to this area. A number of haploid zoospores are then developed and the life cycle is complete.

Plasmopara

A disease known as downy mildew of grapes is caused by a species of *Plasmopara* (Fig. 7) which produces much-branched sporangiophores. The mycelium is freely branched and multinucleate.

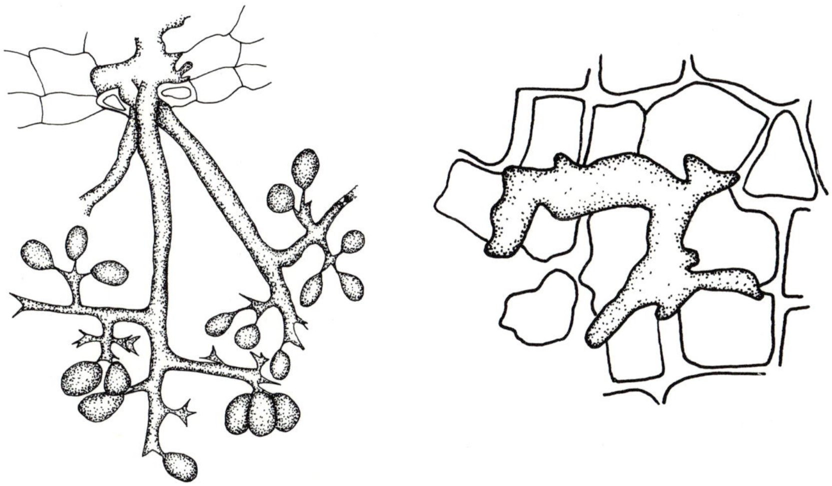

FIGURE 7 *Plasmopara viticola. Left:* Sporangiophores growing through the stomate of a grape leaf. *Right:* Haustorial hyphae in the cells of a grape leaf. (redrawn from Alexopoulos, C. J. 1952. *Introductory Mycology.* Wiley.)

It produces numerous hyphae, some of which form modified **haustoria** that penetrate the walls of the host cells and absorb food from the host's protoplasts.

Once established in a host the mycelium develops the sporangiophore, which emerges from a stomate or through a break in the host's tissue. Several branches then arise at right angles from the main axis of the sporangiophore, and at the tips of tiny branchlets conidiosporangia develop which readily become detached.

The conidiosporangium matures from a small uninucleate structure to one that becomes pyriform and multinucleate as a result of successive nuclear divisions. These nuclei become surrounded with bits of protoplasm and mature into uninucleate zoospores.

If a zoospore develops in the vicinity of susceptible host tissue, it changes from a reniform shape with two laterally attached flagella into a globose quiescent spore. The spore then sends out a germ tube that grows through a stomate, spreading the infection on the same plant or perhaps starting a new infection on a different grape plant.

The Life Cycle of Plasmopara

Asexual cycle	**Sexual cycle**
Mycelium	Mycelium
↓	↙ ↘
Sporangiophore	Oogonium Antheridium
↓	↘ ↙
Conidiosporangia	Egg Sperm
↓	↘ ↙
Zoospores	Zygote
↓	↓
Host infection by germ tube	Host infection by germ tube
	↓
	Sporangium
	↓
	Zoospores

Sexual reproduction occurs by the formation of oogonia and antheridia at the tips of hyphae within the host's tissues. The oogonium is multinucleate at first, but as it matures a central uninucleate egg separates from a peripheral multinucleate periplasm. The adjacent antheridia send forth fertilization tubes which contact the oogonia. Only a single male nucleus migrates through the tube, fusing with the egg nucleus.

Germination of the zygote occurs in spring with the establishment of a germ tube and its penetration through a host's stomate. The multinucleate protoplast which forms in the process of germination then develops a sporangium containing fifty or more zoospores. The zoospores are then liberated through a pore in the sporangial apex and reinfection of the host or infection of a different plant is imminent.

Rhizopus

One of the most universally spread and most widely studied molds is *Rhizopus nigricans*, the common bread mold. The common name may be misleading, for this fungus is by no means limited to bread. It will grow on many plant organs with a fair amount of sugar stored in their cells. It is readily cultivated on the roots of sweet potatoes, the tubers of white potatoes, and numerous fruits. The early growth of the mycelium is white, but as the black sporangia form and the dark spores appear the entire growth changes from white to gray to black.

Germination of a spore is initiated by the absorption of water. A short hypha emerges which soon branches and rebranches prolifically, forming the mycelium. Some hyphae form small rootlike appendages, the **rhizoids.** They serve as anchoring appendages and release enzymes into the host tissues which convert the stored foods into a soluble substance. Some hyphae remain relatively undifferentiated, but they form a connecting link between the clusters of sporangiophores and their rhizoids. These horizontal, connecting hyphae are termed **stolons** and superficially they resemble the runners of a strawberry plant. The sporangiophores which arise from the mycelium in the vicinity of the rhizoids give rise to the black, globose sporangia.

Sexual reproduction in *Rhizopus* (Fig. 8) takes place by the union of like gametes in a manner of conjugation similar to that

FIGURE 8 Life Cycle of *Rhizopus*. A-Portion of mycelium showing nonseptate hyphae, basal rhizoids, and three sporangia; B-liberation of spores; C-germinating spore; D-hyphae of different strains make contact and form progametangia; E-isogametes cut off from tips of progametangia; F-fusion to form zygote forms germ tube with terminal sporangium and spores.

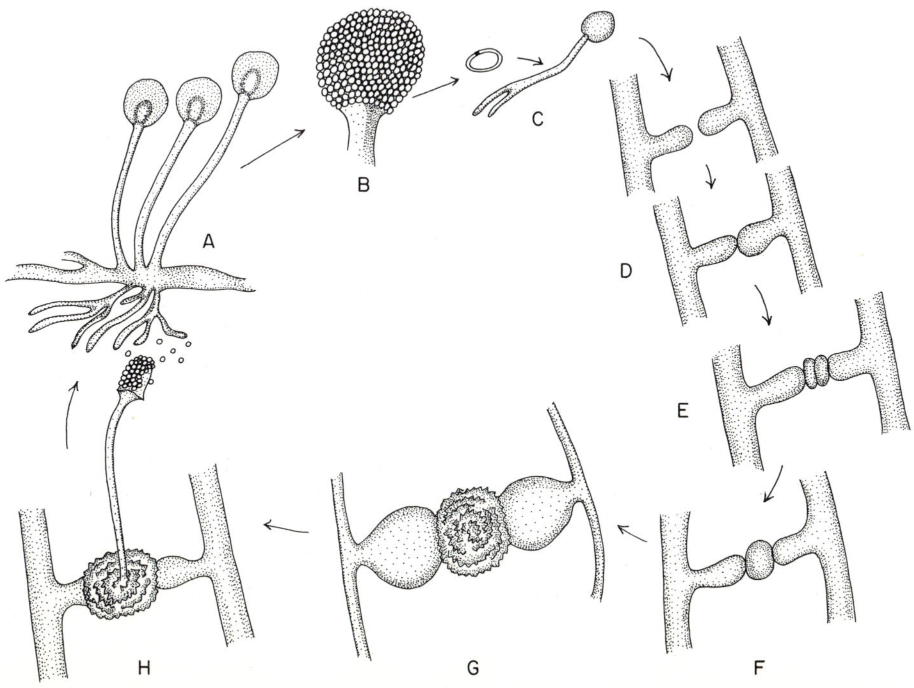

of *Spirogyra*. Conjugation occurs only between plants of different strains. Since distinguishing morphological features which would indicate maleness or femaleness, are lacking, the two different strains are referred to as **plus** and **minus.** In order for gametes to form and union to occur, mycelia of the two strains must be growing together. Since this is not a common occurrence, the usual method of asexual reproduction may continue for countless generations.

When two mycelia of different strains are growing in close proximity the hyphae send out short branches termed **progametangia** towards the opposite partner. As the progametangium develops many nuclei and some cytoplasm migrate into the tip. Soon after the swollen tips make contact with each other, crosswalls form separating each progametangium into two parts. The outer cell becomes the **gametangium** and the basal attached portion is the suspensor. A multinucleate zygote is formed when the gametangia fuse by dissolution of their contacting walls.

The zygote develops a thick, rough wall and often remains in this stage for months. When the zygote germinates it forms a hyphalike germ tube, which produces a terminal sporangium and then the typical spores.

CLASS ASCOMYCETAE

The "sac" fungi include some of our most important fungi, such as the yeasts and *Penicillium*. Some species are parasitic on plants and animals but most are saprophytic, living on leaf mold, decaying logs, or other dead organic matter.

The mycelium as well as the fruiting structures are in general more advanced than those of the phycomycetes. In the ascomycetes the hyphae are septate, and each septum is perforated by a minute central pore which enables protoplasm to move from cell to cell (Fig. 9).

FIGURE 9 Diagram of a septum and pore in cells of *Neurospora*. Drawing was made from an electron micrograph picture in Weisz, P. B. and M. S. Fuller 1962. *The Science of Botany*.

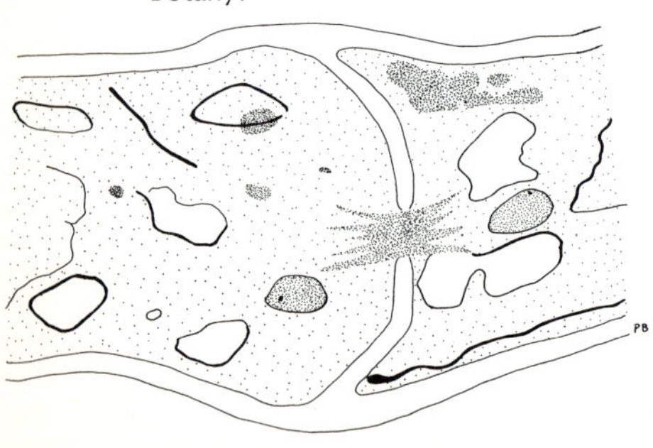

There are usually two types of hyphae, distinguished by their chromosomes. The cells of most hyphae have only one nucleus, and these hyphae are more active vegetatively. In addition to contributing to the greatest mass of the mycelium these hyphae also produce the gametangia. Some hyphae just before nuclear fusion have two nuclei in each hyphal cell and following fusion develop the specialized fruiting structures which give rise to the **ascospores.**

The growth of fungal hyphae, like that of most elongating plant cells, is confined to the tips. This is due mainly to the migration of protoplasm from the older portions of a hypha to its tip.

As a hypha matures, vacuoles are formed in the older cells, forcing the cytoplasm through the hypha towards the place of least resistance, the tip. Although hyphae usually have crosswalls, the division into separate cells is not always complete. The septa are more like diaphragms with an opening in the center that permits relatively free streaming of the cytoplasm. In this way the hyphal tip is constantly supplied with cytoplasm, enabling it to grow at a very rapid rate. It is apparent that a hypha may differentiate into a tip, in which new cell-wall material is laid down, and into an older portion primarily concerned with the production of vacuoles. Actually a growing mycelium has been described as similar to a crawling plasmodium except that the mycelium is continuously laying down its own tubing as it grows.

Cytological studies of the hyphae show that distinct regions can be observed. While the cytoplasm is produced all along the hyphae, the streaming brings building material in the form of a number of amino acids and glycogen to the tips. Here a concentration of enzymes converts the building material into new cell walls. The tips are slightly richer in proteins, and particularly in protein-bound sulfhydryl groups and ribonucleic acid. Glycogen diminishes in quantity at the immediate tip, probably because it is hydrolyzed into sugar, which is used in building the new cell wall.

In the higher ascomycetes the mycelium is often organized into specialized tissue in which the hyphae become so tightly intermeshed as to lose their individuality. Cells of these hyphae become isodiametric, resembling the parenchyma in green plants. Usually these specialized tissues are associated with the fruiting structures, but occasionally vegetative tissues become modified into hardened little bodies **(sclerotia)** that abscise from the main thallus, blow away, and vegetatively propagate a new fungus.

Reproduction

Asexual Reproduction Asexual reproduction in the ascomycetes is usually accomplished by the production of conidia. As in the previous class, conidia are formed at the ends of conidiophores, where they are budded off by constriction. The conidiophores in this group have a wide variety of shapes and forms. Some are long and intricately branched; others are extremely short stalks on globose hyphae or slender elongated structures. Sometimes the conidiophores form with little evident organization. In others, distinctive fruiting bodies appear.

In some ascomycetes hyphal cells may become enveloped by

thick cell walls. These are **chlamydospores** and they may be formed singly or in a short series. This type of spore is considered an overwintering or resting spore, while conidia are less resistant and seldom overwinter.

Sexual Reproduction Sexual reproduction is rare in some ascomycetes but where it occurs it often results in peculiarly shaped fruiting bodies. Fusion occurs when compatible hyphae or their gametangia come in contact. The protoplast of the antheridium migrates into the female organ, referred to in this group as the **ascogonium.** In phycomycetes fusion of the nuclei takes place almost immediately, but in the ascomycetes fusion is delayed. Here

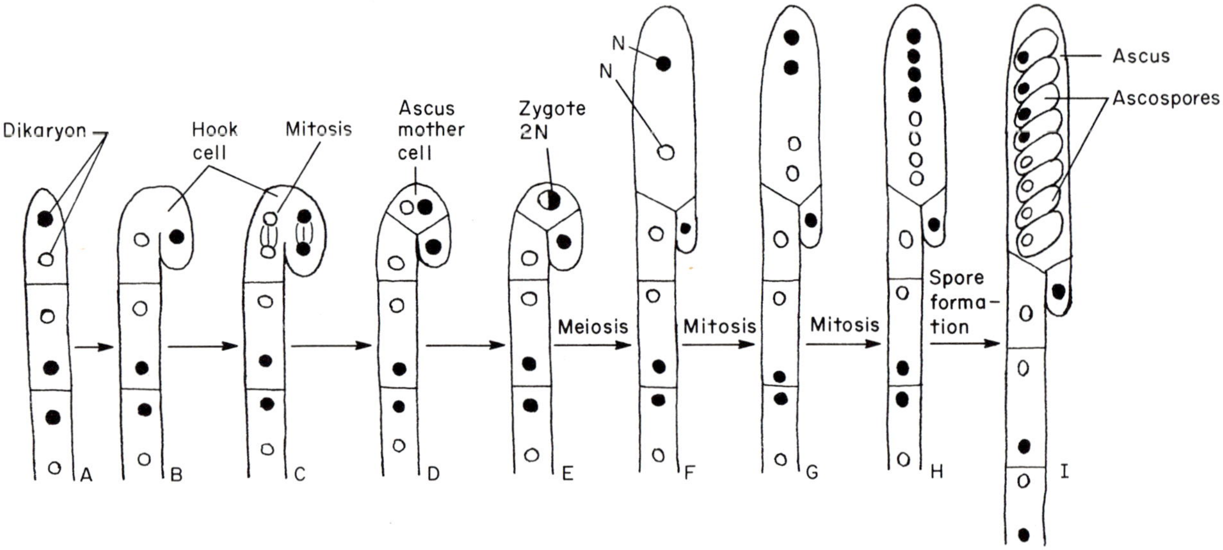

FIGURE 10 Diagrammatic illustration of ascus and ascospore development following union of compatible strains. A-cells with dikaryon stage, B, C-hook formation, D-initiation of ascus, F-fusion (Karyogamy) of dikaryon nuclei to form zygote, G-reduction division to form haploid nuclei; H, I, J-mitotic divisions in the formation of ascospores and ascus. (redrawn from Alexopoulos, C. J. and H. C. Bold, 1967. *Algae and Fungi*. The Macmillan Co., New York.)

the nuclei remain in close association as a functional pair. This stage is known as the **dikaryon** and for a time the nuclei divide separately, producing a number of dikaryon cells. Eventually nuclear fusion takes place and initiates the formation of an **ascus,** the saclike sporangium for which the class has been named. Meiosis occurs in the young ascus soon after fusion, forming four haploid nuclei. These four nuclei divide once more, resulting in eight ascospores in an ascus. Although this is the usual number there may be a variation of from one to hundreds of ascospores in an ascus in certain species. In some species asci are produced singly, but in most they are located in a specialized reproductive structure, the **ascocarp,** formed from female gametangia and associated hyphae (Fig. 10).

In many fungi the ascocarp is the most conspicuous part.

There are three general types of ascocarps, based largely on their morphology.

1. **Apothecium.** This fruiting structure may be cuplike or broadly expanded like a sauce dish or disc (Fig. 11). The asci are tightly arranged in a palisadelike layer within the saucer.

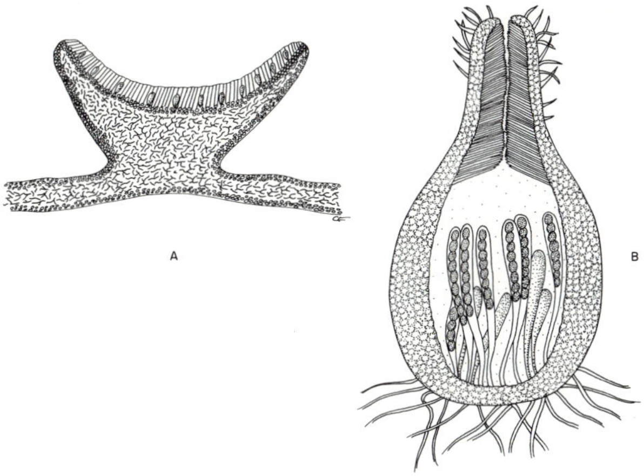

FIGURE 11 Fungi reproductive bodies. A-apothecium. B-Perithecium. B redrawn from Sinnott. 1955. *Botany-Principles and Problems.* 2nd Ed. McGraw-Hill.

2. **Perithecium.** This type is spherical or flask-shaped, is often hollow, and is usually closed except for a small opening at its apex (Fig. 11).
3. **Cleistothecium.** In this spherical fruiting structure the asci are completely enclosed so that the only way the spores can escape is by a break in or disintegration of the ascocarp wall (Fig. 15).

Crozier Formation. One of the unusual features in sexual reproduction and formation of the ascus is known as **crozier formation** (Fig. 12). After the male nucleus has been received by the

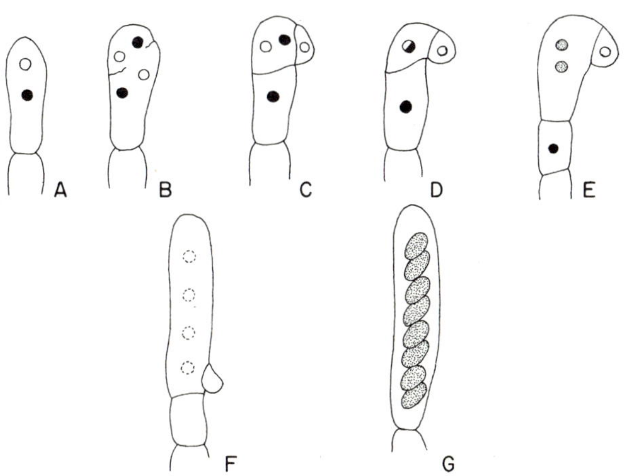

FIGURE 12 Development of an ascus. Following fusion of hyphae from two parents the nuclei in the cells remain in a dicaryon stage for some time. A shows the two nuclei from two different parents. The crozier is forming in C and fusion has occurred in the apical cell in D; In E the fusion nucleus has divided in its first stage in the formation of ascospores. F and G show further stages in enlargement of the ascus and increase in number of the ascospores. Redrawn in part from Sistrom, W. R. *Microbial Life.* 1962. Holt, Rinehart and Winston.

ascogonium, the male and female nuclei approach each other in a pair but do not fuse. Following conjugation the ascogonium produces a number of hyphal protrusions into which pairs of nuclei (the dikaryon stage) migrate. Once in the hyphae, the nuclei undergo independent divisions to form multinucleate ascogenous hyphae. Each hypha is then separated into a number of cells by the formation of septa, each cell containing one or more pairs of nuclei. In many ascomycetes one of the binucleate cells of the ascogenous hypha, usually at the tip, bends over to form a hook or crozier. The two nuclei in the crozier cell divide on a vertical plane so that the two daughter nuclei lying next to each other are of different origin. Septa then form separating the crozier into three cells. Each of the lower cells then contains a single male or female nucleus and the apical cell has the binucleate stage. This is the cell that will become the ascus following fusion.

Since this technique is repeated many times, it results in a clustering of asci within a small area. The various types of ascocarps defined above have their asci arranged in a layer termed the **hymenium.** Generally sterile hairs (paraphyses) are distributed among the asci and form a portion of the hymenial layer.

The complete life cycle of a more advanced member of the ascomycetes may be summarized as follows:

Asexual Cycle

Mycelium
↓
Conidiophore
↓
Conidia
↓
Mycelium

Sexual Cycle

Mycelium	Mycelium
↓	↓
Antheridium	Ascogonium
↓	↓
Male Nuclei	Female Nuclei
↘	↙

Dikaryon in Ascogonium
↓
Dikaryon in Ascogenous hyphae
↓
Crozier
↓
Fusion of Nuclei in hyphae
↓
Development of asci
↓
Meiosis
↓
Mitosis

↓
Ascospore
↓
Mycelium

There are some twelve thousand species of ascomycetes identified at the present time. We will consider a few representative types.

Yeasts

Most yeasts are species of the genus *Saccharomyces* and range from about two to a hundred microns in diameter (Fig. 13). They

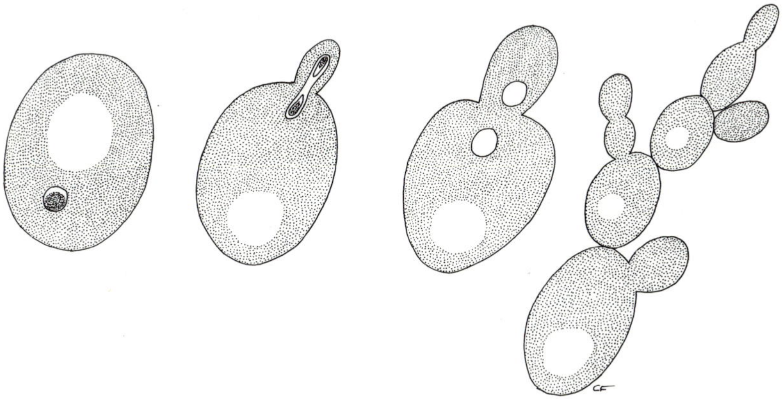

FIGURE 13 Yeast cells in the process of budding.

are usually unicellular, but they have a habit of budding so rapidly that they form clusters of mature and immature cells before completely separating into a pseudomycelial mass. Furthermore, yeasts occasionally produce ascospores as a result of a union of two vegetative cells. The resulting zygote forms a spherical ascus which will ultimately contain from one to eight ascospores.

The economic importance of yeast is based largely on the growth habits of the organisms. They have the ability to ferment carbohydrates partially in the presence of free oxygen and partially where free oxygen is unavailable. Sugars are oxidized in the respiratory processes of the yeast cell to form a simple organic acid with a release of energy in the process. If the supply of oxygen is deficient the organic acid is changed to carbon dioxide and alcohol, and it is this process that is so important to the baking and brewing industries. In the bakery the carbon dioxide is the important product and in the brewery the important product is alcohol.

The equation that describes the net result is

$$C_6H_{12}O_6 \rightarrow 2C_2H_5OH + 2CO_2.$$

This equation indicates that sugar is the substrate, the process is anaerobic, and the products are ethyl alcohol and carbon dioxide.

Most yeasts can ferment the monosaccharides glucose and fructose as in the equation above, as well as the disaccharides sucrose and maltose. The polysaccharides starch and cellulose are not fermented directly by yeasts and therefore require preliminary hydrolysis to change the more complex carbohydrate to a monosaccharide or disaccharide. This is accomplished in the brewing industry by permitting grain to undergo partial germination, which allows the starch-digesting enzymes in the seed to convert the starch into maltose so that yeast can handle the fermentation process.

Aspergillus

Aspergillus (Fig. 14) is frequently called the red or black mold because these colors are often present in its mycelium. However, do not be misled by this characteristic, as the author has on at least one occasion, for *Aspergillus* may be blue and look very much like *Penicillium* macroscopically.

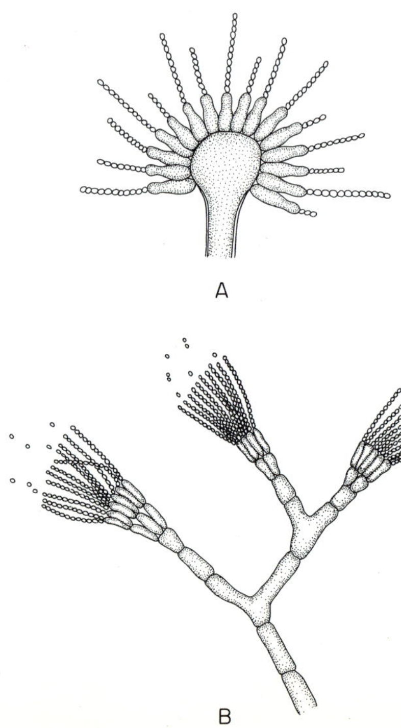

FIGURE 14 A-*Aspergillus* showing globose vesicle, bottle-shaped sterigmata and chains of conidia; B-*Penicillium* slender branched conidiophores with clustered sterigmata and chains of conidia.

The distinguishing feature of *Aspergillus* is the conidiophore. The vegetative hypha which produce the conidiophores terminate in a globose tip, termed the **vesicle.** It is multinucleate and from its surface arise a large number of little stalks or **sterigmata.** These may be in one or two layers depending on the species. From their tips spherical conidia arise in chains and are continually shed as new conidia form below.

Sexual reproduction results in formation of a number of asci in a soft cleistothecium. Usually eight ascospores are formed in each ascus but they are soon liberated into the cleistothecium, for the asci break down soon after the formation of the ascospores.

Penicillium

The genus *Penicillium* (Fig. 14) is usually called the blue mold but like *Aspergillus* it may sometimes be red. Sexually *Penicillium* reproduces much like *Aspergillus,* but the wall of the cleistothecium is firmer and more definitely formed. The ascospores, as seen in edge view, look like pulley wheels; in face view they are round to star-shaped and the walls may be smooth or sculptured.

The most obvious morphological difference between the two genera is the structure of the condiophores. In *Penicillium* the conidiophores are slender and may be simple or branched several times. The sterigmata which arise from the upper branches of the conidiophores produce the usual chains of spores.

Economically *Penicillium* is well known for its production of **penicillin,** the first of a developing line of antibiotics. *P. notatum* was the first species used in medicine, then *P. chrysogenum* was found to be more productive. Two other well-known species, *P. roqueforti* and *P. camemberti,* have been used for perhaps hundreds of years in the manufacture and aging of such cheeses as Roquefort, Camembert, Gorgonzola, and blue.

Powdery Mildews

"Powdery mildew" is a general term for a number of parasitic ascomycetes which form rather cobwebby mycelia over the surfaces of the host tissue. The conidiophores and conidia are usually whitish in color. Frequently found scattered with the white mycelia are tiny spherical dark bodies, the cleistothecia, which have anchoring appendages of rather unusual form (Fig. 15).

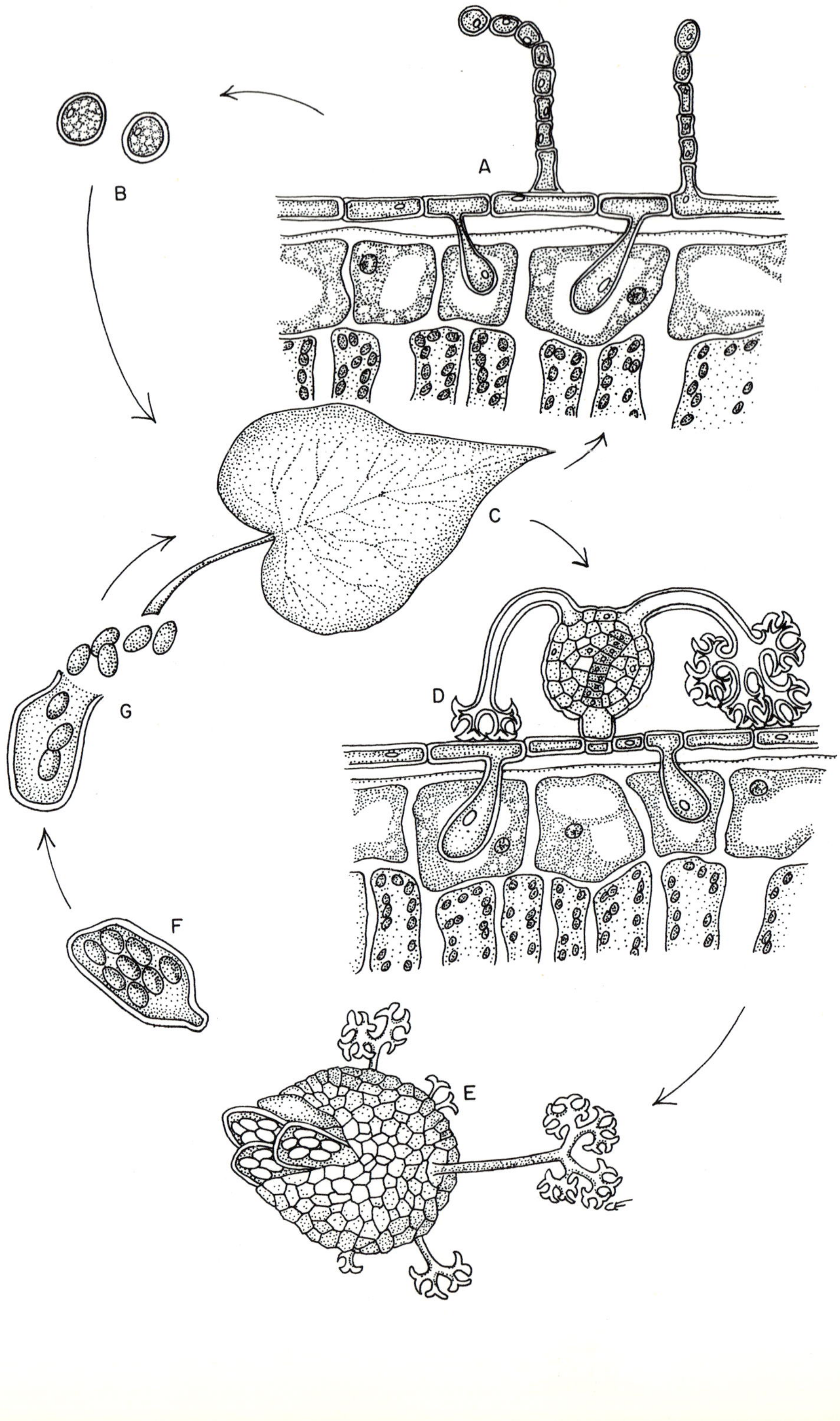

FIGURE 15 Life cycle of *Microsphaera alni* on lilac. A-Conidia produced on upper surface of lilac leaf, young haustoria have developed in epidermal and subepidermal cells; B-conidia disseminated for new or reinfection; C-lilac leaf; D-young cleistothecium developing on leaf; note branched appendages; E-mature cleistothecium split open to show asci; F-ascus; G-discharge of ascospores for new or reinfection. Redrawn from *Laboratory Outlines*, H. Freeman, 1963.

Microsphaera alni is the fungus which causes powdery mildew of lilac; *Uncinula necator* is the cause of powdery mildew of grape. There are many others. In most cases they infect the leaves and young stem tissue, using the host's food which might otherwise be used in flowering and fruiting. The fungus usually does not kill the host, even though it may remain for years, but it impairs the health and reduces the size and quality of the crop.

Neurospora

Neurospora is a ubiquitous mold that was at one time known as *Monilia sitophila*. It has long been known as a bakery pest, since it usually withstands bread-baking temperatures. The conidia are found almost everywhere and get into bread dough while it is being prepared. These spores can tolerate a temperature of 120° to 125° C for an hour or more in dry air but are killed in boiling water in a few minutes. During the process of baking, the temperature deep inside a loaf of bread does not exceed 95° to 98°C. Consequently many spores survive the baking and then may begin the breakdown of the bread.

Neurospora (Fig. 16) has been used widely for a number of years in mycological and genetic experiments. It is selected for

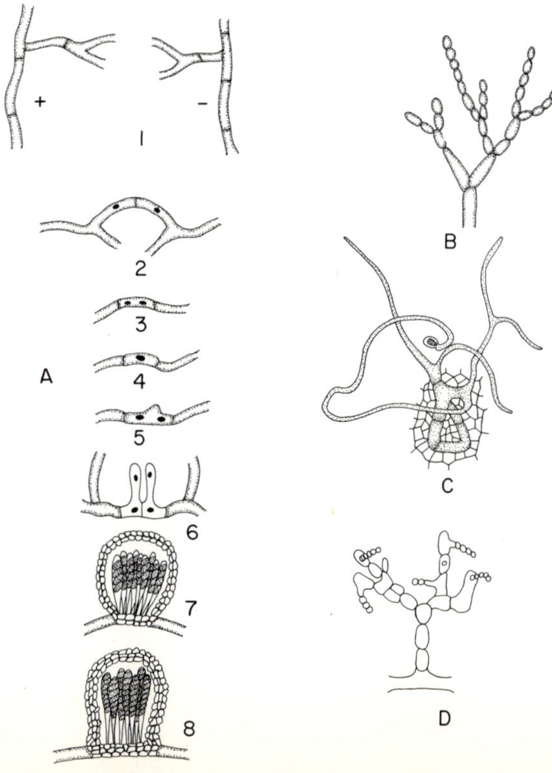

FIGURE 16 *Neurospora*
A. Fusion of the hyphae from plus (+) and minus (—) strains: (1) the + and — strain are growing close together. (2) union of hyphae. (3) dikaryon stage with + and — nuclei in same cell. (4) fusion of nuclei. (5) division of fusion nucleus and initiation of ascus. (6) further division of nuclei and two asci developing. (7) young perithecium formed with asci in four spored stage. (8) mature perthecium with asci containing 8 ascospores.
B. Conidiophore with mature conidia.
C. An ascogonium with trichogynes and one of the filaments has a microconidium attached.
D. A microconidiophore with developing microconidia. C and D from Alexopoulos, C. J. 1952. *Introductory Mycology*. John Wiley & Sons, Inc.

this type of work because it is easy to grow both asexually by conidia or fragmentation and sexually where the ascocarp produced is a perithecium. *Neurospora* hybridizes readily, it produces an ascus with, usually, eight large linearly arranged ascospores, and it has few chromosomes.

When certain strains of *N. sitophila* are mated, the ascospores will be arranged in different ways in the ascus (Fig. 17). One mating will result in pairs of spores of one sex alternating with pairs of the other sex; occasionally the spores alternate singly, and occasionally there may be four of one sex at the top of the ascus with the four of the opposite sex at the bottom. The particular distribution of the spores depends upon whether the factors for sex were segregated during the first or second nuclear divisions in the ascus.

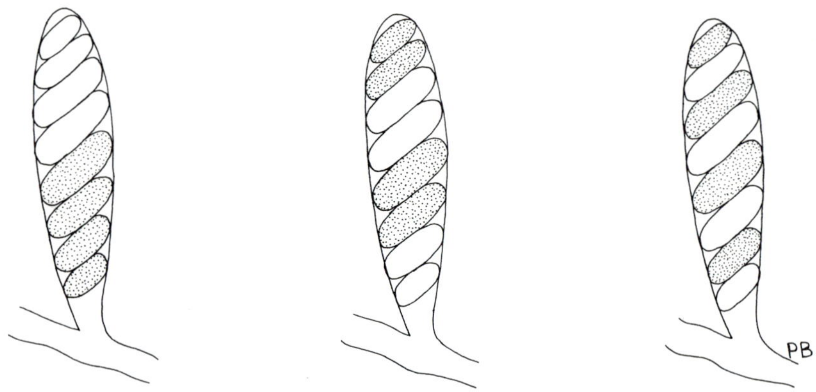

FIGURE 17 Asci of *Neurospora* showing how the ascospores with different genes may be arranged in an ascus.

One of the advantages in using this type of fungus in genetic experiments is that both nuclear fusion and reduction division occur in the ascus. In *Neurospora* the large ascospores are tightly packed in the slender ascus so that by careful micromanipulation the spores can be removed individually, separately cultured, and their resultant characteristics observed.

Beadle and Tatum exposed *Neurospora* to X rays and obtained a number of mutant forms which varied in a specific physiological trait from the wild type. It was found that the trait was due to a single gene which segregated in typical Mendelian fashion. More than twenty different mutants have been found, each lacking the ability to produce some substance such as various vitamins, amino acids, or other compounds necessary for growth. It was concluded that mutant genes prevent the fungus from producing a vital chemical. Some mutating genes in humans have been shown to have similar effects.

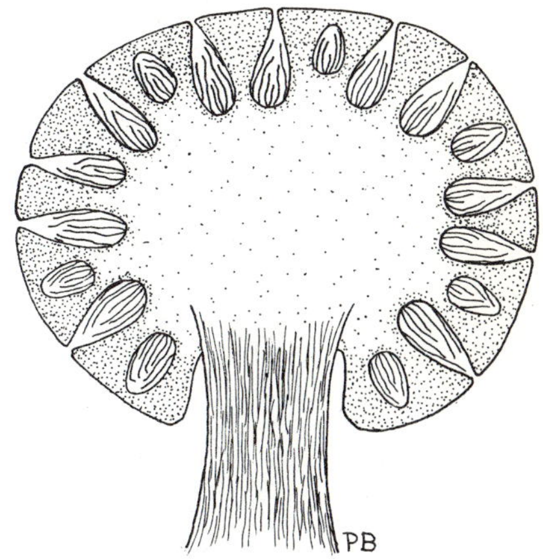

FIGURE 18 *Claviceps purpurea.* A diagram of the stroma with imbedded perithecia.

Claviceps

Ergot of rye is the common name given to a growth by *C. purpurea* (Fig. 18) which replaces the tissue in the rye grain with its own tissue. The substance formed by the fungus is a drug given the name **ergot.** The drug is used in medicine, but when taken in excess it can be lethal. There are many recorded cases where people have died from eating rye bread made from flour containing this fungus.

Morchella

One of the largest of the ascomycetes is a spongelike mushroom commonly called a morel (Fig. 19). In this genus the ascocarp is an apothecium which is spongelike as the result of a variously ridged and reticulate fruiting body. The species of *Morchella* are among our tastiest mushrooms, but they are seldom seen in the market since they are not easily cultured and are very seasonal in the wild state. Another genus, *Gyromitra,* is called a false morel since it has the same spongelike fructification, but it is considered poisonous. Some people have reported eating them without ill effects, but fatalities have also been recorded among others who have eaten species of *Gyromitra*.

FIGURE 19 *Morchella.* An edible member of the ascomycetes. Courtesy Carolina Biological Supply Company.

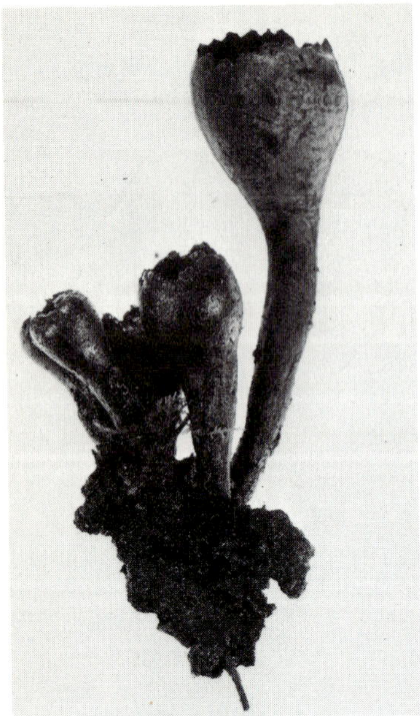

FIGURE 20 *Urnula*—a cluster of the fruiting bodies. Courtesy Carolina Biological Supply Company.

Related to the morels are a variety of cup fungi such as *Peziza* and *Urnula* (Fig. 19-20) which contribute to the interest and color of the flora in moist forests.

Tubur

The members of the genus *Tubur* are commonly called **truffles;** they are widely recommended by epicures as the tastiest of all mushrooms. *Tubur* has a small, globose ascocarp which grows underground and consequently is rarely seen. Several species of truffles have been found growing in the United States, particularly in the Northwest Pacific coast states. Commercially, however, most truffles are obtained from Central Europe, where pigs and dogs are trained to find them.

Plant diseases caused by members of the Ascomycete group include the Chestnut blight and the Dutch elm disease.

CLASS BASIDIOMYCETAE

The club fungi are probably more familiar to most people than any other fungus group, since the most conspicuous fungi are in this class. Included are the usual mushrooms of field, forest (Fig. 21), and grocery store, shelf or bracket fungi that grow on the sides of trees, and puffballs that may reach a size of three feet in diameter. Less conspicuous but of far greater economic importance, in a negative way, are the rusts and smuts. These are microscopic in size in most stages, but their virulent parasitic nature costs us millions of dollars in crop loss every year.

FIGURE 21 *Russula,* a mushroom emerging through the forest litter. *Left:* Only the cap is visible. *Right:* A few hours later.

The basidiomycetes, like the ascomycetes, have a septate mycelium, but there is a difference in the behavior of the nuclei. In the ascomycetes the vegetative mycelium is chiefly uninucleate in the individual cells, and this is the dominant condition. The dikaryon stage is found mostly in the ascogenous hyphae and is comparatively short-lived. The basidiomycetes, on the other hand, usually have a limited uninucleate stage and a dominant dikaryon mycelium.

The distinguishing feature of the basidiomycetes is the production of a *basidium* (Fig. 22). Usually this is a club-shaped structure and it produces *basidiospores* on its apex externally rather

FIGURE 22 Development of a basidium. A-hypa with dikaryon stage. B-apical cell with nuclei from two different plants; C. D. E. stages in fusion of nuclei. F-division of fusion nucleus. G-formation of four nuclei which migrate through the sterigmata to enter the developed basidiospores in H.

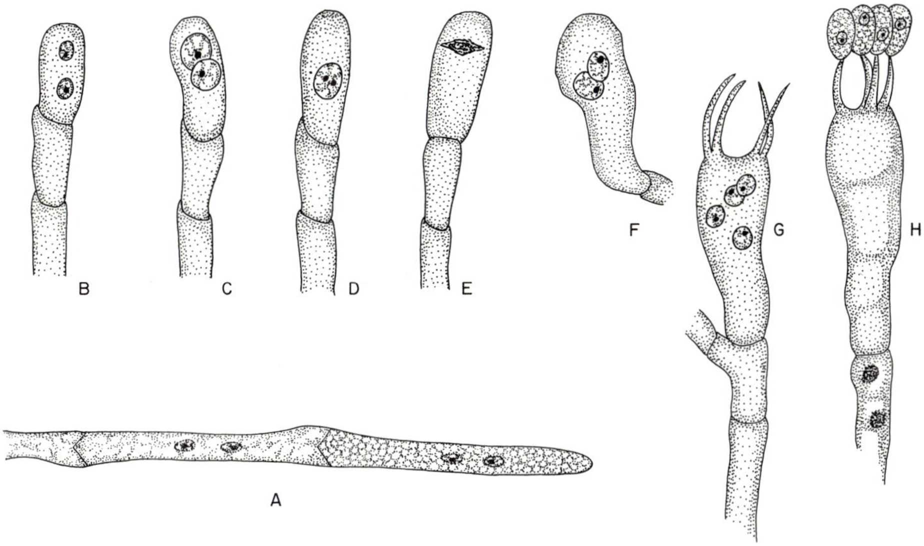

than internally as in an ascus. Generally these spores are uninucleate and haploid, and, like ascospores, they arise following meiosis. The basidium has an origin homologous to that of the ascus in that both are the result of fertilization.

In the ascomycetes the initial ascus cell is formed in the crozier at the apex of the ascogenous hyphae; in the basidiomycetes a clamp connection develops along the lateral wall of a binucleate cell through which a daughter nucleus migrates, to unite ultimately with the daughter nucleus in the adjacent cell. The entire process occurs as follows (Fig. 23): When a binucleate cell is ready to divide, a short tube, similar in appearance to the early progametangium of the phycomycetes or the conjugating tube of *Spirogyra* or

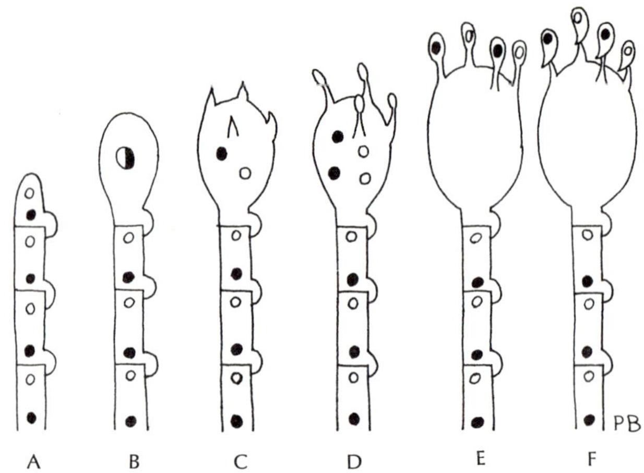

FIGURE 23 Diagrammatic illustration of development of basidium and basidiospores. A-dikaryon stage and clamp connections established, B-karyogamy resulting in 2N zygote nucleus; C, D meiosis resulting in four haploid nuclei in developing basidial cell; E migration of nuclei to sterigmata and into developing basidiospores; F mature basidium and basidiospores. (redrawn from Alexopoulos, C. J. and H. C. Bold, 1967 *Algae and Fungi.* The Macmillan Co., New York.)

Zygnema, is extended from the wall of the hyphae about midway between the two nuclei in the mother cell. The two nuclei then divide. The main axis of the one dividing nucleus is oriented in such a way that the one daughter nucleus can readily migrate into the newly formed tube and the other remains behind. The second of the original two nuclei divides with its orientation along the main axis of the cell at the same time. The one daughter nucleus of this division migrates to the apical end of the cell along with the daughter nucleus of the other parent nucleus. The tube in the meantime has been growing forward and bending down; upon making contact with the hyphal wall its tip dissolves away, forming an open union at the point of contact. This is now the clamp connection between two parts of the same cell. However, at this time two septa are formed. One cuts off the clamp from its original point of origin and the other forms vertically under the bridge separating the new daughter cells. In each cell there is one daughter nucleus from each of the original pair since the nucleus which had migrated into the clamp continues on through to pair with the remaining daughter nucleus of that division.

Although clamp connections have not been observed in all basidiomycetes, they are quite common in most. Since they are generally found at the base of a basidium, where there has been a fusion of daughter nuclei and then meiosis similar to nuclear behavior in an ascus, we consider the entire process as homologous to ascus formation.

Following the clamp development the basidium arises from the terminal cell of the binucleate hypha separated from the remainder of the hypha by the septum at the formation of the clamp.

The basidium gradually enlarges into the club-shaped form it ultimately attains; meanwhile the two nuclei within the young basidium fuse. Soon after this zygote nucleus is formed meiosis occurs resulting in four haploid nuclei. In the later stages of basidium development, four (usually) sterigmata emerge from the apex. The four nuclei migrate to the sterigmata and squeeze through the narrow sterigmatal opening forming basidiospores on the tops of the sterigmata.

Basidiospores upon germination usually give rise to germ tubes which develop into a mycelium with uninucleate cells. In some species the spores may be produced by budding of both basidiospores and mycelial hyphae. Some species like the rusts differentiate conidia as specialized spores termed **uredospores.**

Asexual reproduction may also take place by fragmentation, or the hyphae may break up into specialized unicellular sections called **oidia. Oidiospores** may be formed also on short hyphal branches (Fig. 24) which form the oidia successively from tip to base of the specialized hyphae.

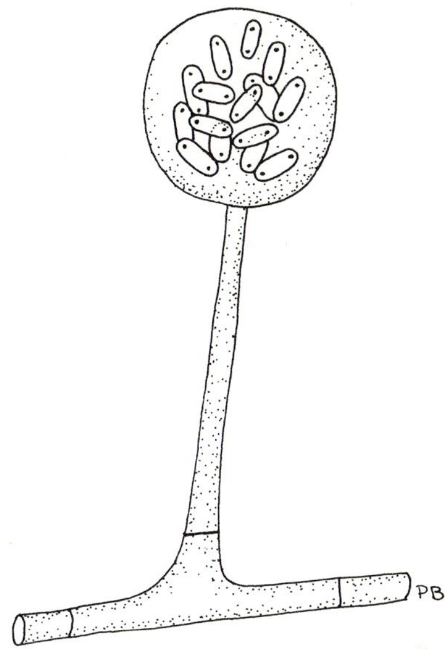

FIGURE 24 Oidiophore with enclosed oidia and septate hypha typical of most Basidiomycetes. (after Alexopoulos, C. J. 1952. *Introductory Mycology*. Wiley.)

In the basidiomycetes the mycelium is usually made up of uninucleate cells, but this phase of the life cycle is brief. In many species a very simple conjugation takes place between two uninucleate vegetative hyphae. The walls dissolve at the point of contact of two hyphae and the nucleus of one cell passes through the opening into the other cell making it binucleate. This binucleate cell then gives rise to the binucleate mycelium by successive divisions. Some species achieve the dikaryon mycelium by uninucleate oidia carried to vegetative hyphae by wind, water, or insects which then discharge their contents into the hyphal cell making it binucleate and in that way initiating the dikaryon stage in the formation of a new mycelium.

Basidiospores in some species are binucleate. In these fungi the germ tube that is formed is binucleate and continues to develop a binucleate mycelium. Fusion takes place as usual in the formation of the basidium, and meiosis occurs as usual, but instead of the four basidiospores only two are formed and each receives a pair of the haploid nuclei.

Rusts

There are over seven thousand species of rusts. All are obligate parasites and many are **heteroecious,** i.e., they require two or more hosts to complete their life cycles.

Puccinia graminis is the name given wheat rust, probably our most virulent and widespread fungus parasite. The wheat rust cycle (Fig. 25) begins very early in the spring when the overwintering

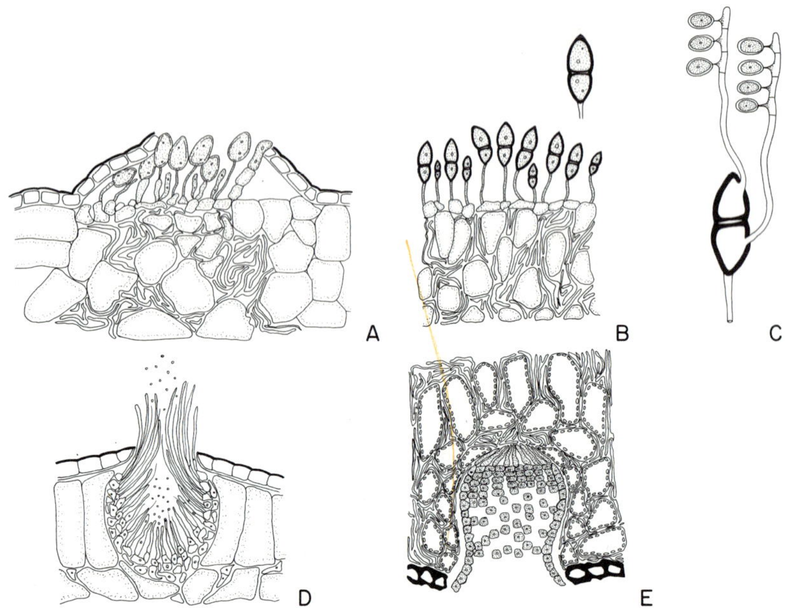

FIGURE 25 Spore stages in the cycle of wheat rust. A-Uredospores, the reinfecting short cycle spore on wheat; B-Teliospores, the overwintering spore; C-Teliospore germinating to form a basidium and basidiospores from each cell. D-Spermatia formed on upper surface of American barberry bush; E-Aeciospores formed on lower surface of barberry. Between each spore stage a mycelium with many hyphae is formed.

teliospores germinate in the soil or mulch. The two cells of these spores may each produce modified hyphae and form four haploid nuclei in each as the result of reduction division. The nuclei become equally spaced in the hyphae, septa form, and each cellular segment develops lateral sterigmata from which the basidiospores arise.

The basidiospores are carried about by air currents, and if by chance some should land on the leaves of the common (American) barberry bush they germinate and germ tubes will grow into the tissues of the barberry leaf.

In the barberry leaf the hyphae which form from the basidiospores are uninucleate. If the hyphae formed in the leaf have developed from spores of different strains, which we may designate as + and −, there will be fusion. Normally this fusion takes place between small specialized cells called **spermatia** produced from hyphae in a flask-shaped structure called the **spermagonium** on the upper surface of the leaf, and small hyphae produced on the leaf by the opposite rust strain.

After fusion of the spermatia and the receptive hyphae, the dikaryon stage is achieved and the next stage or **aecium** is initiated.

The aecium or cup stage forms on the lower surface of the leaf when numerous hyphae become massed to create the cup. Within the cup, chains of **aeciospores** are formed, the binucleate aeciospores being carried off by wind or insects to a wheat plant. Here the aeciospores germinate, forming a delicate mycelium with its imbedded haustoria, and soon form the red pustules on the wheat plant within which the binucleate **uredospores** arise.

Uredospores and their mycelia may grow for many generations on wheat. Reinfection of wheat fields from uredospores may continue throughout the wheat season. This is the short cycle of wheat rust and is the principal way that the rust does its damage, since nearly all common barberry bushes have been removed from the American scene.

Late in the season the uredo stage is changed to the telial stage and the hyphae now form the two-celled teliospores, which are capable of overwintering in cold climates. In most areas this ends the cycle, but in warm winter climates uredospores will overwinter and reinfect young wheat plants when they germinate in the spring.

White pine blister rust, *Cronartium ribicola,* has as its hosts the white pine tree and currants and gooseberries. Control of this disease has greater promise of success, since the short-cycle repeating uredo stage is on the currants and gooseberries, which are considered less desirable than white pine. The white pine has the spermogonial and aecial stages.

The cedar apple rust, *Gymnosporangium juniperi-virginianae,* forms its spermogonial and aecial stages on apple trees and its telial stage on the cedar tree. The uredo stage is lacking. Since both types of trees are desirable, no eradication program has been developed to break the cycle.

Smuts

The members of this group produce a grayish growth in the grains of many cereals, causing a hypertrophied enlargement of the grain which appears very sooty (Fig. 26). The smuts often produce spindle-shaped or sickle-shaped conidia instead of uredospores. They also often form chlamydospores instead of the typical teliospores. A further difference is observed in some smuts where the promycelium (basidium) developing from the teliospore (chlamydospore) does not form basidiospores. Instead the hyphae germinate directly from the cells of the promycelium (Fig. 27). In other

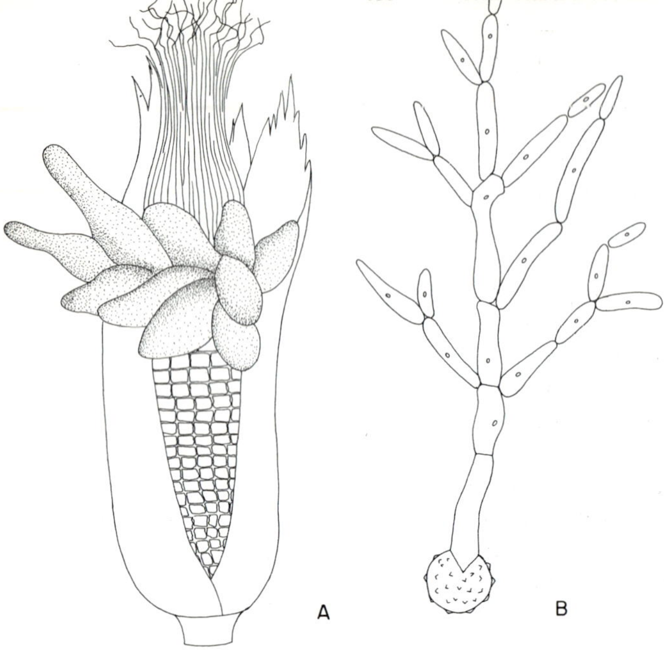

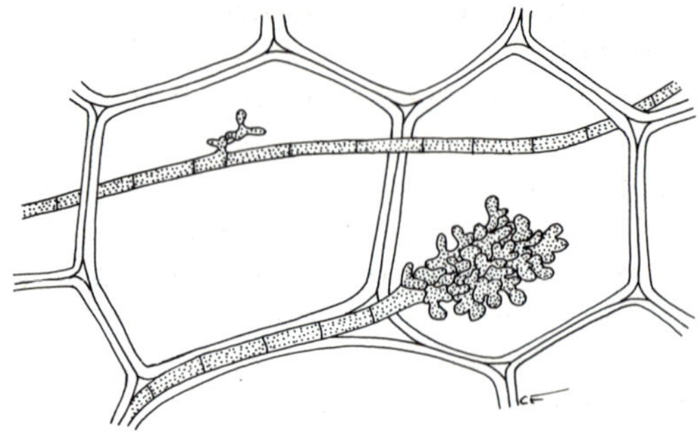

FIGURE 26 Corn smut. A-Note hypertrophied grains; B-Germination of chlamydospore.

FIGURE 27 Corn smut, fungal hyphae growing through host cells; one hyphae has developed modified haustoria.

FIGURE 28 Chlamydospore showing germination to form the filamentous basidium and some liberated basidiospores.

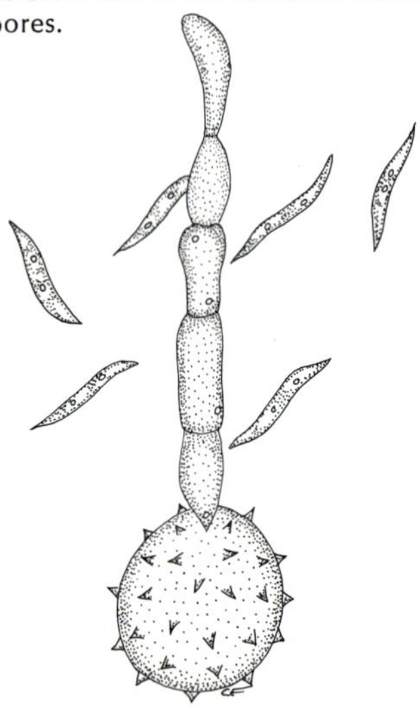

smuts basidiospores are formed (Fig. 28), and when two compatible spores are close to each other they send out copulation tubes uniting them and the tube into an H-shaped structure. Fusion occurs by the migration of one protoplast through the tube into the other basidiospore.

Mushrooms and Toadstools

While the terms above have been used for certain ascomycetes and basidiomycetes, the type of fungus that most people recognize as mushrooms is the gill type or agaric with the typical cap and stalk (Fig. 29).

The part of the mushroom we usually see is the "fruiting" body, or **basidiocarp.** It consists of a cap or **pileus** and a stem or **stipe,** and in some cases the stem develops from a cuplike base, the **volva.** In some mushrooms a very prominent **annulus** or **veil** is found around the midpoint of the stipe where the tissue was left as the cap expanded. The underside of the cap is made up of radiating **gills** which are composed chiefly of the hymenium with its thousands of basidia (Fig. 30a).

The life cycle of the typical mushroom begins when two basidiospores of plus and minus strains fall together on a suitable substrate. The hyphae formed from the spores fuse and the dikaryon stage is formed. After the mycelium has developed for some

FIGURE 29 Habitat pictures of two different species of *Lactarius*.

FIGURE 30a. Stages in the development of a gill type mushroom. A-Mature stage showing expanded cap with radiating gills, stipe with annulus, cup-shaped volva at base, hyphae. The poisonous *Amanita* looks like this illustration. B-section through cap showing gills, in vertical section. C-microscopic section through gill showing basidia on gill with basidiospores; stages and nuclear behavior in development of basidium and basidiospores; E-mature basidium with sterigmata and spores.

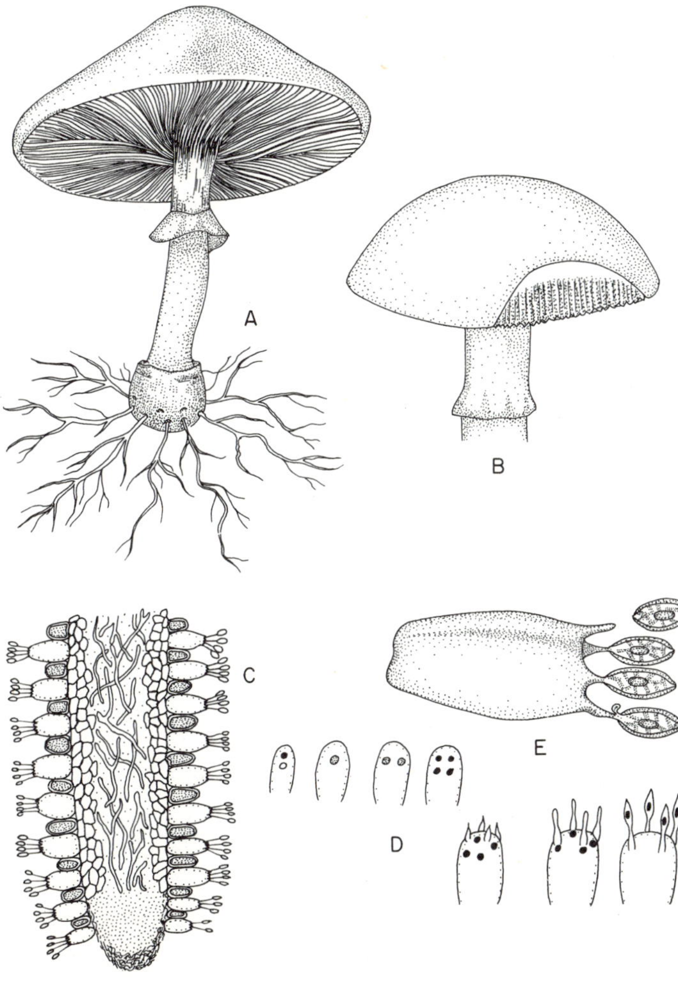

FIGURE 30b. *Amanita verna.* Note the cuplike volva at the base of the stipe and the annulus or veil about halfway up the stipe. Photo by H. L. Dean.

time, the hyphae begin to form a small spherical mass of closely packed hyphae. In this stage the young mushroom is called a **button.** In some species, like those of the deadly **Amanita,** the entire button is surrounded by a thin tissue known as the **universal veil.** As the basidiocarp enlarges the veil is torn apart, leaving a portion on the edge of the cap and a portion underground around the base of the stipe forming the volva. Part of the veil remains around the stipe as the cap pulls away from this structure (Fig. 30b).

Outline of the Life Cycle of a Gill Mushroom

Basidiospore ⟶ Hypha
Basidiospore ⟶ Hypha
Hypha + Hypha ⟶ Fusion
Fusion
|
Dikaryon Stage
|
Mycelium
|
Button
|
Basidiocarp
|
Gills
|
Basidium

A good rule to follow in collecting mushrooms for food is to leave the ones with a veil and volva alone. The poisonous ones have these two structures and a few edible ones do too. However, most edible mushrooms lack a volva, a veil, or both, so you can be "reasonably" safe if you follow this rule. A better procedure to follow would be to do your collecting in the vegetable market if you cannot readily identify mushrooms.

In addition to the gill type mushrooms, there are some we call the **polypores.** One genus, *Boletus,* has a large stipe and cap, but instead of radiating gills the underside of the cap has numerous tiny pores within which the basidia are borne on the ramifying hymenium.

The bracket or shelf fungi are also mostly polypore types. The largest of these is *Fomes aplanatus* (Fig. 31). It is a perennial species with a white, flat lower surface, and is about as hard as the wood on which it is growing. Each year a new layer of basidia-lined pores is added to those of the early growth, increasing the size of the fungus.

Most puffballs are in the genera *Calvatia* (Fig. 32) or *Lycoperdon*. In these fungi the basidia are produced within a closed, globose fruiting body. The outer wall is the **peridium;** in *Calvatia* this consists of one layer; in *Lycoperdon* it has two.

Another, small group of puffballs is the earthstars (Fig. 33) in the genus *Geastrum*. They have several peridial layers of which the outer ones split radially and peel back forming an expanded, flat, star-like structure with the tiny globose puff ball left in the center.

Bird's nest fungi are unique little organisms in which the basidiocarp forms like a little cone-shaped nest or basket. Nestled at the base of the basket are the little eggs or **peridioles** which contain the numerous basidia and basidiospores.

Finally there is a group of basidiomycetes justly named the **stinkhorns.** The odor is much like the sulfurous odor of rotten eggs. The two most common genera are *Phallus* and *Dictyophora*. Although the writer has on occasion been very much aware of the presence of *Phallus,* the specimen in Figure 34 did not have an odor, at least at the time the picture was taken. When the mycelium has formed it develops a soft, round, globose basidiocarp which looks like a potato but is very soft. As the mushroom elongates a spongy pileus emerges on a stalk, leaving a loose volva at the base. The outer part of the pileus is made up of basidia, basidiospores, and considerable sterile tissue, called the **gleba.** The volva remains at the surface of the ground, supporting the phalloid aerial portion.

FIGURE 31 The shelf fungus, *Fomes,* growing at the base and side of a spruce tree.

FIGURE 32 Two puffballs, *Calvatia gigantea,* growing in an open meadow.

FIGURE 33 *Geastrum.* Earthstar type puffballs with the outer peridial layers curved back.

FIGURE 34 *Phallus.* Stages of development. At left are two soft globose young basidiocarps which when cut in two lengthwise show the young stipe and cap in formation. In the center are two nearly mature plants with cone-like pileus, stipe, and basal volva.

Relationships of the Fungi

The three classes of fungi have certain characteristics in common, but their differences have led mycologists into several different paths in attempting to explain their origins and evolution. Some contend that the phycomycetes arose from some algal ancester by loss of chlorophyll and the acceptance of a saprophytic or parasitic existence. The types of oogonia and antheridia and even the filamentous mycelia suggest this. Other phylogenists prefer to think of the phycomycetes as coming in part from flagellate ancestors and in part from non-flagellate primitive groups like the early slime molds. Certainly the heterogeneity of the members shows good reason to suspect a polyphyletic origin. Acceptance of many origins for this group as well as the other Thallophytes would point up the theories of parallel lines of evolution and the convergence in development of organs and tissues in the different groups.

It seems reasonable to assume that the ascomycetes and basidiomycetes have a relationship to the higher phycomycetes. The oogonium and the ascogonium seem to have a similar origin, and the ascus and basidium are certainly homologous structures. Conidia are produced in all groups; the mycelium is likewise present in all but differs significantly in its degree of specialization and differentiation. The fungi are a distinct group and while they are usually studied in connection with plants, some of the more primitive are often considered in the animal kingdom. Some mycologists are also of the opinion that the organic world should be divided into three major categories: Plant, Animal, and Fungus.

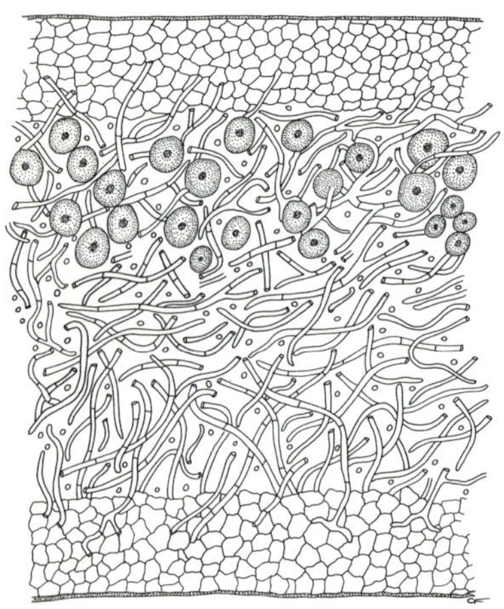

FIGURE 35 Section through the thallus of a lichen showing fungal hyphae and circular unicellular algae.

Lichens

One of the biotic world's strange associations is the intimate relationship of a fungus and an alga which grow so closely intertwined as to form a distinct thallus (Fig. 35). These are the lichens. They are considered excellent examples of a symbiotic relationship, but the fungus has the dominant growth and the alga is relegated to a subservient or slave position. The fungus provides the place and anchorage; the algal cells lie within the common thallus photosynthesizing food and having it absorbed by the fungus. The fungus probably also absorbs most of the water initially, and perhaps mineral nutrients, some of which filter through to the alga.

Most lichens have an ascomycete as the fungus member, and usually a unicellular member of the green or blue-green algae is the associated organism. There are three principal categories in which lichens are classified:

1. **Crustose lichens** (Fig. 36) live closely compressed to the surface of rocks, trees, and other substrates. They are so firmly anchored to the substrate that the easiest way to collect them is to bring the rock or substrate along too.
2. **Foliose lichens** (Fig. 36) have a leaflike thallus which looks very much like a liverwort except for color. They grow in much moister areas than the crustose lichens.
3. **Fruticose lichens** are larger and rather freely branched, and often hang pendulously from tree branches, resembling Spanish moss (Fig. 37).

A

B

FIGURE 36 Lichens. A-The circular light patches on the rocks are crustose lichens; the foliar plants which look like liverworts are foliose lichens. B-The elongated structures are the fruticose lichen, *Cladonia*, growing on a spruce log.

FIGURE 37 The fruticose lichen, *Usnea*, hanging from the branches of a spruce tree.

Reproduction by the algal component is usually limited to simple fission within the fungus mycelium, but for the complete union vegetative reproduction may occur by fragmentation. Some lichens also form very small outgrowths, called **soredia,** on their upper surface. A soredium is made up of one or more algal cells surrounded by a number of fungal hyphae. These structures are blown or washed away to where they may grow into a new thallus.

The fungus partner may also reproduce sexually in the usual fusion method of ascomycete hyphae. The sexual body produced is usually an open apothecium with the asci and ascospores typical of this type of fungus. In some lichens the asci are formed in a perthecium imbedded in the thallus. The ascospores ultimately are blown away and then begin to develop the hyphae and mycelium. If algal cells are in the vicinity and come in contact with the fungus a symbiotic union is established.

Although we may pay little attention to lichens, they are pioneers in ecological succession. The crustose forms become established on bare rock. Then by the chemical products of their growth and the slight penetration of their hyphae the bare rock is gradually decomposed and disintegrated into soil particles. Thus lichens are preparing a substrate for larger plants.

Lichens have been used as a source of food for man and also as a source of dye, but their chief benefits are indirect. Reindeer moss, a lichen of the Arctic Circle, has always been a major food for caribou. In 1963 it was discovered that radioactive fallout was being absorbed into the spongy tissue of the reindeer moss. This fallout included the dangerous cesium-137. Caribou eating the lichen transferred this radioactive material from their stomachs to their flesh and bones. Eskimos in the area then ate the caribou and were gradually transferring the cesium to their tissues. Thus far no serious symptoms have been observed among the Eskimos, but the United States Atomic Energy Commission is running regular tests.

Supplementary Readings

Alexopoulos, C. J., *Introductory Mycology,* 2nd ed. John Wiley & Sons, New York, 1959.

Beadle, G. W., The Genes of Men and Molds, *Scientific American,* September 1948.

Bracker, C. E., Jr., and E. E. Butler, The Ultrastructure and Development of Septa in Hyphae of *Rhizoctonia solani, Mycologia* 55:627–632, 1963.

Christensen, C. M., *The Molds and Man*. University of Minnesota Press, Minneapolis, 1951.

Cochrane, V. W., *Physiology of the Fungi*. John Wiley & Sons, New York, 1958.

Fitzpatrick, H. M., *The Lower Fungi: Phycomycetes,* McGraw-Hill, New York, 1930.

Gray, W. D., *The Relation of Fungi to Human Affairs*. Holt, Rinehart and Winston, New York, 1959.

Ingold, C. T., Spore Liberation in Higher Fungi, *Endeavor* 16(62) 78–83, 1957.

Large, Ernest C., *The Advance of the Fungi*. Holt, Rinehart and Winston, New York, 1940.

Northern, H. T., *Introductory Plant Science,* 2nd ed. Ronald Press, New York, 1958.

Ramsbottom, Jr., *Mushrooms and Toadstools*. Collins, Glasgow, 1953.

Sparrow, F. K., *Aquatic Phycomycetes*. University of Michigan Press, Ann Arbor, 1960.

Woodham-Smith, C. B., *The Great Hunger: Ireland*. Harper and Row, New York, 1962.

Questions for Discussion

1. What evidence indicates that the basidiomycetes descended from the ascomycetes?
2. Many crop plants, such as wheat and corn, are attacked by parasitic fungi which infect the stems and leaves. How does this affect the crop?
3. Why cannot all fungi requiring alternate hosts to complete their life cycles be eliminated?
4. If the spores produced in a giant puff ball two feet in diameter are ten microns in diameter, how many spores would be produced in this one puffball?

5. Describe the different types of sporocarps produced by the basidiomycetes.
6. What would be gained by eliminating the uredospore stage in wheat rust?
7. What advice should you give a novice who has decided to go out in the woods to collect mushrooms for dinner?
8. Why are fungi more restricted in their habitat than higher plants?
9. Prepare an evolutionary tree of the various fungus groups and be prepared to defend your reasons for placing each group in the position selected.
10. Why do you suppose aeciospores will not germinate on the common barberry and teliospores will not parasitize wheat, even though each was produced on the plants it does not parasitize?
11. Compare the development of a basidium with that of an ascus.
12. How do certain fungi form "fairy rings" in certain areas?
13. Some subterranean fungi cause the grass to be greener in certain spots than other areas where the fungus is not growing. What effect does the fungus have on the grass?
14. Why are crustose lichens considered pioneers in some habitats?
15. How do certain fruticose lichens kill leaves and branches of trees without penetrating into the tree's tissues?
16. Why isn't the radioactive cesium-137 washed out of the lichen tissues above the Arctic Circle?

20

The Amphibious Plants

Subkingdom Embryophyta
Phylum Bryophyta
Class Hepaticae
Class Anthocerotae

Members of the Thallophyta are essentially aquatic organisms, but the groups we are now about to consider are virtually all terrestrial. A few members of the subkingdom Embryophyta are aquatic; these are species which apparently returned to water after their ancestors had become terrestrial and developed enough of the tissues and physiological characteristics to make it possible for them to live on land.

Introduction of the Embryo

The Embryophyta are comprised of two phyla, the **Bryophyta** and the **Tracheophyta.** As the subkingdom name indicates, these are plants with **embryos.** Following germination of the zygote a mass of cells are formed which may exhibit little differentiation, or they may be conspicuously differentiated into distinct organs (Fig. 1). In the Bryophyta the embryo matures into a sporophyte which remains dependent on the gametophyte. In this group the embryo has little in the way of tissue differentiation or organ development.

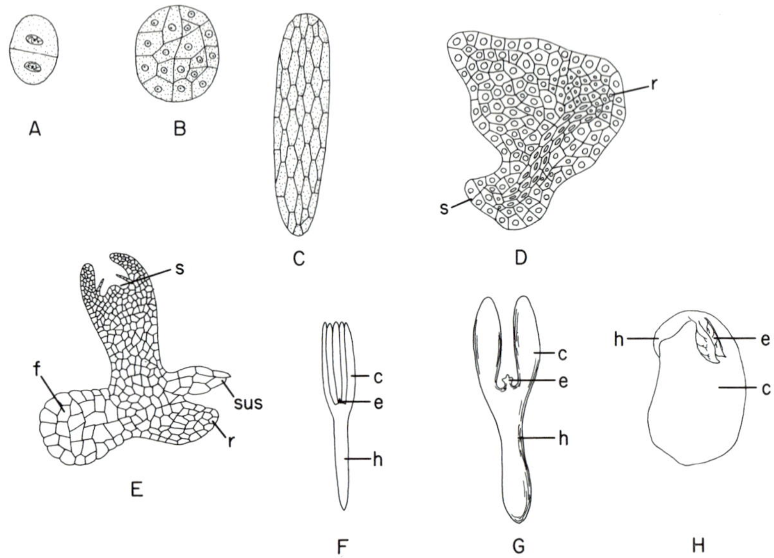

FIGURE 1 Embryos in the plant kingdom. A-*Riccia,* two celled stage. B-Riccia later stage. C-moss. D-fern. E-*Selaginella.* F-pine. G-*Capsella.* H-lima bean: c-cotyledone, f-foot, e-epicotyl, h-hypocotyl, r-root growing point, s-stem growing point, sus-suspensor.

In fact, even the mature sporophyte never becomes a prominent part of the plant; the emphasis is still on the gametophyte as it was in the members of most Thallophyte groups. The embryo is there, and it is developing into a sporophyte, but in the Bryophyta the sporophyte still does not become independent.

The Tracheophyta, on the other hand, were able to take advantage of genetic changes which occurred in them relatively early. The embryo formed in this group developed a stem, a leaf, and a root and although it was at first dependent on an independent gametophyte it was photosynthetic and it had an active growing region which almost immediately began to produce the roots, leaves, and stem structure necessary for successful terrestrial existence. The embryos of all plants, like those of all animals, do not look like the mature organism they are going to develop into. In fact, there is a great similarity among the embryos of all plants just as there is similarity among the embryos of all animals. It is the big differences in the nucleic acids in the arrangement of the molecules which guide differentiation and growth into the mature organism.

The origin of a new individual organism from the fertilized egg involves differentiation at both the cell and cell population levels. In the algae, where there is little tissue differentiation, the modification is mainly within the cell where there is considerable variation in the organelles (such as chloroplasts, nuclei, cytoplasmic content, and wall development). In the higher plants the formation of the embryo brings with it differentiations at all levels of complexity. As mentioned above, the embryo begins with a mass of generalized cells which in the Bryophyta differentiate into a **foot,** a **seta,** and a **sporangium,** and little else. These structures retain their identity and mature into the adult sporophyte.

In the Tracheophyta we shall see that the embryo starts out the same way but soon differentiates a meristematic stem apex, a leaf, and a root. Only the meristem region persists throughout the life of the plant; the initial leaf and root are gradually replaced by the adult roots and leaves. In nature many different patterns of development have been produced, but from an evolutionary view the development of the embryo has been quite conservative with remarkable sameness throughout the plant kingdom. However, with the development of vascular tissues an element of experiment and diversity has been introduced. This has resulted in considerable variation in the adult sporophyte in the Tracheophyta.

The Bryophyte Gametophyte

Members of this phylum are often described as the most primitive of terrestrial plants. They are amphibious in nature since they are dependent on water to a greater extent than most members of the Embryophyta to keep them moist and to provide a means for the sperm to reach the egg. They have retained these and other algal characteristics, consequently we find them growing, for the most part, along the edge of streams and water bodies or in heavily shaded forest areas where they are in little danger of dying out. Bryophytes never developed the corticating tissues which are so prominent in the higher successful terrestrial plants.

Within the narrow confines of their environment, Bryophytes are successful but, compared with the Tracheophyta, liverworts and mosses never have been able to occupy large areas of the land. They lack what we may refer to in ecology as an **area of dominance.** They never have become **climax** plants in large geographical areas; they are never the dominant plants in an **association.**

The emphasis on development in the Bryophyta has been on the gametophyte generation; the major portion of the Bryophyte body is haploid, just as are those of most members of the Thallophyta. The thallus developed by most members of the Bryophyta has a fair degree of differentiation but the tissues necessary for large growth were never formed. Vascular tissues for the conduction of water and nutrients were never developed. They receive their water and nutrient supplies by means of capillarity and osmosis. They never developed mechanical tissues for support. Body shape is maintained largely by cell turgor. Secondary meristems to give thickness and additional corticating cells were never formed. The thallus is essentially a thin-walled, more or less porous structure which could not long survive without moisture.

The greatest advance the Bryophytes have made has been in the formation of multicellular sex organs in which a sterile jacket of cells surrounds the fertile cells. The **archegonium** (Fig. 2) is a

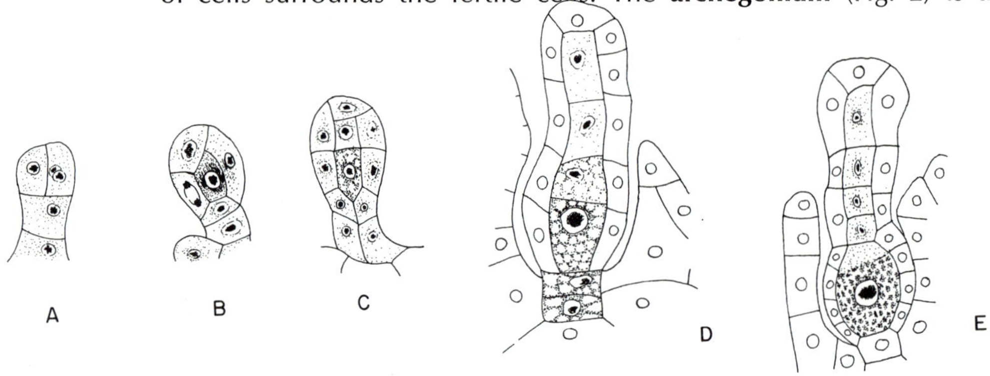

FIGURE 2 Development of the archegonium in *Marchantia* beginning with a two celled stage. In B the first central fertile cell is formed; C has two fertile cells, in D the egg cell and ventral canal cell are differentiated and in E four neck canal cells are present (from Hill, J. B. and L. O. Overholts, H. W. Popp, and Alvin R. Grove, Jr. 1960, Botany. McGraw-Hill Book Co.)

flask-shaped structure in which a single **egg,** a **ventral canal cell,** and associated **neck canal cells** are enclosed in a jacket of sterile cells. The basal expanded portion of the archegonium, the **venter,** is usually deeply imbedded in the thallus but the neck is freely exposed where it may be readily reached by the swimming sperms. After fertilization within the archegonium the embryo sporophyte develops within the gametophytic tissue. The sporophyte may develop almost completely enclosed by gametophyte tissue or remain attached by only the foot, but it never leaves the gametophyte.

The antheridium (Fig. 3) is spherical or oval in shape and also has a protective layer of sterile cells enclosing the spermatogenous cells. The latter are cuboidal at maturity and each gives rise to a pair of biflagellate sperms (Fig. 4). The antheridia are also rather deeply imbedded in thallus tissue and often are open to the exterior only by means of a tiny pore through which the sperms are released.

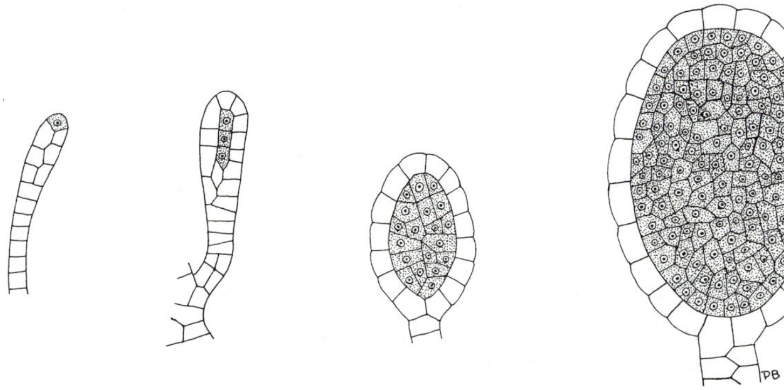

FIGURE 3 Stages in the development of an antheridium in *Porella*.

FIGURE 4 Biflagellate sperm of *Marchantia*.

FIGURE 5 Section of *Marchantia* thallus showing stomate with its four tiers of cells arranged chimneylike; the loosely organized upper portion of the thallus; the more compact lower portion; the unicellular rhizoids and multicellular scales. Redrawn from *Laboratory Outlines*.

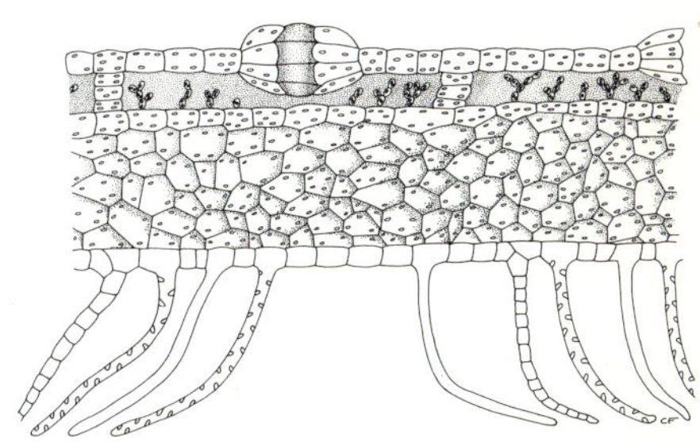

CLASS HEPATICAE

The liverworts have gametophytes of two general types. The most typical is the broad, flat, dichotomously branched thallus which grows dorsiventrally along the surface of the ground. The second includes the leafy-stemmed liverworts, which have two rows of tiny blades arising from a slender axis. The blades arise on opposite sides of the axis, giving the plants a flat appearance. They grow oriented in a prostrate position but may be supported above the ground by nearby vegetation. Most of the leafy liverworts also produce a third rudimentary row of leaf-like protrusions on the ventral side or underside of the gametophyte.

The thallus of liverworts has very little tissue differentiation, but it has developed distinct upper and lower surfaces, which gives it a dorsiventral orientation (Fig. 5). The upper epidermis contains multicellular structures, analogous to stomates, which open into air chambers within the loosely arranged spongy layer between the epidermal tissues. The lower epidermis is not as distinct as the upper. From these cells arise **rhizoids** which look like transparent cylindrical hairs, and **scales** which are somewhat flattened multicellular structures. Rhizoids and scales serve as anchoring and absorbing organs. They provide capillary surfaces in the movement of solvents to various parts of the thallus. Chlorenchyma cells are found just beneath the upper epidermis; those cells nearer the lower epidermis have fewer chloroplasts.

Evolutionary Development

The invasion of the land (it seems to have been only a mild evolutionary skirmish) was accomplished by the development of this dorsiventral thallus. In all evolutionary processes the changes are gradual, the intervening time between changes enormous, and the immediate progress difficult to discern. Even in so small a group as the liverworts it is obvious that there was no single line of development from order to order or from genus to genus; instead there were many little alterations resulting in one plant with a more highly organized epidermis, in another a more compact thallus, in another more highly specialized stomates, and in another more strategically located sex organs, and still others were more successful in sporophyte development. As dozens of different modifications occurred, new species and genera appeared. Adaptations evolved gradually, first one feature emerging and then another. It would be difficult to conceive that all changes evolved together and simultaneously. Whatever their mode of evolution the changes that occurred and persisted all led in the same direction: toward a plant better suited for life on land.

The evolutionary changes that occur in organisms are spontaneous independent alterations in the germ plasm as a result of gene and chromosome rearrangements. Occasionally these alterations favor survival in the organism and are inherited from generation to generation. Gradually other innovations appeared, and these along with the old resulted in sufficient modifications to create new species. By continued propagation, additional genetic alterations, and cross-fertilization, other species were established. Some plants were favored for survival by the new characteristics. Others undoubtedly received fewer favorable genes which did not provide them with adaptations to meet the stresses of their environment; these organisms perished. An environment is demanding on its inhabitants; only organisms that adapt survive. Evolution through the behavior of the genes is the way adaptation is brought about.

Classification

The class Hepaticae is usually divided into three major orders: (1) the Marchantiales, (2) the Sphaerocarpales and (3) the Jungermanniales. We will consider a few genera of the first and third orders, but will omit the Sphaerocarpales since it is a very small order of only three genera and about thirty species. These species are of the thallose type and similar in many characteristics to the Marchantiales.

ORDER MARCHANTIALES

Riccia is one of the simplest of all liverworts because it develops a sporophyte which consists of little more than a sporangium. Both foot and seta are lacking (Fig. 6). The thallus is usually rather fleshy, dichotomously lobed and fan-shaped. Internally the cells are loosely organized with much intercellular space. The air chambers formed in the upper epidermis layer are not so highly specialized nor made up of so many cells as those in the other members of the order.

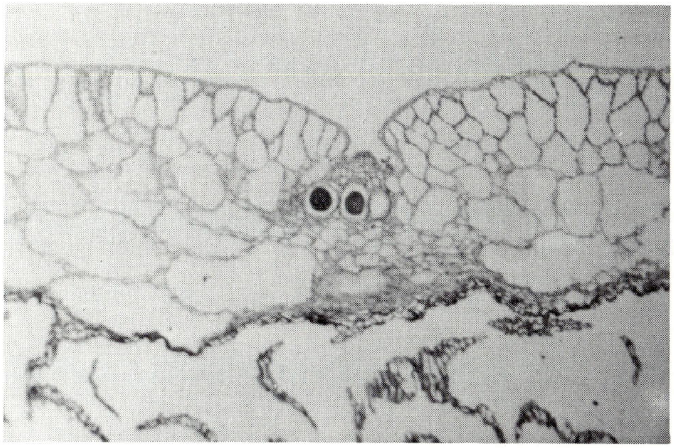

A

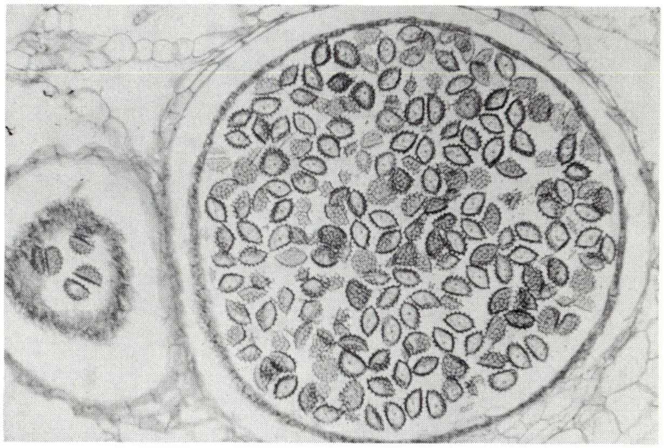

B

FIGURE 6 Cross section of the thallus of *Riccia* showing imbedded sporophyte. This is the simplest sporophyte in the liverworts consisting of nothing more than a sporangium and associated cells. The foot and seta are lacking in the genus. A: two sporophytes imbedded in the thallus. B: a section through a sporophyte with spores.

In the center of the thallus and its lobes is a fissure-like midrib within which the archegonia and antheridia are located. As the thallus grows the antheridia are produced first, followed a short time later by the archegonia. Consequently in the mature thallus the antheridia are towards the posterior portion and the archegonia nearer the anterior.

The sex organs arise from superficial cells but during their formation the thallus tissue grows around them, leaving only the tips of the antheridia and the necks of the archegonia free in the fissure. Following fertilization the zygote undergoes a series of divisions forming a spherical mass of sporophytic tissue.

Early in the enlargement of the developing sporophyte the haploid cells of the surrounding venter undergo periclinal divisions forming an enclosing wall two cells thick. These cells then keep pace with the growing sporophyte by additional increase in size and numbers. The sporophytic tissue is separated by periclinal divisions into an outer layer and an inner mass of cells. After a period of growth the inner cells round up, separate from each other, and become **spore mother cells.** These cells then divide meiotically, forming tetrads of haploid **spores.** The outer cells become the sporangial wall, while the venter wall remains as a protective tissue around the entire sporangium. In many bryophytes the enlarged archegonial wall is termed the **calyptra** and becomes a specialized cover. In *Riccia* this tissue remains buried in thallus tissue.

The spores of *Riccia* are liberated when the thallus around them disintegrates; there is no other method of spore dissemination in this genus.

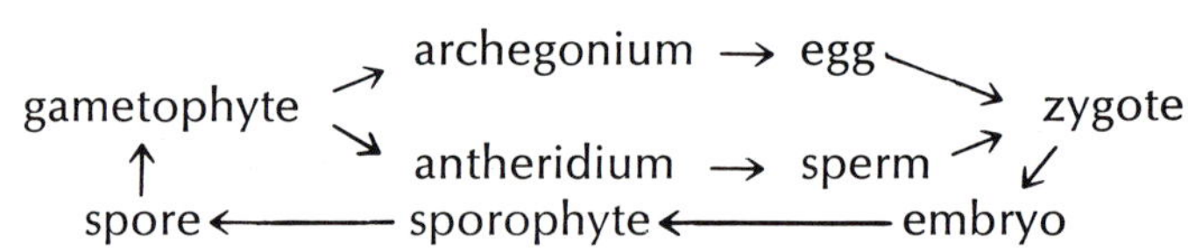

Generalized Outline of Liverwort Life Cycle

Marchantia is probably our most commonly studied liverwort and is frequently found growing on moist substrates in nature. It is several times larger in body surface area than *Riccia* but of about the same thickness. The thalli are rather elongated and covered by a well-differentiated upper epidermis marked by polygonal areas. Within the center of each polygonal area is a small pore which opens into a large air chamber below. Surrounding the pore is an arrangement of cells forming a multicellular border like a chimney around the opening. In *Marchantia* as in most other thallose liverworts the airchamber (Fig. 5) is so organized as to have both upper and lower openings, a double diffusion mechanism which may control the entrance and exit of gases at the bottom and top of the chamber.

The ventral surface, like that of the other thallose bryophytes, is abundantly supplied with scales and rhizoids. The scales are usually arranged in four or more longitudinal rows. They are multicellular, brownish in color, membranous in texture, and often somewhat papillate. Rhizoids are usually unicellular and may be either smooth or marked with tiny protuberances.

Asexual reproduction in *Marchantia* and in many other liverworts is usually accomplished by fragmentation. As the forking thallus grows forward from divisions of its apical cells, the older portions die and decompose, isolating the newly formed branches into separate plants. *Marchantia* and a few other genera also reproduce asexually by specialized bodies termed **gemmae** (Fig. 7). They are formed on the dorsal surface in specialized bodies termed **cupules.** The gemmae arise in the bottom of the cup and are attached there by small stalks. Each gemma is broadly two-lobed and in longitudinal section they are spindle-shaped. On each side a gemma has a notch which partially encloses a meristematic cell or cells. Once a gemma has been washed from the cupule new thalli may grow from its lateral growing points. Although new thalli may grow from both notches once one has started to grow, the other side seems to be inhibited and does not produce a thallus. The gemmae contain chloroplasts and may live independently for brief periods before developing into a new thallus.

Sexual reproduction in *Marchantia* occurs in a union of the oogamous gametes from two different plants, since the species are dioecious. The antheridia are formed in a disclike apical appendage borne on a stalk. The entire structure is termed an **antheridiophore** (Fig. 8). The antheridia develop in individual cavities which open to the exterior at the top (Fig. 9). The oldest antheridia are near the center of the disc and the youngest towards the periphery.

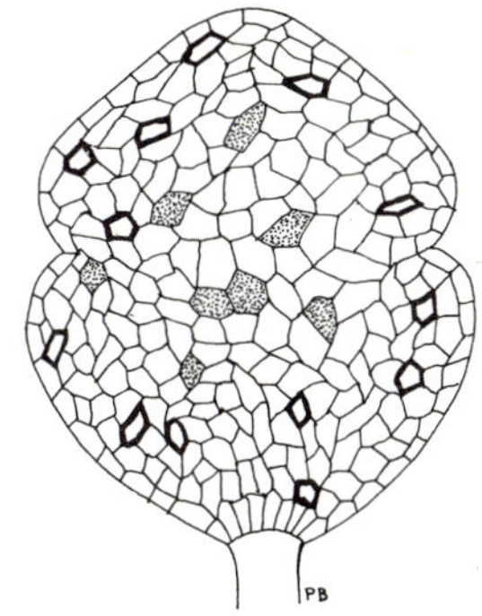

FIGURE 7 A single gemma of *Marchantia* in dorsal view.

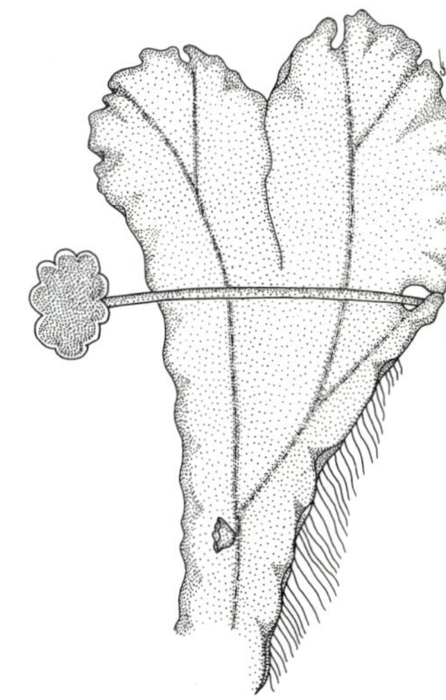

FIGURE 8 *Marchantia*-Male plant showing antheridiophore and one cupule on thallus.

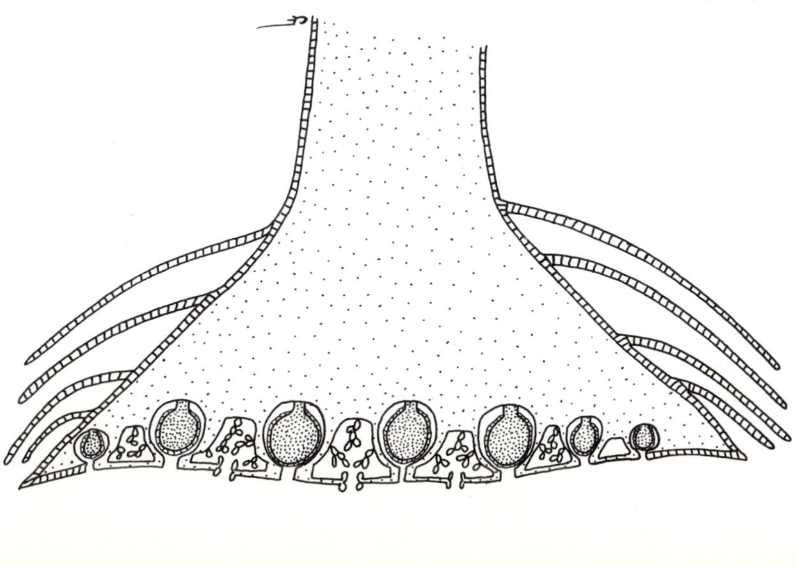

FIGURE 9 Section through an antheridial disc of *Marchantia* showing enclosed atheridia.

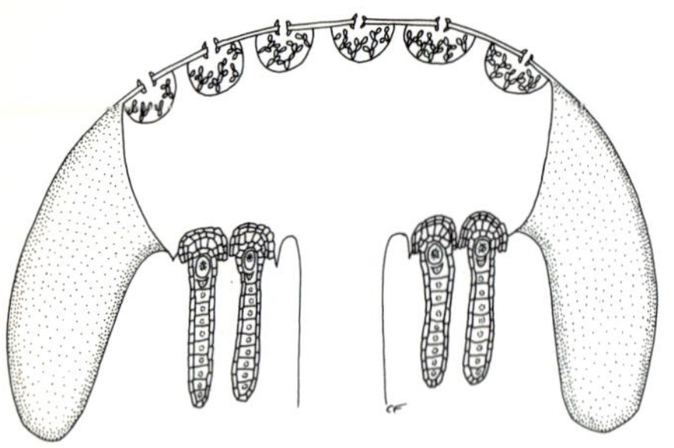

FIGURE 10 Section through the archegoniophore apex with four archegonia.

Sperms appear in the cubical chambers within each antheridium. A single sperm is an elongated, slightly spiral band which tapers to a point at the anterior end. Attached to this end are two flagella which are pointed backwards over the main axis of the sperm.

The archegonia arise on the megagametophytic thallus in a specialized disc. This structure is easily distinguished from that of the male since it has fingers radiating out from the central disc. The entire structure with its supporting stalk is an **archegoniophore** (Fig. 10).

Although the archegonia appear to arise and hang down from the ventral surface of the archegonial disc, they actually arise from an inturned portion of the dorsal surface in such a way that the younger archegonia are towards the central stalk region and the older ones toward the margin. Before the stalk elongates and while at least some of the archegonia are virtually touching the thallus surface, sperms enter the archegonia and fertilize those eggs that

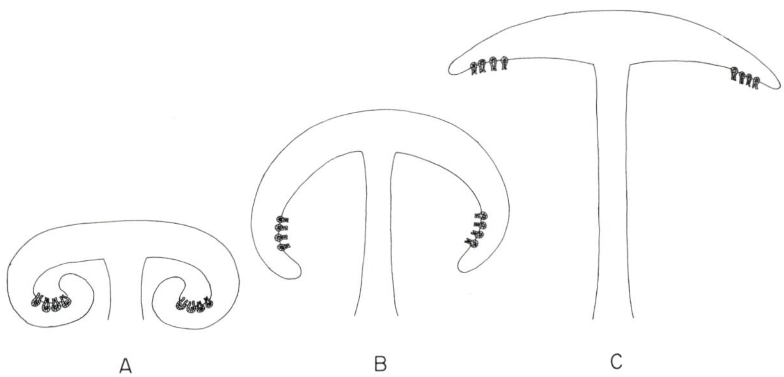

FIGURE 11 Diagrammatic illustration of archegoniophore of *Marchantia* showing in A the inverted position of the archegonia at time of fertilization and in B and C elongation of the stalk and final position of the archegonia as the sporophyte develops.

have matured. Transfer of the sperms requires a continuous water column or passage which is easily broken once the archegoniophore has reached full stature.

When a zygote is formed it begins to develop into the embryo (Fig. 12), which then matures into the sporophyte and remains suspended from the archegonial ray. The jacket of the archegonium enlarges, forming a temporary membranous sac around the embryo. As the sporophyte elongates it pushes through the somewhat enlarged archegonial jacket and the surrounding gametophyte tissue.

Growth of the sporophyte in *Marchantia* (Fig. 13) is marked by much greater differentiation than in *Riccia*. In *Marchantia* the basal cells differentiate into an anchoring and absorbing structure, the foot, while those at the apical end differentiate into an inner

FIGURE 12 *Marchantia*-development of the sporophyte. A-zygote, B-two-celled stage, C-undifferentiated multicellular stage, D-early differentiation of cells into f-foot, s-seta, sp-sporangium.

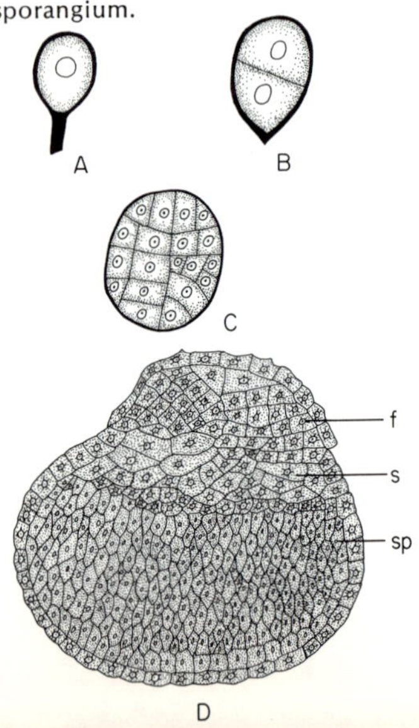

mass of sporogenous tissue, derived from an **endothecium;** an outer layer, the **amphithecium,** forms the sporangial wall or **capsule.** Between the capsule and foot the cells divide and elongate forming a stalk or **seta.** Within the sporangium the sporogenous cells are arranged in rows of four to eight spore mother cells, each of which gives rise to a tetrad of spores by meiosis. Between the tiers of potential spore mother cells are other sporogenous cells which become greatly elongated and tapered at each end. Spiral thickenings, the result of secretions, are on the inner surfaces of sporogenous cell walls, following which their protoplasts disintegrate forming springlike rods. These are **elaters.** The spiral thickenings give elaters a hygroscopic property in the rather violent discharge of spores once the capsule breaks open. Springlike in their actions, the elaters often hold the spores together in clusters and catapult others some distance from the parent plants. An outline of the liverwort life cycle is given below.

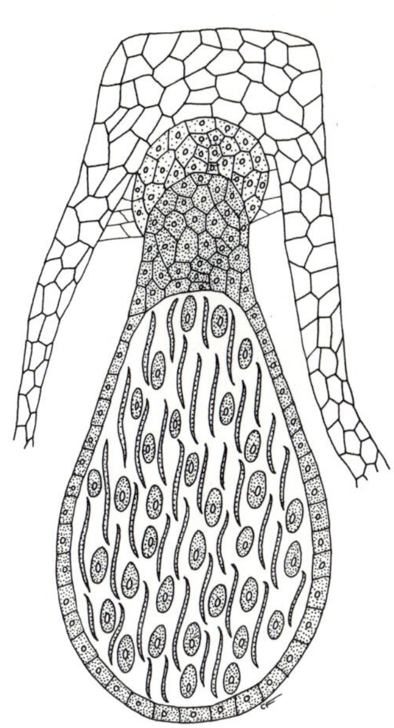

FIGURE 13 *Marchantia* sporophyte with imbedded foot, seta, and sporangium.

Riccia

Thallus
(Gametophyte)
/ \
Archegonium Antheridium
\ /
Egg Sperm
\ /
Zygote
|
Embryo sporophyte
|
Sporangium
|
Spore mother cells
|
Spores

Marchantia

Female thallus (Megagametophyte) | Male thallus (Microgametophyte)
| |
Archegoniophore | Antheridiophore
\ /
Archegonium Antheridium
\ /
Egg Sperm
\ /
Zygote
|
Embryo sporophyte
|
Mature sporophyte
(foot, seta capsule)
|
Sporangium
|
Spore mother cells
|
Spores

The Life Cycles of Riccia, a Simple Liverwort, and Marchantia, a More Advanced Genus

Conocephalum looks very much like *Marchantia,* but the central stomate pores of the latter are in diamond-shaped areas while in *Conocephalum* they are oblong-hexagonal near the margin and broader hexagonal towards the center. Gemmae are not produced

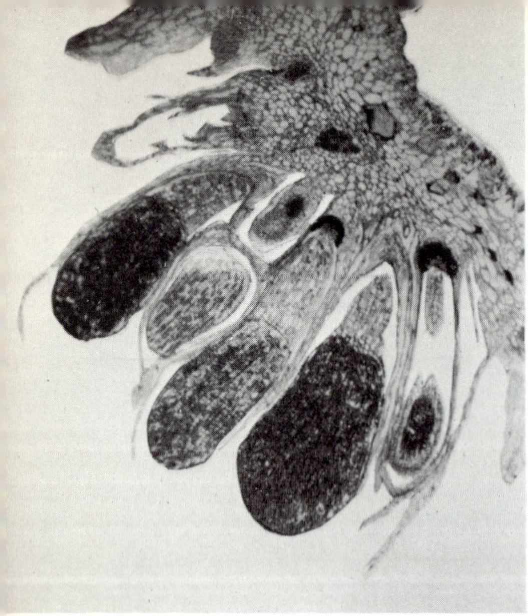

FIGURE 14 Sporophytes of *Conocephalum;* note elongated capsule and setae and relatively small foot.

in *Conocephalum,* the antheridial receptacles are sessile, and the female archegoniophore is more conical. The sporophytes (Fig. 14) in the two genera are similar except that in *Conocephalum* the foot is smaller and the capsule is much more elongate.

Order Jungermanniales

Members of the Jungermanniales are far more numerous in numbers of species than those of the Marchantiales. Most of them inhabit soils of the tropics and subtropics, but many are also found on moist substrates in more temperate latitudes.

Because of the large number of species the Jungermanniales are divided into two series based mainly on the manner in which the archegonia are differentiated. **Anacrogynae** is the term used for those species in which the archegonia originate from a cell or cells behind the apical cell. This places the archegonia, and therefore the sporophytes, on the dorsal surface of the vegetative body. Most of the anacrogynous Jungermanniales are thallose but a few are leafy. The second series in the order are the **acrogynae.** In this group the archegonia arise from the apical cell or one of its segments. Thus the archegonia and the sporophytes occupy a terminal position on the fruiting thallus. The acrogynous members are nearly all leafy and in numbers of species far exceed those in the anacrogynous series.

FIGURE 15 *Pellia.* Branched thallus with dehisced sporangium on the sporophyte at upper left and unopened sporangium on the young sporophyte at right.

Anacrogynae

Pellia (Fig. 15) is one of the best known genera in the Anacrogynae series. The gametophyte is dichotomously branched with several forks per plant, and the thallus is dorsiventral as in the Marchantiales. Rhizoids arise in the thallus tissue. Their position can be easily ascertained by the slightly raised conical areas, below which are the antheridia. Sperms escaping from the antheridia emerge through a tiny pore at the apex of each cone.

Archegonia usually arise in groups of four to twelve along the midline of the dorsal surface. They are produced in cuplike or tubular cavities and are more elongated than those of *Marchantia.* In *Pellia* the number of neck canal cells may be as high as eighteen.

The sporophyte of *Pellia* (Fig. 16) may reach a height of ten centimeters, almost entirely attributable to its very long seta. In its early growth, in the venter, the zygote divides transversely forming two cells. The lower usually develops into an anchoring appendage, while the upper continues to divide, forming a row of cells before vertical walls appear to give thickness to the sporophyte. The lowest cell in the initial row continues growth and division, forming an anchoring and absorbing foot, while the upper cells form a capsule. The cells between the capsule and the foot divide rapidly and undergo extensive elongation, carrying the ripening sporangium several inches above the thallus. In the early development of the capsule, sporogenous cells toward the periphery of the capsule elongate and ultimately form elaters.

When the capsule is mature and dry it splits open along four thin lines from apex to base, which throws the four valves outward violently discharging the spores. Shortly before the spores are shed, the nucleus and cytoplasm of each undergo a few divisions and form a small amount of prothallial tissue. This is a type of "head start" program exhibited by other members of the Bryophyta as well. The spores have thus started on thallus development before they reach the habitat where they may continue growth into a thallus.

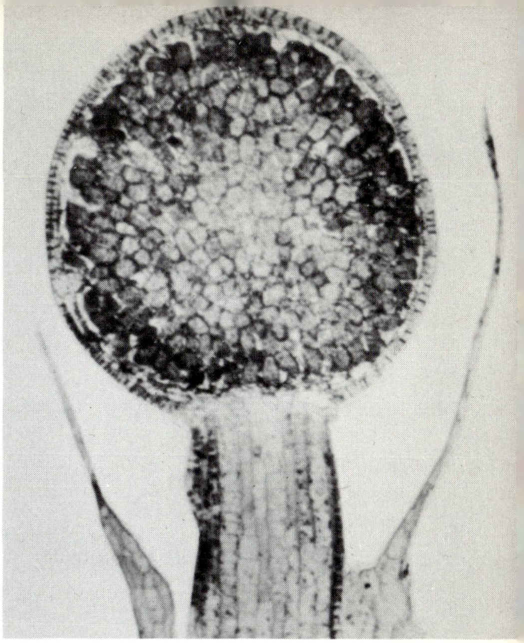

FIGURE 16 *Pellia,* portion of sporophyte showing seta and globose capsule.

Acrogynae

The reproductive organs of the leafy liverworts are essentially the same as those of the anacrogynous series, but the vegetative parts strongly resemble those of the true mosses except that the leaflike organs of the liverworts lack a midrib. The leaves are flattened into two rows on either side of the main axis, and are frequently cleft or lobed. In some species the blades overlap, but in all the characteristic bilateral symmetry of the liverworts, in contrast to the radial symmetry of the true mosses, is apparent.

A few of the acrogynous species have rhizoids arising along the length of the main axis, but for the most part rhizoids are not numerous. Water moves fairly easily by capillarity along the main axis from leaf to leaf and branch to branch.

Porella (Fig. 17) is a typical member of this group. It has a much branched thallus composed of a central stalk bearing numerous tiny blades. Many species occur throughout much of the central, northern, and eastern part of the United States. The plants form thick mats on logs, rocks, and rich soil wherever there is protective cover from the direct rays of the sun. The plants have many short branches arising from the ventral surface of the main axis in a manner of pinnate division; tertiary branches may also be produced in the same manner.

FIGURE 17 *Porella*. Diagram of a vegetative portion showing three young branches and the two rows of blades.

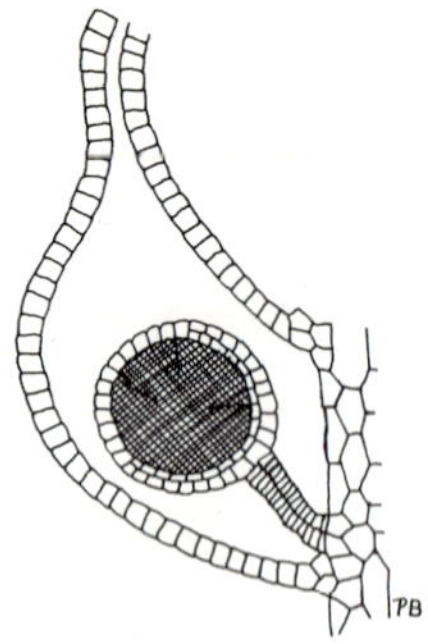

FIGURE 18 *Porella*-Antheridium.

In the heterothallic species the male plants are smaller. Antheridia arise on short stalks in the axils of a few of the blades (Fig. 18). The antheridium arises from a superficial initial cell which first divides transversely, forming a basal cell that becomes the stalk while the outer develops into the globose antheridium. As the antheridium matures the wall becomes two and three cells thick towards the base but remains single-celled at the apex. The spermatogenous cells formed within gradually mature into the motile sperms.

The archegonia (Fig. 19), which arise from a cell close to the apex, occur at the ends of very short branches. Normally several archegonia are produced together, but following fertilization of

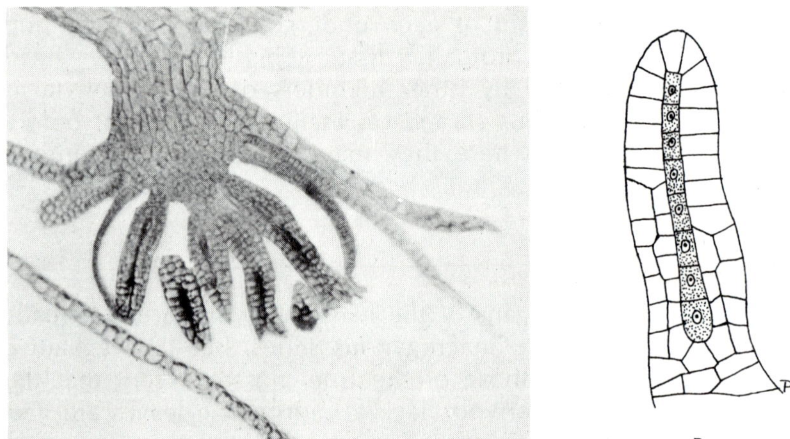

FIGURE 19 A: Photomicrograph of female branch of *Porella* showing a number of archegonia. B: *Porella* archegonium.

the egg in one, development of the others is inhibited. In certain respects the development of the archegonia is like that of a female moss plant. The archegonia are clustered, they are more cylindrical than flask-shaped, and usually only one develops a sporophyte. After an egg has been fertilized the archegonium is surrounded by a saclike overgrowth which is termed the **perianth.**

After fertilization the zygote undergoes a series of divisions to form a foot, seta, and capsule. The foot is bullet-shaped, the seta is quite short, and the capsule is relatively large and globose (Fig. 20). The wall of the capsule may be several layers thick, especially at the base, but the cells in general are rather irregularly thickened in four areas so that the capsule separates in quarters in response to changes in the humidity and the action of the capsule's hygroscopic properties. The spores are also aided in their dissemination by elaters, which also respond to changes in the humidity.

FIGURE 20 *Porella.* Young sporophyte still imbedded in gametophyte tissue.

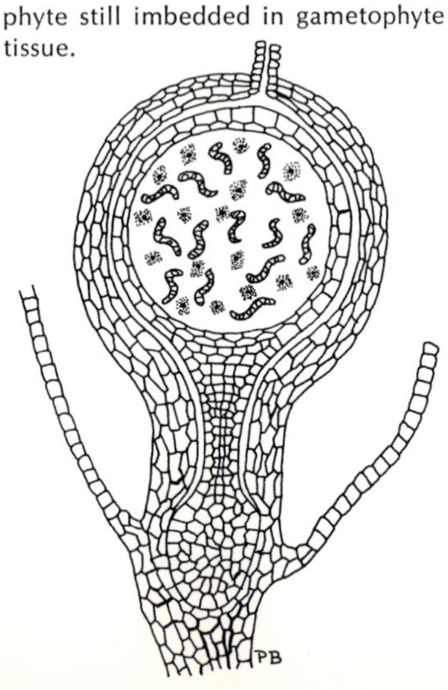

The elongation of the seta is much less pronounced in *Porella* than in such genera as *Pellia*. However, it does elongate and furthermore it grows upward and around the base of the developing capsule. The latter is therefore protected by the perianth, the calyptra, a portion of the seta, and the ensheathing apical leaf-blades.

The acrogynous Jungermanniales have achieved a greater degree of specialization in the gametophyte body, and there is a greater amount of sterile tissue in the capsule. As in the other members of the phylum the sporophyte consists of foot, seta, and capsule. The capsule is either globose or cylindrical and usually dehisces along four lines. The spore mother cells are conspicuously four-lobed, and at maturation elaters are formed along with the spores. The seta is more or less prominent, but it always remains undifferentiated in tissue organization. It is the opinion of botanists that the members of this group have the characteristics, both in gametophyte and sporophyte, to be ancestral to the mosses.

CLASS ANTHOCEROTAE

Order Anthocerotales

If the members of this class and order were classified according to their gametophytes they would be placed in the Marchantiales next to *Riccia,* for the thallus they form is one of the simplest known in the Bryophyta. However, the sporophyte is considered the most advanced in the entire group, thus making it necessary to place them in a separate and distinct class.

The genus *Anthoceros* (Fig. 21), which means "flower horn," is one of the four genera in the entire order. It is found in typical

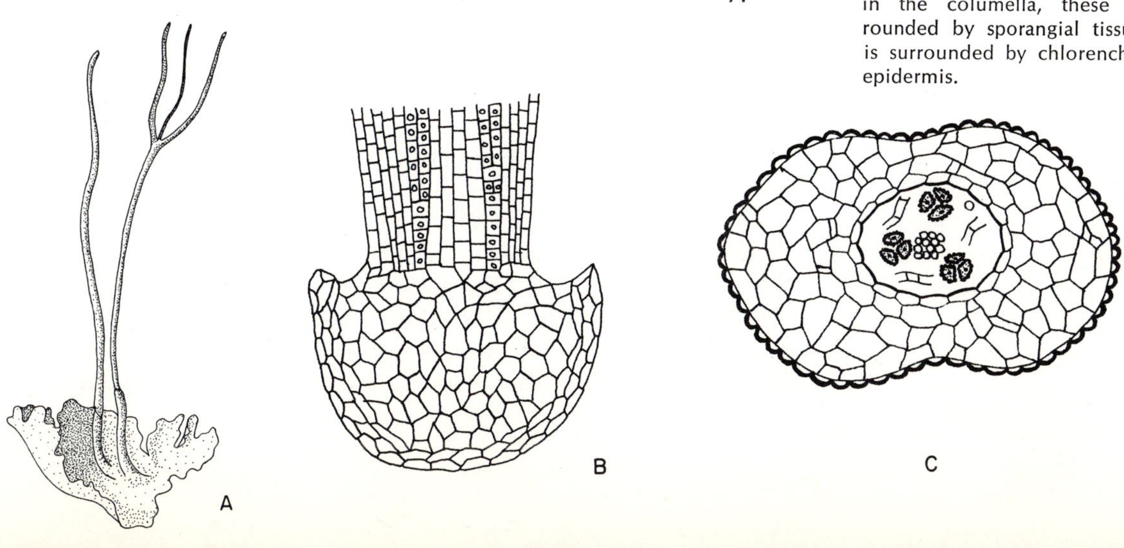

FIGURE 21 *Anthoceros.* A-Thallus with two sporophytes; the one on the right is dehiscing apically to release the spores which ripen first at the apex. The central strand in this sporophyte is the columella. B-Foot of the sporophyte and differentiation of tissues longitudinally. C-A cross section of the mature sporophyte. The central few cells are in the columella, these are surrounded by sporangial tissue which is surrounded by chlorenchyma and epidermis.

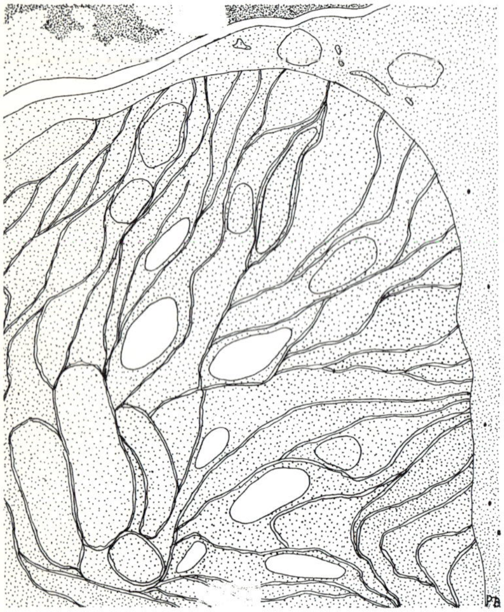

FIGURE 22 A portion of a chloroplast (cross section) in the sporophyte of a liverwort. Drawn from an electron micrograph in Bonner, James and J. E. Varner, 1965, *Plant Biochemistry*, Academic Press.

liverwort habitats wherever the climate is not too cold and moisture is adequate. Most species of the four genera live in the tropics and subtropics, but some species of *Anthoceros* and also some species of *Notothylas* live in temperate areas.

The gametophyte of *Anthoceros* is often less than one-half inch in diameter, rather irregularly dichotomously branched, orbicular in shape, and dorsiventral in orientation. Each cell in the thallus usually has one large chloroplast with a single pyrenoid, a characteristic of many algae (Fig. 22). Anchorage to the substrate is gained by the production of smooth-walled rhizoids from the ventral surface but scales are absent. Within the thallus the amount of air spaces is considerably reduced giving the thallus a more compact structure than those of most members of the Marchantiales.

Vegetative reproduction may occur by fragmentation but gemmae are not produced. In some species tiny tuberlike growths arise on the thallus. These tiny growths may then separate from the main body and give rise to a new thallus. This tuberlike growth is analogous to the soredia of lichens but the similarity is superficial.

The sex organs of *Anthoceros* are produced in small grooves on the dorsal surface imbedded in thallus tissue. Usually both archegonia and antheridia are produced on the same thallus, but in some species the sexes are separate. In either case the sex organs arise from superficial initial cells just back of the marginal apical cell.

Usually several eggs are fertilized on a single thallus and these zygotes then continue growth into mature sporophytes. The sporophyte consists of a bulbous foot (Fig. 21) and a columnar capsule. Early in its development the capsule is divided into two regions. The inner differentiates into a central axis of sterile tissue, the **columella,** and the outer forms the sporogenous tissue.

When mature, a cross section of this cylindrical sporophyte reveals a central columella surrounded by a cylinder of sporogenous tissue. Surrounding the sporogenous tissue are several layers of cortical chlorenchymatous cells which in turn are covered by an epidermis.

As the sporophyte matures the spores towards the apex ripen first, with a progressive maturation from the top to the bottom of the columnar capsule. The latter dehisces along two lines separating the structure into two valves (Fig. 21). As the two valves peel back, the ripened spores are liberated from around the central columella, leaving the latter projecting from between the two halves. Germination of the spores in dioecious species apparently results in an equal production of male and female gametophytes, since it has been reported that two spores of a tetrad have male chromosome determiners and two have the complementary female genotype.

The most significant advance *Anthoceros* has made over the preceding groups is in the greater amount of sterile tissue formed in the sporophyte and the fact that it has an intercalary meristem just above the foot which pushes new growth upward. The earlier hepatics had capsules which were nearly all sporogenus tissue, but in the Anthocerotales the sporophyte is well supplied with chlorenchymatous tissue. Although the chlorophyll present is enough for the plant to carry on an independent existence, it never became so highly modified that it could live independently since the sporophyte never developed roots or rootlike organs. The sporophyte always remained dependent on and attached to the gametophyte. To be a successful independent land plant the sporophyte of *Anthoceros* needed vascular tissue instead of a columella and absorbing organs which could be rooted directly into the soil; these structures never evolved in this group.

Supplementary Readings and Questions for Discussion for this chapter follow Chapter 21.

21

The Gregarious Mosses

Phylum Bryophyta

Class Musci

Mosses are among our most interesting plants but it requires a hand lens and a prostrate position to appreciate them and the world in which they live. They are a very gregarious group, growing side by side in dense mats but never occupying large areas. Living with them are numerous worms, insects, snails, fungi, algae, and bacteria—a microcosm that is quite different from the habitat of the larger forest organisms just a few feet away. Although mosses are not of great economic importance themselves, they help to build the soil, store water, retard erosion by decreasing surface runoff, and provide food and shelter for numerous microorganisms.

The most important of the mosses economically is *Sphagnum*. The accumulated remains of *Sphagnum* species in northern latitudes build peat, which serves as a fuel and as an absorbent packing material for trees, shrubs, and herbs which must be kept moist in shipment.

General Characteristics

The gametophyte body of the typical moss emerges from a germinating spore much like a growing algal filament. This structure, the **protonema** (Fig. 1), grows horizontally on the moist soil surface, branching frequently and giving rise to many rhizoids on its ventral surface and tiny **buds (bulbils)** on its dorsal surface (Fig. 2). These buds gradually elongate and develop into vertical

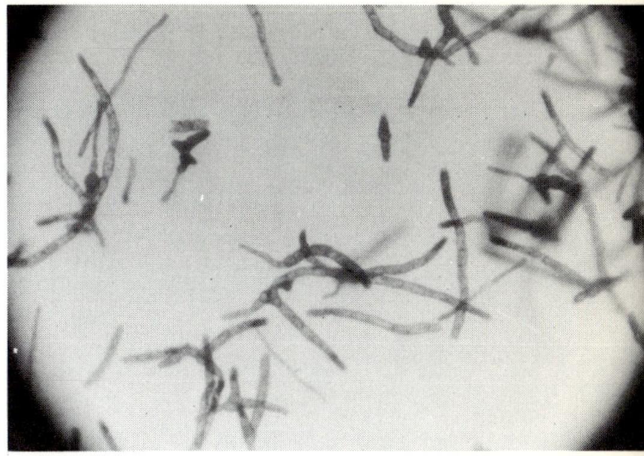

FIGURE 1 The formation of moss protonema from germination spores.

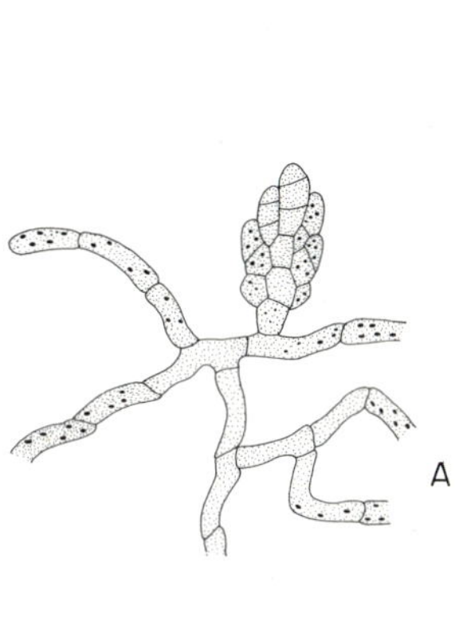

FIGURE 2 A-Protonema giving rise to bulbil which will develop into the leafy gametophyte in B. A sporophyte is also shown growing from the gametophyte. The calyptra covers the capsule.

branches bearing the minute leaves. Technically neither the "buds" nor the "leaves" are organs of this type as defined for the vascular plants, but they do resemble them and they have essentially the same function.

The growing point of the vertical axis continues to add stature to the growing gametophyte until the maximum size for the species is reached. This may be only a few millimeters or several inches in height. At maturity the gametophyte begins to produce archegonia and/or antheridia in its apex.

Mosses may be monoecious or dioecious but in both cases the sex organs arise in much the same manner. The antheridia are usually very much elongated and are borne on short stalks that arise from the sterile tissue of the shoot. As usual, the antheridia have a sterile jacket enclosing a large number of cubical spermatogenous cells. Each cell at maturity gives rise to one or two minute flagellated sperms. When ripe the antheridium absorbs a great deal of water, causing it to burst and free the sperm. Water held between the moss leaves and between the adjacent plants provides the necessary medium for sperm to swim to nearby plants where they may fuse with the egg.

Archegonia may arise in the same apex with antheridia or they may appear alone in the dioecious species. Like the antheridia the moss archegonia are usually very much more elongated than those of the liverworts and the egg is surrounded by one or more layers of sterile cells. At the time the egg is ready for fertilization the ventral canal cell and neck canal cells disintegrate into a mucilaginous substance from which sugars diffuse into the surrounding film of water, attracting the sperms. Apparently sugar is in greater concentration in the venter near the egg, and this attracts the sperms down the neck and into the venter where a male gamete may reach the egg and transform it into a zygote.

As in all members of the Embryophyta, the zygote begins at once to form the embryo sporophyte. The first phase of differentiation begins with the bulbous foot establishing a basal anchorage and absorbing organ in the gametophyte apex. With continued growth the seta elongates and the tip begins to increase in size and finally differentiates into the capsule.

From the standpoint of specialization, the moss capsule is the most complex organ found in the Bryophyta. But like the volvocine line of evolution in the green algae, which ends with the very complex *Volvox* colony, the capsule of the moss never gave rise to anything else in the evolutionary series.

The capsule in true mosses is usually composed of a central **columella,** surrounded by a cylinder of nutritive tissue, the **tapetum,** then a cylinder of sporogenous cells, followed by another cylinder of tapetal cells (Fig. 3). In many mosses only one tapetal

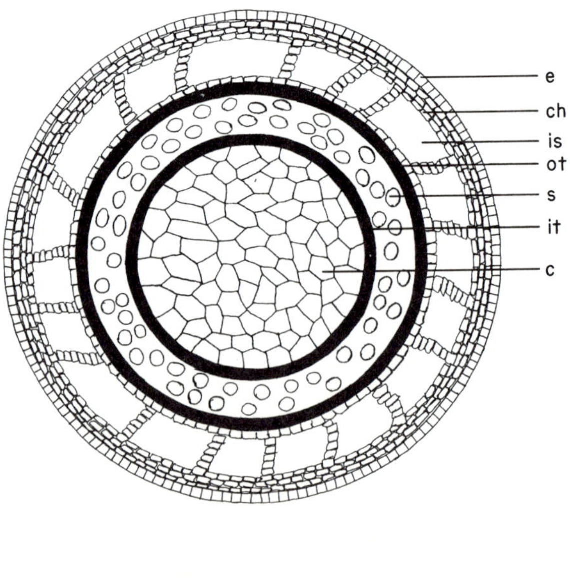

FIGURE 3 Cross section of a moss capsule, such as that of *Polytrichum,* showing e-epidermis; ch-chlorenchyma; is-intercellular space; ot-outer tapetum; s-sporogenous tissue; it-inner tapetum; c-columella.

cylinder is formed, but others have the two layers as described here. Surrounding the outer tapetum many mosses have an area of chlorenchymatous tissue which is rather loosely organized owing to the large amount of intercellular space. Outside this chlorenchymatous area is an epidermis which usually contains a number of small stomates.

One of the most specialized structures in the moss capsule is found at the apex. Usually there is an **operculum** or lid which covers the cylindrically arranged tissues. Just below and enclosed by the lid is a **peristome** (Fig. 4), which is made up of a number of filamentous teeth that open and close in response to changes in the humidity after the operculum is removed. The peristome is another of many highly specialized structures in plants which aid in spore dissemination. These are the tissues, all diploid, which make up the moss capsule, but in addition some mosses have a loose-fitting hood covering the entire capsule. This, the **calyptra,** is formed by enlargement of the archegonial jacket and is therefore haploid in chromosome count.

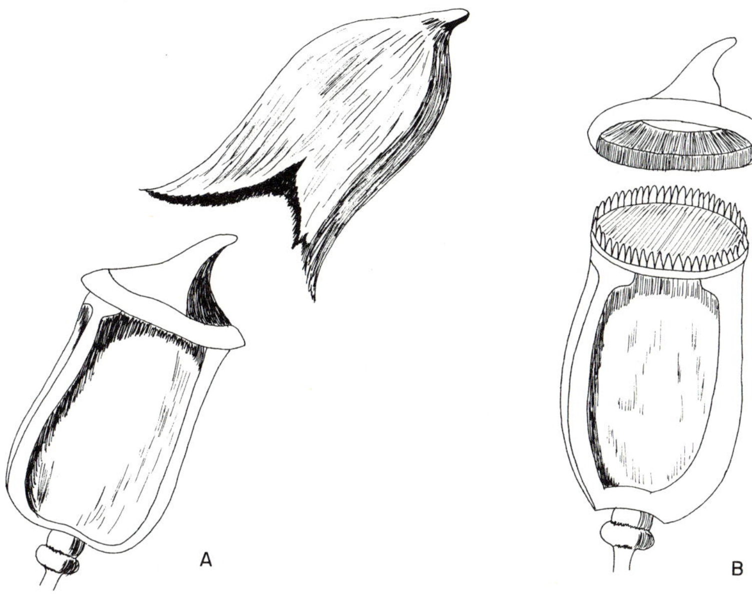

FIGURE 4 Moss *(Polytrichum)* capsule. A-capsule with Calyptra removed; B-operculum removed exposing teeth of the peristome.

The preceding description of the typical moss is mainly about members of the order **Bryales,** the so-called true mosses. Like the class Hepaticae, the Musci are generally divided into three orders. In addition to the Bryales we also have the **Sphagnales** and the

Andreaeales. We will discuss in detail a few members of the first two and omit the Andreaeales except to say that they are the granite mosses and are usually all classed in the single genus *Andreaea.* They live on rocky substrates, and on the basis of their characteristics occupy an evolutionary position about midway between the Sphagnales and Bryales and probably were derived from the leafy Jungermanniales.

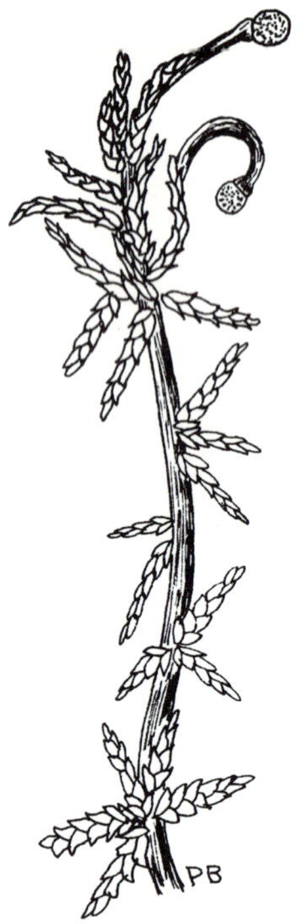

FIGURE 5 *Sphagnum* showing branching gametophyte and two attached sporophytes.

ORDER SPHAGNALES, THE PEAT MOSSES

General Description

An indication that the peat mosses have arisen from the liverworts is exhibited in the germination of the spore. Instead of the filamentous protonema typical of the other mosses a flat, greenish thallus, one cell layer in thickness and bearing numerous septate rhizoids on its ventral surface, is formed. After this thalloid protonema is formed, the members of the genus *Sphagnum* form the leafy shoot from the protonema and the latter dies. As the shoot develops numerous branches arise laterally, but the main gametophyte body continues to grow year after year. The plants grow closely matted together, forming a vegetative cover over the water and creating the quaking bogs in the north temperate latitudes and beyond.

The plants produce two types of branches, which are described as **excurrent** and **decurrent** (Fig. 5). The latter hang pendulously and are usually twisted about the main and secondary axis. The excurrent branches project outward and grow horizontally from the main axis.

Each branch has an apical cell with three cutting faces which gives rise to three rows of spirally arranged leaves. As the leaves begin to form all cells are of about the same size and shape, but as they mature some become greatly enlarged, rhomboid in shape, and hyaline. These cells lose their protoplasm and develop spiral or reticulate markings in the wall with rather large circular pores. The cells are known for their capacity for imbibing and retaining large quantities of water, which is the reason *Sphagnum* is used in packing materials for shipment where water retention is necessary.

Surrounding the large hyaline cells are small but elongated chlorophyllous cells that carry on the usual metabolic functions of the leaf (Fig. 6). The combination of small green cells and large

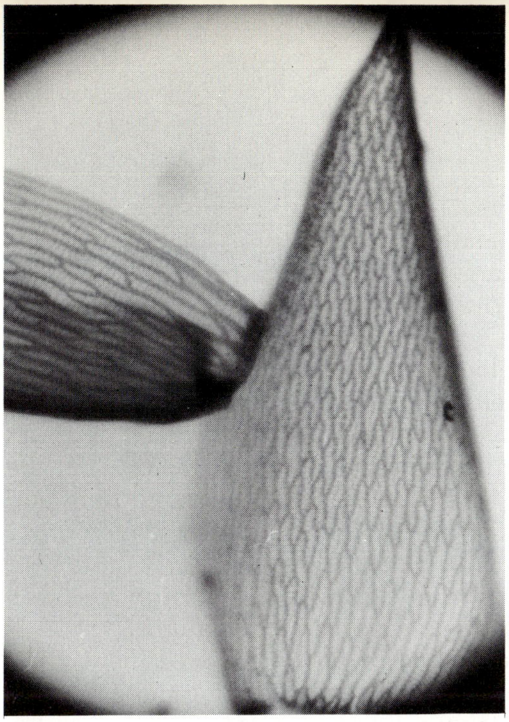

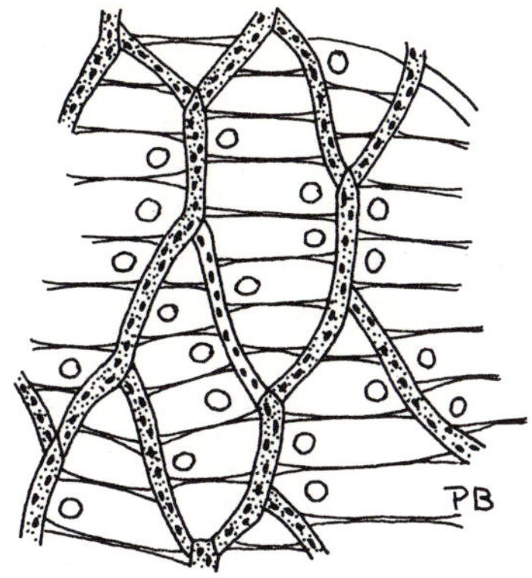

FIGURE 6 A-Cells from the leaf of *Sphagnum*. The large cells are hyaline with markings that appear as cross bars, the small circles are pores. B-The small stippled cells are chlorenchymatous.

hyaline units gives the species of *Sphagnum* their characteristic light green color.

The stem of *Sphagnum* is differentiated into (1) a central axis of thin-walled cells forming a pithlike storage region (Fig. 7); (2) a cylinder of elongated cells, usually about two or three cells thick, with slightly thickened walls; (3) a cortical region of very much enlarged, thin-walled, hyaline cells capable of substantial water storage; and (4) an irregularly formed and poorly differentiated epidermis. The stem serves as a good storage organ, but its primary function seems to be asexual reproduction, in which older plant parts die and new growth continues on the terminal portions.

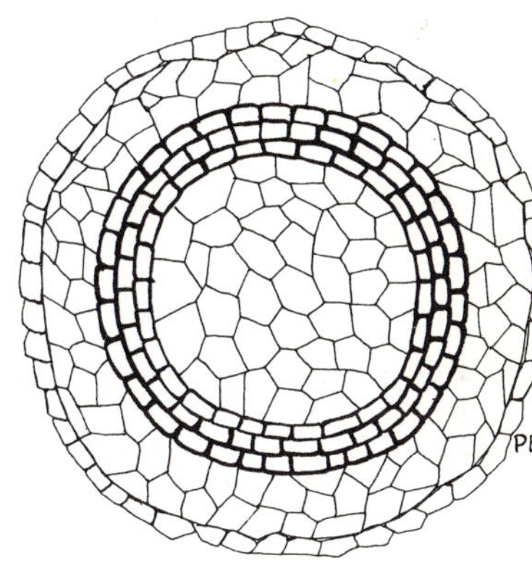

FIGURE 7 *Sphagnum* cross section of branch; note thin-walled central area, surrounded by three rows of thick-walled cells, surrounded by several rows of large thin-walled cells and a covering of irregularly formed epidermal cells.

Sex Organs

The archegonia arise as they do in the acrogynous members of the Jungermanniales, from the apical cell or in segments of it when more than one archegonium is produced on a branch. The archegonia usually contain eight neck canal cells, are relatively long, and have a long supporting stalk. Antheridia also arise from an apical cell, and usually only one is produced in axils of leaves

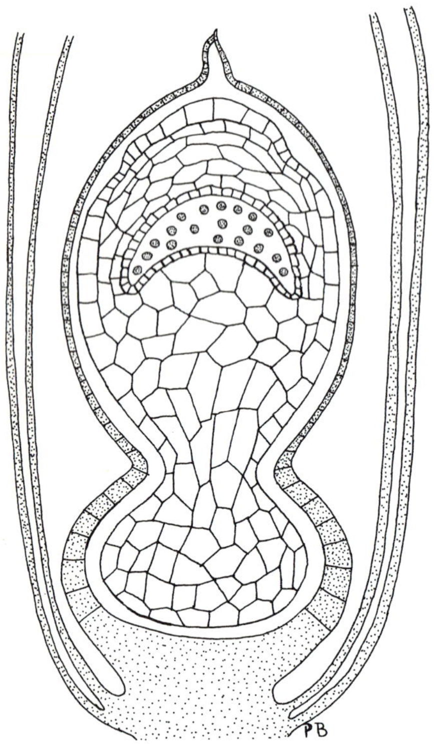

FIGURE 8 *Sphagnum* sporophyte. The sporophyte is the cellular portion drawn in the center. The stippled portion is gametophyte. The latter forms a short stalk and apex in which the sporophyte foot is imbedded and which is termed the pseudopodium. The stippled cellular layer immediately around the sporophyte is the calyptra.

on the short lateral branches. The antheridia at maturity are globose, supported on comparatively long stalks, and at maturity dehisce irregularly to discharge the biflagellate sperms.

The sporophyte has characteristics of both the liverworts and the true mosses and even a characteristic or two of the Anthocerotales. In its initial development the first embryonic cells form a filament of about eight cells. The apical half of this filament differentiates into the capsular portion and the basal half becomes the foot and stalk (Fig. 8). The foot develops most rapidly, but the seta never becomes more than a neck between the capsule and foot. Perhaps in compensation for this omission, the gametophyte develops a **pseudopodium** in which the foot rests. This stalk of haploid tissue elongates rapidly, elevating the sporophyte well above the remainder of the vegetative gametophyte.

Growth of the capsule is marked by a very prominent columella which forms as it does in *Anthoceros*. Capping the columella is the dome-shaped sporogenous tissue, a tapetal layer over the spore-bearing region, and several layers of chlorophyllous cells (Fig. 8). At the apex an operculum differentiates which dehisces when the spores are matured. The spore mother cells give rise only to spores. Elaters are not differentiated, nor is a peristome formed beneath the lid as it is in true mosses. The *Sphagnum* capsule, therefore, has most of the tissues found in mosses but it is a globose structure more closely resembling the shape of liverwort capsules.

ORDER BRYALES, THE TRUE MOSSES

About 90 percent of all species of mosses are members of this order. They occupy nearly all types of terrestrial habitats where it is possible for members of the Bryophyta to survive except those highly acid bogs dominated almost exclusively by *Sphagnum*.

The Gametophyte

The spore germinates, forming an algalike protonema which usually branches freely. It is septate, contains numerous elliptical chloroplasts, and produces many tiny buds (Fig. 2). From the buds leafy shoots develop in close proximity to each other, establishing a velvety greenish mat over the surface of the soil. In a few mosses the protonema is the most prominent portion of the gametophyte, but in most the leafy shoots become the most conspicuous part.

In some, which rarely produce sporophytes, the protonemata break up into small pieces, and may be distributed by wind or water or other agencies and carry on a vegetative propagation of the species. Some mosses also reproduce vegetatively by the production of **brood bodies** which arise adventitiously in the axils of leaves, or on the leaf surface, or in specialized little cupules much as in the production of gemmae in *Marchantia.*

The shoots develop from **gametophore** initial cells within the tiny buds. The initial cell has three cutting faces as in *Sphagnum* and gives rise to three (rarely two) vertical rows of spirally and often imbricated leaves. The individual leaves may be fairly broad with a rounded apex, or they may be almost like threads. Some species have blades which are only one cell thick except for the midrib or **costa;** however, most moss blades are two, three, or more cell layers thick (Fig. 9). The shapes of the individual cells in the leaf are often used taxonomically since within the species they are consistantly linear, squarish, rectangular, diamond-shaped, hexagonal, etc.

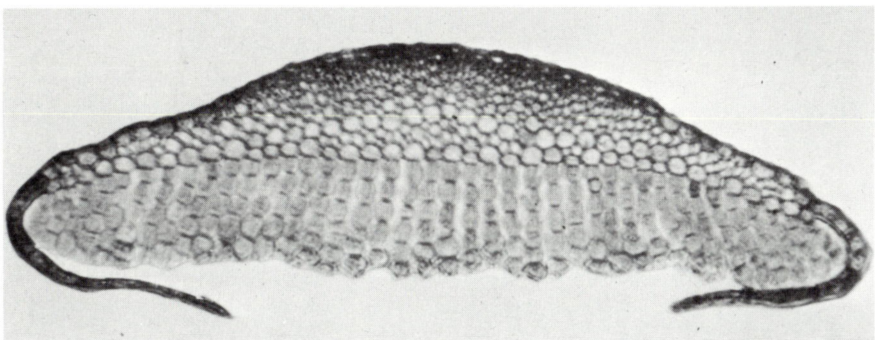

FIGURE 9 Cross section of moss leaf.

The central axis of the leafy shoot is differentiated into three regions: (1) a rather large central area composed mostly of thin-walled elongated cells; (2) a surrounding cortex of thin-walled cells; (3) an epidermal layer in which the cell walls are somewhat thickened on the outer surface. These stemlike structures usually contain plastids, provide some support, may reproduce vegetatively, and provide an external surface where water may move by capillarity, but mostly they produce sex organs in their apices.

Sexual Reproduction

Generally the archegonia and antheridia are produced in the apex of the main axis, but in some species of *Mnium,* and a few others, the sex organs appear on special lateral branches. Monoecious mosses like some species of *Mnium* and *Polytrichum* have both antheridia and archegonia in the same apex, while others like *Funaria* have separate male and female branches (Fig. 10). Dioecious mosses, as in other species of *Polytrichum* and *Mnium*, produce only antheridia or archegonia in their apices.

The archegonia arise from superficial cells and become very much elongated structures in comparison to those of liverworts. In other respects they are similar. Antheridia also arise from superficial papillate cells, and also become very much elongated. Within the antheridial jacket each of the minute cubical cells gives rise to one or two spiral biflagellate sperms.

Dehiscence of the antheridium occurs in the presence of water as the sperms are splashed by falling rain drops or perhaps swim to the apex of another moss plant where they may find a receptive archegonium.

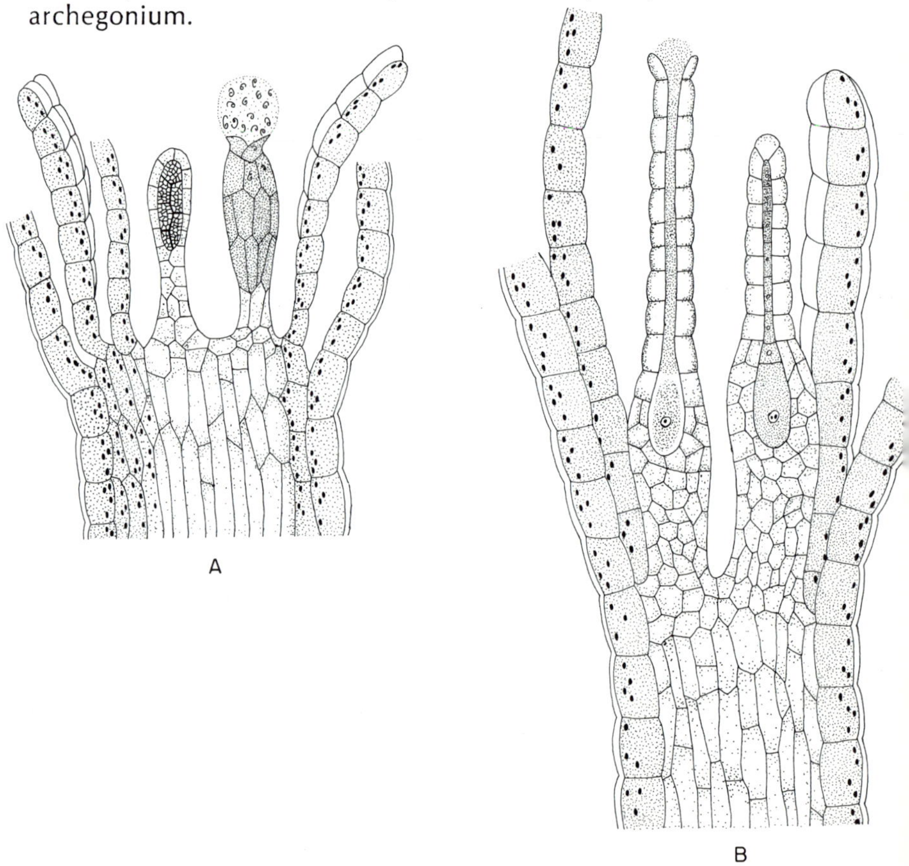

FIGURE 10 A-Antheridial tip showing two antheridia, one in section and the other discharging sperms. B-Archegonial tip with two archegonia; the one on the left is open following disintegration of neck canal and ventral canal cells while the one on the right shows neck cells intact.

The Sporophyte

As pointed out earlier in this chapter, the sporophyte of the Bryales and especially their capsules are the most complex in the Bryophyta from the standpoint of tissue differentiation. As in nearly all members of the phylum initial growth results in the establishment of a foot deeply imbedded in gametophytic tissue (Fig. 11). The upper portion of the developing embryo differentiates into two regions. In contrast to the Sphagnales the inner region of the Bryales gives rise to the sporogenous tissue and also to the central columella, while all other tissues in the capsule are derived from the outer region. When mature the moss capsule (Fig. 12) has the following parts, although one or more tissues such as the tapetum or calyptra may be absent in some:

Inner region:	Columella Sporogenous tissue
Outer region:	Epidermis Chlorenchyma Tapetum Peristome Operculum
Archegonial jacket:	Calyptra

The most unique structure of the moss capsule as mentioned earlier is the peristome (Fig. 4). It is formed from a layer of cells differentiated near the apex of the capsule. The cells develop into filamentous teeth with broad bases and pointed tips. At first these teeth are connected laterally, but as the capsule ripens the connecting membranes split producing the individual teeth. The number of teeth varies from four to sixty-four in different species. Some have a single row of teeth; others have two rows. In those mosses with two rows the inner row acts as a sieve, permitting the escape of a few spores at a time, while the outer row acts as a protective roof, preventing moisture from entering the capsule in wet weather. The teeth are variously marked and banded and are a fascinating mechanism in regulating the dissemination of spores. By carefully removing the operculum with a needle one may gently blow on the peristome, and the teeth will almost instantly respond to the changes in the humidity caused by one's breath. Once the spores are liberated and fall on a suitable substrate the life cycle may begin with germination of the spore.

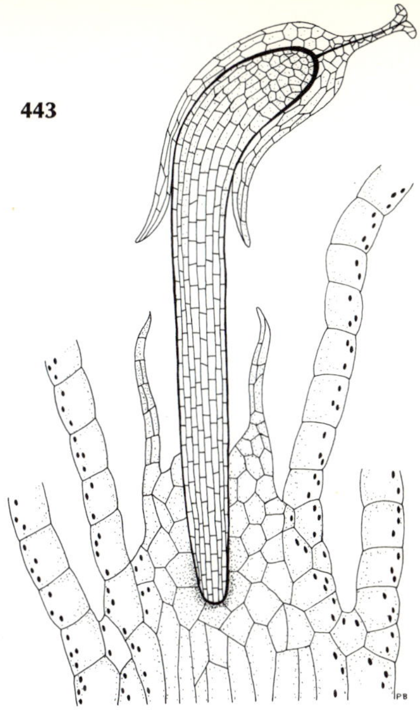

FIGURE 11 Young sporophyte of moss, *Polytrichum,* showing establishment of foot in apex of gametophyte and elongation of the embryo. Archegonium at apex of young sporophyte is developing into the calyptra.

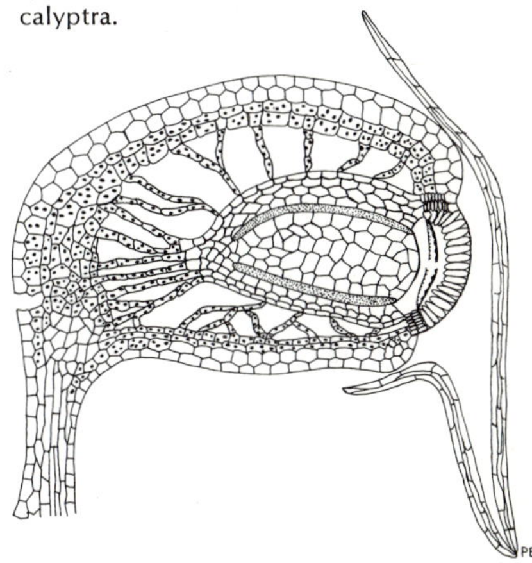

FIGURE 12 A young moss capsule in longitudinal section. The central columella is shown by isodiametric cells; the stippled area is young sporogenous tissue. A tapetum is formed immediately outside the sporogenous tissue. Chlorenchyma tissue is loosely arranged with many air spaces; and this is surrounded by epidermis and hypodermal layers. At the apex the operculum and peristome are seen in section; covering the tip of the capsule is a section of the loosely fitting calyptra. See also Fig. 3.

```
            Spore
              |
          Protonema
              |
            Bulbil
              |
         Leafy Shoot
          /       \
  Archegonia   Antheridia
          \       /
          Egg   Sperm
            \   /
           Zygote
              |
           Embryo
              |
     Mature Sporophyte
   (Foot, Seta, Capsule)
              |
         Sporangium
              |
     Spore Mother Cells
              |
            R D
```

Life Cycle of a Typical Moss

The World of Mosses

While *Polytrichum* and *Mnium* are probably the two most commonly studied genera in botany classes, there are almost six hundred other genera, and many have very fascinating characteristics. The genus *Dicranum* has an elongated capsule with the operculum extending out like a beak so that one doesn't need too much imagination to visualize this capsule as a miniature egret's head. *Tortula,* the twisted moss, not only has variously twisted and curved leaves but the teeth are also twisted around each other like the fibers of a rope. The cord moss, *Funaria,* received its name because the long slender seta is hygroscopic and when dry twists tortuously. *Schistostega* is called the luminous moss because one can frequently see a golden-green glow of reflected light emanating from the smooth surfaces of its fluorescent leaves. Anyone with a hand lens and an interest in nature can spend many fascinating hours examining and identifying the members of the moss world.

Relationships of the Bryophyta

It was pointed out in the preceding chapter that the simplest sporophyte in the Bryophyta, in fact, in the Embryophyta, is in *Riccia,* which consists of a sporangium but lacks both foot and seta. Yet *Riccia* has a reasonably well-developed gametophyte, while other liverworts may have a thallus with less differentiation but with a sporophyte that has the usual foot, seta, and capsule characteristic of the bryophytes. We cannot say, therefore, that *Riccia* is the most primitive member of this phylum any more than we can pick out one moss and say it is the most advanced of all members of the Bryophyta. If we had to pick one genus that seems to have the greatest potential for developing into a Tracheophyte we would have to go back to *Anthoceros*. This genus had species with very simple gametophytes and a sporophyte with the tissues that could make it a free living plant, but it never became an independent structure and it probably never gave rise to anything beyond another species of *Anthoceros*.

In a phylogenetic interpretation of the group we may visualize the Bryophyta arising from some thallose green alga. Once the earliest of liverworts were formed, others appeared through mutation. It is very likely that four major lines of development appeared very early from liverwort ancestors. The Musci seem to have followed three lines of development, coming from the same primitive ancestors that gave rise to the acrogynous Jungermanniales outlined below:

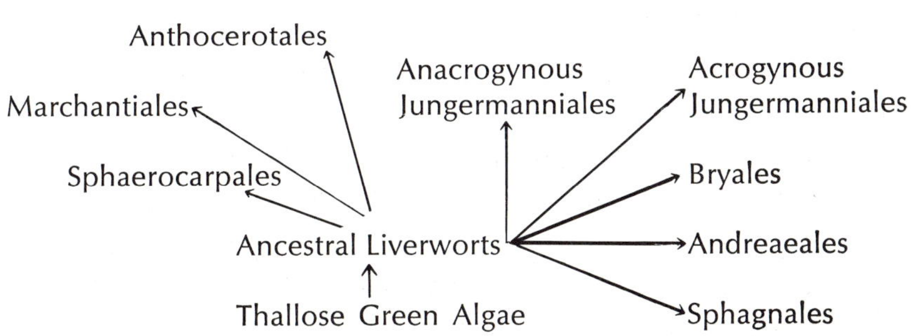

While this is no more than a hypothetical outline, it is based on the prominence of the thallose gametophyte which is prevalent throughout the phylum. The change from this condition was partially achieved by the differentiation of the thallus into a main axis with two-ranked bladelets and finally the development of an apical cell with three cutting faces, producing a gametophyte with three-ranked leaves.

The sporophyte appeared as a simple embryo which persisted throughout the phylum and gradually matured into a structure with an absorbing foot, an elongated seta, and a variously modified capsule. However, it failed to develop vascular tissue, it never differentiated roots or rootlike absorbing organs, and it remained as a dependent organ on the gametophyte. The Bryophyta remained amphibious and retained haploid cells as the dominant phase of their life cycle. To achieve greatness in the plant world, as in the world of animals, organisms had to cast their lot with dominant diploid tissues and independence. These are pinnacles the Bryophyta never attained.

Supplementary Reading

Bower, F. O., *Primitive Land Plants.* Hafner Publishing Company, New York, 1959.

Bower, F. O., *The Origin of a Land Flora.* Hafner Publishing Company, New York, 1959.

Conard, H. S., *How to Know the Mosses.* H. E. Jacques, Mt. Pleasant, Iowa, 1944.

Conard, H. S., *How to Know the Mosses and Liverworts.* H. E. Jacques, Mt. Pleasant, Iowa, 1956.

Proskauer, J., Studies in the Anthocerotales, VI, *Phytomorphology* 10:1–19, 1960.

Smith, G. M., *Cryptogamic Botany,* Vol. II: *Bryophytes* and *Pteridophytes.* McGraw-Hill, New York, 1955.

Steere, W. C., Bryology, *A Century of Progress in the Natural Sciences,* 1853–1953. California Academy of Sciences, San Francisco, 1955.

Wardlaw, C. W., *Embryogenesis in Plants,* John Wiley & Sons, New York, 1955.

Watson, E. V., *The Structure and Life of Bryophytes.* Hutchinson University Library, London, 1964.

Questions for Discussion

1. Structurally, how is the internal tissue development of *Marchantia* similar to the angiosperm leaf? How is it different?
2. What advances in evolutionary development have the liverworts made over the thallophytes?
3. In what ways may we compare liverworts in their development and environmental requirements to the amphibians of the animal kingdom?
4. What additional characteristics would the liverworts need to develop into large terrestrial plants?
5. Why are leafy liverworts considered members of the Hepaticae rather than classified as mosses?
6. Suggest reasons why mosses and liverworts have remained small and insignificant plants.
7. What advances have the archegonia and antheridia of the Bryophyta made over the sex organs of the Thallophyta?
8. From what group of liverworts would you expect the mosses to have arisen? Explain.
9. How does the alternation of generations in the thallophytes differ from that in the bryophytes?
10. What special methods of spore dissemination do we find in the Bryophyta?
11. How is water transported in mosses?
12. In what ways are the leaves of *Sphagnum* different from those of other mosses?
13. How is the capsule of a moss different from the sporophyte of *Marchantia* or *Anthoceros?*
14. Why do we believe the early vascular plants did not arise from members of the Bryophyta?

22

The Lower Vascular Plants

Subkingdom Embryophyta
Phylum Tracheophyta
Subphylum Psilopsida
Subphylum Lycopsida
Subphylum Sphenopsida

The second phylum of the subkingdom Embryophyta is Tracheophyta, of which three subphyla will be considered in this chapter. The fourth, the **Pteropsida,** is comprised of three separate and distinct classes each of which will be discussed in separate chapters. From the standpoint of numbers of species in all of the above subphyla, as well as their morphological differences, we are justified in making this allotment of space.

In earlier systems of classification the Filicineae, the first class of the Pteropsida, along with the Psilopsida, Lycopsida, and Sphenopsida, were placed in a division called the **Pteridophyta** while the remaining two classes of the Pteropsida, the **Gymnospermae** and the **Angiospermae,** were placed in a separate division —the **Spermatophyta** or seed plants. The change in classification has been due largely to a greater emphasis on the vascular system, a feature of all Tracheophytes, and less emphasis has been given to the production of the seed, found only in the gymnosperms and angiosperms. Vascular plants are still frequently termed spermatophytes or pteridophytes depending on whether or not they produce seeds, but most taxonomists now consider these names invalid, although they are quite descriptive.

Subphylum Psilopsida

There is but one class, the **Psilophytineae,** comprised of two orders, the **Psilophytales** and the **Psilotales.** We may with comparative safety say that the members of the first order were our earliest vascular plants. All members of the Psilophytales have been extinct for millions of years, but while they were here (they probably lived on earth for over fifty million years) they gave rise to several lines of vascular plants which in turn gave rise to others now living. Numerous fossils of this order are found in rocks of the Devonian epoch (Fig. 1), when they apparently flourished, and many also have been collected from the even earlier Silurian period (see Table 1).

FIGURE 1 A museum restoration of a Devonian forest. Note the conspicuous circinate vernation of the primitive psilopsids and larger ferns, and the whorled habit of the early sphenopsids. (Courtesy Field Museum of Natural History.)

Table 22-1 Geologic Time Chart of the Dominant Organisms and Probable Climatic Conditions under Which They Lived

ERAS	PERIODS, YEARS AGO		DOMINANT ORGANISMS	ADVANCES IN PLANT LIFE	CLIMATE AND GEOLOGIC EVENTS
		Recent	Age of herbs and man	Dominance of herbs	Modern seasons
	QUATERNARY	Pleistocene 0-1,000,000		Extinction of many trees; increase in number of herbs	Periodic glaciation
		Pliocene 1,000,000-12,000,000		Increasing restriction of plant distribution and of forests; rise of herbs	Continued cooling of climate with temperate zones appearing; Cascades, Coast Ranges
CENOZOIC began 60,000,000 years ago	LATE TERTIARY	Miocene 12,000,000-28,000,000	Age of angiosperms, mammals, and birds	Restriction of distribution of plants; retreat of polar floras; forest reduction	Climate greatly changed, cool and semiarid; Himalayas, Alps
		Oligocene 28,000,000-40,000,000		World-wide distribution of tropical forests	Climate warm humid; Pyrenees
	EARLY TERTIARY	Eocene and Paleocene 40,000,000-60,000,000		Modernization of flowering plants; luxuriant forests of angiosperms even in arctic regions	Climate cool, semiarid, then warm, humid; mountain glaciers
		Upper Cretaceous		Angiosperms dominant; gymnosperms dwindling	Climate fluctuating; Rocky Mountains and Andes
	LATE MESOZOIC 60,000,000-130,000,000	Middle Cretaceous		Rapid development of angiosperms	Climate fluctuating

Table 22-1 (Continued)

ERAS	PERIODS, YEARS AGO		DOMINANT ORGANISMS	ADVANCES IN PLANT LIFE	CLIMATE AND GEOLOGIC EVENTS
MESOZOIC began 185,000,000 years ago		Lower Cretaceous	Age of higher gymnosperms and reptiles	Rise of angiosperms; conifers and cycads still dominant	Climate very warm
		Jurassic 130,000,000-155,000,000		First known angiosperms; conifers and cycads dominant; cordaites disappear	Climate warm; Sierras, great continental seas in western North America
	EARLY MESOZOIC	Triassic 155,000,000-185,000,000		Floras not luxuriant; higher gymnosperms increase, including cycads, conifers, and ginkgoes	Climate warm, semiarid, mostly temperate
		Permian 185,000,000-210,000,000		Dwindling of ancient groups; first cycads and conifers	Climate dry with periodic glaciation; Appalachians, Urals, drainage of seas from continents
	LATE PALEOZOIC	Pennsylvanian (Upper Carboniferous) 210,000,000-235,000,000	Age of lycopods, seed ferns, and amphibians	Dominant calamites, lepidodendrons, ferns, seed ferns, cordaites; extensive coal formation	Period of crustal unrest; alternation of marine and terrestrial conditions
PALEOZOIC began 520,000,000 years ago		Mississippian (Lower Carboniferous) 235,000,000-265,000,000		Dominant lycopods, horsetails, and seed ferns; early coal deposits	Widespread shallow seas on North American continent; Acadian Mountains
		Devonian 265,000,000-		Early land plants; Psilop-	

Table 22-1 (Continued)

ERAS		PERIODS, YEARS AGO	DOMINANT ORGANISMS	ADVANCES IN PLANT LIFE	CLIMATE AND GEOLOGIC EVENTS
		320,000,000		sida, Rhynia, etc., primitive Lycopsida, Sphenopsida, horsetails, ferns, and seed ferns; first forests	
	MIDDLE PALEOZOIC		Age of early land plants and fishes		
		Silurian 320,000,000-360,000,000		First-known land plants; algae dominant; Rhynia	
		Ordovician 360,000,000-440,000,000		Possible rise of land plants; marine algae dominant	
	EARLY PALEOZOIC	Cambrian 440,000,000-520,000,000	Age of algae and higher invertebrates	Algae, especially marine forms	Narrow seas within the borders of North America; Climate warm, uniform over earth
PROTEROZOIC began 2,100,000,000 years ago	520,000,000-2,100,000,000		Age of primitive marine invertebrates	Bacteria and algae	Grand Canyon, younger Laurentian rocks, chiefly sedimentary; glaciation
ARCHEOZOIC began 3,250,000,000 years ago			Age of unicellular life	No fossils found; all organisms probably unicellular or very simple	Older Laurentian rocks, mostly igneous or metamorphosed; few sedimentary

Estimates of time based on the U.S. Geological Survey Chart of 1956; descriptive data in part from Fuller and Tippo, **General Botany** (Holt, 1954).

Rhynia was probably the most primitive vascular plant (Fig. 2). It was only about eight to twenty inches tall, with a naked, dichotomously branched stem, and lacked both roots and leaves. The aerial shoot arose from a horizontal rhizome which bore rhizoids but not roots. At the apex of the branching aerial stems tiny sporangia developed. The tissues apparently arose from an apical meristem and are therefore primary. No cambium was present, nor were secondary tissues developed.

The vascular system, which is the chief reason we have for placing *Rhynia* in this phylum, is called a **protostele** (Fig. 3) and consists of a solid central strand of xylem surrounded by a narrow zone of phloem. The xylem elements were all tracheids and in *Rhynia* they had only annular markings, while in one other genus at least—*Asteroxylon*—the tracheids were often marked by spiral and scalariform thickening. The phloem cells were elongated with oblique end walls. Surrounding the phloem was a thick cortical region made up in part of chlorophyllous and storage parenchyma cells. Covering the cortex was a simple epidermis containing stomates and covered with a waxy cuticle.

Other genera in the Psilophytales included *Psilophyton,* a slightly larger plant than *Rhynia; Asteroxylon,* which had a protostele that was star-shaped in cross section, and *Horneophyton,* in which the sporogenous tissue was dome-shaped and covered a central columella much like the capsule of *Sphagnum.*

Both *Psilophyton* and *Asteroxylon* had leaflike **emergences**

FIGURE 2 *Rhynia;* one of the simplest and possibly first of the vascular land plants. The horizontal rhizome produced vertical stems and rhizoids but neither roots nor leaves. The sporangia were apical.

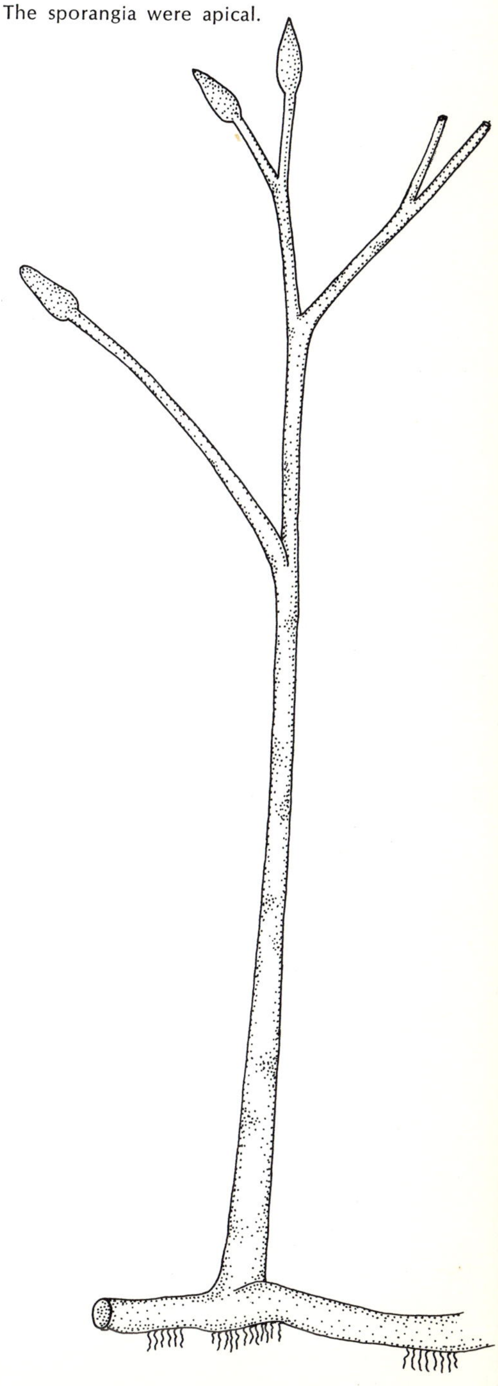

FIGURE 3 A protostele, typical of the stem of *Rhynia*. e-epidermis, c-cortex, p-phloem, x-xylem.

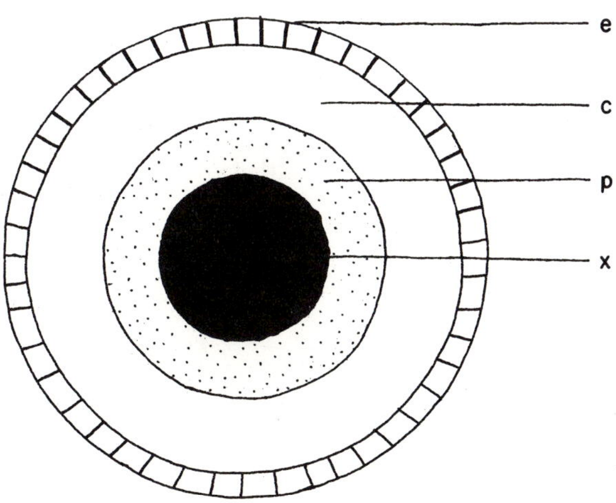

covering their aerial stems. In *Psilophyton* they were like spines, and in *Asteroxylon* they were somewhat broader and mosslike.

The leaflike appendages of these plants were not connected with the vascular system of the stem and had no xylem or phloem elements. They were simply superficial outgrowths of the stem. It is the opinion of some botanists that the first leaves appeared this way and later vascular strands differentiated, connecting the leaf to the vascular system of the stem. This is known as the **enation theory** of leaf origin (Fig. 4).

Just how organisms made the transition from simplicity to complexity is still largely a matter of speculation, but considering the characteristics of these early members of the Psilopsida it is possible to see how some of the organs might have become modified as a result of genetic mutations.

It is generally believed that the early ancestors of the first land plants were green algae with a dichotomous manner of branching and siphonaceous thallus. They became more than algae when the first vascular elements appeared, but at first they had only horizontal stems. Gradually, as the tips curved upward and the dichotomies continued, the xylem elements kept pace and with their thick walls gave needed support and conduction for the enlarging plant.

FIGURE 4 Diagrammatic illustration of the change in leaf structure as discussed in the Enation Theory. The leaves first appeared as emergences, then vascular tissue began to leave the stem's central cylinder, and finally continued on into the leaf in C.

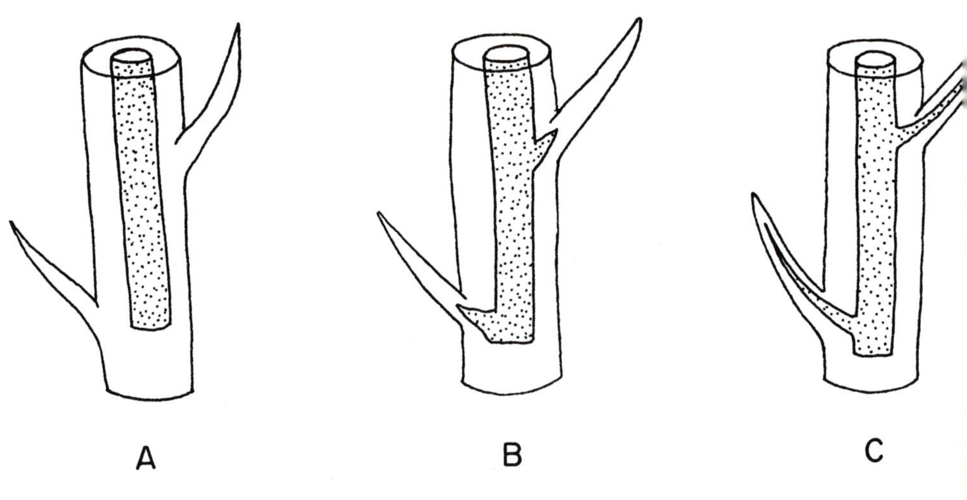

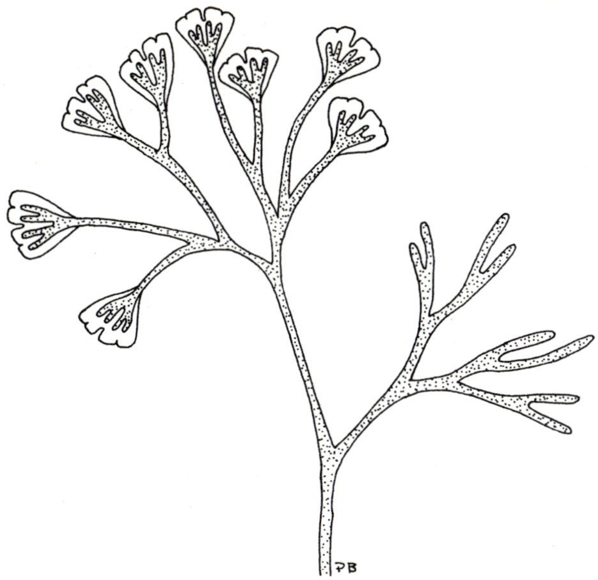

FIGURE 5 Telome theory of leaf origin. In the most primitive vascular plants stems only were present; then gradually by fusion of the other portions the blades developed by expansion as in the branches on the left of the drawing.

Another theory concerning the modification of this early sporophyte body is known as the **telome theory.** Again starting with a simple stem, we may refer to the ultimate branching dichotomies as **telomes** and the internodes below the forking branches as **mesomes.** According to this theory foliar leaves developed by a fusion of one or more terminal dichotomous branches to become a simple lateral leaf (Fig. 5). The stem branches formed the veins within the leaf, and expansion of other tissues (cortex and epidermis) between the veins gave the structure a blade, the mesome forming a petiole. Evidence supporting this view is found in the open venation of many fossil ferns, as well as in some now living, and in some members of the Sphenopsida.

Finally, we have a theory (which is accepted in part by telome theorists) that leaves arose by increased sterilization of potentially sporogenous tissue. This is such a very common situation in higher plants, where stamens and other parts may be differentiated into petals (as in double petunias, roses, etc.), that it appears quite likely that this may have happened many times in the past with a flat leaf as the end result.

It is so easy to show examples of each of the above theories (enation, telome, sterilization) that it is quite probable that leaves arose in different groups in different ways. Genetic changes through mutation, polyploidy, and cross-fertilization result in many variations and, given time (the Psilopsida had about fifty million years), organogenesis was bound to occur.

ORDER PSILOTALES

The two surviving genera of the Psilopsida are *Psilotum* (Figs. 6, 7) and *Tmesipteris* (Fig. 7). The latter grows as an epiphyte in land areas of the South Pacific as far north as the Philippine Islands, but more commonly in New Zealand, Australia, and nearby South Pacific Islands. *Psilotum* lives pretty much on the other side of the world from *Tmesipteris,* in Florida, islands of the Caribbean, and elsewhere, and is more terrestrial in habitat.

Geologically the members of the Psilotales are of much more recent origin than the Psilophytales, but they are so similar morphologically that it would be difficult to imagine that they arose independently. It is quite possible that the fossil evidence of their close relationship has not been uncovered.

FIGURE 6 *Psilotum.* The plant is all stem except for the very small bracts and sporangia.

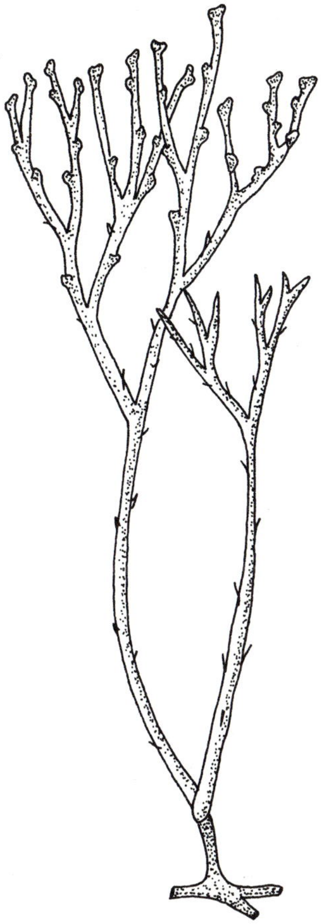

A

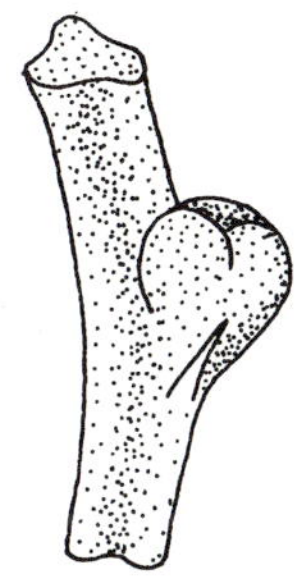

B

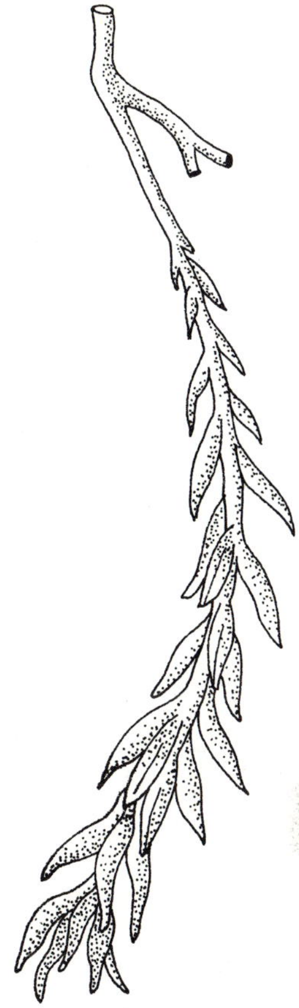

C

FIGURE 7 A-*Psilotum,* B-the sporangium of *Psilotum,* C-*Tmesipteris.*

General Characteristics

The sporophyte of *Psilotum* is dichotomously branched, with numerous aerial shoots arising from a subterranean rhizome. Roots are lacking, and the leaves are tiny emergences spirally arranged on the aerial stems.

Stem growth is maintained by an apical cell with three cutting faces which gives rise to the **primary meristem.** The central cells formed from the meristem differentiate into a protostele which is star-shaped like that of *Asteroxylon.* The xylem elements are tracheids with annular and spiral thickening.

Maturation of the xylem is **exarch** (Chapter 9, Fig. 8); that is, lignification of the xylem elements takes place first in the periphery of the xylem, then the inner cells gradually become lignified as they mature. The cells of the phloem are arranged in a very narrow

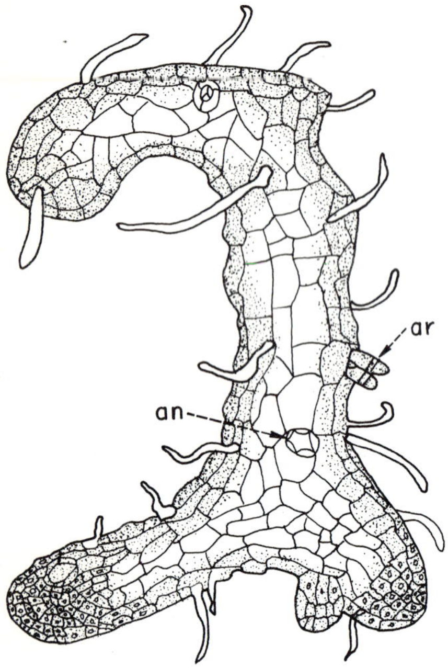

FIGURE 8 The gametophyte of *Psilotum*. This small structure is cylindrical with antheridia (an) and archegonia (ar) imbedded in the surface. Rhizoids are also scattered over the surface.

and often irregular band around the xylem. At maturity they are well marked by sieve areas in both lateral and end walls.

Cambium is lacking in the stem of *Psilotum* but surrounding the phloem is a well-developed **endodermis.** The surrounding cortex has many parenchymatous cells of a storage nature in the inner portion, but the peripheral cortical cells are chlorenchymatous and there are also patches of sclerenchyma in this area. The surrounding epidermis is heavily cutinized and has numerous stomates.

Sporangia of both *Psilotum* and *Tmesipteris* appear as short-stalked axillary bodies along the main axis rather than in a terminal position as in the Psilophytales. The mature sporangium of *Psilotum* is trilobed, while that of *Tmesipteris* is bilobed. The spore mother cells give rise to tetrads of spores by meiosis.

Gametophytes

Germination of the spores results in a cylindrical, rather imperfectly dichotomously branched gametophyte (Fig. 8). They may be as much as twenty millimeters in length, yellowish in color, mostly subterranean, and well supplied with rhizoids. Many of the cells in the prothallus contain endophytic fungi which probably aid the little plants to absorb foods, since the gametophyte lives as a saprophyte. The gametophytes are monoecious, both archegonia and antheridia being formed on the same plant. The archegonia are more like those of other vascular plants than those of the Bryophyta. This is also true of the antheridia and especially of the sperms, which are multiflagellate instead of biflagellate as they were in the Bryophyta.

Growth of the embryo is significantly different from that of the other vascular plants. The embryo is much more primitive in that it does not form a leaf or root but does produce a foot and rhizome. After the rhizome is differentiated an aerial shoot is produced by it, and the foot remains imbedded in the gametophyte and disintegrates it as the gametophyte is separated from the independent sporophyte.

Fossil remains of the gametophytes of members of the Psilophytales have not been found, but it is probable that they were very much like those of the Psilotales. The characteristics of the sporophytes of the two orders are so similar that we may assume the surviving two genera are living remnants of a Devonian flora that has come down to the present with very little change from ancient times.

Subphylum Lycopsida

Members of the Lycopsida can be found as far back as the Silurian epoch, and they were very abundant during the Devonian, but like most members of the Psilopsida many of the members of the Lycopsida became extinct (Fig. 9). Now we have but four surviving genera. They probably descended from early members of the Psilopsida, but the Lycopsida often become much larger and achieved a higher degree of differentiation.

FIGURE 9 *Lycopodites,* a fossil member of the Lycopodiales. Note similarity to present day species of *Selaginella.*

Characteristics of the Sporophyte

Roots appear for the first time in the Lycopsida. In general the roots were dichotomously branched and arose from rhizomes or came from dichotomous branches of the vertical stems. The leaves were more than enations and even the smallest leaves had at least a small vein.

The vascular system in the roots is a protostele, but the stelar arrangement of the stems was often more than a simple protostele. Here we have for the first time a **siphonostele,** which is a vascular cylinder with a central **pith.** Where phloem is developed externally only to the xylem, as it is in most higher plants, it is an **ectophloic siphonostele;** where phloem develops both external and internal to the xylem it is termed an **amphiphloic siphonostele** or a **solenostele** (Fig. 10).

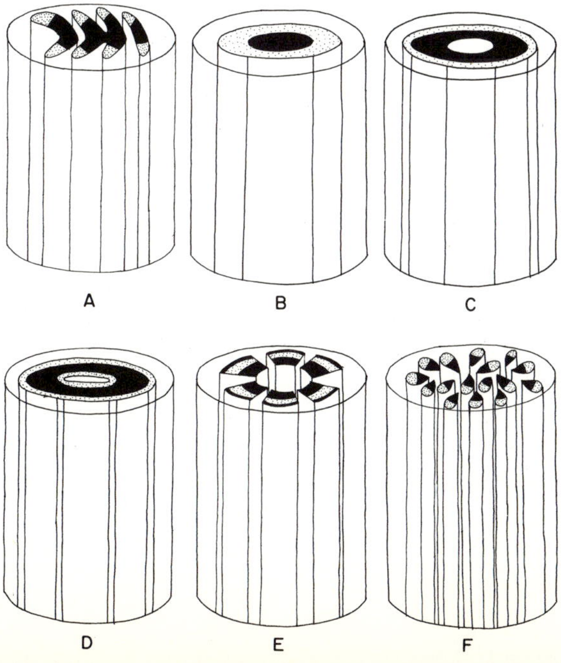

FIGURE 10 A comparison diagrammatically of the major stele types found in plants of the various groups. A. plectostele B. protostele C. ectophloic siphonostele D. amphiphloic siphonostele E. dictyostele F. tactostele. In all drawings the phloem is stippled and the xylem is solid black.

In some of the Lycopsida and in other lower vascular groups the solenostele is broken up into separate vascular strands of xylem surrounded by phloem. This is known as a **dictyostele.** In the higher plants, where the ectophloic siphonostele is broken into separate strands with phloem only on the outer portion of the separate bundles, we term the stem's vascular arrangement a **eustele.**

Where branches occur in stems with siphonosteles and the vascular tissue is continuous, the connecting vascular strand is a **branch trace.** In the vascular cylinder there is often a gap immediately above where the trace departed from the stem. This is a **branch gap.** Branch gaps are present in all plants possessing pith, but they are not present in stems with protosteles. Similarly, **leaf gaps** may occur where vascular strands leave the branches to enter leaves. Members of the Psilopsida and Lycopsida do not have leaf gaps since they have mostly protosteles. In the few members of the Lycopsida where leaf traces do enter the leaves the amount of vascular tissue connecting the stems to the leaves is so small that no noticeable gap appears.

In the Lycopsida the protostele is the dominant stelar arrangement. Here it may be a simple **haplostele,** without radiating points, or it may be lobed or star-shaped in cross section and therefore called an **actinostele,** or the protostele may consist of separate plates of tissue with phloem appearing between and around the xylem, in which case it is a **plectostele.**

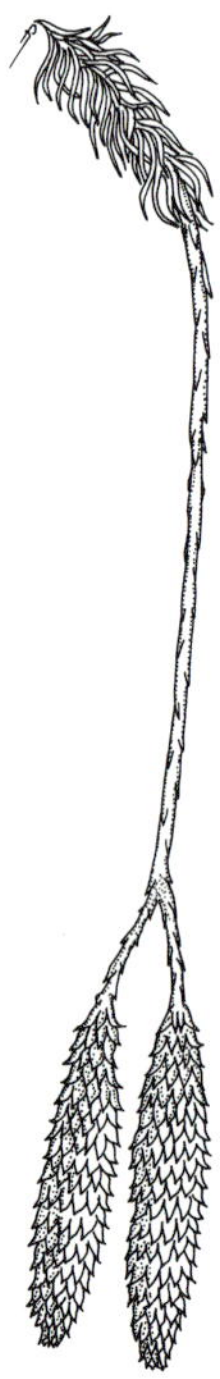

FIGURE 11 Branch of *Lycopodium* showing two strobili.

ORDER LYCOPODIALES

There are two living genera, *Lycopodium* and *Phylloglossum,* still surviving. The latter is found only in Australia, New Zealand, and adjacent islands. It is a small plant with a tuberous base and a few spirally arranged leaves arising from the base of the vertical stem.

Lycopodium is found in most parts of the world, from tropical habitats to the cool temperate latitudes, particularly in acid soils. The stems are usually dichotomously branched and copiously covered with small moss-like leaves. The underground rhizomes are often covered with leaves and give rise to adventitious roots at many of the nodes. Arising from the apex of the aerial branches is a strobilus (Fig. 11), which is usually well differentiated from the vegetative portion of the stem. The sporophylls making up the strobilus are less leaflike than those of *Selaginella* in the next order. On the surface of each sporophyll is a kidney-shaped sporangium in which the spore mother cells give rise to the spores by meiosis (Fig. 12). Both *Lycopodium* and *Phylloglossum* are classed as **homosporous** since the spores they produce are alike.

Upon germination the spores produce a prothallus that varies from about two to eighteen millimeters in length depending on the species. As in many of the lower vascular plants, the prothalli become infected with mycorrhizal fungi which apparently have an endophytic and symbiotic relationship with the prothallus.

The gametophyte is monoecious, both antheridia and archegonia developing in the periphery of the expanded apex. The number of neck canal cells in the archegonia vary from one to sixteen in the different species. Antheridia are globose and produce numerous curved, biflagellate, mosslike sperms.

After fertilization the zygote divides to form an outer **suspensor** cell and an inner embryonic cell. The suspensor cell may not divide again, but usually it does form a few more cells and elongates pushing the developing embryo deeper into the prothallial tissue. The embryonic cell forms two plates of four cells each, very much like the formation of the embryo in seed plants. The quadrant nearest the suspensor gives rise to the foot. The lower quadrant gives rise to the stem from the two cells on one side; the first leaf and root develop from the two cells on the other side. As the embryo develops, food is absorbed by the foot from the prothallus until the primary leaf is pushed upward into the light. After the embryo is well established, the gametophyte disintegrates and, as in higher plants, the primary root and leaf also disappear as the permanent roots and leaves develop.

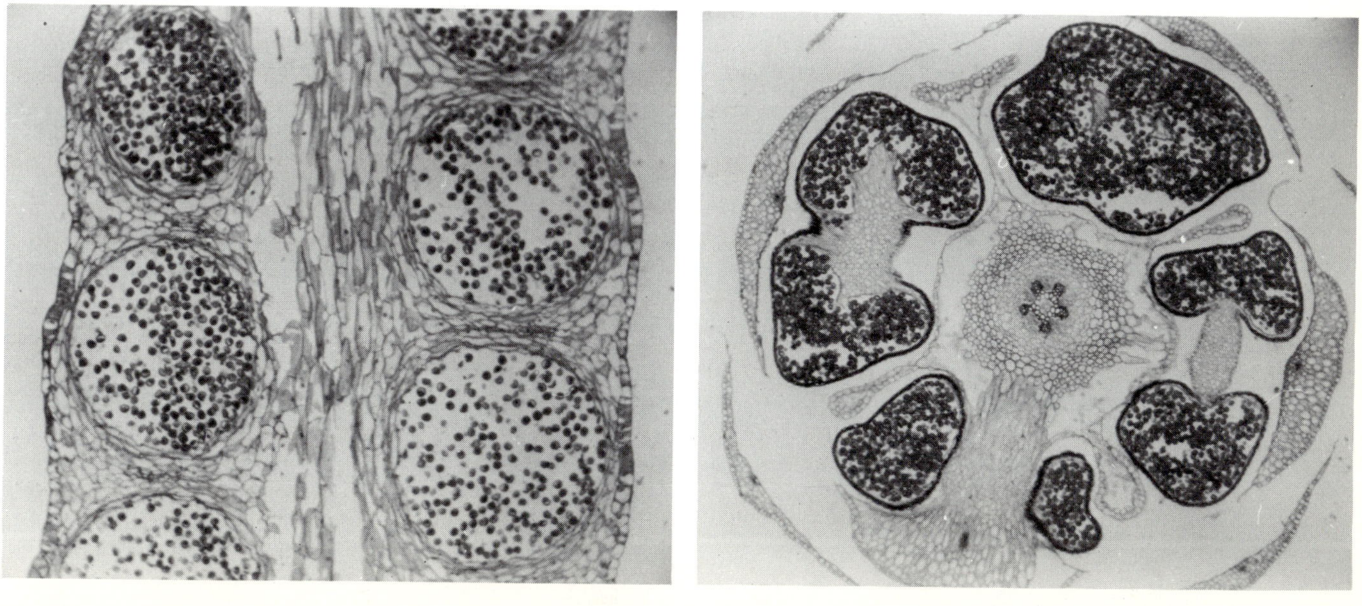

FIGURE 12 Sections through the strobilus of *Lycopodium*. A. Longitudinal section of a portion of the strobilus showing spores in the sporangia. B. Cross section of strobilus showing the central vascular area surrounded by the sporangia and sporophylls.

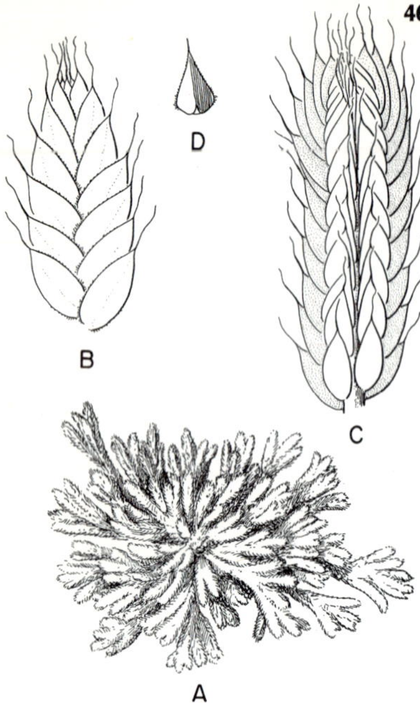

FIGURE 13 "Ressurection plant." a-expanded; b-ventral view of stem tip. c-dorsal view; d-dorsal view of sporophyll. From Dittmer, H. J., E. F. Castetter and Ora Clark, 1954. The Ferns and Fern Allies of New Mexico, University of New Mexico Press.

FIGURE 14 *Selaginella neomeicana* a-Entire plant; b-vegetative leaf; c-strobilus; d-sporophyll dorsal surface; e-ventral surface showing abortive megasporangium. From Dittmer, H. J., E. F. Castetter, and Ora Clark, 1954. The Ferns and Fern Allies of New Mexico. University of New Mexico Press.

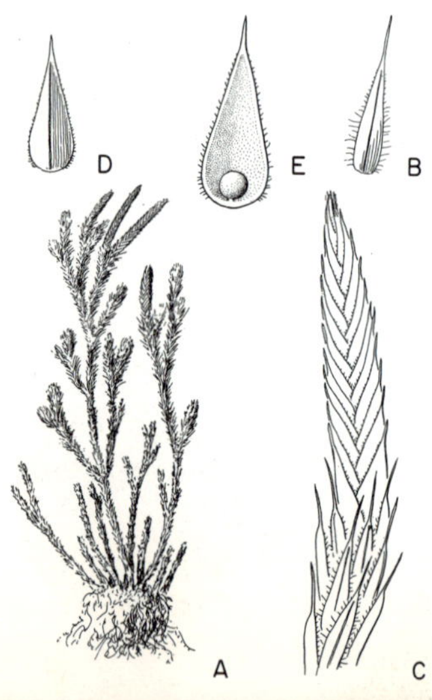

ORDER SELAGINELLALES

Selaginella is the only extant genus in this order, but there are approximately seven hundred species. They live in a wide variety of habitats. Several species live in desert areas and occupy soils that are much more alkaline than *Lycopodium* can endure. Two species, *S. lepidophylla* and *S. pilifera* (Fig. 13), commonly called "resurrection plants," live in the dry deserts of the Southwest where they curl into balls when dry but rapidly uncurl and spread outward when water is available.

Description

Species of *Selaginella* are commonly called "small club mosses" because they are usually smaller and more delicate in texture than those of *Lycopodium*. The vegetative leaves are mosslike (Fig. 14), usually arranged in four rows, and often of different sizes within the rows. A small flaplike outgrowth, the **ligule,** arises from the basal portion of the adaxial surface of each leaf but is prominent only during early development of the leaf. Each leaf is connected to the vascular system of the stem by a single leaf trace. The mesophyll cells have relatively large but few chloroplasts, and stomates are present in both upper and lower epidermis.

The stem grows by means of an apical cell or by a group of apical meristematic cells. From the meristem a central stele and its adjacent tissues differentiate. The vascular arrangement is usually one of several types of protosteles, but in some species an amphiphloic siphonostele or even a dictyostele may be present.

Surrounding the vascular bundle (or bundles) there is usually a single row of pericyclic cells. Extending outward from this area are air spaces traversed by elongated cells, the **trabeculae,** which arise from the endodermis or first row of cortical cells. The cortex itself is quite thick and made up of large parenchyma cells and some hypodermal sclerenchyma beneath the epidermis.

Roots of *Selaginella* are chiefly adventitious in origin, as they are in *Lycopodium,* and may come from either the underground rhizomes or the aerial stem.

The strobili produced on the tips of branches are often difficult to distinguish from the upper portion of the stem since the vegetative leaves are of about the same size and appearance as the sporophylls. The latter can be readily determined, however, by noting the sporangia hidden in their axes. This is the outstanding difference between *Lycopodium* and *Selaginella,* for while the former is homosporous, *Selaginella* is heterosporous (Fig. 15). Note the similarity in the life cycles of these two genera outlined below.

The Life Cycles of LYCOPODIUM, a Homosporous Club Moss, and SELAGINELLA, which is Heterosporous

Lycopodium

Sporophyte (roots, stem, leaves)
→ Strobilus
→ Sporophylls
→ Sporangium
→ Spore mother cells
→ Spores
→ Gametophyte (thallus)
→ Archegonium → Egg
→ Antheridium → Sperm
Egg + Sperm → Zygote
→ Embryo

Selaginella

Sporophyte (roots, stem, leaves)
→ Strobilus
→ Megasporophyll → Megasporangium → Megaspore mother cell → Megaspore → Megagametophyte → Archegonium → Egg
→ Microsporophyll → Microsporangium → Microspore mother cell → Microspore → Microgametophyte → Antheridium → Sperm
Egg + Sperm → Zygote
→ Embryo

In the strobilus of *Selaginella* there are both **microsporophylls,** which produce **microsporangia** on their surfaces, and **megasporophylls** which produce **megasporangia.** Each megasporangium contains a **megaspore mother cell** which produces four large **megaspores** (Fig. 16). The microsporangia contain **microspore mother cells** which give rise to numerous small **microspores.**

Heterospory is a characteristic of all seed plants, but in the lower vascular plants it is found in only a few genera. Although it is generally conceded that heterospory gave rise to the seed habit and in *Selaginella* fertilization and initial development of the embryo also occur within the megaspore (Fig. 17), it is quite probable that heterospory arose independently a number of times in plant evolution. Generally heterospory did not continue to evolve seeds, but the time finally came when it did and once the seed habit developed it was never replaced.

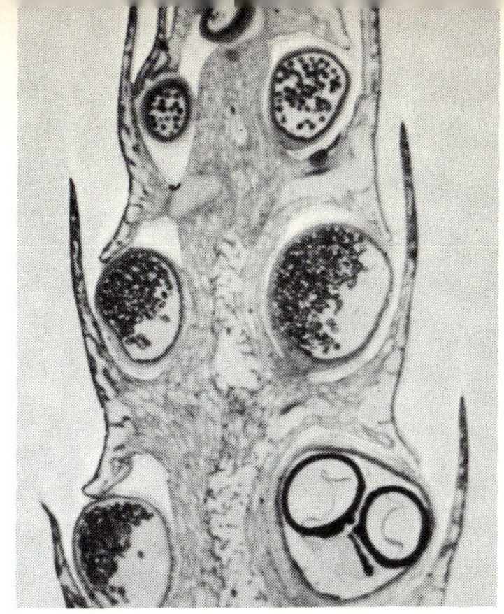

FIGURE 15 Longitudinal section through *Selaginella* strobilus showing five microsporangia and one megasporangia in lower right.

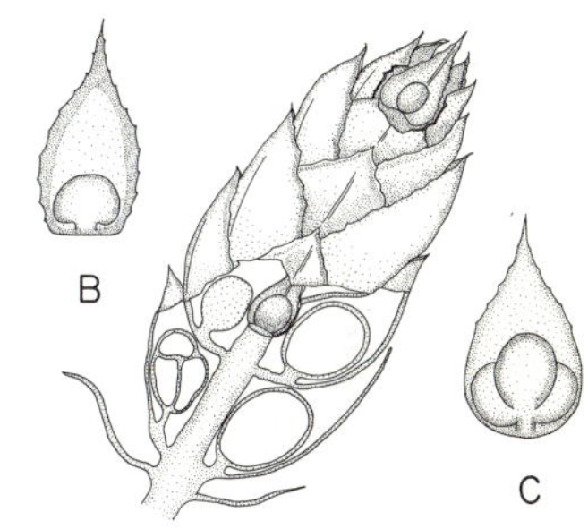

FIGURE 16 *Selaginella.* A. Section through strobilus showing two open microsporangia on right and a section through a megasporangium at left. B. Microsporophyll with microsporangium C. Megasporophyll with mature megasporangium showing three of the four megaspores.

FIGURE 17 *Selaginella* sporeling. The primary leaf and root have been formed but the embryo sporophyte still remains attached by its foot and suspensor to the prothallus in the megaspore.

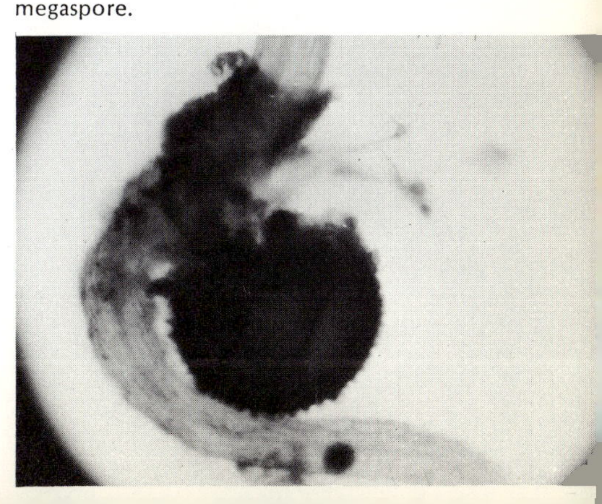

Gametophytes

How close *Selaginella* is to producing seeds is best understood by a consideration of their gametophytes, which of course are dioecious since they develop from separate spores. The male gametophyte develops entirely within the microspore wall and begins with a small prothallial cell and a second larger cell which enlarges to form an antheridium. The male gametophyte is usually shed from the microsporangium when the gametophyte has developed an antheridium made up of a sterile jacket of eight cells enclosing four primary spermatogenous cells. At maturity the antheridium will form 128 to 256 sperm primordial cells, each of which will give rise to a single biflagellate sperm. In some species the antheridium is retained within the microspore wall and the sperms are not liberated until the spore wall is ruptured and the antheridial jacket cells disintegrate. The single prothallial cell, which represents the remaining tissue of a free-living gametophyte, disintegrates with the sterile antheridial wall tissues. This is very similar to the behavior of the few gametophyte cells in the pollen grain of seed plants.

The female gametophyte, although not as much reduced in size as the male, develops within the megaspore wall, at least partially, while the megaspore is still within the megasporangium. The bulk of the female gametophyte is made up of vegetative cells well stocked with a food supply which may be used by the young embryo sporophyte as it develops. The female gametophyte begins to develop by free nuclear division (as in gymnosperms), but walls begin to form around the nuclei at one end of the spore and as the gametophyte enlarges it begins to protrude from the ruptured spore wall. Several archegonia are formed near this opening, and occasionally rhizoids form from some of these gametophytic cells and also protrude. The archegonium of *Selaginella* usually has a very short neck with but a single neck canal cell; otherwise its development is similar to that of *Lycopodium*.

The gametophytes are rarely exposed to light, although chlorophyll is present in some. They are, therefore, mostly dependent on the spore and sporangium for food. Thus the gametophyte, which was the dominant generation in the Bryophyta, has in *Selaginella* dwindled in magnitude until it has lost its independence and has become a rather inconspicuous part of the life cycle.

The similarity of the megaspore and its contents to a seed is exhibited by the embryo which develops within the spore walls. It develops like that of many seed plants with its primary leaf, primary root, and meristematic stem. The stored food which will be used by the developing embryo is megagametophytic tissue as it is in gymnosperms, but its only protective covering is the megaspore wall, whereas seed plants develop a seed coat formed from an integument and the megasporangium wall.

ORDER LEPIDODENDRALES

The giants of the Lycopsida were members of this order. They arose in the late Devonian epoch and flourished for millions of years. On the basis of fossil remains it is estimated that some reached heights of over one hundred feet. Some were homosporous but apparently most were heterosporous, and some paleobotanists believe that they succeeded in producing seeds. During the Carboniferous era vast forests of these giant club mosses covered many sections of the earth, and their remains formed many of the extensive coal deposits we mine today (Fig. 18.)

Two of the most noted genera were *Lepidodendron* (Fig. 19) and *Sigillaria* (Fig. 20). Sigillaria was a tall, columnar tree with little or no branching; the leaves were about three feet long and were produced in dense clusters at the apex. Its common name was "bottle brush tree" because of the straight trunk and position of the leaves. The leaves were deciduous and left vertical rows of leaf scars which were squarish or hexagonal in outline and contained a central vascular bundle scar.

Lepidodendron was a much more highly branched genus with numerous dichotomies forming a more or less flat-topped tree. The leaves were much shorter than those of *Sigillaria* and were arranged in a steep spiral. When shed, the leaves left diamond-shaped leaf scars with the usual centrally located vascular bundle scar.

FIGURE 18 A Carboniferous forest scene as restored in the museum. Included are specimens of *Sigillaria, Lepidodendron, Calamites, Sphenophyllum* and some ferns. (Courtesy of Field Museum of Natural History.)

FIGURE 19 *Lepidodendron,* a museum restored specimen. (Courtesy Field Museum of Natural History.)

FIGURE 20 *Sigillaria,* a museum restored specimen. (Courtesy Field Museum of Natural History.)

Internally the members of these two genera produced numerous scalariform tracheids in both primary and secondary wood. Surrounding the very ample cortex was an extensive barklike periderm made up of thick, fibrous, suberized cells. The cork cambium in these plants gave rise to the periderm cells mostly on its internal side and not, as is more commonly true in plants living today, on its external side. The vascular arrangement varied from a protostele to an ectophloic siphonostele, more commonly the latter. The xylem was exarch and thickening of the stem, which sometimes reached a diameter of six feet, was due to an active cambium.

ORDER ISOETALES

The fourth living member of the Lycopsida is *Isoetes* (Fig. 21), a genus with some sixty species. They are small herbaceous plants which look like small tufts of grass, or perhaps more like clumps of chives, and are commonly called quillworts. They are mostly hydrophytic plants inhabiting temperate regions.

The stem of *Isoetes* is a short, tuberous, unbranched, two-lobed cormlike structure, which gives rise to spirally arranged leaves and similar sporophylls (Fig. 22). The vascular pattern is that of a flattened protostele with numerous parenchyma cells scattered throughout the xylem. Surrounding the xylem strands are narrow

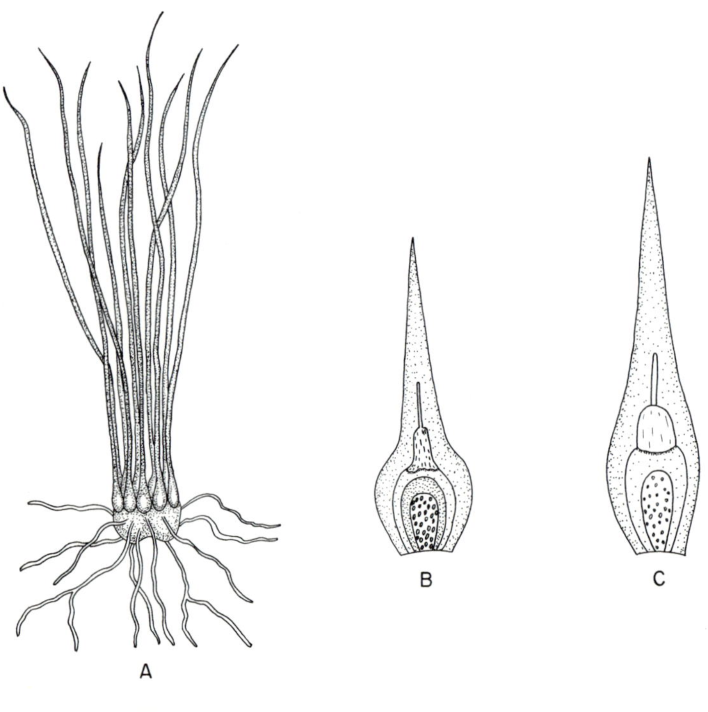

FIGURE 21 *Isoetes*. A-Habit; B-Sporophyll showing adaxial surface with megasporangium; C-Sporophyll, adaxial surface with microsporangium. The tiny elongated structure at the apex of each sporangium is the ligule.

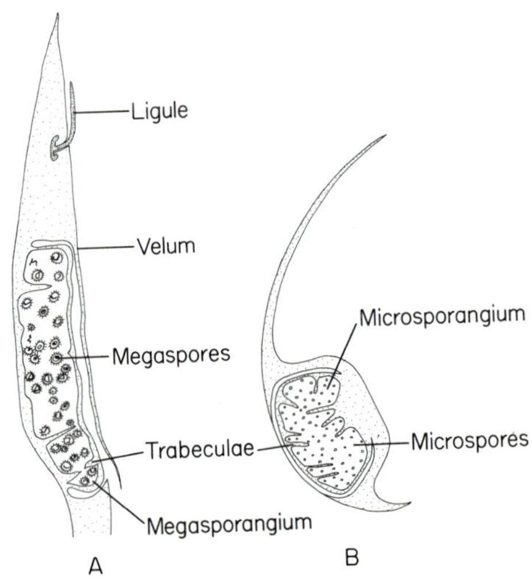

FIGURE 22 *Isoetes*
A-lateral view of megasporophyll
B-lateral view of microsporophyll

bands of phloem. Outside the phloem is a ring of meristematic tissue which functions as a cambium. This meristematic tissue gives rise to cortical cells, a few xylem elements with reticulate and scalariform thickening, and other thin-walled cells resembling secondary phloem. Most of the tissue in the stem at maturity is cortical; there is no well-defined pericycle, endodermis, or periderm.

Isoetes is heterosporous, with megasporangia arising from the stem on the adaxial surface of the outer leaves and microsporangia from the adaxial surface of the inner sporophylls. The most centrally located leaves are usually sterile. The sporangia are much larger than those of either *Lycopodium* or *Selaginella.* Arising just above the sporangia is a small ligule which unites with other superficial cells to form a protective flap, the **velum,** over the sporangia. The latter are partially chambered by a series of sterile plates, the **trabeculae,** which extend inward from the sporangial walls.

The megasporangia produce many spore mother cells, most of which disintegrate. The remainder form tetrads of megaspores by reduction division. The microsporangia also produce numerous spore mother cells, nearly all of which divide to form microspores.

The gametophytes produced are similar to those developed by *Selaginella,* although they are smaller. The male gametophyte consists of but a single prothallial cell and a single antheridial cell at first. The latter develops into an antheridium consisting of a sterile jacket of four cells enclosing four spermatogenous cells, each of which gives rise to a single coiled multiflagellate sperm. *Isoetes* has one of the smallest antheridia among vascular plants, and it produces the smallest number of sperms.

The megaspore is uninucleate when shed, but soon a series of free nuclear divisions occurs. Walls develop in the apical end of the spore, which is determined by the triradiate crack exposing the developing megagametophyte. In this region several small archegonia are differentiated and rhizoids appear. In most respects this phase of development is like that of *Selaginella.*

The embryo forms very much as do those of the other members of the Lycopsida except that a suspensor is lacking. Soon after the primary root and leaf are formed the meristematic stem area begins to form the permanent vegetative organs of the plant.

ORDER PLEUROMEIALES

Pleuromeia (Fig. 23) is the only genus of the extinct order. The plants were about six feet tall with an unbranched trunk about four inches in diameter. *Pleuromeia* resembled *Sigillaria* in its columnar habit and apical cluster of long grasslike leaves. The plants were heterosporous and represent a group midway between the giant members of the Lepidodendrales and the very much reduced Isoetales.

FIGURE 23 *Pleuromeia.* This genus had species with unbranched trunks with enlarged basal rhizophores, roots, slender leaves, and a terminal cone. (after Hirmer.)

Summary of Characteristics

It is probable that the Lycopsida arose from one or more members of the Psilopsida at a very early time, probably in the Silurian. They attained a fair degree of organ development with true roots, leaves, strobili, heterospory, cambiumlike tissue, secondary tissues, the retention of the gametophyte and the embryo sporophyte within the megaspore wall, and many variations in the vascular system, including a few siphonosteles. Although their type of heterospory led them almost to the seed, it is doubtful that they developed beyond this point and so they probably did not give rise to any of our present-day seed plants.

Subphylum Sphenopsida

The members of this group were among the most unique plants that ever lived. Our vegetation would be much more colorful if some of the larger members had survived through the ages and remained to populate our forests. Their outstanding characteristic was the whorled arrangement of their stems and leaves, which made them quite different in appearance from the members of other groups. The sporophytes had true roots, stems, and leaves although the leaves were often very much reduced in size. Both branch and leaf traces were present but only branch gaps. Protosteles prevailed in all roots and in some stems, although siphonosteles were more common in the latter. Sporangia were borne in distinct, compact strobili on **sporangiophores** rather than on leaflike sporophylls.

The Sphenopsida has but one class, the Equisetineae, in which there are four orders only one of which has a single surviving genus.

ORDER HYENIALES

One genus, *Hyenia* (Fig. 24), in this order resembled *Asteroxylon* of the Psilopsida in that its aerial stems were well-covered with very small leaves. The plants were relatively small with slender, dichotomously branched aerial stems and rhizomes. The leaves were bilobed and whorled on the stems. Sporangia were borne on sporangiophores in open, loosely organized strobili.

Calamophyton was a somewhat more advanced genus with similar leaves and more pronounced whorling of leaves, sporangiophores, and stems. Both genera were apparently homosporous.

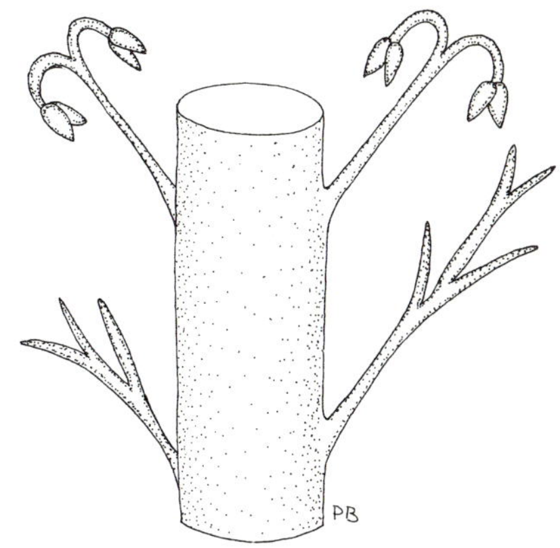

FIGURE 24 *Hyenia.* A portion of the stem; note similarity of sporangia to those of the early psilopsids. (After Benson.)

ORDER SPHENOPHYLLALES

Most of the members of this order were comparatively small prostrate or climbing plants that ranged from the Devonian to the Triassic epoch. They had slender, fluted stems with long internodes and branches arising from some of the nodes (Fig. 25). The leaves were wedgeshaped, more or less lobed, and arranged at the nodes in whorls of three, six, or nine. Internally the vascular arrangement was a protostele arranged in a triarch pattern in some and a haplostele in others.

Sphenophyllum is the best known genus, since numerous fossil remains of its many species have been found (Fig. 26). All of the genera were apparently homosporous.

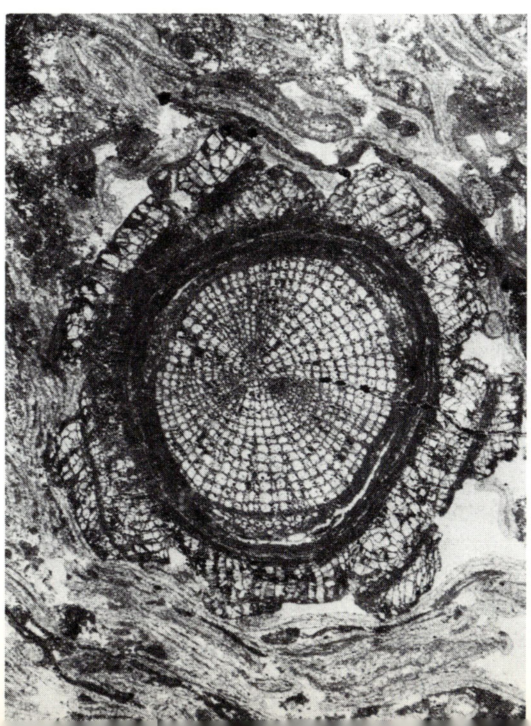

FIGURE 25 *Sphenophyllum.* A picture of a museum model. (Courtesy Field Museum of Natural History.)

FIGURE 26 *Sphenophyllum.* A cross section of the fossilized stem. The central area is xylem, surrounded by compressed layers of phloem endodermis, and fragmentary cortex.

ORDER CALAMITALES

The largest members of the Sphenopsida were in this order. Some members of the genus *Calamites* (Fig. 27) reached sixty feet in height. The stems often reached a basal diameter of three feet, although they were hollow as in the present-day *Equisetum* (Fig. 28). The vascular anatomy in the stem was the most advanced of

FIGURE 27 A section of a stem of a fossilized calamite on the left and leaves on the right. (Courtesy Field Museum of Natural History).

FIGURE 28 *Calamites.* Note nodes and internodes, vertical ribbing and similarity to present-day *Equisetum.*

all in the three subphyla we have considered. Most of them had ectophloic siphonosteles with the primary elements occurring in separate collateral bundles around the hollow center. Maturation of the tracheids was mostly endarch; in a few it seems to have been mesarch, while in the plants discussed in these three subphyla thus far it was always exarch. Secondary tissues were produced by an active cambium and a phellogen layer outside the cortical region formed a periderm.

The strobili of the calamites were made up of whorled sporangiophores, each of which developed four sporangia. Most of the calamites were homosporous, although in a few species spores of two sizes have been found. Nothing is known of the gametophytes in the fossil Sphenopsida, but in the homosporous ones, at least, they were probably similar to those now produced by the modern day *Equisetum*.

ORDER EQUISETALES

Equisetales is the only order in this subphylum with living species, and they are all in the genus *Equisetum*. They are commonly called horsetails and scouring rushes, since the vegetative portion of some species is supposed to resemble a horse's tail, and in early days the silica-impregnated stems were used to scour blackened kettles. They grow in tropical, subtropical, and temperate latitudes usually in quite moist habitats. A few tropical species are over ten feet tall, but most are around two or three feet in height. In all the stems are fluted and all have very small bract-like leaves.

Internally the stem is quite porous, with a central cavity in the internodes and two cylinders of canals in the cellular layers. The epidermis is made up of cells with very thick and heavily silicified walls. On the slopes of the ridges the epidermis has many stomates. Beneath this tissue, particularly in the ridges, are small clusters of sclerenchyma fibers. The cortical tissue, which is chlorenchymatous, forms a continuous band around the stem but is most abundant beneath the grooves. Within the cortex is the first cylinder of canals, termed **vallecular canals,** which are located beneath each groove of the fluted stem. Alternating with these canals and internal to them is the ring of **carinal canals.** These canals are positioned beneath the ridges and are associated with the vascular bundles (Fig. 29). The vascular tissue is essentially a siphonostele

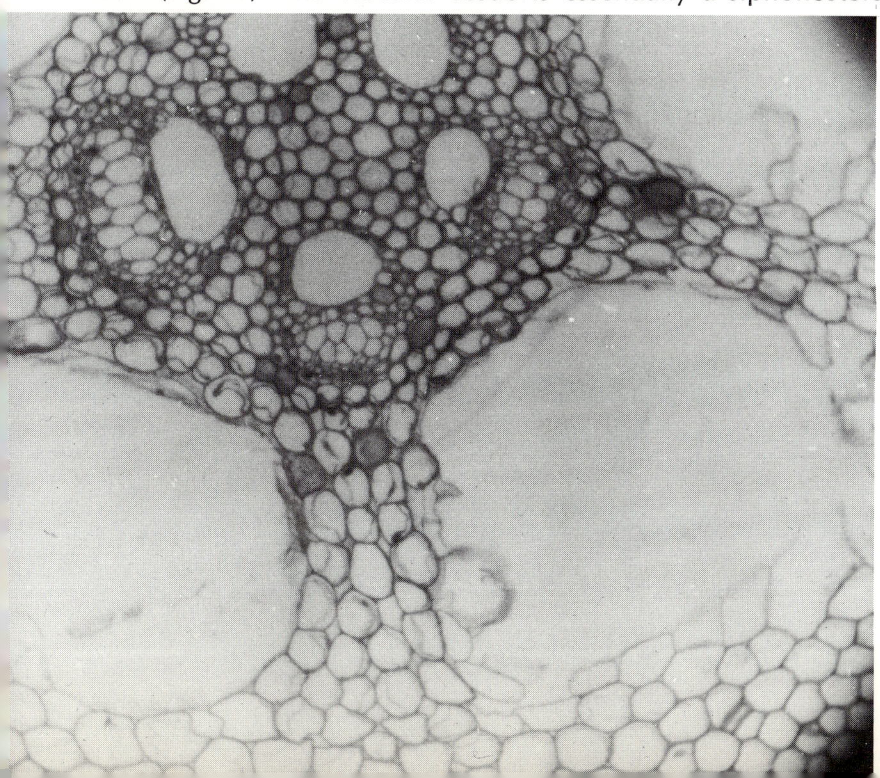

FIGURE 29 Portion of *Equisetum* rhizome in cross section. The large air spaces in the cortex are vallecular canals; alternating with them in the central pith area are the five smaller carinal canals. Immediately outside the carinal canals are the vascular bundles. Just outside the vascular bundles is a partially suberized endodermal layer.

The tissues in an aerial stem are the same except that the center is hollow and the outer layers of the cortex are chlorenchymatous.

but is broken up as the result of branch gaps in some species. In others it is a distinct ectophloic siphonostele. Usually there is an endodermis just outside the vascular cylinder; in some species there is also an inner endodermis formed just around the pith, when the pith is present.

The root systems of the Equisetales are of adventitious origin and fibrous in texture. The epidermal cells of the roots are well-supplied with an abundance of large root hairs.

Some species exhibit a type of **dimorphism** in that fertile shoots arise from the underground rhizomes, usually earlier and independently of the vegetative shoots (Fig. 30). One species, *E. arvense,* produces an unbranched, achlorophyllous fertile shoot in early spring, then after the spores are shed the much branched chlorophyllous vegetative shoot is formed on the same rhizome. Other species have just one type of branch with the strobilus produced at the apex of green shoots (Fig. 31).

Within the rather compact strobilus, sporangia arise in whorls on the underside of umbrellalike sporangiophores (Fig. 32). The sporangia number from about five to ten in the species of *Equisetum*. This is a larger number than are found in the fossil members of the Sphenopsida. The species of *Equisetum* are all homosporous. The spores contain numerous chloroplasts and in addition to the usual wall there is an outer layer that splits in a regular pattern to form two ribbonlike bands arranged spirally around the spore and attached in their central area to one portion of the spore wall (Fig. 32). The bands are hygroscopic structures like elaters. When spores are shed they are frequently entangled by the bands so that they fall in clumps. This means that the gametophytes will develop close together, which favors fertilization.

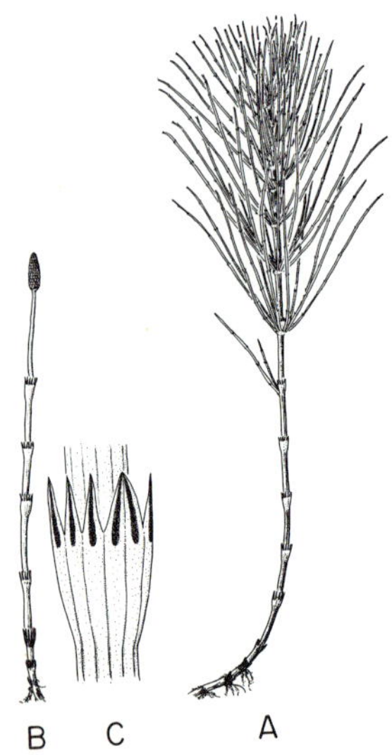

FIGURE 30 *Equisetum arvense,* a dimorphic species. a. Vegetative shoot which develops from the rhizome after the fertile shoot (b) has discharged its spores from the apical strobilus. c. node of fertile shoot. The fertile shoot in this species does not develop chorophyll.

FIGURE 31 *Equisetum hiemale.* In this species only one type of aerial stem is formed. The strobilus forming at the apex of the green unbranched shoot. b. node. The number, length, color, and shape of the leaf bracts at the nodes is a diagnostic characteristic in the identification of species of *Equisetum.*

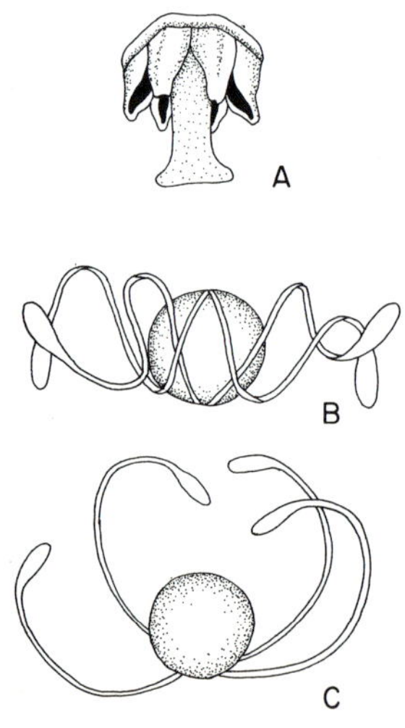

FIGURE 32 *Equisetum.* A. Sporoangiophore with sporangia B, C. spores with elaters derived from the exine layer.

The prothalli are usually less than one-half inch long, are variously lobed, and contain chloroplasts. Generally the prothalli, which are monoecious, develop their sex organs at different times, ensuring cross-fertilization. Studies have indicated that antheridia appear only on those prothalli which are crowded or inadequately supplied with food. When mature the antheridia contain numerous large, coiled, multiflagellate sperms.

Archegonia, like the antheridia, arise from superficial cells and at maturity consist of four vertical rows of neck cells containing one or two neck canal cells. The venter is imbedded in thallus tissue. Following fertilization the zygote forms a quadrant of cells. The two inner cells form the foot and root, and the two upper or outer cells give rise to the first stem and leaf. Both root and stem develop apical meristems very early and soon establish the young plant as an independent sporophyte.

Summary

The Sphenopsida, like the Lycopsida, probably arose from some member or members of the Psilophytales during the Devonian era. The two subphyla apparently had parallel lines of development. In the Sphenopsida the whorled arrangement of sporangia, leaves, and branches was the central theme of development which persisted throughout the subphylum. Lycopsida members tended to be more variable, often reached extremes in size, and had many more woody members, a variety of vascular systems, and much more emphasis on heterospory and its possibilities in the evolution of the seed.

Supplementary readings for this chapter follow Chapter 23.

Questions for Discussion

1. What group of algae probably gave rise to the early vascular plants?
2. Describe the different types of sperms found in the Thallophyta, Bryophyta, and early tracheophytes.

3. What advantages does heterospory have over homospory in the development of land plants?
4. Many of the relatively early vascular plants were arborescent. What advantages and what disadvantages does the arborescent habit have compared with the herbaceous habit?
5. Species of *Equisetum* have stems with many cylindrical cavities. What does this suggest about their habitat?
6. Compare and contrast the different types of steles in the Psilopsida, Lycopsida, and Sphenopsida.
7. Of what advantage is it to a plant stem to have a siphonostele rather than a protostele?
8. How do members of the Sphenopsida differ essentially from members of the Lycopsida?
9. Would you expect to find *Lycopodium* and *Sphagnum* living in the same habitat? Explain.
10. Compare and contrast the gametophytes of these early land plants with those of the liverworts.
11. Compare and contrast the sporophytes of these early land plants with those of the Bryophyta.

23

The Ferns

Tracheophyta Subphylum Pteropsida
Class Filicineae

The Pteropsida include the largest number of species of any of the comparable groups in the Tracheophyta. In fact, there are probably more species of Pteropsida now living than all of the three other subphyla both living and extinct.

Class Filicineae

The class Filicineae, exceeded in numbers by only the Angiospermae, has today approximately 285 living genera with about 9,000 species; in addition there are many species now extinct. The class is divided into five orders with living species in all except one, Coenopteridales. During the Carboniferous era the ferns, giant club mosses, and giant horsetails formed the dominant vegetation, but most of the present-day ferns are of more recent origin.

FIGURE 1 Circinate vernation in fern leaves. Note the small coils within the larger. The smaller ones are the coiled pinnules which have the same type of vernation as the larger ones.

FIGURE 2 Frond of *Pteridium aquilinum*. A-In this compound leaf the leaf divisions are pinnae and the alternate segments are pinnules. B-A pinnule showing the marginal sori; C-a section through the margin showing a minute inner indusium and an outer false indusium formed by the reflexed margin of the pinnule.

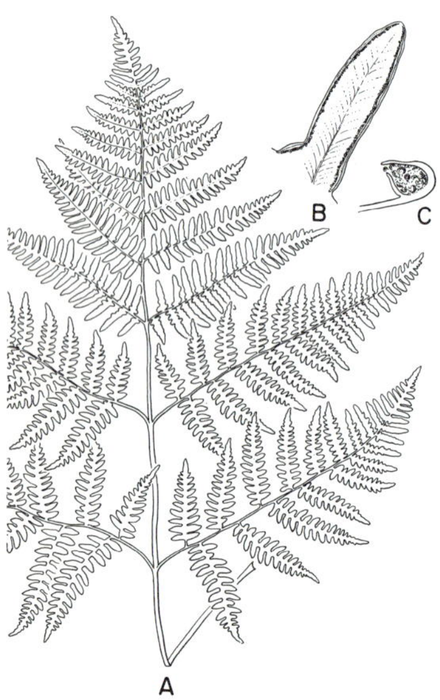

Most ferns live in tropical and subtropical habitats, but a fairly large number live in moist temperate latitudes and a few are xerophytes living in the desert habitats of the North American Southwest and elsewhere.

The typical fern has an underground rhizome which may be slender and grow horizontally considerable distances, or the rhizome may consist of nothing more than a short, stubby vertical axis. Most ferns have leaves **(fronds)** arising directly from the rhizome, but some produce an aerial stem that may extend upward forty feet or more as in tree ferns of the tropics and subtropics. The leaves of most ferns (by no means all) develop by a type of uncoiling process which we term **circinate vernation** (Fig. 1). Leaves may be simple, without any lobing, or they may be divided and compound. When compound, the leaf or frond will have divisions which are called **pinnae** and if these are further divided the ultimate segments are **pinnules** (Fig. 2).

Venation in the leaf and its segments is usually dichotomous (Fig. 3). Internally a leaf may have a typical mesophyll with palisade and spongy layer (Fig. 4), but many ferns are lacking either or both of these layers. The extreme in leaf cell organization is reached in a few species where we find only an upper and a lower epidermis, although the epidermal cells have protuberances resembling palisade tissue extended inward but still just a part of the epidermal cells.

FIGURE 3 Photomicrograph showing dichotomous venation in a fern leaf.

FIGURE 4 Cross section of fern leaflet showing palisade layers, spongy mesophyll, small vein and young indusium before appearance of sporangia.

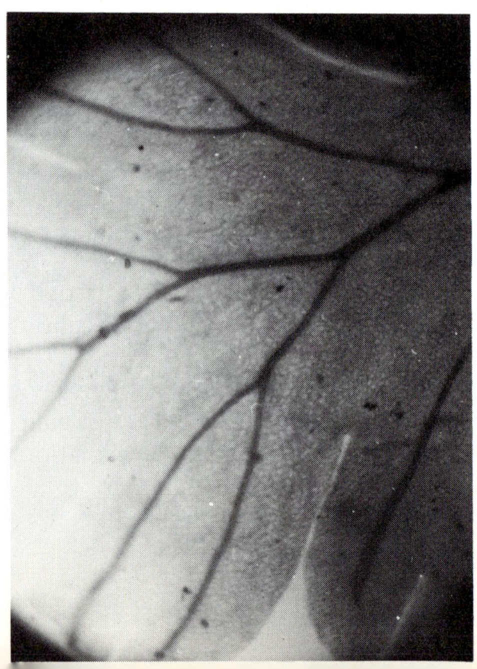

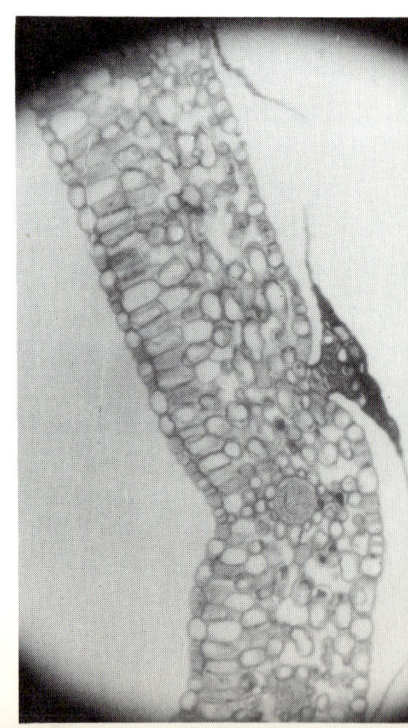

Sporangia are not borne in strobili, but usually appear in clusters on the underside of the leaf (Fig. 5). These clustered bodies are termed **sori** (**sorus,** singular). A few ferns have the sporangia compactly grouped in a fruiting structure termed a **synangium** (Fig. 6) and a few have the sporangia borne in loose clusters on a fertile segment of the frond like small berries on a stem.

Ferns have a greater variety of vascular systems than any other group. The rhizomes are often protostelic with every variation possible; some have siphonosteles, the amphiphloic type being the most common. Most large ferns and those with short stubby bases and many leaves have dictyosteles often elaborately developed. Leaf traces and leaf gaps account for most of the breaks in the vascular cylinders where dictyosteles are common, and branch traces and branch gaps (Fig. 7) are present in all stems with siphonosteles.

The root system of the typical fern is adventitious in origin and fibrous. As is true of most roots, these are protostelic but serve the usual functions of anchorage and absorption.

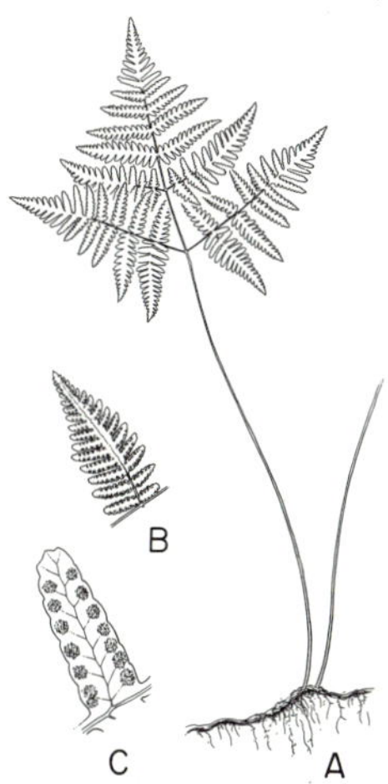

FIGURE 5 *Gymnocarpium dryopteris,* A-frond and horizontal rhizome; B-pinna; C-pinnule, underside showing sori and clustered sporangia. In this fern an indusium is lacking.

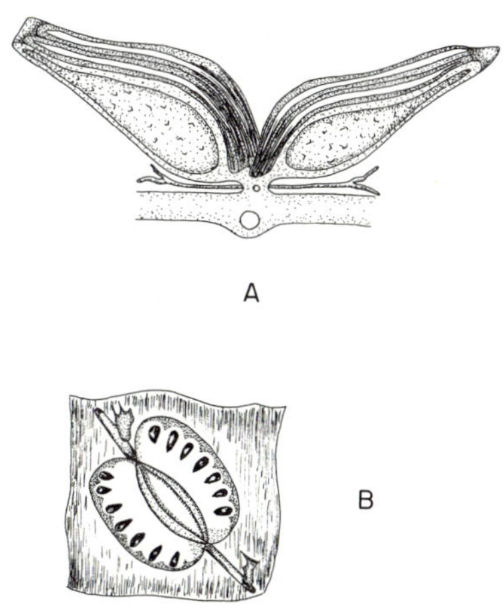

FIGURE 6 *Synangium.* A. lateral view showing walls enclosing the sporangia. B. open view from above showing the two rows of sporangial chambers. In both diagrams the sporangium is divided into two halves as the result of splitting down the middle.

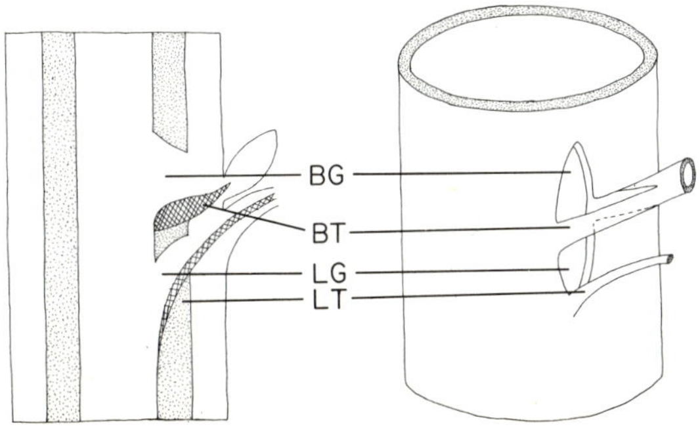

FIGURE 7 Leaf and branch traces and gaps; BG-branch gap; BT-branch trace; LG-leaf gap; LT-leaf trace. (After Eames and MacDaniels.)

ORDER COENOPTERIDALES

The members of this extinct group were probably derived from the Psilophytales and resembled them, for the most part, in that they were mostly stem and were dichotomously branched. The vascular system was protostelic, and the sporangia were homosporous and were borne either at the terminus of branches or along the margins of leaflike segments. Generally the sporangia were enclosed in a synangium-like structure, but some were free and borne singly. The sporangia were fernlike in that they developed a rather imperfect annulus which aided in the liberation of the spores. The members of this order apparently arose during the Devonian and disappeared in the Permian era.

ORDER OPHIOGLOSSALES

There are but three living genera of Ophioglossales, the most primitive order of living ferns. *Helminthostachys* is a prostrate or trailing plant found only in tropical Asia and Polynesia. The other two, *Ophioglossum* (Fig. 8) and *Botrychium* (Fig. 9), are found in temperate North America, Eurasia, and the Orient.

Ophioglossum, commonly called "adder's tongue" because of the shape of the leaf, which is elongate oval and entire, is usually only about four to six inches tall although one tropical epiphytic species has a leaf four feet long. *Botrychium,* known as "grape fern" because of the clustered sporangia on the fertile portion of the frond, has sterile, pinnately divided leaf segments on the lower portion.

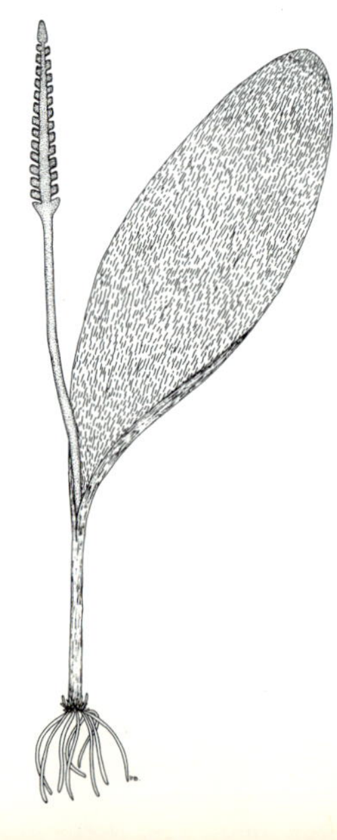

FIGURE 8 *Ophioglossum.* This fern has a fertile spike with two rows of sporangia at its apex and an elliptical sterile blade.

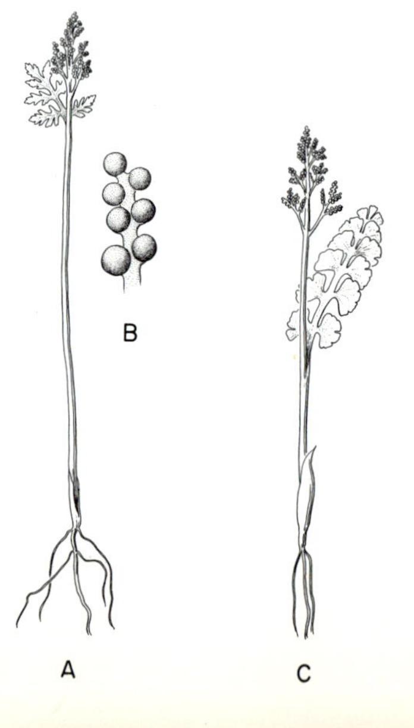

FIGURE 9 *A-Botrychium lanceolatum* showing dissected sterile blade and fertile portion. B-fertile segment showing double row of sporangia. C. *B. lunaria.*

Fertile Stalks

The most distinctive feature of the Ophioglossales is the complete separation of the sporogenous tissue from the vegetative portion of the leaf. In their origin the sporangiophores conform to foliage leaf segments and may be regarded as homologous to them. Spores are borne in sporangia that are on a specialized stalk arising at the base of the leaf segment but elongating well above the blade. In *Ophioglossum* the fertile stalk is unbranched, but species of the two other genera have compounded fertile stalks.

Leaf Development

Vernation of the leaf is erect rather than circinate, and the young leaf develops from a primordium that is protected by the sheathing leaf base of the preceding leaf. Venation is net or reticulate in *Ophioglossum* but of the open dichotomous type in the other genera.

Vascular Patterns

The stems, which develop from an apical cell and its derivatives, usually have an epidermis, a cortex, and a central stele. The stele is protostelic in early growth, but in some species it becomes an ectophloic siphonostele and in others it becomes dictyostelic as the result of large leaf gaps.

GAMETOPHYTES

Although prothalli have not been studied extensively in this order, they are mostly rather compact bodies somewhat like those of *Lycopodium* and *Equisetum*. Most of their nutrition is saprophytic but some chlorophyll is found in a few prothalli. Although there is some variation in the types of gametophytes and the positions of the antheridia and archegonia, they all seem to be homothallic. The antheridia produce many large, multiflagellate sperms and the archegonium has but a single binucleate neck canal cell, a large ventral canal cell, and an egg. Both archegonia and antheridia are well imbedded in the prothallus.

Chromosomes and Polyploidy

In addition to the many variations in vascular development, sporangial types, and other features in the Filicineae, an unusual variation in numbers of chromosomes has also been observed. Some ferns have very few chromosomes, but in *Ophioglossum vulgatum* diploid cells have over five hundred chromosomes and *O. petiolatum* has been reported to have over a thousand, the largest number of chromosomes reported in diploid cells of any vascular plant.

Klekowski and Baker in *Science,* July 15, 1966, report that 96 percent of the homosporous Pteridophyta have a gametic chromosome number greater than twenty seven, whereas 90 percent of the heterosporous ones possess a gametic chromosome number less than twenty-eight. These figures, particularly those of the former, are considerably greater than the chromosome numbers for seed plants. It therefore seems that polyploidy is a means by which homosporous Pteridophyta may achieve and maintain genetic variability in spite of the production of homozygous sporophytes. Since most of the homosporous Pteridophyta are monoecious and homothallic, polyploidy provides a means of variation.

ORDER MARATTIALES

The Marattiales is an ancient group of ferns that goes back to the Carboniferous era. Today there are seven genera and some one hundred species, found exclusively in tropical and subtropical areas.

In general, the Marattiales have short, erect stems with a pair of persistent fleshy stipules surrounding each leaf base. The leaves develop circinately and become large compound fronds. The sporangia are clustered in synangia and arranged in rows on the lower surfaces of fronds (sporophylls).

The spores germinate into dorsiventral thalli that are one to two inches long and cordate or ribbon-shaped. The thalli are somewhat like those of the Filicales although usually much larger. The sex organs are imbedded much as are those of the Ophioglossales and are of about the same size and content.

Following fertilization the embryo begins as a two-celled structure. The inner of these two cells gives rise to the apex of the shoot and the first leaf while the foot is formed from the outer cell, which is towards the archegonial neck. After considerable cell division in the shoot area a root arises from this portion of the embryo.

The vascular anatomy of the stem is quite complex since it is developed in a stem large in diameter but short in length. At first it is a siphonostele, but with the production of many large leaves and by girth expansion the vascular cylinder becomes dissected into a complex dictyostele consisting of two or more concentric cylinders of vascular bundles.

The largest genus in this order, *Angiopteris,* has species distributed throughout the Far East, particularly in the South Sea Islands. *Marattia* (Fig. 10) is world-wide in its tropical distribution, while *Danaea* is limited to tropical America.

FIGURE 10 A diagram of *Marattia,* a tropical fern. (After Coulter, Barnes, and Cowles.)

ORDER FILICALES

By far the largest number of individuals and also of different species of lower vascular plants are members of the Filicales. They were most abundant during the Carboniferous era, sometimes called the "age of ferns," but many species were abundant in the late Devonian and the remains of some have been taken from middle Devonian strata. Although the number of ferns declined in the Jurassic era and soon after, we still have over 8,500 surviving species in this order. Most ferns live in warm, moist climates but there are hundreds of species living in temperate habitats and some have the same enduring characteristics of the *Selaginella* resurrection plant and live in desert areas, where they become dormant during dry periods and quickly revive with moisture.

The various types of vascular systems are more commonly studied in members of the Filicales than in any other group since they have nearly all types in their rhizomes. Genera such as *Gleichenia* (Fig. 11) and *Trichomanes* have protosteles; *Pteridium* (Fig. 12) and *Dicksonia* have dictyosteles; and *Adiantum* (Fig. 13) and *Dennstaedia* have amphiphloic siphonosteles.

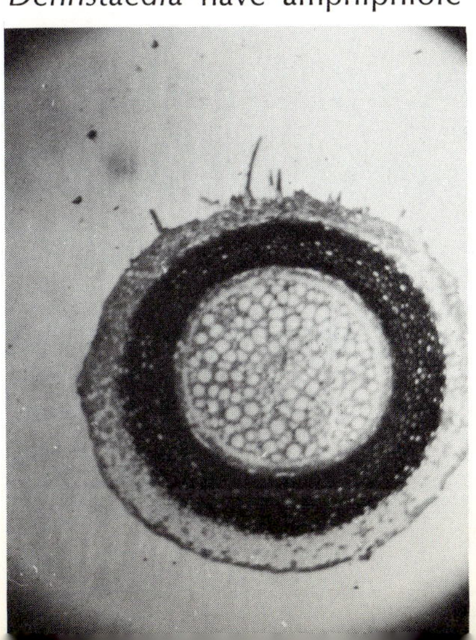

FIGURE 11 A cross section of the rhizome of *Gleichenia* showing a protostele. The large central area is composed of large xylem elements with some undifferentiated parenchyma. A thin layer of phloem surrounds the xylem. The deeply stained ring (about 10 cell layers deep) is sclerenchymatous mechanical tissue followed by the cortical parenchyma and epidermis.

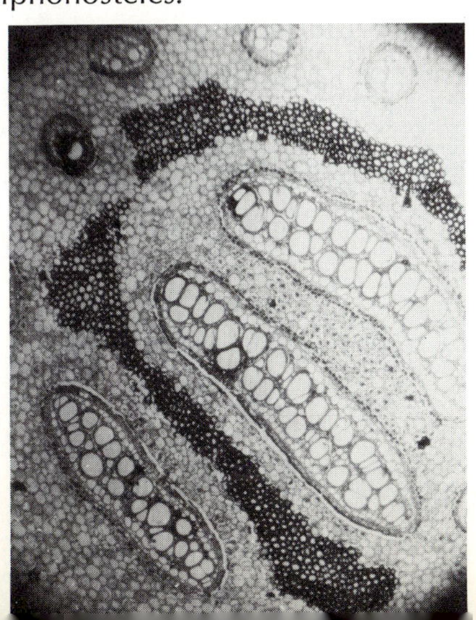

FIGURE 12 Portion of the cross section of the rhizome of *Pteridium* showing a dictyostele. The two elongated vascular bundles in the center are surrounded by two regions of deeply stained mechanical tissue. Surrounding this tissue are a number of large and small vascular bundles. Each vascular bundle has a central xylem of large tracheids surrounded by phloem.

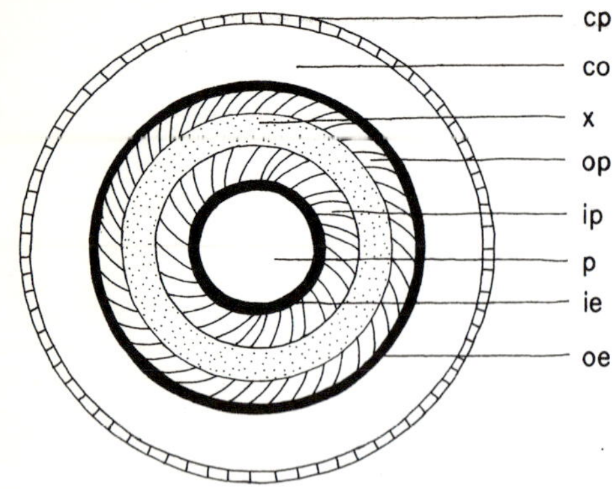

FIGURE 13 The amphiphloic siphonostele of *Adiantum*. ep-epidermis; co-cortex; x-xylem; op-outer phloem; p-pith; ie-inner endodermis; oe-outer endodermis.

Leaves

Most ferns have horizontal rhizomes with the leaves coming off this structure and its branches (Fig. 14). Some ferns, particularly the desert species, have short, stubby, erect rhizomes (Fig. 15) and some have long, erect aerial stems with large apical fronds. These are the tree ferns (Fig. 16) found mostly in the Southern Hemisphere.

There is great variety in the shape, sizes, and internal anatomy of the fronds. *Asplenium rhizophyllum*, the "walking fern," has very much elongated lanceolate leaves and reproduces asexually by rooting at the tips of the leaves where they contact the soil. Another fern, *Notholaena standleyi*, has a palmately lobed leaf which closes up like a fist in dry weather. Most ferns, however, are pinnately lobed or pinnately compound. In all cases the leaves of members of the Filicales develop circinately. In the usual herbaceous fern the young apical coil is less than a half inch wide, but in tree ferns the coiled tip is about the size, and looks like the scroll, of a bass fiddle.

Earlier in the chapter it was mentioned that the mesophyll may be present or absent. Haberlandt has descriptively called those leaves in which the mesophyll is nothing more than internal lobes of the epidermal cells as having "armpalisade structures." The chloroplasts are located in these protuberances and stomates are usually found only in the lower epidermis.

FIGURE 14 *Polypodium vulgare*, showing horizontal rhizome (a) and sori in (b).
FIGURE 15 *Dryopteris filix-mas*. A-Entire plant showing short stubby rhizome; B-pinnule with sori; C-dorsal view of indusium; D-lateral view of indusium showing marginal attachment.
FIGURE 16 A tree fern jungle in Hawaii, Volcanoes National Park. (Courtesy National Park Service.)

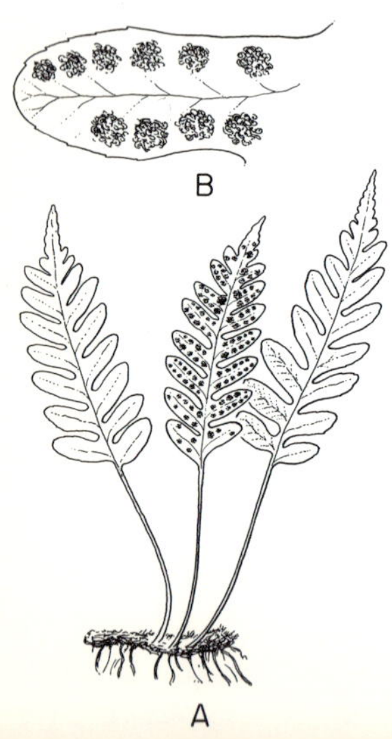

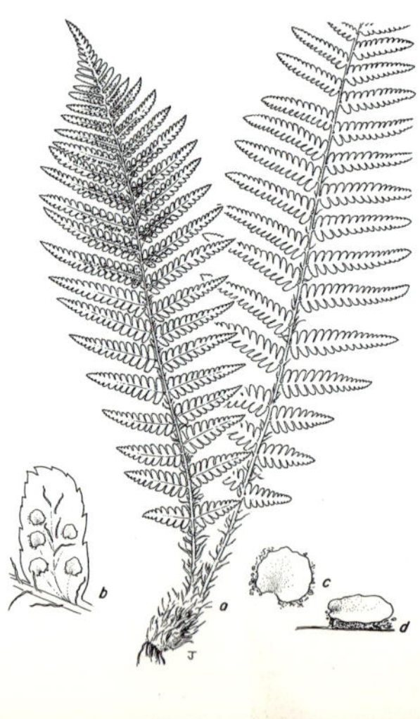

Sporangia

The sori usually are formed separately on the lower surface of the leaf, but many ferns have sporangia in continuous rows along the margin of the leaf. The manner in which sporangia are borne is quite specific and therefore provides a good diagnostic characteristic in taxonomy. Some ferns have a flaplike growth, the **indusium** (Fig. 18), covering the sporangia. In others, where the sporangia are covered by the inrolled margin of the leaf (Fig. 17) or by leaf flaps at the terminus of the leaf, we speak of it as a **false indusium** (Figs. 18, 19). The indusium may be umbrella-shaped, crescent-shaped, or even broken up into short segments growing among the sporangia, looking very much like paraphyses and probably serving the same function.

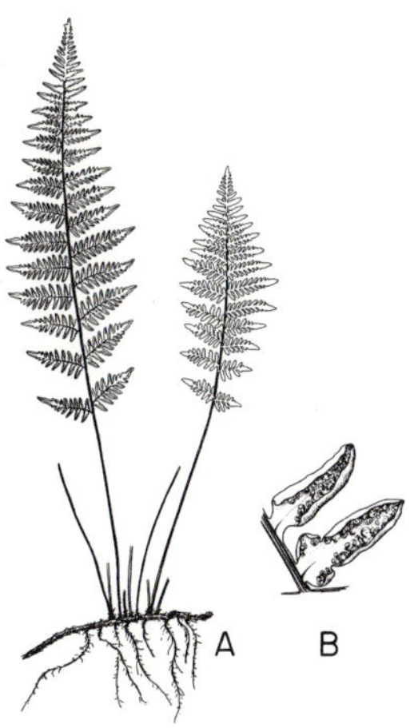

FIGURE 17 *Cheilanthes alabamensis.* A-plant; B-pinnules showing false indusium formed by inrolled leaf margin.

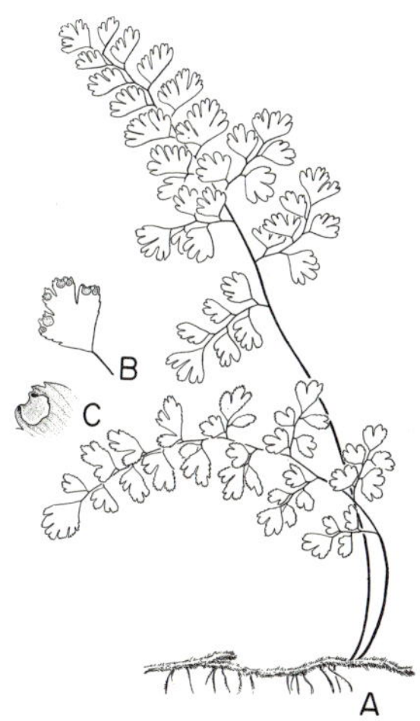

FIGURE 18 *Adiantum capillus-veneris.* A-entire plant; B-pinnule, often rather fan-shaped; C-note the false indusium formed by marginal leaf flap.

FIGURE 19 Pinnule tips of the fern *Hymenophyllum* showing marginal false indusium with enclosed sporangia produced as they are in *Adiantum*, Figure 18.

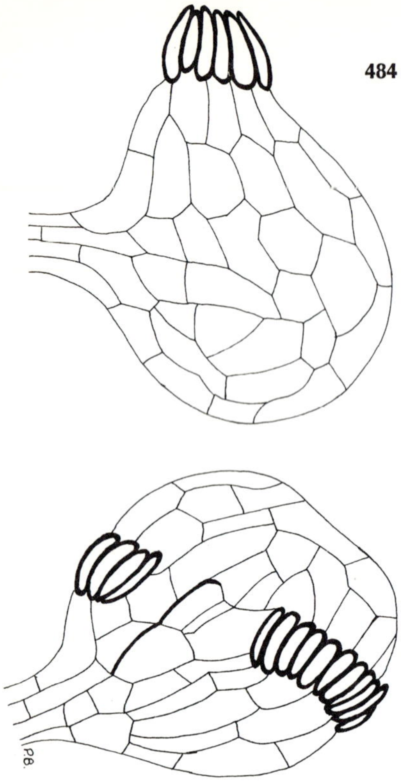

FIGURE 20 Fern sporangia. The thick walled cells show the variations in position of the annulus in two different ferns.

Sporangia also have a good deal of variation in the Filicales. In nearly all the wall is one cell layer in thickness but a specialized row of cells with very much thickened inner walls differentiates as a ring or **annulus** around the sporangium, like a tire around a sphere (Fig. 20). In many ferns dehiscence is accomplished by a vertical slit along the side of the sporangium. In others dehiscence is determined by the position of the annulus. The latter does not form a complete ring about the sporangium but usually leaves a number of thin-walled cells, the lip cells along the axis of orientation of the annulus. With maturity of the spores the annulus becomes a highly hygroscopic structure. Shortening of the annulus from drying causes the thin-walled lip cells to pull apart, rupturing the cells of the sporangium (Fig. 21). As the annulus pulls back

FIGURE 21 Diagram of a fern sporangium showing thin-walled lip cells, dehiscence and dissemination of spores.

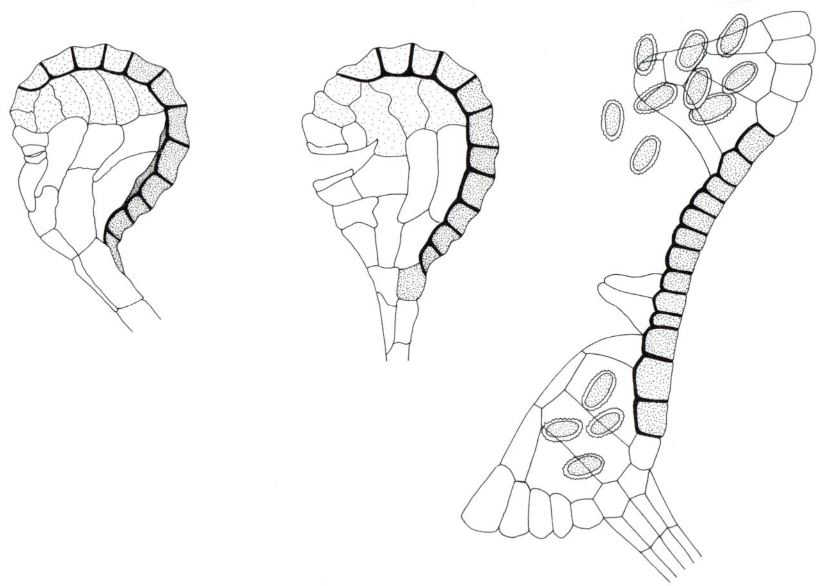

the sporangium is opened; the annulus then explodes forward like a spring throwing out a number of spores. This process may be repeated a number of times, as a result of alternate drying and moistening, before all spores are discharged.

The position of the annulus varies in different ferns. It may be in a horizontal position around the sporangium, or it may be vertical or oblique, or it may be at the very top like a cap (Fig. 20).

Gametophytes

When discharged, the spores may fall in a favorably moist location where they germinate forming a small green prothallus. This little, more or less cordate-shaped gametophyte is usually only a millimeter or two in diameter, and is anchored by unicellular rhizoids (Fig. 22). Both sex organs are usually developed on the ventral side of the same prothallus but they vary in position in different fern species. The antheridia usually form first before the prothallus has reached maturity. As the prothallus grows from the apical cell, the antheridia are gradually moved backward and the archegonia, which then begin to appear, are located nearer the notch and median portion of the prothallus.

The antheridia are small, spherical and usually give rise to thirty-two multiflagellate sperms (Fig. 23). The archegonia are quite small and generally have a single binucleate neck canal cell, although more primitive members of the Filicales generally have two distinct neck canal cells. In both types the eggs and ventral canal cells are present.

There may be many archegonia and consequently many eggs developed on a single prothallus, but once a zygote is formed in an archegonium further union of eggs and sperms on a prothallus is inhibited, or at least only one zygote goes on forming an embryo. A few exceptions have been reported in nutrient studies of prothallus culture but it is obviously a very rare situation for a prothallus to have multiple embryos.

The development of the embryo (Fig. 24) is very much like that described for other members of the subphylum. As in seed plants the first leaf is quite different in appearance from the mature types (Fig. 25) and the embryo and prothallus disappear when the permanent leaves and roots are established.

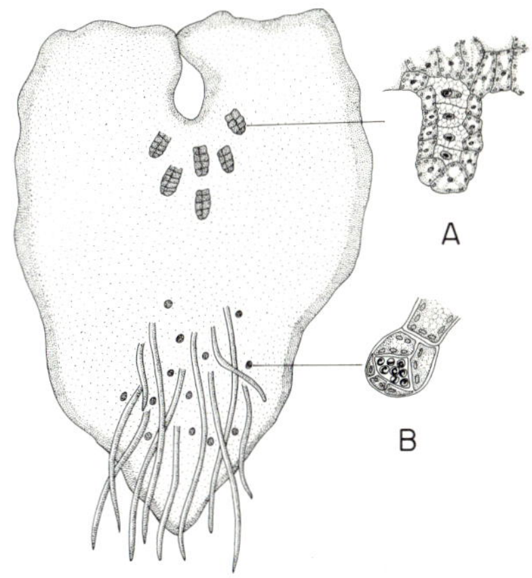

FIGURE 22 Underside of fern prothallus showing rhizoids and A-archegonium; B-antheridium.

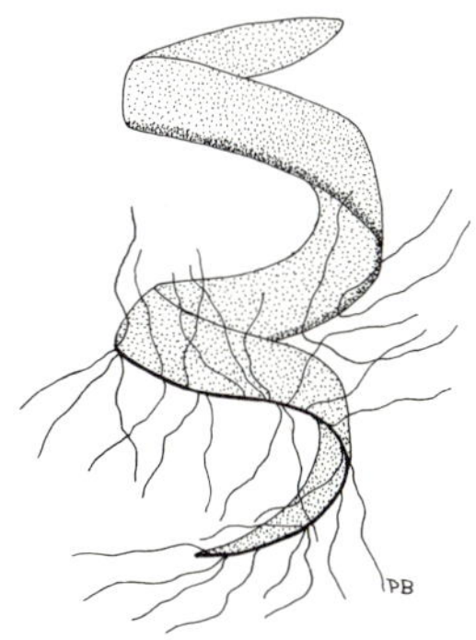

FIGURE 23 Fern sperm.

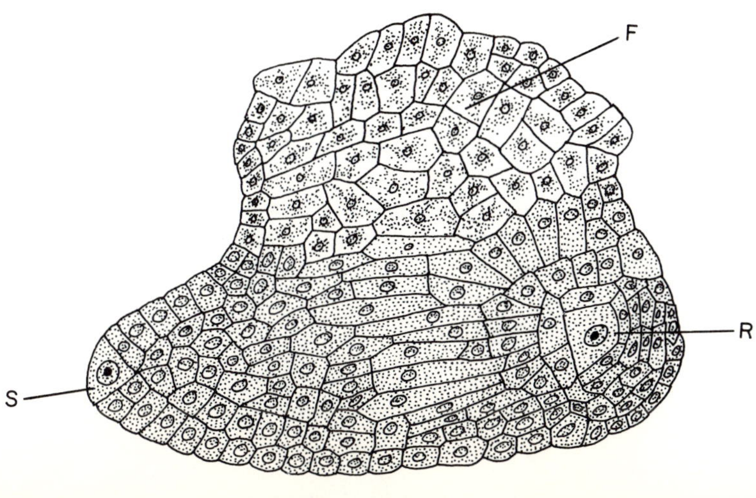

FIGURE 24 Young embryo of fern showing growing points of S-stem; R-root; F-foot. Redrawn from Chamberlain-*Elements of Plant Sciences.* McGraw-Hill.

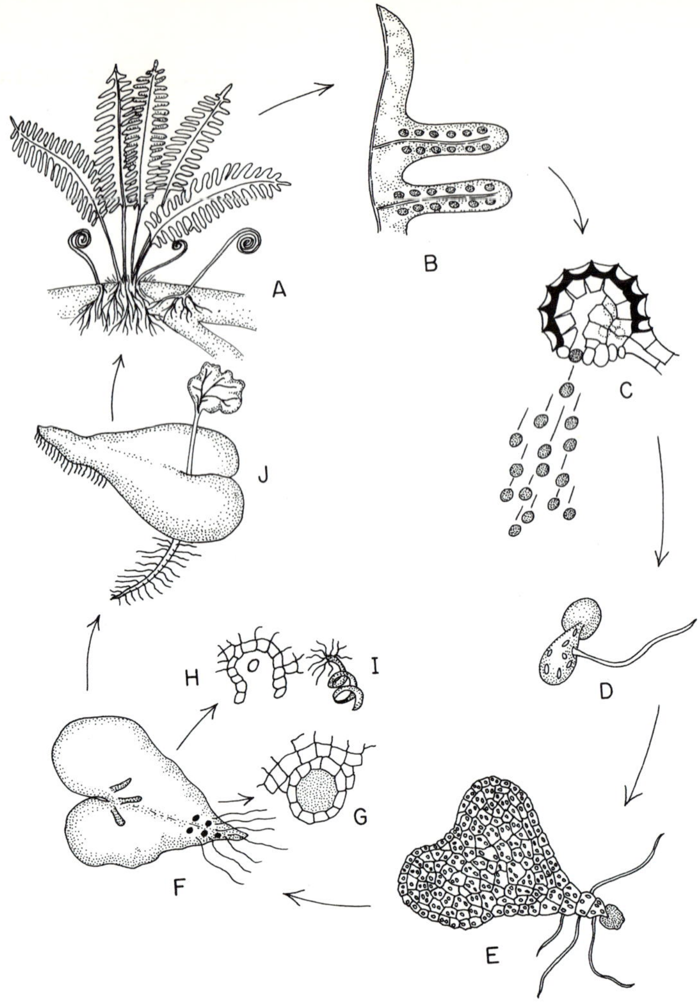

FIGURE 25 Life cycle of a fern. A-Mature sporophyte; B-portion of pinnule; C-sporangium; D-germinating spore; E-young prothallus; F-prothallus with archegonia and antheridia. G-antheridium, H-archegonium; I-sperm; J-embryo sporophyte developing on prothallus.

Apogamy

A phenomenon observed in some ferns, and in other plants as well, is the development of a sporophyte from a gametophyte in which fertilization has not taken place. This is **apogamy** and it has been induced in some ferns by withholding water from the prothallus. It has been demonstrated in a number of desert species and it is possible that most xerophytic ferns are capable of this type of reproduction. In some of these species at least, the chromosome numbers of both gametophyte and sporophyte generations are the same, the sex organs in these species are imperfect, and some sporangia are abortive.

Classification of the Filicales

The Filicales are divided into seven distinct families and vary in characteristics from the primitive Osmundaceae, which have more characteristics in common with the earlier orders than any of the other Filicales families, to the Cyathaceae, the tree fern family, found only in tropical and subtropical habitats. The Polypodiaceae is by far the largest family, with some six thousand species. Members of this family live wherever ferns are capable of surviving. Species of the genera *Cheilanthes* and *Notholaena* usually live in dry habitats, while those of such genera as *Adiantum, Polypodium,* and *Pteridium* live in moist, temperate habitats. A comparison of the three major homosporous orders of the Filicineae is given in Table 23-1.

Table 23-1

Ophioglossales	**Marattiales**	**Filicales**
Gametophyte	*Gametophyte*	*Gametophyte*
Without chlorophyll	With chlorophyll	With chlorophyll
With endophytic fungus	With endophytic fungus	Usually without endophytic fungus
Cylindrical in shape	Strongly dorsiventral	Strongly dorsiventral
Sex organs on all surfaces	Sex organs on ventral surface, in some antheridia may be on dorsal surface also	Sex organs on ventral surface
Homosporous	Homosporous	Homosporous
Sporophyte	*Sporophyte*	*Sporophyte*
Sporangia borne on fertile portion of spike	Sporangia borne on leafy sporophylls either free or in synangia	Sporangia borne in sori or in rolled leaf margins
Erect vernation	Circinate vernation	Circinate vernation
Leaf tissue fleshy but not highly differentiated	Leaf tissue well differentiated	Leaf tissue usually well differentiated
Stipules absent	Stipules present	Stipules usually absent
Vascular anatomy of stem varies from siphonostele to simple dictyostele	Stem vascular anatomy varies from siphonostele to complex dictyostele	Stem anatomy varies from a protostele to a siphonostele to a complex dictyostele

Hydropteridales

Most individuals who are familiar with ferns would have a difficult time recognizing the "water ferns" as members of the class Filicineae. For the most part they live exclusively in water or in very muddy substrates. Instead of sporophylls or sporangiophores the spores are produced in **sporocarps,** specialized globose bodies which develop on short stalks from the rhizomes and contain both microspores and megaspores.

Description

Marsilea (Fig. 26) is the largest genus, with some sixty-five species, and the individual plants are also larger than those of the other genera to be described here. The leaf of *Marsilea,* with four broad terminal leaflets at the apex of a long petiole, looks very much like a four-leafed clover. The leaves are produced in two rows on the horizontal rhizome and arise circinately.

The mature stem has an amphiphloic siphonostele vascular system (Fig. 27). Both inside and outside the two phloem layers

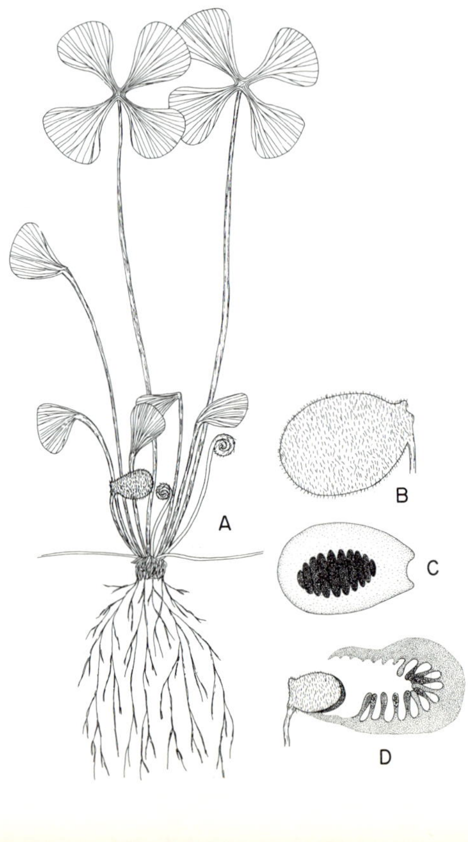

FIGURE 26 *Marsilea*. A-Plant showing horizontal rhizome, leaves, circinate vernation; B-sporocarp; C-section through sporocarp showing sporangia; D-germinating sporocarp with the swollen gelatinous material carrying the sporangia out of the sporocarp. Redrawn from Benson, L. 1957. *Plant Classification.* D. C. Heath Co.

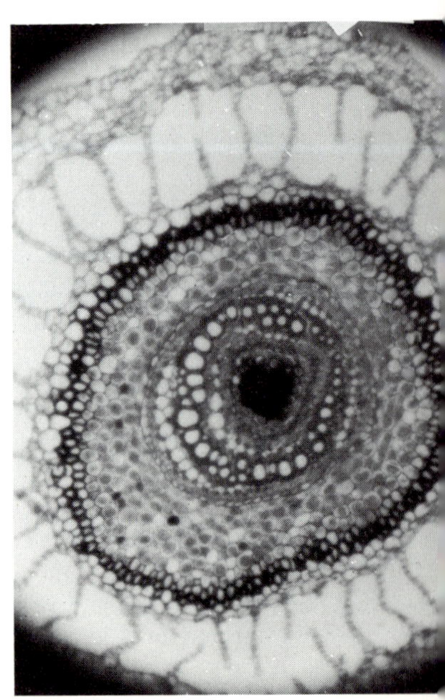

FIGURE 27 *Marsilea* rhizome in cross section. The very dark center is a pith with very thick-walled sclerenchymatous cells; outside of the pith is a single ring of endodermal cells; next is a ring of pericycle cells; then a ring of phloem; next is a double ring of large xylem cells which in turn are surrounded by phloem, pericycle, and endodermis. About five or six rows of storage parenchyma cells are next followed by one to two rings of heavy walled and deeply stained sclerenchyma cells. This is followed by a very spongy cortex layer with many large air-spaces. More compactly arranged cortical cells are next with some hypodermal cells enclosed by the epidermis. This is an example of an amphiphloic sphonostele in an aquatic plant.

are single rows of pericyclic cells, and these are in turn covered by endodermal layers one cell thick. The cortical region is made up of many rows of parenchymatous cells that are frequently interrupted by large air spaces. A slightly cutinized epidermis covers the cortex.

The roots also have considerable parenchyma cortical tissue, and are protostelic with the exarch xylem formed in a monarch or diarch pattern.

The sporocarps arise singly or in clusters from the nodes at the base of the petioles. Within the sporocarps sori, looking like tiny clubs, are developed in two rows. In *Marsilea* the microsporangia and megasporangia arise in the same cavity of the sporocarp, but in *Azolla* and *Salvinia* they are borne in separate sporocarps. Liberation of the microspores and megaspores in *Marsilea* is accomplished by a swelling and splitting of the sporocarp (Fig. 28). The two rows of sori then float out attached to a gelatinous filament like clothespins on a clothesline. The sporangia then burst, liberating the microspores and megaspores.

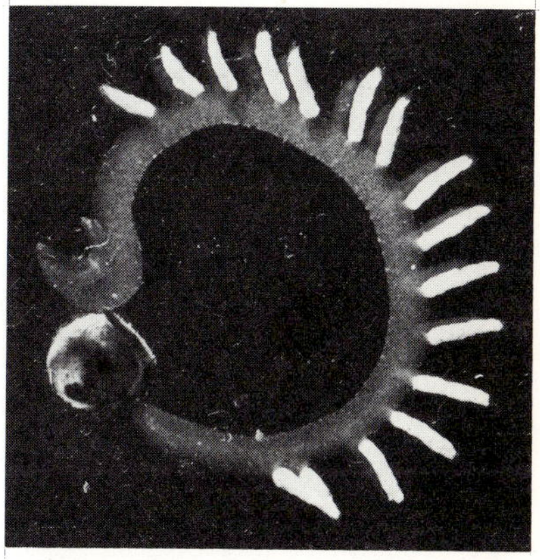

FIGURE 28 *Marsilea*. Germinating sporocarp with sporangia carried on an expanding mucilaginous ring. (Courtesy Carolina Biological Supply Company.)

Gametophytes

The microgametophyte develops within the microspore and consists of a small prothallial cell and a much larger cell. The latter divides successively to form an antheridium. Within the antheridial jacket sixteen spermatogenous cells form and each gives rise to a corkscrew-like multiflagellate sperm.

The megaspore develops a triradiate crack as in *Selaginella*, and in this vicinity a large cell and a smaller one are differentiated from the cytoplasm of the megaspore. The larger cell forms nutritive tissue, but the smaller is the archegonial initial and develops an archegonium consisting of a venter and a neck two cells long.

Like *Selaginella* and *Isoetes* the embryo begins its development within the megaspore wall. The zygote forms a quadrant with the two cells nearest the neck of the archegonium giving rise to the first leaf and root, and the two lower cells forming the stem and foot of the embryonic sporophyte.

Salvinia and *Azolla* are both floating ferns, and both are quite small. *Salvinia* (Fig. 29) has a slender stem bearing two kinds of leaves. The dorsal leaves are produced in rows and are flat, oval, or oblong, and float on the surface of the water, often raised like tiny sails. The ventral leaves are in two rows, are very much dissected, and hang down in the water resembling roots.

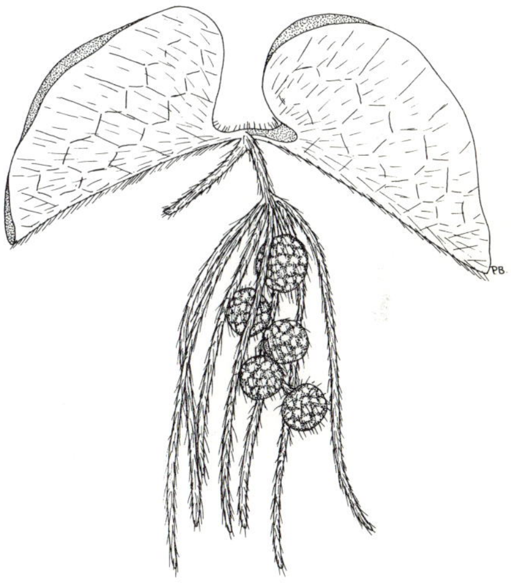

FIGURE 29 *Salvinia*. The root-like organs are morphologically dissected leaves; the globular sporocarps may contain either microsporangia or megasporangia. Redrawn from Benson, L. 1957. *Plant Classification*. D. C. Heath Co.

Azolla is similar to a small seed plant called duckweed in the way it forms dense mats on the surface of the water (Fig. 30). *Azolla* has a much-branched, pinnately divided stem covered with two rows of minute crowded leaves. The latter rarely exceed two millimeters in overall length. Very small, slender roots develop from the lower surface of the tiny rhizome.

Both *Salvinia* and *Azolla* develop separate sporocarps, but in *Salvinia* they are the same size while in *Azolla* the microsporocarp is considerably larger than the megasporocarp. In most respects the development of the sporogenous tissue is similar to that of *Marsilea;* however, there is some variation in numbers of cells in the different spore-bearing organs and gametophytic tissue produced. These differences are listed below in the comparative life cycles of the three major genera of the Hydropteridales.

Comparison of the Life Cycles of Water Ferns

Marsilea

Zygote
Embryo (foot, root, stem, leaf)
Sporophyte (root, stem, leaves)
Sporocarp
Sori

- Microsporangium → Microspore Mother Cell → Microspores (32-64) → Microgametophyte → Antheridia (one) → Sperm (16)
- Megasporangium → Megaspore Mother Cell → Megaspores (32-64) → Megagametophyte → Archegonium (one) → Egg

Salvinia

Zygote
Embryo (foot, stem, leaf)
Sporophyte (stem, leaves)

- Microsporocarp → Microsorus → Microsporangium → Microspore Mother Cell (16) → Microspores (64) → Microgametophyte → Antheridia (two) → Sperm (4 in each antheridium)
- Megasporocarp (same size) → Megasorus → Megasporangium → Megaspore Mother Cell (8) → Megaspore (1 matures) → Megagametophyte (free nuclei) → Archegonia (several) → Egg

Azolla

Zygote
Embryo (foot, root, stem, leaf)
Sporophyte (root, stem, leaves)

- Microsporocarp (large) → Microsorus → Microsporangium → Microspore Mother Cell (16) → Microspores (64) → Microgametophyte → Antheridium (one) → Sperm (8)
- Megasporocarp (small) → Megasorus → Megasporangium → Megaspore Mot[illegible] Cell (8) → Megaspore (1 matures) → Megagametophy[illegible] (free nuclei) → Archegonia (several) → Egg

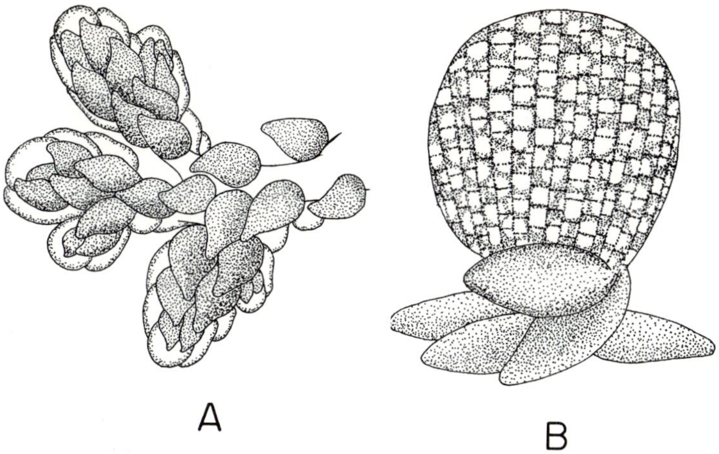

FIGURE 30 *Azolla* A-Growth habit. B-A very much enlarged globose microsporocarp with oval megasporocarps below it. B-Redrawn from Benson, L. 1957. *Plant Classification,* D. C. Heath Co.

Summary of the Filicineae

It is virtually impossible to summarize the characteristics of this class except to point out the wide variation that exists vegetatively and reproductively in the group. Leaves vary in size from the one-eighth-inch size of *Azolla* to the fifteen-foot length of the fronds of tree ferns. Stems vary just as much in size and have practically every type of vascular arrangement found in the plant kingdom. The fertile portions may be sporangiophores with separate and loosely arranged sporangia or they may be modified or unmodified sporophylls or more specialized sporocarps. The vast majority of the ferns are homosporous, but the water ferns are heterosporous.

All Filicineae produce a prothallus, but this may be as much as two inches in length or it may be reduced to a few cells within the spore walls. The prothalli may be independent and green or achlorophyllous within the spore. The embryo may develop externally on the independent prothallus or it may develop in early stages within the megaspore.

With so many differences, it is not surprising that experts are of the opinion that once the Coenopteridales were derived from the Psilophytales ferns of the Ophioglossales, Marattiales and Filicales took separate roads of development. Furthermore, the members of the Filicales show several evolutionary series with one family finally giving rise to the Hydropteridales through a reduction of parts.

Supplementary Readings

Andrews, H. N., Jr., *Studies in Paleobotany.* John Wiley & Sons, New York, 1961.

Bold, H. C., *Morphology of Plants,* 2nd ed. Harper, New York, 1967.

Bower, F. O., *Primitive Land Plants.* Macmillan, London, 1935.

Cronquist, A., *Introductory Botany.* Harper, New York, 1961.

Delevoryas, T., *Morphology and Evolution of Fossil Plants.* Holt, Rinehart and Winston, New York, 1962.

Dittmer, H. J., *Phylogeny and Form in the Plant Kingdom.* Van Nostrand Reinhold, New York, 1964.

Dittmer, H. J., E. F. Castetter and Ora M. Clark, *The Ferns and Fern Allies of New Mexico.* University of New Mexico Press, Albuquerque, 1954.

Foster, A. S., and E. M. Gifford, Jr., *Comparative Morphology of Vascular Plants.* Freeman, San Francisco, 1959.

Haberlandt, G. F. J., *Physiological Plant Anatomy.* Macmillan Book Co., London, 1914.

Manton, I., *Problems of Cytology and Evolution in the Pteridophyta.* Cambridge University Press, London, 1950.

Smith, G. M., *Cryptogamic Botany, Vol. II,* 2nd ed. McGraw-Hill, New York, 1955.

Sporne, K. R., *The Morphology of Pteridophytes.* Hutchinson & Co., London, 1962.

Wardlaw, C. W., *Phylogeny and Morphogenesis.* Macmillan, London, 1952.

White, R. A., Tracheary Elements of the Ferns, *Amer. Jour. Bot.* 50: 447–454; 514–522, 1963.

Questions for Discussion

1. Why should there be so much variation in the steles and other organs in ferns compared to those of other groups?

2. What is meant by vernation?
3. How do the gametophytes of most ferns compare with those of the tracheophytes in the preceding chapter?
4. Why do not the trunks of tree ferns make good lumber?
5. The chromosome number in ferns varies from some of the smallest figures known to the very largest. What does this indicate about the plasticity or stableness of the germ plasm of ferns?
6. What are the characteristics of eusporangiate and leptosporangiate ferns?
7. How do the leaves of ferns differ from those of angiosperms?
8. Where are spores borne in different types of ferns?
9. Is embryonic development in *Selaginella* more like that of an angiosperm, a fern, or *Marchantia?*
10. Why would we expect to find apogamy more common in xerophytic ferns than in mesophytic species?
11. Why would you expect to find polycyclic steles more common in short vertical rhizomes than in those ferns with long horizontal rhizomes?
12. Why do we consider the sporophyte of tracheophytes and bryophytes as developing from an embryo but not the sporophytes in the thallophytes?
13. Why are the members of the Hydropteridales considered more advanced than the members of the Filicales?
14. Suggest reasons why ferns are much more numerous today than members of the Psilopsida, Lycopsida, and Sphenopsida.
15. Of what advantage is it to a fern plant to have its archegonia and antheridia mature at different times?

24

The Primitive Seed Plants

Class Gymnospermae
Subclass Cycadophytae
Subclass Coniferophytae

The Gymnosperms are a major portion of the world's mountain flora and cover thousands of square miles of land in other areas as well, yet there are only about 60 genera with about 650 living species.

Like the more primitive members of the Tracheophyta, all members of the Gymnospermae have a vascular system composed of xylem and phloem. However, in the earlier groups secondary xylem and phloem were not formed in large quantity since cambium was often not formed, or when present did not grow as extensively as it does now in most gymnosperms. In fact, vascular cambium has reached its greatest potential in present-day conifers. Some of the Coniferophytae, such as members of species *Sequoia gigantea,* have basal diameters of over thirty feet and reach heights of approximately three hundred feet. Practically all of this bulk is due to the cambium which has been adding xylem and phloem continuously for over three thousand years. Very few plants can approach this growth achievement from the standpoint of either age and continuous growth or the amount of material produced.

The term gymnosperm (Greek *gymnos,* naked; *sperma,* seed) has a very literal translation and was adopted because the ovules which mature into seeds are borne in an exposed position on the megasporophylls in the female cone. The angiosperms, on the other hand, have their ovules borne within an enclosed ovary. This is by no means the only distinction between these two classes, nor is it the most obvious. Most gymnosperms still produce archegonia; some still have flagellated sperm, which for all practical purposes are still aquatic; nearly all gymnosperms depend upon tracheids for water conduction; and stored food for the developing embryo is still gametophytic tissue.

The gymnosperms apparently arose during the Paleozoic era but became much more prominent during the Mesozoic. At that time a number of evolutionary lines were established. Some flourished for a time and became extinct, but others continued on to the present day (Fig. 1).

The class is usually divided into two subclasses, the **Cycadophytae** and the **Coniferophytae.** Of the three orders in the former, two are known only by fossil members while the third has only a few surviving genera and species.

FIGURE 1 (Left) A petrified tree trunk of the now extinct *Araucarioxylon* in Petrified Forest National Park in northeastern Arizona. The basal portion of this uprooted tree is about six feet in diameter. (Right) Parts of petrified tree trunks as they appear scattered over the ground in Petrified Forest National Park. The park encompasses 94,000 acres but only a few areas are as densely covered with large specimens as this. They are of Jurassic age.

Subclass Cycadophytae

ORDER CYCADOFILICALES

The extinct order Cycadofilicales is sometimes known as the **Pteridospermae** since its members are commonly called seed ferns. They were fernlike (Fig. 2) in appearance and produced their seeds exposed on modified leaves rather than in cones. Their antiquity goes back to the Devonian era when they probably evolved from psilophyte stock through a very ancient fossil fern such as *Archaeopteris* (Fig. 3).

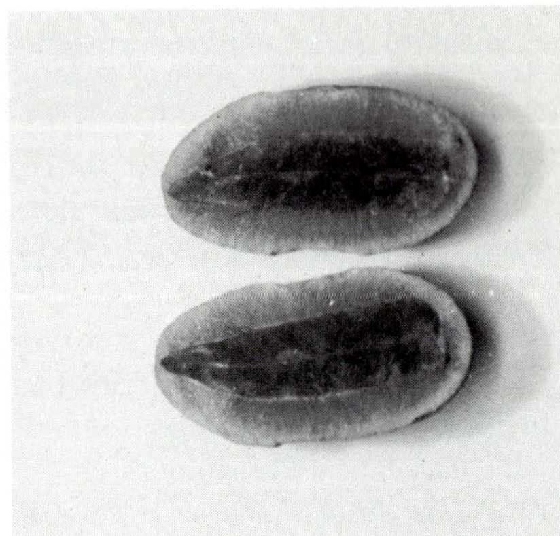

FIGURE 2 (Left) Fossil of a pinnule of *Neuropteris scheuchezeria.* Mazon Creek, Illinois. (Right) *Neuropteris rarinervis,* Mazon Creek, Illinois. Portion of frond with pinnae.

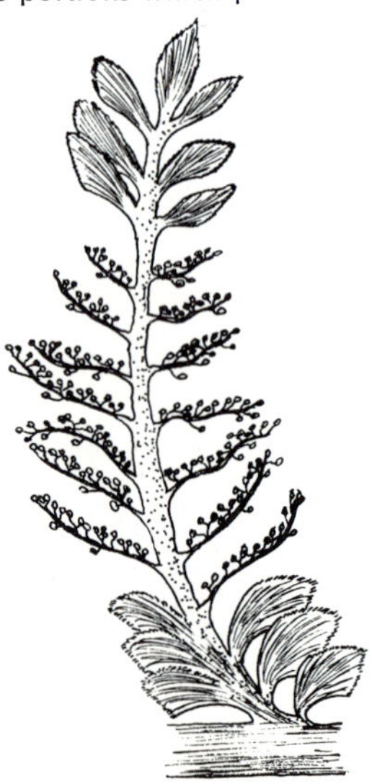

FIGURE 3 *Archaeopteris.* A fossil seed fern showing a compound leaf with foliage blades and modified fertile portions which produce seeds.

Both microsporangia and megasporangia were produced on pinnae of the compound leaves. The microsporangia produced possibly the first wind-blown pollen in the evolution of the plant kingdom. The megasporangia produced the megagametophytic tissue, including archegonia and ultimately the seed. The Pteridospermae apparently became extinct during the Jurassic era.

ORDER BENNETTITALES

This order, occasionally referred to as the **Cycadeoidales** since they more closely resembled plants of the next order, probably evolved from the Cycadofilicales during the late Paleozoic era, reached world-wide distribution in the early Mesozoic, then became extinct in the Cretaceous. Their abundance was so great that the Mesozoic era has often been called the Age of Cycads. Numerous fossils have been found in the Great Plains, particularly in

Nebraska, South Dakota, and Wyoming. The U. S. government recognized them in establishing Fossil Cycad National Monument.

In *Bennettites* the megasporophylls were aggregated into a distinct apical megastrobilus, although the megasporophylls were foliaceous and loosely organized. The microsporophylls still resembled the fronds of the Cycadofilicales and formed a leaflike ring around the central strobilus.

One genus, *Cycadeoidea* (Fig. 4) had a tuberous, unbranched trunk which grew to about twelve feet in height and formed a crown of pinnate cycadlike leaves at its apex. The stem had xylem which developed in an endarch manner and ultimately formed a siphonostele. The central pith was quite large, as was the cortex, and the latter was covered with a fairly thick periderm.

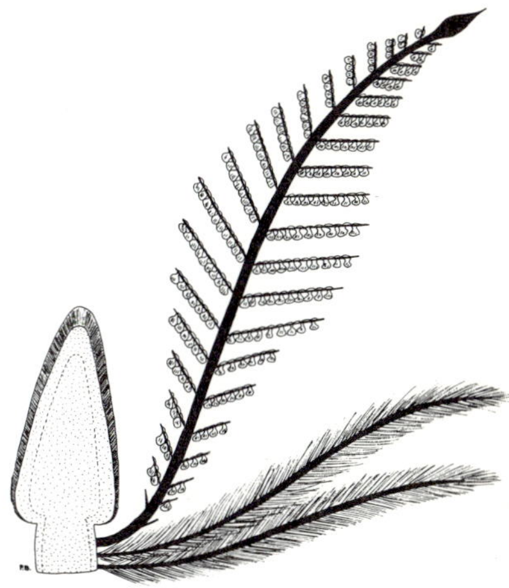

FIGURE 4 *Cycadeoidea,* one of the primitive members of the Bennettitales. The central turbinate structure is the ovulate cone. The expanded leafy structure is a microsporophyll and surrounding the entire structure are hairy bracts. (After Wieland.)

ORDER CYCADALES

All living cycads, of which there are nine genera and approximately a hundred species, live in tropical or subtropical countries. Only one genus, *Zamia* (Fig. 5), lives in the continental United States, mainly in Florida and adjacent Gulf states. It is also found in Mexico and in Central and South America. The others are relatively isolated and live in comparatively small groups in such places as Australia, Africa, Cuba, and Mexico. The cycads are a very ancient group and, although a few are declining in numbers, most are in no immediate danger of extinction.

The general appearance of a cycad is roughly like that of a palm tree and one, *Cycas revoluta,* is called sago palm (Fig. 6).

FIGURE 5 *Zamia florida.* The pinnate leaves are leathery.

FIGURE 6 *Cycas revoluta,* the Sago palm.

Cycads vary in size from the small, tuberous stemmed *Zamia florida* to the *Macrozamia hopei* of Australia, which reaches a height of sixty feet. They grow very slowly but, judged by the rate of growth of a few specimens, some have been estimated to be over a thousand years old.

The Stem

The stem of a cycad differentiates from a single group of apical initial cells. They at first give rise to an inner mass of thin-walled cells arranged in a conical or dome-shaped body and an outer mantle of small, deeply stained cells. The outer mantle gives rise to the foliar structures by a localized acceleration of growth just below the shoot apex. The inner dome-shaped body gradually differentiates into the pith, the provascular tissues, and the cortex.

At maturity cycad stems have a very large pith and cortex with a relatively narrow cylinder of vascular tissue in between. The endarch xylem usually consists of tracheids, wood parenchyma, and rays. This is surrounded by phloem, making the vascular system a siphonostele.

The types of xylem elements are usually scalariform; some have spiral markings. In considering the evolution of xylem development we usually think of the sequence as from narrow spiral to narrow scalariform, then wide scalariform, and finally pitted tracheids. On this basis the long, narrow scalariform type most commonly found in cyads relegates this group to a rather low position in seed plants.

Secondary growth by means of a vascular cambium occurs in many species but the amount of tissue produced is relatively small. Furthermore, the new growth is not always laid down in distinct annual rings. Definite concentric rings can be distinguished in some stems but this seems to be associated with the development of new crowns of leaves at the stem apex and may be biennial or extended over even longer periods.

FIGURE 7 *Cycas* showing young circinately coiled leaves at the base of old leaves.

Leaves

The leaves are all pinnate with the exception of *Bowenia,* in which they are bipinnate. In *Cycas* the leaves develop circinately, but in all others they are moderately circinate (Fig. 7) to erect. The leaves resemble those of ferns not only in gross morphology but also in their usually dichotomous venation. However, the leaf texture is leathery with a fairly thick epidermis and hypodermis; internally the leaves are quite compact with a large mesophyll.

Roots

The roots of most cycads are quite fleshy and begin as a primary root which develops into a tap root. As the plants grow year after year a number of adventitious roots arise from the base of the plant and become large and fleshy forming a fascicled root system. The fleshy character of the root is due to a large cortex whose cells increase greatly in size, rather than number, as the root matures. Contributing to root diameter is a many-layered pericycle and a distinct endodermis. The epidermis becomes suberized quite early and beneath it there are usually several suberized hypodermal layers.

The protoxylem usually begins at seven or eight scattered points in *Cycas revoluta* which then gradually merge to form a diarch protostele. Maturation of the primary xylem is principally mesarch. Cambium originates in the parenchyma between the xylem and phloem. In older roots cork cambium is initiated in the pericycle or in the outer cortical layers. This meristem produces a periderm which is mostly phellem with very little phelloderm.

Reproduction

The strobili produced by members of this order vary in size, but a few develop cones that are believed to be the largest that ever existed. The **megastrobilus** of *Macrozamia dennisoni* has been described as two feet in length and over a foot in diameter at the base and having a fresh weight of fifty to seventy pounds; weights up to ninety pounds have been recorded for *Encephalartos* (Fig. 8) of Africa.

FIGURE 8 Two cycads: *Dioon* on the left, and at right *Encephalartos*. (Courtesy Chicago Natural History Museum.)

FIGURE 9 *Zamia.* The megastrobilus is on the left and the microstrobilus is on the right.

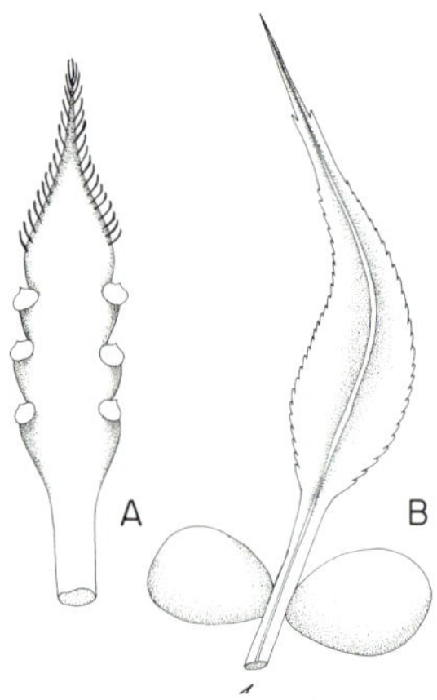

FIGURE 10 Megasporophylls of A-*Cycas revoluta* and B-*Cycad circinalis.* Note position of seeds on the margins. (After Coulter, J. M. and C. J. Chamberlain, 1910. *Morphology of Gymnosperms.* The University of Chicago Press.)

Generally the cones are compact, and the megasporophylls and even the microsporophylls are more like mega- and microsporangiophores rather than leaflike (Fig. 9). However, in *Cycas* the cones are made up of leaflike, leathery sporophylls with usually six to eight large ovules formed along the margins of the megasporophylls (Fig. 10). Most of the other genera have two ovules on the inner surface of the sporophylls.

The ovules of all genera are large. Those of *Zamia* are about one-half inch long. The **integument** (outside covering) is made up of three layers consisting of inner and outer fleshy layers separated by a stony middle layer. Within the integument is a megasporangium of many cells. One of these cells becomes a megaspore mother cell, which divides to form four megaspores arranged linearly. Three of these degenerate, but the innermost greatly enlarges to form the megagametophyte containing numerous nuclei as the result of free nuclear divisions. This is the general pattern of development of the megagametophyte in gymnosperms. Gradually the nuclei towards the periphery of the gametophyte (embryo sac) become enclosed by walls.

Archegonia (Fig. 11), numbering from three to ten, begin to form near the **micropylar** end of the gametophyte, but consist only of two neck cells and a large saclike structure within the gametophytic tissue. Within the saclike base of the archegonium a single large egg is finally differentiated.

The microstrobili of cycads, while much larger than those of most higher gymnosperms, are considerably smaller than the megastrobili. The **microsporangia,** borne on the abaxial surface of the microsporophylls (Fig. 12), arise from hypodermal initials. Microspore mother cells within the microsporangia give rise to the microspores. When the microspores are first formed they are uninucleate, but a division results in the formation of a pollen grain with

FIGURE 11 Microphotograph of one archegonium of *Zamia.* Note two neck cells and large saclike venter surrounded by endosperm.

FIGURE 12 *Zamia.* Longitudinal section through a microsporophyll showing the early development of two clusters of microsporangia.

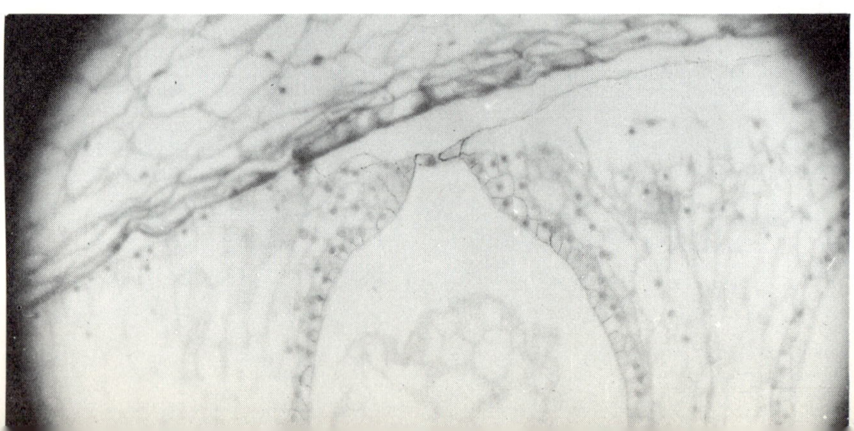

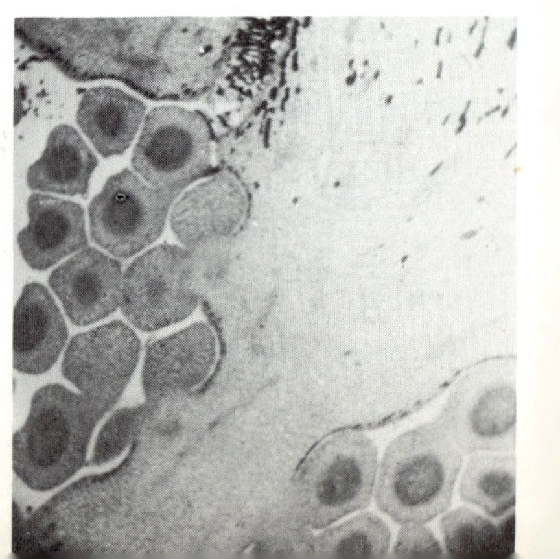

a small **prothallial cell** and a large cell which then divides, forming a **generative cell** and a **tube cell.** At this stage the microsporophylls separate, exposing the young microgametophytes within the spore wall to the wind.

Now for the first time we have reached wind-blown **pollen** and the transfer of the male gamete to the female organ without the necessity of water. At the time the pollen is shed, the megasporophylls open slightly and pollen sifts in between the megasporophylls and lodges near the apex of the megasporangium. In this tissue a few peripheral cells disintegrate, forming droplets of protoplasmic material. This opens a small cavity known as the **pollen chamber.** When pollen grains come in contact with this fluid they germinate, forming pollen tubes which grow down through the micropyle, between the integumentary layers, to the pollen chamber.

Once the pollen tube is formed the generative cell and prothallial cell move down the tube. The generative cell divides in the tube to form the **stalk** and **body cell** and the latter soon divides into two microgametes.

Each sperm is covered by a spiral band (Fig. 13) and for all practical purposes is still a swimming gamete. Sperm travel down the pollen tube, which grows to the apex of an archegonium where the tube bursts, liberating the sperm. Here a sperm may enter the archegonium and fertilize the egg.

Pollination to fertilization is a slow process in gymnosperms; it takes about five months in *Zamia,* and in other gymnosperms the intervening period is often much longer. Since there are several archegonia and often excess sperms, it is not unusual for several eggs to be fertilized. It is very rare for more than one zygote in an ovule to continue development into an embryo.

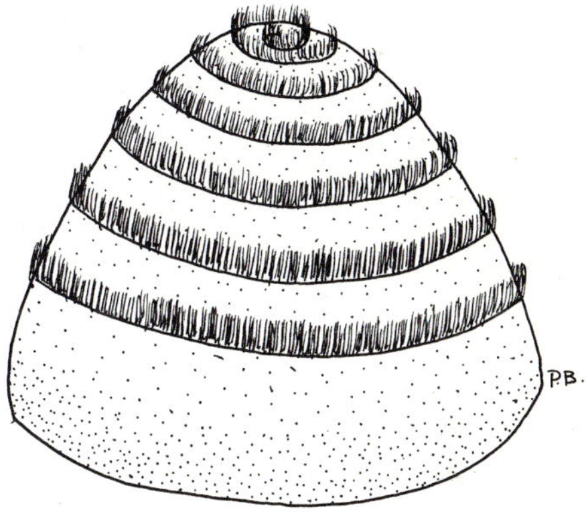

FIGURE 13 Sperm of cycad.

Seed

In most gymnosperms there is a period of free nuclear division in the embryo before wall formation begins. In *Zamia,* wall formation is confined to the nuclei in the basal portion of the fertilized egg; these cells then form the **proembryo.** As growth continues a mass of tissue is pushed out of the base of the egg and differentiates into three parts:

1. An upper portion which absorbs food for the developing embryo.
2. A middle zone of elongating cells which form the massive **suspensor.**

3. The most basal group of meristemetic cells which make up the true embryo.

The embryo developing then at the end of a suspensor differentiates into two **cotyledons** (Fig. 14), a tiny epicotyl between the cotyledons, and a large hypocotyl at the opposite end. In most gymnosperms these are the parts differentiated in the embryo, but in the conifers the number of cotyledons is usually considerably more than two and in one cycad, *Ceratozamia,* there is but one.

As the embryo develops the entire ovule greatly increases in size. The integument, which remains three-layered, becomes the seed coat, and stored food accumulates in cells of the embryo sac. In gymnosperms stored food is in the haploid cells of the megagametophyte, while in angiosperms the stored food is usually present in triploid cells as the result of divisions from a triple fusion nucleus.

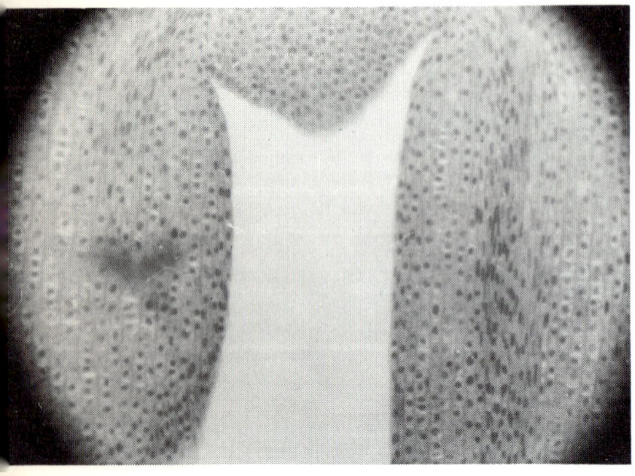

FIGURE 14 Apex of *Zamia* embryo showing two large cotyledons on either side and central epicotyl.

Subclass Coniferophytae

The largest, most numerous, and most successful members of gymnosperms are Coniferophytae. Of the four orders included, only one is without living members.

ORDER CORDAITALES

The members of the extinct order Cordaitales of arborescent plants arose during the Devonian era and were very abundant in the Carboniferous, but apparently died out during the Triassic era over 160 million years ago.

Cordaites

Cordaites (Fig. 15), the dominant genus in the order, had a tall, slender, branched stem with clusters of elongated leaves. Internally the stem had an endarch siphonostele with spiral or reticulate tracheids. These formed in the secondary wood and were often pitted. Cambial activity was quite limited, with the result that few annual rings are obvious.

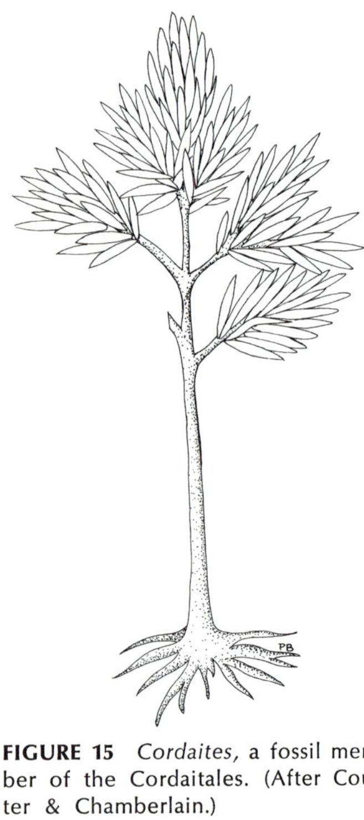

FIGURE 15 *Cordaites,* a fossil member of the Cordaitales. (After Coulter & Chamberlain.)

The leaves were quite distinct when compared with those of conifers or cycads. The leaves were usually up to three feet long but never more than eight inches in width. They were quite leathery and xeromorphic in structure, with dichotomous venation. Little is known of the roots, but they were dichotomously branched and in primary development had a protostele that varied from diarch to tetrarch.

Reproduction

The Cordaitales were either monoecious or dioecious, but none had bisporangiate strobili. Both male and female strobili were of about the same size. Each microsporophyll bore from one to five microsporangia. The pollen grains were multicellular internally and produced sperms which were probably similar to those of the cycads and therefore of the swimming type.

Ovules were borne on short spur shoots in the axil of a bract. Only two integumentary layers were present in this group and the megasporangial wall was completely free of the integumentary wall. At the micropylar end of the ovule a prominent beak was present which separated the two archegonia and enclosed a pollen chamber.

The Cordaitales seem to have a parallel history to that of the Cycadofilicales. They were both very ancient groups and probably were derived from similar heterosporous ancestors. While the Cycadofilicales gave rise to the cycadophytean line the Cordaitales were evolving the Coniferophytean line.

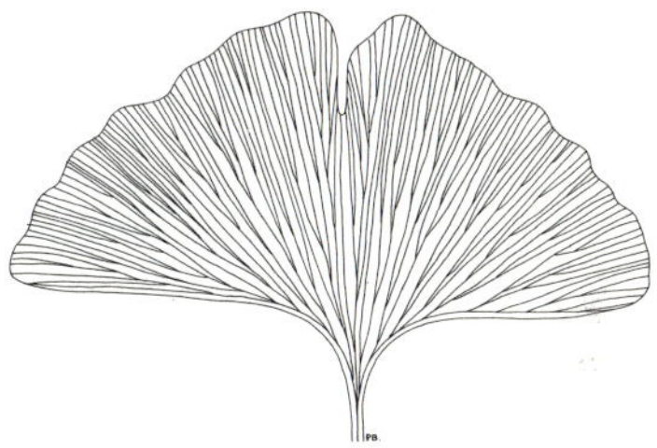

FIGURE 16 Diagram of fan-shaped leaf of *Ginkgo biloba*. Note dichotomous venation.

ORDER GINKGOALES

During the Mesozoic era this order flourished with at least fifteen genera. Now all are extinct with the exception of *Ginkgo biloba*. However, it is in no danger of extinction for it has been planted in most temperate regions and is now world-wide in its distribution, although if it still exists in the wild state it is only in the forests of southeastern China.

The unique features of this very ancient species are its dichotomously veined, fan-shaped leaves (Fig. 16), the retention of motile sperms, and terminally borne seeds. The *Ginkgo* is a deciduous tree which has a prominent excurrent manner of branching when young, but as it matures this changes to a more expanded crown.

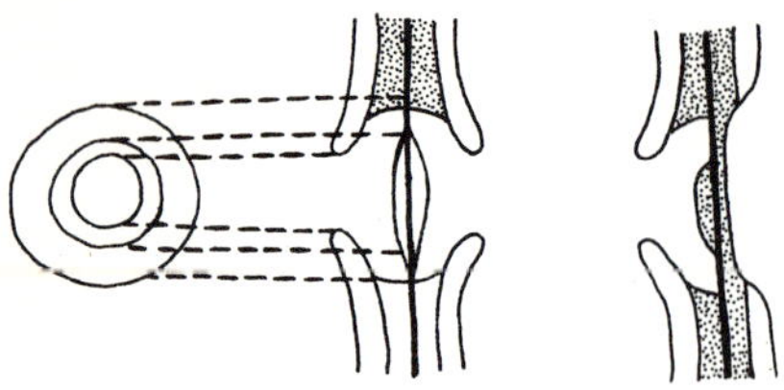

FIGURE 17 A-Full-bordered pit showing central membrane with tylose in pore. B-Half-bordered pit. (After Weis & Fuller.)

FIGURE 18 *Ginkgo* leaves from the same branch showing some variation in lobing.

Internally the stem is an endarch ectophloic siphonostele in which the protoxylem has spirally marked tracheids pitted with prominent pit borders (Fig. 17). Annual rings are much more prominent than in the other gymnosperm groups already described, but are not as well-defined as those in trees of the Coniferales. The pith is comparatively small, as is the cortex, but the vascular rings are broad.

The fan-shaped leaves (Fig. 18) vary in their degree of lobing and are similar in shape and venation to the maidenhair fern, *Adiantum;* consequently the *Ginkgo* has been given the common name "maidenhair tree."

Ginkgo biloba is dioecious. The male plant produces a loose, pendulous microsporangiate strobilus (Fig. 19) that superficially resembles the male catkin of the cottonwood tree. The slender microsporophylls produce at the tip two to four microsporangia within which the microspores are formed.

The microgametophyte within the spore wall consists of two prothallial cells, a generative cell, and a tube cell. After the pollen grain has reached the megastrobilus, the pollen tube develops and the generative cell divides to form the stalk and body cell. The body cell then divides into two large, turbinate, multiflagellate sperms.

The female reproductive organ looks more like a large globose ovule (Fig. 20) than a strobilus. It consists of a long stalk with two

FIGURE 19 Microstrobilus of *Ginkgo*. (After Coulter and Chamberlain.)
FIGURE 20 Megastrobili of *Ginkgo*. (After Coulter and Chamberlain.)

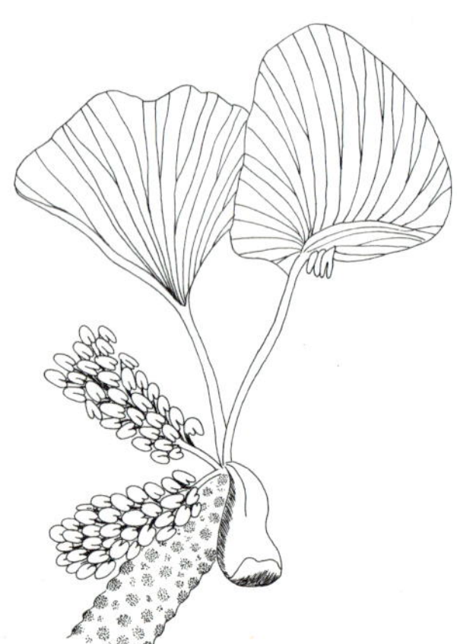

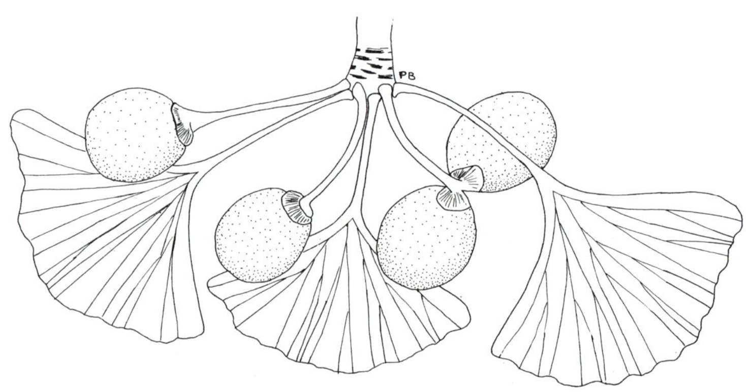

or three ovules, but usually only one persists to become the seed. The megasporangium **(nucellus)** has a large pollen chamber and a prominent apical beak (Fig. 21). Megasporogenesis results in a linear arrangement of four megaspores, the basal one enlarging to form the megagametophyte. Two or three archegonia develop at the micropylar end, and the remainder of the megagametophyte has an abundance of free nuclear divisions before walls begin to form in the peripheral area separating the nuclei. Surrounding the megasporangium three integumentary layers, consisting of an inner fleshy layer, a reduced middle stony layer, and an outer fleshy layer, form a heavy coating over the developing seed.

After fertilization there is another period of free nuclear division, this time in the embryo before walls form. The embryo ultimately consists of a short hypocotyl, two cotyledons, and a short epicotyl (Fig. 22). Upon germination the root developing from the tip of the hypocotyl is established first. The cotyledons remain within the seed, supplying food to the main axis as it develops. The first leaves that form are typically bilobed since they are true leaves rather than "seed leaves" as generally seen in the higher seed plants.

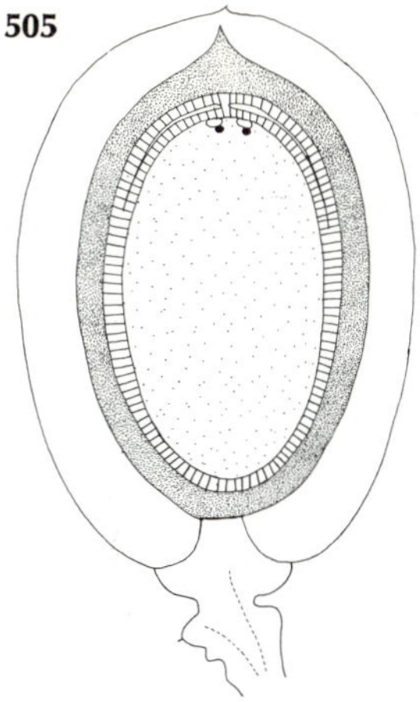

FIGURE 21 *Ginkgo* ovule, the thick outer layer is fleshy, next to the stippled layer is the middle stony region, then the inner fleshy layer fused in part with the nucellus is cross-hatched; two archegonia are represented by small black dots with archegonial chambers just above them. The remainder of the interior is gametophyte tissue which becomes endosperm. A prominent beak is shown at the apex. (After Coulter and Chamberlain.)

ORDER CONIFERALES

When gymnosperms are mentioned most of us in the Northern Hemisphere think immediately of pine, spruce, fir, and juniper, all members of the order Coniferales. They, and a good many other species, are the most prominent gymnosperms, occupying thousands of square miles of the earth's surface. For the most part they live in shallow soils of our mountain ranges and in the more distant latitudes where they have less competition from the highly competitive and expanding angiosperms. Many conifers are now extinct, and a few like the Monterey Cypress (Fig. 23) and giant sequoias

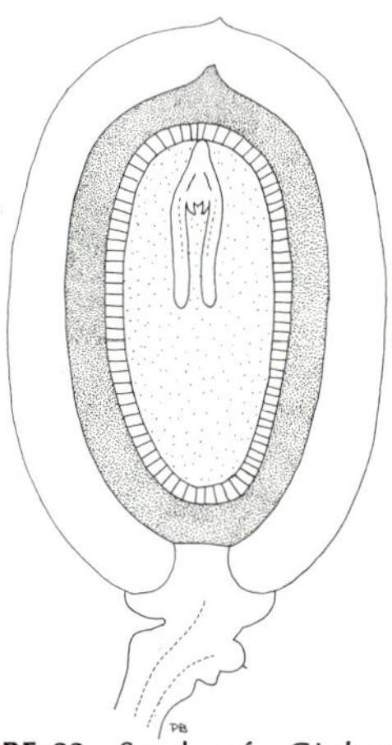

FIGURE 22 Seed of *Ginkgo*. The central embryo consists of two cotyledons, a small central epicotyl and a short hypocotyl. The surrounding gametophyte becomes stored food. The inner cross-hatched layer which was fleshy and rather watery in the ovule becomes dry and papery in the seed. (After Coulter and Chamberlain.)

FIGURE 23 Montery Cypress—a declining species in central, coastal California.

of the Pacific coast are making a last stand in their restricted habitats, but for the most part the surviving conifers are successful plants and dominate the forests in which they have become climax species.

Most conifers are monoecious with ovulate and staminate cones produced on the same plant (Fig. 24). Generally the strobili form as distinct cones, but a few such as the juniper bear their seeds in structures that are more like berries than cones. The leaves are usually needlelike, although those of *Juniperus* and *Cupressus* are usually scalelike and those of *Podocarpus* and *Dacrydium* of the Southern Hemisphere are often broad and flat.

The sporophytes of many Conifers reach the greatest size of all plants in the world. Some members of the genus *Sequoia* attain heights of over three hundred feet with basal diameters of over thirty feet. Most of the plants in the order exhibit a strong apical dominance resulting in beautiful, excurrently branched trees.

Most conifers are evergreen, but a few such as those in the genera *Larix* and *Taxodium* are deciduous. The leaves (Fig. 25) are well-adapted to xeric environments with their heavily cutinized epidermis, sunken stomates, thickwalled hypodermal layers, compact mesophyll composed of rosette shaped cells, resin ducts, and rather viscous protoplasm. The few veins with their prominent tracheids are surrounded by thick-walled endodermal cells.

FIGURE 24 Cones of the western yellow pine, *Pinus ponderosa;* left is a cluster of staminate cones, right are two female cones.

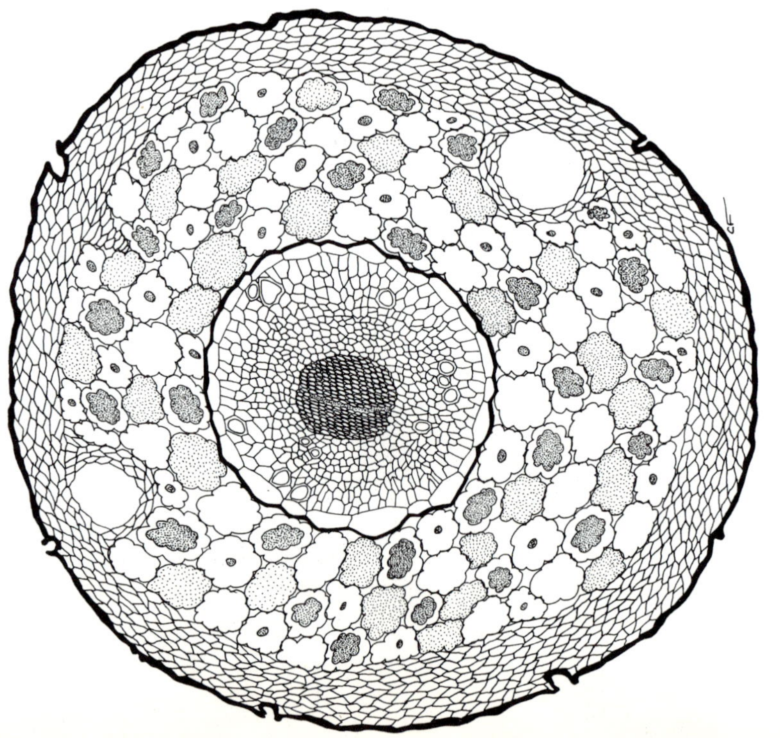

FIGURE 25 Diagram of a typical conifer leaf in cross section. The epidermis is heavily cutinized and has sunken stomates; next there are several hypodermal layers, within which resin ducts may be present as in drawing, then the mesophyll which consists of rosette-shaped cells, then a ring of heavily suberized endodermal cells. In the very center is a large vascular bundle surrounded by the thin-walled transfusion cells.

The stems of many Conifers are excellent examples of the longevity of vascular cambium. In a tree such as the General Sherman *(Sequoia gigantea)* the cambium has been actively producing xylem and phloem for over three thousand years and now has produced approximately six hundred thousand board feet of lumber, all in one tree. In contrast to cycads and other gymnosperms the stems of conifers have very little pith or cortex but massive vascular cylinders in distinct annual rings (Fig. 26). The xylem consists of tracheids which are usually prominently bordered, and some parenchyma cells form rays between the radial rows of tracheids.

Towards the outer portion of the narrow cortex a cork cambium differentiates, which gives rise to outer layers of suberized cells. As the stem increases in girth this outer bark becomes furrowed and wrinkled (Fig. 27), which is quite distinctive in the different species.

FIGURE 26 Section of a pine stem about sixty years old. Note the large amount of heartwood to the right and the smaller amount of sapwood, which is much lighter in color.

FIGURE 27 Section through a western yellow pine stem showing rhytidome of bark. Note the layering of cork cells in the bark.

FIGURE 28 Cross section of pine root approximately two years old. Note three resin ducts in central first-year xylem and several more in second-year xylem. The ring of phloem is not clearly differentiated but an active cambium is giving rise to radial rows of phloem and new xylem. A periderm is beginning to form in the outer two or three layers.

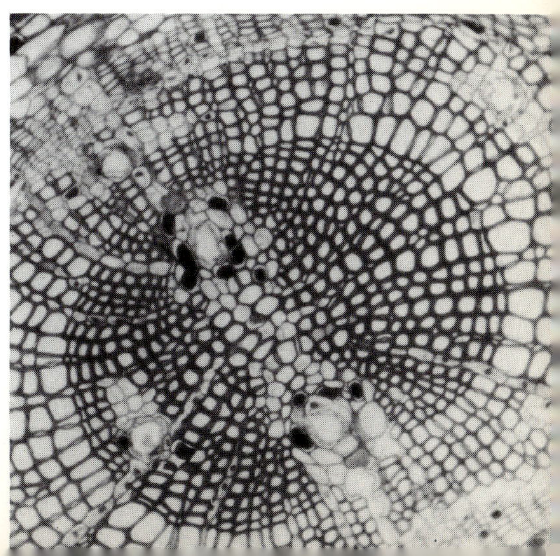

Roots

Most conifers have a very prominent tap root, but numerous laterals develop as the tree matures. Mycorrhizae are present in and on the roots of most conifers and these, along with root hairs, increase the absorptive capacity of the subterranean plant parts. The number of protoxylem arms or points is variable in different species but the primary xylem is exarch and the tissues are organized in about the same way that they are in angiosperms and most other gymnosperms (Fig. 28).

FIGURE 29 The pinyon pine, *Pinus edulis*. This is a short-leaved, two-needle pine. At each stem tip new growth is extended out for two or three inches and just back of the new stem tip are clusters of small staminate cones. The cones are only about seven or eight millimeters long.

Microstrobilus

The **staminate** cone is usually much smaller than the ovulate cone at maturity and is made up of microsporophylls consisting of two microsporangia arising from the **lower** surface of each sporophyll (Fig. 30). In most conifers the number of microsporangia is two per microsporophyll, but in *Agathis* or *Araucaria* they may number as many as fifteen. The microsporangia produce microspores which are usually winged.

The microspore (Fig. 31) develops its gametophyte in the same manner as that described for *Ginkgo*.

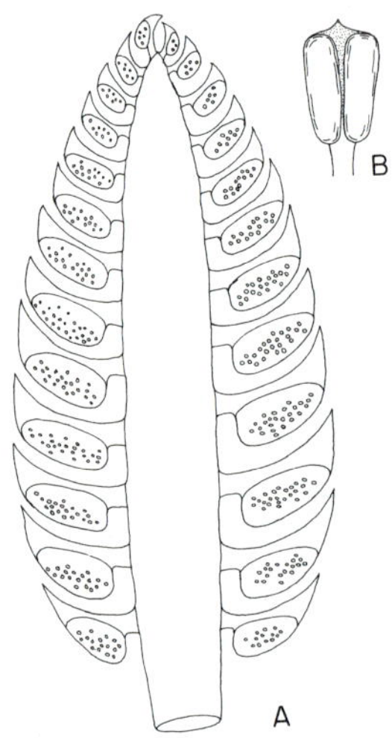

FIGURE 30 A: Longitudinal section of microstrobilis of *Pinus* showing microsporangia in lateral view. B: Diagram of a single microsporophyll with its two microsporangia.

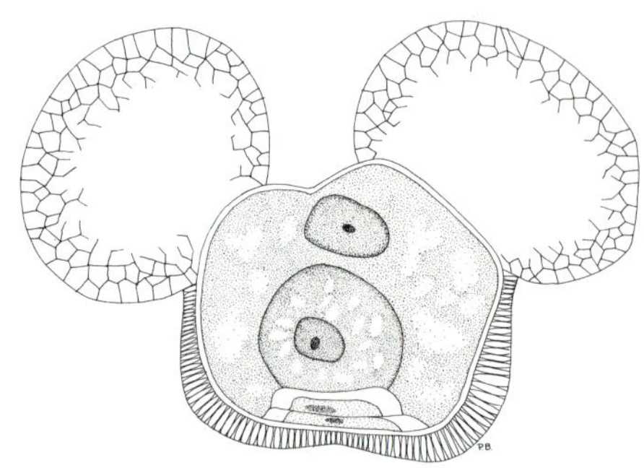

FIGURE 31 Pollen grain of pine showing two reduced prothallial cells, a large central body cell which will divide to form two sperms and a tube cell which will form the pollen tube. (After Brown.)

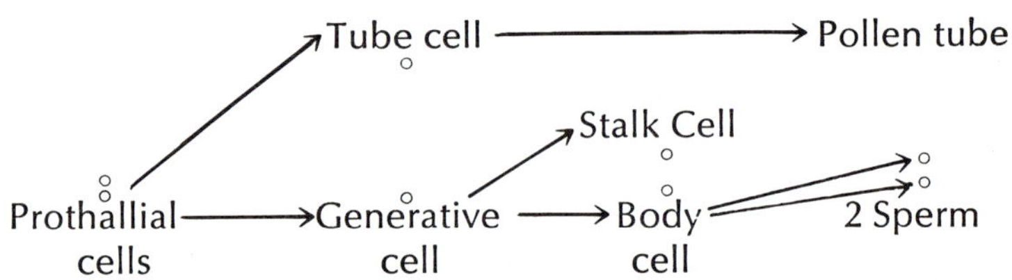

The big difference is in the two sperm, which vary in size and now for the first time are nonflagellate. The other nuclei or cells (tube nucleus, stalk cell, and extra sperm) are discharged into the archegonium along with the larger sperm but only the latter unites with the nucleus in the egg; the others remain within the periphery of the cytoplasm of the egg and eventually disintegrate.

Megastrobilus

The ovulate cone (Figs. 32, 33), at maturity, is usually larger and more persistent than the staminate cone. The latter usually only lasts a few weeks and after the pollen is shed the remnants of the cone fall from the tree. A female cone is on the tree for at least a year and in some species may persist for years, much of which time it is still in the process of developing the product of pollination and fertilization.

FIGURE 32 Megastrobili of the ponderosa pine. Note tiny spine on underside of megasporophylls. All members of the genus *Pinus* have a small spinous process similar to this.
FIGURE 33 Female cone of *Pinus edulis*. This cone is two years old and has already shed its seeds.

The megasporophylls usually develop two ovules (Fig. 34) which consist of an integument of three layers and an inner megasporangium. The megaspore mother cell gives rise to four megaspores in most conifers and it is usually the most deeply imbedded one (farthest from micropyle) that gives rise to the megagametophyte. Several archegonia are usually developed at the micropylar end but the pollen chamber and nucellar beak of earlier gymnosperms are lacking (Fig. 35).

FIGURE 34 A: Megasporophyll of *Abies*. B: Megasporophyll of *Pinus*. Note extensive winglike extensions which have developed on the ovules.

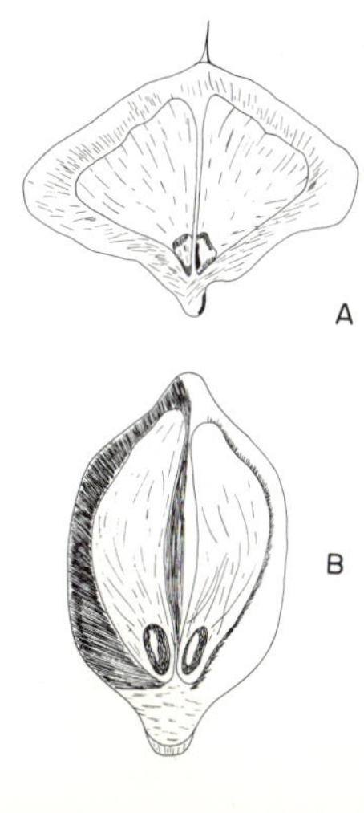

The Embryo

Embryonic development in the conifers is quite variable in the number of free nuclei produced following egg fertilization, but in

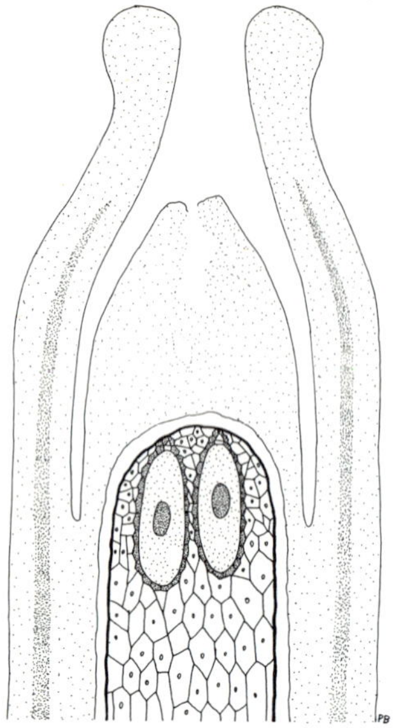

FIGURE 35 Section through young ovule of *Pinus* just prior to fertilization. Two archegonia are shown in the gametophyte tissue. A large double integumentary layer surrounds the nucellar tissue. (After Torrey.)

all there are fewer free nuclei than in the cycads or *Ginkgo*. In *Pinus* the fertilized egg gives rise to a **proembryo** composed of sixteen cells in four tiers of four cells each. The lowest four cells each give rise to an embryo; the next four form the **primary suspensor** (Fig. 36) cells. Those cells of the third tier may also develop into a dozen or more cells which then disintegrate. The fourth tier is used up in the same way.

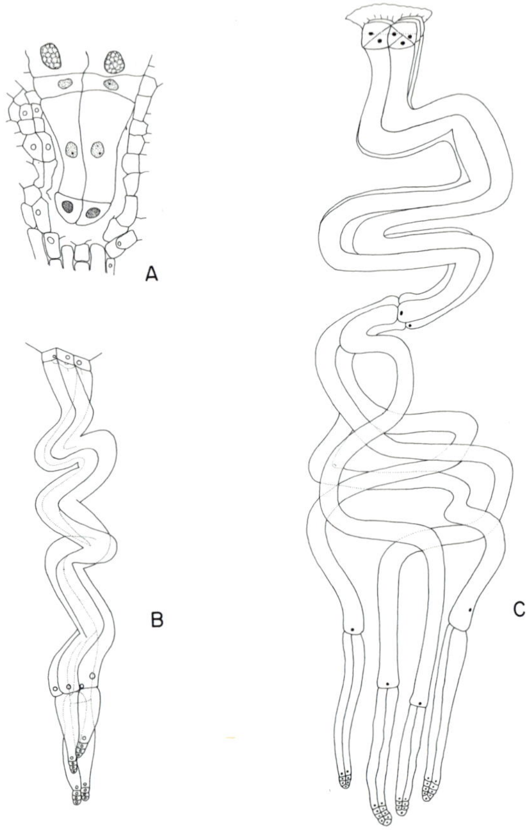

FIGURE 36 Development of four embryos in *Pinus*. A: Two tiers of four cells each are shown. The two lowest will become embryos, the next two will become suspensors, and the next two in each tier will become additional embryos or more commonly simply increase in number for a short time and then become food for the developing embryo(s). B: Now four developing embryos are shown at the very tip. The suspensor cells have greatly elongated and have divided once to form secondary suspensor cells. C: The four embryos are increasing in size and the suspensors have divided once more. Usually all but one embryo will abort. (After Brown.)

In pine, therefore, four embryos begin development in the gametophyte. Usually the cells of the proembryo nearest the initial suspensor cells divide again forming **secondary suspensor** cells. This type of division may be repeated several times. The most conspicuous feature at this stage of development is the great elongation of the suspensor cells which thrusts the developing embryos deep into the megagametophytic tissue. Elongation of the suspensors causes them to coil and twist in the small space of the embryo sac.

Generally the largest embryo formed continues on in growth while the others abort, but this is one way several embryos may be formed in a single pine seed. When the embryo is mature it consists of a very short epicotyl, surrounded by a whorl of cotyledons, and a rather prominent hypocotyl. The number of cotyledons may vary. For example, in our own laboratories on the seedlings from a number of *Pinus ponderosa* seeds collected from a single tree we found that the number of cotyledons varied from seven to eleven (Fig. 37). The redwood, *Sequoia sempervirens,* consistently has four cotyledons per seedling, but in the entire order two is the most common number, as it is in most gymnosperms and dicotyledonous angiosperms. The stored food remains in the megagametophyte tissue, the seed coat is formed from the two inner integumentary layers, and the outer layer is absorbed. In conclusion, note the diagram of the pine life cycle in Figure 38 and the general characteristics of six orders of gymnosperms in the table below.

FIGURE 37 Germination of pine seed. A: Two seeds still attached to megasporophyll; B: longitudinal section through seed showing enclosed embryo; C: splitting of seed coat with enlargement of embryo; D: emergence of radicle and hypocotyl; E: portion of hypocotyl now in soil and cotyledons leaving seed; F: seedling becoming erect and carrying seed coat on cotyledon tips; G: seed coat discarded, roots established and cotyledons expanding and by now green in the light. (Courtesy of St. Regis Paper Co.)

FIGURE 38 Life cycle of *Pinus*. A: sporophyte; B: cluster of staminate cones; C: one staminate cone in section; D: microsporophyll with two microsporangia; E: pollen grain; F: pollen tube; G: megastrobilus (actually much larger than staminate cone); H: section through megastrobilus; I: lateral view of megasporophyll; J: megasporophyll at time of pollination; L: detail of ovule showing two archegonia and developing pollen tube; M: seed with enclosed embryo; N: pine seedling following seed germination. (Redrawn from *Laboratory Outlines,* W. H. Freeman and Co.)

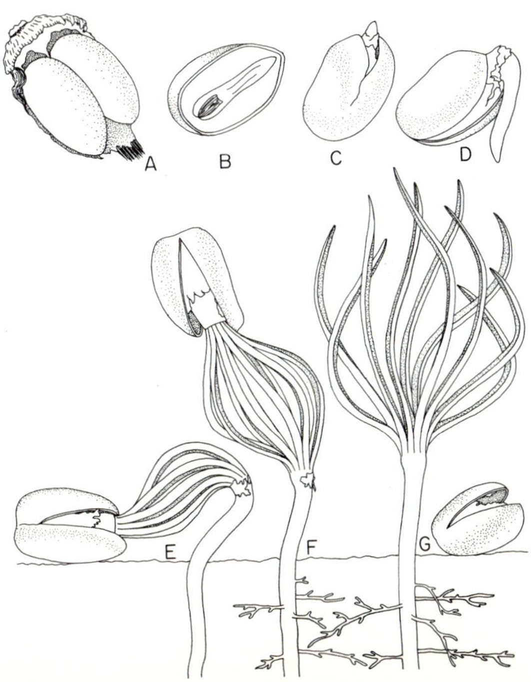

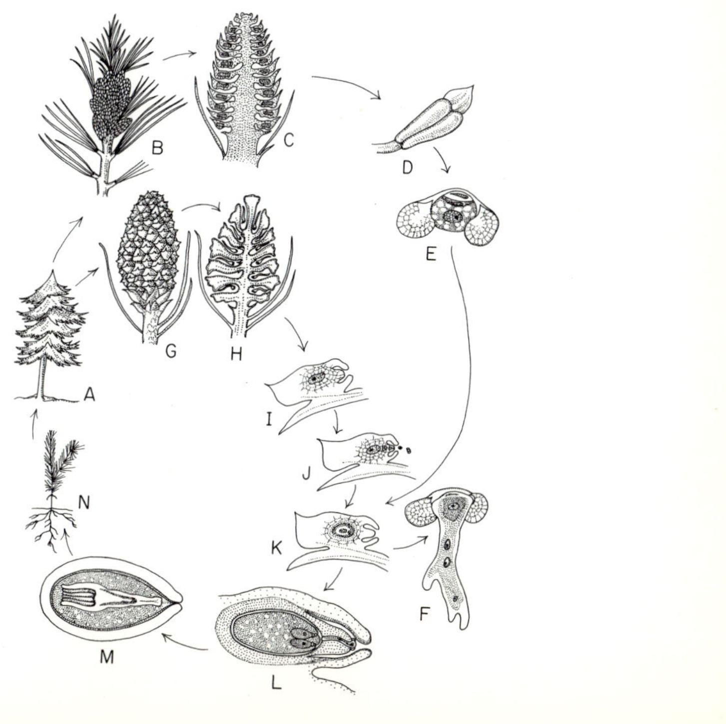

Summary of Six Orders of the Gymnospermae

Cycadofilicales	Bennettitales	Cycadales
1. *Status of members:* All extinct Fernlike	1. *Status of members:* All extinct Cycadlike	1. *Status of members:* Nine surviving genera
2. *Leaves:* Compound, fernlike Venation mostly dichotomous Probably circinate	2. *Leaves:* Pinnately compound Venation probably dichotomous Probably circinate	2. *Leaves:* Pinnate to compound Venation dichotomous Partly circinate in development
3. *Stem:* Short fernlike Siphonostele-probably also dictyostele Large pith Xylem mesarch Cambium weak Some periderm	3. *Stem:* Often large tuberous and small treelike Siphonostele—also dictyostele Large pith and cortex Xylem endarch Cambium limited Periderm	3. *Stem:* Tuberous to treelike Ectophloic siphonostele Large pith and cortex Xylem endarch Cambium fairly limited Periderm
4. *Cones:* Lacking	4. *Cones:* Ovulate cone present Staminate cone lacking Microsporangia borne on vegetative parts or bisporangiate strobilus	4. *Cones:* Ovulate cones very large Staminate cones smaller
5. *Seeds:* Exposed on leaves	5. *Seeds:* In a strobilus Ovules large 3-layered integument	5. *Seeds:* In a strobilus Ovules large 3 integumentary layers Sperm multiflagellate Embryo with 2 cotyledons

Cordaitales	Ginkgoales	Coniferales
1. *Status of members:* All extinct	1. *Status of members:* One surviving species	1. *Status of members:* About 600 living species
2. *Leaves:* Elongate simple; Erect development; Venation dichotomous	2. *Leaves:* Fan-shaped-bilobed; Dichotomous venation	2. *Leaves:* Needlelike, scalelike, bladelike; Venation variable
3. *Stem:* Slender tree; Ectophloic siphonostele; Large pith and cortex; Xylem endarch; Cambium limited; Some periderm	3. *Stem:* Large tree; Ectophloic siphonostele; Pith and cortex smaller; Xylem endarch; Cambium fairly prominent; Periderm	3. *Stem:* Usually large trees; Ectophloic siphonostele; Pith and cortex small; Xylem endarch; Cambium prominent; Prominent periderm
4. *Cones:* Strobili of same size	4. *Cones:* Ovulate cone berrylike; Staminate cone similar to a catkin	4. *Cones:* Ovulate cones larger than staminate; both vary in size
5. *Seeds:* In a strobilus; Embryo not observed	5. *Seeds:* In a berrylike strobilus; Ovule—3 thick integumentary layers; Sperm multiflagellate; Large pollen chamber; Large nucellar beak; Embryo with two cotyledons	5. *Seeds:* In a strobilus of varying types; Ovule—3 Integumentary layers; Seed—2 Integumentary layers; Pollen chamber absent; Nucellar beak absent; Sperms nonflagellate; Embryo often with over two cotyledons

ORDER GNETALES

For convenience the three genera in the order Gnetales are placed with the gymnosperms, but in many respects they could be considered a primitive order in the angiosperms. A summary of their characteristics will be given here.

Ephedra (Fig. 39) is a very common genus in areas of the southwestern United States; other species are found in such distant places as India and China. *Gnetum* inhabits moist tropical forests of Africa, Asia, and South America. *Welwitschia* is one of the rarest and most bizarre plants in the world; it is restricted to a few isolated desert areas in western South Africa.

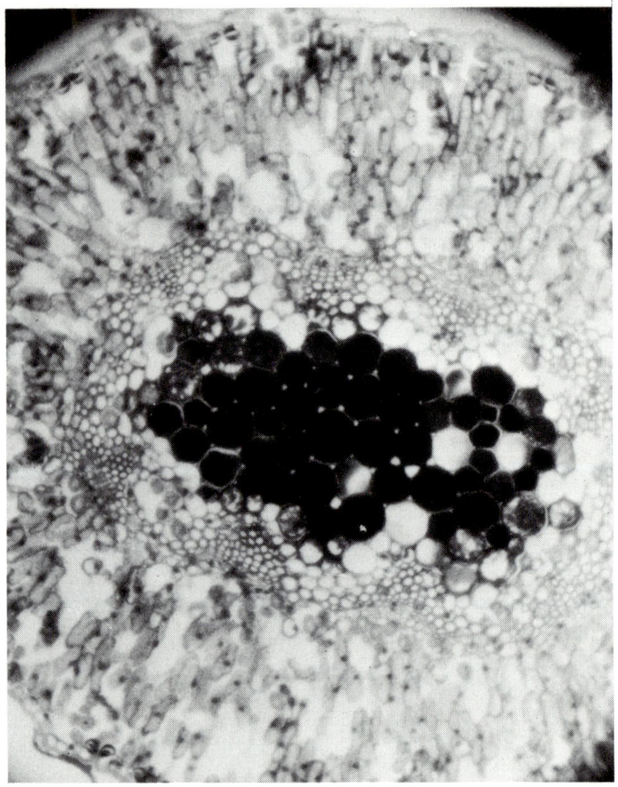

FIGURE 39 *Left: Ephedra trifurca,* one of the largest species of the genus *Ephedra*. *Right:* Cross section of *Ephedra* stem approximately two years old. The central dark cells are in the pith; surrounding the pith are nine discrete vascular bundles, but some interfascicular xylem has already differentiated between the bundles. Note the elongated and loosely organized cells in the cortex.

All three genera produce both tracheids and some tracheal tubes. While the presence of tracheal tubes indicates a relationship to the angiosperms, the tubes are not quite like those of the flowering group. Archegonia are present in *Ephedra* but are absent in both *Gnetum* and *Welwitschia*. Furthermore there is free nuclear division in the development of the embryo in *Ephedra,* but it is lacking in the other two genera. In all three the stored food in the seed is haploid, and the embryos always have two cotyledons. A pollen chamber is present in *Ephedra* but reduced or absent in the other two genera. The male gametophyte develops like that of the conifers in *Ephedra* but it is reduced to four cells (as in angiosperms) in *Gnetum* and *Welwitschia*.

There are other features which may be characteristic of either gymnosperms or angiosperms. It is quite probable that the Gnetales arose from a cycad or cordaite ancestor but at most they are only remotely related to any living gymnosperms. It is equally difficult to show a direct relationship to any living angiosperm. Geologically we have found fossils of them only as far back as the Tertiary era, yet their morphology indicates that they should be much older. Certainly the Gnetales are an interesting order; perhaps someday we will be able to learn how they evolved and where they came from.

Supplementary Reading

Cycads

Andrews, H. N., Jr., *Studies in Paleobotany*. McGraw-Hill, New York, 1961.

Arnold, C. A., Origin and Relationships of the Cycads, *Phytomorphology*, 3:51–65, 1953.

Chamberlain, C. J., *The Living Cycads*. University of Chicago Press, Chicago, 1919 (reprint, 1960, Hafner, New York).

Chamberlain, C. J., *Gymnosperms: Structure and Evolution*. University of Chicago Press, Chicago, 1935 (reprint, 1965, Dover, New York).

Darrah, W. C., *Principles of Paleobotany*, 2nd ed. Ronald, New York, 1960.

Foster, A. S., and E. M. Gifford, Jr., *Comparative Morphology of Vascular Plants*. Freeman, San Francisco, 1959.

Wieland, G. R., *American Fossil Cycads*, Vols. I, II. Carnegie Institution of Washington, no. 34, 1906, 1916.

Cordaites, Ginkgo, Conifers, Gnetales

Andrews, H. N., Jr., *Studies in Paleobotany.* John Wiley & Sons, New York, 1961.

Chamberlain, C. J., *Gymnosperms: Structure and Evolution.* University of Chicago Press, Chicago, 1935.

Darrah, William C., *Principles of Paleobotany,* 2nd ed. Ronald, New York, 1960.

Foster, A. B., and E. M. Gifford, Jr., *Comparative Morphology of Vascular Plants.* Freeman, San Francisco, 1959.

Wardlaw, C. W., *Embryogenesis in Plants.* John Wiley & Sons, New York, 1955.

Questions for Discussion

1. Is it advantageous for a seed to be polyembryonous as occurs in a few gymnosperms?
2. What are the sperms like in the various groups of gymnosperms?
3. Staminate cones remain for only a few weeks on conifers while the ovulate cones last for several years. Explain.
4. To what structures in ferns are the embryo sac, the pollen grain, and the megasporophylls of conifers comparable?
5. Of what advantage is it to the species to have wind-blown pollen?
6. Why are cycads considered very primitive plants?
7. Why do we say cambium has reached its greatest stage of development in the conifers?
8. Why have seed plants become the most successful terrestrial plants?
9. Compare and contrast the internal organization of a conifer leaf with that of an angiosperm.
10. What changes have taken place in the development of the archegonium in gymnosperms?
11. Superficially tree ferns resemble cycads; what are their differences?
12. Diagram the life cycle of a pine and compare it with that of a fern, and that of *Selaginella*.

25

The Advanced Seed Plants

Class Angiospermae

The class of Angiosperms is the largest in numbers of species and the most advanced group in the plant kingdom. They rose to a position of dominance during the latter part of the Mesozoic era, but it seems likely that they have had a longer and more extensive pre-cretaceous history than has thus far been revealed by the fossil record. It is possible that the habitat they occupied was not conducive to ready fossilization and preservation, for comparatively few angiosperms live in the swamps and bogs frequently occupied by the lower vascular plants and gymnosperms. In any event, our present knowledge of pre-cretaceous angiosperms indicates that they were apparently fewer in number of both species and individuals than gymnosperms at that time.

A number of theories have been proposed concerning the possible ancestors of the angiosperms. These theories suggest that they may be found in such orders as the Cycadofilicales, Coniferales, Bennettitales, Cycadales, Cordaitales, or Gnetales. Even the ferns have been proposed as possible ancestors. Some years ago a group of fossil plants, the Claytoniales, was discovered in Middle Jurassic rocks. At the time they were discovered they were believed to be in the direct line of angiosperm descent; continued investigations indicated that they had a relationship with the Cycadofilicales but were as remotely removed from the angiosperms as the other orders listed above.

Comparative morphological studies now indicate that the remote ancestor of the angiosperms came from early fern or very primitive gymnosperm stock which has been given the convenient name **protoangiosperm** to represent a hypothetical ancestor.

With the emergence of the angiosperms the reproductive organ changed from a cone to a flower. The parts of the two structures are very much the same, but these organs have become highly modified in the flower. Consequently the evolution of the flower is one of the major considerations in studies of angiosperm evolution.

One of the most strongly supported theories, termed the *Ranalian* theory, is that the flower arose from a compound bisporangiate strobilus in which the upper and inner sporophylls had been reduced to whorls of carpels surrounded by whorls of stamens (Fig. 1). The outer and lower sporophylls of the strobilus remained sterile and vegetative; these sporophylls became the sepals and petals.

These early flowers still had a strobiloid (Fig. 2) appearance with their numerous pistils mounted on a conical receptacle ringed by many stamens, petals, and sepals.

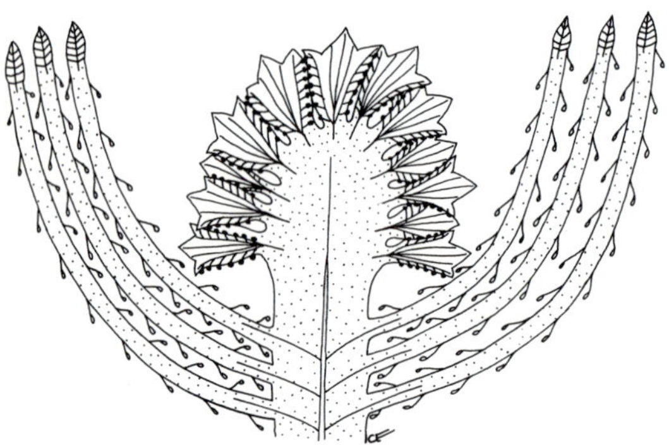

FIGURE 1 An interpretation of a primitive flower. The central axis is made up of megasporophylls with marginal ovules; the surrounding structures are microsporophylls and have marginal anthers.

FIGURE 2 Strobiloid flower of Magnolia, a primitive flower. Note central cluster of carpels (megasporophylls) and stamens (microsporophylls) drawn expanded horizontally. The stamens are not clearly differentiated into a stalklike filament and apical anther. (After Brown.)

Since the first appearance of the flower, according to the Ranalian theory, the central axis has been shortened into a flat disc supporting the floral parts (Fig. 3). These parts have become greatly reduced in number and size and in some cases have been reduced to floral bracts or lost entirely.

FIGURE 3 Compact Ranalian strobiloid flower. The numerous stamens and carpels are clustered in the center.

Classification of the Angiospermae

The angiosperms are divided into two subclasses, the **Dicotyledonae** and the **Monocotyledonae.** As the names indicate, the division is based on the number of cotyledons in the embryo, but this characteristic is not so obvious as other, more easily ascertainable ones. The members of the Dicotyledonae usually have their floral parts in fours or fives or multiples of these numbers. Their leaves are usually net-veined, the stems are usually eusteles with a prominent cambium, and they may have persistent tap roots or a fibrous root system.

The order Ranales, which includes such families as the Magnoliaceae, Ranunculaceae, Degeneriaceae, and Winteraceae, is considered the most primitive order among the dicotyledons. In this group the flowers are usually strobiloid in appearance, with numerous pistils clustered in a spiral arrangement in the center. Numerous stamens surround the pistils, and the stamens in turn are surrounded by showy petals and sepals. In some flowers the sepals are clearly differentiated in structure from the petals; in others they are not. The stamens in the most primitive flowers resemble scale leaves, with their microsporangia (anthers) sunk near the middle of the sporophyll. As the flowers have evolved, the stamen has become a distant cylindrical filament with its anther a lobed apical structure. The megasporophylls (carpels) have changed from an open, flattish organ to a more or less cylindrical organ with its ovules enclosed by union of the edges of one or more carpels. Of course many more modifications have taken place, such as further evolution of the seed and the development of a fruit from the ripening of the combined megasporophylls (ovary or pistil). These modifications will be considered later.

The Dicotyledonae are generally considered to be the more primitive of the two subclasses of angiosperms, but the groups have had parallel lines of development throughout much of their history. Although the monocotyledons may be considered more advanced, they apparently arose long before the dicotyledons had reached their more highly evolved families. The greatest similarities in the two classes can be seen in the two most primitive orders: the Ranales described in part above and the Alismales of the Monocotyledonae.

The flower structures of both have the same general strobiloid appearance. In some of the Ranales, as in the Magnoliaceae, the flower parts are in multiples of three, a common characteristic of monocotyledonous plants. Furthermore the root system of the Ranales is fibrous as in the Alismales and most other monocotyledons, and the conspicuous differentiation of sieve tubes and companion cells in the primary phloem of the Ranales is similar to that of monocotyledons.

The outstanding differences exhibited by the Alismales include its parallel-veined leaves, scattered vascular bundles in the stems, lack of cambium and therefore lack of secondary thickening, and only one cotyledon.

Origin of the Angiosperms

It is quite likely that the angiosperms had a polyphyletic origin with certain groups arising from one or more classes of the early vascular plants. The strongest evidence we have in support of a monophyletic origin is the predominance of the eight-nucleate embryo sac and the prevalence of double fertilization in the formation of the embryo and endosperm. Even these characteristics are not constant, and nearly all other angiosperm characteristics are duplicated or have their counterparts in earlier groups.

Because of their common habit of growth, linear leaves, and heterospory it has been suggested that the monocotyledons descended from the Isoetales. From a geological standpoint this theory is untenable since the Isoetales are of more recent origin than many angiosperms. Furthermore, *Isoetes* seems to have followed the usual evolutionary scheme of development of the vascular plants: the arborescent body preceded and is more primitive than the herbaceous stock.

The seed ferns have been suggested as progenitors of the flowering plants but scalariform pitted tracheids in secondary xylem, found abundantly in the wood of primitive angiosperms, are absent in the Cycadofilicales.

It has also been suggested that the angiosperms were derived from the Coniferales because of the prevalence of wind-blown pollen in many arborescent flowering plants. Now it is conceded that these arborescent angiosperms are much more advanced than was at first supposed and their flower parts have been essentially reduced in size and number.

The Gnetales have the largest number of angiosperm characteristics among the various orders of gymnosperms, but like the Isoetales, the Gnetales are of much more recent origin than the early angiosperms according to the fossil evidence.

One theory which has not been completely abandoned is that the earliest angiosperms came from the Bennettitales. The evidence is based on the possession of a bisporangiate strobilus found in the Bennettitales. The relationship, if any, seems to be quite superficial since the megasporophyll with its terminal ovule formed in the Bennettitales is quite different from the carpel of the angiosperms with its ovules on the adaxial surface. Furthermore the Bennettitales microsporophyll is a large, pinnately compound organ quite different from the staminate appendage of angiosperms. There are other conspicuous differences in wood structure and other vegetative characteristics, leading most botanists to think of the bennettitalean cone and the angiosperm flower as having had parallel lines of development rather than a direct relationship.

Theories of angiosperm origin still strongly favor the development of the flower from a compound bisporangiate strobilus. According to the Ranalian theory the main axis of the strobilus has been shortened into a more or less flat disc forming a receptacle supporting the floral parts. Proponents of this theory are of the opinion that the apical and inner sporophylls were reduced to whorls of carpels surrounded by whorls of stamens, petals, and sepals. Present-day flowers which approach this arrangement are in the order Ranales and include the magnolias, buttercups, and such primitive families as the Degeneriaceae and Winteraceae.

Characteristics of angiosperms which tend to classify a species as primitive or advanced are listed below. However, no one group has all primitive characteristics and no one group has all advanced ones. There is a quantitative distinction, and some groups do have more of the recognized primitive characteristics than the others. This is largely the determining factor in classifying a family as primitive or advanced.

Primitive	**Advanced**
Floral Characteristics	
Flowers regular	Flowers irregular
Flowers perfect	Unisexual
Flowers hypogynous	Perigynous to epigynous
Floral parts of each kind many	Floral parts of each type few
Floral parts spirally arranged	Floral parts whorled

Primitive	Advanced
Floral Characteristics	
Perianth undifferentiated	Corolla and calyx distinct or absent
Floral parts separate	Floral parts united
Stamens not differentiated into distinct filament and anther	Definite filament and anther
Carpels merely folded	Carpel fully closed
Carpels separate	Carpels united
Endosperm copious	Endosperm none
Cotyledons several	Cotyledons two or one
Pollen separate	Pollen massed
Vegetative Characteristics	
Arborescent and woody	Herbaceous
Cambium active	Cambium lacking
Long-lived	Short-lived
Vascular bundles in cylinder enclosing pith	Vascular bundles scattered
Vessels none	Vessels present
Chlorophyll present	Chlorophyll none
Evergreen	Deciduous
Leaves simple	Leaves compound
Stipules present	Stipules absent

Significance of Fruit

In the angiosperms the gametophytes develop, as they did in gymnosperms, within protective tissues, but those in the angiosperms are further protected by an ovary wall. The embryo, therefore, is protected not only by the seed coat but by the ovary wall (fruit coat) as well. The fruit is a major evolutionary advance in the success of angiosperms. The production of this organ and its associated tissues, which aid in seed dispersal, are considered prominent features in establishing angiosperms as the dominant plant group today.

Furthermore, double-fertilization resulting in the formation of polyploid endosperm provides additional storage tissue which is considered an advance over the haploid tissue of gymnosperms.

Within the ovule of the angiosperm only one egg is present. This decreases the chances for polyembryony, and ensures a greater food supply for the single developing embryo, since it is not competing for space and stored food within the seed.

Genetic Variation

From our earlier discussion of genetics it is apparent that chromosomal variation is the method of evolution. This may be the result of gene mutation, polyploidy, or hybridization.

Since the genes contained in the chromosomes are largely responsible for the morphological characteristics and physiological activities of an organism, any change in the complement of genes may alter one or more factors in the organism's development. It is well-known that genes mutate at different rates. A high mutation rate results in more rapid distribution of the gene throughout a population; the distribution of this mutated gene depends upon the number of interbreeding individuals in a population. Furthermore, if the changes which occur modify the structure and metabolism of the new individual in a desirable way, so that it has greater survival value in the environment, it is likely to persist.

Changes in the geographic features, such as the elevation of mountain ranges, the formation of broad valleys and bodies of water, the advancement and recession of glaciers, alterations in climate and moisture, and drought, change the environment. Those plants best able to survive in the new environment gradually replace those less well adapted.

Evolution, therefore, results from changes in the geographical features, the climatic conditions, the variations which occur in the organisms, and the adaptations they are able to make. It is a slow creative process but whether it is the result of gene mutations, the change in complete chromosome complements as in polyploidy, or the fortuitous combination of genes through hybridization, evolution occurs as a natural, persistent process.

Characteristics of a Few Angiosperm Families

DICOTYLEDONAE

Primitive Families

Winteraceae The members of this family are woody plants occurring chiefly in the South Pacific, especially in Australia and New Guinea. The leaves are simple, entire, and alternately arranged. The flowers are bisexual; the sepals, which number two

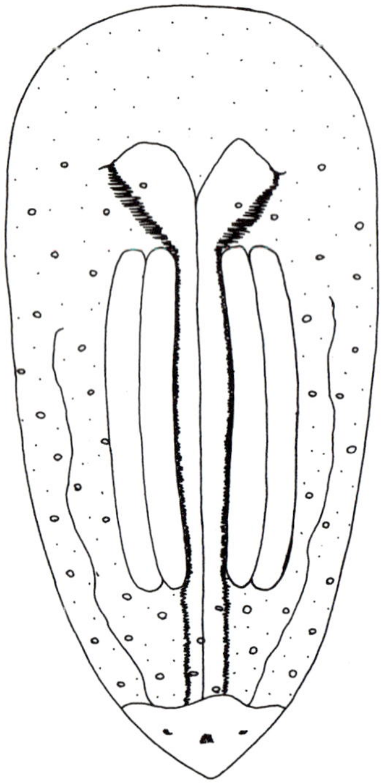

FIGURE 4 Petaloid stamen of *Degeneria* showing two anthers (microsporangia) attached to petal-like stamen but lacking stalk-like filament. (After Benson.)

to six, usually are coalescent and form a cap that falls when the flower opens. The stamens are numerous, variable in shape, and usually not differentiated into filament and anther. The carpels are also variable, they are open with no style or stigma differentiation, and with ovules on the ventral surfaces of the leaflike carpels. Along with these primitive characteristics, the wood is without xylem vessels. There are about ninety species in six genera in this family and most of them have the primitive characteristics listed above.

Degeneriaceae This family is comprised of but a single species, *Degeneria vitiensis,* and was described in 1942 for the first time from plants collected in the Fiji Islands. It has essentially the same characteristics as the Winteraceae: Woody, leaves simple, entire, alternate; flowers bisexual, three sepals, about two petals. Stamens number about 20, and are broad and leaflike with four slender pollen chambers imbedded on the dorsal surface of the microsporophyll (Fig. 4). The single carpel is folded and open on the ventral side; it does not have either a style or stigma differentiation.

Magnoliaceae Some members of the **Magnoliaceae** are fairly abundant in the United States, especially species of *Magnolia* and *Liriodendron* (tulip tree). The species are trees or shrubs with simple, usually entire and alternate leaves. The flowers are bisexual, large, and solitary. Sepals and petals are mostly in multiples of three. Stamens are numerous in multiples of three or four, spirally arranged, and narrowly leaflike. Carpels are also numerous and spirally arranged on an elongated axis with a more or less distinct stigma and style. The characteristics of this family are somewhat more advanced than those of the two families described above, but they still have enough in common to be included along with a number of other families in the order Ranales.

There are other plant families with numerous primitive characteristics, but these three should be enough to emphasize what is meant by "primitive" in plant phylogeny.

Intermediate Families

Cruciferae Commonly called the mustard family, the many species of the Cruciferae are herbs or small shrubs. The leaves are simple or pinnate, occasionally opposite but usually alternately arranged. The flowers are bisexual and radially symmetrical with usually six stamens, the four inner ones usually longer than the outer two (Fig. 5). They are distinctly differentiated into a filament and anther, the latter almost always having two pollen chambers. The single carpel is considered a two-chambered organ divided by a false partition. The fruit is a silique (or silicle) and dehiscent at maturity. Many of our garden vegetables such as cabbage, cauliflower, and rutabaga are members of this family.

Malvaceae Usually termed the Mallow or Cotton family, the species, like those in the preceding family, have some primitive and some advanced characteristics. The plants are mostly herbs, occasionally shrubs or trees, with simple and sometimes palmately lobed leaves. The flowers are almost always bisexual and radially symmetrical, with five sepals and five petals which are separate but usually coalescent at the base. The numerous stamens have distinct anthers with one pollen chamber which splits lengthwise. The ovary, which becomes a capsule in fruit, is usually divided into six to many (rarely three to five) fused carpels. Some of the members are cotton, okra, hollyhock and hibiscus (Fig. 6).

There are dozens of other families which could be described as intermediate types. Two very large families, the Rosaceae and Leguminosae, might be described as upper intermediate in phylogeny since they have some species that have a number of advanced characteristics. For example, the sweet pea in the Leguminosae has a zygomorphic corolla, with partially united petals and stamens. The floral parts are arranged in multiples of five, and the leaves are pinnately compound.

These two families have a total of over 15,000 species and consequently have so many individual species differences that we will not list their numerous variations here.

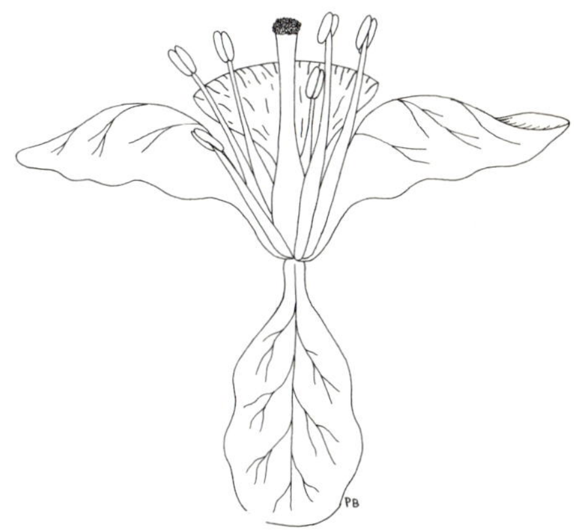

FIGURE 5 A flower of the Cruciferae. Note the 6 stamens (four of one length and two shorter) and four separate petals. (After Benson.)

FIGURE 6 *Hibiscus,* a member of the Malvaceae. Note the elongated style around which the numerous stamens are borne on a tube.

Advanced Families

The most advanced families in the Dicotyledonae are in the order Asterales. The members do not have characteristics that are all advanced, but they do have a preponderance of such characteristics as: few floral parts (usually five); the parts are usually united; the flowers are often epigynous and zygomorphic; the leaves are usually lobed or compound; and the ovary is compound with few and fused carpels. Some of our most prominent advanced families include the **Cucurbitaceae** (Gourd family), **Caprifoliaceae,** (Honeysuckle family), the **Campanulaceae** (Bellflower family), and the **Compositae** (Sunflower family).

Detailed descriptions of these families and others can be found in the books listed at the end of this chapter.

MONOCOTYLEDONAE

Primitive Families

Alismaceae The members of this family, considered the most primitive in the monocotyledonae, are aquatic or marshland herbs. The leaves are chiefly basal, simple, with net venation except for the main veins which run parallel. Flowers are mostly bisexual although one genus, *Sagittaria,* is dioecious (Fig. 7). The sepals and petals are separate and in threes; the stamens are usually numerous, and the ovary superior, with a distinct and persistent style, usually becoming an achene in fruit.

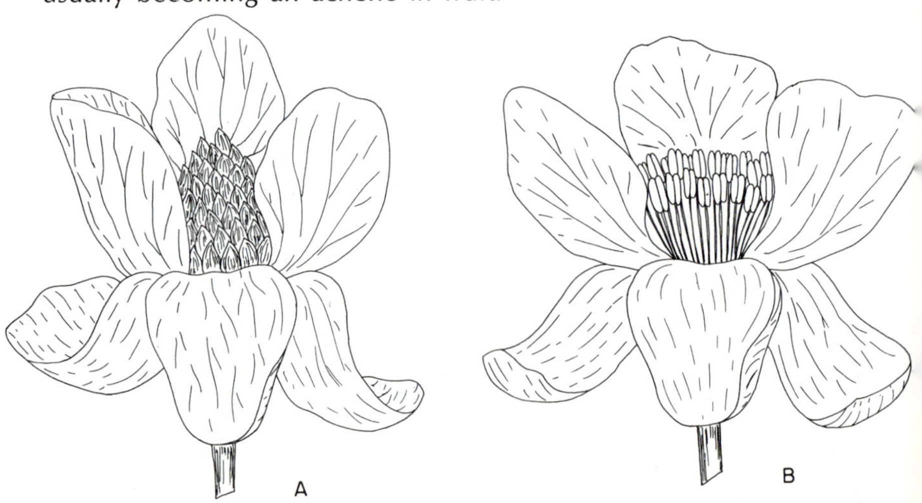

FIGURE 7 Strobiloid flowers of *Sagittaria latifolia.* A-female flower with many pistils; B-male flower with many stamens. (After Benson.)

Although the characteristics of the Alismaceae are in general quite primitive, the members have a greater differentiation of some floral parts than those in the Ranales. It is quite probable that the monocotyledons arose after the dicotyledons had developed the flower. Very likely the monocotyledons branched off very early from a Ranalian ancestor. The similarities in parts and morphology would indicate that the Alismales made the initial break when one examines and compares them with the other orders of monocotyledons.

Liliaceae This very large family (about three thousand species) is comprised chiefly of herbaceous members although a few, such as members of the genus *Yucca*, are woody. Many also develop fleshy vegetative structures such as bulbs, corms, and rhizomes. The leaves are usually alternate or whorled or (very rarely) opposite. Flowers are radially symmetrical and usually bisexual (a few plants are dioecious), and the floral parts are in multiples of three. The ovary is usually hypogenous and develops into a capsule or, occasionally, a berry. Some of the members are the lily, onion, Joshua tree, and bear-grass.

Intermediate Families

Araceae The arum family is characterized by a large white spathe (Fig. 8) which encloses a spadix. The flowers are somewhat imbedded on this large central axis. They may be bisexual or unisexual. If the latter, then the upper flowers on the axis are staminate and the lower either perfect or in some pistillate. The flowers

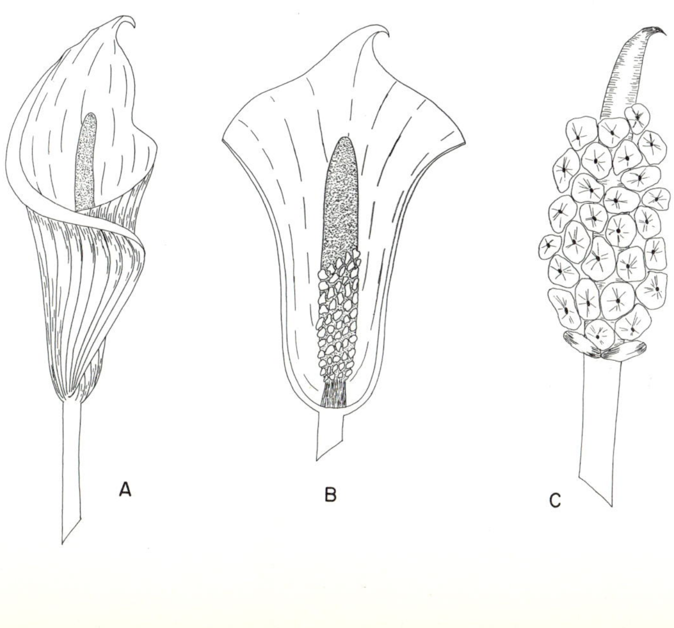

FIGURE 8 The calla, a member of the Araceae. A-the spathe is closed in the flower. B-opened flower showing central column of the spadix with numerous flowers. C-spadix in fruit. (After Benson.)

are radially symmetrical, and hypogenous but in the unisexual species the perianth may be lacking. The stamens number from one to eight and carpels one to many. The ovary is one-to-many-chambered and becomes a berry. These usually develop towards the base of the spadix. Included in the family are Jack-in-the-pulpit, skunk-cabbage, the garden white calla, and philodendron.

Gramineae The grass family is described as an intermediate family in the phylogeny of the monocotyledons here, but some botanists would consider it very much advanced since the floral parts are greatly reduced and modified. The plants are herbaceous except for some giant reeds and bamboo, which in tropical forests may be over sixty feet tall and have a basal diameter of over twelve inches. The stems **(culms)** are usually hollow except at the nodes and often arise from underground rhizomes. The leaves are generally basal or alternate or both. They are chiefly linear and two-ranked. Flowers are usually bisexual, occasionally unisexual, and inconspicuously bilaterally symmetrical. The flowers are usually borne in spikes or spiklets (Fig. 9). Each flower of a spiklet is enclosed in a lower bract, the lemma, and an upper bract, the palea. The spiklet as a whole is subtended by one or two glumes, and each flower is termed a **floret** some flowers have extra, sterile lemmas or paleas. The perianth is sometimes represented by hyaline scales; stamens usually number three, carpels are three but the ovary is one-chambered and develops into a caryopsis (grain) in most species. This family, which includes rice, wheat, sugar cane, and most other cereals, is our plant family of greatest economic importance.

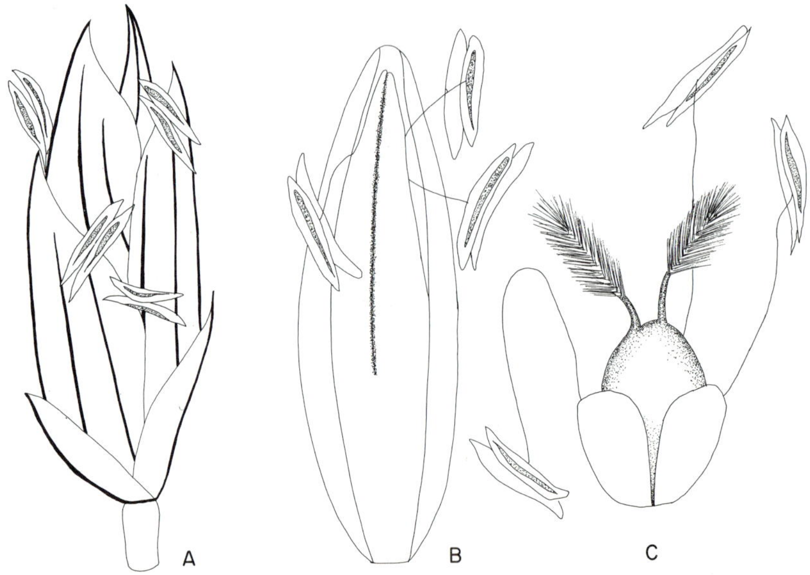

FIGURE 9 Inflorescence of a grass. A-spike, B-spiklet, C-floret with three stamens, central ovary, subtended by two scales. (Redrawn from Benson, L. 1957. *Plant Classification.* D. C. Heath.)

Advanced Family

Orchidaceae The orchid family is usually considered the most advanced in the monocotyledonae. It has some features that are unique to this group and rarely found elsewhere. The pollen is a waxy mass imbedded in a structure known as the **pollinium** and transported by insects as a unit. This feature occurs also in the milkweed family, the Asclepiadaceae, of the Dicotyledonae but in almost no other group. Orchid seeds are exceedingly small and the embryo is not differentiated into a cotyledon, or a hypocotyl, or an epicotyl. The twisting of the flower (about 180°) results in an inversion of flower parts (termed resupinate) (Fig. 10). The corolla is bilaterally symmetrical, the stamens are usually one or two, and they are fused to the style forming a sexual structure called a **gynandrium.** Many of the members are saprophytic and virtually all of them start as saprophytes with germination. The characters are in general advanced and the unique features, if we may consider them advanced also, really set the Orchidaceae considerably ahead of the other plant families in evolutionary development.

FIGURE 10 Irregular flower of an orchid. Note 3 sepals, 3 petals, 2 stamens and twisted perianth tube which encloses the inferior ovary. (Redrawn from Benson, L. 1957. *Plant Classification*. D.C. Heath.)

In many respects there has been a parallel development in the evolution of the Dicotyledonae and Monocotyledonae. Once the two groups had separated there was a tendency in both to go from flowers which were strobiloid, with many and distinct floral parts, to flowers which had flattened or greatly reduced receptacles with few and often coalesced floral parts. The end results appear quite different but close examination of the floral parts indicates that reduction of the sexual parts has been the general rule.

Supplementary Reading

Arber, A., The Interpretation of the Flower, a Study of Some Aspects of Morphological Thought, *Biological Review* 12:157–184, 1937.

Axelrod, D. I., A Theory of Angiosperm Evolution, *Evolution* 6:29–60, 1952.

Bailey, I. W., Origin of the Angiosperms, *Jour. Arnold Arb.* 30:64–70, 1949.

Benson, L., *Plant Classification*. D.C. Health and Co., Boston, 1957.

Cheadle, V. I., Independent Origin of Vessels in the Monocotyledons and Dicotyledons, *Phytomorphology* 3:23–44, 1953.

Constance, L., The Systemetics of Angiosperms. A Century of Progress in the Natural Sciences: 1855–1953, *Calif. Acad. Science* 405–483, 1955.

Dittmer, H. J., *Phylogeny and Form in the Plant Kingdom*. D. Van Nostrand, Princeton, N. J., 1964.

Eames, A. J., *Morphology of the Angiosperms*. McGraw-Hill, New York, 1961.

Esau, K., *Plant Anatomy*. John Wiley & Sons, New York, 1953.

Foster, A., and E. M. Gifford, Jr. *Comparative Morphology of Vascular Plants*. Freeman, San Francisco, 1959.

Maheshwari, P., *An Introduction to the Embryology of Angiosperms*. McGraw-Hill, New York, 1950.

Scott, R. A., E. S. Barghoorm, and E. B. Leopold, How Old Are the Angiosperms? *Amer. Jour. Sci.* 258:284–299, 1960.

Questions for Discussion

1. What are the evidences that the angiosperms arose from gymnosperm ancestors?
2. How has pollen changed in the angiosperms from the type found in gymnosperms?
3. Why are certain primitive families of angiosperms limited to a few isolated world habitats while most others have representatives in many areas?
4. Why is a flower considered a more efficient reproductive organ than a strobilus?
5. Discuss ten different primitive and advanced characteristics found in angiosperms.
6. Gymnosperms are practically all woody while the big majority of angiosperms are herbaceous. What does this suggest about the evolution of terrestrial plants?
7. Compare the parts of an ovulate gymnosperm cone with those of an ovary.
8. Why do not gymnosperms produce fruits?

9. The parts of a flower are often considered modified leaves. What evidences do we have for this opinion?
10. There are many angiosperms living in fresh-water habitats but few living in the ocean. Explain.
11. Compare the seeds of angiosperms with those of gymnosperms.
12. What structures have become modified into petals in double petunias, roses, etc.?
13. Why are zygomorphic flowers considered of more recent evolutionary origin than the actinomorphic types?
14. How are new plant varieties and species produced?

26

Evolution: A Biological Principle

Evolution has been defined as the development of organisms through time. Through natural selection it is the most unifying concept in biology, for it applies to organisms as well as to molecules, to plants as well as to animals. Essentially evolution involves the development of an entity, in the course of time, from a simple to a more complex state as a result of gradual change.

We are primarily interested in **organic evolution** since we are dealing with living organisms, but evolution may also be **inorganic.** The origin, development, and positioning of the galaxies is **cosmic evolution.** In studies of human culture one is frequently introduced to the evolution of automobiles, furniture, clothing, philosophy, religion, and virtually everything which has developed as a result of human inventiveness and imagination. This we classify as **cultural evolution.**

Diversity and Unity

The concept of organic evolution has been with us since at least the time of Aristotle, for most scientists have sought an explanation of the mechanisms which cause change in all organisms. It has been obvious to all that there is great diversity among the two million or more species now living and the millions of species that are now extinct. It has also been reasonably obvious, to all biologists at least, that there is unity in all organisms. That is, the biological systems of all organisms exhibit the same basic fundamental life processes and are composed of the same basic molecules.

Man has been aware of diversity for a much longer period than he has been aware of unity; in fact, it has been only in the most recent years that we have been able to prove unity beyond the usual, well-known physiological processes of respiration, digestion, etc.

Now a new dimension has been added to our concept of organic evolution through molecular biology. Molecular biology has also revealed to us the unity of all matter. We know that all materials are formed of fundamental substances in which two or more particles are combined to form atoms; these in turn are combined to form molecules and they in turn form the basis of all nonliving and living systems. The nonliving systems are composed of groups of one kind of molecule or perhaps a combination of several, in which case they are mixtures. Living systems are mixtures of very large and complex molecules organized and functioning in a coordinated manner. Both living and nonliving systems may be composed of the same basic atoms, but only in the living systems have these elements achieved the status of self-reproduction and the ability to use lesser molecules in growth.

The early evolution of the atoms and molecules to form simple and then complex substances was at first inorganic. Ultimately, when the molecules had developed sufficiently to carry on growth and self-replication (and therefore the processes of life), we attained organic evolution.

Through millions of years these atoms and molecules had been drawn together with different elements and molecules combining and recombining. It seems only logical that with all the time available and all the elements present, and with all the combinations possible, the critical quantities of carbon, hydrogen, oxygen, and nitrogen should eventually be linked in such a way that life was possible.

In the substances which make up all living systems we have unity, not only in what they are chemically but also in what they do physiologically. It is how these molecules are put together which determines diversity. The nature of all the steps (atom→molecules→macromolecules→subcellular particles→cells) which led to the formation of life is conjectural, but it seems that once the right conditions were present the formation of complex organic substances and life was an inevitable consequence.

Cosmic Origin

What evolution needed was time and inorganic substances. Both were always here. Early in this book we mentioned the number of galaxies that we can see with powerful telescopes and the numbers of planets and suns that make up these galaxies. We also mentioned that they were formed from great clouds of cosmic dust by molecular and gravitational attraction. We mentioned how our sun and its planetary system were probably formed and how life is believed to have originated from these inorganic substances, but let us look at these concepts again.

Origin of Organic Molecules

Of all the elements which are found within the earth, four (carbon, hydrogen, oxygen, and nitrogen) make up about 99 percent of the living and nonliving parts of all organisms. However, before we can produce the organism we must first produce organic molecules. At first it was thought that this would be impossible since organic molecules were being formed only by living organisms. In 1828 a German scientist succeeded in synthesizing the organic molecule, urea. Since then, it has been generally conceded that the generating of organic molecules was a possibility.

Now we recognize the fact that organic molecules are produced in nature by ordinary processes. When water is dripped on iron carbide, acetylene gas (C_2H_2) is produced. This is a highly reactive organic molecule which can react with other molecules, forming more complex ones. In geological processes, such as the eruption of volcanoes, iron and other metal carbides are blown out and spread over the surface of the earth where they may combine with water and form the highly reactive organic molecules. Two distinguished scientists, Haldane of Great Britain and Oparin of Russia, used this information to visualize the origin of life in the sea before it was salty and when the atmosphere was virtually without oxygen.

Synthetic Organic Molecules

In 1951 Stanley Miller succeeded in producing complex organic molecules by passing an electric charge through a mixture of water vapor, ammonia, hydrogen, and methane. He found that when the electric spark was shut off, more than half of the carbon which had entered the reaction as methane (CH_4) became a mixture of much more complex molecules. Analysis of the products disclosed them to be minute quantities of amino acids—the basic components of proteins and the principal organic substances of all living things. Since the substances used in the experiments are universally accepted as the most likely to have been present in the early atmosphere and since an energy source simulating lightning can be used in the laboratory, it is quite possible that organic molecules, which may give rise to living organisms, could be formed in nature with the same substances and energy sources (lightning and cosmic radiation, including the ultraviolet) available. Later experiments of the same type have confirmed these results (Fig. 1).

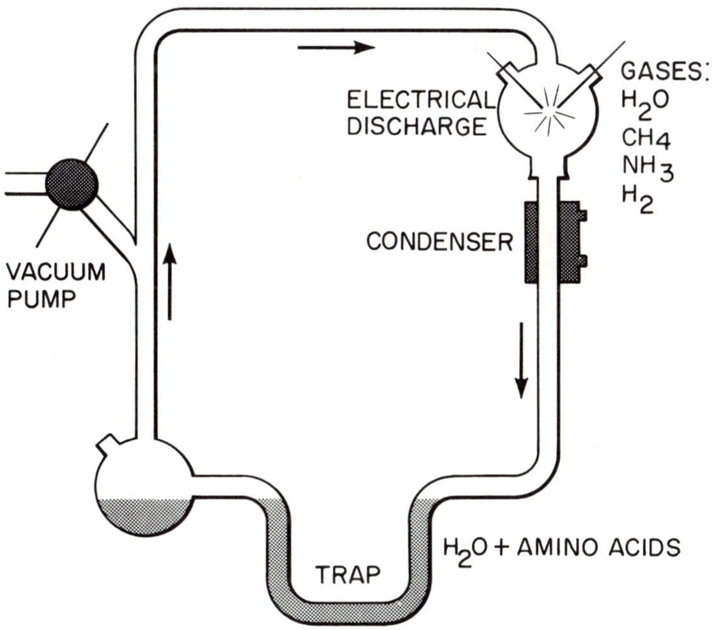

FIGURE 1

Most scientists are of the opinion that life arose like this in the sea where water and all necessary elements were available and where the first organic molecules could come in contact with others, thereby forming many aggregates of molecules. Perhaps the first were not DNA or even a type of nucleic acid, but with the many combinations possible, it probably was not too long before a self-reproducing compound like DNA was formed.

A question frequently raised against this theory was why can we not do the same thing now? That is, take these organic molecules, which we can produce, and build them into a living organism.

One day we might be able to create the same environment that these primordial organisms were formed in. For the present we can say these conditions no longer exist. Now our world is heavily populated with all types of microorganisms which quickly gobble up any organic molecules formed. Furthermore, many of these microorganisms of decay cause the newly formed molecules to rot in a short time, but three billion years ago these microorganisms had not yet been formed. Another thing which destroys these molecules is oxidation. Now oxygen is found almost everywhere; but when these first were formed, there was no free oxygen. Atmospheric oxygen began to appear after there were living organisms capable of photosynthesis.

The difference between chemical evolution and organic evolution lies in the capacity of the product to reproduce itself. As more and more complex substances were formed as the result of increased energy reception from the sun and the combination of more and more chemical substances, substances began to grow much in the form of crystals. This was not yet life. Once the property of self-duplication was acquired, probably through the union of chemical substances to form a nucleic acid similar to or identical with the DNA found in the chromosomes of present-day organisms, the ability of self-duplication was at hand and life was present. There were perhaps many such combinations and some of the self-duplicating molecules were more abundant and more successful at reproduction than others. These had the greatest potential for survival and it was, therefore, at this level that natural selection was initiated. This is an exclusive characteristic of organic evolution and was first observed as a characteristic of higher organisms many years before molecular biology was described.

The Status of Evolution

The vast majority of scientists, and virtually all biologists, accept as fact that organisms have evolved. Of course it is impossible to watch an organism evolve into another, for the major steps in the evolution of organisms require millions of years to complete. Nevertheless, it is possible to duplicate in the laboratory, in the garden, or even among native organisms some of the evolutionary processes known to have taken place in nature.

How this may be done and the various mechanisms of evolution will be described a bit later, but first we might consider some of the historical background which led to our modern evolutionary concepts.

Greek Concepts

As early as the sixth century B.C. the Greek philosophers considered the possibility of relationship among organisms. Although they did not see evolution as we do today, they described a type of succession in living organisms. They looked upon succession as being from the "less nearly perfect" to the "more nearly perfect" rather than from the simple to the complex. By the fourth century B.C. Aristotle had described several hundred different species and continued with an elaboration of the earlier evolutionary philosophies, but for many centuries following his time scientific concepts were forced into stagnation because of the power of the Church and religious dogma.

It was not until the seventeenth century, when Redi proved that maggots would not form spontaneously in meat if flies were prevented from laying their eggs on meat, that spontaneous generation was discredited as the manner in which species were created.

Lamarck

In the eighteenth and nineteenth centuries there was a good deal of straightforward discussion of the origin and evolution of organisms. Finally, in 1809, Lamarck not only asserted emphatically that the theory of organic evolution was true but also explained how the process must have occurred. He based his explanation on the idea of **use and disuse** of parts and the **inheritance of acquired characteristics.** He stated that during the lifetime of the individual organism any part of its body which is used a great deal will develop in size and functional capacity, while those parts which are not used will gradually deteriorate and atrophy. Furthermore, he stated that these acquired characteristics will be inherited by offspring. This in itself is not evolution, but according to Lamarck such traits provide a start from which evolutionary processes may follow.

Plausible as the Lamarckian theory seems, few modern biologists believe in it—at least as it was proposed. Acquired characteristics which affect only the body tissue of the organism are not transmitted to their offspring, but if the external stimulus affects the characteristics of the genes of the reproductive cells in some way, then the effect may be inherited. Knowledge of the cell, the nucleus, the chromosomes, and the genes was to come long after Lamarck died, but his theory did much to spread the idea of evolution.

Darwin and Wallace

The cliché that the thoughts of great minds often follow the same channels was true, at least, in the case of the **concept of natural selection.** It was first proposed by Charles R. Darwin (1809–1882) and Alfred R. Wallace (1823–1913) through their papers presented at the 1858 meeting of the Linnaean Society of London. Darwin had had an opportunity to spend five years traveling around the world on a British naval vessel, H.M.S. *Beagle*. He had made extensive collections and observations and spent many years studying the data before he was ready to expose his ideas in print. Shortly before the Linnaean Society meeting Darwin received Wallace's manuscript through the mail. In the contents, Darwin immediately recognized his own theory. He promptly arranged for the joint reading of Wallace's essay and one of his own, through the geologist Lyell and botanist Hooker. While we may consider the general concept of natural selection as the Darwin-Wallace theory, Wallace readily gave credit to Darwin because of his earlier papers and for his famous book *The Origin of Species by Means of Natural Selection,* which was published in 1859.

In general, the Darwin-Wallace concept of how evolution occurs through natural selection may be summarized as follows:

1. *Overproduction*
   a. Every species of organism produces far more offspring than can ever survive to become productive individuals.
   b. Because the numbers of individuals in a species are relatively constant, there must be a high death rate.
2. *Variation*
   The individuals of any species and different species vary in their characteristics.
3. *Adaptation*
   Those species with characteristics that naturally fit their en-

vironment will be favored over those with less fit characteristics.

4. *Survival of the fittest*
   a. Those individuals with characteristics best fitted for survival and inheritance will pass these characteristics on to the next and succeeding generations in greater numbers than those characteristics which are less fit.
   b. Succeeding generations will continue to exhibit and improve the adaptations received from their ancestors.

Both Darwin and Wallace were impressed with the effect of the environment on the individual. They believed that over a period of time, following a long series of generations, the influence of the environment would eventually produce many new favorable variations which would ultimately produce a new species.

One of the several principles upon which Darwin, in particular, built this theory of natural selection is that of the "geometric ratio of increase." He stated that all living species have a potential reproductive ratio of more than one to one; i.e., if all reproductive units function and all offspring survive, the second generation will outnumber the first. In bacteria and other simple fission organisms with a reproductive ratio of only two, the number of individuals increase tremendously under conditions which favor growth, reproduction, and survival. An extreme in overproduction is the potential of spore production in puffball fungi. It is estimated that if every spore from one giant puffball about three feet in diameter were to grow into another giant puffball, by the third generation they would produce a mass equal in volume to the earth. All plants and animals have great potential for increase, although not as great as that of a puffball. As a result, any small population invading a new fertile locality will rapidly increase in numbers until it has reached the saturation point beyond which limitations in the food supply and other natural checks will make further increase impossible. The potential geometric ratio of increase and the natural checks of the environment have, therefore, arrived at an approximate equilibrium which is maintained until some outside agency upsets the environment. At any given time most species in most localities have attained this equilibrum, so that the surviving population of the second generation is essentially equal to that of the first. The maintenance of this equilibrium necessitates that of the thousands of individuals that start life each generation only a few hundred (usually less) actually survive to maturity. The majority are eliminated during the early stages of life. In other words in most natural situations "death is the rule and life is the exception." For the puffball not one spore in a million actually goes on to produce another puffball.

Darwin also built his theory on the struggle for existence. He pointed out that, with rare and at best only temporary exceptions, life is essentially a struggle. In part this struggle is the competition between organisms that inevitably results from overproduction. Since space and food supply are limited, organisms compete with others of the same species. The available light, water, and soil salts are usually present in limited quantities. The more vigorous, or perhaps better positioned, plants secure these essentials, while other plants are crowded out and die. Competition is interspecific as well as intraspecific, and competition between plants and animals may also occur.

Green plants are frequently the victims of bacteria, fungi, and animals; carnivorous animals are dependent upon herbivorous animals. Frequently there is a food chain of many links starting with the green plant and ending with the dominant carnivore of the region. Wherever one link of the chain forms the next, there is usually a life-and-death competition. In most natural situations these intricate interdependencies have been established, and an "equilibrium of species" has been approximated with about the same number of individuals of each species reaching the adult condition in each generation. But this state of equilibrium does not mean a state of tranquility. The equilibrium is not static but dynamic; it involves a relatively continuous interplay of natural forces that are roughly balanced in their effects.

Competition with other organisms is not, however, the only thing that enters into the struggle for existence. Even where no other organisms are on hand to compete with them, plants may be restricted in their growth and reproduction by the purely physical and chemical limitations of the environment. Temperature often decides the fate of organisms as do light and the physical and chemical composition of the substratum. The struggle for existence is, then, with problems of adjustment to the nonliving forces of nature as well as with living competitors.

Darwin also based his theory on the "universality of variation." We are well aware of the individual differences which we readily see in other humans and if we were that familiar with two plants from the same seed pod or two dogs from the same litter, the individual differences would be just as apparent. Of course the more obvious differences which exist between different species are usually thought of in connection with the evolutionary process, and it is the cause for these variations which we are most concerned about in studies of the origin of species.

Darwin summed up the significance of variation by saying that certain organisms by their characteristics were more fit than were others. By "fit" he meant that they were better adapted to their environment. He simply pointed out that the fit stand a better chance for survival in the struggle for existence than the unfit. The successful are able to reproduce and continue the species while the unsuccessful perish usually before they reach the reproductive period. Thus the inevitable operation of environmental forces leads to a certain amount of natural selection, a selection of the more fit in the population to perpetuate the species.

The Modern Theory of Evolution

Darwin wrote that "any variation which is not inherited is unimportant to us." Essentially this summed up the weakness of Lamarckianism, and at the same time it opened the door for Mendelism. Although Darwin and Mendel were contemporaries, Mendel's work was apparently never observed by Darwin. If it had been, Darwin could undoubtedly have delved much more deeply into the role of genetics in evolution. Mendel's work demonstrated a mechanism which provides for stability as well as variability in hereditary factors. Mendel showed that characteristics were independently inherited, were assorted during gamete formation, and exhibited a phenomenon of dominance and recessiveness.

Now we have greatly elaborated on both Darwin's and Mendel's work and have come to recognize evolution through five basic processes:

1. Gene mutation
2. Changes in chromosome structure and number
3. Genetic recombination and hybridization
4. Natural selection
5. Reproductive isolation

Genetic variability (which is necessary if change and evolution are to occur) is provided through the first three. The last two processes provide for the development of populations as well as species in acceptable environmental situations. The migration of individuals from one population to another increases the amount of genetic material available to the population and the probability of hybridization within the groups may further alter the way in which natural selection guides the course of evolution (Fig. 2).

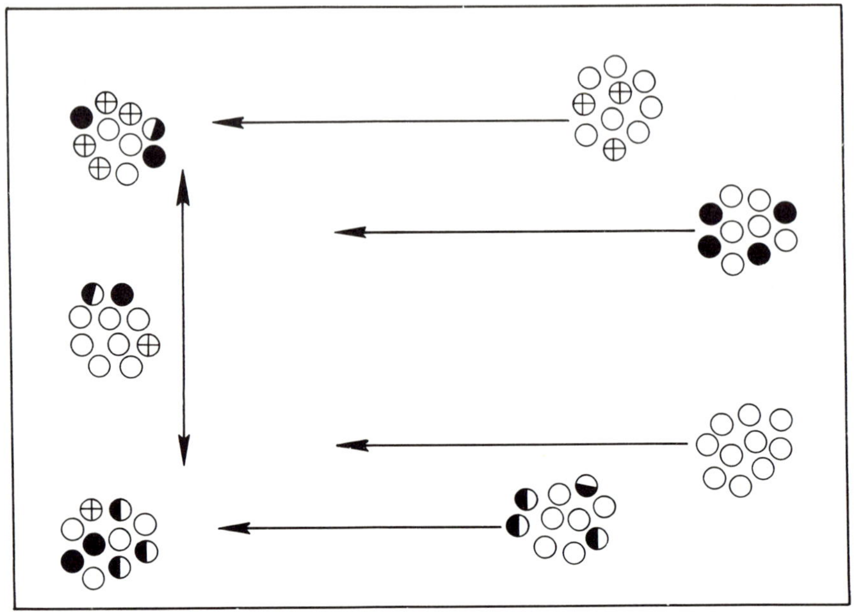

FIGURE 2

Genetic Drift

The introduction of new genes into a population, usually from contiguous populations, in the form of foreign pollen or seed is often termed **gene flow** or **genetic drift.** Thus hereditary material present in a small part of a population may spread in time to a large segment of the population, altering to a degree the genetic content of the population. With the production of many sexual generations and the shuffling and reshuffling of the genes in the population, a gene pool is established providing genetic materials which may be used in many different combinations by the interbreeding members of the community. Evolution operates essentially by the appearance of inheritable variations as a result of sexual recombination and mutation, and these variations are then spread through a population by interbreeding members in successive generations (Fig. 3).

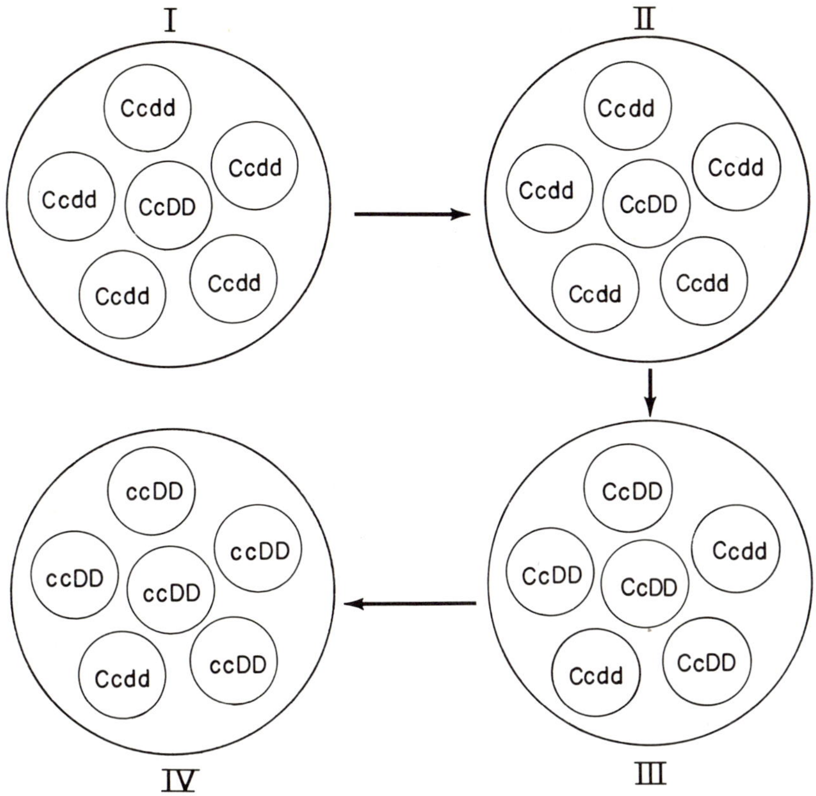

FIGURE 3

The Hardy-Weinberg Law

Since evolution may be characterized by a progressive change of gene frequencies, it should be possible to measure the speed of evolution. Of course, there may be other limiting factors such as the adaptive and survival value of a certain characteristic and of the organism receiving the characteristic. Generally, well-adapted organisms contribute most to the gene pools, especially if they have a high reproductive ratio.

Some years ago a mathematical corallary was worked out independently by a British mathematician, G. H. Hardy, and a German geneticist by the name of G. Weinberg. Now known as the Hardy-Weinberg law it demonstrates that the original variability in a population will be maintained in the absence of forces such as selection which tends to decrease the variability, or of mutation which tends to increase variability.

For example, we may consider a population composed of a large number of individuals in which breeding is at random. If in the population there is a pair of genes, A and a, with a frequency of p and q respectively so that $p + q = 1$, the two genes could then combine at random and produce genotypes with the following frequencies:

$$\begin{aligned} \text{AA} &: p \times p = p^2 \\ \text{Aa} &: p \times q = pq \\ \text{aA} &: q \times p = qp \\ \text{aa} &: q \times q = q^2 \end{aligned}$$

Since Aa and aA are genetically identical, their combined frequencies are $2 \times pq$ or $2pq$. Since the two alleles, A and a, occur in a population in equal frequency the result of random matings between the possible genotypes of AA, Aa, and aa will continue them at frequencies found in the F_2 progeny of the typical monohybrid cross of AA × aa, or 25 percent aa, 50 percent Aa and 25 percent AA.

However, if the frequencies of the alleles are unequal the genotypes may be calculated by binomial expansion. Since the alleles A and a are represented by the symbols p and q, then the genotypes AA, Aa, and aa will occur in the frequencies of p^2, $2pq$, and q^2.

The frequency of A and a gametes produced by the population, without selection or mutation, is therefore the same as the parental gametes provided mating is at random and the genotypes have equal adaptive values. Evolution therefore depends upon unequal or differential adaptive values of genotypes, on deviations from random matings, or on both factors operating together. Since we know that mating is usually nonrandom, that mutations do occur, and genes are being continually reshuffled by sexual recombination, individual variations within a population inevitably occur.

Speciation by Isolation

Occasionally new species originate when single individuals or relatively few become isolated from the remainder of the population. This may be due to geographic barriers or to reproductive incompatibility to the remainder of the population. Following reproductive isolation, mutations may appear due to breaks in the chromosomes, rearrangements of the linear order of the genes in the chromosomes, or perhaps to changes in the number of entire complements of chromosomes. Individuals with the new chromosome complements may mate with the ancestral types, but because the chromosomes often fail to pair at meiosis, the progeny are frequently sterile. In plants which are self-fertile such changes in the chromosomes will not interfere with their fertility; consequently new species may be produced this way. According to chromosome counts made of many plant species thus far it appears that more new species have arisen by polyploidy than by chromosome inversions or translocations and other mutations.

When interbreeding between subgroups of a population becomes less frequent and the hybrids are commonly sterile, the individual groups become new species. Land masses which become separated from each other by physical barriers such as the rising of mountain ranges, by the invasion of seas, rivers or glaciation, provide isolated habitats which ultimately develop species peculiar to that area. Occasionally the isolating barrier breaks down and the isolated species may again be in contact. As a result two isolated groups may again interbreed. However, it is possible that each may have mutated to such a degree that they no longer are compatible genetically and any hybrids produced will be sterile.

New species may also arise when two groups have been growing in different habitats. The different soil conditions, light exposure, and temperature of the two groups may cause pollen to mature at different times, and consequently cause them to breed at different times. Such groups would, therefore, be unable to breed even if now grown together.

Hybridization

Since relatively few plant species will interbreed and produce fertile offspring, it seems unlikely that many species have originated this way. Nevertheless it is a method frequently used by plant breeders to establish new combinations of desirable characters in progeny which are economically useful to us. Hybrid corn is an example.

If species with different chromosome numbers are crossed, the offspring are usually sterile. Occasionally the chromosomes of the sterile hybrid will double; then meiosis can take place normally, and normal fertile eggs and sperms can be produced. Such a phenomenon has apparently happened in many plants. One example is tobacco. There are two wild South American species, *Nicotiana sylvestris* and *Nicotiana otophora*. The hybridization of these two species resulted in a sterile hybrid, but doubling of its chromosomes has produced *Nicotiana tabacum,* the most widely grown species of tobacco today.

Another species of grass arose as a result of hybridization in which the hybrid received complete sets of chromosomes from each parent. *Spartina stricta* is a species that grows in England, while *Spartina alterniflora* is a species native to the New England states. As a result of commerce, some of the seeds from the New England species were transported to England. Soon *S. alterniflora* was growing near *S. stricta*. Hybrids resulted from natural crosses between the species, and one of them proved to be an outstanding type with characteristics more desirable than either of its parents. It was found to breed true, was given the same *Spartina townsendii,* and recognized as a distinct species.

Upon cytological examination it was found that the hybrid *S. townsendii* has 126 chromosomes, *S. alterniflora* has 70, and *S. stricta* has 56. In this case the fusing gametes of each parent carried complete sets of chromosomes; the hybrid produced is a polyploid.

Summary of the Principles of Evolution

1. Life originated from inorganic sources through the interaction and combination of such substances as carbon, hydrogen, oxygen, and nitrogen.
2. Physical factors in the environment apparently influenced the interaction of the substances to form complex molecules.
3. Complex molecules united and became increasingly diversified, forming primitive heterotrophs.
4. Heterotrophs eventually mutated into autotrophs initiating the appearance of the first plants.
5. Diversity as a result of chromosome and gene alterations in mutations, polyploidy, and hybridization resulted in the establishment of numerous species.
6. Evolution occurs by populations rather than by individuals, through nonrandom reproduction, natural selection, and genetic drift.

Evidences for Evolution

Much of the earlier discussion presented in this chapter emphasized the role of the genes and chromosomes in evolution through their biochemical changes, resulting in variation. This type of information was not available to the early scientists who strongly supported evolution. Yet most of them must have been as firmly convinced of the evolutionary concept as we are today. The earlier proponents depended to a very large extent on the fossil record, comparative morphology and anatomy, comparative embryology, comparative physiology, biogeography, and taxonomy.

Paleontology, the Study of the Fossil Record

The study of ancient organisms from their fossil remains is the only direct way we have of knowing exactly what these organisms were like. Undoubtedly thousands of species were never fossilized, and some were fossilized so imperfectly as to leave little for us to observe. Yet thousands of remains have been found and at different levels in the earth's crust which gives us a reasonably clear picture of what life was like hundreds of thousands and even hundreds of millions of years ago.

Fossils are of several types:

1. Entire specimens or parts of the organism, which have been preserved in acid bogs, tar pits, and elsewhere where specimens can occasionally be found.
2. A cast or impression of the organism shows the size, shape, and often very minute details of the surface characteristics of the organ or organism.
3. Petrifactions are specimens which have had the plant parts, often on a cell for cell basis, replaced by various minerals. Numerous petrified trees and other species have been preserved in this way.
4. Compressions consist of carbonized plant material in much the same shape as the original but generally much compressed.

Geologists have worked out a comprehensive time table showing the probable time rocks in various layers of the earth's crust were laid down. These layers have been fairly successfully dated in time by radioactive substances. The disintegration rate of certain substances is quite accurately known, as well as the end products of disintegration. Radium, for example, is known to disintegrate into lead in a specific period of time. Occasionally a rock is found which is composed of both radium and lead. If we assume the entire rock was at one time radium, then the amount of lead present makes it possible to calculate the age of the rock by determining how long it took the lead to form.

Another method, but which can be used for dating material no older than 50,000 years, is known as radiocarbon dating. In this case a measurement is made of the amount of Carbon-14 present in the specimen. In approximately 5568 $\pm$ 30 years half of the Carbon-14 will have disappeared, and half of the remainder will be gone in another 5568 years. Consequently, by measuring the Carbon-14 in the specimen and comparing this with the amount of Carbon-14 in a modern specimen, a date can be established for the death of the fossil specimen.

In general the deeper a rock stratum is located the more ancient are the remains found in it. As a result of world-wide collection of fossils and the recording of the strata in which they were found, a great deal of information has accumulated concerning the time certain species lived, when they disappeared, and the types of organisms that followed them. Consequently, we know that the first vascular plants arose during the Silurian era (see the geologic time chart on pp. 450–452) and became very abundant during the Devonian. Many of these early species became extinct long before the first angiosperms apparently arose in the Mesozoic era. Few, if any, of the vascular species in the Devonian era have survived to the present day.

Comparative Morphology

It is quite probable that some concept of organic evolution would have developed even if there had never been any fossils found. It is true that the fossil record was very influential in the establishment of early evolutionary concepts, but anatomical studies of the structure and major organs of both animals and plants also show relationships which may be used as evidence of evolution. In animals similarities in the circulatory, excretory, skeleton, muscular, and other systems are quite evident. The vascular plants show similar relationships in leaf structure, stem, root, flower, and vascular development. Homologous organs have the same origin and have basically the same relationship to adjacent organs and tissues. In the higher plants we see that roots and the other organs mentioned above are similar in their origin in the various groups of ferns, gymnosperms, and angiosperms. Analogous structures have a similar function but a different origin. For example, the blades of a moss have the same general characteristics as the leaves of an oak tree, but they have different origins. The stipe of a kelp is similar in function and appearance to the stem of a pine, but again they have different origins. Such analogous structures indicate remote relationships, while the homologous organs show closer relationships. Comparative morphology and anatomy is a very strong point for evolution for it shows the unity of organisms on a macroscopic scale.

Comparative Embryology

Similarity in the development of the embryo for many groups of organisms is readily seen in animals. While not as obvious in plants, the same general plan is followed in development of the embryo with little variation. All embryos begin as a zygote which produces an undifferentiated mass of cells. In the bryophytes this mass of tissue elongates and differentiates into an absorbing foot, a seta, and a sporangium. In the tracheophytes there is somewhat greater differentiation with the zygote first forming the typical mass of undifferentiated cells, then differentiating an absorbing foot. In most vascular plants a suspensor, of usually a few cells, and then the main body of the embryo are formed consisting of cotyledon(s), epicotyl, and hypocotyl.

While there is variation in the development of these parts in the various plant groups, all organs differentiating in the embryo are homologous to those of the other groups.

In organogenesis of the stem tip and root tip, there is also great similarity in the differentiation of the various tissues from these growing points. The embryological evidence indicates relationship among organisms is due to the similarities of development and differentiation.

Comparative Physiology

Throughout much of the discussion in this book, the unity of all organisms was emphasized because of the similarities in the nucleic acids, amino acids, enzymes, protein formation and the various physiological processes. Basically respiration is the same in all organisms for it is essentially a surface phenomenon of oxidation and diffusion. In all plants, whether blue-green algae or orchids, photosynthesis is the same. In all organisms energy cycles are much the same, the foods utilized and the end products obtained are the same.

Variations in the proteins formed in different organisms also indicate a close or distant relationship. The more similar the proteins the closer the relationship of the organisms; this is also confirmed by studies of their morphology and embryology, which bear out the chemical indication of relationship or the lack of it.

Biogeography

It is clear to us today that the plants and animals of different geographical localities vary, but even Aristotle was aware that there was some variation in the organisms in the small area of which he was acquainted. The extent of this variation was finally made clear by Darwin following his extensive voyage. He was impressed by the various species of animals and plants he found on the isolated Galapagos Islands, six hundred miles west of Ecuador, and those of the Cape Verde Islands, about four hundred miles west of Dakor. While the species on these islands were quite different, Darwin noted that those of the Galapagos were similar to those found on the mainland of South America, and those of the Cape Verde Islands were similar to those of the African mainland. But Darwin also noted that the species living in the two groups were quite dissimilar from each other.

Certain of the animal and plant species had been able to migrate voluntarily and involuntarily to these islands, or they were isolated when cut off from the mainland by sea barriers in the remote past. In any case, they were able to develop as separate groups and gradually evolved characteristics somewhat different from those of the mainland and the common ancestor they had descended from.

Throughout much of the world the same types of observations may be made for the species in a given locality. Centers of origin have been well established for many of our crop plants as well as for others of less economic value. The geographic distribution of plants (and animals) and the various life zones are discussed in the next chapter.

Taxonomy

The science of taxonomy or systematic botany in which the classification of plants was described is discussed in Chapter 14. Organisms were placed in their various groups according to our opinion of their relationships. Plants placed in the same species are there because the individuals are similar in their morphology, embryology, physiology; and because they may interbreed. Those species placed in the same genus have many characteristics in common but vary more than those within the same species. In plants, more than in animals, individuals within the same genus but of different species will occasionally interbreed. This characteristic alone then does not indicate that they should all be in the same species, but it does indicate a certain genetic compatibility which probably means they are more closely related to each other than to those of other species within the same genus with which they do not mate.

Higher and broader groups such as families, orders, classes, and phyla have basic structural patterns in which the members of each group will have more in common with its own members than with those of the other groups. Yet as distantly related as the members of different phyla may be, they are all structurally the same for each has the cell as its basic unit upon which its body plan was formed.

Supplementary Reading

Andrews, H. N., Evolutionary Trends in Early Vascular Plants, *Cold Spring Harbor Symp.* 24:217–235, 1959.

Calvin, M., Chemical Evolution and the Origin of Life, *Amer. Scientist* 44:248–263, 1956.

Davis, P. H., and V. H. Heywood, *Principles of Angiosperm Taxonomy*. Oliver and Boyd Ltd., Edinburgh, 1963.

Hamilton, T. H., *Processes and Pattern in Evolution*. The Macmillan Co., New York, 1967.

Oparin, A. I., *The Origin of Life,* 3rd ed. Academic Press, New York, 1957.

Savage, J. M., *Evolution.* Holt, Rinehart, and Winston, New York, 1963.

Solbrig, O. T., *Evolution and Systematics.* The Macmillan Co., New York, 1966.

Sporne, K. R., The Phylogenetic Classification of Angiosperms, *Biol. Rev.* 31:1–29, 1956.

Stebbins, G. L., Chromosomal Variation and Evolution, *Science* 152:1463–1469, 1966.

Stebbins, G. L., *Processes of Organic Evolution.* Prentice-Hall, Inc., Englewood Cliffs, New Jersey, 1966.

Stewart, W. N., More About the Origin of Vascular Plants, *Plant Sci. Bull.* 6:1–5, 1960.

Wald, G., Theories of the Origin of Life, *Frontiers of Biology.* Houghton Mifflin Co., Boston, 1962.

Questions for Discussion

1. What are the general lines of descent which probably took place in the evolutionary development of flagellated organisms?

2. How has Darwin's theory been improved and elaborated in recent years?

3. What evidence do we have that certain cultivated vegetables, which we raise extensively today, are descendants of wild species of relatively little economic value?

4. How may life have originated and developed into the great mass of living species, if at one time there was nothing but dust and space?

5. Explain the significance of the following in evolution: Macromolecule, Time, Radiation, Variation, Molecular Attraction, Mutation, Polyploidy, Hybridization, and Adaptation.

6. What would you suggest as the reasons why the species of fungi and angiosperms are so much more numerous than the species of other groups?

7. What are the reasons many species have become extinct?

8. How is it that certain species such as the *Metasequoia* and *Ginkgo,* considered "living fossils," have not become extinct?

9. In the Great Plains regions trees are found only in isolated areas. Why do not trees take over the grasslands?
10. Explain the survival of cacti in desert habitats and gymnosperms on many mountain tops.
11. What explanation can you give for the incomplete fossil record displayed in the many sedimentary deposits?
12. Is evolution a well-planned, organized process or development by trial and error? Explain.

27

Ecology: Biologic Interrelationships

In the preceding chapter we discussed the evolutionary descent of organisms and some of the adaptations acquired by certain organisms living in specific habitats. Furthermore in earlier chapters the characteristics of many species were described. In many cases the characteristics were described in relation to the environment to illustrate how these characteristics made the organisms better adapted to their environments. As one observes different organisms such as moths, caterpillars, other insects, and other animals, as well as plants, he is usually amazed by the remarkable adaptations the organisms have acquired through evolution by natural selection which serve them so well for the environment in which they live.

How the organism lives in its environment and has adapted itself to the physical factors of cold, heat, drought, excess moisture, light, and dark, as well as to the other species which live with it, is the subject of the study of ecology. All organisms, whether an individual or a population, depend directly on the physical factors of their environment.

Controlling the Environment

Within the last few years we have suddenly come to the realization that the domination man has had over the major environments on earth has been with complete disregard to any long-range consequences. We have successfully controlled the reproduction and populations of all species and brought them in a subservient position to man. We have manipulated most environments to serve our own purposes but have failed to realize what we were doing to them. We have designed a great many habitats to increase our food supplies and to provide the raw materials for our industrial and domestic needs. Unfortunately we have been extracting more than we have been returning and what we have been returning has resulted in a poisoning of the environment to the extent that the physical base on which we live is losing its productivity. While many species of animals and plants are decreasing in numbers because of lack of space and pollution, man, the only species capable of living wherever life is possible, has been increasing at a rate that promises nothing but destruction. How soon we recognize the limitations of our environment and the steps that must be taken to live within the earth's productive capacity will determine how serious the overpopulation problem will become.

THE PHYSICAL ENVIRONMENT

On a world-wide basis the physical environment consists of land, water and air. For thousands of years man did not know the extent of any of them. Now the inventory is complete.

1. The **land base** on which we live is comprised of about 110 billion acres of which only about 4 billion acres is arable food-producing land. The most optimistic individuals agree that even by using our most imaginative and utopian land reclamation and development programs the amount of food-producing land cannot be extended beyond 6 billion acres.

Since we know it takes about two acres to produce enough food and fiber to support one human per year at a *reasonable* economic level, such as in the United States, we can see that we have already exceeded the carrying capacity of the land available. At present we have a population of close to 3.5 billion people. If we allow the human population to double in the next thirty years, as predicted, the best we can hope for is survival at minimum subsistence levels. Furthermore the decline in diet will not only be quantitative it will also be qualitative. As more land is used for crops for human food less can be used for our domesticated animals which provide much of the proteins in our diet.

This might be brought home to us in the United States in this way. Our major crop is corn. For each person now living in the United States we raise close to a thousand pounds of corn, but per capita consumption of corn in the forms of cereals, meals, syrup, etc., is only about thirty pounds. About 85 percent of the corn crop is used as animal food and the remainder is processed. In other words, at the present time we eat most of our corn as animal protein, but the time may come when we can no longer afford so expensive a diet and will have to remove the hog or chicken as the middleman and eat the corn directly.

2. Our **water base** seems as inexhaustible as any substance on earth since about three fourths of the earth is covered with water. Yet the inventory shows us that 97.2 percent of all the water is in the oceans where it is too salty to drink or use for irrigation. Two percent of our water lies frozen in glaciers and icecaps in the remote polar regions. Only 0.8 percent of all the water on earth is immediately available to us, and this includes our lakes, streams, ground water and daily exchange between the earth and atmosphere in precipitation and evaporation.

Although water is in short supply in many areas, it is adequate for us at present. However, the problem is not how much water we use but what we do with it afterwards. Much of our sewage and industrial wastes are pumped back into rivers or lakes in a toxic or untreated form. This sewage then flows down stream or in many cases remains locked as a toxic pollutant in lakes. Soon the toxic effects are observed in the death and disappearance of aquatic wildlife and the value of the lake as a water source or recreational area is lost.

Pure water is becoming so valuable that some cities are considering the possibility of having their own closed systems in which sewage water will be purified and sent back into the domestic supply to be used over and over. Some cities have this system now but at present they use the purified water only for irrigation.

Many individuals who minimize the problem of human overpopulation and the limitations of the environment optimistically proclaim the oceans as our salvation. We do have about 200 desalination plants in operation around the world but thus far the cost is so great the water produced can only be used for immediate domestic use; it is much too expensive to use for irrigation. When the proposed and approved nuclear powered desalination plant is in full operation in Long Beach California* it will produce 150,000,000 gallons of fresh water daily, enough for 2,000,000 people. Each day 23,000 tons of salt will be extracted from the 150,000,000 gallons of water. At present much of this salt is put back in the ocean and therefore not accumulated in massive piles. However, over a period of years what will this amount of salt returned to the offshore waters do to the microflora and microfauna which cannot tolerate changes in the salt content of their environment?

Although we can expect an increase in the amount of food we obtain from the oceans, the ocean is no panacea. Only in the waters of estuaries, the continental shelf and offshore comparative shallow waters can we expect to increase the productivity by fertilizing the waters, increasing the plankton and thereby making fish more abundant. The oceans will not be as important a source of increased food production as the arable land we now have in cultivation. This seems to offer us the most favorable prospects but again the limitations are all too clear.

3. The **air base** is also an inexhaustible resource. We used to think of the atmosphere as a well-balanced gaseous entity in which there was an exchange of oxygen and carbon dioxide between plants and animals. Now we are seeking methods to control the pollution in the form of carbon monoxide, sulfur dioxides, radioactive substances, insecticides, herbicides and dozens of other substances which modern civilization is discharging into the atmosphere in ever increasing quantities. Although many factories and car manufacturers are building devices to reduce the pollutants they place in the atmosphere by as much as 50 to 65 per cent, the increase in population means more cars on the highways, more planes in the sky, more industrial plants to provide electricity and all the products we use daily. The net result is still more atmospheric pollution.

There is also considerable concern about the possible decrease in the amount of oxygen in the atmosphere because of the elimination of many square miles of green plants for the sake of thousands of miles of highways, and new and larger cities, none of which add oxygen to the atmosphere.

For the past decade we have been transferring land in the United States, at the rate of about 2,000,000 acres per year, from open, rural into highways, urban development, and industry. This transferred land includes cropland, pasture land, forest and desert. With this gradual elimination of green plants from our environ-

* This project has now been abandoned.

ment we are decreasing our potential for food production and oxygen.

This then is our physical environment: land, water and air. It seems incredible that with such an abundance of these basic units everywhere we have to be concerned about them. It is not that we are depleting the earth of them, it is simply through overuse and misuse we are polluting and poisoning the only environment we have in which to live.

THE BIOTIC ENVIRONMENT

There are many limiting factors in an environment which govern the establishment of a species in a certain habitat other than the purely physical. These are the many interrelationships such as competition, predation, parasitism, and symbiosis which are regulatory for the establishment of any given species. Those organisms living in any given region are termed its biotic community. Each community is comprised of few to many populations. A population in ecology is defined as a group of organisms of the same species living in an area.

The Population

For the sake of definition, let us assume that a population is more than one individual. This does not seem to be a very profound statement until we begin defining the individual. This also seems simple enough but in plant (and some animal) populations we often have what appears to be a large number of plants but all are connected by subterranean roots or rhizomes and have arisen vegetatively from one original plant. We speak of these as **clones.** The wild morning glory, purple night shade, bindweed, bananas, and many others form a considerable population in this way. In some cases the newly arisen plant has no root system of its own but is completely dependent upon the root system of the original plant, often many feet away. A clone is usually thought of as the offspring of a single individual. This may be by vegetative propagation from rhizomes or runners with all plants connected as above, or the progeny may have been produced by cuttings or tubers or other plant parts from one individual, and many generations may have intervened between the life of the original and its progeny. Genetically they are all identical (Fig. 1) but from a population standpoint they are in competition with each other for survival. Ecologically the members of a community are in competition with

FIGURE 1 Diagrammatic illustration of asexual reproduction, where the parent cell simply splits in two. The chromosome complement remains the same and there is no variation in the offspring.

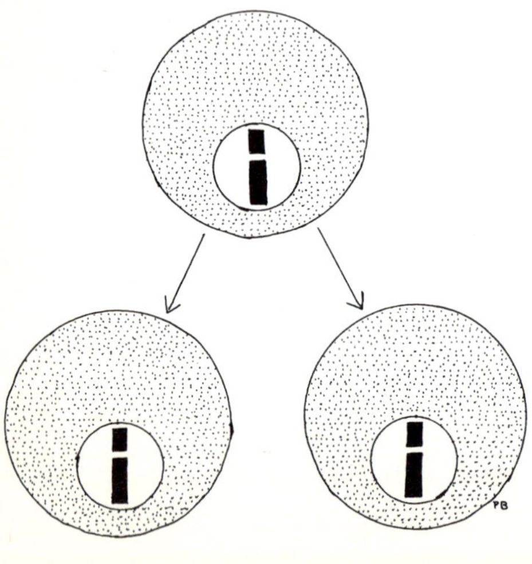

each other for the raw materials of life. These members may be those of the same clone, members of the same species, or of different species. Not only is the plant competing against other plants but it may also be competing against animals. The relationship is certainly not all deleterious for many organisms are successful because they carry on a mutually beneficial relationship, which we have come to call **symbiosis,** with other organisms. Many plants have endotrophic or ectotrophic fungi growing in or around their roots which aid them in the breakdown and uptake of inorganic salts. Microorganisms such as bacteria, protozoans, and the Actinomycetes are often of great benefit in changing the chemistry and physics of the soil for the betterment of plant growth. Animals may be helped in a similar manner, not only by microorganisms which make their external environment more suitable but also improve their digestion and assimilation by being present as an intestinal floral population in the animal's digestive tract.

The Ecosystem

The study of a population reveals that it is an aggregation of organisms associated in the utilization of energy and raw materials. Individuals and populations do not live alone but in an association with many different species of plants and animals. The relationship is one that has many subpopulations in which there may be a microflora and a microfauna within the soil each virtually living on the others. But this association could not have become established or remain long in an active state if it were not for the larger organisms which contribute their remains and at the same time, in a living state, offer protection and improved habitat for the smaller organisms.

At the soil surface, in a forest of minimum light penetration, is a Liliputian world of lichens, liverworts, and mosses with a few small, invading, herbaceous angiosperms. Within this association of flora numerous small animals such as insects, the larvae of insects, nematodes, earthworms, etc., are actively at home.

Offering some protection for the small but numerous organisms in the soil are scattered clusters of shrubs and larger herbs which have a certain interdependence with the microflora and fauna. Towering above the forest cover are trees which may be all of one species such as in a forest of Western Yellow Pines, or of a few species, such as a Beech-Maple association, or a tropical forest where there is a much greater variety in the species population (Fig. 2).

FIGURE 2 A mixed species habitat of a subtropical forest.

The various groups of organisms do not live in a haphazard arrangement but become carefully populated and associated on the basis of their dependence on each other. Each population unit is a biological system in which the raw materials of nature are available. Some of the organisms are capable of building foods and protoplasm from these inorganic substances but others must await a natural synthesis of food materials by other organisms in order to survive. Obviously such a natural community is a high level biological system. This we call the **ecosystem.** It may be as small as a tiny aquarium or as large as a continent. However, an ecosystem operates as a unit. This unit may be somewhat variable from year to year and season to season. The ecosystem may have many subsystems within it and the lines of demarcation are not always easy to determine. In the ecosystem, however, we must always have available the raw materials and energy that the biological community needs for survival.

The transformation of solar energy to chemical energy by plants as outlined below initiates the formation and ultimate utilization of food.

Solar energy assimilated by plants	→	Chemical energy formed by autotrophic organisms in growth and reproduction	+	Heat energy of respiration
Chemical energy eaten by heterotrophic organisms	→	Chemical energy assimilated by heterotrophic organisms	+	Chemical energy disposed of through waste products of heterotrophic organisms
Chemical energy assimilated by heterotrophic organisms	→	Chemical energy of growth, reproduction, and waste products of heterotrophic organisms	+	Heat energy of respiration

In the above steps of energy flow there is a continuous use of the products formed and in many cases a temporary storage of the energy in both the autotrophic and heterotrophic organisms. At each step there is a transfer of energy and heat is evolved; the series shows that in any ecosystem solar energy is the source and heat energy is its ultimate fate.

The effective operation of the various ecosystems which are described in the pages to follow depends upon the transfer of energy from the sun to the organism either directly or indirectly.

The quantity of solar energy entering the earth's atmosphere is mostly dissipated in the form of heat. Only about 1 to 5 percent is used in photosynthesis and is transformed into the chemical energy of plant tissues. In ecosystems such as a desert, where the plants grow rather far apart, much more of the sun's rays are lost than in a rich tropical rainforest where the plants grow close together and frequently grow in layers as undershrubs, shrubs and trees. Plants growing in shallow ponds and lakes also absorb much more of the solar energy directly than those living in oceans and deep lakes in proportion to their area of growth.

Our interest in the ecosystem is then not only of what types of organisms it is composed but of how efficiently they may be in absorbing solar energy and transforming this energy into other types.

Billings has divided the biological part of an ecosystem into **trophic levels** based on how far the original energy has come through the community.

Trophic level 1(T_1) is the green vegetation. This is the producer level for it is at this level that the radiant energy of the sun is stored in photosynthesis, primary foods are formed and oxygen is released for other organisms.

T_2 consists of the **herbivores** which range in size from the microorganisms to elephants. These organisms derive their food and energy from those of T_1 and are classified as **consumers.**

T_3 are those **carnivores** which live strictly on the herbivores of T_2. They are also consumers.

T_4 are those **carnivores** which live not only on the herbivores of T_2 but frequently eat carnivores of T_3. These animals include insects and spiders, shrews, weasels, some birds, and some reptiles. Added to this category might be the human and some other animals which are **omnivores** eating both plants and animals. These are also consumers.

T_5 is a large category which includes insects, insect larvae, fungi, bacteria, some protozoans, and many other small organisms that use dead plant and animal remains as food. These are the **decomposers** which break down organic substances releasing many compounds and elements back into the environment.

In the past relatively little interest had been shown in the order of decomposition or the identity of the decomposers except to generalize with a blanket identification of simply bacteria and

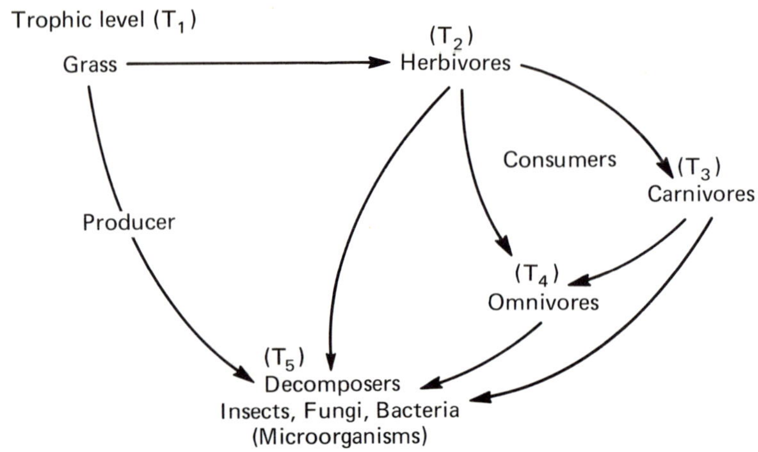

FIGURE 3 Trophic levels in the ecosystem; each level is dependent on those preceding but in T_5 we find the organisms which can live on each other as well as on those organisms in all other levels.

fungi. In 1965 an interested scientist, Dr. Jerry A. Payne, described an extensive study of decomposition in still-born pigs. He found that insects and insect larvae played a very significant role in the early stages and that different groups of insects moved in at different stages in the process providing a definite ecological succession of decomposition (Fig. 3).

Carrion exposed to insects and all other microorganisms underwent six stages of decomposition: fresh, bloated, active decay, advanced decay, and remains. Most of the decomposition took place within six days with only some dried remains left for bacteria and fungi to finish over the next several weeks period. Carrion kept free of insects decomposed very slowly, retaining its form for many months. Five stages were recognized for the protected carrion: Fresh, bloating and decomposition, flaccidity and dehydration, mummy, and desiccation and disintegration. It is therefore apparent that the bacteria and fungi are the final decomposers of organic remains but the initial stages are often started by larger organisms which in turn add their remains for the action of the microorganisms.

The efficiency with which organisms convert solar energy to the chemical energy of plant protoplasm and this in turn to the protoplasm of heterotrophs gives us a perspective of the role certain key species play in the energy flow within ecosystems. Since 95 to 99 percent of the solar energy is immediately lost from plants in the form of heat reflection or unabsorbed transmitted light, only a very small percentage is transformed into the chemical energy of plant tissues through absorption.

The autotrophs, which are then eaten by herbivores, have only about ten percent of their protoplasm assimilated and transformed into herbivore flesh, the remainder is lost as feces and heat. Carnivores may assimilate as much as 75 percent of the herbivore flesh eaten, although 30 percent is a more realistic figure, and then only through the early stages of growth or until adulthood is reached following which the efficiency of assimilation declines to 4 or 5 percent.

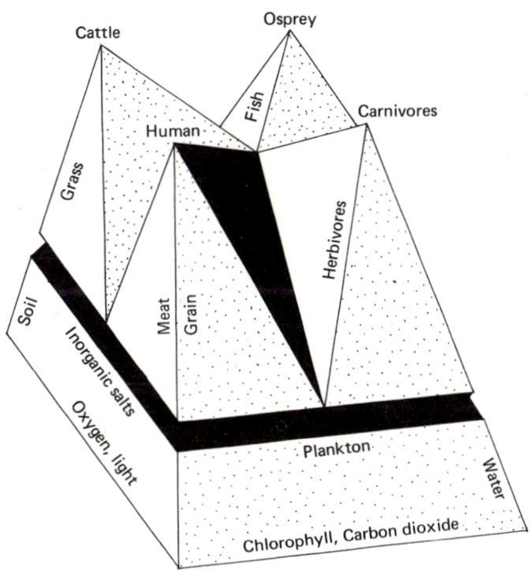

FIGURE 4 A food pyramid which shows four major lines of development, each of which uses some or all of the substances in the pyramid base. Each point of the pyramid has an organism at its pinnacle which is dependent on the organisms or food immediately below it.

The effective operation of all ecological systems depends then upon the transfer of energy from organism to organism and constitutes the programming of the many food chains and more complex food "webs" which are more than isolated linear systems involving but a few species (Fig. 4).

The decomposers constitute a more heterogeneous collection of organisms than those of the grazing chains. These microorganisms of the detritus food chains are estimated to account for as much as 90 percent of the energy flow through an ecosystem. Studies made of these organisms indicate the bacteria and fungi are much more effective in completing their role of energy transfer if earthworms, larvae and other soil invertebrates are present to aid the microorganisms in the initial food breakdowns.

The decomposers are also most active where the organic remains are most plentiful and the temperature and moisture are most suitable for their action. Normally this would be in tropical rain forests where the production of vegetation may be twenty five to thirty five tons of vegetation per year. In temperate regions the most productive communities produce about fifteen to twenty five tons of vegetation per year while a desert will often produce much less than one tenth of that figure. Consequently the animal populations will be limited accordingly and the microorganisms will have a similar reduction in quantity.

FIGURE 5 World distribution of primary production in grams of dry matter produced per square meter per day in the major ecosystems. The illustration shows that only a relatively small part of the biosphere is naturally fertile. (Redrawn from E. P. Odum, *Fundamentals of Ecology*, 2nd ed. 1959. Saunders, Philadelphia.)

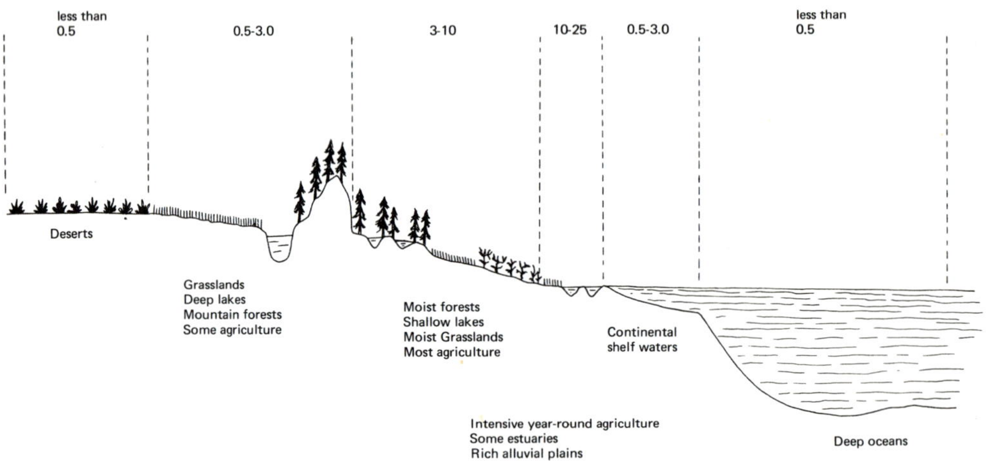

Thus we find that an ecosystem is an environment involving many organisms and many overlapping energy cycles. Ecologists often use the term **biotic community** to describe the members associated in a particular place. Certain plants usually dominate a given habitat in association with others. In the short grass prairies buffalo grass *(Buchloe)* and blue grama grass *(Bouteloua)* are usually living together in the same plant community, while in the deciduous forest ecosystem one of the plant communities would be the beech and maple and another would be oak and hickory. Each community would have its own more prominent shrubs and herbs and each would have its own species of microorganisms and animals living together as an associated biotic community (Fig. 5).

Until the organisms in an environment become relatively stabilized there is an orderly process of change in which one community replaces another in area. We speak of this as **succession.** It begins with a **pioneer stage** which may continue for many years until the **edaphic factors** (soil conditions) have so changed that other organisms are able to live there. The new organisms coming in are described as **invaders.** There may be a good many sequential stages with a different set of invaders coming in each time until the environment has been stabilized and the **climax** community has been established. At this stage the changes are minor and succession has stopped. The climax organisms will then live here permanently unless some drastic changes occur such as might happen as the result of a forest fire, flooding, the draining of a lake, severe climatic changes, or the eruption of a volcano.

FIGURE 6 Everglades National Park in Florida. It is essentially a hydrosere wherein live numerous plants adapted to a wet habitat. Many aquatic animals including alligators and the cormorant in the picture also live here.

Hydrosere All lakes and ponds are gradually being filled as the result of soil being carried in by streams, rivulets, and runoff following a rain. Furthermore the organisms which live in and around the lake add their remains as they die and are succeeded by others. We called this type of habitat a **hydrosere** (Fig. 6). It is usually characterized by several concentric zones. The middle of the lake is open water and relatively deep. The organisms which live there are essentially plankton but it also consists of many larger aquatic animals capable of swimming about freely. Surrounding the open water is a relatively broad band of vegetation composed of some free floating angiosperms, such as the water hyacinth in the south or more commonly water lilies which have their leaves and flowers on the surface of the water but with their roots anchored in the soil many feet below. The depth of the water will determine how extensive an area these plants will occupy.

Near the shore, in but a few inches of water, will be cattails and large sedges which are rooted in the soil but with most of the plant parts out of water. Surrounding the emerged aquatics is usually a mixture of sedges, grasses, and willows which grow very well in water saturated soils. Farther back and away from the lake will be the climax plants typical of that particular area.

This type of aquatic habitat will persist for thousands of years in large deep lakes, while a small shallow pond may be transformed from an aquatic habitat to a climax beech-maple community within a human lifetime. As the sediment and organic remains accumulate in the lake the water becomes more shallow and each vegetation zone tends to occupy more area towards the center. Gradually the environment changes from a lake to a swamp to a grassland to a forest or to whatever that particular habitat will support as the climax community.

Xerosere One of the oldest of biotic successions is the xerosere (Fig. 7), which begins with bare rock. In areas where the parent rock is on the surface or new rock has been brought in such as by a lava flow, we would see that the first plants to appear are

FIGURE 7 A xerosere in Southwestern Arizona inhabited by the organ pipe cactus, creosote bush, ocotillo and many other xeric plants.

crustose lichens. They will persist for many years, but as the result of their growth bits of sand and dust accumulate around them. The plants die and thereby add a little humus to the accumulating soil material. By their respiratory products and those of bacteria and fungi chemical action from organic acids begins to decompose and disintegrate some of the parent rock. After a number of years larger foliose lichens are able to grow in the very shallow soil base provided by the crustose lichens. Mosses may then follow the lichens, and then with an accumulation of more soil, with better water-holding capacities, some annual short-lived herbs become the invaders followed in turn by grasses or even shrubs and trees until the climax plants have finally succeeded. The process can be hastened where the rocks are broken and form numerous crevices. In these crevices wind-blown soil accumulates, often to a considerable depth. Here shrubs and even trees are able to grow. They hasten the break-up of the rocks and reduce the number of invasions necessary to turn the area over to the climax species.

The succession of plant growth in a xerosere is shown briefly below. As the bare rock is weathered into soil, plants of different species become established according to their tolerance for the habitat and their ability to compete with the already established species.

Bare rock → disintegration and decomposition → soil
crustose lichens → foliose lichens → moss → annual herbs → grass

We have now briefly discussed the manner in which communities are developed, the general types of organisms involved, and the general order of succession in at least two ecological habitats. Now let us look at some of the large ecosystems as they appear today. There are a number of ways in which these ecosystems may be classified and there are many ways in which they may be subdivided. This chapter is intended only as an introduction to ecology; much more extensive treatment can be found in books written specifically for ecology and plant geography.

Major Ecosystems

Oceans, lakes, and **ponds** vary considerably in the habitats they provide for the organisms which are flourishing there. Depth of the water is one determining factor but the salts they contain, the abundance of oxygen and carbon dioxide, the temperature of the water, and the biota itself are all limiting factors in determining the aquatic population.

There is usually a distinct zonation of organisms in lakes based largely on the depth of water. The **littoral zone** is located around the edge of the water body and contains the rooted vegetation along with the animals that live on the plants and microorganisms in this zone. Open water is termed the **limnetic zone** which is occupied by millions of free floating organisms termed **plankton.** In all bodies of water with a rich microflora and microfauna there are numerous crustaceans and fishes forming a typical aquatic food chain. In relatively deep lakes of temperate regions there is a distinct thermal stratification during the year due to a differential heating and cooling.

The upper portion is the **epilimnion,** and the lower is the **hypolimnion.** When there is a distinct separation between the two zones, due to temperature, exchange of materials is greatly retarded. Oxygen may often be of insufficient quantity in the hypolimnion and nutrients in the epilimnion may be inadequate. This thermal stratification is most pronounced in summer and winter while in spring and fall, with changing temperatures, the entire body of water has about the same temperature and there is a mixing of water and its contents.

The seasonal growth of plankton is closely associated with the rise and fall of temperatures within the epilimnion in many large lakes in midsummer. If a lake has a surface temperature of about 72° F, with the water clear and plenty of light, oxygen, and carbon dioxide, algae will be abundant. As the wind blows, the water cools at the surface and this cooler water gradually sinks. Water is a unique substance in that it becomes denser and heavier as it cools to a temperature of 39.2° F (4° C) when it becomes less dense and lighter, until finally when it turns to ice the water is at its lightest weight per unit of volume. As the heavier water sinks plankton is carried with it, and thus there comes about a circulation of the free-floating microorganisms, nutrients and gases.

Much of the ocean area is virtually sterile from the standpoint of food supply to man. Nevertheless the vast stretches of the sea and extreme depths of water provide a considerable energy flow because of size, but there is very little energy exchange per unit of area. The physical factors of waves, currents, tides, pressures, temperatures, light penetration, and salinity determine the growth and location of the biological communities which in turn modify many of the physical factors.

Along the shores and within the intertidal zone is found the most diverse and richest areas of flora and fauna found anywhere. Out to a depth of about fifty feet brown and red algae often grow like a veritable underwater forest. In this area, which may extend seaward for miles, fish and other aquatic animals are extremely abundant and in this area many sea animals give birth to their young. After living and feeding in this fertile zone for perhaps years many of the fishes and other animals leave for the open water. In the rock crevices and water holes within the intertidal zone the rock algae, crustaceans, molluscs and many other small species live an abundant life.

Temperature Zones

In addition to vertical changes in temperature in water it has also been observed that there is a habitat distribution of algal flora along the Pacific Coast of North America. In general it is divided into four zones. A majority of the species are found restricted to any one of these zones; a few are found in two zones; and fewer still are capable of growth in three zones. The brown algae, especially the giant kelps, are much more abundant in the two northern zones, while red algae become more prominent as we go southward. These zones are:

1. The **Boreal Zone,** which extends southward from the Arctic as far south as the Strait of Juan de Fuca, just south of Vancouver Island in Canada.

2. The **Temperate Zone** extends southward from the Strait of Juan de Fuca to Point Conception, Santa Barbara County in Central California.
3. The **Subtropical Zone** extends south from Point Conception to Magdalena Bay in Baja California, Mexico.
4. The **Tropical Zone** extends south from Magdalena Bay to the Isthmus of Panama.

Although we might expect to find a similar distribution of marine algae in coastal waters of the Southern Hemisphere the distribution of marine algae is not so clear cut and there are far fewer species in these waters. *Macrocystis* and several other genera are found on both sides of the equator and on the stormy coasts of Chile, Australia and New Zealand but the greatest number and largest seaweeds are found in the Arctic ocean waters which have been called the "Sea of the Laminariaceae."

Light Relations

Light from the sun is a very important factor in the distribution of land plants but it is especially important in the vertical distribution of aquatic plants. Quantity of light is probably a more limiting factor to land plants than it is to those inhabiting water since land plants make about equal use of the red and blue colors of the spectrum. Water has a light-filtering action which greatly decreases the quantity of light as well as altering the amount of penetration of certain wavelengths. In clear water approximately one-third of the light is lost within the top three feet; at a depth of sixteen feet, three-fourths of the light is lost. So far as quality is concerned an alga growing thirty feet beneath the surface of the water receives only 10 percent of the blue rays and only one percent of the red rays of full sunlight. The amount of light actually available may be considerably less than this for the total amount of light that penetrates the surface of the water depends upon (1) the intensity of the light; (2) the direction of the rays, which varies with the time of day, the time of year, and the latitude; and (3) the turbulence and the turbidity of the water.

For algae growing in the intertidal belt the total amount of light seems to be of greater importance than its quality. It has been shown for *Fucus* that if the amount of light is reduced by one-half the plant dies. Algae growing in deep waters, chiefly red algae, are quite efficient photosynthetically with only blue.

So far as the aquatic medium is concerned we might say that life is fairly uniform and often without the hazards associated with terrestrial life. Temperature does not have the extreme daily or seasonal changes we find in most land areas. While light, in general, is much less than on land, aquatic plants have become either much more efficient in the absorption of the wavelengths that penetrate the water or require less light to carry on their activities. So far as nutrients are concerned they are mostly in solution and fairly uniformly dissolved throughout the water and readily available to the organisms which they surround. Oxygen and carbon dioxide are readily available and are generally in adequate quantity. In permanent waters desiccation is not a problem as it is on land; very few plants living in this type of habitat have developed corticating tissues. The development of vascular and mechanical tissues are not of great necessity to plants in water since the aquatic medium brings substances to nearly all parts of the plant and the water gives buoyancy to plant-parts. Although there is competition in every habitat it seems reasonably safe to say that life is easiest for the organism and the population in this type of ecosystem than that in the desert, the tundra, the temperate forest or even in a tropical forest.

DESERTS

In contrast to the aquatic ecosystem the desert presents a habitat which generally is inadequately supplied with moisture, and sparsely covered with vegetation. The arid lands are often classified as warm deserts if located in tropical or subtropical areas, or cold deserts in regions where the winters are quite severe.

Many of us have some knowledge of our southwestern United States deserts (Fig. 8) where the rainfall is often considerably less

FIGURE 8 Vegetation of the warm Southern Arizona desert. The large cylindrical cacti are the giant saguaro, the very much branched tree-like cactus in center is the jumping cholla, *Opuntia fulgida,* the small slender cactus at lower right is *O. leptocaulis* and the much branched species at lower left is *O. spinosior.* To the upper right and left center is the slender cane plant *Fouquieria spendens* (ocotillo).

than ten inches per year, and where the dominant vegetation consists of many species of cacti, including the organ pipe *(Lemaireocereus thurberi)* and giant saguaro *(Carnegiea gigantea)*, shrubs like the creosote bush *(Larrea tridentata)* and ocotillo *(Fouquieria splendens)*, the jumping cholla *(Opuntia fulgida)* and some trees of low stature like the palo verde *(Cercidium microphyllum)*. Few of the species may be considered of high palatability to cattle and consequently the warm deserts are poor quality range land.

Life on the desert is not all bad; humans often prefer this type of habitat over all others. Economically the desert needs help in order to make it productive, but the clear air, warm winters, widespread availability of air conditioners, and desire for outdoor living have brought thousands of people to our southwestern desert habitats.

Irrigation systems and deep wells have made the desert, which was nearly barren, some of our most productive agricultural regions in the United States. The Imperial Valley of Southern California has proved so productive, through the channeling of water from the Colorado River, that two crops can often be raised on the same acreage year after year (Fig. 9). Southern Arizona has had a

FIGURE 9 A field of broccoli in Southern California raised under irrigation. The photograph was taken December 30 and is the second crop of the year, alternating with cotton.

similar history and industries which do not require large amounts of water for their operation have been moving into the southwest for many years. Modern agricultural methods and modern living aids have made the desert habitable and so pleasant that a sizeable segment of our population is now living in these regions.

The first civilizations were established in desert areas as early as ten thousand years ago. At first these early civilizations of the Tigris-Euphrates and the Nile river valleys had an easy time, but as the populations increased and more sheep and goats were raised in the hills, overgrazing and erosion led to serious soil depletion and loss of fertility, particularly in the Tigris-Euphrates Valley. The Nile river, which flowed through many hundreds of miles of comparatively flat land, was much more dependable. Population in this valley has continued to increase for thousands of years although many groups have had to move to more productive regions in the Nile Valley from time to time. Now through the benefit of modern engineering, the construction of dams and the extensive use of fertilizers, this great desert basin has become a dependable productive area and we no longer think of it only as the cradle of civilization.

In our own warm deserts of the southwest much the same story can be told of early peoples moving into our desert river valleys. Occasionally the water in the stream beds disappeared and occasionally the people became too numerous for the water and soil to support. As a result many archaeological sites have been left by the homes and towns the Indians were forced to abandon.

Today these hot southern deserts provide a substrate for cotton, sugar beets, grapes, and many truck crops. The native plants are well adapted to the hot sun, the daily extremes in temperature, to minimum quantities of moisture and to the highly saline soils. However, the introduced species could never have survived without human help.

In the United States the cold deserts range from about central New Mexico to the Canadian border as intermittent strips in the Great Plains east of the Rocky Mountains. A second larger area is west of the Rockies, for the most part, and east of the Cascade-Sierra Nevada range. It is known as the Great Basin and covers portions of Utah, Oregon, California and Wyoming. The winters are quite severe but the snowfall is generally light and here also the precipitation is often less than ten inches a year.

The principal vegetation is sagebrush *(Artemisia tridentata)* and a sparse growth of grasses including bunch wheat grass *(Agropyron spicatum)*. Many annuals and perennials dot the Great Basin landscape from year to year depending upon the moisture available and the plants' tolerance for salts which often are heavily concentrated in the soils of the playas and shallow lake beds.

Like the warm deserts these plains are used for limited grazing and agriculture where supplemental water through wells or surface irrigation can be obtained. Because of the severe winters the cold deserts have fewer human inhabitants than the warm deserts.

THE CHAPARRAL OR MAQUIS

This ecosystem might be described by a number of different names and often is. The vegetation is dominated by scrubby, evergreen, broadleafed species. Many of the trees are live oak species, usually of rather low stature while over much of the area spiney shrubs, grasses, and some herbaceous weeds abound. In portions of Southwestern United States, particularly in the hills of Southern California, this type of vegetation covers many square miles and is called **chaparral.** The same type of vegetation found in Southern Europe and Northern Africa around the Mediterranean Sea is called **maquis.**

The climate is warm, almost subtropical but characterized by long, hot dry summers and short mild, wet winters. This type of climate is usually referred to as Mediterranean and while we may refer to this ecosystem by that name the true Mediterranean zone is coastal and has more precipitation than that found in hills of the chaparral (Fig. 10).

While the area is not highly productive in plants or animals, from a biological consideration, it is an area that is becoming highly populated with humans because of the mild climate and nearness to large population centers. Where water from wells or streams is available specialized agriculture, such as the raising of olives, citrus, and other warm weather fruit crops, is profitable.

FIGURE 10 A chaparral habitat best shown by the shrubby vegetation on the hills. The large Washington palms growing in the stream bed are not part of the typical chaparral vegetation but grow in this warm climate where water is abundant.

TUNDRAS

More forbidding than the desert ecosystem is the barren, windy, freezing habitats of the Arctic and Antarctic. Tundra soils are usually sterile since they form largely from boggy peats and sandy gravels. They may be either wet or dry but are underlain with permafrost. During the brief arctic summers there is some surface thawing but for the most part the habitat supports very little vegetation.

The vegetation which exists consists of surface lichens (reindeer moss) and a variety of low growing herbaceous perennials (grasses and sedges) and a few dwarf shrubs such as miniature willows and birches. The few plants that tolerate the tundra conditions are able to store enough food in the short growing season to put forth new growth the following year but dwarfed, stunted plants are the general rule (Fig. 11). The number of species is limited to a few hundred while an equal area in the tropics has over 100,000.

FIGURE 11 A mountain tundra habitat. The scrubby spruce in the background is developed because of wind and complete snow coverage several months each year.

Animal life is usually limited to migratory birds and a transient population of herbivores like caribou, reindeer, and musk ox as well as lemmings and a few other small herbivores. The carnivore population of foxes, wolves, weasels, polar bear, and owls has a quite variable number depending on the number of herbivores available for food which in turn are limited by the amount of forage present for them (Fig. 12).

Man has explored and exploited the tundras to a limited extent but he has less effect on this ecosystem than probably any other. The Eskimo and the Lapps of northern Europe have more or less adapted themselves to this type of habitat by domesticating herds of reindeer and being good enough hunters to live off the animal herds and the marine life that abounds in the arctic seas.

Undoubtedly man will move in ever increasing numbers to the arctic and with each generation there will probably be improved accommodations, but it is expected that the economy will probably be developed in the fields of mining and perhaps in marine biology more than in any other area.

FIGURE 12 North and Central America showing principal vegetation areas.

Alpine vegetation is often quite similar to that of the tundras and in the high mountain elevations of the northern latitudes it is referred to as arctic-alpine or alpine tundra. In most countries the arctic-alpine areas are limited to the peaks of mountains and therefore are more likely to be a few acres in size rather than thousands of square miles as in the tundra (Fig. 13). The arctic-alpine is mostly above timber line, but the few trees that survive there are dwarfed and stunted by wind, ice, and snow. Herbs are limited to sedges, grasses, and a few high mountain species like sedum and columbine, which find protected alcoves behind rocks. The animals are usually transient and move up and down the mountain as the seasons progress.

FIGURE 13 A rugged arctic-alpine habitat in northern United States. Wind, snow and ice limit the vegetation to herbs and dwarfed woody species.

FIGURE 14 A western boreal forest with numerous aspens and some evergreens.

BOREAL FOREST

Extending southward from the subarctic latitudes to the southern Canadian border and down into northern Montana, Minnesota, Michigan, and the northern New England States is a vast coniferous forest which we have come to call the "North Woods" or the **boreal forest.** On the North American continent the dominant trees are the white spruce *(Picea glauca)*, the balsam fir *(Abies balsamea)*, and the jack pine *(Pinus banksiana)*. Scattered among the coniferous trees are pockets of a few broad leafed species such as the quaking aspen *(Populus tremuloides)* (Fig. 14), the balsam poplar *(Populus balsamifera)*, and the canoe birch *(Betula papyrifera)*.

In the Euroasiatic area the same type of forest spreads across the country from Norway and Sweden through Finland, Russia and

Siberia. The same general types of coniferous trees are found there but they are of different species since each population evolved independently, separated by the oceans.

Soils of the boreal forest are covered by thick layers of twigs, leaves, and cones which decay very slowly due to the slow action of bacteria and fungi in this cold climate. Decomposition ultimately results in the formation of humic acids which percolate downward with falling rain, leaching the calcium and other bases from the upper layers and depositing them in lower strata.

Frequently this results in hardpan layers from several inches below the soil surface to several feet in depth. Consequently many trees in this area are shallow rooted and the hardpan layer often retards drainage contributing to the building up of acid soils.

Numerous lakes, formed as the result of glaciation (Fig. 15), gradually fill in with peat produced by sphagnum moss and other hydrophytic species. In many areas this growth produces bogs with a mat of floating debris until sufficient vegetation fills in the entire lake to form the peat bogs. This material is excavated and used in packing by nurseries for plants in shipment. Solidification and continued decomposition result in peat which can be cut up into bricks or chunks and used as fuel for inhabitants of the northern climates.

The boreal forest is the home of most of our furbearing animals including mink, beaver, fox, and ermine as well as our largest animals such as moose, elk, deer, and bear. This forest is where

FIGURE 15 Lake Louise in Canada. The mountains in the background have perennial glaciers which supply the lake below.

the trapper and hunter are making their last stands, at least from a commercial standpoint. The principal industries are lumbering and mining. If it were not for those forests our paper industry with its tremendous requirements for pulp would probably have to curtail production.

The outlook for the future use of the boreal forest is bright for in spite of low temperatures it has a high productive potential. With the development of more roads and better transportation greater numbers of people will be moving into the area, more industries and agriculture will be established and last but not least, recreation will be greatly expanded.

DECIDUOUS FOREST

Throughout much of the temperate latitudes of the world deciduous forests were once luxuriant. In continental North America this deciduous forest extended southward from the New England States to southeastern Texas and the southern states, practically to the Gulf of Mexico. The forests extended westward from New York and the Appalachian Mountain range to the Mississippi river, which it crossed over to cover portions of Iowa, Missouri, eastern Kansas, eastern Oklahoma, and a good deal of eastern Texas.

Throughout much of this midwestern area the dominant trees were oak of many species. The large deciduous oaks such as *Quercus alba* and *Quercus rubra* gradually gave way to other species southward and finally were largely replaced by "live" oaks in the southern states. In most of the northern portions many species of maples *(Acer)* were common, along with beech *(Fagus)*, buckeye *(Aesculus)*, walnut *(Juglans)*, chestnut *(Castanea)*, and hickories *(Carya)*. Further south magnolias, persimmons, tulip trees, and others were common. In general the deciduous forest is characterized by having a relatively large number of tree species when compared to the northern coniferous forests.

The soils are quite deep, fertile, and well drained. The rainfall, which usually comes at a seasonal time of greatest benefit to crop production, is usually over 30 inches per year.

In profile the forest has an overstory or crown of deciduous trees which often are over a hundred feet in height. Many of the luxuriant stands have a secondary tree stratum of smaller species which grow to a height of twenty to thirty feet. Beneath the tree stands, there is usually a scattered or occasionally a thick shrub layer from one to eight feet in height. Covering the forest floor is a layer of herbaceous annuals and perennials intermingled with lichens and moss which develop on the moist humus and fallen logs.

Actually a discussion of the deciduous forest in the Northern Hemisphere should be stated in the past tense. For several million years this forest extended throughout western Europe, eastern Asia, and eastern North America. Civilization and the expanding population, within a few hundred years, cleared most of the trees for lumber, fuel and especially for the benefit of agricultural crops. Within the past century most European countries became aware of what was happening to their forests and began a program of forest conservation to save the few remaining islands of trees. Now the forests of western Europe are, in general, well managed and have become a productive segment of the peoples' economy with little danger of further destruction. Virtually the same story can be told of forest depletion in the eastern Asiatic area, only the losses were greater with the recovery later and slower.

The deciduous forests of eastern North America were the most extensive and luxuriant and lasted longer. But with the migration of Europeans to this country the economy gradually changed from the American Indian practice of hunting, fishing and the gathering of fruits, nuts and wild plants to an agricultural program. In order to live in a smaller locale the early settlers had to grow crops which required the clearing of forest land. Fortunately the power saw had not been invented when the land was being cleared or there probably would not be any deciduous trees left in most of the area. Fortunately also, there was a vast grassland to the west. While many people settled in the forests and gradually cleared acreage for farms another segment of the population moved into the great prairies where they began raising cattle and plowed the land for grain crops.

The deciduous forest lands now contain some of our most fertile soils and produce our richest agricultural crops. Most of the corn and a good deal of the other cereal grains are now raised in these soils along with thousands of acres of truck crops and numerous orchards of apples, peaches, cherries, etc.

The original animal population was made up largely of deer, bear, red squirrels, turkey, rabbits, and numerous other species of mammals and birds. In many areas the larger animals were completely exterminated but now, where the proper forest cover remains or has been restored, restocking of game animals has been successful. With reasonable care the deciduous forest, as it is today, can be expected to be at least as productive in the future as it now is.

TEMPERATE RAIN FOREST

Along the coasts of several continents, where strong westerly winds bring in heavy rain storms, the precipitation is over 100 inches per year. These rain forests are usually found in latitudes between 40 and 35 degrees. As the heavily laden rain clouds blow in from the ocean they are forced higher by the coastal mountains which causes a natural cooling and condensation, resulting in heavy rainfall.

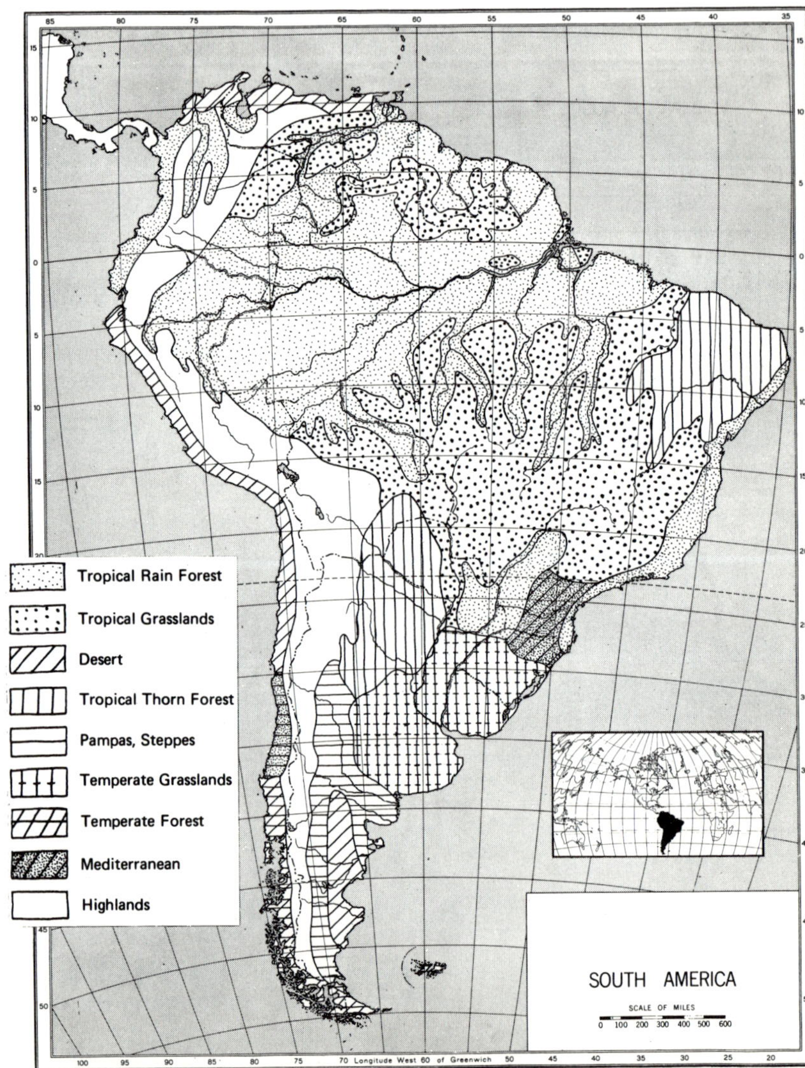

FIGURE 16 South America, major vegetation areas.

Strips of coastal rain forest are found in southern Chile, the west coasts of New Zealand and Tasmania, southeastern Australia, and along the northwestern coast of North America from about northern California, to Oregon, Washington, Canada, and Alaska (Fig. 17). The rainfall on the Olympic Peninsula in Washington in this ecosystem is the highest in the Continental United States, measuring about 140 inches annually. However, on the southwest coast of the South Island of New Zealand the precipitation has been measured at close to 350 inches per year. With this amount of rainfall the forests are literally dripping although there are often a few

FIGURE 17 Temperate rain forest along our northern California Coast. The largest trees are redwood, Douglas fir and some western hemlock.

weeks each year when the winds are bringing in very little moisture. During these times even the rain forests are subject to fire and frequently burn because of the dry mosses, epiphytes, and the highly flammable resins which are concentrated in the coniferous tree tissues.

The dominant species in the coastal rain forest in the United States is the Douglas Fir *(Pseudotsuga menziesii).* This tree is the major species in the entire Northwest Pacific forest which is the most important lumber forest in the United States. Other species include redwood *(Sequoia sempervirens)* (Fig. 18), Sitka spruce

FIGURE 18 A redwood forest in northern California, near the southern portion of the temperate rain forest.

(Picea sitchensis), western hemlock *(Tsuga heterophylla)*, and arborvitae *(Thuja plicata)*. Trees in the Southern Hemisphere temperate rain forests are also chiefly gymnosperms but belong mainly to the family Podocarpaceae of which two principal genera are *Dacrydium* and *Podocarpus*.

In addition to the gymnosperms a number of angiosperm broad-leafed species are found in these forests. Numerous epiphytes ornament the branches, trunks and fallen logs giving the forest a verdant covering of ferns, mosses, liverworts, and foliose lichens. Decomposition is relatively slow since the cool climate retards the action of the decomposers. As a result the forest floor is covered to a considerable depth with humus, fallen leaves, branches and logs.

The big game animal population is rather sparse partly due to the thickness of the undergrowth and partly because most of the plants found here are not of high forage value.

The principal uses man has made of these forests has been in lumbering and recreation. The temperate rain forests of the United States are largely under the control and protection of our U.S. Forest Service and are mostly part of the National Park and State Park systems.

FIGURE 19 Epiphytes growing on a mahogany tree. In the crotch are ferns and on the side is a bromeliad.

FIGURE 20 Lianas and epiphytes on trees in the tropical rain forest.

TROPICAL RAIN FOREST

The richest forests in the world are those in the tropics where the rainfall is heavy, the temperature is always warm, and the soils are relatively good, although sometimes severely leached of nutrients in local areas. Here the limiting factor for growth is usually light for there is such a concentration of rich flora, with many species of trees towering over 150 feet in height, that with their broad leaves forming a dense mosaic pattern they cut off nearly all direct light to the forest floor. Beneath the dense canopy of lofty leaves and branches grow many shade tolerant species of shorter trees and shrubs and numerous herbaceous species. Covering many of the larger plants are numerous epiphytes (Fig. 19) such as orchids and bromeliads, in contrast to the moss and liverwort epiphytes of the temperate rain forests. Hanging from the branches of even the highest trees are **lianas** (Fig. 20), which are vines rooted in the ground but with their flowers and leaves a hundred or more feet above.

The most extensive tropical rain forest is in the great Amazon basin of South America; others are found in the East Indies of tropical Asia and in portions of Africa. These areas have the most extensive stands of hardwood forests anywhere in the world. One of the unique features of the tropical rain forest is its large number of species. In very few localities would we find many square miles dominated by one or two species. Generally we find over a hundred species per square mile all closely intermingled by their root systems in the soil and by their spreading branches and leaves in the rooflike canopy above (Fig. 21).

With such a luxuriant flora it is not surprising that the fauna should be equally abundant. Numerous of the smaller animals including monkeys, birds, amphibians, reptiles, and insects rarely ever come down to the forest floor but remain arboreal all of their lives. Many biological niches, where a few species may form a small interdependent but integrated unit, are frequently observed in this large ecosystem (Fig. 22).

FIGURE 21 Tropical forests supply an abundance of water to the areas below. This picture is of Dunn River Falls in Jamaica.

FIGURE 22 Tropical forest in Hawaii, note the vines on the trunks of the dicotyledonous trees and rich growth on the forest floor. (Courtesy Bernice P. Bishop Museum, Baker Collection.)

The formation of all metabolic products is more abundant in the rain forest than anywhere else. The greatest activity is probably the least obvious for it is here that the decomposers are most active. The role of bacteria, fungi, termites and other insects have an important role in the continued growth of the forest. The tremendous amount of subterranean plant growth removes many of the inorganic nutrients from the soil storing it in the standing crop. Microorganisms then decompose the dead plants and animals, releasing the inorganic nutrients and returning them to the soil and atmosphere for new growth (Fig. 23).

FIGURE 23 Forest of monocotyledonous species and some ferns in Hawaii. (Courtesy Bernice P. Bishop Museum, Baker Collection.)

The tropical rain forest is contributing more and more each year to the economy of man. Where once it was too formidable and dense a habitat to penetrate more than temporarily, modern machinery and equipment, and new drugs and medicines have made the jungle habitable. In most tropical countries lumbering is still slow but ever increasing quantities of highly valuable hardwoods are being removed from the forests of the Philippines, Central America, and elsewhere. In nearly every tropical region man is clearing areas for bananas, coconuts (Fig. 24), and other important tropical plants which now can be quickly brought to the highly populated cities of the temperate regions. Agriculture, which cleared the temperate deciduous forest, is also the force exerting its pressure on the tropical populations to move deeper into the forest and clear more land.

FIGURE 24 A coconut-banana plantation in Jamaica. The coconut trees remain productive for many years but the bananas must be replanted or at least thinned out every year for maximum production.

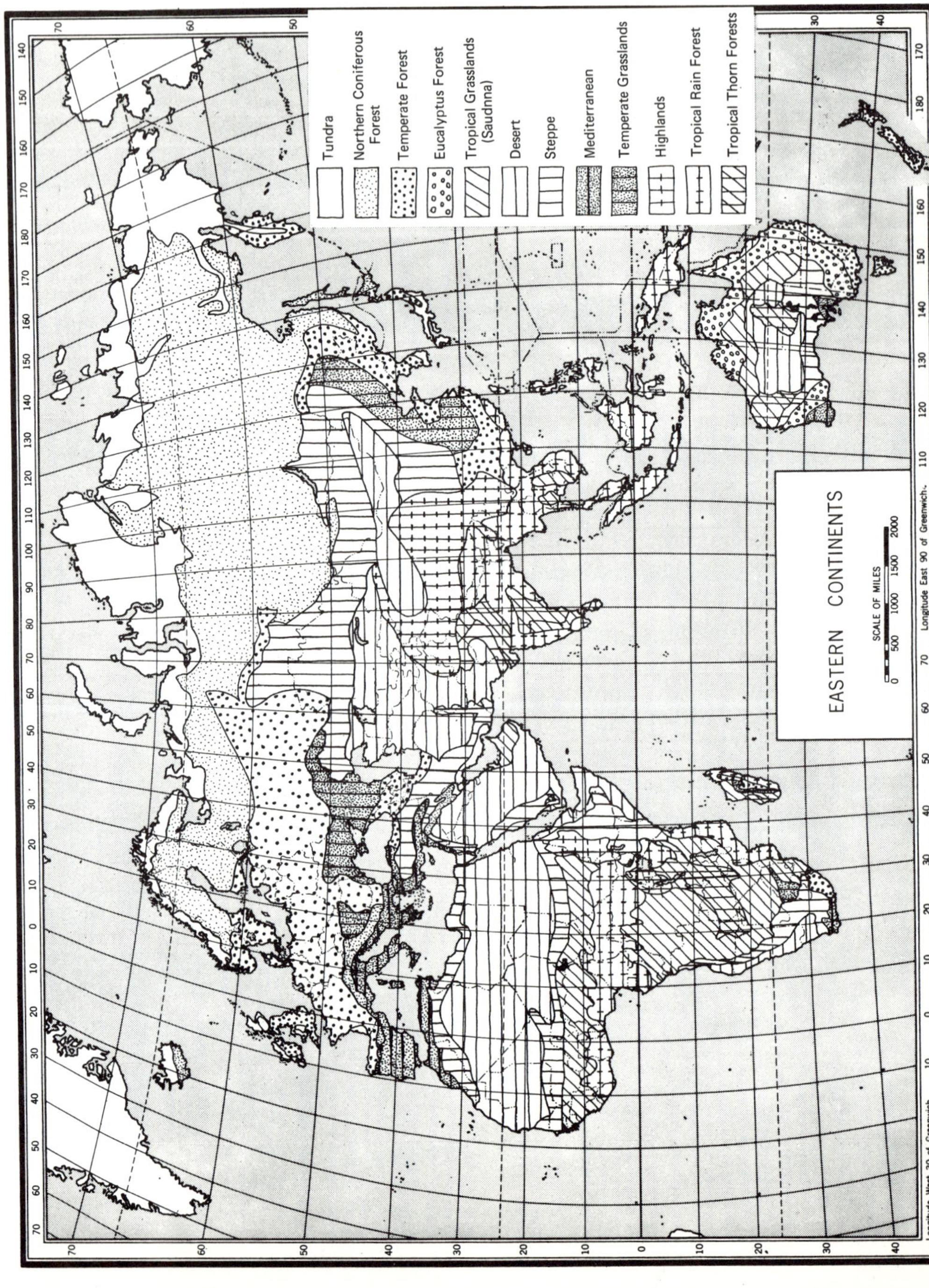

FIGURE 25 The Eastern Continents showing principal vegetation areas.

THE SAVANNA

Bordering the tropical rain forests in many regions are grasslands dotted with trees or occasionally with clusters of open forest. These are the savannas. They are apparently the product of seasonal droughts as well as heavy rainfall and are always warm. The grasses are tall and of high palatability for herbivores, and consequently this habitat is the home of numerous species of antelope, wildebeast, gnu, and buffalo. These animals still roam in large herds in several areas, particularily in Africa, India, and some parts of South America. Along with the large herbivores are many carnivores such as the lions of Africa and tigers of India. The trees, which sparsely dot the landscape, are often legumes such as the *Acacia* and *Mimosa*. Usually flat-topped, the arboreal species provide some protection for herbivorous species but in general the tall grasses provide cover and protection not only for the herbivores but for the preying carnivores as well.

While the amount of food produced is not as great as that in the tropical rain forest it is of higher quality. The climate is more suitable to man and as a result he has made extensive inroads into the savannas. Clearing of the land for the development of agriculture has been easier, the seasonal drought has limited the growth of parasitic organisms such as the disease causing protozoans, their carriers the Tsetse fly and mosquito, and numerous flukes, flatworms, and nematodes. Control of tropical diseases has therefore, been easier than in the tropical forest. Now with the requirement for more land and more agriculture man is rapidly moving into the savanna. In fact the concern of conservationists for the protection and preservation of the wild animal populations of Africa and elsewhere has become so great that requests and help have been offered these areas to set up National Parks for the protection of large game animals. Some success has been achieved in this direction but much more must be accomplished before the natural ecosystem of the savanna can be preserved in even a small way in its natural state (Fig. 25).

THE GRASSLANDS

In many of the temperate regions where rainfall is not heavy, grasses dominate the landscape. Trees are nonexistent except along the waterways or where they have been planted and given a little extra moisture, as around homesites. The most extensive grasslands are the Eurasian steppes, the Argentine Pampas, and the great North American Prairies.

Where the rainfall is 15 inches or more annually, as in Nebraska, the Dakotas, Kansas and adjacent states, tall grasses such as the *Andropogons* and *Sorghastrums* are dominant. This vast area from about Central Iowa to the Rocky mountains, and from the Canadian border southward to about the latitude of Northern New Mexico, is known as the **tall-grass prairies** and is the richest of our grasslands. To the south and west of this area, for the most part, lie our **short-grass prairies,** characterized by *Bouteloa gracilis* and other low growing species including those of grama and the buffalo grasses (Fig. 26). This area usually has between ten and fifteen inches of rain annually but in some areas of the Southwest it may be less. The short grass prairies cannot support as large a native animal population as those of the tall grasses but like them were at one time the natural habitat for the American bison, antelope, many smaller herbivorous species and numerous coyotes, foxes, wolves and other carnivores.

FIGURE 26 A short-grass prairie consisting of species of such native grasses as grama *(Bouteloa),* and buffalo *(Buchloe).* Lovegrass *(Erogrostis)* has also been planted on this range to help restore the range which some ten years ago was heavily overgrazed but now is in reasonably good shape.

The grasslands have, in general, withstood the encroachment of trees because of the lack of moisture and the fact that range fires which seldom do little permanent damage to the grasses burn out young trees. In many of these areas, as the rich prairie soils of Illinois, Iowa, Kansas, and Nebraska, introduced cereal grains have largely replaced native grasses of the range lands, changing them to some of our most productive agricultural areas. Unfortunately in other areas, where the rainfall and soil are adequate for the wild grasses, plowing of the soil has resulted in dust bowls and crop failure. Misuse of the land in much of western Kansas, Oklahoma, western Texas and eastern Colorado, during drought periods a few decades ago, resulted in the "Terrible Thirties"—a time when economic depression and unwise cultivation caused thousands of families to leave these marginal lands and seek a livelihood elsewhere. Fortunately we have been equal to the challenge. The establishment of the Soil Conservation Service, qualified help from agricultural agents, state and federal aid, and seeding regulations brought these soils under control. Now after one human generation most of this land is again in production and through wise use we can avoid the catastrophe of the thirties.

Both the tall-grass prairies and those of short grass are more or less sparsely inhabited. The determining factor seems to be the amount of rainfall and therefore the amount of grass. Few places are so windswept as to be uninhabitable. However, it is a region of marginal productivity. If left in range land and not overgrazed the grasslands are a great asset, if overgrazed or forced into production beyond its capabilities it can be destroyed. We have changed the vegetative population over much of the eastern area from range grasses to cereal grains with phenomenal success. We have also overgrazed, and still are doing so, much of the southwestern portion. We had killed off virtually every wild bison by the late 1800's, and forced the Pronghorn antelope out of much of the area. Now the antelope has been restored in limited numbers throughout much of the area and the bison is allowed to continue life in the protection of a few government refuges and national parks.

THE MONTANE FORESTS

One characteristic of mountains is that they are usually covered with trees. Generally these trees are of the coniferous species, but in some sections of the Appalachian Mountains and in those

FIGURE 27 A mountain forest. While the dominant tree here is the Englemann spruce a fire some thirty years ago destroyed the spruce. Now through succession the dominant tree is the aspen but small spruce are coming in to the area and in about fifteen more years the spruce will have replaced the aspen and again restored the climax community.

mountains of the eastern Asiatic area we find deciduous forests occupying large areas of the mountainside (Fig. 27). Mountains generally have very shallow soils since much soil that is formed from the disintegration of parent rock material and decomposition of organic matter is easily eroded from the surface by wind and water and carried to the valleys below. Most angiosperms require better soil conditions than those provided in the mountains yet most gymnosperms are capable of surviving under minimal quantities of soil nutrients and rocky substrates.

Wind is often a limiting factor for plant growth, but gymnosperms have the shape and plasticity or flexible tissue structure which can more readily cope with it. The spire or pyramidal shape of the conifer sheds more snow than the flat-topped shape of the broad-leafed species. Conifers can bend with much less breakage when covered with ice and snow than can angiosperms. Furthermore the resinous materials and greater viscosity of the protoplasm in the cells of gymnosperms give them greater tolerance to the desiccating conditions caused by high winds.

Water is quite variable in mountains at different elevations, depending on the latitude and nearness to coastal waters. Water-laden clouds, bringing in moisture from the sea or other land areas, usually drop their water in the higher portions while the lower elevations get less precipitation and more runoff (Fig. 28).

FIGURE 28 Mt. Ranier in the State of Washington. While vegetation is lacking on this permanently ice- and snow-covered peak the value of mountains such as this one in supplying water to the rivers and valleys below is virtually incalculable.

Mountains have a distinct zonal distribution of flora and therefore of fauna as well. However, most of the birds and larger animals move up and down into the different floral zones with the seasons unless they are restricted to a specific food found only in one zone. The location of a mountain so far as latitude is concerned, its exposure, that is, north or south, and its degree of slope determine the type of vegetation and the altitudinal range each zone will have.

If we examine a mountain range in Southwestern United States which has a maximum elevation of over 13,000 feet we would have an altitudinal distribution of flora with about six different vegetation zones (Fig. 29). Assuming we could start at about 1500 feet

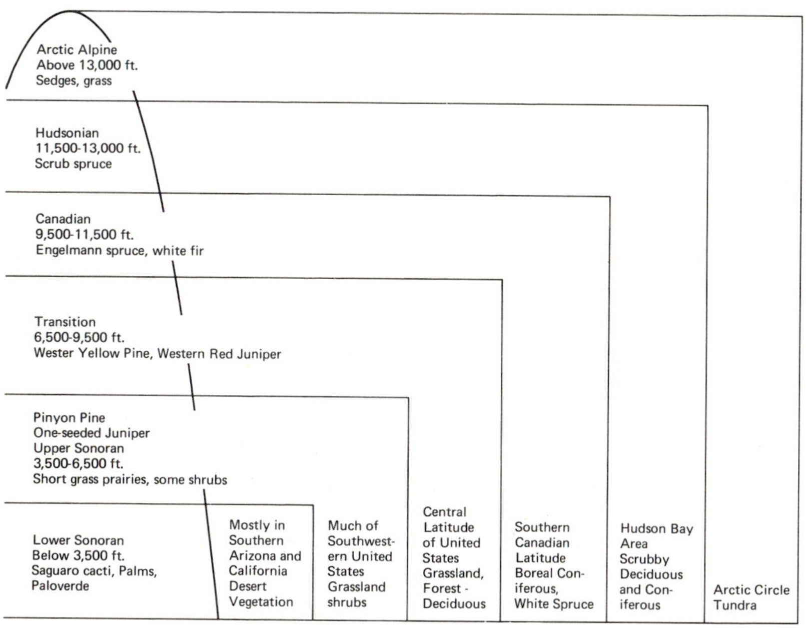

FIGURE 29 Diagram of a mountain profile from southwestern United States showing the type of vegetation at various elevations and extended to the right to show the latitudinal distribution of vegetation in comparison. See text for further explanation.

above sea level we would be in the warm desert in which the giant saguaro, palo verde, a few grasses and some spiney shrubs formed the dominant vegetation. This type of flora would cover many hundreds of square miles and would inhabit the mountain side to an elevation of about 3500 feet. The term given this zone is the **Lower Sonoran** and is based on the type of vegetation generally found in the lower elevations in the state of Sonora, Mexico. The amount of rainfall is usually less than 10 inches per year and supports only xerophytic plants except around the waterways where palms and other trees are able to grow.

From about 3500 feet to about 6500 or even 7000 feet, depending upon the slope gradient and exposure, we have the **Upper Sonoran Zone.** Here the rainfall is somewhat heavier, often up to fifteen inches a year, and the winters are much more severe, with many days of subfreezing temperatures. This zone in the southwestern United States covers many thousands of square miles, is the largest in total area, and is the most important zone for range land, dry farming, and agricultural soils where irrigation water is available. Where the Upper Sonoran blends with the Lower Sonoran short grasses, numerous cacti, and creosote bush are abundant but in the higher elevations the one seeded junipers and pinyon pines are dominant.

Above the Upper Sonoran from about 6500 or 7000 feet to about 9500 feet ± 500 feet we have a zone termed **Transition.** In our Southwestern mountains it is the most important lumbering area and is the zone where the Western Yellow Pine *(Pinus ponderosa)* is the principal species, although the Douglas Fir *(Pseudotsuga menziesii)* and several other species are often present. Throughout much of Central United States this zone covers the prairies and deciduous forest lands described in earlier sections. The elevation is of course much less than that listed for the mountains. In the plains states the elevation may be less than 1,000 feet and the area extends about as far north as the Canadian border.

The **Canadian** zone extends from about 9500 to about 11,500 ± 500 feet. The dominant tree is the Englemann Spruce *(Picea englemanni)* although the Colorado blue spruce *(Picea pungens)* is found locally in some abundance. Frequently in the Canadian Zone of our Southwestern mountains the white fir *(Abies concolor)* and the white-barked quaking aspen *(Populus tremuloides)* are found in dense stands where the spruce has not become established. The Canadian Zone is named after the boreal coniferous zone which extends from southern Canada almost to the Arctic circle. In both areas spruce is the dominant tree and like the Canadian boreal forest the animals found in this zone in the mountains are chiefly deer, elk, bear, and many smaller animals.

The **Hudsonian Zone** in our southwestern mountains is named after the Hudson Bay Area of Canada because it has the same type of wind swept gnarled and crooked trees. In these southern mountains the Hudsonian is largely above timberline. Those trees which are found there are chiefly spruce but they are dwarfed, stunted, and exhibit the one-side flagged effect typical of wind swept trees. The altitude is from about 11,500 to 13,000 ± 500 feet.

The **Arctic-Alpine zone** in our Southwestern mountains is but a few acres in extent and is found only on the highest peaks. It is above timberline, very rocky, and windswept with low vegetation of mostly sedges and grasses. The vegetation is similar to that of a tundra but the soil is not underlain with permafrost although snow and ice persist for about ten months of the year.

The zonal vegetation which has been outlined for a mountain in a specific locality can be generalized for other areas as well, and the same latitudinal comparisons can be made. The chief difference would be in the direction of travel we might take. If we went north into Colorado, Wyoming or Montana, we would note that the Lower Sonoran with its desert vegetation was nonexistent. We would see typical vegetation of the Upper Sonoran in some of the drier, lower elevations but its total acreage would become smaller as we traveled north. By the time we had reached the Canadian border this zone would occupy large areas and both the Hudsonian and Arctic-Alpine Zones would be present in large areas and thousands of feet lower in elevation than in our Southwest.

If we had traveled southward, by the time we traveled about a thousand miles south of the Mexican border the rainfall would have increased to such an extent that the lower elevations would be covered with a dense subtropical forest (Fig. 30). Above this

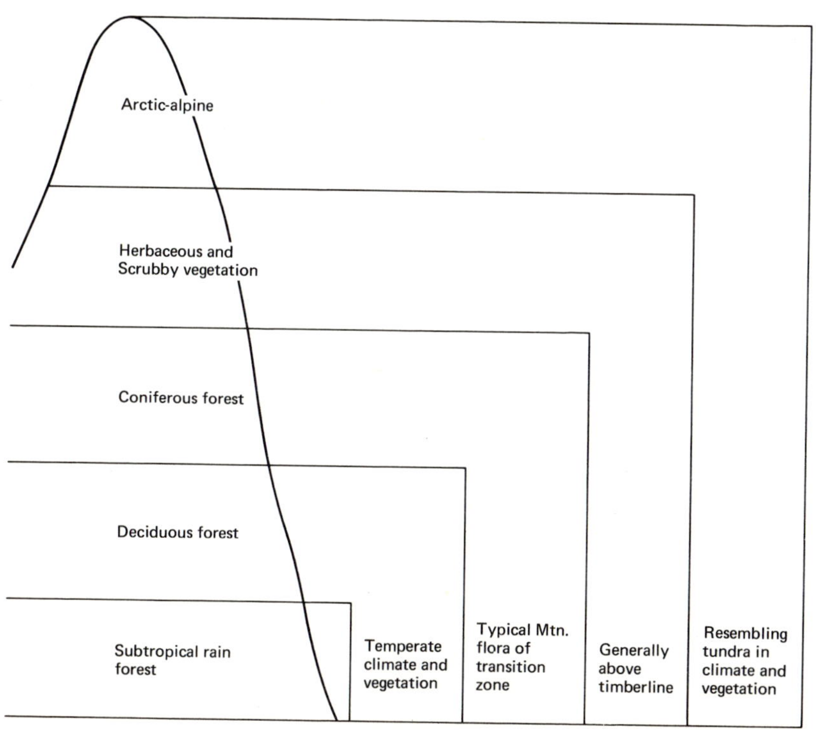

FIGURE 30 A chart similar to figure 29, but the extra rainfall of the subtropical forest has resulted in a replacement of the desert and short grass prairies of the Sonoran Zones with tropical forest and deciduous forest.

zone would be broadleafed deciduous forests of fewer species than below and the tops of the mountains would be covered with a coniferous forest of pines (mostly white) resembling our Transition Zone. In only the highest mountains would we find habitats typical of the zones described for the Hudsonian and arctic-alpine.

An Artificial Ecosystem

The various types of ecosystems which have been described are in general natural systems most of which can be found on any continent. We can also create ecosystems artificially in an aquarium, a terrarium, or a garden spot. To be self-containing and self-supporting, they must have the same four basic components—producers, consumers, decomposers, and the inorganic substances light, air, water, and minerals.

Man has now undertaken the problem of creating a closed ecosystem to be used in space craft for travel to distant planets, and also submarine capsules for living over extended periods in the oceanic depths. At present man takes his oxygen, food, and

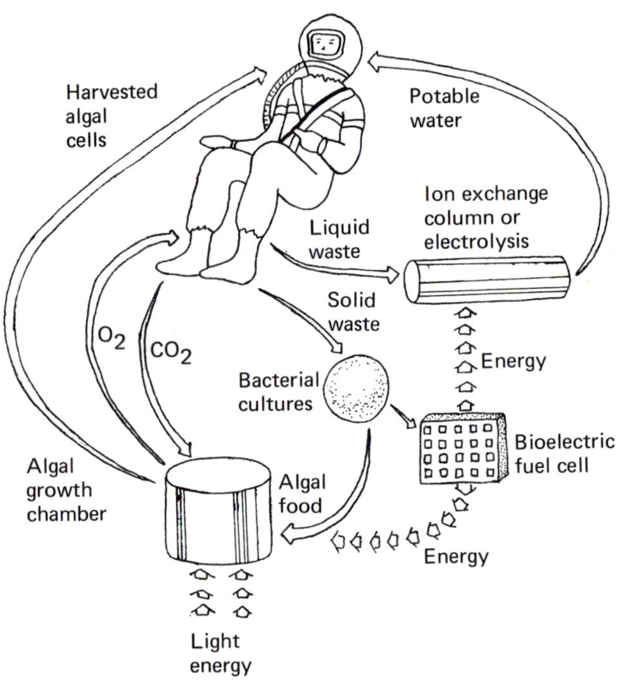

FIGURE 31 A closed ecological system. The regeneration of all products is necessary to sustain life on long-term space flights. The energy produced by the fuel cell is distributed to several points. (Redrawn from Lenhoff, E. S. 1966. *Tools of Biology*. Macmillan, New York.)

all necessary supplies along. However, in the future if we are to travel to distant planets, which will require years, even "light" years in space, our only hope for survival is in a completely balanced unit. In this type of craft plants, probably algae, will have to be grown to provide a continuous supply of oxygen, and they in turn will be provided with carbon dioxide from the humans in the craft. Food will also also be provided by the growth of the algae which will be changed by enzymes to a more palatable food and the algae will in turn be provided with nutrients from the waste materials of the human travelers (Fig. 31).

This is only a small part of the complete system which will have to be worked out before we can submerge for great lengths of time or travel to distant planets. Undoubtedly we will eventually be successful in creating and maintaining our own closed ecosystem, and very likely it will require the same basic components as all natural ecosystems.

Supplementary Reading

Billings, W. D., *Plants and the Ecosystem*. Wadsworth Publishing Co., Belmont, California, 1964.

Cain, S. A., and G. M. de O. Castro, *Manual of Vegetation Analysis*. Harper and Row, Inc., New York, 1959.

Carson, R. L., *Silent Spring*. Houghton Mifflin Co., Boston, 1962.

Dasmann, R. F., *The Last Horizon*. The Macmillan Co., New York, 1963.

Daubenmire, R. F., *Plants and Environment,* 2nd ed. John Wiley & Sons, New York, 1959.

Gates, D. M., *Energy Exchange in the Biosphere*. Harper and Row, New York, 1962.

Odum, E. P., *Ecology*. Holt, Rinehart and Winston, Inc., New York, 1963.

Phillipson, *Ecological Energetics*. St. Martins Press, New York, 1966.

Polunin, N., *Introduction to Plant Geography*. McGraw-Hill, New York, 1960.

Sears, P. B., *Where There is Life*. Dell Publishing Co., New York, 1962.

Thomas, W. L., Jr. (editor), *Man's Role in Changing the Face of the Earth*. University of Chicago Press, Chicago, 1956.

Woodwell, G. M., The Ecological Effects of Radiation, *Scientific American* 208:40–49, 1963.

Future Environments of North America. Nat. History Press, 1966.

Questions for Discussion

1. Civilizations have invariably developed and risen to higher levels in agricultural areas. Explain.

2. In what continental areas may we expect greater improvement in economic status for the people living there? Why?

3. What is the role each of the organisms plays at the different trophic levels and how does this activity improve the entire ecosystem?

4. If our hemoglobin molecule mutated to the extent that it became chlorophyll would it be possible for us to become photoautotrophic organisms? Explain on the basis of caloric and energy requirements.

5. Over a hundred years ago the early settlers in the American Southwest saw and described the area as made up of thousands of square miles of grassy prairies. Now much of this area is covered with creosote bush and other obnoxious weeds. What have been the causes for this change in vegetation?

6. Why have grain crop plants been more important to the establishment of a sedentary life than apples, berries, or even potatoes?

7. How do you account for the fact that conifers are more successful in mountainous areas, grasses on prairies, and hardwood trees in the tropics?

8. Define ecosystem, niche, association, climax, and biome in relation to ecology.

9. What do we mean by "indicator" plants?

10. How do you account for the fact that coniferous forests of the northern United States and Canada are similar in species content to that found in the mountains of the southwestern United States?

11. What is the order of succession by which a plant community is established?

12. What are some of the barriers which impede the distribution of plants?

13. If an area in central Illinois had been used consistently for the raising of corn for many years but was removed from cultivation and human direction for a hundred years, what do you think the vegetation would probably be like after that time?

14. What would probably be the future of a shallow lake in central Kansas?

15. How does light affect the distribution of marine algae?

16. Why have warm desert areas played such an important role in the development of our early civilizations?

17. Why is most of the big game population on the North American continent found in the coniferous forest areas rather than in the deciduous forests?

18. Why do tropical forests have so many different species while northern coniferous forests have relatively few species but many individuals?

19. How do grasslands differ from those of deciduous forests?

20. Diagram the altitudinal distribution of plants in a mountain area.

28

Plant Products of Economic Importance

To the average person the field of applied science, which deals with the economic aspects of a particular subject, is of greater interest than the study of pure science. The uses that we have made of the various plants and their products have been intimately bound to our culture and the comforts we have enjoyed. Plants have also had a profound influence on the development of civilizations and the locations of population centers since the origin of mankind.

In recent years the importance of greater food production and other materials of plant origin has become increasingly apparent. Consequently a greater knowledge of agricultural production, medicinal products, industrial uses of plants and strategic materials, which can be obtained from plants and their by-products, has become imperative. While many plants have been in use for food and fiber for thousands of years, and many still remain our most important products, new plant products in the form of drugs and new industrial uses are being discovered every year. In times of national and international strife the need of an adequate food supply and the many products required by our modern civilizations become of paramount importance. Recognizing that these political crises seem to be always with us, the necessity of improved and increased quantities of food production, oils, fats, insecticides, rubber, drugs, wood products and many others are of the greatest urgency to maintain an adequate supply for domestic use as well as for export to the people of less fortunate countries.

No more than an outline of the various products we obtain from plants will be given here but it will show our complete dependence on the members of the plant kingdom. In the final analysis all animals are dependent directly or indirectly on plants.

Food Plants

If we had lived millions of years ago, such as during the Carboniferous, very few of the plants we now use would have been available to us. Then we would have had to eat algae, the club mosses, and other lower vascular plants, for angiosperms were not present. Fruits and seeds, which are the most important vegetal part of our diet, had not yet evolved.

The most important of the plant families from which we derive our greatest amount of food is the **Gramineae,** or grass family. There are many members of this family used as food and forage but the most important to humans are rice, wheat, maize, oats, rye, and barley. Those of lesser value, some of which are used as pasture grasses, include millet (Fig. 1), sorghum, timothy, bluegrass, wild wheat and rye grasses, and grama and buffalo range grasses. An unusual grass used for centuries by the American Indians, and today an expensive food often used in stuffing wild fowl, is wild rice. In Table 28-1 the major members of the Gramineae used in this country for food are listed.

You will undoubtedly notice that corn exceeds all other plants in acreage, total production, and value, and yet most of you will be immediately conscious of the fact that you eat relatively little corn. Most of the corn produced goes into animal food, especially pork, and in the United States the per capita consumption of pork is 63.8 pounds, which is chiefly the way we consume our corn. Only about 15 percent (500,000,000 bushels) of the corn produced is not fed on farms. This relatively small percentage is used in making alcohol, corn sugar and syrup, laundry starch, oil, and many related products.

On a world-wide basis rice is the most important grain consumed by humans; yet the United States exports a large percentage

FIGURE 1 A field of millet, one of the most drought resistant of the grain crops.

Crop	Acreage	Quantity	Value	Leading States in Production
Maize (Corn)	60,550,000	4,100,700,000 bushels	$4,490,480,000	Iowa, Illinois, Indiana, Minnesota
Wheat	49,200,000	1,290,500,000 bushels	1,775,000,000	Kansas, North Dakota, Oklahoma, Montana
Oats	20,420,000	880,900,000 bushels	552,000,000	Minnesota, Iowa, North Dakota, Kansas
Sorghum (for grain)	11,930,000	490,253,000 bushels	519,760,000	Texas, Kansas, Nebraska, California
Barley	10,670,000	403,072,000 bushels	373,800,000	North Dakota, California, Montana, Idaho
Rice	1,786,000	73,113,000 CWT.	360,706,000	Texas, Arkansas, Louisiana, California
Rye	1,725,000	33,472,000 bushels	35,029,000	North Dakota, South Dakota, Kansas, Nebraska
Sugar Cane	686,000	25,180 Short tons	108,658,000	Hawaii, Louisiana, Florida

Table 28-1 Grasses of Major Value in the United States. The statistical data presented in this table was taken from the *Statistical Abstract of the United States 1965*. U.S. Bureau of Census (86th edition), Washington, D. C.

of its crop since our per capita consumption of rice is only 6.8 pounds. Our most important grain in the United States, in fact, the food used in greatest quantity of all, is wheat.

On an average basis the yearly per capita consumption of wheat is about 116 pounds. In comparison, we eat about 99 to 100 pounds of beef, 99 pounds of white potatoes, about 180 pounds of all vegetables combined and about 130 pounds of all fruits fresh, dried, canned, and frozen per person per year. All other foods are consumed in much smaller quantities with the exception of refined sugar which we use at the rate of 96 to 98 pounds per person per year. In 1930 our per capita consumption of sugar was 110 pounds. Since then the amount of sugar consumed has gradually been going down.

The history of our food consumption is very much like that in other countries. That is, when we are in recession and our per capita income is low the amount of expensive foods we eat declines and the amount of cheap foods consumed rises. For example

Crop	Quantity Produced	Value of Crop	Leading Producing States
Apples	3,370,000 tons	$248,000,000	Washington, New York, Michigan, California
Grapes	3,500,000 tons	221,800,000	California, New York, Washington
Peaches	1,400,000 tons	156,850,000	California, Michigan, Georgia, Pennyslvania, New Jersey, South Carolina
Prunes (fresh)	490,000 tons	44,500,000	California, Oregon, Washington, Michigan
Pears	674,000 tons	65,750,000	California, Oregon, Washington
Plums	128,000 tons	18,000,000	California, Oregon
Cherries (sour)	254,000 tons	22,400,000	Michigan, New York
Cherries (sweet)	115,000 tons	33,260,000	California, Oregon
Oranges and Tangerines	117,400,000 boxes	550,000,000	Florida, California, Arizona, Texas (In Florida and Texas a box weighs 90 pounds, in California and Arizona a box weighs 75 pounds.)
Grapefruit	40,600,000 boxes	106,550,000	Florida, California, Texas, Arizona (In Arizona and California a box weighs 65 pounds; in Texas 80 pounds; in Florida 85 pounds.)
Lemons	13,800,000 boxes	34,000,000	California, Arizona (Average weight is 76 pounds per box.)
Pineapples	1,010,000 tons	38,400,000	Hawaii
Watermelons	1,420,000 tons	45,100,000	Florida, Texas
Cantaloupes	642,000 tons	64,000,000	California, Arizona
Strawberries	275,000 tons	110,000,000	California, Oregon, Washington, Michigan

Table 28-2 Most of the leading fruit crops and approximate amounts of the fruit produced on a yearly basis. There is considerably more variation from year to year in the production of these crops than the cereal crops in Table 28-1, since many of these crops may be damaged by frost. For the same reason the leading states frequently change position, first one state being the major producer then another. These data were taken from the *Statistical Abstract of the United States 1965.* U. S. Bureau of Census (86th edition), Washington, D. C.

in 1930, following the 1929 stock market decline we ate only 48.9 pounds of beef per person and in that year we ate an average of 171 pounds of wheat in the form of flour per person. Since that time our economy has improved; we now eat twice as much beef and about sixty percent less wheat.

Although grains are our "staff of life," either directly or indirectly, many other botanical products provide us with a varied diet. In examining the lists of fruits and vegetables in Table 28-2 and Table 28-3 remember that you are looking at the products of a wealthy nation which has not quite reached the population saturation point. There is no reason why every person in the United States should not have an adequate, healthy, well-balanced diet. This is a statement that can be made for very few other countries and at the rate of population increase in this country it is a statement we may not be able to make for long here.

Several plant families stand out in the production of foods besides the Gramineae. The Leguminosae is the family in which peas, beans, soybeans, peanuts, clover, alfalfa, lentils and many other crop plants are classified. A tropical leguminous tree, the tamarind, has recently been introduced into Southern United States and elsewhere. The fruit is that of a typical legume but it is quite pulpy, very sweet and highly nutritious. Its origin was apparently India but like many desirable plants it is now being cultivated in many other areas.

On page 606, a number of major plant families are listed with some of the crop plants found in them. Several dozen others could be listed with many more products that are obtained from them. Many of the families not listed have tropical or subtropical species which will probably be of greater importance in the future when we have introduced people of temperate climates to the many unusual and piquant flavors of such tropical fruits as the custard apple, cherimoya, carambola, akee, papaya, sapodilla and many others (Figs. 2, 3, 4, 5).

Several of the families have mostly fruits as the principal products; others have roots, stems, leaves, or even flowers as the edible portions. The average person speaks quite freely of fruits and vegetables as a portion of his diet. Yet if a student in a botany class asks the difference between a fruit and a vegetable, the instructor can define one with ease but not the other. As defined earlier in the discussion of the flower and its parts we defined a fruit as a ripened ovary. Fruit, therefore, has a definite botanical meaning but a vegetable is really just a category on the menu. For a vege-

FIGURE 2 Lemon.
FIGURE 3 Avocado. This large variety is raised much more commonly in Florida than elsewhere.

FIGURE 4 Young dates forming in the apex of the date palm.
FIGURE 5 Pineapple plantation in Hawaii. (Courtesy Bernice P. Bishop Museum, Baker Collection.)

table we may eat roots (sweet potato, parsnip), stems (white potato, kohlrabi), leaf petiole (celery, rubarb), flower (cauliflower, broccoli), leaf blade (spinach, Swiss chard), apical bud (cabbage, artichoke), axillary bud (Brussels sprouts, garlic), fruits (okra, corn), and seeds (peas, beans).

Crop	Acreage	Quantity	Value	Major States
Tomatoes	430,000	5,570,000 tons (not including) Hawaii)	$315,000,000	California, Florida, New Jersey, South Carolina
Beans (Dry edible)	1,450,000	18,000,000 CWT	135,000,000	Michigan, California, Idaho, Colorado
Potatoes (white)	1,350,000	250,000,000 CWT	780,000,000	Idaho, Maine, California, New Jersey
Sweet Potatoes	190,000	16,000,000 CWT	75,000,000	Louisiana, North Carolina, Alabama, Texas
Peanuts	1,397,000	2,167,100,000 lbs.	242,648,000	Georgia, North Carolina, Alabama, Texas
Soy Beans	30,738,000	700,000,000 Bu.	1,835,596,000	Illinois, Iowa, Indiana, Arkansas
Flaxseed	2,831,000	24,408,000 Bu.	68,597,000	North Dakota, South Dakota, Minnesota, Texas
Cottonseed	14,060,000	6,333,000 tons	296,618,000	Texas, Mississippi, California, Arkansas
Cotton lint	14,060,000	15,356,600 tons	2,356,600,000	Texas, Mississippi, California, Arkansas
Sugar Beets	1,398,000	23,184,000 tons	271,253,000	California, Idaho, Colorado, Nebraska
Lettuce	215,000	1,900,000 tons (including Alaska and Hawaii)	170,000,000	California, Arizona, Colorado, Texas

Table 28-3 Some vegetable and miscellaneous major crops produced in the United States. These data were taken from the *Statistical Abstract of the United States 1965*. U. S. Bureau of Census (86th edition), Washington, D. C.

The various food products (Figs. 6, 7) are almost endless, yet if you were to think of the ones you eat with regularity I am sure you would be amazed to see how few of all the world's plant food products are a part of your diet.

FIGURE 6 Harvesting and boxing head lettuce. The workers are "braceros," imported from Mexico to help in the harvesting of crops, a common practice in Southwestern United States when crops must be harvested by hand quickly.
FIGURE 7 Red chili peppers laid out to dry. The bank in the center of the picture separating the two cotton fields is several hundred yards long and covered with peppers. In the background the hills have been scraped free of native vegetation and the peppers are spread out on the slopes to dry. This crop is worth a good many thousands of dollars.

Sugars

Sugar is a food product but most of that which we eat is in the form of candy, desserts, sweetening agents, or syrup. However, sugar as we know it is a relatively recent product historically. Prior to the use of sugar cane and beet sugar, people depended upon an animal product, honey, as a sweetening agent. In this country the American Indians also found that they could extract a sweet substance from certain maple trees and therefore discovered maple sugar.

We extract very little sugar from other plants even though all green plants produce it. We do make some syrups and molasses from sorghums and a palm sugar from several members of the Palmaceae but that is virtually all.

Major Plant Families and the Crops Which Are Classified in Them

Gramineae Maize, Rice, Wheat, Oats, Barley, Millet, Sorghum, and numerous other grasses

Leguminosae Beans (lima, kidney, etc.), peas, peanut, soybean, alfalfa, clover, lentils, cowpeas, tamarind

Rosaceae Apple, pears, plum, cherry, peach, almond, apricot, raspberries, strawberries, blackberries, nectarine, loquat, quince, and others

Vitaceae Grapes

Ericaceae Blueberry, cranberry, and huckleberry

Rutaceae Oranges and all other citrus fruits

Moraceae Mulberry, Breadfruit, and fig

Bromelidaceae Pineapple

Musaceae Banana

Euphorbiaceae Manioc (tapioca)

Chenopodiaceae Beet, spinach, Swiss chard, and other leafy vegetables

Umbelliferae Carrots, celery, parsnip, and parsley

Cruciferae Cabbage, mustard, radish, turnip, rutabaga, kohlrabi, and others

Solanaceae White potato, tomato, green pepper, egg plant, and others

Starches

Most of the starch we use, either as a food or industrially, comes from corn and some other cereal grasses. Other sources include the potato, manioc of the tropics, and to a very limited extent arrowroot starch from several different families, and sago starch from *Zamia florida* of the Cycadaceae.

Beverage Plants

Of the nonalcoholic beverages produced readily from plants, coffee (*Coffea arabica* of the Rubiaceae) and tea (*Thea sinensis* of the Theaceae) are by far the most significant. The history of both of these plants is very colorful but it is not possible to discuss their effect upon the several different countries, where they are now raised in quantity, here. In South America, especially in Argentina and Paraguay, a tea is made from a plant, *Ilex paraguayensis,* of the Aquifoliaceae. The tea is called maté or yerba maté and looks very much like the tea of oriental origin but has a much stronger flavor.

Another beverage, as well as other sweet products, is obtained from *Theobroma cacao* of the Sterculiaceae. The products are chocolate and cocoa. Of lesser importance are teas and soft drinks made from various flavoring extracts such as sarsaparilla and sassafras.

Alcoholic beverages are made from almost any fermentable materials but the most widely used are obtained from the grains. Liqueurs and cordials are made from the distilled fermented products of numerous fruits such as cherries, peaches, apricots, apples and many others. Wines, of course, are principally a product of the fruit of the grapevine. While sake is a fermented product from steamed rice; pulque from fermented agave leaves; rums from sugar cane and sorghums; whiskies, vodka, and gin are the distilled alcoholic products of fermented grains.

FIGURE 8 Cell from the endosperm of castor beans with several oil drops.

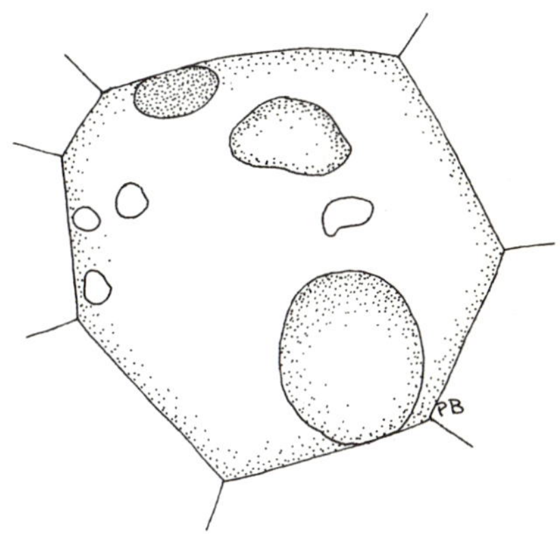

Vegetable Oils

In recent years we have been hearing much of the saturated and unsaturated oils in the foods we eat. The unsaturated oils, generally considered to be less active in the formation of cholesterol in our circulatory systems, come mainly from the oils of plants (Fig. 8).

The most widely used vegetable cooking oils and those used in margarine come chiefly from maize, cottonseed, olive, peanut, soybean, and safflower oils. Some idea of the amount of oil extracted from these plants may be obtained by checking the tables which include maize, peanuts and cottonseed. For example, the number of pounds of peanuts produced in the United States is over two billion pounds per year. Since our per capita consumption of peanuts is about five pounds per year the remainder is used to extract cooking oils and the dried cake is made into animal foods and for other industrial uses including fiber board.

Oils are also extracted from castor beans of the Euphorbiaceae, from almonds, sunflower seeds, linseed, and various nuts (Fig. 9). In addition highly flavored and aromatic cooking oils are made from the seeds of anise, sesame, fennel, dill, caraway and other scented seeds.

FIGURE 9 A grove of young pecan trees. A grain forage crop is raised between the trees for additional revenue, to cover the soil, and to retard the growth of weeds.

Waxes

Closely related to the oils are waxes. The majority are at least in part animal products but the best, so far as durability is concerned, is carnauba wax obtained as an exudate from the leaves of a Brazilian palm named *Copernicia cerifera.* Aside from some waxes, made in part from the oils mentioned above, we have one called candelilla obtained from a shrub, *Euphorbia antisyphilitica,* which grows in the Big Bend Country of Texas and Mexico.

Pectins

These substances are found in nearly all plants as the cementing substance in cell walls. It is especially abundant in young growing tips and in green fruits. It is obtained chiefly from citrus and green apple residues but many other plants including guava, carambola, soursop, surinam cherry, and manioc are now being widely used for the extraction of pectin. It is used in the manufacture of jellies, candies, to increase viscosity in tomato juice, for the preparation of various medicinals, and in the tobacco industry.

Gums

Another cell exudate, usually produced in small quantity, is a gum used in sizing, pharmaceuticals, inks, cosmetics, etc. A number of gums such as gum arabic, tragacanth, carob gum and mesquite gum come from the Leguminosae family. One medicinal product used chiefly as an intestinal conditioner is obtained from the gum of the seeds of *Plantago psyllium.* The gum is made up principally of the sugars mannose and galactose.

There are many other minor gums used but closely related to them are agar and alginates obtained from certain red and brown algae mentioned in earlier chapters. The agars and alginates are used and produced in much greater quantity than the gums mentioned above.

Resins

Most of the resins are produced by members of the Pinaceae (Fig. 10). In the United States *Pinus palustris* and *P. caribea* are the principal trees from which the resins are extracted to form the basis of our large naval stores industry. Fossil amber is also a resin product of the Pinaceae deposited many years ago. Some members of the Leguminosae also produce fine resins in the form of Balsam of Peru, Congo Copal, and Copaiba balsam.

Damar resins are obtained from several genera of the Dipterocarpaceae, and frankincense and myrrh of biblical fame are resins obtained from two different genera in the Burseraceae. Very colorful laquers are made from resins obtained from *Rhus verniciflua* of the Anacardiaceae by the Chinese and Japanese for their highly laquered and colorful paintings.

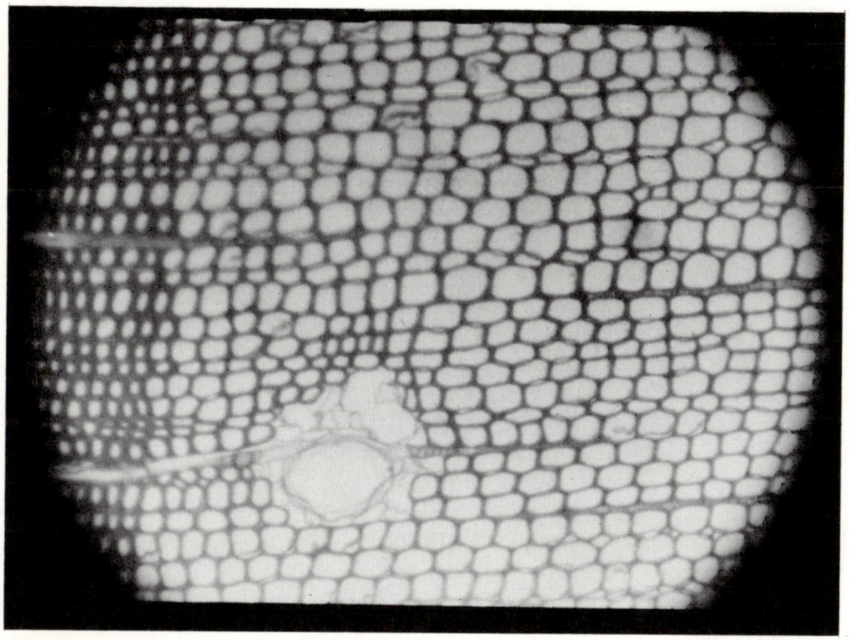

FIGURE 10 Cross section of pine stem showing about one annual ring. Note resin duct in spring wood.

Perfumes and Flavors

Perhaps these two terms do not seem too closely related but many of the plant products obtained from the highly aromatic plants used in the preparation of these products are used for both.

For example ginger is used as a spice and in the manufacture of exotic oriental perfumes. Some 500 tons of lemons are used annually in the manufacture of extracts for flavoring and perfumes. Other flavoring substances from the mint family include basil, lavender, sage, sweet marjoram, and rosemary. In Table 28-4 many plant families and the flavor plants produced in them are listed.

The number of different flowers from which perfume oils could be extracted is quite large. Commercially most of the essences are derived from such flowers as those of rose, carnations, gardenias, hyacinths, lilacs, jasmine, orange, tuberose, geranium and a grass, *Cymbopogon nardus,* from which we get an oil, citronella, used in making perfumes, flavoring, menthol and other pharmaceuticals.

Family	**Plant name**	**Product**
Labiatae	*Mentha piperata*	peppermint
	Melissa officinalis	balm
	Salvia officinale	sage
Rosaceae	*Prunus amygdalus*	oil of almond
Compositae	*Artemisia dracunculus*	tarragon
Orchidaceae	*Vanilla* spp.	vanilla
Zingiberaceae	*Zingiber officinale*	ginger
Liliaceae	*Allium sativa*	garlic
	Allium cepa	onion
	Smilax spp.	sarsaparilla
Cruciferae	*Rorippa aromoracia*	horseradish
	Brassica alba	yellow mustard
	Brassica nigra	black mustard
Lauraceae	*Cinnamomum zeylanicum*	cinnamon
Myristicaceae	*Myristica fragrans*	nutmeg
Myrtaceae	*Eugenia caryophyllata*	cloves
	Pimenta officinale	allspice
	Eucalyptus spp.	Eucalyptus oil
Moraceae	*Humulus lupulus*	hops
Umbelliferae	*Apium graveolens*	celery
	Coriander sativum	coriander
	Angelica archangelica	angelica oil
	Foeniculum vulgare	fennel
	Anethum graveolens	dill
	Apium sativum	parsley
	Pimpinella anisum	anise
	Carum carvi	caraway

Table 28-5 Flavorings used in cooking and the production of many pharmaceuticals

Although perfume is big business, it is not the highly specialized industry it was years ago before the advent of the bath tub, hot running water, and scented soaps. However, in recent years our Internal Revenue Department has been making an annual tax collection of approximately $180,000,000 from the sale of toilet preparations, which is, of course, the cosmetic industry.

Vegetable Tannins and Dyes

Tannin is a dye used in the preparation of leathers, in the paint industry, and many others. We use thousands of tons of the extract which in this country is obtained from the wood of many oak trees, from the wood of hemlock, and some is still obtained from the wood of the American Chestnut, although the living trees of this species are nearly all gone. Most of our tannin is, however, imported. Probably the most important foreign source is from the wood of the quebracho tree of South America and the mangrove of Central America and elsewhere. A number of other plants are used for specific types of tannin such as a Sicilian sumac, *Rhus coriaria,* which produces a light colored tannin used in preparing leather for gloves.

The dye industry has moved largely into the use of aniline and other synthetics. However, a number of plants still are used for highly prized dyes. The legume family has *Indigo tinctoria* and *Haematoxylon campechianum* from which we get dyes of the same name as the generic epithet of each. Other dyes obtained from plants are woad from a member of the Cruciferae, madder from the Rubiaceae, saffron from the elongated stigmas of *Crocus sativus* in the Iridaceae, and safflower from *Carthamus tinctorius* of the Compositae. There are a few others but the ones listed are those with which most of us are familiar.

Medicinals

Hundreds of plants have been used as the source of drugs, many quite ineffectively, but a good many have been so successful no adequate substitute has been found for them. One of the most effective, discovered by the Andean Indians, hundreds of years ago, and now used on a world wide basis, is quinine obtained from the bark of several species of *Cinchona* in the Rubiaceae. Ipecac also comes from *Cephaelis ipecacuana* in the same family. Although several synthetic drugs to supplement quinine have been developed in the last twenty-five years certain forms of malaria can be controlled better by the natural product than any other.

One of the great drugs to benefit mankind, but which unfortunately too frequently falls into illegal channels, is opium. The product is an exudate of the fruit of *Papaver somniferum* of the poppy family. It is quite common for botanists to have specimens of illegally raised plants, such as marijuana or peyote, brought in by representatives of the law for proper identification. Although this has happened many times to the writer only once has a member of the Bureau of Narcotics or any other law enforcement agency brought in a specimen of *Papaver somniferum* for the confirmation of its identity. While marijuana is often raised illegally the opium poppy is rarely raised in this country for illegal or any other purpose. Most of our legal opium is imported from the Mediterranean area and apparently most of our illegal opium is brought in from the orient.

Cocaine is a product of the Andes, of the species *Erythroxylon coca* in the Erythroxyllaceae. One of the hallucination producing narcotics is peyote from a member of the Cactaceae, *Lophophora williamsii;* and marijuana or hashish is a product of the hemp plant, *Cannabis sativa* of the Moraceae.

The family Solonaceae has a number of species which produce such alkaloids as atropine, hyoscyamine and scopalamine. These drugs and several others are obtained from *Atropa belladonna*. Several species of the jimson weed, *Datura,* also produce these same drugs and several others.

One of the most widely used heart regulating drugs is digitalis, a glycoside formed by *Digitalis purpurea* of the Scrophulariaceae. Another drug which may affect the heart is strychnine from *Strychnos nux-vomica* of the Loganiaceae. This species is native to the orient but other species, *S. taxifera* and *S. guianensis,* which grow in South America, are the source of a similar drug called curare, applied by natives on arrows and blowgun darts.

Since the effectiveness of certain antibiotics derived from members of the Ascomycetes *(Penicillium),* and Actinomycetes *(Streptomyces)* are reasonably well known, they will not be considered again here. It should be emphasized, however, that some of the drugs obtained from these and related fungi have greatly broadened the effectiveness of drugs in medicine and we are discovering more of this type every year.

Of less specific effect than the drugs mentioned above, but still used to increase the comfort and alleviate the discomfort of many individuals are such pharmaceuticals as camphor, hoarhound, pennyroyal, witch hazel, bay rum and some related industrial oils as cedarwood oil, santonin oil, and cubeb oil.

Insecticides

Closely related to the drugs are two members of the Leguminosae, *Lonchocarpus nicou* of the Amazon region, and *Derrii elliptica* of the Far East, which produce **rotenone.** *Chrysanthemum coccineum* and several other species of the Compositae produce **pyrethrum.** Both of these products are very effective insecticides but perhaps the most widely used of all is **nicotine sulfate** made from a member of the Solonaceae, *Nicotiana tabacum.* In 1964 the farmers of the United States marketed 2,206,000,000 pounds of tobacco. This was almost 23 percent of the total produced throughout the world. We exported almost thirty percent of our crop and although most of the remainder was used in the manufacture of fumitory and masticatory products, hundreds of thousands of pounds were used in the manufacture of nicotine sulfate.

FIGURE 11 Cluster of bast fibers from the phloem region of flax.

Fibers

Except for cotton, little commercial fiber is now raised in the United States. In the extraction of fiber a good deal of labor is required; consequently we find it cheaper to import the many fibers we need than to raise them.

Fibers are classified as **soft** if they are obtained from the bast fibers of the phloem or pericyclic tissue just outside the phloem. Flax *(Linum usitatissimum)* produces a soft fiber known as linen (Fig. 11). While we raise a good deal of flax, we raise it for the seed for linseed oil rather than the fiber. Other soft fibers are ramie, jute, and hemp.

Hard fibers are obtained chiefly from the large leaves of monocotyledonous plants where the entire vascular bundle is combed out for the fiber. They are very coarse and used mostly for rope. Examples are abacá or Manila hemp, from *Musa textilis,* a banana plant; sisal from the leaves of *Agave,* and New Zealand hemp from a member of the lily family. There are many others, including a species of *Sorghum,* used in making brooms.

The most important **surface** fiber, in fact the most important fiber plant is cotton (Fig. 12). The amount produced in this country is listed in Table 28-3. The fiber is an elongation of the epidermal cells of the seed coat and may be obtained from several species of the genus *Gossypium* in the Malvaceae. While most plants cannot be raised satisfactorily for two products, like flax for both seeds and fiber, or the pineapple plant for both fruit and a hard fiber, cotton is an exception. The lint is first removed from the seed, then the oil is extracted from the seed, and finally the dried seed cake can be used as animal food, or as a fertilizer, or for some other industrial products.

FIGURE 12 Cotton before the first picking. Many of the bolls have not yet opened.
FIGURE 13 Mechanical cotton picker at work.

Other surface fibers are **kapok** from a tropical member of the Bombacaceae and **coir** from the husk of the fruit of the coconut palm. The latter is an extremely coarse fiber used in making door mats and similar products.

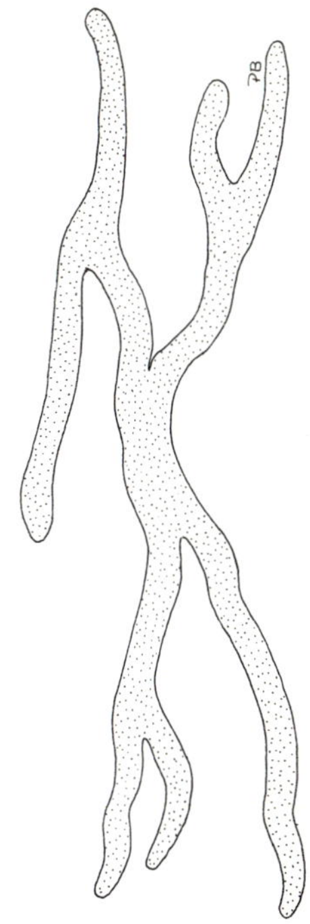

FIGURE 14 Diagram of anastomosing latex tubule.

Latex Products

The Euphorbiaceae is one of the most versatile families in the variety of products obtained from it. Its two most important products are manioc from *Manihot glaziovii,* and rubber, *Hevea brasiliensis.* While rubber can be made from the latex exudate of plants from several other families the commercial source is chiefly *H. brasiliensis.* In recent years this has been about 2,400,000,000 tons of raw rubber annually. Now most of the rubber is obtained from the Far East and Africa but the tree is a native of Brazil. Disease in the trees and lack of labor forced the transfer of the rubber industry from this hemisphere (Fig. 14).

Many synthetics have been developed to replace rubber but none thus far have proved as versatile in the various products that can be obtained from it.

One additional latex product should be mentioned. This is chicle from *Achras zapota* of the Sapotoceae. It is used almost exclusively for chewing gum and the United States is the greatest importer and user of this questionable product.

Wood

In Chapter 8 a considerable amount of space was given to softwoods and hardwoods, their differences, characteristics, and uses. It would seem appropriate to discuss at great length the importance of wood to our general economy for we use it in various forms everyday. However, since a good deal about wood and forests has been discussed earlier I will close this chapter with an imposing list of characteristics of wood possessed by no other product, natural or manufactured. The source of this material has been lost to me but I am sure you will appreciate the value of wood after reading it.

Could you name a substance that has the following properties?

1. As strong, weight for weight as the average metal.
2. Absorbs shock that will shatter all ceramic materials.
3. Can be worked into various shapes by simple tools.
4. Can be fastened with ordinary nails and screws.
5. Can be produced in very large sizes, in fact into almost any size.
6. In durability, it will last indefinitely if protected from moisture and even in wet places can be made to last indefinitely by simple chemical treatment.
7. Possesses a variety of beautiful patterns.
8. Appearance can be altered by a variety of stains or paints.
9. Will not rust, and resists the action of strong acids and other chemicals.
10. Excellent for sound and heat insulating material.
11. Can be converted into a pulpy mass from which paper, fiber board, rayon fabric, cellophane film, photographic film, and plastics can be made.
12. Can serve as a raw material in the manufacture of explosives, methyl and ethyl alcohol, acetic, oxalic and other acids, industrial solvents, lubricants, tars and aromatic substances such as vanilla extract.
13. Can be used as a soil conditioner.
14. Can be produced in unlimited quantities and at a reasonable price.
15. Molasses can be made from it to feed livestock; steaks made of proteins grown on sugars derived from this substance may be obtained and a party may be enlivened by serving whiskey brewed from this product and aged and stored in it.

Supplementary Reading

Baker, Herbert G., *Plants and Civilization*. Wadsworth Publishing Co., Inc., Belmont, Cal., 1965.

Hayward, Herman E., *The Structure of Economic Plants*. The Macmillan Co., New York, 1948.

Hill, Albert F., *Economic Botany*, 2nd ed. McGraw-Hill, New York, 1952.

Schery, Robert W., *Plants for Man*. Prentice-Hall Inc., Englewood Cliffs, N. J., 1961.

Questions for Discussion

1. May we be expected to improve the quantity and quality of plants of economic importance for centuries to come? Explain.
2. How are new varieties developed?
3. How may we increase the production of a given crop?
4. How could we make plants, such as weeds of little direct nutritional value and low palatability, useful for human or even herbivorous animal consumption?
5. List twenty-five plants, the products we obtain from them, and the tissues or organs which produce these products.
6. List ten drugs which are obtained from plants.
7. In raising a garden why do you find that weeds often do much better than the desirable plants?
8. Which tissues in a plant more commonly form food products; fibers; drugs?
9. Why are some plants of economic importance planted in the early spring or even the preceding fall, while others are not planted until late spring or summer?
10. How would the type of transportation and storage which was prevalent fifty years, one hundred years, and two hundred years ago have affected your diet if you had lived at that time?

Glossary

abaxial. Situated on the side away from the stem or main axis, as a sporangium.

aberration. Deviating from the normal; as a chromosome aberration is a change from the usual chromosome complement or arrangement.

abscission layer. A zone of cells formed at the base of a leaf or other plant part.

absorption. Taking in a substance, as in the assimilation of digested foods; part of the osmotic phenomena.

acetic acid. A compound, CH_3COOH, found in vinegar and other substances.

acetone. A volatile, fragrant, liquid ketone, CH_3COCH_3, frequently used as a solvent for organic compounds.

acetylation. The act or process of forming acetate from the breakdown of fatty acids and carbohydrates and subsequently oxidized to carbon dioxide and water.

acid-fast (of tissues, cells, bacteria). Able to take and hold an acid stain.

acid soils. Soils with a pH of less than 7, as contrasted with alkaline soils, which have a pH of more than 7.

acrasin. A hormonal substance secreted by *Dictyostelium*.

acrogynae. Liverworts of the Jungermanniales in which the apical cell ultimately becomes an archegonium.

acropetal succession. Order of development in which the youngest structures are at the apex and the oldest at the base.

actinomorphic. Manifesting radial symmetry, as flowers that have the floral parts in each series of equal size and position.

actinostele. A protostele in which the xylem arms are star-shaped in cross section.

active transport. The movement of substances through membranes by means of an energic coupling such as ATP.

adaptation. A change of form or habit in an organism that makes it better able to thrive and reproduce in its environment.

adaxial. Situated on the side towards the stem or main axis, as a sporangium.

adenine. A component of nucleotides and the nucleic acids.

adenosine diphosphate (ADP) and triphosphate (ATP). Phosphorylated compounds used in cellular energy cycles such as respiration and photosynthesis.

adsorption. The adhesion of molecules in solution to solid surfaces such as to the surface of root hairs.

adventitious. Arising at an unusual place, as roots arising from stems or leaves.

aeciospore. The spore produced in the aecial stage develops on the barberry leaf.

aecium. A spore stage in rusts; in wheat rust the aecial stage develops on the barberry leaf.

aerial stem. A stem which grows above the soil, as opposed to a rhizome.

aerobic. Living or active in the presence of free oxygen.

agar. A mucilaginous substance obtained from red algae and used as a culture medium.

agaric. Any fungus in the family Agaricaceae.

aggregate fruit. A fruit developed in a cluster from the separate ovaries of a single flower.

agronomist. One who studies field crops and their products.

akinete. A nonmotile, thick-walled resting spore in which the vegetative wall of the cell becomes the wall of the spore.

alanine. A white crystalline amino acid with the formula $C_3H_7NO_2$.

albinism. Lacking pigmentation, as a plant without chlorophyll.

albumen. Nutritive parenchyma, as the storage tissue in cotyledons.

aldose. A sugar (as xylose, glucose, mannose) containing one aldehyde group per molecule—contrasted with ketose.

algology. The study of algae.

algin. A mucilaginous product of certain brown algae.

alkaline. Having a pH of over 7.0; soils which contain a large amount of calcium or sodium carbonates.

alkaloid. An alkaline nitrogenous substance usually poisonous to animals.

allele. Either of a pair of alternative, contrasting hereditary characteristics, situated at the same locus in homologous chromosomes.

allergy. Hypersensitivity to certain substances such as bacteria, viruses, and pollen.

allopolyploid. An individual with more than 2 sets of chromosomes derived from different species.

alpha ray. An alpha particle moving at high speed (as in radioactive emission).

alpine. High altitude environment.

alternate leaves. Leaves each of which arises singly at a node; also alternate buds.

alternation of generations. A pattern in which two distinct plant bodies or organs produced in the same life cycle alternate with each other, as the gametophytic generation produces gametes and the sporophytic generation arising from the zygote produces the spores which produce the gametophytic plant.

amino acid. An organic acid with the amino radical (NH_2); one of the components of protein.

ammonium. A strongly basic radical (NH_4).

amoeboid. Resembling an amoeba.

amorphous matrix. The intercellular substance, usually lacking definite form and shape, in which certain unicellular algae or bacteria are often imbedded.

amphibious. Capable of living at least at some time during the life span in both water and on land.

amphicribral. (Of a vascular bundle) having phloem surrounding the xylem.

amphigastrium. One of the reduced ventral leaves of a leafy liverwort.

amphiphloic. Having phloem both internal and external to the xylem.

amphithecium. The outer layer of cells that in some plants develops into the young sporophyte.

amphitropous. Having the ovule inverted but with the attachment near the middle of one side.

anabolism. A constructive, or building phase of metabolism.

anacrogynae. Liverworts in which the archegonia develop from behind the apical cell and never from the apical cell itself.

analogous. Of similar function and often similar in appearance, but not of similar origin.

anaphase. A stage in mitosis in which the divided chromosomes move towards the opposite poles of the cell.

anatomy. The division of morphology concerned with the internal structure of organisms.

anatropous. (Of ovules) having the main body inverted and straight, with the micropyle bent downward parallel with the funiculus.

androsporangium. A sporangium which produces androspores.

androspore. A zoospore that produces a male filament, as in *Oedogonium*.

angstrom. A unit of length equal to one ten-thousandth of a micron; used in expressing the length of light waves.

anisogamous. Characterized by gametes which are morphologically similar, but differ in size.

annual. A plant which completes its growth and dies within one year.

annual ring. The yearly growth of xylem formed by cambium.

annulus. The ring found on the stipe of certain mushrooms.

anther. The portion of a stamen which produces pollen.

antheridiophore. The structure on which an antheridium is borne.

antheridium. A sperm-producing structure in the thallophytes, bryophytes, and lower tracheophytes.

anthocyanin. The blue or red pigment found in plants.

antibiotic. A substance in the tissues of an organism which combats pathogenic organisms.

anticlinal. Formed at right angles or perpendicular to the outside surface, as a cell wall.

antigen. A substance, as bacteria or foreign protein, which when introduced into the body stimulates the production of antibodies.

antipodal. At the opposite end, as the nuclei or cells in the embryo sac which are at the end opposite the egg.

antitoxin. An antibody or substance that neutralizes poisons.

apetalous. Without petals, as a flower; an apetalous flower may have sepals.

apical growth. Growth occurring at the tip of an organ, as the meristem region of a root or stem tip.

aplanospore. A nonmobile spore in which the wall of the spore is not derived from the wall of the vegetative cell in which it is formed.

apogamy. The development of a sporophyte from gametophytic tissue without sexual fusion.

apomixis. The setting of seed without fertilization.

apophysis. An enlargement at the base of the capsule of some mosses.

apospory. The formation of a gametophyte directly from sporophyte tissue, rather than from a spore.

apothecium. A cuplike or disclike base for spore-bearing structures in lichens and ascomucetes.

aquatic. Living in water.

arboreal. Treelike.

arborescent. Treelike in size and form.

arboretum. A place where trees and shrubs are cultivated mainly for educational or scientific purposes.

Archaeozoic. The earliest era in geological history during which life is believed to have originated.

archegoniophore. Structure on which archegonia are borne.

area of dominance. The region occupied by an organism so as to restrict or exclude organisms of other types.

arm palisade. The fingerlike projections of the inner walls of the epidermis of certain leaves.

ascocarp. A fruiting structure in which asci are borne.

ascogenous hyphae. Hyphae that bear asci.

ascospore. A spore borne in an ascus.

ascus. The saclike cell in which ascospores are produced.
asexual. Not sexual; reproducing without the benefit of gametes.
assimilative. Capable of the absorption of nutrients into the body.
atactostele. A structure in which the vascular bundles appear scattered throughout the parenchymatous ground tissue.
atom. The smallest particle of an element.
atomic energy. Energy liberated by changes in the nucleus of an atom; as by fission of a heavy nucleus, or condensation of light nuclei into heavier ones with an accompanying loss of mass.
aureomycin. An antibiotic drug.
autosomic chromosomes. Chromosomes other than the sex chromosomes.
autotrophic. Able to synthesize food from inorganic materials, as green plants.
auxiliary cell. Any of a group of cells produced in certain red algae in early formation of the cystocarp.
auxin. A plant hormone or growth-regulating substance.
auxospore. A reproductive cell in diatoms formed by the union of the gametes or by asexual development.
axil. The upper angle where a leaf or twig comes off the main axis of the stem.
axile placentation. Position of the placentae (and ovules) on the axis of an ovary with more than one carpel.
axillary bud. The bud borne in the axil of a leaf.
axoneme. The central filament of a flagellum.

bacillus. A rod-shaped bacterium.
bacteria. Microorganisms of the phylum Schizomycophyta.
bacteriochlorophyll. A substance produced by certain bacteria which enables them to manufacture their own food; the substance is similar to the chlorophyll of higher plants.
bacteriocide. Anything that destroys bacteria chemically as a strong salt solution.
bacteriology. The study of bacteria.
bark. In woody stems the tissue, outside the functional phloem, derived chiefly from cork cambium and outer nonfunctional phloem layers.
barometer. An instrument for determining atmospheric pressure and thereby predicting probable weather changes.
basal cell. The lowermost cell of the suspensor; or the lower of two cells formed following initial division in the development of a gametangium or sporangium.
basalt. A dark-gray to black, fine to coarse grained, igneous rock.
basidiocarp. The fruiting structure of a mushroom; chiefly the stipe and cap.
basidiospore. Spore produced on a basidium in the Basidiomycetes.
basidium. The club-shaped structure bearing basidiospores, usually four, as the result of karyogamy and meiosis.
basipetal. Order of development in which the youngest structures are at the base of the developing axis and the oldest are at the apex.
bast. Phloem; especially the fibrous cells in the phloem.
berry. A simple fruit, the entire pericarp of which is fleshy.
beta ray. A beta particle; a stream of beta particles.
bibulous. Capable of absorbing liquids as a blotter absorbs ink or water.
bicollateral. Having the phloem both internal and external to the xylem, as a vascular bundle.
biennial. A plant which produces seed during its second year of life and then dies.
biflagellate. Having two flagella, as a gamete.
bilateral symmetry. The condition of having distinct and similar right and left sides, as in the flowers of sweetpea and snapdragon.
binomial nomenclature. The system of naming that makes use of two names, the generic and specific names, for each type of organism.
biochemistry. A study of the chemical compounds and processes occurring in organisms.
biome. A large, natural assemblage of associated plants and animals, extending over a large region of the earth's surface.
biophysics. A study of minute physical processes, including radiation, and their relation to organisms.
biosphere. The various habitats of the lithosphere, hydrosphere, and atmosphere within which plants and animals live.
biotic. Relating to life and therefore biology.
biotin. A member of the vitamin B complex.
bisexual. Producing both eggs and sperms; (a flower) bearing both stamens and pistil(s).
bisporangiate. Having two kinds of sporangia in the same structure, as a strobilus of *Selaginella*.
bivalent. Having a valence of two; applied also to homologous chromosomes when two are associated in synapsis.
black alkali. An alkaline soil containing sodium carbonates.
blade. The expanded portion of a leaf.
blepharoplast. The granular neuromotor apparatus at the base of a flagellum.
blight. A widespread and destructive plant disease.
blue-green algae. Any member of the Cyanophyta.
bond. The means by which atoms are held together in a molecule; indicated in formulas by a line (—) or a dot (·).
border parenchyma. The cylinders of thin-walled cells surrounding the individual veins in a leaf.
boreal. Pertaining to the northern forests.
boron. A nonmetallic element; occurring in borax and boric acid.

botany. The science dealing with the study of plants.
botulinism. Poisoning by the toxin produced by the bacillus *Clostridium botulinum.*
bract. A modified leaf associated with a flower or inflorescence.
branch gap. An interruption in the vascular tissue of a stem at the point where a branch trace arises.
brittlewort. Common name for members of the Charales.
brood bodies. Specialized, asexual reproductive bodies formed in some plant groups.
bud. A small axillary protuberance containing rudimentary foliage or floral parts or both.
budding. A method of asexual reproduction by a small outgrowth from the parent cell.
bud scale. A specialized protective leaf of a bud.
bulb. A short, usually globose underground stem, bearing mainly fleshy, food-storing scale leaves; essentially a subterranean bud.
bulbil. An asexual body produced on the protonema of some mosses, and on the leaves of certain ferns; a small bulb.
bundle scars. Scars left in leaf scars at the time of leaf fall by the breaking of vascular bundles passing from a stem into a petiole.
bundle sheath. The tissue, usually of thick-walled cells, surrounding the veins in a leaf or stem.
button. An early stage in the development of a basidiocarp.

calcium. A silver-white soft metallic element commonly found as a compound in white alkaline soils.
calyptra. In the sporophyte of a moss, a sheath which covers the capsule; the tissue develops from a portion of the archegonium.
calyptrogen. The histogen in a growing tip of the root from which the root cap is developed.
calyptrogen-dermatogen. The outer histogen in certain young roots.
calyx. Collective term for the sepals of a flower.
cambium. Layer of meristematic cells between xylem and phloem tissues.
Cambrian. The earliest period of the Paleozoic era.
campylotropous. Having the ovule turned so that the base and apex are close together.
cap cell. One or more cells formed at the apex of a specialized reproductive structure; liberation of the gametes or spores usually occurs when the cap cells disintegrate or separate.
capillarity. Movement of water and solvent in soil and in plant tissues by molecular attraction.
capillitium. A delicate, branched network in the sporangia of slime fungi.
caprifig. The fig of the Smyrna type which provides the pollen for pollination of the female flowers. A process commonly called caprification.
capsule. A simple, dehiscent, dry fruit composed of two or more carpels; the spore case of a moss or liverwort; a slimy outer wall layer of bacteria.
carbohydrate. Any of certain foods composed of carbon, hydrogen, and oxygen in the ratio of 2 to 1. Examples: sugars, starches.
carbon. An element occurring native, as the diamond and as graphite; one of the constituents of all organic compounds.
Carboniferous period. A period of the Paleozoic era characterized by the formation of great coal beds.
carboxydismutase. An enzyme capable of splitting certain carboxyl organic compounds.
carinal canal. In the stem of the Equisetales, canals which develop beneath the stem ridges in the stele.
carnivore. A flesh eating animal; also applied to insectivorous plants.
carotene. A reddish crystalline hydrocarbon ($C_{40}H_{56}$) found in various plant organs.
carotenoid. Any of a group of yellow, orange, red, or purplish pigments in plants.
carotin. A reddish-orange pigment found in plant cells, made up of carbon and hydrogen atoms.
carpel. A floral organ which bears and encloses ovules; a megasporophyll.
carpogonium. In red algae, the basal portion of the procarp containing the egg.
carpospore. In red algae, one of the nonmotile spores produced on short filaments arising from the carpogonium following fertilization.
caruncle. An outgrowth of fleshy tissue at or near the hilum of a seed.
caryopsis. A fruit characteristic of grasses; the fruit is dry, one-seeded, indehiscent, with the seed coat and pericarp completely united.
casparian strip. A secondary thickening in many endodermal cells in the form of a continuous band or strip on the radial and transverse walls.
catabolism. A destructive or breakdown process of metabolism.
catalyze. The process, catalysis, which accelerates a reaction, such as the action of an enzyme in digestion.
cell. The unit of structure of plants and animals. The essential feature of a cell is its living protoplasm, surrounded in plant cells by a wall.
cell division. A process of division whereby cells reproduce.
cell sap. The liquid in the vacuoles of plant cells.
cell theory. Theory that all organisms are composed of cells.
cellulose. The carbohydrate substance found abundantly in most plant cells.

cell wall. The outer covering of a cell; in plants it is usually rigid and contains a high quantity of cellulose.

Cenozoic. An era of geological history extending from the beginning of the Tertiary to the present.

centimeter. A linear unit of measure in the metric system; there are ten millimeters in a centimeter and 100 centimeters in a meter.

central cell. In development of an archegonium, the cell which gives rise to the primary neck canal cell and a primary ventral cell.

central placentation. The placenta is borne on the axis of an ovary made up of but one carpel.

centripetal. Developing or maturing from the outside toward the center, as maturation of metaxylem.

centromere. A small body or constriction on a chromosome where the latter is attached to a spindle fiber.

cesium. A soft silvery metallic element, the most electropositive of the elements.

chalaza. The portion of an ovule just below the place where the integuments are united to the nucellus.

chaparral. An environment comprised principally of thorny shrubs or dwarf trees.

chemoautotrophic. Organisms capable of utilizing the energy released in chemical reactions involving inorganic compounds in synthesizing their food and protoplasm.

chemosynthesis. The synthesis of food in certain bacteria through energy released by such chemical reactions as the oxidation of certain inorganic compounds.

chemotropism. A growth movement in response to a chemical stimulus.

chiasma. The twisting and fusion with possible crossing over at the point of contact of chromosomes.

chlamydospore. A thick-walled resting spore produced in certain fungi; in the smuts it is a transformed vegetative cell.

chlorenchyma. Parenchyma tissue containing chloroplasts.

chlorophyll. The green coloring substance in plants.

chlorophyllous. Containing chlorophyll.

chloroplast. A specialized structure in cytoplasm containing chlorophyll.

chlorosis. A yellowing of green plants due to a loss of chlorophyll.

cholesterol. A white, fatty, crystalline alcohol ($C_{27}H_{45}OH + H_2O$).

chromatid. A chromosome half, formed in the prophase of mitosis by a longitudinal splitting of a chromosome.

chromatin. The granular, deep-staining protoplasmic material in the cell nucleus.

chromatophore. A chlorophyll containing body in the cells of the more primitive algae.

chromoplast. A plastid containing red, orange, or yellow pigments.

chromosomal aberrations. Changes resulting from breaks in the chromosomes and abnormal fusion of the broken ends, causing a change in structure.

chromosome. A nuclear body of a definite structure and number formed from chromatin and bearing hereditary determiners or genes.

chromosphere. A ruddy gaseous layer of incandescent hydrogen, helium, etc., surrounding the sun and between it and the corona (the latter is defined as a luminous envelope surrounding the sun, beyond the chromosphere).

circinate. Manifesting an unrolling development, like fern leaves.

circumnutation. A movement of rotation in growing stem tips.

cladophylls. Stems which resemble thick leaves, as pads of certain cacti.

clamp connection. The connection, tubular in form, between two cells in the Basidiomycetes, providing a pathway for nuclei to achieve the dikaryon phase.

class. A plant group which is immediately above an order in status but below a phylum.

cleistothecium. A closed ascocarp.

climax. A relatively permanent biome or association which maintains itself with little change in a given region so long as there are no major changes in environmental conditions.

clone. Any of the asexual offspring of an individual which is identical with the individual's genetic makeup.

club mosses. Members of the Lycopsida.

coalescent. To grow together or combine into one body.

coccus. A spherical bacterium.

coenocyte. The plant body in certain algae and fungi in which the vegetative filament is multinucleate and divided into separate cells.

coenzyme. A substance necessary for the activity of an enzyme.

cofactors. A coenzyme (see above).

cohesion. Molecular attraction by which the particles of a body are united throughout the mass; as the substances in a xylem element.

coir. A stiff fiber extracted from the husk of a coconut.

coleoptile. In some seed plants, a protective sheath over the young stem tip.

coleorhiza. In some seed plants, a protective sheath over the root tip of an embryo.

collateral bundles. Vascular bundles in which the phloem is external to the xylem.

collenchyma. A strengthening tissue composed of cells with walls usually thickened at the angles of the walls.

colloid. A substance in which finely divided particles, larger than in a solution but smaller than in a suspension, are dispersed through a continuous medium which may be liquid, solid, or gas.

colony. A group of similar organisms living together in close association; usually a group of associated unicellular organisms among which there are no marked structural differences and little or no division of labor.

columella. A sterile central axis in the sporangia of certain fungi; also, the sterile central tissue in the sporophytes of liverworts and mosses.

columnar. The type of growth chiefly in palm trees where the leaves are apical and the trunk forms a tall unbranched column.

compact body. A region where the cells in tissues are arranged close together with a minimum of intercellular space.

companion cell. One of the elongated cells adjoining sieve tubes in phloem tissue of most plants.

complete flower. A flower which bears four types of floral organs; sepals, petals, stamens, and carpels.

compound leaf. A leaf whose blade is subdivided into several distinct parts.

conceptacle. In algae such as *Fucus* a cavity with an apical pore and containing the reproductive bodies.

condiment. A seasoning for foods.

cone. A specialized branch bearing an aggregation of sporophylls.

conidiophore. A structure which bears conidiosporangia.

conidium. A conidiospore.

coniferous. Having cones as reproductive structures.

conjunctivitis. An inflammation of the mucous membrane which lines the inner surface of the eyelid.

conjugation. The fusion of two similar gametes, as in the Conjugales.

contractile vacuole. One or more spherical bodies forming in the body of certain primitive algae which are considered regulatory in water control and possibly play a role in the elimination of waste materials.

convergent evolution. Evolution of unrelated or distantly related groups of organisms along similar lines, resulting in the development of similar traits or features in the unrelated groups.

coprophilous. Living on a dung substrate, as certain fungi.

cordate. Heart-shaped.

cork. A suberized tissue formed on the stem and root surfaces (sometimes on other parts) chiefly of woody plants from the cork cambium.

corm. A short, often globose, upright, underground stem which stores food. It differs from a bulb in that the latter consists chiefly of fleshy storage leaves growing from a small stem, whereas a corm is chiefly stem tissue.

corolla. Collectively, the petals of a flower.

corona. Crown.

corpus. The histogen forming the portion of the growing tip; often, all the undifferentiated tissue within the tunica.

cortex. Tissue lying between the epidermis and pericycle of a stem or root.

corticating tissue. A protective tissue which retards loss of water in transpiration. In some plants it may be a multicellular layer; in other plants it may be only a cutinized epidermal layer.

cosmos. The universe conceived as an orderly and harmonious system.

cotyledon. A food-digesting and food-storing part of any embryo; a seed leaf.

cover cell. In the development of an archegonium, one of the two cells formed by division of the primary axial cell.

crenate. The margin of a leaf cut into small rounded scallops.

Cretaceous. Geological period of the Mesozoic era represented by deposits frequently made up of rocks containing chalk.

cross fertilization. The union of gametes from two different plants or animals.

crossover. In meiosis the exchange of parts between two homologous chromatids.

cross pollination. The transfer of pollen from a stamen to the stigma of a flower on another plant.

crozier. The apex of a young fern leaf; a formation in the Ascomycetes.

crustose. Manifesting a crustlike habit of growth, as do certain lichens.

cucurbit. A member of the gourd or pumpkin family, cucurbitaceae.

culm. The jointed stem of a grass.

cuneate. Wedge-shaped, as an apical cell with four cutting faces.

cupule. The asexual body on the thallus of certain liverworts which produce gemmae.

curare. The extract of vine *(Strycnos toxifera)* used as an arrow poison by South American Indians; in medicine it is used as an antispasmodic.

cuticle. A waxy layer formed on outer walls of epidermal cells and derived from them.

cutinization. Process of impregnating cell walls with cutin.

cuttings. A section of a plant, usually a stem, used in propagation.

cyme. A flower cluster in which the central apex ceases growth early but lateral branches keep forming towards the base and grow upwards and outwards, forming a flat-topped cluster.

cyst. Various spore-like structures produced by certain plant groups.

cystidium. A large sterile structure in the hymenium of a Basidiomycete.

cystocarp. In the red algae, a spore-producing body formed by the carpogonium.
cytology. The science of cell structure and function.
cytopharynx. A minute portion of a unicellular organism which may function in the intake of solid foods.
cytoplasm. All protoplasm in a cell except the nucleus.
cytosine. An amino acid; a cleavage product of nucleic acids; $C_4H_6N_3O$.
cytostome. The mouth of a unicellular organism.

daughter cell. A cell newly formed as a result of the division of a parent cell (mother cell).
day-neutral. Plants which bloom under either short or long photoperiods.
deciduous. (Of plant parts) falling off within a year of the time of production; producing leaves which fall regularly each year, as trees other than evergreens.
dehiscent. Splitting open along more or less definite lines at maturity, as a fruit pod or moss capsule.
deliquescent. To divide apically into many branches, as the broadly flat topped branching of an elm tree and many other deciduous species.
dendrochronology. The study of tree rings in relation to history.
dendroid. An organism with a branching habit, resembling a tree.
dentate. A margin of a leaf with sharp-pointed obtuse teeth directed outward.
deoxyribonucleotide. A five carbon sugar with one oxygen atom less than ribose.
dermatogen. The outermost histogen of a growing root or stem tip, which differentiates into the epidermis.
dermatitus. An inflammation of the skin.
desert. An arid region receiving less than fifteen inches annual rainfall.
deuterium. The heavy hydrogen isotope having a symbol of H^2 or D.
Devonian. The geological period following the Silurian and preceding the Carboniferous.
diaminoimelic acid. An organic acid found only in certain bacteria and blue-green algae.
diarch. Having two arms, as the xylem in a stele.
diastase. A starch-digesting enzyme.
diatomaceous earth. The soil residue formed from the siliceous valves of diatoms.
dichotomy. A type of branching in plants in which the main axis or stem forks repeatedly into two branches, as in certain seaweeds, liverworts, etc.
dicotyledonous. Having two cotyledons or seed leaves.
dictysostele. A stele consisting of separate vascular bundles, or of a network of bundles, produced by overlapping leaf gaps.
differentiation. The formation of tissues from the ground or lateral meristems.
diffusion. The movement of molecules from a region of greater concentration to a region of lesser concentration; ultimately reaching equilibrium.
digestion. The transformation of insoluble or complex foods into soluble or simpler substances, through the action of enzymes.
dihybrid. The offspring of parents differing in two characteristics; a genetic problem involving two pairs of homologous chromosomes.
dimorphism. Having two forms or types, as in the life cycle of certain algae with different gametophytic and sporophytic generations.
dinoflagellate. A member of the class Dinophyceae.
dioecious. Having the male and female organs on separate plants (or animals).
diplococcus. Bacteria that occur in groups of two.
diploid. Having the double chromosome number, as is characteristic of cells of the sporophyte.
diplotene. The stage of the meiotic prophase immediately following pachytene during which the homologous chromosomes tend to repel one another.
disaccharide. A double sugar such as sucrose or lactose.
disease. An abnormal condition of sufficient duration to produce marked derangements in the structure or activities of an organism or its parts.
DNA. Deoxyribonucleic acid; the essential constituent of chromatin.
dominance. A genetic principle, established by Mendel, that when genes productive of two contrasting characters are brought together in a hybrid as the result of a cross, only one character (dominant) may appear in the hybrid, while the other is latent (recessive); the role of the dominant character.
dominant. In genetics, the hereditary character resulting from a gene with greater biochemical activity than another termed recessive.
dormancy. A period of reduced physiological activity occurring in seeds and buds; resting period.
dorsal. The upper or outer surface of a plant organ. In a leaf, the outer surface in position of development is the dorsal surface even though at maturity it is usually the under surface when expanded.
dorsiventral. Having an upper surface structurally different from the lower surface. Many leaves, fern prothalli, and liverwort thalli exhibit this type of differentiation.
double fertilization. The process in angiosperms in which one sperm unites with the egg and another sperm unites with the fusion nucleus to form the primary endosperm nucleus.

drug. A substance, often of plant origin, used in the treatment of disease.

drupe. A simple fleshy fruit in which the endocarp (inner wall of ovary) becomes hard and stony and encloses one or two seeds.

ecology. The division of biology dealing with the relation between organisms and their environment.

ecosystem. A biotic environment of small or great area in which the organisms live together as an orderly organized unit.

ectophloic. Having outer phloem only.

ectotrophic. Referring to fungi and bacteria which live near or on the outer surfaces of roots and root hairs.

edaphic. Pertaining to the soil. Roots are edaphic structures since they grow in soil.

egg. Female gamete.

elater. Elongated structure which is produced in the sporangium of liverworts and on the spores of horsetails; it aids in spore dispersal.

elaterophore. Structure to which the elaters are attached.

electron. A constitutent of an atom; they constitute cathode and beta rays, and are emitted by hot bodies; the part of an atom having a negative charge.

element. Any one of a limited number of distinct varieties of matter which singly or in combination compose substances of all kinds.

elongation. Applied particularly to the portion of the stem tip or root tip in which the cells lengthen much more rapidly and extensively than elsewhere.

embryo. A young sporophytic plant before the beginning of its rapid growth; the "germ" of a seed.

embryogeny. Development of the embryo.

embryology. The study of the embryo.

embryo sac. The structure containing the embryo, usually restricted in meaning to the megagametophyte.

emergency. An outgrowth of epidermal or cortical tissue, as the spine of a cactus.

emulsoid. A colloidal system in which both phases are liquids.

enation. Outgrowth from a previously smooth surface, as the outgrowth of leaf bracts from a stem surface.

endarch. Where the first-formed xylem elements arise near the inner limit of the procambial strands.

endocarp. Inner layer of fruit wall.

endodermis. A layer of cells just inside the cortex, most conspicuous in roots.

endogenous. Having an internal origin.

endophyte. Growing within the cells of a plant host.

endoplasmic reticulum. The portion of cytoplasm which forms a channeling network for ready diffusion of solvents.

endosperm. A food storage tissue in seeds.

endospore. A spore formed within a cell.

endothecium. Inner tissue of sporophyte which gives rise to spores in some cases.

endotrophic. Feeding internally, as a fungus living inside the tissues of a host.

energy. Capacity for doing work; especially the generation of heat and chemical activity.

enzyme. An organic catalyst, manufactured by living protoplasm and controlling digestion and other physiological processes.

Eocene. The earliest period of the tertiary division in the Cenozoic era.

eon. A grand division of geological time.

epicotyl. The portion of an embryo axis above the attachment of the cotyledons. The growing apex of the epicotyl forms the young stem of a plant.

epicure. One who is fastidious in his food tastes.

epidermis. The surface layer of roots, stems, leaves, and other organs.

epigyny. A condition in which the ovary is imbedded in the receptacle, so that the other floral organs seem to originate at a point above the ovary.

epilimnion. The layer of water that overlies the thermocline of a lake and is subject to the action of wind.

epiphyte. A plant which gains physical support, by growing on it.

epistasis. The suppression of the effect of a gene by a nonallelic gene.

epitheca. The larger, coverlike valve of a diatom.

epithet. The part of a taxonomic name designating a species or lesser division of a genus.

epizoic. Growing on an animal, like algae on the back of a turtle.

ER. *see* endoplasmic reticulum

ergosterol. A sterol first prepared from ergot.

ergot. A disease of rye and other cereals due to a fungus; the fungus is used as a drug.

ester. A compound formed by the replacement of acid hydrogen by a hydrocarbon radical.

ethyl acetate. A colorless liquid, $CH_3CO_2C_2H_5$, used as a solvent, in organic synthesis, and in flavoring.

eusporangiate. Characterized by the development of a sporangium from the inner segment of the initial cell.

eustele. Vascular stem structure with either collateral or bicollateral bundles arranged in a cylinder, with anastomosing vascular strands.

evergreen. A plant with leaves persisting more than a year; or leaves, not abscissing seasonally.

evolution. The history of development of a race or species or larger biological group; phylogeny.

exarch. (Of xylem) characterized by a centripetal development, that is, toward the center of the stem.
excurrent. In trees, strong apical dominance resulting in an unbranched main stem as in conifers.
exine. Outer wall of a spore or pollen grain.
exocarp. Outermost layer of a fruit wall.
exogenous. Growing from or on the outside, or by addition to the exterior.
exudate. A substance secreted by a gland, or pore, or incision.
eyespot. The eyelike spot of pigment in the interior region of the body of certain unicellular algae, occasionally found in flagellated reproductive bodies.

facultative parasite. An organism capable of living either as a parasite on another living organism or of growing on dead organic matter.
family. A taxonomic category ranking between certain genera and an order. A family is composed of a group of genera.
fascicle. A cluster of leaves, as the leaves of *Pinus.*
fat. A type of food consisting of carbon, hydrogen, and oxygen in which the amount of oxygen is very small compared to carbohydrates.
fatty acid. A saturated acid ($C_nH_{2n}O_2$) such as stearic or palmitic acid; an end product of fat digestion.
fermentation. The conversion of organic substances into simpler substances under the influence of enzymes with little or no oxygen involved.
fertile. Having the potentiality of reproduction.
fertile spike. A strobilus or inflorescence with reproductive structures.
fertilization. The essential feature of sexual reproduction; the fusion of one gamete with another.
fertilization tube. A tube originating from one cell, usually a male gametangium, and connecting with another cell, usually a female gametangium, through which the gametes (nuclei) are transferred.
fiber. Elongated, thick-walled cells in various plant tissues.
fibrous. Containing fibers, generally stringy in texture.
fibrous root. A root system with numerous fine roots, as in grasses.
filament. A colony of cells in linear series, as in certain algae; the stalk of a stamen.
filterable. Capable of being filtered, as viruses passing through a filter.
filtrate. To filter; the liquid or substance which has been passed through a filter.
fission. Reproduction of a unicellular organism by division into two equal cells.
flagellum. A long whip-like appendage of a cell, usually playing a role in locomotion.
floret. A small flower such as found in members of the composite and grass families.
floridean starch. A carbohydrate reserve food of red algae.
florigen. The name suggested for growth regulators concerned with flowering.
flower. The characteristic reproductive structure of angiosperms, interpreted morphologically as a specialized branch system.
flower bud. A bud which grows into a flower or flowers.
fodder. Plant material eaten by herbivorous animals.
foliaceous. Leaflike; having an abundance of leaves.
foliar. Leaflike; consisting of leaves.
foliose. Leafy.
follicle. A simple, dry, dehiscent fruit, producing several to numerous seeds and composed of one carpel, which splits along one side.
foot. In mosses, liverworts, ferns, and other plant groups, a portion of a young sporophyte which attaches the sporophyte to the gametophyte and often absorbs food from the latter.
forage crop. A field crop raised specifically for domesticated herbivorous animals.
forest. A type of biome in which the dominant plants are trees.
formaldehyde (formalin). A colorless gas, HCHO, with a strong odor. Aqueous solutions are used as a preservative or disinfectant.
form genus. A name applied to plant parts of certain fossils until a more complete identification is made.
fossil. An impression, a trace, or remains of a plant or animal from a former geologic age.
fragmentation. Reproduction by a segment of the main plant, a form of asexual reproduction.
frankincense. An incense resin from the Burseraceae family.
frond. A very large leaf, especially a fern or fernlike leaf.
fructification. Any fungal structure which contains or produces spores.
fructose. Fruit sugar; a monosaccharide such as levulose.
fruit. A matured ovary, or cluster of matured ovaries.
frustule. The siliceous shell of a diatom, consisting of two valves, one overlapping the other.
fruticose. Having the form of a shrub; branching freely, as certain lichens.
fucoid. Resembling *Fucus* in appearance; having a flat, dichotomously branched thallus.
fucoxanthin. The brown pigment of certain algae.
fungicide. A substance that destroys fungi or inhibits the growth of fungus spores or hyphae.
fungi imperfecti. A group of fungi in which the sexual phase of the life cycle has not been observed and therefore cannot be perfectly classified.

fungus. A thallophyte without chlorophyll, usually filamentous. True fungi belong to the phylum Eumycophyta.

funiculus. The stalk which attaches an ovule to the inner portions of an ovary.

fusion. Blending together as the union of gametes; the nuclei of chemical elements unite with a release of energy, as in a hydrogen bomb.

fusion nucleus. The nucleus resulting from the union of two others, particularly of the two polar nuclei in angiosperms.

galactose. A white crystalline monosaccharide.

galaxy. An organized celestial assemblage, such as the Milky Way.

gall. Abnormal tissue enlargement due chiefly to infectious microorganisms or insects.

gametangium. A structure which contains or bears gametes.

gamete. A sex cell; an egg or sperm.

gametogenesis. The formation of gametes.

gametophore. A structure or stalk bearing gametes.

gametophyte. A plant producing gametes.

gamma rays. Radiation similar to X rays but of shorter wavelength, emitted during some nuclear transformations.

Geiger counter. A mechanical device with a needlelike electrode for the detection of ionizing particles such as cosmic ray particles and radioactivity.

gel. A colloid in which a liquid is dispersed in a solid phase; a jelly-like colloid.

gemma. An asexual or vegetative outgrowth of a parent body, capable of growing into a new individual directly, as in liverworts.

gene. A hereditary determiner, located on a chromosome.

generative cell. The cell or nucleus in a pollen tube which upon division forms two sperms.

genetics. The science of heredity.

genotype. The genetic constitution of an organism; the particular set of genes in the cells of an organism.

genotypic selection. Selection of progeny in a pedigree based on the known genotypes of the parents.

genus. A taxonomic category which includes one or more species. The genus name is the first name in the binomial.

geotropism. A growth movement in response to gravity.

germicide. A substance which destroys germs (microorganisms).

germination. The growth or sprouting of a zygote, spore, or seed to form a new plant.

germling. Early development of a plant from germination of zygote in certain Thallophytes.

germ tube. The tube formed by germination of a pollen grain through which the sperms are transferred.

gibberellin. A growth regulator which causes elongation of the internode in certain plants.

gills. The thin fruiting plates on the underside of the cap of a mushroom.

glaciation. Ice sheet; scoring or erosion of rocks by movement of glaciers.

gleba. The inner portion of the basidiocarp in the Gasteromycetes.

globule. (Formerly) The antheridium of the Charales.

glucose (dextrose). A monosaccharide sugar.

glume. One of the two empty bracts at the base of the spikelet in grasses.

glycerol. A sweet, sirupy trihydroxy alcohol ($C_3H_5(OH)_3$); a breakdown product of fat digestion.

glycine. An amino acid, $NH_2CH_2CO_2H$.

glycogen. A carbohydrate reserve related to starch and dextrin.

glycoside. A compound, which yields sugar following hydrolysis.

glycolysis. Decomposition of glucose or glycogen by hydrolysis.

glycophyte. A term occasionally used to describe plants which live in the usual fertile soils, as opposed to halophytes.

golgi structures (bodies). A cytoplasm component believed to be the site where high energy compounds from the mitochondria collect.

gonimoblast. One of many short filaments arising from the fertilized carpogonium and giving rise to carpospores in the cystocarp.

graft. The union of separate plants, the stock and scion, resulting in permanent fusion.

grafting. The joining of two plant parts, usually stems, so that their tissues grow together.

grain. The fruit of a grass, especially of a cereal grass.

gram. A metric unit, defined as one thousandth of a kilogram.

gramstain. A differential stain, utilizing iodine, alcohol, and gentian violet, for bacteria.

grana. The minute, discrete, green bodies imbedded in a chloroplast.

granitic rocks. Igneous rocks containing quartz, feldspar, and mica.

graphite. Soft, black native carbon of metallic luster.

gravitational water. Water which moves downward in soils in response to gravity.

ground meristem. Undifferentiated isodiametric cells in stem and root tips.

growth regulator. A substance active in controlling growth and development in plants, referred to as auxins or hormones.

guanine. A base, $C_5H_5N_5O$, originally found in guano, but best known now as one of the four bases in the DNA and RNA macromolecules.

guard cell. One of the paired epidermal, chlorophyllous cells which enclose a stomate.

gullet. A tubelike structure leading to the interior of such flagellated organisms as *Euglena.*

gum. A colloidal, glutinous substance exuded by or extracted from certain plants.

gynandrium. Having the sexual parts, as in orchids, united into a column.

habitat. The natural abode of an organism; the particular location where it normally grows or lives, such as the desert or plains.

halophytes. A plant which grows naturally in soil or water impregnated with salts, such as brakish water along seacoasts and alkaline deserts.

haploid. Characterized by cells with the single or reduced number of chromosomes found in mature gametes, in contrast to the diploid number found in sporophytic cells.

haplostele. A protostelic arrangement in which the central xylem is circular in cross section.

hardiness. Hardy quality or state; physical vigor; robustness; endurance.

hardwood. The wood of angiosperms such as oak, cherry, maple.

haustorium. An absorbing organ originating from a hypha of a parasite and penetrating into the tissues of a host.

head. A type of inflorescence in which numerous small flowers are densely crowded upon a usually disc-shaped common receptacle as in a sunflower.

heartwood. The hard central part of the trunk of a dicotyledonous tree.

helium. An inert, colorless, gaseous element, first observed spectroscopically in the sun's atmosphere. Next to hydrogen it is the lightest of gases and because of its noninflammability is separated from natural gas for use in dirigibles, etc.

heme (haem). A blood substance oxidizing to haematin; $C_{34}H_{32}O_4N_4Fe$.

hemicellulose. A carbohydrate found in many plant cell walls.

hemoglobin. The respiratory pigment in the red corpuscles of vertebrates, a compound of hematin and globin.

herb. A plant which does not develop much woody tissue. The tops die down to the ground each year.

herbaceous. A plant with a minimum of woody tissue that remains soft and succulent.

herbarium. A collection of dried plants, usually mounted and classified; also the room or building where the collection is kept.

herbivore. An animal feeding chiefly on herbage, especially any of a group (Herbivora) of mammals, mostly feeding on herbage.

heredity. The tendency of an organism to resemble its parents; the transmission of genetic characteristics from parents to progeny.

hesperidium. A type of berry with a distinct rind formed from the fused mesocarp and exocarp and an inner juicy endocarp, such as an orange.

heterocyst. An enlarged resting cell of certain members of the Cyanophyta. Usually it is transparent and thick-walled.

heteroecious. Requiring two hosts to complete the life cycle, as a parasite.

heterogamous. Reproducing by a union (heterogamy) of unlike gametes, one of which is larger than the other.

heteromorphic. Manifesting a type of alternation of generations in which the gametophytic plant is distinctly different from the sporophytic plant.

heterosis. Increased vigor or capacity for growth often displayed by crossbred animals or plants.

heterosporous. Producing spores of two types.

heterothallism. The condition of a species in which the individuals are self-sterile and therefore two compatible thalli are required for sexual reproduction, even though both male and female organs may occur on a single individual.

heterotrophic. Obtaining food from organic matter, and incapable of photosynthesis, in contrast to autrophic.

heterozygous. Containing one or more gene pairs made up of dissimilar members.

hexagonal. Six-sided, as a cell.

hilum. 1) The mark at the point of attachment of an ovule forming the "eyes" as of a bean. 2) The nucleus of a starch grain.

histidine. A crystalline basic amino acid, $C_6H_9N_3O_2$, formed in the splitting of proteins.

histogen. Tissue layers in early development, especially stem or root tips before differentiation into the primary tissues.

histology. 1) That branch of science which deals with the minute structure of animal and vegetable tissues. 2) The tissue structure or organization, as of an organism.

holdfast. The basal attaching cell or organ of an alga or fungus, often composed of rhizoids.

holozoic. Ingulfing solid food particles; obtaining food in the manner of animals.

homologous. Of similar origin, but not necessarily of similar function, as limbs in different species.

homosporous. Producing spores of a single type.

homothallism. The type of plants in which sexual reproduction takes place in a single thallus. The plants are self-compatible.

hormogonium. In certain blue-green algae a portion of the filament that becomes detached because of dead cells or heterocysts, but continues to grow into a new filament.

hormone. A substance, esp. a specific organic product of the cells of one part, as of the adrenal glands, transported in the body fluid or the sap of an organism and producing a specific effect on the activity of cells remote from its source; an internal secretion.

host. A plant or animal which affords nourishment to a parasite, usually within its tissue.

humus. A brown or black material formed by the partial decomposition of vegetable or animal matter in or on the soil; the organic portion of the soil.

hybrid. The offspring of the union of a male of one race, variety, species, genus, etc., with the female of another; a crossbred animal or plant.

hybridization. The production of progeny by interbreeding plants of different types or species.

hybrid vigor. *see* heterosis.

hydrocarbon. A compound containing only hydrogen and carbon, as acetylene, benezene, etc.

hydrogen. The hydrogen atom is the simplest of all atoms; the ordinary isotope (H^1) consists of a single proton in the nucleus and a single electron outside of the nucleus.

hydrolysis. A chemical process of decomposition involving the addition of the elements of water.

hydrophyte. A plant which lives in water or on exceedingly wet soils.

hydrophytic. Of a plant which grows in water or in saturated soil.

hydrotropism. A tropism in which water or water vapor constitutes the orienting factor, as in many plant roots.

hygroscopic. Absorbing or attracting moisture from the air with resultant swelling or moving.

hygroscopic moisture. The very fine film of water adhering to soil particles in an air dry soil; this water is unavailable to plants.

hymenium. A layer of cells from which the asci and basidia arise.

hyperplasia. Excessive increase in number of cells in an organ or tissue.

hypertrophy. Excessive enlargement of cells in an organ or tissue.

hypha. One of the branches or filaments of a mycelium.

hypocotyl. The portion of an embryo axis below the attachment of the cotyledons. The growing tip of a hypocotyl forms the primary root of a seedling.

hypodermis. Tissue beneath the epidermis, often considered a part of the epidermal tissue.

hypogeny. The situation in a flower in which the ovary is above and free of the receptacle and in which the sepals, petals, and stamens are attached below the ovary.

hypolimnion. The water between the thermocline and bottom of lakes; the opposite of epilimnion.

hypotheca. The lower or inside valve (frustule) of a diatom.

hypothesis. A tentative theory or supposition provisionally adopted to explain certain facts and to guide the investigation of others.

IAA. Indoleacetic acid.

imbibition. The property in solid substances of taking in liquids, as in absorption.

imbricate. Overlapping in a regular pattern, as the leaves of a moss or bracts of an involucre.

imperfect flower. A flower which bears stamens or carpels, but not both.

impression. A type of fossil; an imprint of a plant or animal structure in rock.

inbreeding. Breeding or mating of closely related individuals or stocks, to preserve or fix favorable, or eliminate unfavorable characters.

incomplete dominance. In genetics the lack of a dominant characteristic in homologous chromosomes; usually a blending of the factors.

incomplete flower. A flower which lacks one or more of the four kinds (sepals, petals, stamens, carpels) of floral organs.

indehiscent. Not splitting open along regular seams or sutures, as in a fruit.

independent units. In genetics each characteristic may be independent of any other; one of the points of Mendelism.

indigenous. Native to a region, opposed to introduced; as a plant species.

indusium. The membranous cover of a fern sorus.

inferior ovary. *see* epigeny.

inflorescence. A flower cluster; the pattern in which flowers are arranged on an axis.

infrared. Pertaining to or designating those rays lying just beyond the red end of the visible spectrum, such as are emitted by a hot nonincandescent body. Their wavelengths are longer than those of visible light and shorter than those of radio waves.

insecticide. An agent or preparation for repelling or destroying insects.

integument. The coat of an ovule.

intercalary. Occurring between cells or regions of a different kind; as an intercalary meristem, opposed to apical meristem.

intercellular. Between cells.

internode. The length of stem between two nodes or joints.

interphase. In cell growth and division the period between telophase and prophase; commonly referred to as a resting period.

intine. The inner lining of the cell wall of a pollen grain or spore.

inversion. Conversion of a substance showing dextrorotation into one showing levorotation, or vice versa. Thus, cane sugar, which is dextrorotatory yields by hydrolysis or inversion a levorotatory mixture of dextrose and levulose; the reversal of genes on a chromosome.

invertase. One of several enzymes capable of digesting double sugars.

invertebrate. An animal without a backbone or spinal column.

in vitro. In glass; hence, in a test tube or beaker.

involucre. Plant-enveloping tissue; as a collective term, the floral bracts of a head-type inflorescence.

iodine. A nonmetallic element of the halogen group, isolated as a shining, blackish-gray crystalline solid of peculiar odor.

ion. An electrically charged atom or group of atoms.

ionization. To convert or be converted (wholly or partly) into ions.

ionosphere. That part of earth's atmosphere beginning at an altitude of about 25 miles and extending outward 250 miles or more. Contains electrically charged particles.

irregular flower. *see* zygomorphic.

iron. A silver-white metallic element found almost universally in combined forms; it constitutes about 5% of the earth's crust.

isogamete. A gamete with no apparent morphological difference from the gamete with which it unites; a like gamete.

isogamous. Reproducing by union of gametes of equal size and structure.

isomorphic. Manifesting a type of alternation of generations in which gametophytic and sporophytic plants are similar in appearance vegetatively.

isotopes. Any two or more forms of the same element, occupying the same position in the periodic table (that is, having the same atomic number), closely similar in chemical behavior, but distinguishable by radioactive transformations, differences in atomic weight, etc.

isthmus. A passage or channel connecting two similar halves or cavities, as in the desmids.

Jurassic. The mid-Mesozoic geological period following the Triassic and preceding the Cretaceous.

karyogamy. The fusion of two nuclei.

kelp. Large brown seaweeds of the Laminariales.

ketone. A compound containing the carbonyl group (CO) doubly united with carbon, as with two hydrocarbon radicals, with a single bivalent radical, or with derivatives of these.

ketose. *see* ketone.

kilogram. A unit of mass and weight; equal to 1,000 grams or 2.2046 pounds.

kilometer. One thousand meters or 3,280.0 feet or about five-eighths of a mile.

kinetic. Of, pertaining to, or due to motion or action.

Krebs cycle. A series of oxidation-reduction reactions whereby, step by step, hydrogen and carbon dioxide are liberated from a series of organic acids through the action of various enzymes in the aerobic oxidation of pyruvic acid in respiration or similar metabolic activities.

lactic acid. A colorless sirupy acid, $CH_3CH(OH)CO_2H$.

lactobacillus. Any of genus *(Lactobacillus)* of lactic-acid forming bacteria.

lamella. The layer (middle lamella) of intercellular material, composed chiefly of pectins, that cements together adjacent cells.

lamina. The broadly expanded portion of a leaf, the blade.

laminarin. A dextrinlike carbohydrate forming the food reserve in the blade of brown algae.

lateral bud. An axillary bud. It may be a branch or flower bud.

lateral conjugation. The union of gametes of two adjacent cells in the same filament.

latex. A milky, usually white, fluid found in certain cells in some families of seed plants. The latex contains various gum resins, fats, wax, and often a complex mixture of other substances.

layering. A method of propagation often used in vines and shrubs whereby a branch is bent down, covered with soil and allowed to root, then cut off.

leaflet. One of the several blades of a compound leaf.

leaf scar. A scar left on a twig following the fall of a leaf.

leaf trace. A branch of the vascular tissues of a stem, extending out into a leaf.

lecithin. One of several complex nitrogenous substances; used especially for its phosphorous content.

legume. A simple, dry dehiscent, one-carpeled fruit splitting along two seams (sutures); also, a member of the legume family, the plants of which have such fruits.

lemma. In grasses, the lower of the two bracts enclosing the flower.

lenticel. In woody stems (and other plant parts), a pore surrounded by many cells through which exchange of gases occurs. In woody stems, lenticels occur in bark.

leptosporangiate. Characterized by the development of a sporangium from a single cell in which the outer segment is formed by division of a superficial cell.

leptotene. Early stage of the prophase of meiosis where chromatin is in the form of fine threads.

lethal. Fatal or pertaining to death; in genetics one or more genes may occur in a chromosome complement, such as a factor for albinism in plants, which carry a lethal factor.

leucoplast. A colorless plastid.

leucosin. A proteinlike food reserve in certain algae.

lichen. Any of several plant forms consisting of a combination of an alga and a fungus in such close association as to form what appears to be a single individual.

lid. The cap (operculum) of a moss capsule.

lignification. The impregnation of cell walls with lignin, most commonly in cells of woody tissue.

lignin. A substance often associated with cellulose in the cell walls of wood.

ligule. The membranous appendage projecting from the leaf sheath in grasses, etc.

limestone. A sedimentary rock which may be impregnated with fossils. The rocks are usually high in calcium carbonates.

limnetic. Living in or pertaining to lakes, etc; zone of deep water between surface and compensation depth (limit of depth below which plants lose more by respiration than they gain by photosynthesis).

linkage. The tendency of certain genes to be inherited together because of their location in the same chromosome.

lipase. An enzyme that digests fats into fatty acids and glycerol.

lip cell. One of the cells (usually two) formed along the line of dehiscence in a fern sporangium.

lipid. Any of a group of substances comprising the fats and other esters that possess analogous properties; they are insoluble in water but soluble in fat solvents.

littoral. A sea shore.

liverwort. A member of the class Hepaticae, having a flat, lobed thallus.

lobed. Divided by clefts, as a maple leaf.

locule. A cavity in a sporangium or ovary.

loculicidal. A fruit capsule which dehisces by means of a dorsal suture.

long day. A photoperiod expression in which a plant blooms when the length of the continuous light period is over 12 hours.

lorica. A protective covering in certain flagellated organisms.

luminescence. An emission of light; to shine or glisten.

lysis. The dissolution or breaking down of cells by lysins, such as an antibody.

macandrous. Applied to algae such as *Oedogonium* in which the antheridia appear on long or regular-sized filaments.

macromolecule. Large organic molecules such as that of DNA in which a number of substances are combined in the molecule.

magnesium. A silver-white metallic element; the element forms a part of the chlorophyll molecule.

main axis. The central or major line of growth, as the primary root-stem axis.

manganese. A grayish-white metal with a reddish tinge; it is one of the important minor elements in plant growth.

manila hemp. A bananalike plant.

mannitol. A white crystalline alcohol stored in certain plants.

mannose. A monosaccharide sugar obtained by the oxidation of mannitol.

maquis. An environment usually described as a chaparral.

marginal plants. Plants able to survive on minimal substrates in relatively poor environmental conditions. Rarely would plants of this type become dominant in any large area.

marl. A sedimentary deposit composed principally of clay and carbonates.

matter. That which is considered to constitute the substance of the physical universe; anything which occupies space.

maturation. The process in growth of becoming mature.

mechanical tissue. Supporting tissue, usually fibrous or woody.

medulla. A central tissue; pith.

medullary ray. An extension of pith cells between the vascular bundles.

megagamete. A term used to define the egg or large gamete in heterogamous and oogamous plants.

megagametophyte. A female gametophyte resulting from the growth of a megaspore and producing female gametes or eggs; the embryo sac of angiosperms.

megasorus. A sorus containing only megasporangia, and therefore producing only megaspores.

megasporangium. A sporangium containing only megaspores.

megaspore. A spore which gives rise to the megagametophyte.

megaspore mother cell. A cell which gives rise by meiosis to megaspores.

megasporophyll. A leaf or similar structure which bears megaspores.

megastrobilus. A cone or strobilus bearing megasporangia.

meiosis. Reduction division; a process in which a pair of nuclear divisions occurs successively, reducing the number of chromosomes by one-half.

meristem. A mass of growing cells, capable of frequent cell division.

mesarch. Characterized by primary xylem in which maturation of the metaxylem is both centripetal and centrifugal, that is, both toward the center and the outside of a stem or other organ.

mesocarp. Middle layer of a fruit wall.

mesome. In the telome theory, the internodes (main axis) below the telomes (dichotomies).
mesophyll. The chlorophyllous leaf tissues between the epidermal layers.
mesophyte. A plant that grows under medium conditions of moisture.
mesosphere. A layer of atmosphere located above the ionosphere generally more than 250 miles above the earth's surface.
Mesozoic. The geological era between the Paleozoic and Cenozoic.
metabolism. The sum total of the chemical transformations occurring in the body of a living organism.
metaphase. The stage in mitosis at which the chromosomes lie at the equator of the spindle.
metaxylem. Xylem cells formed towards the close of the growth period of the primary xylem.
meter. The equivalent of 39.37 inches.
methane. A gaseous hydrocarbon, CH_4, odorless and inflammable; a product of decomposition in marshes.
microbiology. The study of minute organisms; a broader term than bacteriology.
microcosm. A little world, as the habitat of mosses and their associated organisms.
microfibrils. The extremely small threadlike particles found in cell walls.
microgametophyte. A male gametophyte resulting from the growth and development of a microspore.
micron. One-thousandth part of a millimeter; approximately 25,000 in an inch.
micropyle. A small opening; especially, the tiny opening in the integuments of an ovule through which the pollen tube grows.
microsorus. A sorus containing only microsporangia and therefore producing only microspores.
microsporangium. A sporangium producing only microspores.
microspore. The spore which gives rise to the microgametophyte; in seed plants, a pollen grain.
microspore mother cell. The cell which, by reduction division, gives rise to microspores.
microsporocarp. A structure which gives rise to or contains microsporangia.
microsporophyll. A leaf, or similar structure, which bears microsporangia.
middle lamella. The middle layer of a cell wall; the original membrane, consisting of pectic material, between two adjacent protoplasts and remaining as a cementing layer between adjacent cell walls.
milligram. One-thousandth part of a gram.
milliliter. One-thousandth part of a liter.
millimeter. One-thousandth part of a meter; approximately 25 in an inch.
millimicron. One-thousandth of a micron.
Miocene. A period of the Tertiary, preceding the Pliocene.
mitochondria. Specifically formed bodies in the cytoplasm concerned with respiratory activities of the cell.
mitosis. The series of complex changes through which a nucleus ordinarily passes when it divides.
molecule. A unit of matter; the smallest particle of a chemical compound with the characteristic properties of the compound. A molecule is composed of atoms.
monarch. A protostele in which the central xylem is circular and without arms.
monera. An organism which lacks an organized nucleus, such as bacteria or blue-green algae.
monocotyledonous. Having a single cotyledon.
monoecious. Having male and female organs or staminate and pistillate flowers on the same plant.
monophyletic. A single line of descent in plant evolution.
monoploid. Or haploid; one set of chromosomes, as the *n* number of chromosomes in a gamete.
monosaccharide. A simple sugar having 5, 6, or less carbon atoms.
monosiphonous. Consisting of a single filament.
morphology. The study of form and structure in biology.
mosaic disease. A diseased condition in plants, often caused by viruses, in which different parts of the leaf show unequal development of chlorophyll, resulting in a mottled appearance.
mucilage. Any of certain gelatinous substances found in plants such as seaweeds.
multicellular. Many-celled.
multiflagellate. Having many flagella.
multiple factors. In genetics where the inheritance of certain characters, such as size or seed color, depends upon the collective action of a group of two or more similarly acting genes.
multiple fruit. A cluster of matured ovaries produced by separate flowers; a compound fruit.
mutation. A sudden, heritable change produced in the offspring of a parent organism as a result of an alternation in a gene or chromosome, or of an increase in chromosome number; also the process by which such changes occur.
mycelium. The mass of hyphae forming the vegetative body of certain fungi.
mycology. The branch of botany dealing with the study of fungi.
mycorrhiza. A symbiotic association of a fungus with the roots of higher plants.

nannandrous. In algae, such as *Oedogonium,* which have antheridia produced on special dwarf filaments.

natural selection. The natural process tending to cause the "survival of the fittest" (that is the survival of those organisms best adapted to a given environment). Darwin's theory.

neck canal cell. A cell or cells within the archegonial neck.

necrosis. Death of tissue or organ.

nematode. A member of a class of round worms; often parasitic in plants and animals.

neomycin. An antibiotic obtained from *Streptomyces fradiae.*

net venation. A scheme of vein arrangement in leaves, in which the veins branch frequently, forming a network.

neuromotor. A mechanism in lower organisms which controls the action of the flagella, similar to the nerve mechanisms of higher animals.

neurotoxin. A poison affecting the nervous system.

neutron. A particle having neither a positive nor negative charge.

niche. Microhabitat; the place or status of an organism in its environment.

nitrogen. A colorless gaseous element, tasteless and odorless, comprising about four-fifths of the atmosphere; a constituent of all living tissues.

nitrogen fixation. The conversion of free gaseous (atmospheric) nitrogen into a combined form.

node. A point on a stem from which a leaf and bud arise.

nucellus. The central cellular body of an ovule containing the embryo sac.

nuclear fission. The splitting of an atomic nucleus, as by neutron bombardment with the release of great quantities of energy.

nuclear fusion. The union of atomic nuclei to form heavier nuclei resulting in the release of enormous quantities of energy when certain light elements are united, such as heavy hydrogen to form helium.

nucleic acid. Any of a group of acids, combinations of a sugar or a derivative of a sugar with a phosphoric acid and a base found in nuclear material.

nuclide. An atom characterized by the constitution of its nucleus and hence by the number of protons; the number of neutrons determines the energy content.

nucleolus. A small, usually spherical body found within the nucleus of most kinds of cells.

nucleotide. The basic unit of nucleic acid, consisting of a sugar molecule (deoxyribose or ribose), a phosphate, and an organic base (cytosine, guanine, adenine, thymine, or uracil).

nucleus. A usually spherical or ovoid protoplasmic body found in most cells and considered to be a directive center of many protoplasmic activities, including the transmission of hereditary characteristics.

nut. An indehiscent, dry, one-seeded, hard-walled fruit, produced from a compound ovary.

obligate parasite. An organism which can only obtain its food from living protoplasm.

oidia. Short, cylindrical, thin-walled parts of hyphae of certain Basidiomycetes.

oidiospore. A spore formed by segmentation of short hyphal branches.

oidium. A segment of a hypha which is sporelike in function.

oil. A fat that is liquid at ordinary temperatures.

Oligocene. A period of the Tertiary.

omnivore. An organism that eats both plant and animal foods.

ontogeny. The life history, or development, of an individual, as opposed to that (phylogeny) of the race.

oogamus. Characterized by a type of sexual reproduction involving a large non-flagellated egg and a small sperm.

operculum. The lid covering the apex of a moss capsule.

opposite. Bearing two leaves or two buds at a node on opposite sides of a stem.

order. A category in the classification of plants, ranking between a family and a class. An order is composed of families.

Ordovician. The period following the Cambrian.

organ. Any of the major parts of a plant body—leaf, stem, root, floral organs. An organ is composed of tissues.

organelles. The small structural units of cells such as mitochondria.

organism. An individual plant or animal.

organogenesis. The origin or development of organs in a plant or animal.

orthotropous. Straight and symmetrical, with the chalaza at the base and the micropyle at the opposite end, as an ovary.

osmosis. Diffusion through a differentially or selectively permeable membrane.

ostiole. The pore or opening of a fruiting body.

outbreeding. Breeding or mating of individuals which are relatively unrelated.

ovary. The basal, enlarged portion of a pistil, within which seeds develop.

ovule. A structure consisting of a female gametophyte, nucellus, and integuments which develops after fertilization into a seed.

oxidation. The process of combining oxygen in a chemical reaction, such as in burning or in respiration.

oxygen. An element occurring free as a colorless, tasteless odorless gas in the atmosphere; the most abundant element in the earth's surface since it is combined with many other substances including water and various rocks.

pachytene. A prophase stage in meiosis during which homologous chromosomes are associated as bivalents.

palea. In grasses, the upper bract that subtends the flower.

paleobotany. The division of paleontology concerned with the study of fossil plants.

Paleocene. A geological period.

Paleozoic. The oldest geological era having abundant fossils. It includes divisions from the Cambrian to the Permian.

palisade tissue. A leaf tissue composed of long cylindrical chlorophyllous cells, as the portion of mesophyll lying below and at right angles to the upper epidermis.

palmate. Having the lobes, divisions, or leaflets radiating from a common center.

palmately veined. A type of net venation in which the main veins of a leaf blade branch out from the apex of the petiole like the fingers of a hand.

palmella stage. A reproductive phase of the Volvocales in which the protoplast of the parent cell undergoes changes to form a number of cells within a common matrix.

panicle. A compound inflorescence with several main branches, each of which bears pediceled flowers arranged along its axis; in grass panicles, the flowers are borne in spikelets on the pedicel.

pantothenic acid. One of the water-soluble B group of vitamins.

papillate. Manifesting small nipplelike projections.

parallel evolution. Evolution in a similar direction in different groups of organisms.

parallel venation. A type of leaf venation, in which principal veins are parallel to each other and to the longitudinal axis of the leaf.

paramylum. The starchlike reserve food in the Euglenophyta.

paraphysis. A sterile hair or filament usually associated with the sporangia and gametangia of lower plants.

parasite. An organism living on or in another organism and subsisting by getting its food at the latter's expense.

parenchyma. A tissue composed of thin-walled, often isodiametric cells, which often store food and usually retain meristematic potentialities.

parietal placentation. Having the placenta and therefore the seeds developing along the inner walls of the ovary.

parthenogenesis. The development of an egg into a new individual without fertilization by a sperm.

pathogen. An organism causing a disease.

pathology. The study of diseases and their treatment.

pectinate. In flagella, having fine hairs arising from the side of the main axis.

pedicel. The stalk of an individual flower in an inflorescence.

pedigree. An ancestral line; a geneological tree.

peduncle. The stalk of a solitary flower, or the main stalk of an inflorescence.

pellicle. The thin, firm outer membrane of certain unicellular organisms such as *Euglena.*

penicillin. A powerful, nontoxic, anti-infective substance found in the mold fungus *Penicillium.*

pentaploid. A polyploid with five sets of chromosomes.

pepo. A large, fleshy, succulent berry, as a watermelon.

perennial. A plant which lives for more than two years.

perfect flower. A flower which bears both stamens and carpels.

perianth. The floral envelope, i.e., calyx and corolla.

periblem. A group of embryonic cells (histogen) in a root or stem tip that gives rise to the cortex.

pericarp. The wall of a ripened ovary (fruit).

pericentral cell. In the red algae, the cell which ultimately gives rise to the carpogonium and the auxiliary cells.

periclinal. Parallel to the outer suffAce, as the periclinal walls in cell division.

pericycle. One or more layers of cells forming the outermost tissue of the stele.

pericyclic fibers. Thick-walled, elongated cells arising in the pericycle.

periderm. The outer tissues of a stem, usually consisting of cork cambium and the cork.

peridiole. *see* peridium.

peridium. The outer coat or wall of the fruiting body of certain fungi; the outer covering of the basidiocarp of puffballs.

perigynous. Having a flower structure in which the stamens and perianth parts are borne on a cup or disc formed by the receptacle surrounding the ovary.

perimedullary. Surrounding a central core, as the tissues around the pith.

periplast. The outer cell membrane, especially of such organisms as *Euglena.*

perisperm. The nutritive tissue in a seed formed outside the embryo sac. It usually comes from the nucellus.

peristome. A ring of teeth surrounding the opening of a moss capsule and by its hygroscopic movements scattering the spores.

perithecium. A spherical or pear-shaped ascocarp opening by means of a terminal pore.

permafrost. Certain soils in the high latitudes (and altitudes) which remain permanently frozen 12 months of the year.

permeable. Capable of being passed through, as a cell membrane which allows water and solutes to pass through.

Permian. The most recent period of the Paleozoic. The Permian follows the Carboniferous and precedes the Triassic.

petal. One of the divisions of a corolla.

petiolate. Having a petiole.

petiole. A leaf stalk.

petiolule. A division of a petiole (a leaflet stalk).

phalloid. Resembling a mushroom of the genus *Phallus,* named for a fancied likeness to the human penis.
phellem. Cork or cells of the bark.
phelloderm. Secondary tissue formed by cork cambium or phellogen on its inner surface.
phellogen. The cambium (cork) that produces cork tissue.
phenotype. The appearance or visible characteristic of an organism, as opposed to its genotype.
phenotypic selection. Selection of organisms for mating based on their immediate physical appearance.
phloem. A conducting tissue, consisting of sieve tubes and usually also of companion cells, fibers, and parenchyma. The chief function of phloem is food conduction.
phospholipid. A fat or oil in which the molecule is combined with some phosphorus containing substance.
phosphorus. A nonmetallic, poisonous, active element of the nitrogen group, usually obtained as a waxy crystalline solid with a disagreeable odor.
photoautotrophic. Able to synthesize food in the presence of light, as a plant.
photonasty. Response of plants to diffuse light stimuli; or to variations in illumination.
photoperiodism. Growth and development responses of plants to differing lengths of light.
photosphere. A sphere of light; the luminous envelope of the sun.
photosynthesis. The manufacture of carbohydrate food from carbon dioxide and water in the presence of chlorophyll, utilizing light energy and releasing oxygen; photosynthesis occurs in certain bacteria, involving the participation of pigments other than chlorophyll.
phycobilin. Chromoproteins of seaweeds, as phycocyanin or phycoerythrin.
phycocyanin. A blue-green pigment associated with chlorophyll in the Cyanophyta.
phycoerythrin. A red pigment associated with chlorophyll in the Rhodophyta.
phycology. The study of algae.
phyllotaxis. The arrangement of leaves on a stem or axis.
phylogeny. The evolutionary development of a race of related individuals.
phylum. A large taxonomic unit, subdivided into closely related classes.
physiology. The branch of biology dealing with the processes, activities, and functions of the organism.
phytochrome. Chromoproteins of plants, pigments of certain plants.
pileus. The cap or umbrella-shaped portion of a mushroom.
pinna. A leaflet of a fern leaf; the primary division of a pinnate leaf.
pinnate. Featherlike; having the leaflets of the primary division of a leaf arranged on each side of an axis.
pinnately veined. Manifesting a net venation in which the secondary veins branch out in parallel fashion from the single midrib of a leaf blade.
pinnule. One of the ultimate divisions of a bipinnate leaf.
pistil. The ovule-producing part of a flower, consisting of a carpel (simple pistil); or of two or more partly or wholly fused carpels (compound pistil).
pistillate flower. A flower which bears a pistil but no stamens.
pit. A thin place in a cell wall. A simple pit has no overarching wall; a bordered pit has such a wall.
pith. The parenchyma (usually) tissue occupying the central portion of a stem inside the xylem.
placenta. A small mass of ovary tissue to which the seeds are attached.
placentation. The arrangement of placentae within an ovary.
plankton. Free-swimming or floating aquatic organisms.
plasma membrane. A thin membrane that encloses the protoplasm of all living cells.
plasmodesma. Protoplasmic connections between cells.
plasmodium. The mass of naked multinucleate protoplasm usually having amoeboid movement which is the vegetative stage of a myxomycete.
plasmogamy. Fusion of cytoplasm without nuclear fission.
plastid. An organized body in the cytoplasm of many plant cells, usually pigmented and concerned with the metabolic activities in the plant.
plectostele. A specialized type of protostele in which the xylem seems to consist of separate plates with phloem in and around the xylem areas in cross section.
plerome. A group of embryonic cells (histogen) in the tips of stems and roots that gives rise to the stele.
pliocene. The latest period of the Tertiary.
plurilocular. Having many cavities (locules) as in certain sporangia.
polarity. Morphological and physiological direction in plants, largely based on a gradient of metabolic activities.
polar nucleus. One of two free nuclei in an embryo sac that unites to form the fusion nucleus.
pollen. A microspore; the fertilizing structure of higher plants.
pollen chamber. A cavity developed at the apex of the nucellus in certain gymnosperms.
pollen sac. One of the pollen-containing cavities in an anther; a microsporangium in higher plants.
pollen tube. A tube which is formed by a pollen grain and which transports sperms to the eggs in ovules.
pollination. The transfer of pollen grains from a stamen or microsporophyll to a stigma or ovule.
pollinium. Massed pollen.
polyarch. Characterized by a primary xylem with many points or arms radiating from the central cells.

polycotyledonous. Having many cotyledons, as in certain gymnosperms.
polyembryonous. Having many embryos.
polypeptide. Any of the simple nonprotein combinations of several amino acid molecules.
polyphyletic. Derived from more than one ancestral stock; showing several lines of descent.
polyploid. Having more than two sets of chromosomes.
polypore. Any large number of members of the Agaricales whose basidia are borne in pores instead of gills.
polysaccharide. A carbohydrate, such as starch or cellulose, containing more than three monosaccharide units per molecule.
polysiphonous. Composed of many cylindrical filaments.
pome. A fruit of the apple or pear type, in which the true fruits (core sections) are surrounded by an enlarged, fleshy floral tube and receptacle.
primary. Pertaining to the first or highest; as primary xylem is the first xylem to form.
primary axis. The main or central branch, as the trunk of a tree, or tap root.
primary endosperm nucleus. The nucleus (later a cell) formed by the union of polar nuclei with a sperm, usually a triploid structure.
primary neck cell. The first cell differentiated in development of the neck of an archegonium. It will give rise by successive divisions to the remaining neck cells.
primary root. The root which develops directly from the hypocotyl of an embryo.
primordium. The rudiment or beginning of a part.
procarp. The female sex organ of the red algae, consisting of a basal carpogonium and a trichogyne.
proembryo. A group of cells, derived by division of the fertilized egg, which will differentiate into the suspensor and embryo.
progamete. The enlarged initial cell before division, separating the basal cell from the gamete, as in *Rhizopus*.
progeny. Offspring or descendants.
promeristem. Undifferentiated cells in growing regions which will shortly become meristem.
promycelium. The germ tube arising from the teliospore of rust.
prophase. An early stage in mitosis in which the chromosomes become distinct and organized and in which their longitudinal splitting begins.
prostrate habit. The tendency in many plants to grow horizontally along the ground, as the dorsiventral body of the liverwort.
Proterozoic. A division of geological history from the Archeozoic to the Paleozoic.
prothallus (prothallium). The gametophyte of ferns and similar plants.
protista. A group of unicellular organisms, including both plant and animal species.
protoangiosperm. The term given to the precursor of the flowering plants; a hypothetical designation.
proton. Nucleus of the atom of the light isotope of hydrogen and probably all other atoms.
protonema. The filamentous stage in the gametophyte of a moss.
protophloem. The first-formed cells in the primary phloem of a root, stem, or other plant organ.
protoplasm. The living material found within the cell walls of all organisms.
protostele. A stele with a solid core of xylem, thus lacking pith, and with a surrounding mass of phloem.
protovirus. A macromolecule, believed to be the ancestor of the virus.
protoxylem. The first-formed cells in the primary xylem of a root, stem, or other part of the plant.
pseudoparenchyma. Fungus tissue in which the hyphae have lost their filamentous identity and become isodiametric.
pseudoperianth. An involucral envelope enclosing certain fruiting structures, but not composed of sepals or petals.
pseudoplasmodium. An aggregation of amoeboid cells forming the initial stage of fruiting of the *Acrasieae*.
pseudopodium. In certain mosses, the stalk developed from the apex of the gametophyte.
pustule. A surface blister formed by the rupture of surface tissues in certain plant diseases.
pyrenoid. A starch-forming body in the chloroplast of many algae.
pyriform. Pear-shaped.
pyrimidine. A base that is a component of nucleosides and nucleotides.

quadrant. In growth, initial division of the zygote. It often results in four cells grouped together.
quadriflagellate. Having four flagella.
quadrilateral. Having four sides.
quantum. An elemental unit of energy.
quarter-sawed. A term used to describe wood sawed along a radial plane so as to expose the radial surface.
Quaternary. The later division of the Cenozoic era down to the present.
quillwort. The common name for plants of the genus *Isoetes*.
quiescent. At rest, latent.
quinine. An alkaloid ($C_{20}H_{24}N_2O_2$) drug obtained from the bark of the cinchona tree.

rachis. In compound leaves, the extension of the petiole bearing the leaflets.

radial symmetry. Symmetry such that the half parts of an organ or structure are always equal to each other in size and shape when cut from any angle; having uniform parts which radiate out from a central axis.

radiant energy. Energy arising from and transmitted by light.

radical. A fundamental constituent of a compound; a group of atoms replaceable by a single atom or remaining unchanged during a series of reactions, and therefore regarded as playing the same role as a single atom.

radicle. The part of the embryo that develops into the primary root.

radioactivity. The process whereby certain elements or isotopes (chiefly strontium, cesium, thorium, etc.) spontaneously emit particles and/or rays by the distintegration of the nuclei of their atoms.

radium. A radioactive metallic element found in pitchblende and other uranium rocks.

raffinase. An enzyme which digests certain complex sugars, as raffinose.

raphe. In diatoms, a longitudinal slit extending along the median line of a valve.

ray. The corolla of a marginal flower of a composite inflorescence; also, a vascular ray.

recapitulation. The repetition of ancestral stages in the development of an individual.

receptacle. Usually applied to the apex of the pedicel or peduncle from which the flower parts arise.

recessive. One of a pair of contrasting characters which is masked by the dominant character when both are present.

reciprocal translocation. Exchange of chromosome segments between two chromosomes.

reduction division. *see* meiosis.

region of elongation. The area just back of the apical meristem in a root or stem tip in which the cells are very much elongated in comparison with their diameters.

region of maturation. The area in a root or stem tip in which the various tissues have become differentiated from the ground tissue.

regular. *see* actinomorphic.

resin. One of the sticky to brittle plant products derived from essential oils and often possessing marked odors. Resin is used in varnishes, incense, medicines, etc.

resin duct. A cell, tube, or cylinder containing resin.

respiration. A chemical oxidative process whereby living protoplasm breaks down certain organic substances with the release of energy.

reticulate. Netlike, as the chloroplasts of certain algae, or as spore walls.

retting. To soak or expose to moisture as the treatment of stems to remove the fibers for commercial use.

rhizoid. In some fungi, mosses, liverworts, one of the hairlike appendages which penetrate the soil or another substrate, anchoring the plant and absorbing water and other substances.

rhizome. A horizontal underground stem, often enlarged by food storage.

rhizomorph. Resembling a root.

rhizophore. A specialized root-bearing organ.

rhizoplast. A fiber connected to the base of the flagella.

rhytidome. The furrowed, wrinkled outer bark of stems and roots.

rib meristem. A type of primary meristem which contributes to the growth in length of young internodes. In stem apices it is between the central mother cell zone and the maturing pith.

riboflavin. Vitamin B_2; important in oxidation processes.

ribonucleotide. Any of various nucleic acids that yield ribose as one product of hydrolysis.

ribulose. A five carbon sugar.

rickets. A human disease due to a deficiency of calcium deposition in bones; the disease often responds to treatment by vitamin D and sunlight.

rickettsiae. A very small microorganism having both bacterial and protozoan characteristics, and causes such disease as typhus.

ring cell. A cell in the annulus of a sporangium.

ring-porous wood. Wood in which the pores (vessels) of one part of an annual ring are distinctly different in size or number (or both) from those in the other part of the ring.

RNA. Ribonucleic acid.

rockweed. A common name for a number of brown algae, such as *Fucus, Desmarestia,* etc., which live in the intertidal zone on rocky substrates.

root. A major vegetative organ of a plant, usually subterranean and exhibiting dominant apical growth.

root cap. A thimblelike mass of cells which fits over the apical meristem of a root and protects it.

root hair. An elongated projection of an epidermal root cell.

root pressure. The pressure of sap developed in a root when absorption is active and transpiration relatively low.

rosette. A circular cluster of leaves or other organs; as the arrangement of leaves on a dandelion.

runner. A slender prostrate stem (stolon) which gives rise to roots and aerial portions at various places (nodes) along its length to vegetatively propagate the plant.

rust. An infectious parasite of the Uredinales which produces reddish or brownish spots on infected host tissue.

safranin. A synthetic dye, usually red.

samara. A dry indehiscent, one-seeded, winged fruit, such as that of elm or maple.

saprophyte. A plant living on dead organic material.

sapwood. The outer annual rings of a tree; the cells are physiologically functional in conduction.
sarcina. Certain bacteria which grow in cubical masses.
scalariform. Ladderlike; having bars or markings like the rungs of a ladder.
scales. The structures arising, along with rhizoids, from the ventral surface of certain members of the Bryophyta, usually multicellular.
schiff base. A class of bases obtained typically by condensation of an aldehyde or ketone, used chiefly in organic synthesis.
scion. In grafting, the portion which is inserted into the stock; it may be a bud or twig.
sclerenchyma. A strengthening tissue composed of thick-walled, elongated cells (fibers) or shorter cells (stone cells).
sclerotium. In certain fungi, a resting stage consisting of a mass of the mycelium or plasmodium used as an asexual method of reproduction.
scutellum. In monocotyledons, a thick disclike structure between the main axis of the embryo and endosperm; commonly considered the cotyledon.
secondary root. A branch of a primary root.
secondary tissues. Tissues arising from a lateral or secondary meristem, such as the cambium or cork cambium, thereby increasing the diameter of a stem or root.
sedentary. Settled, not migrating, as applied to a population.
sedimentary rock. Rock formed from materials which settle out of water and then undergo compression.
seed. The characteristic reproductive structure of seed plants, consisting of an embryo, enclosed by a seed coat, and a food-storage tissue called endosperm. In some species, the endosperm is absorbed by the embryo before the seed reaches maturity.
segmental interchange. Exchange of portions of chromosomes during meiosis.
selenium. A nonmetallic element resembling sulfur; excessive quantities taken up by certain plants makes them poisonous to cattle.
seminal. Arising from the seed.
sepal. One of the divisions of a calyx; one of the outermost floral organs.
septate. Partitioned, as with crosswalls.
septicidal. A capsule dehiscing at the partitions thus separating the fruit into its carpels.
sessile. Lacking a stalk or petiole.
seta. The stalk that supports the capsule in Bryophytes; a slender bristlelike structure, as the hairlike tip of a moss leaf.
sex chromosome. Sex-determining chromosome.
sexual. Characterized by a reproduction involving fusion of two compatible nuclei.
shoot. A young stem with its leaves; a young branch, twig, or sprout.
short day plant. A plant blooming only when the continuous dark period is 12 or more hours long.
shrub. A woody plant with several stems arising from the root system.
sieve plate. The perforated end wall of a sieve tube.
sieve tube. A cell of phloem tissue; an elongated living cell or linear series of cells in which the end walls at least are perforated.
silicle. A short broad silique, characteristic of the mustard family.
silique. The elongated, two carpeled capsule of the mustard family.
Silurian. An early Paleozoic period following the Ordovician and preceding the Devonian.
simple. Not divided into parts, as a simple leaf opposed to a compound leaf.
siphonostele. A stele with a central pith.
slime mold. A member of the class *Myxomycetae*.
smut. An infectious parasite of the Ustilaginelales, which produces masses of black spores on the host tissue.
sodium. A soft, waxy, sliver-white metallic element; the element is toxic to most plants.
softwood. The wood of gymnosperms; a wood lacking vessels.
soil fire. The rapid oxidation of organic matter in soils due to the metabolic activities of microorganisms.
solar. Pertaining to the sun; radiation of light and heat.
solenostele. An amphiphloic siphonostele. The xylem tissue has both internal and external phloem in the vascular cylinder.
somatic. Pertaining to the body of an organism as distinguished from its reproductive organs.
soredium. A special reproductive body of lichens, consisting of a few algal cells surrounded by fungal hyphae.
sorus. A cluster of sporangia.
spadix. A fleshy spike or head with small, often imperfect flowers, commonly surrounded by an enveloping sheath called the spathe.
spathe. The enveloping sheath of a spadix.
species. The basic category in plant (or animal) classification. The members of a given species are usually sexually compatible.
spectrum. The wavelengths of light brought to focus so that the component waves are arranged in the order of their wavelengths.
sperm. A male gamete.
spermagonium. A flask-shaped structure in rust fungi, producing spermatia.
spermatiophore. A specialized hypha that bears spermatia.

spermatium. In red algae, a naked nonmotile gamete; a tiny sporelike cell in rusts which may function as a gamete.
spermatogenous. Sperm-producing.
spike. An inflorescence consisting of sessile flowers on an elongated axis; a strobilus in certain vascular plants.
spindle. The ovoid mass of fiberlike structures formed in a cell during mitosis.
spirillum. A bacterium that is spiral or curved in shape.
spirochaete. Spirally undulating microorganisms, as the causative organism of syphilis.
spongy tissue. A leaf tissue composed of loosely packed, chlorophyllous cells of diverse form.
sporangiophore. A stalk or hypha bearing sporangia.
sporangium. A cell or organ containing spores.
spore. A reproductive body, usually one-celled, capable of developing into a new individual.
spore mother cell. A cell, which, by cell divisions, typically produces four spores.
spore print. The pattern left on paper by spores after placing the cap of a mushroom on the paper.
sporocarp. A specialized structure of water ferns in which sporangia are borne
sporogenous. Producing spores; reproducing by means of spores.
sporophyll. A spore-bearing leaf.
sporophyte. The plant capable of bearing spores, having the diploid number of chromosomes.
springwood. The earliest cells formed in an annual wood ring.
stalk cell. In conifers, one of the cells formed by the division of the generative cell.
stamen. In a flower, the part on which the microspores are borne; the microsporophyll.
staminate flower. A flower which bears stamens but no carpels.
staphylococcus. Any of certain bacteria whose spherical cells are arranged in irregular clumps, resembling bunches of grapes.
starch. A white, complex, water-insoluble carbohydrate which is a common storage food in many plants.
stele. A collective name for the vascular and closely associated tissues in stems, roots, and other plant parts; xylem, phloem, pericycle, endodermis, and pith.
stem. The ascending, usually branched axis of a plant, externally consisting of nodes and internodes.
sterigma. A slender tip arising from a basidium and bearing a basidiospore.
sterile. Having the reproductive organs undeveloped, as vegetative tissue, opposed to fertile (reproductive) tissue.
sterol. Any of a class of solid alcohols, as cholesterol.
stigma. The part of a pistil, usually the apex, which receives pollen and upon which pollen grains germinate.
stinkhorn. A slender phalloid type of mushroom, with an offensive odor.
stipe. A stem or stalk; the supporting stalk of a mushroom.
stipule. An appendage borne at the base of certain leaves, usually scalelike or leaflike.
stock. The plant on which a scion is grafted.
stolon. A slender branch or runner which grows as a prostrate stem, giving rise to new roots and aerial shoots or leaves in vegetative propagation.
stomate. A small mouthlike opening in the epidermis of a leaf or young stem, usually bounded by a pair of guard cells.
strata. (In geology) layers of sedimentary rock; (in biology) layers of tissue.
stratosphere. An upper portion of the atmosphere above 7 miles, more or less depending on latitude, season, and weather.
streptococcus. Any of several spherical bacteria typically arranged in chains.
streptomycin. An antibiotic obtained from *Streptomyces griseus*.
strobilus. A conelike collection of sporophylls borne on a stem axis.
stroma. A compact, cushionlike mass of mycelium in which perithecia are produced.
strontium. An alkaline-earth metal occurring naturally only in combination; radioactive.
style. In flowers, a cylindrical structure which rises from the top of an ovary and through which pollen tubes grow.
subepidermal. Beneath the epidermis.
suberin. A fatty or waxy substance.
substrate. A substance acted upon by an enzyme.
subterranean. Developing or living beneath the surface of the ground.
subtropical. Bordering on the tropical, the geographic region between the temperate and tropical area.
succession. The sequence of organisms in any area from the original pioneer organisms to the climax plants.
succulent. Having juicy tissues, as cacti; fresh.
sucrose. A double sugar, as cane or beet sugar.
sulfur. A nonmetallic element; a common constituent of proteins.
summerwood. The cells formed late in the season in an annual ring.
superior. Growing or attached above a part or organ.
superior ovary. *see* hypogenous.
superphosphate. An acid phosphate made from mineral phosphates by treatment with sulfuric acid.
suspensoid. A colloid in which the dispersed phase is of solid particles.
suspensor. In many seed plants and other vascular plants, a portion of the embryo which often elongates, pushing

the main portion of the embryo deep in gametophytic tissue.

swarm spore. A motile, naked spore as in the Myxomycetes.

symbiont. An organism in an association of symbiosis.

symbiosis. The intimate association of two dissimilar organisms to their mutual advantage, as an alga and fungus in close association form a lichen.

synangium. A group of sporangia developed as a single structure typical of the fruiting structure of certain ferns.

synapsis. The pairing of homologous chromosomes in meiosis.

synconium. The inverted multiple fruit of the Smyrna fig.

synergid. Either of two small cells lying near the egg at the micropylar end of the embryo sac in an ovule.

synthesis. The process of forming a compound by the union of simple compounds or their elements.

systematic botany. Taxonomy; the identification and classification of plants.

tannin. A bitter, astringent organic substance found in the bark and other plant tissues.

tapetum. One or more layers of nutritive cells surrounding the sporogenous tissue in a young sporangium.

tap root. A root system in which there is a primary root distinctly larger and more conspicuous than any of its branches.

taxon. A name.

taxonomy. The science of the classification and arrangement, according to relationships, of living organisms.

teliospore. A thick-walled spore, usually two-celled, in rusts that produces the basidium, commonly called the overwintering spore.

telome. A morphological unit consisting of two stalks or a stalk and a sporangium or two sporangia.

telophase. The final stage in mitosis in which the daughter nuclei are organized into two nets of chromosomes and processes leading to the formation of a new cell wall separating the daughter nuclei are initiated.

tendril. A slender, coiling organ which aids in the support of plant stems. A tendril may be a modified stem leaf, leaflet, or stipule.

terramycin. An antibiotic obtained from *Streptomyces rimosus.*

terrestrial. An inhabitant of the earth; growing on or in the soil.

tertiary. The geological period which precedes the Quaternary.

testa. The hard outer portion of the seed coat.

tetrad. A group of four cells (usually spores) produced by meiotic division of a mother cell; also, a group of four chromosomes formed at meiosis.

tetrahedral. Having four sides or parts like a pyramid.

tetrasporangium. A sporangium in which the spores are borne in tetrads as in the Dictyotales.

tetraspore. In certain algae, one of the nonmotile spores produced in groups of four.

tetrasporine line of evolution. An evolutionary line in the green algae, beginning with a flagellate ancestor, including chiefly the various filamentous species of the Ulotrichales and ending with a thallus plant such as *Coleochaete.*

thallus. A relatively simple plant body lacking roots, stems, and leaves.

theory of relativity. Einstein's theory on the motions of the planets and the measurements of the velocity of light.

thermocline. A layer of water in a thermally stratified lake or other body of water separating an upper, warmer, lighter oxygen-rich zone from a lower, colder, heavier oxygen-poor zone.

thermonasty. Plant movements (nasties) in response to temperature.

thermonuclear energy. The result of changes in the nucleus of atoms of low atomic weight (as hydrogen) that require a high temperature for their inception, as a hydrogen bomb.

thiamin. Vitamin B_1—the antiberiberi vitamin.

thigmotropism. A growth movement in response to the stimulus of contact.

thorium. A radioactive element which may be isolated as a heavy, gray metal which is difficult to fuse.

thorn. A sharp-pointed aborted branch, generally considered as of endogenous origin, as opposed to the more superficial spine.

thymine. A crystalline pyrimidine base $C_5H_{14}N_2O_5$ that is obtained by hydrolysis of deoxyribonucleic acid; 5-methyl-uracil.

tinsel. A term when applied to flagella refers to a type of flagellum with lateral fibrils.

tissue. An aggregation of cells, usually of similar structure, which perform the same or related functions.

tonoplast. The cytoplasmic membrane bordering a vacuole.

toxin. Poisonous secretion of a living organism or cell.

TPN. Triphosphonucleotide, a coenzyme derived from the vitamin nicotinic acid and functioning in dehydrogenation reactions.

trabecula. A plate of cells bridging or separating an intercellular space, as the separating or partitioning cell plates formed in certain sporangia.

tracheal tube. A xylem vessel; a xylem element lacking end walls, typical of angiosperms.

tracheid. A type of nonperforate conducting and strengthening cell in xylem tissue; of elongated, tapering form and with pitted walls.

transfusion tissue. The tissue in gymnosperm leaves between the vascular bundles and endodermis. It is believed to function in lateral conduction of substances from the vascular elements to the mesophyll.

translocation. The movement of materials in solution from one part of the plant to another; in genetics the transfer of one or more genes to anther chromosome.

transpiration. The loss of water in the form of vapor from living plants.

triarch. Having a primary xylem with three points in cross section.

Triassic. The geological period that constitutes the earliest principal division of the Mesozoic era.

trichogyne. A hairlike extension of the carpogonium in red algae; also, a hairlike extension of the ascogonium in the ascomycetes.

trichome. A plant hair or hairlike appendage.

triglycerides. A triester of glycerol with one, two or three acids.

triploid. Having three times the haploid chromosome number, as the primary endosperm nucleus in certain angiosperms.

tritium. The hydrogen isotope of mass number 3.

tropism. A bending movement of a cylindrical organ or other cylindrical structure, caused by differences in growth rate in different parts of the organ and induced by external stimuli.

troposphere. That portion of atmosphere below the stratosphere in which temperatures gradually decrease with altitude, clouds form in it, and convection is active.

trypsin. A proteolytic enzyme, i.e., for the digestion of certain proteins.

tube nucleus. One of the nuclei in a pollen tube, influencing the growth and behavior of the tube and not involved in sexual fusion.

tuber. Enlarged, fleshy, underground stem, commonly borne at the end of a rhizome.

tunica. The outer growth zone in a stem tip, composed of an outer growth zone and the central embryonic growth zone.

tunica corpus. The apical histogen of a stem tip consisting of an outer growth zone and the central embryonic growth zone.

turgor. The distention of a cell resulting from its water content.

tyrosinase (tyrosine). An enzyme capable of digesting certain amino acids.

undulate. Having a wavy or curved pattern.

uniflagellate. Having one flagellum.

unilocular. Having a single cavity or locule.

uniseriate. Having a single row of cells or parts, as a single-row filamentous alga.

unisexual. (Of an organism) producing eggs or sperm, but not both; (of a flower) bearing stamens or pistils, but not both.

universal veil. A thin veillike membrane covering certain young mushrooms.

uracil. A compound $C_4H_4N_2O_2$ obtained by hydrolysis of ribonucleic acid or by reaction of urea with ethyl-formyl-acetate; 2,4-dihydroxyl-pyrimidine.

urea. A very soluble, crystalline, nitrogenous compound, $CO(NH_2)_2$.

uredospore. The red, unicellular summer spores of the rust fungi.

urgot. A drug obtained from the fungus ergot (of rye).

utricle. Modified club-shaped branch in *Codium*.

vacuole. A cavity in the protoplasm of a cell commonly filled with cell sap.

valence. The degree of combining power of an element as shown by the number of atomic weights of hydrogen or the like, with which the atomic weight of the element will combine, or for which it can be substituted.

valine. A crystalline amino acid essential in nutrition of man, $(CH_3)_2CHCH(NH_2)COOH$.

vallecular canal. In the Equisetales, one of the canals in the cortex of the stem lying beneath the stem grooves.

valve. One of the two halves of a diatom.

vascular. Conducting tissue, as xylem and phloem.

vascular bundle. A strand of conducting and strengthening tissues (xylem and phloem) in a plant organ.

vascular bundle scar. The scar, usually centrally located, in a leaf scar left by the broken end of the vascular bundles.

vascular ray. A ribbonlike aggregrate of cells extending radially in stems through the xylem and often into the phloem.

vascular tissue. Conducting tissue; xylem and phloem.

vector. An organism, such as an insect, which is a carrier of disease-causing organisms.

vegetative propagation. A method of asexual reproduction involving portions of a plant as the rooting of cuttings.

veil. The annulus or ring of fungus tissue left on the stipe when the cap pulls away in growth.

vein. One of the vascular bundles of a leaf.

velamen. The thick, spongy integument covering the aerial roots of epiphytic orchids.

velum. A membranous partition or veil, as the membrane covering the gills of immature mushrooms.

venation. The arrangement of the vascular bundles (veins) in a leaf.

venter. The enlarged base of an archegonium within which an egg develops.

ventral canal cell. The cell at the apex of the venter derived from the primary cell which divides to give rise to the egg and ventral canal cell.

ventral surface. The inner surface of an organ. In leaves of higher plants it becomes the upper surface, in liverworts the lower surface.

vermiform. Worm-shaped; long and slender.

vermifuge. A medicine used to expel worms from the intestines.

vernation. The distribution of the young foliage leaves within the bud.

vesicle. A thin bubblelike sac in which zoospores or aplanospores are formed or enclosed.

vessel. A conducting tube in xylem tissue.

viability. The ability to live and grow; applied usually to seeds and spores.

virulent. Highly infectious.

virus. A minute substance which may or may not be living, but is infectious.

volva. The cuplike base of some mushrooms in which the stipe is borne.

volvocine line of evolution. A development series of flagellated green algae, beginning with *Chlamydomonas* and ending with *Volvox*.

wax. Any of a class of substances of plant or animal origin, containing esters and often free fatty acids, free alcohols, and higher hydrocarbons.

whip graft. A graft in which the scion is one or more slender shoots inserted into the stock.

whip lash. A flagellum of the smooth type.

white alkali. A white crust of soils, chiefly of calcium and magnesium salts.

whorl. A circle of similar parts arising from the same part of an axis.

wilt. A plant disease in which water conduction is impaired or stopped, causing the plant to lose turgor and droop.

wood. Technically, xylem; popularly, the hard, fibrous xylem of trees and shrubs.

xanthophyll. A light-yellow pigment, usually associated with chlorophyll; it differs from carotin chiefly by containing oxygen as well as carbon and hydrogen.

xeric. Having a desert or semidesert habitat; living under conditions of minimum water.

xeromorphic. Having tissues and organs modified for successful life in an arid habitat.

xerophyte. A plant adapted to growth under dry conditions.

xylem. A complex plant vascular tissue, composed of such cells as tracheids, vessels, wood fibers, ray cells, and parenchyma cells; wood.

zinc. A bluish-white crystalline metallic element, considered a minor element for plant growth but most useful in hydroponics.

zoosporangium. A sporangium which produces zoospores.

zoospore. A naked spore which is equipped with flagella.

zygomorphic. Having a flower with bilateral symmetry, the parts arranged symmetrically on either side of a median longitudinal axis.

zygospore. A cell formed by fusion of two similar gametes.

zygote. A cell formed by the fusion of two gametes.

zygotene. Pairing threads; in the prophase stage of meiosis where spireme threads are uniting in pairs.

Index*

* The terms of the glossary, since they are in alphabetical order, are not listed in the Index. Those numbers in italics refer to pages where there is an illustration.